Lecture Notes in Computer Science 5069

Commenced Publication in 1973
Founding and Former Series Editors:
Gerhard Goos, Juris Hartmanis, and Jan van Leeuwen

T0189854

Bertram Ludäscher   Nikos Mamoulis (Eds.)

Scientific and Statistical Database Management

20th International Conference, SSDBM 2008
Hong Kong, China, July 9-11, 2008
Proceedings

 Springer

Volume Editors

Bertram Ludäscher
University of California
One Shields Avenue, Davis, CA 95616, USA
E-mail: ludaesch@ucdavis.edu

Nikos Mamoulis
University of Hong Kong
Pokfulam Road, Hong Kong
E-mail: nikos@cs.hku.hk

Library of Congress Control Number: 2008929534

CR Subject Classification (1998): H.2.8, H, G.3, I.3.5, E.1-2, J.1-3

LNCS Sublibrary: SL 3 – Information Systems and Application,
incl. Internet/Web and HCI

ISSN   0302-9743
ISBN-10   3-540-69497-8 Springer Berlin Heidelberg New York
ISBN-13   978-3-540-69497-7 Springer Berlin Heidelberg New York

Springer is a part of Springer Science+Business Media

springer.com

© Springer-Verlag Berlin Heidelberg 2008

Typesetting: Camera-ready by author, data conversion by Scientific Publishing Services, Chennai, India
Printed on acid-free paper   SPIN: 12324200   06/3180   5 4 3 2 1 0

Preface

The International Conference on Scientific and Statistical Database Management (SSDBM) is an established forum for the exchange of the latest research results on concepts, tools, and techniques for scientific and statistical database applications. The 2008 meeting marked the 20th time that scientific domain experts, databases researchers, practitioners, and developers came together to share their new insights and to discuss in a stimulating environment future research directions.

This volume contains the proceedings of the 20th SSDBM Conference, held in Hong Kong, China, July 9–11, 2008. The conference included 3 keynote talks, 28 long and 7 short papers in 9 sessions, and 8 posters and demonstrations in a single session.

Distinguished members of the community delivered the three keynotes, which were about the past, present, and future management of scientific and statistical data. Alex Szalay, an expert in large-scale scientific data management, discussed "New Challenges in Petascale Scientific Databases," managing huge scientific data repositories, with a focus on particular examples taken from astronomy. Nick Koudas, a leader in semi-structured text management, talked about "Adventures in the Blogosphere," a huge network of textual data (including blogs, social networks, wikis), and BlogScope, a system that collects and analyzes such data. Finally, Per Svensson, a pioneer in database systems development for scientific applications, provided a historical review on "The Evolution of Vertical Database Architectures" from the perspective and performance needs of a scientific or statistical large-scale data analyst user.

The Program Committee, consisting of 37 members, accepted 43 papers (28 long, 7 short, and 8 posters/demos) from a total of 84 submissions. The reviewing process was managed by the EasyChair Conference System, an excellent free conference management system, developed by Andrei Voronkov.

The program and activities of SSDBM 2008 were the result of a large effort by the authors, reviewers, presenters, and organizers. We thank them all for helping to make this conference a success. In particular, we would like to thank Max Egenhofer for his great help in the early stages of the organization. We are grateful to the Department of Computer Science of Hong Kong University, especially to Maria Lam and the student helpers for their great help with the local organization. We believe that SSDBM 2008 continued the successful tradition of the series, providing an interesting program and lively discussions in a pleasant environment.

Davis, California   Bertram Ludäscher
Hong Kong, China   Nikos Mamoulis
April 2008

Conference Organization

General and Program Chairs

Nikos Mamoulis (General Chair)
Bertram Ludäscher (PC Co-chair)
Max Egenhofer (PC Co-chair)

Program Committee

Ken Barker
Elisa Bertino
Claudio Bettini
Shawn Bowers
Gilberto Camara
Lei Chen
Reynold Cheng
Ian Davidson
Jim Frew
Michael Gertz
Carole Goble
Moustafa Hammad
Panos Kalnis
Larry Kerschberg
Martin Kersten
George Kollios
Feifei Li
Xiaosong Ma
Francesco Malvestuto

Mohamed Mokbel
Kyriakos Mouratidis
Mario Nascimento
Silvia Nittel
Moira Norrie
Beng Chin Ooi
Dimitris Papadias
John Pfaltz
Philippe Rigaux
Doron Rotem
Bernhard Seeger
Kurt Stockinger
Yufei Tao
Stijn Vansummeren
Marianne Winslett
Man Lung Yiu
Jeffrey Xu Yu
Xiaofang Zhou

External Reviewers

Spiridon Bakiras
Ilaria Bartolini
Jeffrey Bergamini
Angelo Brayner
Alexander Brodsky
Lijun Chang
Jinchuan Chen
Zhihong Chong
Chi-Yin Chow

Maria Damiani
Ke Deng
Alexandre de Spindler
Cédric du Mouza
Fatima Farag
Conny Franke
Dario Freni
Gabriel Fung
Antoon Goderis

Hoyoung Jeung
Mohamed Khalefa
Jay Kola
Ashish Kundu
Justin Levandoski
Jiangtian Li
Xiang Lian
Erietta Liarou
Jessica Lin
Dan Lin
Heshan Lin
Dekang Lin
Yimin Lin
Aretusa Lopes
Saravanan Muthaiyah
Francesco M. Malvestuto
Alex Markowetz
Sergio Mascetti
Rimma Nehme
Niels Nes
Moira Norrie
Stavros Papadopoulos

Linda Pareschi
Michalis Potamias
Belen Prados Suarez
Daniele Riboni
Carols Rueda
Lefteris Sidirourgos
Shaoxu Song
Kurt Stockinger
Arash Termehchy
Sonny Vaupel
Quang Hieu Vu
Changliang Wang
Leanne Wu
Xike Xie
Jongpil Yoon
Ting Yu
Cammie Zhuang
Charles Zhang
Jianting Zhang
Yang Zhang
Zhe Zhang
Lei Zou

Table of Contents

Searching and Mining Graphs

Data Streams

Scientific Database Applications

Advanced Indexing Methods

Data Mining

Advanced Queries and Uncertain Data

Short Presentations

Poster and Demonstration Papers

New Challenges in Petascale Scientific Databases
(Keynote Talk)

Alexander Szalay

Department of Physics and Astronomy
The Johns Hopkins University
3701 San Martin Drive
Baltimore, MD 21218
szalay@jhu.edu

Abstract. Scientific data is doubling every year. Virtual Observatories are established over every scale of the physical world: from elementary particles to materials, biological systems, environmental observatories, remote sensing, and the universe. These collaborations collect increasing amounts of data, often close to a rate of petabytes per year. Many scientists will soon obtain most of their data from large scientific repositories of data, often stored in the form of databases. The talk will discuss the different requirements for such databases, and discuss user behavior in a few concrete examples taken from astronomy, in particular from the 6 year usage of the Sloan Digital Sky Survey database. Interesting query patterns are emerging, where users create custom "crawlers" to break large queries into many repetitive ones. The trial-and-error behavior of many exploratory projects will be also discussed. The talk will also present various scalable alternatives to large scientific analysis facilities.

About the Speaker. Alexander Szalay is the Alumni Centennial Professor of Astronomy at the Johns Hopkins University. He is also Professor in the Department of Computer Science. He is a cosmologist, working on the statistical measures of the spatial distribution of galaxies and galaxy formation. He was born and educated in Hungary. After graduation he spent postdoctoral periods at UC Berkeley and the University of Chicago, before accepting a faculty position at Johns Hopkins. In 1990 he has been elected to the Hungarian Academy of Sciences as a Corresponding Member. He is the architect for the Science Archive of the Sloan Digital Sky Survey. He is Project Director of the NSF-funded National Virtual Observatory. He has written over 340 papers in various scientific journals, covering areas from theoretical cosmology to observational astronomy, spatial statistics and computer science. In 2003 he was elected as a Fellow of the American Academy of Arts and Sciences. In 2004 he received an Alexander Von Humboldt Prize in Physical Sciences, in 2008 a Microsoft Award for Technical Computing.

B. Ludäscher and Nikos Mamoulis (Eds.): SSDBM 2008, LNCS 5069, p. 1, 2008.

Adventures in the Blogosphere
(Keynote Talk)

Nick Koudas

Department of Computer Science
Bahen Center for Information Technology
University of Toronto
40 St. George Street Rm BA5240
Toronto ON M5S 2E4
koudas@cs.toronto.edu

Abstract. Blogs, social networks, wikis and microblogging are prolif-
erating at unprecedented pace. The numbers reported quantifying user
engagement are profound. In this talk, I will present BlogScope
(www.blogscope.net) a system under development at the University of
Toronto, that aims to collect, process and distill in real time the information in social media. I will present the system, its architecture the difficulties encountered and highlight the various research challenges in building
the various components of the system. I will also present, Grapevine,
BlogScope's sister project that aims to make sense in real time of the
social media space. I will detail areas of research related to the scope of
these projects and present challenges that could be addressed via the utilization of scientific and statistical database techniques. If time permits,
I'll present demos.

About the Speaker. Nick Koudas is an associate professor at the University
of Toronto. He was a principal member of technical staff at AT&T labs and an
adjunct professor at Columbia University. He holds a PhD from the University of
Toronto. His research interests are in data management, managing information
at web scale, indexing, algorithms and information mining.

B. Ludäscher and Nikos Mamoulis (Eds.): SSDBM 2008, LNCS 5069, p. 2, 2008.
© Springer-Verlag Berlin Heidelberg 2008

The Evolution of Vertical Database Architectures – A Historical Review
(Keynote Talk)

Per Svensson

Dept. of Decision Support Systems, Swedish Defence Research Agency,
SE 164 90 Stockholm, Sweden
per.svensson@foi.se

1 Background

My intention in this lecture is to discuss the evolution of key concepts behind today's emerging vertical database architectures. The Cantor project [5, 7] pioneered the analysis and coordinated application of many of these concepts in relational systems, which is one reason why references to this work are a recurring theme in what follows. The other reason is that although the work was duly reported in reasonably well-known conference publications, it has left no trace in citations. Thus, from a strictly evolutionary perspective, Cantor was a dead branch which left no progeny, but from a historical perspective it might still provide useful lessons.

Transposed files as such were used in a number of early non-relational data base systems, mostly intended for statistical or scientific applications. A fairly comprehensive list of such systems was given by the paper [6] which is cited below. One great conceptual step that is now being taken is the realization that the adoption of transposed files opens a whole range of architectural opportunities. By careful combination of these opportunities dramatic performance gains may be provided, in particular of course when systems are used in those statistical and analytical kinds of application for which the concept was originally developed.

So what do these architectural opportunities consist of? Below is a list, however, due to space limitations it is not possible here to give a fair account of all of them:

1. column-wise storage of data, or *fully transposed files*
2. use of ordering
3. use of various kinds of "light-weight" data compression: RLE, minimum byte size, dictionary encoding, differencing
4. dynamically optimized combinations of these and other compression techniques
5. use of run-length encoding (RLE) for columns that are ordered
6. lazy decompression of data
7. use of vectorized method interfaces to reduce call overhead costs
8. special method interfaces for accessing RLE-coded data to enable higher-level search and join operations to work directly on compressed data when available
9. use of B-tree variants or other techniques to store and retrieve variable-length data in columns
10. conjunctive search and join algorithms working directly on ordered, RLE-compressed data

B. Ludäscher and Nikos Mamoulis (Eds.): SSDBM 2008, LNCS 5069, pp. 3–5, 2008.
© Springer-Verlag Berlin Heidelberg 2008

11. use of the vectorized data flow network architecture paradigm and vectorized operations on data streams, to allow efficient query evaluation by interpretation of algebraic expressions rather than by compilation to low-level code.

Today's experiments and analyses are usually better planned and executed than those of the early days. It is therefore at least possible that the current research interest in vertical architectures will result in a better-founded kind of consensus than was achieved, and criticized in [6], see below, in the early 80´s for the standard tabular scheme for storage of relations.

2   The Effects of Modern Processor Architectures

The research group behind the Monet DBMS [9, 10] has made thorough analyses of the effect of modern computer hardware architectures on data base performance. In [9] a detailed discussion is presented of the impact of modern computer architectures, in particular with respect to their use of multi-level cache memories to alleviate the widening gap between DRAM and CPU speeds that has been a characteristic for computer hardware evolution since the late 70's. They show that it is progressively less appropriate to think of the main memory of a computer system as "random access" memory, and that accessing data sequentially also in main memory may provide significant performance advantages.

3   Transposed Files and Decomposed Storage Models

The term transposed file was used in early papers, such as [2,3,4], to denote what is today usually called "vertically fragmented" or "vertically decomposed" data structures [9], "vertical partitioning", "column-oriented" data bases [12] or "column store" data bases [11]. In my opinion, there is a need for more terminological consistency here.

The first published paper on transposed files and related structures that is widely recognized in recent literature is [6]. While the paper notes that some database systems use a fully transposed storage model, in this paper the *fully decomposed storage model* (DSM) is described.

A DSM is a *[fully] transposed storage model with surrogates included.* The authors conclude: "There seems to be a general consensus among the database community that the [conventional] n-ary approach is better. ... Instead, we claim that the consensus opinion is not well founded and that neither is clearly better until a closer analysis is made."

4   Data Compression

Two recent papers on the use of data compression in relational data bases are [8, 12]. The most significant advantages are obtained when combining data compression with ordered, fully transposed file or DSM storage, but there are also approaches which use compression for row-oriented storage schemes. The paper [12] also addresses querying compressed data and is briefly discussed in the lecture. In [7], Cantor's approach to data compression is described. It presupposes the existence of an efficient way to organize attribute sub-files containing varying length data.

5 Conclusions

Based on a literature review, it appears that most of the advantages of vertical storage in databases for analytical purposes have been known and exploited since the early 80's at least. As an early contributor to this technology, the author is happy to see a previous lack of interest at last reverse into what might be seen as a canonical vertical architecture, replacing the previous ill-founded consensus around the "flat file with indexes" approach.

About the Speaker. Per Svensson has been with the Swedish Defence Research Agency (FOI, previously FOA) since 1973 and is a Research Director since 1987. He was an Adjunct Professor of Scientific and Statistical Database Management at the Royal Institute of Technology (KTH), Department of Numerical Analysis and Computer Science, from 1996 to 2002. Previous employments include IBM Sweden and the Royal Institute of Technology (KTH). Dr. Svensson authored or co-authored 30 internationally published scientific papers, as well as about 25 technical reports in Swedish or English. He is the editor and one of the authors of the successful HiTS/ISAC proposal to EU PASR 2005.

References

1. Svensson, P.: Performance evaluation of a prototype relational data base handler for technical and scientific data processing. FOA Rapport C20281-D8. Swedish National Defence Research Institute, Stockholm (1978)
2. Batory, D.S.: On Searching Transposed Files. ACM TODS 4(4), 531–544 (1979)
3. Svensson, P.: On Search Performance for Conjunctive Queries in Compressed, Fully Transposed Ordered Files. In: Proc. VLDB 1979, pp. 155–163 (1979)
4. Turner, M.J., Hammond, R., Cotton, P.: A DBMS for Large Statistical Databases. In: Proc. VLDB 1979, pp. 319–327 (1979)
5. Karasalo, I., Svensson, P.: An overview of Cantor – a new system for data analysis. In: Proc. 2nd SSDBM (1983)
6. Copeland, G.P., Khoshafian, S.N.: A decomposition storage model. In: Proc. 1985 SIGMOD Conf. ACM, New York (1985)
7. Karasalo, I., Svensson, P.: The design of Cantor – a new system for data analysis. In: Proc. 3rd SSDBM (1986)
8. Westmann, T., Kossmann, D., Helmer, S., Moerkotte, G.: The implementation and performance of compressed databases. SIGMOD Record 29(3), 55–67 (2000)
9. Manegold, S., Boncz, P.A., Kersten, M.L.: Optimizing database architecture for the new bottleneck: memory access. VLDB Journal 9(3), 231–246 (2000)
10. Boncz, P., Zukowski, M., Nes, N.: MonetDB/X100: Hyper-pipelining query execution. In: Proc. of 2005 CIDR Conference, VLDB Endowment (2005)
11. Stonebraker, M., et al.: C-Store: A Column-oriented DBMS. In: Proc. VLDB 2005, pp. 553–564 (2005)
12. Abadi, D.J., Madden, S., Ferreira, M.C.: Integrating compression and execution in column-oriented database systems. In: Proc. of the 2006 SIGMOD Conf. ACM, New York (2006)

Linked Bernoulli Synopses: Sampling along Foreign Keys

Rainer Gemulla, Philipp Rösch, and Wolfgang Lehner

Database Technology Group
Technische Universität Dresden, Germany
{gemulla,roesch,lehner}@inf.tu-dresden.de

Abstract. Random sampling is a popular technique for providing fast approximate query answers, especially in data warehouse environments. Compared to other types of synopses, random sampling bears the advantage of retaining the dataset's dimensionality; it also associates probabilistic error bounds with the query results. Most of the available sampling techniques focus on table-level sampling, that is, they produce a sample of only a single database table. Queries that contain joins over multiple tables cannot be answered with such samples because join results on random samples are often small and skewed. On the contrary, schema-level sampling techniques by design support queries containing joins. In this paper, we introduce *Linked Bernoulli Synopses*, a schema-level sampling scheme based upon the well-known *Join Synopses*. Both schemes rely on the idea of maintaining foreign-key integrity in the synopses; they are therefore suited to process queries containing arbitrary foreign-key joins. In contrast to Join Synopses, however, Linked Bernoulli Synopses correlate the sampling processes of the different tables in the database so as to minimize the space overhead, without destroying the uniformity of the individual samples. We also discuss how to compute Linked Bernoulli Synopses which maximize the effective sampling fraction for a given memory budget. The computation of the optimum solution is often computationally prohibitive so that approximate solutions are needed. We propose a simple heuristic approach which is fast and seems to produce close-to-optimum results in practice. We conclude the paper with an evaluation of our methods on both synthetic and real-world datasets.

1 Introduction

With the huge amount of data stored in current data warehouse environments, it is impracticable to execute queries directly on the database when human interaction is involved. This applies to explorative queries in data mining and decision-support tasks, which are used as a precursor to more complex analysis tasks; the main goal is to determine which methods and parts of data that are likely to produce interesting results and which will not. It also applies to OLAP and report design, where approximate query processing is able to significantly increase the responsiveness of the system and therefore the productiveness of

B. Ludäscher and Nikos Mamoulis (Eds.): SSDBM 2008, LNCS 5069, pp. 6–23, 2008.

its users. In all these scenarios, random sampling has proven to be a valuable tool for database summarization. Compared to other types of synopses, random sampling is easy to implement and use, it supports a broad range of queries (including grouping) and it provides probabilistic error bounds.

The main problem with most of the available database sampling schemes is that they focus on only a single table in the database; we refer to these schemes as *table-level sampling schemes*. Queries that contain joins between multiple tables are problematic because joins between random samples in general do not result in a random sample of the join of the respective tables [1,2,3]. To avoid this problem, it is crucial that relationships between multiple tables be known and exploited in the sampling process itself. This is accomplished by *schema-level sampling schemes*: these schemes sample multiple relations at once in such a way that it is possible to use the resulting samples to compute results of queries with joins. In this paper, we restrict attention to foreign-key joins. Such joins are ubiquitous in data warehouse scenarios, where most queries join a fact table with multiple dimension tables along predefined foreign-key paths.

The problem of schema-level sampling becomes tractable when only foreign-key joins are of interest. Indeed, Acharya et al. [2] have shown that it is sufficient to maintain a single sample per table to support any potential foreign-key join. The key idea underlying their Join Synopses is to 1) take a sample of each table, 2) join each sample with all tables to which it has foreign keys and 3) store the joined samples in the synopsis. For example, consider a schema with two tables R_1 and R_2, where R_1 has a foreign key to R_2. Let S_1 and S_2 be uniform random samples of R_1 and R_2, respectively. The Join Synopsis then consists of the two samples $S_1 \bowtie R_2$ and S_2. Observe that—by projection on the attributes of R_1—S_1 can be reconstructed from $S_1 \bowtie R_2$; there is a 1:1 relationship between the tuples in these two relations. The synopsis can be used to answer queries on R_1, on R_2 as well as on $R_1 \bowtie R_2$. To reduce the space requirement of the sampling scheme, [2] also suggest to renormalize the join results. After renormalization, the synopsis contains three "samples": S_1, $R_2 \ltimes S_1$ and S_2. Observe that both $R_2 \ltimes S_1$ and S_2 contain tuples from R_2. If samples S_1 and S_2 have a size of, say, $10,000$ tuples each, the entire synopsis consists of up to $30,000$ tuples—a space overhead of 50%.

In this paper, we develop a new schema-level sampling scheme called *Linked Bernoulli Synopses* (LBS), which reduces the space overhead incurred by Join Synopses. In expectation, the size of LBS is at most as large as the size of the corresponding Join Synopsis; it is often much smaller. The key idea behind LBS is to correlate the sampling processes of S_1 and S_2, while maintaining the uniformity of both samples. Intuitively, given a sample of S_1, we try to reuse as many tuples from $R_2 \ltimes S_1$ as possible for the sample S_2. Our synopses are optimal, that is, it is impossible to find a sampling scheme which produces smaller synopses (with the same sampling fractions for each table) in expectation. For example, when the cardinalities of R_1 and R_2 are equal and there is a 1:1 relationship between both tables (the best case), we require a space budget of $20,000$ tuples to sample $10,000$ tuples from each table; the overhead is reduced to 0%.

We also address the problem of computing a Linked Bernoulli Synopsis which fits into a given space budget. The problem is challenging because all the sampling steps are correlated; changing the size of one sample might change the size of many others. We treat the problem as an optimization problem and show how it can be solved numerically using results from convex optimization. Finding the optimum solution requires time exponential to the number of relations and linear to their size; approximate solutions are therefore key to the practicability of our methods. In fact, we found that a simple heuristic seems to produce near-optimal results in practice, so that it might be unnecessary to run the entire optimization.

The remainder of the paper is structured as follows: In Section 2, we review Join Synopses in more detail and show how their space consumption can be reduced with a simple modification. We then analyze the modified sampling scheme in terms of (expected) space consumption. In Section 3, we introduce and analyze Linked Bernoulli Synopses. Section 4 discusses the problem of allocating the available space to the different tables in the schema. Preliminary results of an evaluation on synthetic and real-world datasets are presented in Section 5. Section 6 gives a brief overview of related work and we conclude the paper with a summary of our results in Section 7.

2   Preliminaries

In this section, we review, discuss and analyze Join Synopses. The results presented in this section drive the design of our Linked Bernoulli Synopsis in Section 3.

2.1   Notation

We start by summarizing the notation used throughout this paper. Let $G = (V, E)$ be a schema graph of a relational database with V being the set of vertices and E being a set of directed edges. In more detail, V is the set of tables in the database, while the set $E \subseteq V \times V$ describes foreign-key relationships. An element $(R_1, R_2) \in E$ with $R_1, R_2 \in V$ represents a foreign-key relationship from R_1 to R_2. R_1 is called parent table or predecessor, while R_2 is called child table or successor. For brevity, we write $R_1 \rightarrow R_2$ whenever $(R_1, R_2) \in E$. Moreover, we will use $\Rightarrow$ to denote the transitive closure over $\rightarrow$; $\nrightarrow$ and $\nRightarrow$ denote the inverse of $\rightarrow$ and $\Rightarrow$, respectively. The function $\mathrm{pk}_R(t)$ determines the primary key of a tuple $t \in R$, and $\mathrm{fk}_{R_1 \rightarrow R_2}(t)$ determines the foreign key of a tuple $t \in R_1$ to table R_2. Thus, when $R_1 \rightarrow R_2$, two tuples $t_1 \in R_1$ and $t_2 \in R_2$ join if $\mathrm{fk}_{R_1 \rightarrow R_2}(t_1) = \mathrm{pk}_{R_2}(t_2)$. In this case, we say that t_1 references t_2.

Figure 1 shows the example database that we will use as our running example throughout the paper. The database consists of three tables A, B and C; foreign keys are defined between $A.FK$ and $B.PK$ as well as $B.FK$ and $C.PK$. The foreign-key relationships are graphically encoded using arrows between matching tuples. Using the notation above, we have $V = \{A, B, C\}$ and $E = \{(A, B), (B, C)\}$.

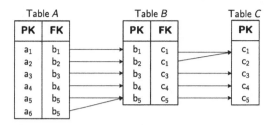

Fig. 1. An example database with 3 tables ($A \to B$ and $B \to C$)

We subsequently assume that the schema graph G is free of cycles, that is, there is no table R with $R \Rightarrow R$. The reason is that—when the schema graph contains cycles—the inclusion of even a single tuple from one of the relations in the cycle might lead to an explosion of the synopsis size. Fortunately, in the setting of data warehouses we are concerned with, cycles in the schema graph do not occur. We also assume that G does not contain multiple edges between two tables; this is not a limitation of our approach but simplifies explanation.

2.2   Join Synopses Revisited

In the following, we take a slightly different view on Join Synopses by incorporating them into the more general concept of schema synopses. In more detail, the *schema synopsis* Ψ_G of a schema graph G consists of a table synopsis for each table in the schema. For brevity, we will omit the schema graph G when referring to the schema synopsis. The *table synopsis* Ψ_R consists of a uniform sample Sample_R and a reference table $\mathrm{RefTable}_R$, both containing items from R. The sample is primarily used for query evaluation, while the reference table is used to maintain foreign-key integrity. In general, the reference table contains all tuples from R which (1) are referenced by a table synopsis of a predecessor of R and (2) are not stored in the sample already. As a matter of notation, we say $t \in \Psi_R$ whenever $t \in \mathrm{Sample}_R \cup \mathrm{RefTable}_R$.

Join Synopses can now be viewed as a special kind of schema synopsis. The algorithm is as follows [2]:

1. Sample each relation independently using Bernoulli sampling. In Bernoulli sampling with sampling rate q, each tuple is included into the sample with probability q and excluded with probability $1 - q$; the process is repeated independently for each tuple. The parameter q essentially controls the desired sample size; see Section 4.
2. Fill the reference tables so that foreign-key integrity is restored. The tables are processed top-down, that is, a table R is processed only after all its predecessors have been processed already. Note that the second step slightly differs from the original Join Synopses because we do not store the same tuple in both the sample and the reference table. For large tables, the resulting space reduction is usually negligible for Join Synopses but sets the foundation for the space reduction possible with Linked Bernoulli Synopses.

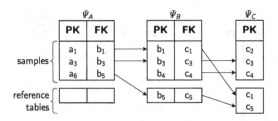

Fig. 2. Join Synopsis with a sampling rate of 50%

Note that both phases of the above algorithm can be interweaved, that is, we can compute both the sample and the reference table of each table in a single scan.

An example for a Join Synopsis of our example database is given in Figure 2. We used a sampling rate of $q = 50\%$ for each table. The reference tables are given below the individual samples. As can be seen, reference tables restore foreign-key integrity. For example, tuple $(a_6, b_5) \in A$ references $(b_5, c_5) \in B$, but the latter is not stored in the sample so that it has been included in the reference table.

2.3 Analysis of Join Synopses

We now analyze the space consumption of the modified Join Synopses described above.[1] Let $R_1, R_2, \ldots, R_k$ be the direct predecessors of a table R. During Join Synopses computation, R is processed after the processing of $R_1, \ldots, R_k$ has completed. Now, let isRef$_{R_i \to R} : R \to \{\text{true}, \text{false}\}$ for $1 \le i \le k$ be a function which evaluates to true if a tuple $t \in R$ has been referenced by Ψ_{R_i} and to false otherwise. Also set

$$\text{isRef}_R(t) = \text{isRef}_{R_1 \to R}(t) \vee \text{isRef}_{R_2 \to R}(t) \vee \cdots \vee \text{isRef}_{R_k \to R}(t),$$

so that isRef$_R(t)$ evaluates to true if t is referenced by any predecessor.

To determine the space consumption of the Join Synopsis algorithm, we view isRef$_R(t)$ as a random function (over the choices made when sampling the predecessors). Our goal is to compute the *reference probability* pRef$_R(t)$ that tuple t is referenced from any predecessor of R. Assume for a moment that pRef$_R(t)$ is known for a tuple $t \in R$. The *selection probability* pSel$_R(t)$ that tuple t is included into the synopsis—either by acception into the sample or by addition to the reference table—is then given by

$$\text{pSel}_R(t) = P(t \in \Psi_R) = P(t \in \text{Sample}_R \vee t \in \text{RefTable}_R)$$
$$= P(t \in \text{Sample}_R) + P(t \notin \text{Sample}_R)P(t \in \text{RefTable}_R \,|\, t \notin \text{Sample}_R)$$
$$= q_R + (1 - q_R)\,\text{pRef}_R(t),$$

where q_R is the sampling fraction used for table R. Here, the final equality follows from the independence of the sampling step and the event that the tuple

---

[1] The analysis is fair with respect to the original algorithms because our modification of the Join Synopsis algorithm leads to a decrease of the space required to store the synopsis.

is being referenced. We now have all the information we need to compute reference probabilities. Suppose that table R has predecessor R' and denote by $\text{pRef}_{R' \to R}(t)$ the probability that a tuple $t \in R$ is referenced by the synopsis of R'. Since the tuples in R' are sampled independently from each other, this probability is given by:

$$\text{pRef}_{R' \to R}(t) = 1 - \prod_{\substack{t' \in R' \\ \text{fk}_{R' \to R}(t') = \text{pk}_R(t)}} (1 - \text{pSel}_{R'}(t')).$$

Note that $\text{pRef}_{R' \to R}(t)$ can be computed incrementally, that is, with only a single scan of R'. To see this, suppose that we have already processed a subset R'_0 of the tuples in R' and let $t^+ \in R' \setminus R'_0$ be the currently processed tuple. If t^+ does not reference t, we can simply ignore it. Otherwise, we can use the following relationship

$$\text{pRef}_{R'_0 \cup \{t^+\} \to R}(t) = 1 - \prod_{\substack{t' \in R'_0 \cup \{t^+\} \\ \text{fk}_{R' \to R}(t') = \text{pk}_R(t)}} (1 - \text{pSel}_{R'}(t'))$$

$$= 1 - \left(1 - \text{pSel}_{R'}(t^+)\right) \prod_{\substack{t' \in R'_0 \\ \text{fk}_{R' \to R}(t') = \text{pk}_R(t)}} (1 - \text{pSel}_{R'}(t'))$$

$$= 1 - \left(1 - \text{pSel}_{R'}(t^+)\right) \left(1 - \text{pRef}_{R'_0 \to R}(t)\right) \tag{1}$$

to update the reference probability. If R' is the sole predecessor of R, then $\text{pRef}_R(t) = \text{pRef}_{R' \to R}(t)$. The discussion of the computation of $\text{pRef}_R(t)$ for tables with multiple predecessors is deferred to Section 3.

Figure 3a shows the selection and reference probabilities for Join Synopses with a sampling fraction of $q = 50\%$ for each table. The reference probabilities are annotated on the arrows, while the selection probabilities are given right next to each tuple (rounded to one digit after decimal point). As can be seen, the selection probabilities—which effectively determine the size of the synopsis— are larger than both the sampling fraction q and the reference probability. The reason is that the sampling steps for each table are performed independently of each other.

Given the sampling fraction q_R, the expected size of the table synopsis of R is given by

$$E[|\Psi_R|] = \sum_{t \in R} \text{pSel}_R(t) = q_R |R| + (1 - q_R) \sum_{t \in R} \text{pRef}_R(t).$$

In expectation, the entire Join Synopsis consists of

$$E[|\Psi|] = \sum_{R \in V} E[|\Psi_R|] \tag{2}$$

tuples by the linearity of expected value. In our example, we have $E[|\Psi_A|] = 3$, $E[|\Psi_B|] \approx 3.88$, $E[|\Psi_C|] \approx 4.16$. The expected total synopsis size $E[|\Psi|]$ is 11.04 tuples.

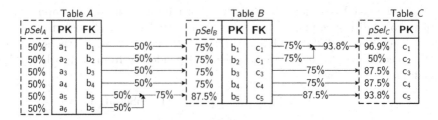

(a) Join Synopsis

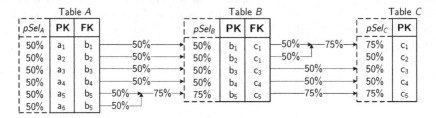

(b) Linked Bernoulli Synopsis

Fig. 3. Reference and selection probabilities for $q = 50\%$. Reference probabilities are annotated on the arrows; tuples with no incoming edges have zero reference probability.

3   Linked Bernoulli Synopses

Linked Bernoulli Synopses are based on Join Synopses, but the synopsis computation is entirely different. The key observation leading to our LBS is that the event of a tuple being referenced by a predecessor already contains some randomness, which can be exploited for sampling. If a tuple is referenced, we *have to* include it into either the sample or the reference table; that is, we have to store it anyway. A tuple in a sample, however, is more valuable because it can be used directly for query answering: larger samples lead to better results. The tuples in the reference tables can be seen as overhead because they are "only" used to preserve foreign-key integrity. Thus, we would like to bias the sample towards referenced tuples, so that the overhead in the reference tables is minimized. LBS perform this biasing in such a way that the resulting sample is still uniform, so that it can be used for query processing as before. In contrast to Join Synopses, however, the samples in LBS are correlated. This correlation does not lead to problems at query time because only one sample is used to answer each query (the sample of the base table of the join) and each sample on its own is a uniform random sample of the respective table.

3.1   Algorithmic Description

We are now ready to describe the computation of Linked Bernoulli Synopses in full detail. The general procedure is as follows:

1. Scan the tables in top-down order, that is, whenever a table R is processed, all its predecessors must have been processed already.
2. For each tuple t, decide whether or not the tuple is included into either the sample or the reference table. This decision is based on (1) the reference probability $\mathrm{pRef}_R(t)$ of the tuple and (2) the fact of whether or not the tuple is referenced by a predecessor ($\mathrm{isRef}_R(t)$ is true). Simultaneously, compute (or update) the reference probabilities for every successor of R using equation (1).

The crux of LBS lies in step 2, where we make use of $\mathrm{pRef}_R(t)$ and $\mathrm{isRef}_R(t)$ to drive the sampling process. We now discuss this step in more detail. The key idea is to compare the reference probability $\mathrm{pRef}_R(t)$ with the desired sampling fraction q_R. For each tuple t, there are three cases:

Case 1: $\mathrm{pRef}_R(t) = q_R$, that is, the reference probability and the sampling fraction are equal. In this case, we add the tuple to the sample if and only if it is referenced. Otherwise the tuple is ignored. It follows immediately that $P(t \in \mathrm{Sample}_R) = q_R$.

Case 2: $\mathrm{pRef}_R(t) < q_R$, that is, the reference probability is smaller than the sampling fraction. We directly add t to the sample whenever it is referenced. When t is not referenced, we add it to the sample with probability $(q_R - \mathrm{pRef}_R(t)) / (1 - \mathrm{pRef}_R(t))$ or ignore it otherwise. The probability that t is included into the sample is given by:

$$P(t \in \mathrm{Sample}_R) = P(\mathrm{isRef}_R(t) = \mathrm{true}) + P(\mathrm{isRef}_R(t) = \mathrm{false}) \frac{q_R - \mathrm{pRef}_R(t)}{1 - \mathrm{pRef}_R(t)}$$

$$= \mathrm{pRef}_R(t) + (1 - \mathrm{pRef}_R(t)) \frac{q_R - \mathrm{pRef}_R(t)}{1 - \mathrm{pRef}_R(t)} = q_R.$$

Case 3: $\mathrm{pRef}_R(t) > q_R$, that is, the reference probability is larger than the sampling fraction. If t is not referenced, we can safely ignore it. If t is referenced, we add it to the sample with probability $q_R / \mathrm{pRef}_R(t)$ or to the reference table otherwise. The probability that t is added to the sample is:

$$P(t \in \mathrm{Sample}_R) = P(\mathrm{isRef}_R(t) = \mathrm{true}) \frac{q_R}{\mathrm{pRef}_R(t)} = \mathrm{pRef}_R(t) \frac{q_R}{\mathrm{pRef}_R(t)} = q_R.$$

This case is the "bad case" because there is a non-zero probability that t is added to the reference table.

To sum up, the tuple is included into the sample with the desired probability of q_R in each of the three cases. Since both the inclusion/exclusion decisions as well as the event of being referenced are independent among the tuples, the algorithm produces a Bernoulli sample with sampling rate q_R. The selection probability $\mathrm{pSel}_R(t)$, that is, the probability that a tuple t is stored in either the sample or the reference table, is given by:

$$\mathrm{pSel}_R(t) = \max\{q, \mathrm{pRef}_R(t)\}.$$

The reference probabilities for LBS are computed from the selection probabilities as before.

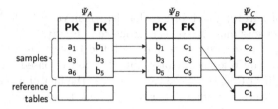

Fig. 4. Linked Bernoulli Synopsis with a sampling rate of 50%

3.2   Example and Analysis

Figure 3b shows the selection and reference probabilities for LBS with a sampling rate of $q = 50\%$ for each table. By inspection, one finds that—for tables B and C—the selection probabilities for LBS are lower than the selection probabilities of Join Synopses; we will formalize this observation below. The expected synopsis sizes are: $E[|\Psi_A|] = 3$, $E[|\Psi_B|] = 2.75$, $E[|\Psi_C|] = 3$ and therefore $E[|\Psi|] = 8.75$ (instead of 11.03).

A potential LBS is shown in Figure 4. Compared to Join Synopses (Figure 2), there is no difference in the sample of table A because A does not have any predecessors (all tuples trivially belong to case 2). When sampling table B, we compare the desired sampling rate of 50% with the reference probabilities given in Figure 3b. Tuples b_1 through b_4 all belong to case 1 above, that is, they are included if and only if they are referenced. In the example, this holds true for only b_1 and b_3; b_2 and b_4 are ignored. Note that—given the sample of table A—this process is entirely deterministic. For tuple b_5, the reference probability of 75% is larger than the desired sampling fraction; this is case 3 and—since b_5 is referenced—it is accepted into the sample with probability 2/3 (as in the example) or in the reference table with probability 1/3. Continuing with table C, both c_1 and c_5 belong to case 3. Both tuples are referenced; c_1 has been added to the reference table (probability of acceptance into sample: $\approx 52\%$) and c_5 to the sample ($\approx 53\%$). Tuples c_3 and c_4 belong to case 1, but only c_3 is referenced and therefore added to the sample. Finally, tuple c_2 belongs to case 2 because it has a reference probability of zero. It is accepted into the sample with a probability of 50%.

We now compare formally the selection probability of a tuple using Join Synopses with the selection probability using LBS when using the same sampling rate q_R for both. Assuming that the reference probabilities are the same for both approaches, we have

$$\mathrm{pSel}_R^{\mathrm{JS}}(t) - \mathrm{pSel}_R^{\mathrm{LBS}}(t) = q + (1-q)\,\mathrm{pRef}_R(t) - \max\{q, \mathrm{pRef}_R(t)\}$$
$$= (1-q)\,\mathrm{pRef}_R(t) - \max\{0, \mathrm{pRef}_R(t) - q\}$$
$$= \begin{cases} (1-q)\,\mathrm{pRef}_R(t) & \mathrm{pRef}_R(t) \le q \\ q(1 - \mathrm{pRef}_R(t)) & \text{otherwise} \end{cases}$$
$$\ge 0.$$

For $0 < q, \mathrm{pRef}_R(t) < 1$ the inequality becomes strict. In general, the reference probabilities for Join Synopses and Linked Bernoulli Synopses will be different.

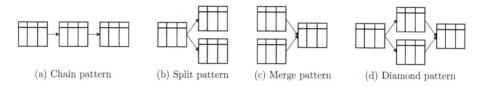

(a) Chain pattern   (b) Split pattern   (c) Merge pattern   (d) Diamond pattern

Fig. 5. Reference patterns

Using the argument above, it is straightforward to show that the reference probability of Linked Bernoulli Synopses is always smaller than or equal to that of Join Synopses.

3.3 Handling Multiple Predecessors

Until now, we have assumed that every table has at most one predecessor. This assumption is clearly too restrictive in practice. In this section, we discuss how to handle tables with multiple predecessors. The key question we are going to answer is how to compute the reference probability $\mathrm{pRef}_R(t)$ from the reference probabilities $\mathrm{pRef}_{R_i \to R}$ (with R_i being a predecessor of R). To that extent, we distinguish the 4 possible patterns shown in Figure 5. As before, we assume that the schema graph is free of cycles.

In a *chain pattern*, each table has at most one predecessor and at most one successor. This is exactly the situation we looked at in the preceding sections. In a *split pattern*, each table has at most one predecessor but arbitrarily many successors. Again, the formulas established in the preceding sections can be used directly. In a *merge pattern*, each table has arbitrarily many predecessors and at most one successor. Fix a table R and denote by $R_1, \ldots, R_k$ the direct predecessors of R. Since the schema graph is free of cycles, tables $R_1, \ldots, R_k$ do not have any common predecessor. They are therefore sampled independently of each other. It follows that the reference probabilities from each of the R_i to table R are independent and thus

$$\mathrm{pRef}_R(t) = 1 - \prod_{i=1}^{k} \left(1 - \mathrm{pRef}_{R_i \to R}(t)\right). \tag{3}$$

Finally, in a *diamond pattern*, at least two of the predecessors $R_1, \ldots, R_k$ of R share a common predecessor, say R_1 and R_2. In this case, the event that a tuple from R is referenced from R_1 and the event that the same tuple is referenced from R_2 are not necessarily independent. As a consequence, equation (3) cannot be used. In the remainder of this section, we have a closer look at the dependencies introduced in diamond patterns and propose possible workarounds.

We will use the example in Figure 6 for illustration purposes. The example shows 4 tables A, B, C and D, which are arranged in a diamond pattern. The sampling rate has been set to 50% for all tables. Suppose that tables A, B and C have been sampled already and that table D is about to be processed. There are two problems which might occur in this setting:

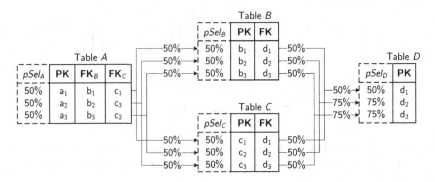

Fig. 6. Reference probabilities for a 50%-sample, dependent references

- *Reference probabilities.* The references to a tuple t might be dependent. In the example, tuple d_1 is referenced whenever either b_1 or c_1 is included into the synopses. The selection probabilities of b_1 and c_1 are 50% each, so that the application of formula (3) indicates that d_1 is referenced with a probability of 75%. However, since in a LBS the samples are correlated, both b_1 and c_1 are included into the sample of tables B and C, respectively, if and only if a_1 has been included in the sample of table A. This event occurs with 50% probability, so that the true reference probability of d_1 is given by $\mathrm{pRef}_D(d_1) = 50\%$.
- *Joint inclusion probabilities.* A more subtle problem is that of joint inclusion probabilities. In the example, both d_2 and d_3 are referenced with a probability of 75%. The references from tables B and C to tuple d_2 are independent, as are the references to tuple d_3. However, if one looks at both d_2 and d_3 simultaneously, one finds that d_2 is referenced whenever d_3 is and vice versa. As a consequence, the sample is biased towards the cases where (1) both d_2 and d_3 are sampled and (2) neither d_2 nor d_3 is sampled. For example, if we ignore the dependencies between the references and proceed as in the previous sections, the joint inclusion probability of d_2 and d_3 is $\approx 44\%$ but should be 25%.

A trivial way of handling the above problems would be to store table D in its entirety. Though simple, this approach is viable in scenarios where table D is very small. For instance, in the schema of the TPC-H benchmark, the only table which is referenced in a diamond pattern is NATION, and this table consists of only 25 tuples. Otherwise, if table D is too large to store it in its entirety, we see two possible solutions: (1) switch back to Join Synopses for table D and all its successors and (2) decide on a per-tuple basis whether to switch back to the Join-Synopses way of sampling or not. The first solution might work well if table D does not have large successors; its main advantage is its simplicity. The second solution is more sophisticated and requires more bookkeeping, but it may reduce the overall space consumption significantly. We omit further details due to lack of space; a detailed description of the second solution can be found in [4].

4 Computing a Synopsis with a Memory Bound

In the previous section, we assumed that the desired sampling rate for each table is given beforehand. In practice, however, it might be difficult to decide on the values of the individual sampling fractions. A more realistic approach is to start with a space budget and to automatically set the sampling fractions so that the space budget is not exceeded and the sample sizes are maximized. To simplify the ongoing discussion, we assume that the space budget is given in number of tuples. That is, for a given budget M, the goal is to find sampling fractions which ensure that $|\Psi| \leq M$ with high probability.

4.1 An Optimization Problem

Suppose that the schema contains tables $R_1, \ldots, R_n$. Denote by $q_1, \ldots, q_n$ the sampling fraction used for each respective table, and let $q = (q_1, \ldots, q_n)$ denote a vector of these sampling fractions. There are many possible choices for q and we have to quantify which choices are considered good and which are not. Suppose that there is a function f so that $f(q) > f(q')$ whenever the sampling rates in q are considered more valuable than those in q'. We can now treat the problem as an optimization problem, that is, we want to find a vector q^* which maximizes the objective function $f(q)$ with respect to the constraint $g(q) \leq M$, where $g(q)$ encodes the (expected) space budget. Using these two functions, the optimization problem can be stated as:

> Maximize
> $$f(q_1, \ldots, q_n)$$
>
> with respect to
> $$0 < q_1, \ldots, q_n \leq 1$$
> $$g(q_1, \ldots, q_n) \leq M.$$

The function f can be derived from workload information or from information about the intended usage of the synopsis. In the absence of such information, a suitable choice for f is the geometric mean of the individual sampling fractions or, even simpler, its n-th power:

$$f_{GEO}(q_1, \ldots, q_n) = q_1 q_2 \cdots q_n.$$

Some insight into this choice for f is given in the next section. In any case, we make the assumption that both f and g are monotonically increasing functions of q. In other words, for any $\Delta q = (\Delta q_1, \ldots, \Delta q_n)$ with $\Delta q_i \geq 0$, we assume that

$$f(q + \Delta q) \geq f(q) \qquad \text{and} \qquad g(q + \Delta q) \geq g(q).$$

This assumption virtually always holds in practice because larger sampling rates lead to larger samples (f) which in turn lead to a larger synopsis size (g).

The monotonicity of f and g introduces structure into the optimization problem, which can be exploited for solving it. In [5], Tuy proposes an outer-approximation algorithm for monotonic optimization called the *polyblock algorithm*. The time complexity of the polyblock algorithm is exponential in the number of tables, so that it can only be used when the number of tables is not too large. But even when the number of tables is small, the polyblock algorithm requires frequent evaluations of the constraint function g. Exact computation of g according to equation (2) is expensive because a table scan of every table in the schema is required. Therefore, for large problem sizes, it is impractical to compute the optimum solution and approximate algorithms are needed.

4.2 A Heuristic Solution

Our heuristic solution is based on two simplifications. First, we do not compute g exactly but make use of a lower bound g_l for which a closed-form expression exists and which can be evaluated quickly without accessing the database. It is easy to see that

$$g_l(q_1, \ldots, q_n) = |R_1|q_1 + \cdots + |R_n|q_n$$

provides the desired lower bound; we simply ignore the size of the reference table. As a consequence, replacing g by g_l will produce oversized synopses. Depending on the data, this may or may not be significant; we show below that the size of the reference tables is often negligible when f_{GEO} is used as the objective function. Second, the optimum solution can be computed analytically for the combination of g_l and f_{GEO}. To see this, consider an n-dimensional hypercube with edges of length $q_1|R_1|, \ldots, q_n|R_n|$. Then, f_{GEO} is proportional to the volume of the hypercube, which in turn is maximized when all edges have equal length. It follows the f_{GEO} is maximized when

$$q_i \propto \frac{1}{|R_i|}$$

for $1 \leq i \leq n$.[2] We refer to this allocation scheme as *equi-size allocation* because—when the reference tables are ignored—the same number of tuples is sampled from every table in expectation. For traditional Join Synopses, the equi-size allocation scheme is known to produce good results [2].

In the following, we argue that the size of the reference tables is often negligible for equi-size allocation. To see this, consider two tables R_1 and R_2 with $R_1 \rightarrow R_2$ and set $r = |R_1|/(|R_1| + |R_2|)$. For a space budget of M tuples and equi-size allocation, we set $q_1 = rM/|R_1|$ and $q_2 = (1 - r)M/|R_2|$. There is no reference table for R_1, so that we focus on R_2. Recall that a tuple $t \in R_2$ is added to the reference table if and only if $\text{pRef}_{R_2}(t) > q_2$ (case 3 in Section 3). Perhaps surprisingly, all tuples from R_2 that are referenced up to $k = |R_1|/|R_2|$ times will not be added to the reference table. To see this, start from the Bernoulli

---

[2] When one of the q_i exceeds 1, we set it to 1 and repeat the process for the remaining sampling fractions.

inequality $(1 + x)^k > 1 + kx$ and set $x = -q_1 = -q_2/k$. It immediately follows that $1 - (1 - q_1)^k < q_2$. The expression on the left hand side is equal to $\text{pRef}_{R_2}(t)$ when t is referenced exactly k times; the inequality also holds when t is referenced fewer than k times. Thus, only tuples which are referenced $k + 1$ or more times have a non-zero chance of being included in the reference tables. There are at most $|R_1|/(k + 1)$ such tuples and only some of them are added to the reference table. Since dependent tables are typically smaller than their parents, the number of tuples in the reference table is expected to be low.

5   Experiments

We ran a variety of experiments in order to evaluate the effectiveness of Linked Bernoulli Synopses. Most of the experiments directly compare Join Synopses (JS) with Linked Bernoulli Synopses (LBS); the main issue we are trying to address is the extent to which LBS are able to reduce the overhead for storing reference tables. We also evaluate how close the equi-size allocation scheme comes to optimum allocation in terms of resulting sample sizes and query accuracy.

5.1   Experimental Setup

We implemented JS and LBS on top of DB2 using Java 1.6. The experiments were conducted on an Athlon AMD XP 3000+ system running Linux with 2 GB of main memory.

We make use of both synthetic and real-world data. The synthetic datasets are based on the TPC-H database of 1GB size. We used a Zipfian distribution for both values (prices, quantities, etc.) and foreign keys. We fixed the skew parameter for values to $z = 0.5$; the skew parameter for foreign keys is modified across the experiments. For our real-world experiments, we make use of the CDBS database[3]. The database contains information about radio and television broadcast services in the United States; only the 14 radio-related tables (without comment tables) were used in our experiments. The sizes of the tables range from $28,000$ to 1.4 million tuples.

5.2   Space Consumption

In a first set of experiments, we evaluated the effectiveness of LBS in comparison to JS on both synthetic and real-world datasets. We computed both synopses for various sampling fractions and datasets and recorded the space overhead required for the reference tables. The space overhead is defined as the size of the reference tables with respect to the size of just the samples, that is, we determine how much space is used for non-sample tuples. We used the equi-size allocation scheme throughout all the experiments.

---

[3] http://www.fcc.gov/mb/cdbs.html

Synthetic Data. Our experiments on synthetic datasets try to determine the key factors that influence the overhead of JS and LBS. We generated several TPC-H datasets with different parameters to examine the impact of skew in the foreign-key columns. We also experimented with varying synopsis sizes. For simplicity, we define the synopsis size as the size of just the sample part of the synopsis with respect to the size of the original tables.

Data skew. In a first experiment, we only consider the orders (O) and customer (C) table of the TPC-H schema. We varied the skew parameter of the foreign keys ($O \rightarrow C$) from 0 (uniformly distributed) to 1 (heavily skewed); each tuple of C is referenced at least once. We used a sampling fraction of 0.55% for O and 5.5% for C; these settings correspond to equi-size allocation with a synopsis size of 1%. The results are shown in Figure 7a. JS have a high overhead on uniformly distributed keys, but the overhead decreases with increasing skew. The reason for this behavior is that, when the skew is low, almost every tuple in the sample of O corresponds to a different customer, which in turn has to be added to the reference table. When the skew is high, however, a large subset of the orders are placed by only a small subset of the customers; the number of distinct foreign keys in the sample of O therefore decreases in expectation. The overhead of LBS is consistently smaller than the overhead of JS. For a skew value of $z = 0$, no reference tables are needed at all; see the discussion at the end of Section 4.2. With increasing skew, some tuples are referenced with a higher probability than their desired sampling rate (case 3), so that they are added to the reference tables from time to time. If the skew increases further, the number of referenced tuples decreases rapidly and the same effect as for JS can be observed.

Number of unreferenced tuples. In the next experiment, we proceeded as before but modified the fraction f of unreferenced customers. A fraction of $f = 40\%$ means that 40% of the customers did not place any order. For the remaining customers, we used a skew parameter of $z = 0.5$. Figure 7d plots the space overhead for various choices of f. For JS, the space overhead decreases as the f increases. The reason is that the number of distinct customers in the sample of O drops as f increases so that less space is required for reference tables. LBS performs better when the values of f are not too extreme. When the value of f increases, so does the space overhead because more and more customers are referenced with probability larger than the sampling fraction (case 3).

Number of tables. We next evaluated the impact of the number of tables. We started with just lineitem and orders and subsequently added customer, part-supp, part and supplier (in this order). The total size of the synopsis was set to 1%. The skew parameter was set to $z = 0.5$. The results are shown in Figure 7b. As can be seen, LBS outperform JS, especially when the number of tables is high. The reason is that an increasing number of tables lead to an increasing number of transitive references, which have to be stored in the reference tables. For Linked Bernoulli Synopses, this effect is reduced to a minimum.

Synopsis size. In a final experiment, we evaluated the impact of the synopsis size when sampling the 6 tables mentioned above. Figure 7c plots the space overhead

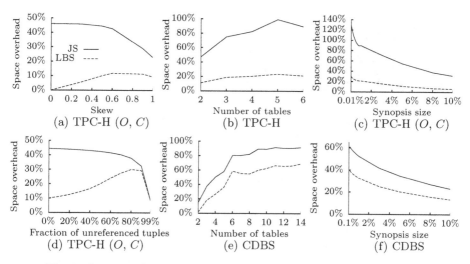

Fig. 7. Space overhead for Linked Bernoulli Synopses and Join Synopses

in dependency of the size of the sample part of the synopsis. As can be seen, the overhead decreases with increasing sample size because more and more tuples qualify for the sample and therefore do not have to be stored in the reference tables. Especially for small synopses, LBS have a significantly smaller overhead.

Real-world Data. We now report our results on the CDBS data. We modified the synopsis size between 0.1% and 10%. As can be seen in Figure 7f, the space overhead decreases with increasing synopsis size for both JS and LBS. The difference between the two is not as dramatic as it has been in the synthetic datasets because the CDBS tables contain large numbers of unreferenced tuples (up to 90%). Figure 7e shows the influence of the number of tables for a synopsis size of 1%. Again, LBS has lower overhead than JS.

5.3   Memory Bounds

In a final experiment, we computed synopses that fit into a prespecified amount of space. We used all tables of the TPC-H database; the nation and region table have been sampled entirely. The foreign-key skew was set to $z = 0.5$ and f_{GEO} was used as objective function. We ran three different combinations: JS with equi-size allocation (JS-ES), LBS with equi-size allocation (LBS-ES), and LBS with polyblock allocation (LBS-PB). In the latter case, we restricted the number of steps of the optimization algorithm to $1,000$; this corresponds to $1 - 2$ days of computation. In contrast, synopsis computation with equi-size allocation is a matter of minutes. The memory bound (samples and reference tables) was varied from 1% to 5%.

Figure 8 plots the objective function f_{GEO} for each combination and memory bound (log plot). The value of the objective function increases with an increasing synopsis size because more space is available for samples. In all cases, the

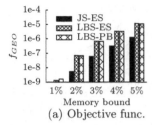

(a) Objective func.

	Q_1	Q_2	Q_3	Q_4
JS	3.51%	3.95%	3.28%	0.18%
LBS	2.69%	3.06%	2.43%	0.14%
	(-23.4%)	(-22.5%)	(-25.9%)	(-22.2%)

(b) Average relative error (1% size)

Fig. 8. Memory-bounded synopses (synthetic data)

LBS-based approaches achieve significantly larger values of the objective function than JS; they are between 8.9 to 20.8 times larger. Comparing LBS-ES and LBS-PB, one finds that both approaches perform similarly. In an additional experiment, we allowed 14,000 optimization steps for LBS-PB with a 1% memory bound (16 days) and found that the resulting value of the objective function was roughly 50% above the one achieved by LBS-ES. Thus, there is room for improvement, but the high computational cost of LBS-PB renders it impractical.

A different view of the results for a 1% memory bound is given in Figure 8b, where we compare the estimation error achieved by both JS and LBS with equi-size allocation. The resulting per-table sample size was 7,407 tuples and 12,296 tuples, respectively. The columns denote the average relative error of the approximate answer for four different queries (over 1,000 independent runs). Q_1 determines the average order value of customers from Germany ($O \bowtie C \bowtie N$), Q_2 the average balance of these customers ($C \bowtie N$), Q_3 the turnover generated by European suppliers ($L \bowtie PS \bowtie S \bowtie N \bowtie R$), and Q_4 computes the average retail price of a part (P). As can be seen in the figure, the increase in sample size for LBS is directly reflected in the precision of the estimates.

6   Related Work

There exists a variety of sampling techniques for approximate query processing. These techniques can be divided into table-level and schema-level sampling schemes.

Table-level sampling. A table-level sample represents a single table (or view) of a database. Most research focuses on sampling techniques which produce good or optimal samples for a specific purpose such as aggregation queries [6,7] or group-by queries [8,9]. It might be possible to combine some of these techniques with the ideas presented in this paper. For example, a combination of LBS with the weighted sampling scheme in [6] requires an appropriate adjustment of the reference probabilities.

Schema-level sampling. Schema-level samples summarize more than one table as well as the relationships between them. The difficulty of joins over random samples is examined in [3,2]. Acharya et al. [2] also propose the Join Synopsis algorithm from which our Linked Bernoulli Synopses have been derived. In contrast to Join Synopses, Linked Bernoulli Synopses correlate the individual samples so that the space consumption of the synopses is minimized.

Other schema-level synopses. Apart from sampling, other synopses have been proposed for approximate query processing over joins. In [10], Spiegel and Polyzotis propose the *Tuple Graph* as a data structure which is able to represent complex relations between tables. *Probabilistic Relational Models* [11] also exhibit statistical dependencies between attributes of multiple tables. Both techniques focus on selectivity estimation of complex queries but are not applicable to approximate query processing.

7   Conclusion

In this paper, we introduced a novel schema-level sampling scheme called Linked Bernoulli Synopses. The scheme computes a uniform sample of every table in the database; foreign-key integrity is maintained for all sampled tuples. Our approach is based on Join Synopses but correlates the sampling processes of the individual tables. As a consequence, the size of the resulting synopsis is significantly reduced without affecting the quality of approximate answers. Indeed, the saved space can be used to store larger samples, which in turn decreases the estimation error.

References

1. Olken, F.: Random Sampling from Databases. Ph.d. thesis, Lawrence Berkeley National Laboratory (1993)
2. Acharya, S., Gibbons, P.B., Poosala, V., Ramaswamy, S.: Join Synopses for Approximate Query Answering. In: SIGMOD, pp. 275–286 (1999)
3. Chaudhuri, S., Motwani, R., Narasayya, V.: On Random Sampling over Joins. In: SIGMOD, pp. 263–274 (1999)
4. Gemulla, R., Rösch, P., Lehner, W.: Linked Bernoulli Synopses: Sampling Along Foreign Keys (Full Version). Technical report (2007),
   http://wwwdb.inf.tu-dresden.de/publications
5. Tuy, H.: Monotonic optimization: Problems and solution approaches. SIAM J. on Optimization 11(2), 464–494 (2000)
6. Chaudhuri, S., Das, G., Datar, M., Narasayya, R.M.V.R.: Overcoming Limitations of Sampling for Aggregation Queries. In: ICDE, pp. 534–544 (2001)
7. Rösch, P., Gemulla, R., Lehner, W.: Designing Random Sample Synopses with Outliers. In: ICDE (2008)
8. Acharya, S., Gibbons, P., Poosala, V.: Congressional Samples for Approximate Answering of Group-By Queries. In: SIGMOD, pp. 487–498 (2000)
9. Babcock, B., Chaudhuri, S., Das, G.: Dynamic Sample Selection for Approximate Query Processing. In: SIGMOD, pp. 539–550 (2003)
10. Spiegel, J., Polyzotis, N.: Graph-Based Synopses for Relational Selectivity Estimation. In: SIGMOD, pp. 205–216 (2006)
11. Getoor, L., Taskar, B., Koller, D.: Selectivity Estimation using Probabilistic Models. In: SIGMOD, pp. 461–472 (2001)

Query Planning for Searching Inter-dependent Deep-Web Databases

Fan Wang[1], Gagan Agrawal[1], and Ruoming Jin[2]

[1] Department of Computer Science and Engineering
Ohio State University, Columbus OH 43210
{wangfa,agrawal}@cse.ohio-state.edu
[2] Department of Computer Science, Kent State University, Kent OH 44242
jin@cs.kent.edu

Abstract. Increasingly, many data sources appear as online databases, hidden behind query forms, thus forming what is referred to as the *deep web*. It is desirable to have systems that can provide a high-level and simple interface for users to query such data sources, and can automate data retrieval from the deep web. However, such systems need to address the following challenges. First, in most cases, no single database can provide all desired data, and therefore, multiple different databases need to be queried for a given user query. Second, due to the dependencies present between the deep-web databases, certain databases must be queried before others. Third, some database may not be available at certain times because of network or hardware problems, and therefore, the query planning should be capable of dealing with unavailable databases and generating alternative plans when the optimal one is not feasible.

This paper considers query planning in the context of a deep-web integration system. We have developed a dynamic query planner to generate an efficient query order based on the database dependencies. Our query planner is able to select the *top K* query plans. We also develop cost models suitable for query planning for deep web mining. Our implementation and evaluation has been made in the context of a bioinformatics system, SNPMiner. We have compared our algorithm with a naive algorithm and the optimal algorithm. We show that for the 30 queries we used, our algorithm outperformed the naive algorithm and obtained very similar results as the optimal algorithm. Our experiments also show the scalability of our system with respect to the number of data sources involved and the number of query terms.

1 Introduction

A recent and emerging trend in data dissemination involves online databases that are hidden behind query forms, thus forming what is referred to as the *deep web* [13]. As compared to the *surface web*, where the HTML pages are static and data is stored as document files, deep web data is stored in databases. *Dynamic* HTML pages are generated only after a user submits a query by filling an online form.

The emergence of the deep-web is posing many new challenges in data integration. Standard search engines like Google are not able to crawl to these web-sites. At the same time, in many domains, manually submitting online queries to numerous query

B. Ludäscher and Nikos Mamoulis (Eds.): SSDBM 2008, LNCS 5069, pp. 24–41, 2008.

forms, keeping track of the obtained results, and combining them together is a tedious and error-prone process. Recently, there has been a lot of work on developing deep web mining systems [6,7,14,15,19,31,38]. Most of these systems focus on query interface integration and schema matching.

A challenge associated with deep web systems, which has not received attention so far, arises because the deep web databases within a specific domain are often not independent, i.e., the output results from one database are needed for querying another database. For a given user query, multiple databases may need to be queried in an *intelligent* order to retrieve all the information desired by a user. Thus, there is a need for techniques that can generate query plans, accounting for dependencies between the data sources, and extracting all information desired by a user.

A specific motivating scenario is as follows. In bioinformatics, Single Nucleotide Polymorphisms (SNPs), seem particularly promising for explaining the genetic contribution to complex diseases [3, 20, 34]. Because over seven million Single Nucleotide Polymorphisms (SNPs) have been reported in public databases, it is desirable to develop methods of sifting through this information. Much information that biological researchers are interested in requires a search across multiple different web databases. No single database can provide all user requested information, and the output of some databases need to be the input for querying another database.

We consider a query that asks for the amino acids occurring at the corresponding position in the orthologous gene of non-human mammals with respect to a particular gene, such as ERCC6. There is no database which takes gene name ERCC6 as input, and outputs the corresponding amino acids in the orthologous gene of non-human mammals. Instead, one needs to execute this query plan. We first need to query on one database, such as SNP500Cancer, to retrieve all SNPs located in gene ERCC6. Second, using the extracted SNP identifier, we query on SNP database, such as dbSNP, to obtain the amino acid position of the SNP. Third, we need to use a sequence database to retrieve the protein sequence of the corresponding SNP. Finally, querying on BLAST database, which is a sequence alignment database, we can obtain the amino acid at the corresponding position in the orthologous gene of non-human mammals.

From the above example, we can clearly see that for a particular query, there are multiple *sub-goals*. These sub-goals are not specified by the user query, because the user may not be familiar with details of the biological databases. The query planner must be able to figure out the *sub-goals*. Furthermore, we can note the strong dependencies between those databases, which constraint the query planning process.

This paper considers query planning in the context of a deep-web integration system. The system is designed to support a very simple and easy to use query interface, where each query comprises a *query key term* and a set of *query target terms* that the user is interested in. The query key term is a name, and the query target terms capture the properties or the kind of information that is desired for this name. We do not need the user to provide us with a formal predicate-like query. In the context of such a system, we develop a dynamic query planner to generate an efficient query order based on the deep web database dependencies. Our query planner is able to select the *top K* query plans. This ensures that when the most efficient query plan is not feasible, for examples, because a database is not available, there are other plans possible.

To summarize, this paper makes the following contributions:

1. We formulate the query planning and optimization problem for deep web databases with dependencies.
2. We design and implement a dynamic query planner to generate the top K query plans based on the user query and database dependencies. This strategy provides alternative plans when the most efficient one is not feasible due to the non-availability of a database.
3. We support query planning for a user-friendly system that requires the user to only include query key terms and a set of target terms of interest. Database schemas are input by an administrator or designer.
4. We develop cost models suitable for query planning for deep web mining.
5. We present an integrated approximate planning algorithm with approximation ratio of $1/2$.
6. We integrate our query planner with a deep web mining tool SNPMiner [37] to develop a domain specific deep web mining system.
7. We evaluate our dynamic query planning algorithm with two other algorithms and show that our algorithm can achieve optimal results for most queries, and furthermore, our system has very good scalability.

The rest of the paper is organized as follows. In Section 2, we formulate the dynamic query planning problem. We describe the details of our dynamic query planner in Section 3. In Section 4, we evaluate the system. We compare our work with related efforts in Section 5 and conclude in Section 6.

2   Problem Formulation

The deep web integration system we target provides a fixed set of candidate terms which can be queried on. These terms are referred to as the *Query Target Terms*. A user selects a subset of the allowable *Query Target Terms* and in addition, specifies a *Query Key Term*. Query target terms specify what type of information the user wants to know about the query key term. From the example in Section 1, we know that for a single query of this nature, several pieces of information may need to be extracted from various databases. Furthermore, there are dependencies between different databases, i.e. information gained from one source may be required to query another source. Our goal is to have a *query planning* strategy that can provide us an efficient and correct query plan to query the relevant databases.

We have designed a dynamic query planner which can generate a set of *Top K* query plans. The query plan with shorter *length*, i.e. the number of databases searched, higher coverage of user request terms, and higher user preference, are considered to have a higher priority. By generating K query plans, there can be back-up plans when the best one is not feasible, for example, because of unavailability of a database.

Formally, the problem we consider can be stated as follows. We are given a universal set $T = \{t_1, t_2, \ldots, t_n\}$, where each t_i is a term that can be requested by a user. We are also given a subset $T' = \{t'_1, t'_2, \ldots, t'_m\}$, $t'_i \in T$, of terms that are actually requested by the user for a given query. We also have a set $D = \{D_1, D_2, \ldots, D_m\}$, where each

D_i is a deep web database, and each D_i covers a set of terms $E_i = \{e_i^1, e_i^2, \ldots, e_i^k\}$, and E_i is a subset of T. Furthermore, each database D_i requires a set of elements $\{r_i^1, r_i^2, \ldots, r_i^k\}$ before it can be queried, where $r_i^j \in T$.

Our goal is to find a query order of the databases $D^* = \{D_1, D_2, \ldots, D_k\}$, which can cover the set T' with the maximal *benefit* and also makes k as small as possible. The benefit is based on a cost function that we can choose. We call it a dynamic query planning problem, because the query order should be selected based on the user specified target terms, and cannot be fixed by the integration system. This is a variant of the famous weighted set cover problem, and can be easily proven as NP-Complete [9].

2.1 Production System Formulation

For discussing our algorithm, it is useful to view the query planning problem as a *production system*. A production system is a model of computation that has proved to be particularly useful in AI, both for implementing search algorithms and for modeling human problem solving [27]. A production system can be represented by four elements, which are a *Working Memory*, a *Target Space*, a set of *Production Rules*, and a *Recognize-Act* control cycle. The working memory contains a description of the current state in a reasoning process. The target space is the description of the aim. If the *working memory* becomes a superset of the *target space*, the problem solving procedure is completed. A production rule is a *condition-action* pair. The *condition* part determines whether the rule can be applied. The *action* part defines the associated problem-solving step. The *working memory* is initialized with the beginning problem description. The current state is matched against the conditions of the production rules. When a production rule is fired, its action is performed, and the working memory is changed accordingly. The process terminates when the content of the working memory becomes a superset of the target state, or no more rules can be fired.

We map our query planning problem into the four elements of a production system as follows. The working memory is comprised of all the data which has already been extracted. Our query plan is generated step by step, and when a database is added into our query plan, the data that can be obtained from this database is considered as stored in the working memory. Initially, the working memory is just the Query Key Term. The target state is a subset of the Query Target Terms selected by the user.

Each online database has one or more underlying query schema. Those schemas specify what the input of the online query form of the database is, and what data can be extracted from the database by using the input terms. The production rules of our system are the database schemas. Note that one database may have multiple schemas. In this case, each schema carries different input elements to retrieve different output results. The database schemas are provided by deep web data source providers and/or a developer creating the query interface.

The terms in working memory are matched against the necessary input set of each production rule. Appropriate rule will be fired according to our rule selection strategy, which will be introduced in Section 3.3. We consider the corresponding database as queried and the output component of the fired rule is added to the working memory. We mark the selected rules as visited to avoid re-visiting the same rule. If either of the following two cases holds, one complete query plan would have been generated. In the first case, the working memory has covered all the elements in the target space, which means that all user requested Query Target Terms have been found. In the second case, there are still some terms in the target state have not been covered by the working memory, but no unvisited rules can cover any more elements in the target space. This means that it is impossible to retrieve all the request terms by using current set of available databases. This normally occurs when some databases are unavailable.

3 Query Planning Approach and Algorithm

Our approach for addressing the problem stated in the previous section is as follows. We first introduce a data structure, *dependency graph*, to capture the database dependencies. Our algorithm is based on this data-structure. Towards the end of this section, we describe the cost or *benefit* model that we use.

3.1 Dependency Graph

As we stated earlier, there are dependencies between online databases. If we want to query database D, we have to query on some other databases in order to extract the necessary input elements of D first. We use the production rule representation of the databases to identify the dependecies between the databases and build a dependency graph of databases to capture the relationship between databases.

Formally, there is a dependency relation DR, $\prec_{DR} \subset 2^D \times D$, where 2^D is the power set of D. If $\{D_i, D_{i+1}, \ldots, D_{i+m}\} \prec_{DR} D_j$, we have to query on data source $D_i, D_{i+1}, \ldots, D_{i+m}$ first in order to obtain the necessary input elements for querying on the data source D_j. Note that there could be multiple combinations of databases that can provide input required for querying a given database.

We use *hypergraph* to represent the dependency relationship. A hypergraph consists of a set of nodes N and a set of hyperarcs. The set of hyperarcs is defined by ordered pairs in which the first element of the pair is a subset of N and the second element is a single node from N. The first element of the pair is called the parent set, and the second element of the pair is called the descendant. If the parent set is not singleton, the elements of the set of parents are called *AND* nodes. In our dependency graph, the nodes are online databases, and hyperarcs represent the dependencies between databases. For a particular hyperarc, the parent nodes of the pair are the databases which must be queried first in order to continue the query on the database represented by the descendent node of the pair.

The dependency graph is constructed using the production rules of each online database. For two databases D_i and D_j, suppose D_i has a set of production rules $R_i = \{r_{i1}, r_{i2}, \ldots, r_{in}\}$ and D_j has a set of production rules $R_j = \{r_{j1}, r_{j2}, \ldots, r_{jm}\}$. If any rule in R_i has an output set which can fully cover any of the rules' input set in R_j, we build an edge between D_i and D_j. In another case, if any rule in R_i has an output set which partially covers any of the rules' input set in R_j, we scan the rules of other databases to find a partner set of databases for D_i together with which can fully cover any of the rules' input set in R_j. If the partner set exists, we build a hyperarc from D_i

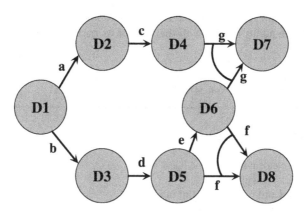

Fig. 1. Dependency Graph Example

and its partner set databases to D_j. If the production rules of a database are updated, our dependency graph can also be updated accordingly. Figure 1 shows an example of hypergraph.

In Figure 1, there are 8 nodes in the hypergraph, and 7 hyperarcs. The first hyperarc is denoted by a. The first element (parent) of the ordered pair of a is node D1, and the second element (descendant) of the pair is node D2. This hyperarc implies that after querying on D1, we can obtain the input elements needed for querying D2. Hyperarcs b to e are similar to hyperarc a. In hyperarc f, the arc connecting the two edges of f shows that this hyperarc is a *2-connector*. The first element of f is a set comprising of D5 and D6, and the second element is D8. This shows in order to query on D8, we need to first query on D5 and D6. Hyperarc g has the same structure as f.

For a node D, the *neighbors* of D are the nodes which have an edge coming from D. The partners of D are the nodes which is connected with D by a hyperarc to a common descendent. For example, in Figure 1, $D5$ has a partner $D6$, as $D5$ and $D6$ connect $D8$.

3.2 Query Planning Algorithm

We now introduce our algorithm. Initially, we define several new terms.

Due to the existence of database dependencies, some databases can be more important because it can link us to other important databases. In Figure 1, suppose we are on node $D1$, and user requested terms can only be obtained by the database $D8$. Using the query key term, we cannot query on $D8$ at the beginning. We must first query on other databases, say $D3$, to gain additional information in order to follow the database dependencies and finally query on $D8$. We call databases $D2$, $D3$, $D4$, $D5$ and $D6$ *hidden nodes*.

In order to determine the hidden nodes, we need to come up with a strategy to find all reachable nodes from a starting node in the dependency graph. This can be done by adapting the breath-first search method to a hypergraph. We use the algorithm *Find_Reachable_Node(DG,s)* to find all reachable nodes from a starting node s in the dependency graph DG. We show the sketch of this algorithm in Algorithm 3.1. In the algorithm, $Q1$ is a queue which stores all reachable nodes from starting node s, $Q2$ stores all reachable nodes from s with the help of its partners, and $PS(t)$ returns the partner set of t.

Algorithm 3.1: FindReachableNodes(DG, s)

Initialize two queues Q1 and Q2
Add s to Q1, and mark s as visited
while Q1 is not empty
  Dequeue the first element in Q1, name it as t
  foreach n which is a neighbor of t
    if $n \in unvisited$ **and** $PS(t) = \Phi$ **and** rules match
      Add n to Q1 and mark n as visited
    else if $n \in unvisited$ **and** $PS(t) \neq \Phi$
    Add n to Q2
  while Q2 is not empty
    foreach $n \in Q2$
      Extract the partner set PS of n
      Denote each partner of n as p
      if $p \in Q1$
        foreach p of n **and** rules match
          Add n to Q1, and remove n from Q2
        Mark n as visited
return ($Q1$)

Next, we introduce a new concept, *Database Necessity*. Each production rule is associated with a set of terms which can be extracted by executing the rule. Some terms can only be provided by one database, while other terms can be provided by multiple databases. If a requested term can only be provided by a single rule, that rule should have a higher priority to be executed. Conversely, if the term can be provided by multiple rules, a lower priority can be assigned to this rule. Based on this idea, each term is associated with a *Database Necessity* value. Formally, for a term t, if K databases can provide it, the database necessity value for t is $\frac{1}{K}$.

As part of our algorithm, we need to make hidden rules partially *visible on the surface*. A hidden but potentially useful rule has the following two properties: (1) It must be executed in order to extract all user requested terms. (2) The necessary input elements of it are hidden, i.e. either they are not in the Query Target Terms or they can only be extracted by rules located in the hidden layer.

We make the hidden rules visible in a bottom-up manner as follows. We scan all the terms in the user provided Query Target Terms, i.e. the initial Target Space. If there is some term in the initial target space with database necessity value of 1, which means only one production rule, say R, can provide this term. This rule R is a hidden rule which must be fired. In order to make rule R visible, we add the necessary input elements of R into the target space to enlarge the target space. Then, we re-scan the newly enlarged target space to find new hidden rules and enlarge the target space further. This procedure continues until there are no more hidden rules of this kind.

Another important issue in our algorithm is the ability to prune similar query plans. The dynamic query planner can generate the top K query plans. When one query plan is generated, the algorithm will trace back from the current point to generate another query plan. It is highly possible that two generated query plans QP_1 and QP_2 use the same set of databases, but differ in the order of querying two or more databases which do not have any dependencies. In this case, we will consider the two query plan QP_1 and QP_2 as the same, and the latter one will be deleted.

Algorithm 3.2: Find_TopK_Query_Plans(PR, WS, TS)

---

while enlargeable(WS)
  Enlarge WS
Initialize queue Q and P
while $size(Q) \leq K$
  if ($\exists e \in TS$ **and** $e \notin WS$)
  and ($\exists r \in PR$ **and** $\exists o \in O(r)$ **and** $o \in TS$)
   Find candidate rule set CR
   foreach $r \in CR$
    Compute benefit score according to benefit model
   Select r_opt, the rule with the highest benefit
   if !$prunable(P, r\_opt)$
    while $r\_opt \neq null$ **and** ($\exists e \in TS$ **and** $e \notin WS$)
    and ($\exists r \in PR$ **and** $\exists o \in O(r)$ **and** $o \in TS$)
     Add r_opt to P, and update WS
     Select next r_opt
    else Empty queue P
   else Add P to Q and re-order Q
  if $size(P) > 0$
   Remove the last rule of P, update WS, trace back
return (Q)

Main Algorithm. Our dynamic query planning algorithm takes the Query Target Terms and Query Key Term, and dynamically generates K query plans which cover as many request terms as possible. At the beginning, we enlarge user provided target space to visualize hidden rules and obtain the enlarged target space. We take the Query Key Term as the initial working memory. Then, the production system begins the recognize-act procedure. Each iteration the system selects an appropriate rule according to a benefit model and updates the current working memory. This procedure terminates when all terms in the target space are covered or no more rules can be fired.

Now, we assume our benefit model can select the *best* rule according to the current working memory and our goal. We will introduce our benefit model in detail in Section 3.3. Algorithm 3.2 shows the sketch of the planning algorithm. In the algorithm, PR is the set of production rules, WS is the working memory, and TS is the target space. Q is a queue to store the top K query plans, and P is a queue to store the rules along one query plan. $O(r)$ returns the output elements of a rule r. $prunable()$ is a function to test whether a candidate query plan can be pruned as we discussed earlier.

Our dynamic query planning algorithm is a greedy algorithm which selects the production rule with the local maximal benefit according to the benefit model. Each greedy algorithm has an approximation ratio which measures the performance of the algorithm. We use $|R|$ to represent the cardinality of the collection of rules R, i.e. the total number of production rules. We have the following result:

Theorem 1. *The approximation algorithm introduced in Algorithm 3.2 has an approximation ratio of* $\frac{|R|+1}{2|R|}$.

The proof is omitted for lack for space.

3.3   Benefit Model

A very important issue in a production system is *rule selection*, i.e., which rule should be executed. We have designed a benefit model to select an appropriate rule at each iteration of the recognize-act cycle. In the algorithm presented earlier in this section, at each step, each rule is scanned and all the rules which can be fired are put into a set called the *candidate rule set*. Then, we compute a benefit score for each of the candidate rules.

We have used four metrics for rule selection, which are Database Availability (DA), Data Coverage (DC), User Preference (UP), and Potential Importance (PI).

Database Availability: A production rule R can be executed if the corresponding database is available. In our implementation, for each rule, we send a message to the database to test the availability of the database. If the database is not available, we just ignore this rule for the current iteration.

Data Coverage: Data coverage measures the percentage of required data that can be provided by a particular rule. Given a rule R_k, the target state TS, and $k-1$ rules $R_1, R_2, \ldots, R_{k-1}$ that have already been selected, we want to compute the data coverage of the current rule R_k with respect to TS. We use the number of Query Target Terms in TS which are also covered by the rule R_k, but have not been extracted by previous rules for this purpose.

User Preference: Some terms can be extracted from multiple databases, and domain users may have preference for certain databases for a particular term. We can assign a user preference value for each term with respect to databases and incorporate user preference into the benefit function. Consider a particular term t, which can be obtained from r databases $D_1, D_2, \ldots, D_r$. A number between 0 and 1 should be assigned to t for each of the r databases as the preference value, such that the r preference values sum up to 1. If t can only be obtained from a single database D, the preference value of t with respect to D is 1 and is 0 for all other databases. The user preference values should be given by a domain expert.

Suppose we are examining the production rule R, which is associated with the database D. The following k terms $UF_1, UF_2, \ldots, UF_k$ have not been found. For each term UF_i, the user preference with respect to database D is UP_i. We use the database necessity value of each term (DN_i for term UF_i) as the weight of its user preference and we compute the weighted sum of all unfound terms as the user preference value of the rule, i.e. the user preference of R is $\sum_{i=1}^{k} DN_i * UP_i$.

Potential Importance: Because of database dependencies, some databases can be more important due to its linking to other important databases. Figure 1 shows an example. In the above case, suppose $D2$ and $D3$ have the same data coverage and user preference. Obviously, $D3$ is potentially more important because $D3$ can help us link to our final target $D8$. As a result, the $D3$ should be assigned a larger benefit value. Based on the above idea, we incorporate potential importance to our benefit function.

Suppose we are considering production rule corresponding to the database D. By using Algorithm 3.1, we find a set of databases $D_{reachable} = \{D_1, D_2, \ldots, D_m\}$, which can be queried by using the data extracted from database D exclusively. We have k term which have not been found, denoted by $UF_1, UF_2, \ldots, UF_k$. For term UF_i, its database necessity value is DN_i, which means the term UF_i can be obtained by $\frac{1}{DN_i}$ number of databases and we denote this set of databases as $NecessaryD_i$. We want to know the number of *Necessary Databases* of UF_i which can be reached by the current rule R. We count the number of databases in $NecessaryD_i$, which are also in the set $D_{reachable}$, i.e. we compute the cardinality of the set $\{d | d \in NecessaryD_i, d \in D_{reachable}\}$. Suppose the cardinality is r_i for term UF_i. The potential importance for UF_i with respect to rule R and corresponding database D is $\frac{r_i * \frac{1}{DN_i}}{|D_{reachable}|} = \frac{r_i}{m * DN_i}$. Finally, the potential importance for the rule R is

$$\sum_{i=1}^{k} \frac{r_i}{m * DN_i}$$

For each candidate rule, a benefit score is computed according to the three metrics, data coverage, user preference and potential importance. The value of the three metrics are closely related to the database necessity values of all unfound terms when a rule is being examined, as a result, if a rule is considered as a candidate multiple times, each time the benefit score must be different, because each time the set of unfound terms is different. As a result, the benefit score of a production rule is dynamically related to the current working space of the production system.

The benefit function of a rule R with respect the current working space WS can be represented as follows:

$$BF(R, WS) = DC * \alpha + UP * \beta + PI * \gamma, \alpha + \beta + \gamma = 1$$

There are three parameters α, β and γ associated with each metric term. These three parameters scale the relative importance of the three metric terms.

3.4 Discussion: System Extendibility

Extendibility is an important issue for any deep web mining system, as new data sources can emerge often. We now briefly describe how a new data source can be integrated with our system. First, we need to represent the database query schemas of the new data source into the form of production rules. Then, a domain expert assigns or changes user preference values for the terms appearing in the newly integrated data sources. We have developed simple algorithms for automatically integrating the new data source into the Dependency Graph and updating the database necessity values. The algorithms proposed are scalable to larger numbers of databases. Furthermore, because the design of our dependency graph and query planning algorithm is based on the inherent characteristics of deep web data sources, such as database dependencies and database schemas, our system is independent of the application domain, i.e., the system can be applied on any domains of application.

4   Performance

This section describes the experiments we conducted to evaluate our algorithm. We ran 30 queries and compared the performance of our algorithm with two other algorithms.

4.1   Experiment Setup

Our evaluation is based on the SNPMiner system [37]. This system integrates the following biological databases: dbSNP[1], Entrez Gene and Protein[2], BLAST[3], SNP500Cancer[4], SeattleSNPs[5], SIFT[6], and BIND[7]. SNPMiner System provides an interface by which users can specify query key terms and query target terms. We use some heuristics to map user requested keywords to appropriate databases. SNPMiner uses Apache Tomcat 6.x to support a web server. After a query plan is executed, all results are returned in the form of HTML files. We have a web page parser to extract relevant data from the files and tabulate the data.

We created 30 queries for our evaluation. Among these 30 queries, 10 are real queries specified by a domain expert we have collaborated with. The remaining 20 queries were generated by randomly selecting query keywords. We also vary the number of terms in each query in order to evaluate the scalability of our algorithm. Table 1 summarizes the statistics for the 30 queries.

Table 1. Experimental Query Statistics

Query ID	Number of Terms
1-8	2-5
9-16	8-12
17-24	17-23
25-28	27-33
29,30	37-43

[1] http://www.ncbi.nlm.nih.gov/projects/SNP
[2] http://www.ncbi.nlm.nih.gov/entrez
[3] http://www.ncbi.nlm.nih.gov/blast/index.shtml
[4] http://snp500cancer.nci.nih.gov/home_1.cfm
[5] http://pga.gs.washington.edu/
[6] http://blocks.fhcrc.org/sift/SIFT.html
[7] http://www.bind.ca

Our evaluation has three parts. First, we compare our production rule algorithm with two other algorithms. Second, we show that enlarging the target space improves the performance of our system significantly. Finally, we evaluate the scalability of our system with respect to the number of databases and the number of query terms. In all our experiments, the three scaling parameters are set as follows: $\alpha = 0.5$, $\beta = 0.3$ and $\gamma = 0.2$.

In comparing planning algorithms or evaluating the impact of an optimization, we use two metrics, which are the *mumber of databases involved in the query plan* and the *actual execution time for the query plan*. We consider a query plan to be good if it can cover all user requested terms using as few databases as possible. A query plan that involves more databases tends to query redundant databases, and cause additional system workload.

4.2   Comparison of Three Planning Algorithms

We compare our Production Rule Algorithm (PRA) with two other algorithms, which are the Naive Algorithm (NA) and the Optimal Algorithm (OA).

Naive Algorithm: As the name suggests, this algorithm does query planning in a naive way. The algorithm selects all production rules which can be queried at each round, until all keywords are covered. This algorithm can quickly find a query plan, but the query plan is likely to have a very low score and a long execution time.

Optimal Algorithm: This algorithm searches the entire space to find the optimal query plan. Because we only had 8 databases for our experiments, we could manually determine the optimal query plan for each query. Such a plan is determined based on the number of databases involved in the query plan and the expected response time of the databases involved. This means that the optimal query plan has the smallest *estimated* execution time, though the measured execution time may not necessarily be the lowest of all plans.

In Figure 2, sub-figures (1a) and (1b) show the the comparison between PRA and NA. In sub-figure (1a), the diamonds are the ratios between the execution time of the query plans generated by PRA and NA, annotated as *ETRatio*. We can see that all diamonds are located below the $ratio = 1$ line, which implies that for each of the 30 queries, the query plan generated by production rule algorithm has a lesser execution time than that of the plan generated by naive algorithm. In the sub-figure (1b), the rectangles are the ratios of the number of databases involved in the query plan generated by PRA and NA, denoted as *DRatio*. We observe that the same pattern, i.e. the query plans generated by the production rule algorithm use fewer data sources.

Sub-figures (2a) and (2b) show the the comparison between PRA and OA. From the sub-figure (2a), we can observe that all the diamonds are distributed closely around the $ratio = 1$ line. This shows that in terms of the execution time of generated query plans, the production rule algorithm has close to the optimal performance. We also observe from the sub-figure (2b) that in terms of the number of databases involved in query plans, the production rule algorithm obtains the optimal result, with an exception of query 11. We examined the query plans generated by PRA, and found that most of the query plans are exactly the same as the optimal query plans. For other cases, we note

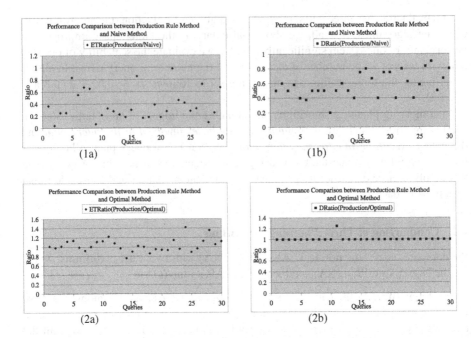

Fig. 2. Comparison among PRA, NA and OA: (1a) Comparison between PRA and NA on Plan Execution Time; (1b) Comparison between PRA and NA on Plan Length;(2a) Comparison between PRA and OA on Plan Execution Time;(2b) Comparison between PRA and OA on Plan Length

that the optimal algorithm uses some databases with lower response time. However, this did not necessarily result in lower actual execution time. We can see from the sub-figure (2a) that some of the execution time with PRA are actually smaller than the execution times of the plans generated by the optimal algorithm.

4.3   Impact of Enlarging Target Space

In this experiment, we compare the number of databases involved and the execution time for different query plans generated using the system without enlarging target space and the system with enlarged target space. We select 8 queries which contains many terms with database necessity value smaller than 1, because these query plan results can better show the usefulness of enlarging the target space. The results are shown in Figure 3.

We have the following observations. From the sub-figure (a) in Figure 3, we can observe that the number of databases involved for most of the query plans is much shorter for the enhanced system than that of the system without enhancement. From the sub-figure (b) in Figure 3, we can also observe that the execution time reduces very significantly for the enhanced system. The above results show that enlarging target space can effectively improve our system to generate query plans with fewer databases and less execution time.

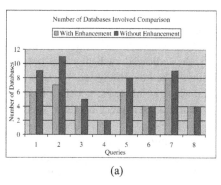
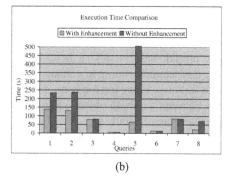

(a)   (b)

Fig. 3. System Enhancement Test: (a) Comparison of Number of Databases Involved; (b) Comparison of Execution Time

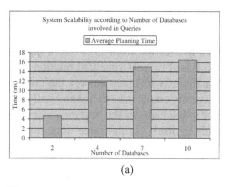
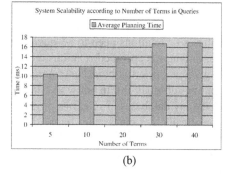

(a)   (b)

Fig. 4. System Scalability Test: (a) System Scalability with respect to Number of Databases Involved; (b) System Scalability with respect to Number of Terms in Queries

4.4 Scalability of Production Rule Algorithm

Our last experiment evaluated how the query planning time scales with increasing number of databases and query terms. From Figure 4, in sub-figure (a), we can observe that in terms of the average planning time, there is a sharp increase in going from 2 data sources to 4 data sources. Then, the planning time increases only moderately with respect to the increase in the number of data sources. In the sub-figure (b), we can see that the average planning time increases very slowly with the increase in the number of terms in the queries. This shows that our system has good scalability.

5 Related Work

We now compare our work with existing work on query planning, deep web mining, and keyword search on relational databases.

Query Planning: There are a number of research efforts on query planning. Raschid and co-workers have developed a navigational-based query planning strategy for mining biological data sources [5,21,22,23,28,36]. They build a source graph representing integrated biological databases and an object graph representing biological objects in each database. The navigational links (hyperlink) between the database objects are assumed to be pre-fetched. Extending their work [23, 28], they allowed each physical link to carry a semantic meaning to enhance their algorithm. The key differences in our work are as follows. First, we focus on deep web database dependencies, not the physical links between database objects. Further, in their work, a user query needs to specify source and target databases.

A lot of work has been done in SQL-based query planning [1,12,18,26,32]. In [18], new SQL operators were introduced to reduce repetitive and unnecessary computations in query planning. In [12,26], a set of pre-defined plans were represented in the form of grammar production rules. For a SQL query, the algorithm first built plans to access individual tables, and then repeatedly refered grammar rules to join plans that were generated earlier. Other work has focused on query planning using database views and multiple databases sharing the same relational schema [1,32].

Much work on query planning is based on the well known *Bucket Algorithm* [10, 11,17,24,25,29,30]. In the above work, they assume that the user query specifies the databases or relations need to be queried, and the task of the work is to find a query order among the specified relations or databases. Based on user specified relations or sub-goals, a bucket is built containing all the databases which can answer the corresponding sub-goal. But in our work, the user query only contains keyword and will not specify any databases or relations of interest. Our system selects the best data sources automatically, i.e. our system figure out sub-goals by itself. At the same time, query planning is performed.

In [35], a query planning algorithm minimizes the query's total running time by optimally exploits parallelism among web services. The main difference between our work and theirs is, they assume that one attribute can only be provided by exactly one data source which is a unrealistic assumption in real application, but we allow the present of data redundancy.

Deep Web Mining: Lately, there has been a lot of work on mining useful information from the deep web [6,7,14,15,19,31,38]. In [19], a database selection algorithm based on attribute co-occurrence graph was proposed. In [31], Nie *et al.* proposed an object-level vertical searching mechanism to handle the disadvantages of document-level retrieval. QUIC [38] was a mining system supporting imprecise queries over incomplete autonomous databases. In [14, 7, 6], Chang *et al* proposed an E-commerce domain deep web mining tool MetaQuerier. MetaQuerier translated user query into several local queries by schema matching. WISE-Integrator [15] was another deep web mining tool similar to MetaQuerier. The key difference in our work is that none of the above systems consider database dependencies.

Keyword Search on Relational Databases: Recently, providing keyword based search over relational databases has attracted a lot of attention [2,4,16,33]. The major technical issue here is to efficiently search several keywords which co-occur in the same row, in a table obtained by joining multiple tables or even databases together, and rank them based on different measures. In addition, the keywords may appear in any attribute or column. This is very different from the problem studied in this paper.

Select-Project-Join Query Optimization: There has been extensive work in query optimization, especially SPJ type query optimization since the early 1970s [8]. A query optimizer needs to generate an efficient execution plan for the given SQL query from a space of possible plans based on a cost estimation technique which is used to measure the cost of each plan in the search space. Our work has some similarities with the above research efforts in that we both do selection as earlier as possible. Two major differences between our work and SPJ query optimization are as follows. First, in traditional query optimization, any join-order is allowed, but for our work, due to deep web properties, the allowable join operations are restricted. Second, in traditional databases, redundant columns seldom occur, so it is impossible to have options to take one project or column from several alternative databases, but redundant data exists in our deep web databases, and we can take different paths. As pointed out above, our problem is different from traditional SPJ query and new techniques are needed.

6   Conclusion

In this paper, we formulated and solved the query planning and optimization problem for deep web databases with dependencies. We have developed a dynamic query planner with an approximation algorithm with a provable approximation ratio of $1/2$. We have also developed cost models to guide the planner. The query planner automatically selects best sub-goals on-the-fly. The K query plans generated by the planner can provide alternative plans when the optimal one is not feasible. Our experiments show that the cost model for query planning is effective. Despite using an approximate algorithm, our planning algorithm outperforms the naive planning algorithm, and obtains the optimal query plans for most experimental queries in terms of both number of databases involved and actual execution time. We also show that our system has good scalability.

References

1. Abiteboul, S., Garcia-Molina, H., Papakonstantinou, Y., Yerneni, R.: Fusion queries over internet databases (1997)
2. Agrawal, S., Chaudhuri, S., Das, G.: Dbxplorer: A system for keyword-based search over relational databases. In: Proceedings of the 18th International Conference on Data Engineering, pp. 5–16 (2002)
3. Brookes, A.J.: The essence of snps. Gene. 234, 177–186 (1999)
4. Bhalotia, G., Hulgeri, A., Nakhe, C., Chakrabarti, S., Sudarshan, S.: Keyword searching and browsing in databases using banks. In: Proceedings of the 18th International Conference on Data Engineering, pp. 431–440 (2002)
5. Bleiholder, J., Khuller, S., Naumann, F., Raschid, L., Wu, Y.: Query planning in the presence of overlapping sources. In: Proceedings of the 10th International Conference on Extending Database Technology, pp. 811–828 (2006)
6. Chang, K., He, B., Zhang, Z.: Toward large scale integration: Building a metaquerier over databases on the web (2005)

7. Chang, K.C.-C., Cho, J.: Accessing the web: From search to integration. In: Proceedings of the 2006 ACM SIGMOD international conference on Management of Data, pp. 804–805 (2006)
8. Chaudhuri, S.: An overview of query optimization in relational systems. In: Proceedings of the seventeenth ACM SIGACT-SIGMOD-SIGART symposium on Principles of database systems, pp. 34–43 (1998)
9. Cormen, T.H., Leiserson, C.E., Rivest, R.L., Stein, C.: Introduction to Algorithms, 2nd edn. The MIT Press, Cambridge (2001)
10. Doan, A., Halevy, A.: Efficiently ordering query plans for data integration. In: Proceedings of the 18th International Conference on Data Engineering, p. 393 (2002)
11. Florescu, D., Levy, A., Manolescu, I.: Query optimization in the presence of limited access patterns. In: Proceedings of the 1999 ACM SIGMOD international conference on Management of Data, pp. 311–322 (1999)
12. Haas, L.M., Kossmann, D., Wimmers, E.L., Yang, J.: Optimizing queries across diverse data sources. In: Proceedings of the 23rd International Conference on Very Large Databases, pp. 276–285 (1997)
13. He, B., Patel, M., Zhang, Z., Chang, K.C.-C.: Accessing the deep web: A survey. Communications of ACM 50, 94–101 (2007)
14. He, B., Zhang, Z., Chang, K.C.-C.: Knocking the door to the deep web: Integrating web query interfaces. In: Proceedings of the 2004 ACM SIGMOD international conference on Management of Data, pp. 913–914 (2004)
15. He, H., Meng, W., Yu, C., Wu, Z.: Automatic integration of web search interfaces with wise_integrator. The international Journal on Very Large Data Bases 12, 256–273 (2004)
16. Hristidis, V., Gravano, L., Papakonstantinou, Y.: Efficient ir-style keyword search over relational databases. In: Proceedings of the 29th international conference on Very large data bases, pp. 850–861 (2003)
17. Ives, A.G., Florescu, D., Friedman, M., Levy, A.: An adaptive query execution system for data integration. In: Proceedings of the 1999 ACM SIGMOD International Conference on Management of Data, pp. 299–310 (1999)
18. Jin, R., Agrawal, G.: A systematic approach for optimizing complex mining tasks on multiple databases. In: Proceedings of the 22nd International Conference on Data Engineering, p. 17 (2006)
19. Kabra, G., Li, C., Chang, K.C.-C.: Query routing: Finding ways in the maze of the deep web. In: Proceedings of the 2005 International Workshop on Challenges in Web Information Retrieval and Integration, pp. 64–73 (2005)
20. Lohnmueller, K.E., Pearce, C.L., Pike, M., Lander, E.S., Hirschhorn, J.N.: Meta-analysis of genetic association studies supports a contribution of common variants to susceptibility to common disease. Nature Genet. 33, 177–182 (2003)
21. Lacroix, Z., Parekh, K., Vidal, M.-E., Cardenas, M., Marquez, N., Raschid, L.: Bionavigation: Using ontologies to express meaningful navigational queries over biological resources. In: Proceedings of the 2005 IEEE Computational Systems Bioinformatics Conference Workshops, pp. 137–138 (2005)
22. Lacroix, Z., Raschid, L., Vidal, M.-E.: Efficient techniques to explore and rank paths in life science data sources. In: Proceedings of the 1st International Workshop on Data Integration in the Life Sciences, pp. 187–202 (2004)
23. Lacroix, Z., Raschid, L., Vidal, M.E.: Semantic model to integrate biological resources. In: Proceedings of the 22nd International Conference on Data Engineering Workshops, p. 63 (2006)
24. Leser, U., Naumann, F.: Query planning with information quality bounds. In: Proceedings of the 4th International Conference on Flexible Query Answering, pp. 85–94 (2000)

25. Levy, A.Y., Rajaraman, A., Ordille, J.J.: Querying heterogeneous information sources using source descriptions. In: Proceedings of the 22nd International Conference on Very Large Databases, pp. 251–262 (1996)
26. Lohman, G.M.: Grammar-like functional rules for representing query optimization alternatives. In: Proceedings of the 1988 ACM SIGMOD International Conference on Management of Data, pp. 18–27 (1988)
27. Luger, G.F.: Artificial intelligence: Structure and Strategies for Complex Problem Solving, 5th edn. Addison-Wesley, Reading (2005)
28. Mihaila, G., Raschid, L., Naumann, F., Vidal, M.E.: A data model and query language to explore enhanced links and paths in life science sources. In: Proceedings of the eighth International Workshop on Web and Databases, pp. 133–138 (2005)
29. Naumann, F., Leser, U., Freytag, J.C.: Quality-driven integration of heterogeneous information systems. In: Proceedings of the 25th International Conference on Very Large Data Bases, pp. 447–458 (1999)
30. Nie, Z., Kambhampati, S.: Joint optimization of cost and coverage of information gathering plans, asu cse tr 01-002. computer science and engg. arizona state university
31. Nie, Z., Wen, J.-R., Ma, W.-Y.: Object-level vertical search. In: Proceedings of the 3rd Biennial Conference on Innovative Data Systems Research, pp. 235–246 (2007)
32. Pottinger, R., Levy, A.: A scalable algorithm for answering queries using views. The international Journal on Very Large Data Bases 10, 182–198 (2001)
33. Sayyadian, M., LeKhac, H., Doan, A., Gravano, L.: Efficient keyword search across heterogeneous relational databases. In: Proceedings of the 23rd International Conference on Data Engineering, pp. 346–355 (2007)
34. Chanock, S.: Candidate genes and single nucleotide polymorphisms (snps) in the study of human disease. Disease Markers 19, 89–98 (2001)
35. Srivastava, U., Munagala, K., Widom, J., Motwani, R.: Query optimization over web services. In: Proceedings of the 32nd international conference on Very Large Data Bases, pp. 355–366 (2006)
36. Vidal, M.-E., Raschid, L., Mestre, J.: Challenges in selecting paths for navigational queries: Trade-off of benefit of path versus cost of plan. In: Proceedings of the 7th International Workshop on the Web and Databases, pp. 61–66 (2004)
37. Wang, F., Agrawal, G., Jin, R., Piontkivska, H.: Snpminer: A domain-specific deep web mining tool. In: Proceedings of the 7th IEEE International Conference on Bioinformatics and Bioengineering, pp. 192–199 (2007)
38. Wolf, G., Khatri, H., Chen, Y., Kambhampati, S.: Quic: A system for handling imprecision and incompleteness in autonomous databases. In: Proceedings of the Third Biennial Conference on Innovative Data Systems Research, pp. 263–268 (2007)

Summarizing Two-Dimensional Data with Skyline-Based Statistical Descriptors

Graham Cormode[1], Flip Korn[1], S. Muthukrishnan[2], and Divesh Srivastava[1]

[1] AT&T Labs — Research
{graham,flip,divesh}@research.att.com
[2] Rutgers University
muthu@cs.rutgers.edu

Abstract. Much real data consists of more than one dimension, such as financial transactions (eg, price $\times$ volume) and IP network flows (eg, duration $\times$ num-Bytes), and capture relationships between the variables. For a single dimension, quantiles are intuitive and robust descriptors. Processing and analyzing such data, particularly in data warehouse or data streaming settings, requires similarly robust and informative statistical descriptors that go beyond one-dimension. Applying quantile methods to summarize a multidimensional distribution along only singleton attributes ignores the rich dependence amongst the variables.

In this paper, we present new skyline-based statistical descriptors for capturing the distributions over pairs of dimensions. They generalize the notion of quantiles in the individual dimensions, and also incorporate properties of the joint distribution. We introduce ϕ-*quantours* and α-*radials*, which are skyline points over subsets of the data, and propose (ϕ, α)-*quantiles*, found from the union of these skylines, as statistical descriptors of two-dimensional distributions. We present efficient online algorithms for tracking (ϕ, α)-quantiles on two-dimensional streams using guaranteed small space. We identify the principal properties of the proposed descriptors and perform extensive experiments with synthetic and real IP traffic data to study the efficiency of our proposed algorithms.

1 Introduction

Much of the data in warehouses and streams contains multiple attributes (those attributes typically having skewed distributions), and analysis of such data often calls for summarizing relationships between attributes via robust statistical descriptors. Quantiles (e.g., percentiles), applied to singleton attributes independently, are more descriptive than just the median, which are in turn more robust than the mean. However, quantiles ignore the rich dependencies that can exist between attributes. In particular, it is important to understand the trade-offs between attribute value combinations as well as comparisons between data points exhibiting similar trade-offs.

Example. The American College Board measures academic performance using the SAT standardized test, which historically consisted of both math and verbal sections, and scores for each are sent to prospective colleges. Each section on its own is a narrow indicator of overall scholastic aptitude (e.g., a student with a top score on the

B. Ludäscher and Nikos Mamoulis (Eds.): SSDBM 2008, LNCS 5069, pp. 42–60, 2008.

math section may have weak verbal skills). Knowing percentiles for sections independently does not reveal the overall dominance of a student. E.g. if 10% of students scored above 1300 on the math section and 10% scored above 1200 on verbal, it does not imply that 10% of students simultaneously achieved these scores on both sections. Ideally, one would like to summarize the distribution of scores having (roughly) the same sectional percentile ratio. So it will be useful to answer questions such as, *"Which students ranked equivalently as well on the math and verbal sections?"* Figure 1 depicts such scores, labeled "0.5-radial" (explained later), using SAT scores for 2244 colleges.[1] Many schools seek balanced candidates; some (e.g., Caltech) have a

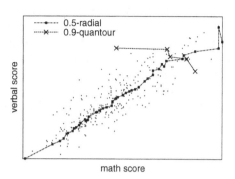

Fig. 1. Radials and quantours on SAT data

preference for mathematical prowess over verbal proficiency. Hence, it is useful to understand the trade-offs between score combinations. Another example question is, *"Which math and verbal score combinations dominate (on both math and verbal scores) 90% of students?"* Figure 1 depicts these scores as the curve labeled "0.9-quantour" on the same data. Using both notions one can obtain the "$(0.9, 0.5)$-quantile": the student dominating 90% of all students and amongst balanced students (along the 0.5-radial); or, equivalently, the balanced student amongst students with 90% dominance (along the 0.9-quantour). ☐

Other examples of 2D data analysis arise in the context of flow size distributions within network service providers (where size is measured in both bytes and duration), and financial transactions (price and volume). The outstanding question is how to define and develop such descriptors for multidimensional data. Such a descriptor should reflect the joint distribution while being composed solely of a representative subset of the data points (which can serve as example records). Furthermore, any given input parameter vector should identify a single point in the distribution (though these mappings need not be unique), so that a range of parameter values yields a fixed number of points. Quantiles provide a simple and robust point descriptors of an arbitrary one-dimensional data set (i.e., distribution). But in many problems, such as indexing, going from one dimension to two induces many new difficulties, both conceptual (what are suitable space partitioning techniques in two dimensions?) and technical (how to define the problem declaratively and find a solution efficiently?). To provide an analogous description of multi-dimensional data, several approaches have been considered in the Statistics and Database literature, the three most popular of which are:

– Quantile-quantile (QQ) plots are a well-established tool in data analysis to compare two one-dimensional distributions [11]. Values from the first distribution form the x-axis of the plot, and values from the second form the y-axis. However, this

---

[1] http://www.ivywest.com/satscore.htm

gives a fundamentally one-dimensional view: when applied to the marginals of two-dimensional data, the resulting pairs will likely not be points from the data set. QQ plots allow for the comparison of one-dimensional distributions but are insufficient to give insight into the joint distribution of multidimensional data.

– Tukey proposed an "onion peeling" technique to order points based on the proximity to the "center" of a data set, which is procedurally defined based on recursively stripping away convex hull layers to determine the *contour depth* of a point [22]. However, since an arbitrary number of points may exist at any given depth, points are not uniquely identified by q, so this technique is not a point descriptor. Other approaches, such as multidimensional equidepth histograms [19] have similar deficiencies and are fundamentally ad hoc in nature.

– Lastly, the skyline operator has been proposed to determine the subset of points not dominated by any other points (e.g., "find the skyline of cheap hotels that are close to the beach") [2] and have been generalized to so-called k-skybands [21], that is, points not dominated by more than k points, typically for some small constant k. The skyline (more generally, k-skyband) may contain an arbitrary number of points from the data (perhaps the entire set), and is thus not a point descriptor. Attempts to address this select a subset of the skyline based on orthogonal criteria, rather than distributional properties, such as subspace dominance [4], additional dimensions [12], or the number of distinct points dominated [18].

Our approach is based on computing skylines of subsets of data based on the "depth" of the data within the points in the dataset. This depth is given by removing points which dominate more than a fixed fraction of the whole data set, and so can access points which are far from the traditional skyline or skybands. We are not aware of any prior generalization of skyline queries in this way. This approach is more robust, and functionally rather than procedurally defined (e.g., in comparison with onion peeling).

Our Approach. We introduce two orthogonal notions: the ϕ-*dominance* of a point, which encodes the fraction of the data set dominated; and the α-*skewness* of a point, which is the ratio of its rank in the y-dimension divided by the sum of its ranks in x and y dimensions. We also introduce the notion of ϕ-*quantours* (short for "quantile-contours"), which is a set of points that dominate at most a ϕ-fraction of the multidimensional points; and the notion of α-*radials*, which are a set of points having an aspect ratio at most α in their marginal ranks. A point in the data set is uniquely identified by supplying values of ϕ and α. Together, we study the notion of (ϕ, α)-quantiles where points satisfy both α-radial as well as ϕ-quantour properties; thus they *simultaneously* capture the notion of being quantiles in *each* of the dimensions as well as in the joint distribution. They generalize the notion of skylines and provide clearly defined point descriptors in multiple dimensions. Parameters ϕ and α are reminiscent of polar coordinates but applied on order statistics rather than values.

Our Contributions. This paper consists of two parts.

– In the first part of the paper, we introduce ϕ-quantours, α-radials, and (ϕ, α)-quantiles as suitable descriptors of multidimensional distributions and describe a simple algorithm for computing them offline; We analyze these properties and illustrate their use in understanding the "local" structure (Section 2) with respect to the overall joint distribution.

- Motivated by the utility of (ϕ, α)-quantiles to understand a distribution, in the second part of the paper we present small-space algorithms for estimating ϕ-dominance and α-skewness at streaming speeds, with provable guarantees. While tracking (ϕ, α)-quantile points over a stream history may be useful for some applications, in many streaming scenarios only newly-arriving points can be acted upon (e.g., due to time criticality). Thus, the online problem we study is the following. Given a range of ϕ-dominance and α-skewness that are of interest (perhaps obtained by offline distributional analysis described in the first part), the (ϕ, α)-values of incoming points are monitored for matches. For example, if an incoming IP flow record exhibits high α-skew at high ϕ-dominance with respect to respective packet and byte counts, it may be a candidate for more thorough inspection.

  Our algorithms take a stream of points in two dimensions and create compact summaries that can answer such (ϕ, α)-quantile queries accurately. We derive these algorithms by constructing a variety of novel combinations of algorithms for quantiles in one dimension, building on prior work. We consider three fundamental approaches to building these combinations: the cross-product approach, the deferred-merge approach and the eager-merge approach. Each of these has different properties in terms of the space required and the amortized cost per point in the stream. From the viewpoint of real life applications, our methods are able to process more than a hundred thousand flows a second when monitoring IP flows on the stream, and as such are suitable for large Internet Service Provider applications.

- Finally, we perform a detailed experimental study of the online algorithms, using real IP network traffic data as well as synthetic data, to study the space and speed efficiency of these algorithms.

2   Preliminaries

We formally define quantile concepts and problems in two dimensions. We first state the definition of quantiles in one dimension, then show how these can extend to higher dimensions. In one dimension, we consider an input of N items. If we sort the input, and pick out a which is the ith in the sorted order, we say that the rank of a, $\mathrm{rank}(a)$, is i; alternatively, a dominates i points. Given an item a (which may or may not be present in the input), its rank is its position within the sorted input.[2] Throughout we use ϵ to denote the permitted tolerance for error. A *one-dimensional quantile query* is, given ϕ, and an error tolerance ϵ to return a so $(\phi - \epsilon)N \leq \mathrm{rank}(a) \leq (\phi + \epsilon)N$. E.g. finding the median corresponds to querying for the $\phi = \frac{1}{2}$-quantile. We also make use of the stronger error guarantee that is the "biased" quantiles requirement which asks to find a so that $\mathrm{rank}(a) \in (1 \pm \epsilon)\phi N$ [7].

Problem Definition and Discussion. In two dimensions, there is no longer a unique single descriptor of the dominance relationships of the points. The input consists of a stream of items: now these items are points in 2-dimensional space, drawn from a

---

[2] Hence the rank of an item which appears multiple times in the input is a range of positions where the same item occurs in the sorted input.

domain of size U, so that each coordinate is in the range $[0 \ldots U - 1]$. Let P be the set of N points.

Definition 1. *The ϕ-dominance of a point $p = (p_x, p_y)$ is the fraction ϕ of points from the input that are ϕ-dominated by p, i.e., $q \in P, (q_x \leq p_x) \wedge (q_y \leq p_y)$. Let $\mathrm{rank}(p)$ be the number of points from the input $q = (q_x, q_y)$ such that p ϕ-dominates q.[3] The ϕ-dominance of p, $\phi(p)$, is $\mathrm{rank}(p)/N$.*

This notion of dominance is reversed from traditional examples of dominance in skyline computations, but this does not materially affect the definition. From this, we can define a skyline-like operator which identifies points with similar dominance that are not themselves dominated.

Definition 2. *Given a set of points P, let P_ϕ be the subset of points such that $P_\phi = \{p \in P | \phi(p) \leq \phi\}$. Define the ϕ-dominance quantile contour, or ϕ-quantour for short, as the skyline of P_ϕ using the ϕ-dominance relation.*

Thus the ϕ-quantour selects those points such that their dominance is at most ϕ, and they are not dominated by any other points with dominance at most ϕ. This definition is carefully chosen so that it is well defined for any $\frac{1}{N} \leq \phi \leq 1$. When $\phi = 1$, this maximal quantour "touches the sky", that is, it is identical to the standard skyline.

Definition 3. *Given a point p, we define its α-skewness as $\alpha(p) = \mathrm{rank}_y(p)/(\mathrm{rank}_x(p) + \mathrm{rank}_y(p))$.[4] Intuitively, this shows how skewed this point is in terms of the ordering of its two dimensions. We similarly set P_α to be the subset of points such that $P_\alpha = \{p \in P | \alpha(p) \leq \alpha\}$. We say a point p α-dominates q if $p_y > q_y$ and $p_x < q_x$ (and hence $\alpha(p) > \alpha(q)$). We define an α-radial based on P_α, as the skyline of P_α, using the α-dominance relation.*

Our definition of α means every point has $0 < \alpha < 1$. Intuitively, $\alpha = \frac{1}{2}$ is 'balanced' between x and y dimensions. If we reflect all points in the line $y = x$ to generate a new point set $P^r = \{p^r = (p_y, p_x) | (p_x, p_y) \in P\}$, then the new $\alpha(p^r) = 1 - \alpha(p)$, showing the symmetry of the definition. Our notions of ϕ-dominance and α-dominance are chosen to ensure that, given any two points p and q, either one ϕ-dominates the other, or one α-dominates the other. From studying the definitions, the dominating point will be the one with maximum y value; if they share the same y value, then it is the one with the greater x value. Thus we can combine these two notions and define a unique point estimator.

Definition 4. *We define the (ϕ, α)-quantile as follows: we take the skyline of $P_\phi \cap P_\alpha$ based on ϕ-dominance and take the skyline of the result with α-dominance. What remains is defined to be the (ϕ, α) quantile.*

Example. Our two-dimensional definitions are illustrated in Figure 2 (points from the same quantours and radials are connected by line segments for illumination). It shows a set of 20 points in two dimensions. For each point we show its (ϕ, α) value:

---

[3] Technically, the rank of a point can be a range when points from the input share the same coordinates. For simplicity of presentation, we avoid further discussion of this issue and treat rank as if it gives a single value (our results hold when it is a range).

[4] Hence the α-skewness is a range for points having an x- or y-value that appears multiple times in the input; again, we treat α as a unique value for simplicity of presentation.

the point marked p with (ϕ, α) = $(0.4, 13/23)$ ϕ-dominates 8 points (including itself), and has rank_y = 13 and rank_x = 10, and so has ϕ = 0.4 and α = $13/23$. It falls on the intersection of $\phi = 0.5$ quantour and the $\alpha = \frac{2}{3}$ radial, and therefore is the unique $(0.5, \frac{2}{3})$ quantile.

For the $(0.5, 0.5)$ quantile, there are several possible points that have $(\phi \leq 0.5, \alpha \leq 0.5)$: these are $P_\alpha \cap P_\phi = \{(0.05, 1/2),$ $(0.1, 3/8),$ $(0.1, 2/9),$ $(0.25, 2/5),$ $(0.2, 3/16),$ $(0.35, 2/7),$ $(0.35, 10/21),$ $(0.45, 9/26)\}$. However, the unique point q at $(0.35, 10/21)$ either ϕ-dominates or α-dominates every other point in $P_\alpha \cap P_\phi$, and is therefore *the* $(0.5, 0.5)$ quantile. As noted next, this is the point with the greatest y rank amongst the set which obey the (ϕ, α) predicates, and it falls on the $\alpha = 0.5$-radial.

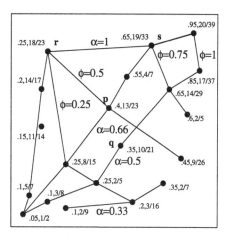

Fig. 2. Sample data set with (ϕ, α) values for each point shown along with ϕ-quantours and α-radials for $\phi = \frac{1}{4}, \frac{1}{2}, \frac{3}{4}$ and $\alpha = \frac{1}{3}, \frac{1}{2}, \frac{2}{3}, 1$.

□

Properties of (ϕ, α)-quantiles, ϕ-quantours and α-radials. One can readily verify the following statements:

1. A unique point from the input is found. This follows since, after taking the ϕ-dominance skyline, we obtain a set of points such that no pair is comparable under ϕ-dominance. Thus, they must all be comparable under α-dominance, and hence there is a unique maximal point.

2. The order of the taking the ϕ-dominance skyline and the α-dominance skyline is unimportant. Moreover, the unique point that is returned is the point in $P_\phi \cap P_\alpha$ which has the greatest y value and, if more than one has this y value, the one amongst them with the greatest x value.

3. The returned point lies on the α-radial or the ϕ-quantour, or possibly both if they intersect. When the maximal points on the α-radial and ϕ-quantour have differing y-values, then the (ϕ, α)-quantile is guaranteed to be on the α-radial.

4. For any two input points p, q, we have $\phi(p) = \phi(q) \wedge \alpha(p) = \alpha(q) \Leftrightarrow p = q$. In other words, all distinct input points have distinct (ϕ, α) values. This is seen by observing that if two points share the same α-value then one must ϕ-dominate the other. So p is *the* $(\phi(p), \alpha(p))$-quantile.

Thus we have a robust definition which selects a unique point p from the input having $\alpha(p) \leq \alpha$ and $\phi(p) \leq \phi$.

Exact Algorithm. We note that one can compute (ϕ, α) quantiles with similar time cost to computing quantiles in one dimension. We omit the details for brevity; the main idea is to scan the data in an appropriate order, and track certain information to rapidly compute the ϕ-dominance of a point. The result we claim is that given $O(N \log N)$ preprocessing, we can find the (ϕ, α)-quantiles (in two-dimensions) in time $O(N)$ per query.

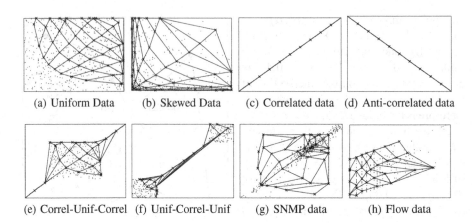

(a) Uniform Data   (b) Skewed Data   (c) Correlated data   (d) Anti-correlated data

(e) Correl-Unif-Correl (f) Unif-Correl-Unif   (g) SNMP data   (h) Flow data

Fig. 3. Radials and quantours on synthetic and real data

Other Dimensionalities. Note that in one-dimension, where the notion of α does not apply, this definition naturally collapses back to the familiar definition of ϕ-quantiles. We can also generalize these definitions to higher dimensions: the same notion of ϕ and ϕ-dominance translate immediately. One can also define $d-1$ new α-dominances between the d dimensions: let $\mathrm{rank}_i(p)$ denote the rank of the projection of p on the ith dimension of d dimensions. We define $\alpha_i(p) = \mathrm{rank}_i(p)/\sum_{j=1}^{d}\mathrm{rank}_j(p)$. Thus, we have $\sum_{i=1}^{d}\alpha_i(p) = 1$, and in two dimensions we recover our original definition. Since our focus is primarily on the two-dimensional case we will only briefly mention extensions to higher dimensions later.

Nature of ϕ-Quantours and α-Radials. Figure 3 plots the ϕ-quantours, α-radials and (ϕ, α) quantiles of several synthetic data sets, for $\phi \in \{0.1, ..., 0.9\}$ and $\alpha \in \{0.1, ..., 0.9\}$. Comparing Figure 3(b) with Figure 3(a) demonstrates how skew affects the angles between radials, causing them to diverge when both x and y have higher skew. In Figure 3(c), every point p has $\alpha(p) = 0.5$, effectively collapsing to 1D quantiles. In Figure 3(d), every point p has $\phi(p) = \frac{1}{N}$ (i.e., every point is on the skyline), effectively collapsing to 1D quantiles along the skyline. Figures 3(e) and 3(f) demonstrate how (ϕ, α) quantiles follow the "shape" of a point cloud on hybrid data sets having (and lacking) regional correlations. In Figure 3(f), the reason why only points above the diagonal merge into the diagonal is due to the definition forcing α-radial points p to have $\alpha(p) \le \alpha$.

Figures 3(g) and 3(h) give examples of quantours and radials of real data sets, plotted in log-log scales. In Figure 3(g), which plots outbound versus inbound traffic volumes, the correlation varies by region of the plot, as indicated by how the radials "bend inwards": at low values there is high correlation (acute quantour

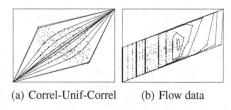

(a) Correl-Unif-Correl   (b) Flow data

Fig. 4. Depth contours on synthetic & real data

angles), at medium values low correlation (obtuse angles), and at high values again high correlation (acute angles). Compare this with Figures 3(a) and 3(b), where there is no correlation between x and y. Hence, inbound and outbound traffic is balanced for medium traffic levels, but the traffic parity doesn't remain when the throughput in either direction is too large or small. In Figure 3(h), which plots flow sizes in bytes versus packets, the distributions of both attributes are skewed, as indicated by the divergence of the radials away from the center. The curvature of the 0.5-radial shows that the relationship between packets and bytes is different for small "balanced" flows than larger ones, perhaps indicating distinct application types. For contrast, Figure 4 shows the data from Figures 3(e) and 3(h) plotted with depth contours. We argue that these plots are less informative, and more idiosyncratic, than the quantour/radial plots: Figure 4 is unable to capture the local information as well as Figure 3(e) and 3(h). Quantile-quantile plots are also unsuitable for these examples: Figures 3(a), 3(c), 3(e) and 3(f), all have the same set of x and y values, and so would be indistinguishable in a QQ-plot.

3   Streaming Algorithms

We now define and solve approximate versions of the problem which will allow us to reduce the amount of space required to answer them. While in one dimension it is possible to give relative error $(1 \pm \epsilon)$ estimates of ϕ-dominance, the same is not true in higher dimensions (we omit the formal information theoretic proof for brevity). Instead, we formalize the requirements as:

Definition 5. *Given a stream of data points in two dimensions, the* approximate (ϕ, α) quantile problem *is to process the stream so that, given any point p, we return an approximation $(\hat{\phi}(p), \hat{\alpha}(p))$ satisfying:*

$$\phi(p) - \epsilon \leq \hat{\phi}(p) \leq \phi(p) + \epsilon \text{ and } (1 - \epsilon)\alpha(p) \leq \hat{\alpha}(p) \leq (1 + \epsilon)\alpha(p).$$

This allows the accurate estimation of ϕ and α values for any point, whether in the input data, or not (e.g. answering "what if" queries, such as how a particular flow would rank amongst the recently observed data). We can draw sample points from the input as quantile points, since our algorithms keep track of a representative set of input points.

3.1   Algorithmic Approach

We provide a selection of algorithms to solve the above approximate problem over streaming data. We separate the two key components of the approximate (ϕ, α) quantile problem: estimating α values and estimating ϕ values. Estimating α is relatively straightforward, since it can be found by combining independent estimations of one dimensional quantiles with appropriate guarantees. Thus, the bulk of our challenge comes in estimating ϕ-dominance of points. Our approach uses techniques from the one-dimensional problem to summarize the two-dimensional data by dividing the data on each axis. We propose three classes of algorithms for combining one-dimensional summaries in order to compute (ϕ, α)-quantiles. These include some algorithms previously proposed for the related but distinct problem of computing rank queries in two-dimensions. Here, we provide a more general framework and demonstrate the different combinations that are possible to combine to form other examples of such classes.

The two-dimensional algorithms retain a subset of points from the input, and additional information to allow the estimation of ϕ and α values of points. We primarily consider algorithms which support several operations: INSERT(x), which takes a new value $x \in [0 \ldots U - 1]$ and updates the summary accordingly; $\hat{\phi}(q)$ which returns the approximate ϕ-dominance of point q; $\hat{\alpha}(q)$ which returns the approximate α-dominance of point q; and COMPRESS, which compacts the data structure.

3.2 Properties of One-Dimensional Algorithms

We first summarize some of the properties of one-dimensional quantile summary algorithms that we use as the building blocks of our two-dimensional algorithms—note that not all algorithms have all properties, which affects which combinations are possible.

- Mergable: an algorithm is considered (strongly) *mergable* if two summaries of different inputs can be combined to create a summary of the union of the two inputs. Our focus is on summaries that can be merged arbitrarily many times and still retain the same asymptotic space bounds. Other summaries are weakly mergable, in that their output can be merged to answer a query on the union of their inputs, but there is no guarantee that the size of the merged summary is less than the sum of sizes of the original summaries.
- Compressing: a quantile algorithm is *compressing* if it stores items from the input and, when the size of the summary is being reduced, tuples are compressed together by summing the counts of particular items or ranges from the input.
- Hierarchical: a quantile algorithm is *hierarchical* if it follows a pre-determined, hierarchical approach to merging: a tree-structure is placed over the domain, and merges only occur between child nodes and their parent.

We briefly outline existing "sample-based" algorithms (which maintain summaries based on selecting points from the stream):

- The Greenwald-Khanna algorithm (GK) [13] retains a set of tuples consisting of an item from the input, a count of how many items have been merged into that tuple, and an upper bound on the rank of the item.
- The Quantile Digest algorithm (QD) [25] retains a set of tuples, where each tuple consists of an item or (dyadic) range of items from the input and a count of how many items have been merged into that tuple.
- Biased Quantiles (BQ) [7] uses Quantile Digest-like data structure with different manipulation routines to estimate the dominance of a point with stronger relative error guarantees, with slightly higher space usage.

QD and BQ do not by default retain points from the input. But it is straightforward to augment them to do so: for every tuple that relates to a range of items, we additionally keep some point from the input that fell in that range, and merge these appropriately. Figure 5 lists the key properties of these algorithms. Randomized methods, such as random sampling, can give similar guarantees, but our focus is on stronger deterministic guarantees, so we do not discuss these. Each of the algorithms we do consider allows us to 1DINSERT a new item, 1DCOMPRESS the structure to compact the size to its theoretical bounds, and 1DQUERY to find the approximate rank of a point.

Method	Space cost	Update time	Compressing	Mergability	Hierarchical
GK [13]	$O(\frac{1}{\epsilon}\log \epsilon N)$	$O(\log(\frac{1}{\epsilon}\log \epsilon N))$	Yes	Weakly	No
QD [25]	$O(\frac{1}{\epsilon}\log U)$	$O(\log \log U)$	Yes	Strongly	Yes
BQ [7]	$O(\frac{1}{\epsilon}\log U \log \epsilon N)$	$O(\log \log U)$	No	Strongly	Yes

Fig. 5. Properties of one-dimensional quantile summary algorithms

Using 1D algorithms to approximate α. Computing $\alpha(p)$ can be carried out with relative error in small space:

Theorem 1. *Using space $O(\frac{1}{\epsilon}\log U \log \epsilon N)$, we can take any point q and compute an approximation $\hat{\alpha}(q)$ such that $(1 - \epsilon)\alpha(q) \leq \hat{\alpha}(q) \leq (1 + \epsilon)\alpha(q)$.*

Proof. Recall that $\alpha(q) = \operatorname{rank}_y(q)/(\operatorname{rank}_x(q) + \operatorname{rank}_y(q))$. By maintaining a biased quantile data summary (BQ) on the x-values and y-values of points independently, we can approximate rank_x and rank_y with $\hat{r}_x$ and $\hat{r}_y$ such that $(1-\epsilon)\operatorname{rank}_x(q) \leq \hat{r}_x(q) \leq (1 + \epsilon)\operatorname{rank}_x(q)$ and $(1 - \epsilon)\operatorname{rank}_y(q) \leq \hat{r}_y(q) \leq (1 + \epsilon)\operatorname{rank}_y(q)$. Thus, finding $\hat{\alpha}(q) = \hat{r}_y(q)/(\hat{r}_x(q) + \hat{r}_y(q))$ ensures that the relative error is between $\frac{1-\epsilon}{1+\epsilon} \geq 1 - 2\epsilon$, and $\frac{1+\epsilon}{1-\epsilon} \leq 1 + 3\epsilon$, for $\epsilon \leq \frac{1}{3}$. We rescale ϵ by a factor 3, which does not affect the asymptotic space costs of the BQ summary, whose space bounds follow from [7].

The next three sections outline three classes of algorithms, and for each give sample instantiations based on combining appropriate one-dimensional algorithms from the list above. For reasons of brevity, we do not give complete proofs of all properties of the algorithms in this presentation, but instead provide an outline of why they hold. The main properties of each algorithm are summarized in Figure 6.

Type	Instance	Space	Amortized time	$\hat{\phi}$-query time
Cross-product	GK×GK	$O(\frac{\log^2 \epsilon N}{\epsilon^2})$	$O(\frac{\log \epsilon N}{\epsilon})$	$O(\frac{\log^2 \epsilon N}{\epsilon^2})$
Deferred-merge	GK ×QD	$O(\frac{\log U \log \epsilon N}{\epsilon^2})$	$O(\log \frac{\log U}{\epsilon})$	$O(\frac{\log U \log \epsilon N}{\epsilon^2})$
Eager-merge	QD×QD	$O(\frac{\log^3 U}{\epsilon})$	$O(\log U \log \log U)$	$O(\frac{\log^3 U}{\epsilon})$

Fig. 6. Comparison of bounds for different instantiations of two-dimensional data structures

3.3 Cross-Product Algorithms

Our first approach to tracking quantile information in multiple dimensions is the cross-product approach, based on keeping one-dimensional compressing sample-based summaries on each dimension independently.

Update Processing in Cross-Product. Each 1D summary consists of a set of items and ranges from each dimension, and we maintain information about the cross-product across dimensions: if we keep information on a set X of points or ranges from the x-dimension, and a set Y on the y-dimension, then we will maintain information on the Cartesian product $X \times Y$. For each cell in this cross-product, we maintain a count of the number of input items associated with the cell, and a point from the input that falls in the cell, if the count is non-zero. The counts must satisfy the property that summing the counts of all cells within a rectangle gives a lower bound on the number of input points within that rectangle.

Periodically, a 1DCOMPRESS operation is run on first X and then Y. When the 1D algorithm merges two tuples together, the corresponding cells from $X \times Y$ are also merged. The count of the merged cell is the sum of the counts of the cells which went to form it. The retained point is chosen arbitrarily from the points of the merged cells. When INSERT is run on a new point $p = (p_x, p_y)$, we update $X \times Y$ accordingly to reflect the changes from the insertion to X and Y. Typically, this means inserting p_x into X and p_y into Y, so we update $X \times Y$ with $\{p_x\} \times Y$, $X \times \{p_y\}$ and (p_x, p_y). We set the count $(p_x, p_y) = 1$ (if (p_x, p_y) already exists in $X \times Y$, we increment its count) and the count of all other new cells to zero (if they already exist, we leave them unchanged).

Estimation and Accuracy on Cross-Product. In order to compute $\hat{\phi}(q)$, we compute the count of all cells dominated by q. The approximation error in our response comes because q may fall within a cell: all cells below and to the left of q contain points that are strictly dominated by q, and all cells above or to the right contain points that are not dominated by q; this leaves the cells containing the x-value of q, and the cells containing the y-value of q. The properties of the one-dimensional structures ensure that this uncertainty is limited to ϵN. The space used by this algorithm depends on the product of the sizes of the one-dimensional summary structures used. Various combinations are possible: GK×GK, QD×QD (which was considered in [26]) or even GK×QD.

Instantiation: Cross-product with GK×GK . Since the GK algorithm typically attains the best space usage on one-dimensional data, it is expected that GK×GK will attain the best space usage of the cross-product algorithms. The space bounds follow from [26], and restated here along with our improved time bounds. Asymptotically, the space usage of GK×GK is bounded by $O(\frac{1}{\epsilon^2} \log^2(\epsilon N))$. The time cost of all cross-product algorithms can be somewhat high. For efficiency, rather than explicitly materializing all cells, our idea is to use a hash table to store only those cells (x, y) in the grid that have a non-zero count. GK (along with other one-dimensional algorithms) adds a new item to the data structure for each INSERT operation. In two dimensions this adds $O(\frac{1}{\epsilon} \log(\epsilon N))$ new cells, but by using the hash table approach, we only have to do $O(1)$ operations since only a single new cell is created with a non-zero count. COMPRESS requires time linear in the (worst case) size of the data structure, $O(\frac{1}{\epsilon^2} \log^2(\epsilon N))$: it consists of compressing each of the one-dimensional data structures independently, and when they merge two of their tuples, merging together the cells associated with those rows/columns. Merging a row takes time linear in its size, and compressing each 1D data structure also takes linear time, so the total time cost of COMPRESS is worst case bounded by the size of the data structure. By performing COMPRESS after every $O(\frac{1}{\epsilon} \log(\epsilon N))$ INSERT operations, the space bounds are preserved, while the amortized time cost is $O(\frac{1}{\epsilon} \log(\epsilon N))$ per update. The time to compute $\hat{\phi}(p)$ is at most linear in the size of the data structure, and to compute the $\hat{\phi}$ of every stored point at most logarithmically longer in the size of the data structure, using the exact algorithm of Section 2.

3.4 Deferred-Merge Algorithms

Our next approach fixes an ordering on the dimensions, and runs a compressing algorithm on the primary (x) dimension with uniform guarantee ϵ_x. It runs multiple instances of (strongly) mergable algorithms on the secondary (y) dimension with parameter ϵ_y. Dominance queries require the merging of multiple secondary data structures to answer, but we defer this merging to query time, hence "deferred-merge". So for the primary axis, we can use either GK or QD; for the secondary axis, we can use QD because of its mergable properties.

Update Processing in Deferred-Merge. For each input point p, we first 1DINSERT p_x into the compressing algorithm on the x-dimension. Instead of just keeping a count of the number of items retained in each tuple, we also keep a second one-dimensional quantile data structure that summarizes the y-values of all points that are summarized in the tuple. So after finding the data structure tuple to insert p_x into, we insert p_y into its associated summary. For compression of the data structure, we first run 1DCOMPRESS on the one-dimensional summary of x-values. If we merge tuples in this structure, then we also merge together their summaries of y-values and 1DCOMPRESS the result.

Estimation and Accuracy on Deferred-Merge. In order to approximate $\phi(q)$, we can query q_x to find the summary containing q_x, and then merge together all summaries of y-values to the left of this, and query the resulting merge structure with q_y. The approximation error from this approach comes from two sources: uncertainty due to the querying on x-values, and uncertainty on y-values. The guarantees of the summaries on x-values ensure that the uncertainty is at most $\epsilon_x N$ points; similarly, posing a query to the merged summary of y-values gives a guarantee depending on ϵ_y. In order to get the required accuracy bounds, we set the parameters ϵ_x and ϵ_y less than or equal to $\frac{\epsilon}{2}$, giving accuracy ϵN or better for each query. Consequently, this algorithm solves the approximate (ϕ, α) quantiles problem. Lastly, we observe that the space bound of merge-based algorithms is at most the product of the space bounds of the algorithms on each axis. We can instantiate the deferred merge algorithms with either GK or QD on the primary (x) axis, but require a strongly mergable algorithm for the secondary (y) axis, which allows us to use GK×QD or QD×QD.

Instantiation: Deferred-merge algorithm with GK×QD. GK is a compressing algorithm that typically achieves the best space in 1D; QD is a mergable algorithm that also has relatively small space cost. The worst case bound on the space needed is the product of the space bounds: $O(\frac{1}{\epsilon^2} \log(\epsilon N) \log U)$. To INSERT a new point, we first insert the p_x into the GK structure, and store p_y along with the inserted points itself as a QD summary of size 1. To COMPRESS the summary, we run the one-dimensional GK-COMPRESS on the GK structure with error parameter $\epsilon_x = \epsilon/2$, and when two tuples in GK are merged, we also merge their corresponding QD summaries (and then run the one-dimensional QD-COMPRESS on the result with error parameter $\epsilon_y = \epsilon/2$). The time to perform the compression is thus worst case bounded by time linear in the total data structure. This can be amortized by running COMPRESS only after every $O(\frac{1}{\epsilon^2} \log(\epsilon N) \log U)$ updates, which retains the asymptotic space bounds, and ensures that the update cost is dominated by the cost of inserting into one GK structure and one

QD structure, which is $O(\log(\frac{1}{\epsilon}) + \log \log U)$. The overall uncertainty in ϕ-dominance queries is at most ϵ.

An important feature of the merge based algorithms is that the size of the data structure is never more than the size of the input, since each input point corresponds to at most one tuple in the summary. This is in contrast to the next methods we consider, which have the potential to represent each point multiple times during the early phases of the algorithm.

3.5   Eager-Merge Algorithms

The class of eager-merge algorithms also combine one-dimensional algorithms. They use a compressing hierarchical algorithm on the primary (x) axis, with uniform guarantee ϵ_x. Rather than demanding that the algorithm on the secondary (y) axis be strongly mergable (as in the deferred-merge case), they eagerly compute the results of merging by inserting the y-dimension of each input point into a summary at each level of the hierarchy, and (weakly) merge appropriate outputs from these structures at query time. The first such algorithm was proposed in [14]. Here, we generalize the description, and give details for completeness.

Update Processing In Eager-Merge. For each materialized node in the primary data structure, we maintain a second data structure on the y-dimensional values of all points allocated to this node *or any of its descendants*, with an accuracy guarantee set to ensure accurate answers. To perform an INSERT operation on a new point, we first find the node in the x-structure corresponding to p_x from the inserted point. We then 1DINSERT p_y into the corresponding y-summary of that node, and also into the y-summaries of every ancestor of the node. When we create a new node based on an input point p, we store p along with that node. COMPRESS takes place firstly over the data structure on the x-axis, and then on each of the data structures covering the y-axis contained within it. We run the 1DCOMPRESS routine over the x structure and when nodes are merged, only their associated counts are updated. The y-summaries corresponding to the deleted children can simply be deleted: they are not merged into their parent, since during insertions, the result of the merge is already (eagerly) computed, by ensuring that every inserted point was put into the y-summaries of every ancestor. After deleting y-summaries, we then perform a COMPRESS on those that remain.

Estimation and Accuracy in Eager-Merge. The product of the x and y space bounds gives a naive space bound, but tighter bounds are obtained using the fact that each point is represented at most once per level. We need to use a hierarchical algorithm such as QD on the primary (x) axis, and a (weakly) mergable algorithm on the secondary axis.

Instantiation: Eager-merge algorithm with QD×QD. Our eager-merge algorithm uses QD as the method on the first dimension, since this method is hierarchical, and also uses QD on the second dimension. (An alternative is QD×GK, but QD×QD yields better worst case space bounds.) Rather than using the number of points within the y-summary as the basis of the threshold for compressing, we use a threshold based on N, the total number of points in the data structure (the same idea was used in [14], though we obtain different bounds due to implementation choices). This is because the overall error guarantee is in terms of ϵN, and to give a tight space bound. Within each y-summary, we set a local error tolerance of $\frac{\epsilon}{2 \log U}$. This is chosen to ensure that when these are summed over $\log U$ different summaries, the total error will be bounded by $\epsilon/2$.

INSERT operations take time $O(\log U \log \log U)$: we perform 1DINSERT operations for $O(\log U)$ nodes in the x-dimensional summary, and each of those can be completed in time $O(\log \log U)$ on the y-dimensional summaries [7]. A COMPRESS operation takes time linear in the size of the data structure, since it reduces to running 1DCOMPRESS on multiple one-dimensional data structures, each of which takes time linear in the size of their substructure. The amortized update time is therefore $O(\log U \log \log U)$.

To answer a ϕ-dominance query given q, we find a set of nodes in the x structure to query, by representing q_x as a union of disjoint ranges from the hierarchy. The binary tree structure of the QD algorithm ensures that there are at most $\log U$ nodes in this set. For each node in the set, we query its corresponding y-summary with q_y and sum the outputs to obtain the estimate of $\text{rank}(q)$: q_x is broken into two y-summaries, one on the first half of the horizontal span, another on the next quarter. We get an accurate count of the number of points within the queried region: summing the accuracy bounds over the at most $\log U$ queries gives error at most $\epsilon/2$. The uncertainty due to the query on the x-axis is also at most $\epsilon/2$, from the accuracy bound on the x-data structure, so the total error is at most ϵ. Since queries probe at most $\log U$ y-summaries, each in time $O(\log U)$, the total time cost is $O(\log^2 U)$ per query after a COMPRESS has updated counts in time $O(\frac{\log^3 U}{\epsilon})$.

3.6 Comparison

We summarize the space and time bounds of various instantiations of the three different approaches in Figure 6. Initially, it is hard to compare them, since the relative asymptotic cost depends on the setting of parameter ϵ, relative to $\log U$. Comparing eager-merge costs to deferred-merge, the space cost trades off roughly a factor $O(\frac{1}{\epsilon})$ for one of $O(\log U)$. This suggests that for very fine accuracy situations $\epsilon \ll \frac{1}{\log U}$, the eager-merge approach will win out. Comparing cross-product to deferred-merge, it seems possible that cross-product methods will use less space, especially since GK has been observed to use closer to $O(\frac{1}{\epsilon})$ space in practice. But the amortized running time of the deferred-merge algorithms are exponentially smaller than those of the cross-product algorithms, which can make a big difference over high speed data streams.

In terms of query time for estimating ϕ-dominance, by default all methods require a linear pass over the whole data structure to answer the query, so those with smaller space have faster queries. But, if many queries are being computed in a batch, then the time cost of eager-merge methods can be improved to $O(\log^2 U)$ by taking advantage of the hierarchical structure of the summary to answer queries much faster. This relies on utilizing stored counts within the data structure that are needed by the INSERT and COMPRESS routines. By recomputing these counts in a linear pass before each computation of $\hat{\phi}(p)$, the running time is much reduced for each query.

4   Experimental Results

In this section, we summarize experiments comparing the three classes of algorithms in Section 3 for answering approximate (ϕ, α) estimation queries: cross-product based on GK×GK; deferred-merge based on GK×QD; and eager-merge based on QD×QD.

In all experiments, we run all three instances with the same accuracy requirements ϵ and observe how their space and time bounds vary while they provide the same accuracy guarantee. We report space usage in terms of the number of tuples and as a function of the number of stream tuples that have arrived; we report performance in terms of the *throughput*, that is, the number of tuples processed per second. These experiments were run on a Linux machine with a 2.8 GHz Pentium CPU, 2 GB RAM and 512K cache. For synthetic data, we generated random uniform data with universe size 2^{32} on each axis independently; we did the same with Zipfian data (skew parameter=1.3); and we generated bivariate Gaussian data sets with varying correlation strengths. We also used real network flow data from a router used in the AT&T Common Backbone and obtained 2D data sets by projecting on the fields (numPkts, numBytes), (duration, numBytes), and (srcIP, destIP).

In addition, we conducted experiments using a live stream of IP packet traffic monitored at a local network interface. The traffic speeds varied throughout the day, with a rate of about 200K TCP packets per second at mid-day. The stream data was monitored using Gigascope, a highly optimized system for monitoring very high speed data streams [8]. Our copy was installed on a FreeBSD machine with a dual Intel Xeon 3.2 GHz CPU and 3.75 GB RAM. The methods were implemented as User-Defined Aggregate Functions (UDAFs) in Gigascope, as explained in [6].

Results on Real and Synthetic Data Sets. Figure 7 shows a space usage and throughput comparison (in log scale) of the different algorithms run with $\epsilon = 0.01$; Figure 8 shows the same with $\epsilon = 0.001$. The algorithms were run longer (up to 10M tuples) in Figure 8 than in Figure 7 (1M tuples), long enough so that the space usage curves can be seen to "level off" and converge to approximately stable values. None of the algorithms

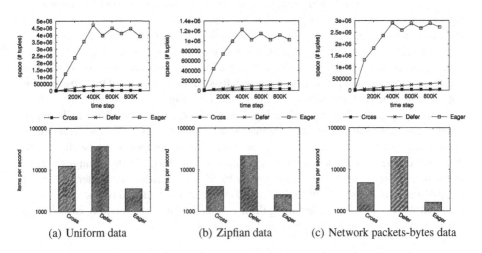

(a) Uniform data   (b) Zipfian data   (c) Network packets-bytes data

Fig. 7. Space usage and throughput ($\epsilon = 0.01$)

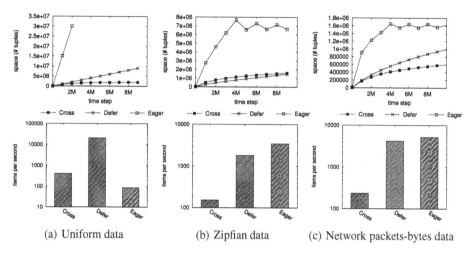

(a) Uniform data   (b) Zipfian data   (c) Network packets-bytes data

Fig. 8. Space usage and throughput ($\epsilon = 0.001$)

was dominant in all cases. Notice that space usage and performance were not always correlated: some cases exhibited a space-time trade-off.

Data skew affected space usage and performance: in general, there was smaller space and faster runtime with increasing skew. Whereas the GK algorithm (in 1D) is impervious to the data values since it only cares about rank-ordering, the universe-based algorithms (eg, QD) benefit from non-uniformity. This can be seen from the space usages in Figures 7(a) and 7(b): the gap between the curves for eager-merge (based on QD) and cross-product (based on GK) narrows significantly. Indeed, real data is often skewed, as is the case with flow data (thought not quite as skewed as the Zipfian data set). Hence, the space usage on this data set in Figure 7(c) was much more similar to that on Zipfian data in Figure 7(b) than uniform data in Figure 7(a). The deferred-merge algorithm, a hybrid of GK and QD approaches, is asymmetric: it is more efficient when the attribute with higher skew is in the y-axis. We observed up to a factor of 5 difference in the space used by deferred-merge by swapping the ordering of the axes.

Another relevant factor is correlation between attributes, as observed using bivariate Gaussian data of differing covariances. This benefits the cross-product approach because, with increasing correlation, cross-product effectively becomes one-dimensional; correlation did not have much impact on the other classes. Real data, such as flow packets and bytes, often exhibit correlations. Skew, which does not benefit cross-product but benefits the other algorithms, appeared to be more significant in our experiments.

In the streaming scenario, throughput often trumps space efficiency.[5] Therefore, deferred-merge is the overall "safest" algorithm to use. In our experiments, it always had higher throughput than cross-product (often 1-2 orders of magnitude) while being competitive with respect to space usage; and it had as good or better performance than eager-merge, with significantly better space usage. In some cases (e.g., Figure 8(a)), the

---

[5] For example, the Gigascope high-level query processor can make use of as much RAM as available.

space usage of eager-merge grew so large that it exceeded the RAM size, causing the system to thrash and resulting in abysmal throughput.

Live Packet Streams. We issued a long-running query to the Gigascope system to maintain per-minute (ϕ, α)-quantiles with $\epsilon = 0.01$, on our methods, over a cumulative window of pairs of approximate flow aggregates on (numPkts, numBytes) grouped by flow. We compared to a 'null' aggregate that just counts updates, which required 85% CPU utilization. Cross-product and eager-merge could not keep up with this stream, but the deferred-merge UDAF achieved processing rates similar to the 'null' aggregate (130-140K per minute) and CPU utilization around 88%.

5   Related Work

In Statistics, there has been significant work on multidimensional quantile descriptors. A good overview can be found in [24], with some specific approaches in [5,9,16]. However, these approaches do not yield point descriptors but algebraic curves. Quantile-quantile (QQ) plots compare the 1D quantiles from each marginal as 2D points. Such plots allow us to compare one-dimensional distributions, but are insufficient to give us full insight into joint distribution of multi-dimensional data.

Similarly, in Computational Geometry, notions of median and other quantiles are procedurally defined in terms of "depths" of point-sets and produce quantile regions [10,15,22,23]. While this may be satisfying for visualization, such region-based quantile descriptors are not point descriptors. Other descriptors such as multidimensional equidepth histograms and one dimensional quantiles on linearized multidimensional data are ad hoc, and have similar deficiencies.

In recent years there has been significant interest in the area of data streams, where the space available for processing is considerably smaller than the input, which is presented in a "one-pass" fashion [1, 20]. As previously noted, there is a wealth of algorithms devoted to the problem of tracking (1D) quantiles in data streams [7,13,25]. For multidimensional data streams, prior work has been scant; it is primarily focused on summaries such as histograms [3,27].

Our three classes of algorithms for combining one-dimensional summaries in order to compute (ϕ, α)-quantiles capture some algorithms previously proposed for the related but distinct problem of computing rank queries in two-dimensions. Algorithms proposed in [14, 26] can be thought of as fitting into our class of "cross-product" and "eager-merging" summaries, respectively. Here, we provide a more general framework and demonstrate the different combinations that are possible to combine to form other examples of such classes. We are not aware of any prior examples which demonstrate the deferred-merging approach that we have proposed. Interestingly, it appears that this class is often the best suited for keeping pace with high streaming data rates.

6   Conclusions

Data in warehouses and streams, such as IP traffic data, are typically multidimensional, and capture relationships between multiple variables. In a variety of applications, one needs simple, statistical point descriptors of such streams. Existing methods use quantiles in single dimensions and therefore miss joint distributional behavior, or give procedural or ad hoc definitions. In this paper, we propose skyline-based statistical descriptors, and introduce ϕ-quantours, α-radials and, in particular (ϕ, α)-*quantiles*. We present fast and small-space streaming algorithms for computing them approximately with guaranteed accuracy by judicious combinations of previously known one-dimensional algorithms. We demonstrate experimentally the efficiency of computing 2D quantiles in data streams with synthetic and real data.

References

1. Babcock, B., Babu, S., Datar, M., Motwani, R., Widom, J.: Models and issues in data stream systems. In: ACM PODS, pp. 1–16 (2002)
2. Borzsonyi, S., Kossmann, D., Stocker, K.: The skyline operator. In: IEEE ICDE, pp. 421–430 (2001)
3. Bruno, N., Chaudhuri, S., Gravano, L.: STHoles: a multidimensional workload-aware histogram. In: ACM SIGMOD (2001)
4. Chan, C.Y., Jagadish, H.V., Tan, K.-L., Tung, A.K.H., Zhang, Z.: On High Dimensional Skylines. In: Ioannidis, Y., Scholl, M.H., Schmidt, J.W., Matthes, F., Hatzopoulos, M., Böhm, K., Kemper, A., Grust, T., Böhm, C. (eds.) EDBT 2006. LNCS, vol. 3896, pp. 478–495. Springer, Heidelberg (2006)
5. Chaudhuri, P.: On a geometric notion of quantiles for multivariate data. Journal of the American Statistical Association 91, 862–872 (1996)
6. Cormode, G., Korn, F., Muthukrishnan, S., Johnson, T., Spatscheck, O., Srivastava, D.: Holistic UDAFs at streaming speeds. In: ACM SIGMOD, pp. 35–46 (2004)
7. Cormode, G., Korn, F., Muthukrishnan, S., Srivastava, D.: Space- and time-efficient deterministic algorithms for biased quantiles over data streams. In: ACM PODS (2006)
8. Cranor, C., Johnson, T., Spatscheck, O., Shkapenyuk, V.: Gigascope: A stream database for network applications. In: ACM SIGMOD, pp. 647–651 (2003)
9. Einmal, J., Mason, D.: Generalized quantile processes. Annals of Statistics 20(2), 1062–1078 (1992)
10. Eppstein, D.: Single point estimators (1999),
    http://www.ics.uci.edu/~eppstein/280/point.html
11. Evans, M., Hastings, N., Peacock, B.: Statistical Distributions, 3rd edn. Wiley, New York (2000)
12. Goncalves, M., Vidal, M.-E.: Top-k Skyline: A Unified Approach. In: OTM Workshops (2005)
13. Greenwald, M., Khanna, S.: Space-efficient online computation of quantile summaries. In: ACM SIGMOD, pp. 58–66 (2001)
14. Hershberger, J., Shrivastava, N., Suri, S., Toth, C.: Adaptive spatial partitioning for multidimensional data streams. In: ISAAC (2004)
15. Johnson, T., Kwok, I., Ng, R.: Fast computation of 2-dimensional depth contours. In: KDD, pp. 224–228 (1998)
16. Koltchinskii, V.I.: M-estimation, convexity and quantiles. Annals of Statistics 25(2), 435–477 (1997)
17. Kumar, A., Sung, M., Xu, J., Wang, J.: Data streaming algorithms for efficient and accurate estimation of flow distribution. In: ACM Sigmetrics (2004)
18. Lin, X., Yuan, Y., Zhang, Q., Zhang, Y.: Selecting Stars: the k Most Representative Skyline Operator. In: IEEE ICDE (2007)

19. Muralikrishna, M., DeWitt, D.: Equi-depth histograms for estimating selectivity factors for multi-dimensional queries. In: ACM SIGMOD (1988)
20. Muthukrishnan, S.: Data streams: Algorithms and applications. In: ACM-SIAM SODA (2003)
21. Papadias, D., Tao, Y., Fu, G., Seeger, B.: Progressive skyline computation in database systems. ACM Transactions on Database Systems 30(1), 41–82 (2005)
22. Preparata, F.P., Shamos, M.I.: Computational Geometry: An Introduction, 2nd edn. Springer, Heidelberg (1985)
23. Rousseeuw, P.J., Leroy, A.M.: Robust Regression and Outlier Detection. Wiley, Chichester (2003)
24. Serfling, R.: Quantile functions for multivariate analysis: approaches and applications. Statistica Neerlandica 56(2), 214–232 (2002)
25. Shrivastava, N., Buragohain, C., Agrawal, D., Suri, S.: Medians and beyond: New aggregation techniques for sensor networks. In: ACM SenSys. (2004)
26. Suri, S., Tóth, C.D., Zhou, Y.: Range counting over multidimensional data streams. In: SoCG (2004)
27. Thaper, N., Indyk, P., Guha, S., Koudas, N.: Dynamic multidimensional histograms. In: ACM SIGMOD, pp. 359–366 (2002)

Query Selectivity Estimation for Uncertain Data

Sarvjeet Singh, Chris Mayfield, Rahul Shah⋆, Sunil Prabhakar,
and Susanne Hambrusch

Department of Computer Science, Purdue University
West Lafayette, IN 47907, USA
{sarvjeet,cmayfiel,rahul,sunil,seh}@cs.purdue.edu

Abstract. Applications requiring the handling of uncertain data have led to the development of database management systems extending the scope of relational databases to include uncertain (probabilistic) data as a native data type. New automatic query optimizations having the ability to estimate the cost of execution of a given query plan, as available in existing databases, need to be developed. For probabilistic data this involves providing selectivity estimations that can handle multiple values for each attribute and also new query types with threshold values. This paper presents novel selectivity estimation functions for uncertain data and shows how these functions can be integrated into PostgreSQL to achieve query optimization for probabilistic queries over uncertain data. The proposed methods are able to handle both attribute- and tuple-uncertainty. Our experimental results show that our algorithms are efficient and give good selectivity estimates with low space-time overhead.

1 Introduction

Recently there has been a surge in interest in managing probabilistic data in relational databases [1,2,3,4,5,6]. This interest is engendered by the needs of numerous applications including scientific data management, data integration, sensor databases, data cleaning, text processing and location-based services. The relational database model has very little support for uncertain data, limited to the use of NULL values. The nature of uncertainty in many applications is such that it is necessary to store alternative values for tuples, or attributes and process probabilistic queries over this data.

Several models have been proposed for extending the scope of relational databases to include uncertain (probabilistic) data as a native data type. These models define new semantics for query processing over uncertain data. The results of these queries are typically probabilistic in nature. Since results with a low probability of occurrence are generally less interesting than higher probability answers, an important new class of *threshold* queries has been identified [7]. These queries return only those answers that have a probability exceeding a threshold. While this thresholding weeds out less relevant answers, it also opens up possibilities for query optimization. There has been some recent work on efficient processing of threshold queries over uncertain data [8]. This work has largely focused on indexing methods to improve query performance.

⋆ Work done while at Purdue University. Current affiliation: Louisiana State University, Baton Rouge, Louisiana, USA.

B. Ludäscher and Nikos Mamoulis (Eds.): SSDBM 2008, LNCS 5069, pp. 61–78, 2008.
ⓒ Springer-Verlag Berlin Heidelberg 2008

The long-term goal for several projects is the development of novel database management systems that natively handle uncertain data. An important step in this direction is the development of automatic query optimization as is available in existing databases. Toward this end, an essential ingredient is the ability to estimate the cost of execution of a given query plan. For probabilistic data this would involve providing selectivity estimates for probabilistic operators. Currently, there is no work on providing such selectivity estimation functions for probabilistic data. With the availability of these estimation functions it is possible to use existing query optimization techniques that are already built into databases to handle the case of probabilistic data.

In this paper we address this problem and develop novel selectivity estimation functions for uncertain data. We also show how these functions can be integrated into PostgreSQL to achieve query optimization for probabilistic queries over uncertain data. Selectivity estimation for uncertain data needs to handle multiple values for each attribute and also novel query types with threshold values. Furthermore, an important type of uncertainty transforms a single attribute value to a continuous distribution – this is especially common in sensor databases [9]. The existing cost estimation methods are therefore not applicable for this domain.

The goal of this paper is to handle selectivity estimation for the two main types of uncertainty that have been proposed in recent work: *tuple uncertainty* [1,2] and *attribute uncertainty* [7]. To demonstrate the effectiveness of our selectivity estimation techniques, we have used an open-source database management system for uncertain data called Orion [3] which is built into PostgreSQL.

The major contributions of this paper are as follows:

– We have developed efficient algorithms for selectivity estimation of probabilistic threshold queries over uncertain data.
– We have implemented these algorithms in a real database system.
– Our experimental results show that the algorithms are efficient and provide good estimates for query selectivities.

The rest of this paper is organized as follows. Section 2 summarizes the related work done in this area. We formally describe the uncertainty model and probabilistic queries in Section 3. Our algorithms for selectivity estimation are presented in Section 4. We present the experimental results in Section 5, and Section 6 concludes this paper.

2 Related Work

There is a rich body of work on selectivity estimation for traditional relational database management systems. Most approaches for selectivity estimation on precise data use histograms. Poosala et al [10] proposed a taxonomy to capture all previously proposed histogram approaches. These approaches are not applicable for uncertain data because both the queries and the underlying data types for uncertain data differ greatly from traditional data and queries.

More recently, there has been a great deal of work on the development of models for representing uncertainty in databases. Two main approaches for modeling uncertain data have emerged in this field: Tuple uncertainty [1,2] and Attribute uncertainty [7].

Similar models have been proposed in moving-object environments [11] and in sensor networks [9]. Several systems that handle such uncertainty in data have been recently proposed (Orion [3], MayBMS [12], Mystiq [13], Trio [14], [4]). This probabilistic modeling of data has also been extended to semi-structured data [15] and XML [16].

Efficient evaluation of probabilistic range queries is discussed in [2,6,7,9,11]. Probabilistic nearest-neighbor queries are presented in [7,17]. An index called Probabilistic Threshold Index was proposed in [18] that can be used to efficiently execute some classes of probabilistic queries.

To best of our knowledge, the issue of selectivity estimation for queries over probabilistic data has not been addressed before.

3   Uncertainty Model

To model the uncertainty present in a data item, a data scheme known as the *Attribute uncertainty model* was proposed in [7]. This scheme assumes that individual attributes, as opposed to complete tuples, are uncertain. The attribute uncertainty model assumes that each data item can be represented by a range of possible values along with the distribution of values over this range. Formally, assume that each tuple of interest consists of an uncertain attribute a. If there are more than one uncertain attributes within the same tuple, they are assumed to be independent of each other. The domain of the uncertain attribute can be continuous (e.g. real-valued) or discrete (e.g. integer). The probabilistic uncertainty of a continuous attribute a consists of two components:

1. **Uncertainty Interval:** The *uncertainty interval* of an item a, denoted by U_a, is an interval $[l_a, r_a]$ where $l_a, r_a \in \Re, r_a \geq l_a$ and $a \in U_a$. The range of R_a of a is defined as $R_a = r_a - l_a$.
2. **Uncertainty pdf:** The *uncertainty pdf* of a, denoted by $f_a(x)$ is a *probability distribution function* (pdf) of a where $f_a(x) = 0$ if $x \notin U_a$.

In addition to the pdf $f_a(x)$, we can also define a *cumulative distribution function* (cdf) $F_a(x)$, which is defined as $F_a(x) = \int_{-\infty}^{x} f_a(x)dx$. Note that, similar to the continuous case, we can also define the pdf and cdf functions in case of a *discrete attribute* by replacing the integral with a sum in the above definitions.

The tuple uncertainty model [1,2,19] assumes that the complete tuple is uncertain. A probability value is attached to each tuple which represents the probability of that tuple being present in the database. In addition, multiple tuples can be grouped together to form an *x-tuple* [1]. The tuples present inside a *x-tuple* are called alternatives and they represent mutually exclusive values for the tuple.

The goal of this paper is to propose estimation solutions that are applicable to both models of uncertainty: attribute and tuple. For our purposes, we are interested in a single attribute at a time, a (real-valued or integer), for which we are estimating the selectivity. Thus, we can ignore the intra-tuple dependencies. We assume that the uncertainty in the data can be captured in terms of attribute uncertainty. In other words, for the attribute in question, we are able to generate a pdf (f_a) and cdf (F_a) for each tuple of the relation. This is directly available from the attribute uncertainty model. For the case of tuple uncertainty, there are two cases to consider. The first is if there are no x-tuples. In this case, each tuple has a probability value associated with it and is independent of any other tuple. For this case, the pdf for each tuple is simply the single attribute value along with the associated tuple probability. In the second case, the x-tuple itself provides multiple alternatives for the given attribute along with associated probabilities. These are collapsed into a single attribute uncertainty (discrete) pdf.

3.1 Operators and Threshold Queries

A number of operators are defined in [8] for comparing uncertain values with both uncertain and certain (precise) values. This paper focuses on selection queries that compare an uncertain value with precise values. For these queries, we present the definitions for comparing uncertain with certain data. Operators between an uncertain value a and a certain value $v \in \Re$ can be defined as:

$$Pr(a < v) = \int_{-\infty}^{v} f_a(x)dx = F_a(v)$$
$$Pr(a > v) = 1 - F_a(v)$$

The set of queries that we consider in the paper are called *Probabilistic Threshold Range Queries* and were proposed in [18]. These queries are a variant of probabilistic queries where only answers with probability values over a certain threshold τ are returned. With this concept, all the operators discussed above can be changed into boolean predicates by adding a probability threshold to them.

4 Selectivity Estimation

In this section we describe various techniques that can be used for estimating the selectivity for a given probabilistic threshold operator.

4.1 Unbounded Range Queries

This approach is based on mapping the uncertain attribute values to a 2-D histogram and estimating the query result size by executing a 2-D box query on the histogram.

To understand the approach, let us consider an unbounded range query Q given by $a <_\tau x_0$, where τ is the probability threshold for the $>$ predicate. This query returns all uncertain items a such that $Pr(a < x_0) > \tau$. In terms of the cumulative distribution function $F_a(x)$, we get the following condition:

$$Pr(a < x_0) > \tau \Leftrightarrow \int_{-\infty}^{x_0} f_a(x)dx > \tau \Leftrightarrow F_a(x_0) > \tau \tag{1}$$

This follows from the definition of pdf and cdf functions.

Let us consider a 2D graph where we plot the cdf function F of all uncertain items. Figure 1 shows an example of this graph. The cdfs for three data items a, b, and c are shown. The range query Q given by Equation 1 can be translated into a (unbounded) box query $x < x_0$ and $y > \tau$ over this 2D plot (the shaded region in Figure 1). Items a and b satisfy the query as they intersect the shaded region.

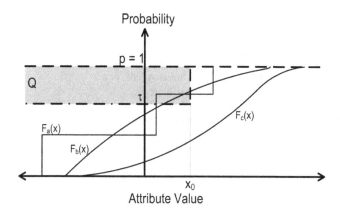

Fig. 1. Example plot for query $Q(x_0, \tau)$

Theorem 1. *All the items whose cdf function $F_a(x)$ lies in the box defined by query Q are part of the result of query Q. That is, $\forall a$, where the cdf function F_a lies in the box defined by query Q, we have $Pr(a < x_0) > \tau$.*

Proof. We observe that for any cdf F_a that lies in the box of query Q, we have $F_a(x) > \tau$ for some $x < x_0$. As F_a is a monotonically increasing function, we can deduce that $F_a(x_0) > F_a(x) > \tau$. Using 1, $P(a < x_0) > \tau$.

Now we state the following theorem without proof:

Theorem 2. *The total number of cdf lines that lie in the query box Q is equal to the number of lines crossing (intersecting) the vertical line-segment given by $\ell : x = x_0, \tau < y \le 1$, which furthermore is equal to the number of lines crossing (intersecting) the horizontal ray $y = \tau, x < x_0$.*

The proof of this theorem follows from basic geometry and the monotonically increasing nature of cdf F.

Now finding all the items whose cdf function lies in the box defined by a query Q is equivalent to finding the total number of intersections of cdf lines with the vertical line-segment ℓ. To efficiently calculate this number we need to develop an approximation of the above technique. For this purpose, we define a *2-D grid* of histogram over the plot region. Given $u_i, 0 \le i < m$ as all the uncertain data items, we define

$$l = \min_i (l_{u_i}), r = \max_i (r_{u_i})$$

where $[l_{u_i}, r_{u_i}]$ is the uncertainty interval of u_i. The plot region is bounded by 0 and 1 in the y (probability) direction and l, r in the x direction. The range R of the histogram is defined as $R = r - l$. The width of the histogram is given by the parameters δ_x and δ_p which represent the size of histogram along x and y (probability) axes respectively. A histogram bucket $H(x, y)$ covers the area given by the box $(x, y, x + \delta_x, y + \delta_p)$. The notations used are summarized in Table 1.

Table 1. Notations

Symbol	Meaning
f_a	Probability distribution function (pdf) of uncertain item a
F_a	Cumulative distribution function (cdf) of a
l_a, r_a	Left and right bounds of a's interval.
R_a	Range of a, $R_a = r_a - l_a$
u_i	All the uncertain data items ($0 \leq i \leq m$)
l, r	Leftmost and rightmost limits of all the uncertain intervals
R	Range of input data, $R = r - l$
δ_x, δ_p	Width of histogram bucket along x and y (probability) axis
H	Histogram structure for cost estimation

Definition 1. *The height of a histogram bucket $H(x, y)$ is the total number of cdf lines of uncertain items intersecting the box $(x, y, x + \delta_x, y + \delta_p)$.*

With this definition, we can now informally define the algorithm for calculating an approximation (upper-bound) of operator selectivity. Using Theorem 2 we see that the sum of individual histograms that cover the vertical line-segment ℓ gives a good approximation of the upper-bound of the result set size. The error in this approximation can be reduced by reducing the size of the histogram buckets. This extra accuracy comes at the cost of increased space overhead for storing the histogram structure.

If a cdf line has a large slope, it can contribute to more than one histogram in a given vertical window. This will result in over-estimation of the result size because the same cdf line will be counted multiple times. To prevent this, we propose a simple fix: If a cdf line intersects multiple (contiguous) histograms in a given vertical window, we only count its contribution in the *topmost* histogram. With this slight change, we will avoid counting the same line multiple times and obtain a tighter upper bound. Note that by adding the contribution of a given cdf line to the topmost histogram, we are guaranteed that there will be no false negatives. The algorithm for constructing this 2-D histogram is presented in Figure 2.

The algorithm presented in Figure 2 takes as input the uncertain data items from an attribute and the parameters δ_x and δ_p defining the width of each histogram inside the structure H. In addition to these values, it also takes the l and r values (defined earlier) which represent the spread of input data values. Depending on the attribute domain, these parameters can be provided by the user or the system can select them by random sampling. For a given uncertain item a, we start counting its contribution from its lower bound l_a and stop when we hit the upper-most bucket in the y-direction (Step 1(ii)). This small optimization saves a lot of computations as this step is repeated for all the input uncertain data items. Note that, for the correctness of our algorithm we *do* need to add the contributions to all the successive top buckets for item a. We take care of this *correction* in step 2 with just one pass over the entire histogram.

Given this histogram structure H, we can easily give an approximation for query result size. Figure 3 shows the algorithm for finding the selectivity estimate for query $Q(x, \tau) = a <_\tau x$.

Input

$u_i, 0 \le i < m$: All the uncertain data items

δ_x, δ_p : Width of histogram along x and y axis

l, r : The left and right bounds for the histogram

Output

H: The histogram structure for the input data

0. Initialize H with all bucket heights $= 0$

1. **for** $a = u_0, u_1 \ldots, u_{m-1}$ **do**

    (i) **let** $x = \lfloor (l_a - l)/\delta_x \rfloor; p = 0$

    (ii) **while** $p < (1 - \delta_p)$

        (a) $p = F_a(l + (x + 1)\delta_x)$

        (b) $H(x, \lfloor p/\delta_p \rfloor)$++

        (c) x++

2. **for** $x = 0, 1, \ldots, \lfloor R/\delta_x \rfloor$

    (i) $H(x, \lfloor 1/\delta_p \rfloor) \mathrel{+}= H(x - 1, \lfloor 1/\delta_p \rfloor)$

3. **return** H

Fig. 2. Algorithm for generating the histogram for unbounded range queries

Input

x_0, τ : Parameters of a query Q

H : Histogram structure

m : Total number of uncertain items

δ_x, δ_p : Width of histogram along x and y axis

l, r : The left and right bounds for the histogram

Output

An estimate (upper-bound) of query selectivity

1. **if** $x_0 < l$ **return 0**

2. **if** $x_0 > r$ **return 1**

3. $x = \lfloor (x_0 - l)/\delta_x \rfloor$

4. **let** $S = 0$

5. **for** $p = \lfloor \tau/\delta_p \rfloor, \ldots, \lfloor 1/\delta_p \rfloor$

    (i) $S = S + H(x, p)$

6. **return** (S/m)

Fig. 3. Algorithm for estimating query selectivity for unbounded range queries

Note that the above discussion applies to $a <_\tau x$ queries only. For unbounded range queries of the form $Q : a >_\tau x$, we have the following result:

$$a >_\tau x \Leftrightarrow Pr(a > x) > \tau \Leftrightarrow F_a(x) < 1 - \tau \qquad (2)$$

Using Equation 2 we can see that if an uncertain item a *does not* satisfy the query $a <_{1-\tau} x$ (i.e. $F_a(x) \not> 1 - \tau$) then it will satisfy the query $a >_\tau x$. The algorithms presented in Figures 2 and 3 can therefore be used for $>_\tau$ queries with slight modifications. The selectivity of $>$ can be calculated by computing the selectivity of $<$ and using the fact that selectivity for $>_\tau$ is 1 - selectivity for $<_{1-\tau}$.

Theorem 3. *The time complexity of algorithm presented in Figure 2 is:*

$$\sum_{i=0}^{m-1} \left(\frac{R_{u_i}}{\delta_x} \right) + O\left(\frac{R}{\delta_x} \right)$$

Proof. The first terms comes from Step (1) in which we go through each item once for each uncertain item. Finally we add up all the contributions in the top histogram buckets in Step (2) which gives us the second term in the above expression.

4.2   General Range Queries

As discussed earlier, a general range query Q is expressed as $Pr(x_1 < a < x_2) > \tau$. This query returns all tuples such that:

$$Pr(x_1 < a < x_2) > \tau \Leftrightarrow \int_{x_1}^{x_2} f_a(x)dx > \tau$$

$$\Leftrightarrow F_a(x_2) - F_a(x_1) > \tau$$

The previous section on unbounded range queries is a special case of the general range query where $x_1 = -\infty$ (or l) or $x_2 = \infty$ (or r).

We can extend the earlier solution to general range queries by adding another dimension to the histogram. In addition to the x-axis and y-axis representing x_2 (end-point of the range query) and the probability threshold τ respectively, we will now have a z-axis representing x_1 (or the beginning of range query).

The theoretical discussion of this selectivity estimation solution is similar to the unbounded case. In place of a 2-D curve, we will now have a 3-D curve for each uncertain item which is given by the function:

$$G_a(x_1, x_2) = \int_{x_1}^{x_2} f_a(x)dx = F_a(x_2) - F_a(x_1) \tag{3}$$

The range query Q will now translate to a box query given by $x < x_2$, $y > \tau$ and $z = x_1$. We can now state the following theorem for the 3-D curve:

Theorem 4. *Each item for which $G_a(x_1, x_2)$ intersects the box defined by query Q is part of the result of query Q. That is, $\forall a$, where the function G_a intersects the box defined by query Q, we have $Pr(x_1 < a < x_2) > \tau$.*

Proof. We observe that for any cdf F_a that lies in the box of query Q, we know that $G_a(x_1, x) > \tau$ for some $x < x_2$. This gives us that $G_a(x_1, x_2) > G_a(x_1, x) > \tau$. Using 3, we have $P(x_1 < a < x_2) > \tau$.

Similar to Theorem 2, we can prove that we can count the total number of items in the result set by counting the total number of intersections of function G_a with the line-segment $x = x_2$, $\tau < y \leq 1$ in the $z = x_1$ plane. The definition and construction of 3-D histogram is similar to the 2-D counterpart and is presented in Figure 4. The algorithm for estimating the answer size for a given query $Q(x_1, x_2, \tau)$ is presented in Figure 5.

Input

$u_i, 0 \le i < m$: All the uncertain items

δ_x, δ_p : Width of histogram along x,z and y axis

l, r : The left and right bounds for the histogram

Output

H: The histogram structure for the input data

0. Initialize H, H_x, H_z, H_{xz} with all bucket heights = 0

1. **for** $a = u_0, u_1 \ldots, u_{m-1}$ **do**

    (i) **let** $x_{min} = \lfloor (l_a - l)/\delta_x \rfloor$, $x_{max} = \lfloor (r_a - l)/\delta_x \rfloor$

    (ii) **for** $z = x_{min}, \ldots, x_{max}$ **do**

        for $x = z, \ldots, x_{max}$ **do**

            (a) $p = G_a(l + z\delta_x, l + (x+1)\delta_x)$

            (b) **if** $(z = x_{min}) \wedge (x = x_{max})$

                $H_{xz}(x, \lfloor p/\delta_p \rfloor, z)$++

            (c) **else if** $(z = x_{min})$

                $H_z(x, \lfloor p/\delta_p \rfloor, z)$++

            (d) **else if** $(x = x_{max})$

                $H_x(x, \lfloor p/\delta_p \rfloor, z)$++

            (e) **else** $H(x, \lfloor p/\delta_p \rfloor, z)$++

2. **let** $x_{max} = \lfloor R/\delta_x \rfloor$

3. **for** $p = 0, \ldots, \lfloor 1/\delta_p \rfloor$

    (a) **for** $x = 0, \ldots, x_{max}$

        for $z = x_{max} - 1, x_{max} - 2, \ldots, 0$

        $H_z(x, p, z) \mathrel{+}= H_z(x, p, z+1)$

    (b) **for** $z = 0, \ldots, x_{max}$

        for $x = 1, 2, \ldots, x_{max}$

        $H_x(x, p, z) \mathrel{+} = H_x(x-1, p, z)$

4. **for** $x = 0, \ldots, x_{max}$

    for $z = x_{max} - 1, x_{max} - 2, \ldots, 0$

    $H_{xz}(x, \lfloor 1/\delta_p \rfloor, z) \mathrel{+}= H_{xz}(x, \lfloor 1/\delta_p \rfloor, z+1)$

5. **for** $z = 0, \ldots, x_{max}$

    for $x = 1, 2, \ldots, x_{max}$

    $H_{xz}(x, \lfloor 1/\delta_p \rfloor, z) \mathrel{+}= H_{xz}(x-1, \lfloor 1/\delta_p \rfloor, z)$

6. **for all** x, z, p

    $H(x, z, p) \mathrel{+}= H_z(x, p, z) + H_x(x, p, z) + H_{xz}(x, p, z)$

7. **return** H

Fig. 4. Algorithm for generating the histogram structure for general range queries

We can apply an optimization similar to the algorithm in Figure 2 by modifying only the local histogram area which is affected by an uncertain item and then propagating the effects globally by adding a post-processing step. This optimization helps in bringing down the running time of the algorithm significantly. To achieve this goal we keep three temporary histogram tables H_x, H_z and H_{xz} along with the main histogram structure H. For an uncertain item a, Step 1 adds the contribution of the item to the main histogram H, along with adding the contributions that are to be propagated globally to the temporary histograms. H_z and H_x store the contribution to the bins corresponding to $z = l_a$ and $x = r_a$ respectively, while H_{xz} stores the contribution to the bin

---

Input

x_1, x_2, τ : Parameters of a query Q

H : Histogram structure

m : Total number of uncertain items

δ_x, δ_p : Width of histogram bucket along x, z and y axis

l, r : The left and right bounds for the histogram

Output

An estimate (upper-bound) of query selectivity

1. if $x_1 < l$ $x_1 = l$

2. if $x_2 > r$ $x_2 = r$

3. let $x = \lfloor (x_2 - l)/\delta_x \rfloor, z = \lfloor (x_1 - l)/\delta_x \rfloor$

4. let $S = 0$

5. for $p = \lfloor \tau/\delta_p \rfloor, \ldots, \lfloor 1/\delta_p \rfloor$

(i) $S = S + H(x, p, z)$

6. **return** (S/m)

Fig. 5. Algorithm for estimating query selectivity for general range queries

corresponding to $z = l_a$ and $x = r_a$. It is easy to see that the local contribution of the item a to H_z needs to be propagated to the plane given by $l_a \leq x < r_a$ and $z < l_a$ as for these values $Pr(z < a < x) = Pr(l_a < a < x)$ (Step 3a). Similarly, H_z needs to be propagated globally to the plane $l_a < z \leq r_a$ and $x > r_a$ as for this plane $Pr(z < a < x) = Pr(z < a < r_a)$ (Step 3b). In a similar fashion, H_{xz} is propagated to $z < l_a$ and $x > r_a$ (Step 4 and 5). Finally, we add all the temporary histograms to the main histogram to get the final histogram structure (Step 6).

Theorem 5. *The time complexity of algorithm presented in Figure 4 is:*

$$\sum_{i=0}^{m-1} \left(\frac{R_{u_i}^2}{2\delta_x^2} \right) + O\left(\frac{R^2}{\delta_x^2 \delta_p} \right)$$

Proof. By counting the number of loops. All the steps in Figure 4, except for Step 1, touch the cells only constant number of times. The number of loops in Step 1 gives the first summation.

4.3   General Range Queries Using Slabs

In Section 4.2 we discussed how the histogram construction technique can be extended to general range queries. While the accuracy of such an estimate is very good, the initial construction time and space trade-off is quadratic in terms of the range of the input data (R). In this section, we present another technique which has, in general, a lower accuracy than the previous technique but better space-time complexity.

In this algorithm, we partition the entire range of input data into slabs. Similar to histograms, the length of a slab is controlled by the input parameter δ_x. Each slab stores estimates of query selectivity for different values of p. A slab with end-points at $x = x_1, x_2$ stores the selectivity of a bounded range query $Q(x_1, x_2, \tau)$ for different values of τ. Once again, the number of divisions (estimates) along the probability axis is controlled by δ_p. Note that, for a query that spans multiple slabs, we cannot just add the contributions of individual slabs. To solve this problem, we have a hierarchy of slabs. The size of slab at the bottom-most level of this hierarchy is exactly δ_x but as we go up the hierarchy the size increases exponentially until we reach the top-most slab, which encompasses the entire input region. At each level of the hierarchy there are two[1] sets of slabs, one starting at the midpoint of the other, so that we can get better estimates. We call these slabs A and B, respectively.

Formally, we have $\log(R/\delta_x)$ hierarchical levels, with each hierarchical level having two sets of slabs $A(i, j, p)$ and $B(i, j, p)$ where $j \leq \lceil \log_2(R/\delta_x) \rceil$.

Definition 2. *The slabs $A(i, j, p)$ and $B(i, j, p)$ cover the regions $\mathcal{R}_1 = [l + 2^j i \delta_x, l + 2^j(i + 1)\delta_x]$ and $\mathcal{R}_2 = [l + 2^j(i + 1/2)\delta_x, l + 2^j(i + 3/2)\delta_x]$ respectively. The height of the slab $A(i, j, p)$ (or $B(i, j, p)$) is given by the number of uncertain items satisfying the bounded query $\mathcal{R}_1$ (or $\mathcal{R}_2$) with probability between $p\delta_p$ and $(p + 1)\delta_p$.*

As mentioned earlier, each of these slabs stores the query answers for different values of query threshold τ. Thus, every $A(i, j)$ or $B(i, j)$ is an array of $\lfloor 1/\delta_p \rfloor$ values. The construction algorithm is presented in Figure 6. In Step 1, for each item, we find the slabs that are affected by the item and add the contribution of the item to the corresponding slabs.

Input
    $u_i, 0 \leq i < m$: All the uncertain items
    δ_x, δ_p : Parameters controlling width of divisions
    l, r : The left and right bounds for the input region
Output
    The slab structure for the input data
0. Initialize A and B with all buckets heights $= 0$
1. **for** $a = u_0, u_1, \ldots, u_{m-1}$ **do**
    (i) **for** $j = 0, 1 \ldots, \lceil \log_2(R/\delta_x) \rceil$ **do**
        (a) let $x_{min} = \lfloor (l_a - l)/(2^j \delta_x) \rfloor$,
            $x_{max} = \lfloor (r_a - l)/(2^j \delta_x) \rfloor$
        (b) **for** $x = x_{min} \ldots x_{max}$ **do**
            (A) let $p = G_a(l + x 2^j \delta_x, l + (x + 1)2^j \delta_x)$,
            (B) $A(x, j, \lfloor p/\delta_p \rfloor)$++
        (c) let $x_{min} = \lfloor (l_a - (l + 2^{j-1}\delta_x))/(2^j \delta_x) \rfloor$,
            $x_{max} = \lfloor (r_a - (l + 2^{j-1}\delta_x))/(2^j \delta_x) \rfloor$
        (d) **for** $x = x_{min} \ldots x_{max}$ **do**
            (A) $p = G_a(l + 2^j(x + 1/2)\delta_x, l + 2^j(x + 3/2)\delta_x)$
            (B) $B(x, j, \lfloor p/\delta_p \rfloor)$++
2. **return** A, B

Fig. 6. Algorithm for generating slabs

---

[1] In general, we can have more than two sets of slabs for each level of hierarchy which will further increase the accuracy of this technique.

Once we have this slab structure, we can get estimates by finding a pair of slabs that contains (over-estimate) and is contained (under-estimate) by the query region. With these estimates, we interpolate the estimates based on the the interval size to get the final estimate. The algorithm for finding the estimate is presented in Figure 7. In the algorithm, Step 1 picks j which corresponds to the slab size just smaller than the query. We have two additional functions *pickLB* and *pickUB*, which given the query limits and a level j, returns the slab that is contained inside and contains the query respectively. If these functions can not find any such slab at level j they return *null*. For $j < 0$, these functions simply return a slab with size 0 and all estimates are set to 0. In the case, these functions find more than one slab which satisfy the conditions of UB (LB) they return the one with minimum (maximum) estimate. This is done in order to get a tighter bound on the final estimate. The details of these functions are omitted due to space considerations. Steps 2 and 3 find the slabs and return them. Once we have a slab T_{LB} that bounds the answer from below and a slab T_{UB} that bounds the answer from above, we find the selectivity estimates of T_{LB} and T_{UB} in Step 6 and then finally in Step 7 we linearly interpolate the estimates based on the size of query and size of the two intervals returned. This gives us an estimate of the query result size.

Lemma 1. *For any query Q, the difference between the levels, from which T_{LB} and T_{UB} are picked up, is at most 2. Thus, the space covered by T_{UB} is at most 4 times that of T_{LB}.*

Proof. It follows from the cases of Figure 7. It remains to show that the *else* cases in Step 2(b) and Step 3(a),(b) are always successful in finding a slab. Note that the size of the slab at level j is less than the query interval. So a slab at level j could fit in the query. If this happens with the A slab being contained, then there is a slab at level $j + 2$ that surely contains the query. This is because, an A slab at level $j + 1$ contains at least one end-point of the query, and hence at level $j + 2$, since an A slab and a B slab extend this A slab at level $j + 1$ in different directions, at least one of the A slabs at level $j + 2$ or B slabs at level $j + 2$ will cover the entire interval. If at level j, the query covers a B slab, then it cuts two consecutive A slabs at level j and hence it is covered in either an A slab or a B slab at level $j + 1$. If the query does not cover any slab at level j, then it again cuts two consecutive A slabs at level j. This means it is covered by a slab at level $j + 1$. Also, it cuts at least one of these A slabs by more than half at the level j. Thus, there is an A slab at level $j - 1$ which is contained in the query.

Theorem 6. *The time complexity of algorithm presented in Figure 6 is:*

$$O\left(\sum_{i=0}^{m-1}\left(\frac{R_{u_i}}{\delta_x}\right) + m\log\left(\frac{R}{\delta_x}\right)\right)$$

Proof. The above result directly follows from the following expression which is the total cost of Step 1.

$$\sum_{i=0}^{m-1}\sum_{j=0}^{\log(R/\delta_x)}\left\lceil\frac{R_{u_i}}{2^j\delta_x}\right\rceil$$

Input

x_1, x_2, τ : Parameters of a query Q

A, B : Slab structure

m : Total number of uncertain items

δ_x, δ_p : Parameters controlling width of divisions

l, r : The left and right bounds for the histogram

Output

An estimate of the query selectivity

1. **let** $j = \lfloor \log_2((x_2 - x_1)/\delta_x) \rfloor$
2. **if** (T = pickLB(x_1, x_2, j)) exists
    (a) $T_{LB} = T$
    (b) **if** $(T = \text{pickUB}(x_1, x_2, j + 1))$ exists
        $T_{UB} = T$
    else $T_{UB} = \text{pickUB}(x_1, x_2, j + 2)$
3. **else**
    (a) $T_{LB} = \text{pickLB}(x_1, x_2, j - 1)$
    (b) $T_{UB} = \text{pickUB}(x_1, x_2, j + 1)$
4. **let** $S_{min} = S_{max} = 0, t_1 = \text{length of } T_{LB}$,
    $t_2 = \text{length of } T_{UB}$
5. **for** $p = \lfloor \tau/\delta_p \rfloor, \ldots, \lfloor 1/\delta_p \rfloor$
    (a) S_{min} += $T_{LB}(p), S_{max}$ += $T_{UB}(p)$
6. $S = S_{min} + (S_{max} - S_{min}) \times (x_2 - x_1 - t_1)/(t_2 - t_1)$
7. **return** (S/m)

Fig. 7. Algorithm for estimating query selectivity using slabs

Similarly, we can also show that the total space overhead is $O(R/\delta_x)$. Both these results are intuitive if we observe that the total cost/space is asymptotically bounded by number of slabs at the bottom-most level as the number of slabs at higher levels decrease exponentially.

5 Experimental Evaluation

We have implemented our statistics collection and selectivity estimation algorithms in *Orion*, a publicly available extension to PostgreSQL that provides native support for uncertain data [3]. To efficiently evaluate the queries discussed in this paper, Orion uses an indexing scheme known as *probabilistic threshold index* (PTI) introduced in [18]. This system not only allows us to validate the accuracy of our methods in a realistic runtime environment, it also gives additional insight into the overall effect our techniques have on query optimization in an industrial-strength DBMS.

5.1 Implementation

PostgreSQL measures the cost of query plans in disk page fetches (for simplicity, all CPU efforts are converted into disk I/Os). The optimizer generally estimates the cost of query plans by calculating the overall selectivity and multiplying it against the estimated cardinality. In the common case of multiple predicates, individual selectivies are multiplied together, except for range queries where the dependence between the lower and upper bounds is simple to evaluate.

Virtually every numeric data type in PostgreSQL shares the same source code for cost estimation. Using this code base, we have built our implementation of the algorithms in Figures 2, 4, 3, and 5. Using the elegant framework PostgreSQL provides for new data management techniques, our implementation extends the functionality of Orion's UNCERTAIN data type by registering the optional callbacks for collecting statistics and estimating selectivity.

5.2 Methodology

To ensure correctness, we ran each experiment on a variety of queries and datasets, and then averaged the results. After populating the database with each test dataset, we first used VACUUM ANALYZE to generate the statistics in advance. The following experiments were conducted on a 1.6 GHz Pentium CPU with 512 MB RAM, running Linux 2.6.17, PostgreSQL 8.1.5, and Orion 0.1. Note that most of the resulting plots show the *relative error* of the selectivity estimates, i.e. the goal is to be as close to 0% as possible.

Synthetic Datasets. Each dataset consists of random "sensor readings," using a schema Readings (rid, value). Without loss of generality, the uncertain values (i.e. reported from the sensors) are floating point numbers ranging from 0 to 1000, and the pdf for each uncertain value is a uniform distribution. The interval sizes are distributed normally, with midpoints distributed uniformly. We refer to our three main datasets as Data-5, Data-50, and Data-100; the numbers correspond to the average width of the uncertain value intervals.

Table 2 summarizes the control variables for the subsequent experiments. In particular, we show that our algorithms perform well without regard to dataset cardinality, and are reasonably robust to query selectivity and probabilistic threshold. In addition, we demonstrate the effect of increased precision as a trade-off between construction time and space versus the resulting accuracy of the selectivity estimates.

Example Query Plan. To illustrate the impact that correct estimates have on query optimization, we present the following example output from PostgreSQL. When no selectivity estimation function is available for a given predicate, PostgreSQL simply returns the default value of 1/3 for estimating unbounded range queries, and 0.005 for general range queries. In practice this estimate favors the use of unclustered indexes, such as PTI [18], to improve I/O performance:

Table 2. Summary of control variables

Variable	Default Value
Cardinality	250,000
Selectivity	2.5 %
Threshold	50 %
Precision	70 bins

```
SELECT * FROM Readings WHERE value < 750;
-------------------------------------------
Bitmap Heap Scan on Readings
  (cost=742.33..4075.67 rows=66667 width=36)
  (actual=20379.348..20824.652 rows=153037)
  Recheck Cond: (value < 750::real)
->   Bitmap Index Scan on pti_value
  (cost=0.00..742.33 rows=66667 width=0)
  (actual=20378.677..20378.677 rows=153K)
  Index Cond: (value < 750::real)
```

With accurate estimates, the optimizer makes the correct decision, namely not to use the available PTI index:

```
(same query as before, but using our algorithms)
-------------------------------------------------
Seq Scan on Readings
  (cost=0.00..5000.00 rows=164333 width=35)
  (actual=83.841..15545.401 rows=153037)
  Filter: (value < 750::real)
```

As shown in this example, accurate selectivity estimation saves the system thousands of disk fetches (i.e. 15545 total cost instead of 20825). In general, incorrect estimates may result in much higher losses of efficiency.

5.3   Results

We now evaluate the accuracy and performance of our cost estimation techniques for unbounded range queries using the 2D histogram structure introduced in Section 4.1 (see Figure 3), and general range queries using the 3D histogram discussed in Section 4.2 (see Figure 5).

Accuracy at Varying Selectivities: The first experiment verifies the accuracy of our algorithms, regardless of query selectivity. Figures 8 and 9 summarize the results using all three synthetic datasets. For clarity, we have only plotted one of them. The x-axis shows the selectivity of the query which was varied from high (1%) to low (100%). The y-axis shows the accuracy of the estimation as a percentage relative to the size of the exact result. Our algorithm significantly outperforms the baseline PostgreSQL estimate. As expected, high selectivity has a slight effect on the accuracy of our methods.

Accuracy at Varying Cardinalities: The next experiment studies the overall scalability of our algorithms, namely the impact of the size of the relation on the accuracy of the estimations. Figures 10 and 11 show the results for three representative queries. The x-axis shows the size of the table in number of tuples which was varied from 50,000 to 800,000. The results show that our approach is unaffected by the size of the dataset. This is in sharp contrast to the baseline PostgreSQL estimator (not shown) which is much more sensitive to the dataset size, particularly for smaller datasets.

Accuracy at Varying Thresholds: Figures 12 and 13 show the impact of query threshold on the accuracy of the estimates. The x-axis shows the threshold probability and the

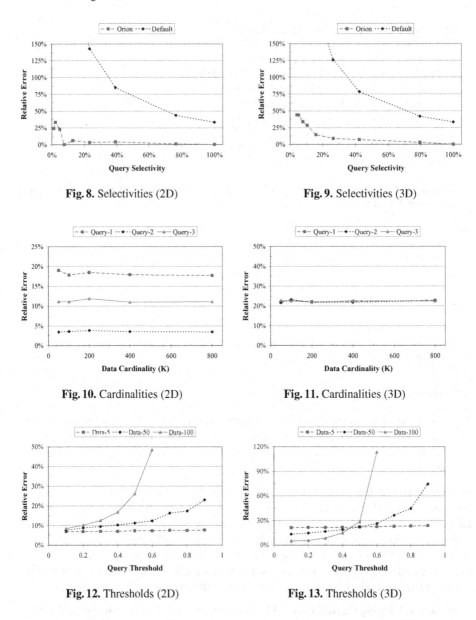

Fig. 8. Selectivities (2D)

Fig. 9. Selectivities (3D)

Fig. 10. Cardinalities (2D)

Fig. 11. Cardinalities (3D)

Fig. 12. Thresholds (2D)

Fig. 13. Thresholds (3D)

y-axis shows the relative accuracy with respect to the correct answer size. Once again, we observe that our algorithm is much more robust than the baseline PostgreSQL estimator (not shown) that simply returns a constant selectivity. Our implementation shows slightly better accuracy for smaller thresholds, in part because larger thresholds result in additional tuples becoming part of the query answer, leading to overestimates. We can see that for highly selective queries, our algorithm is significantly better that the baseline and thus it is more likely to lead the optimizer into choosing a much more efficient plan.

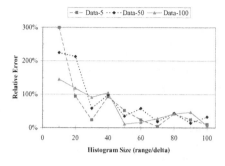

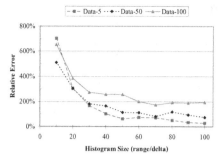

Fig. 14. Precision (2D)   **Fig. 15.** Precision (3D)

Accuracy at Varying Precisions: Next we show the relationship between the size of the histograms and the resulting accuracy. Figures 14 and 15 summarize the results for each dataset. The x-axis shows the number of histogram buckets in each dimension, which was varied from 10 to 100. Clearly, both algorithms perform better with a more detailed histogram. Our algorithm outperforms the baseline for smaller histograms. As expected, we see that after a certain amount (i.e. 70, for these datasets and queries), larger histograms do not provide significant increase in accuracy.

Runtime Performance Overhead: The final set of experiments study the runtime performance of constructing the statistics and estimating the selectivity of a query. We have omitted figures for these findings because of limited space. As expected, the estimation times are constant and almost negligible (on the order of 15 ms). The histogram construction times scale linearly with respect to data cardinality, and grow a little more than linear as the requested number of buckets increases. For the bulk of our experiments, histogram construction only amounted to several hundred milliseconds.

6   Conclusions and Future Work

In this paper, we developed algorithms for computing selectivity estimates of probabilistic queries over uncertain data. The estimation techniques can be applied both to tuple uncertainty and attribute uncertainty models. These techniques were implemented in PostgreSQL and found to provide accurate estimates for uncertain data. The algorithms presented can be further improved by combining them with standard cost estimation techniques such as equi-depth binning and sampling. We showed both theoretically and empirically that our histogram construction algorithms are fast. The experiments show that they give very accurate estimation especially for less selective queries. For more selective queries, the accuracy is not quite as good, but is still much better than the baseline estimator.

Acknowledgements. This work was supported by NSF grants IIS 0242421, IIS 0534702, IIS 0415097, CCF-0621457 and AFOSR award FA9550–06–1–0099.

References

1. Benjelloun, O., Sarma, A., Halevy, A., Widom, J.: ULDBs: Databases with uncertainty and lineage. In: Proceedings of International Conference on Very Large Databases (2006)
2. Dalvi, N., Suciu, D.: Efficient query evaluation on probabilistic databases. In: Proceedings of International Conference on Very Large Databases (2004)
3. Orion (2008), http://orion.cs.purdue.edu/
4. Sen, P., Deshpande, A.: Representing and querying correlated tuples in probabilistic databases. In: Proceedings of IEEE International Conference on Data Engineering (2007)
5. Singh, S., Mayfield, C., Prabhakar, S., Shah, R., Hambrusch, S.: Indexing uncertain categorical data. In: Proceedings of IEEE International Conference on Data Engineering (2007)
6. Tao, Y., Cheng, R., Xiao, X., Ngai, W., Kao, B., Prabhakar, S.: Indexing multi-dimensional uncertain data with arbitrary probability density functions. In: Proceedings of the 31st Very Large Data Bases conference (2005)
7. Cheng, R., Kalashnikov, D.V., Prabhakar, S.: Evaluating probabilistic queries over imprecise data. In: Proceedings of ACM Special Interest Group on Management of Data (2003)
8. Cheng, R., Singh, S., Prabhakar, S., Shah, R., Vitter, J., Xia, Y.: Efficient join processing over uncertain data. In: Proceedings of International Conference on Information and Knowledge Management (2006)
9. Deshpande, A., Guestrin, C., Madden, S., Hellerstein, J., Hong, W.: Model-driven data acquisition in sensor networks. In: Proceedings of International Conference on Very Large Databases (2004)
10. Poosala, V., Ioannidis, Y., Haas, P., Shekita, E.: Improved histograms for selectivity estimation of range predicates. In: Proceedings of ACM Special Interest Group on Management of Data (1996)
11. Pfoser, D., Jensen, C.: Capturing the uncertainty of moving-objects representations. In: Proceedings of International Conference on Scientific and Statistical Database Management (1999)
12. Antova, L., Koch, C., Olteanu, D.: 10^10^6 worlds and beyond: Efficient representation and processing of incomplete information. In: Proceedings of 23rd International Conference on Data Engineering (2007)
13. Boulos, J., Dalvi, N., Mandhani, B., Mathur, S., Re, C., Suciu, D.: Mystiq: A system for finding more answers by using probabilities. In: Proceedings of ACM Special Interest Group on Management of Data (2005)
14. Widom, J.: Trio: A system for integrated management of data, accuracy, and lineage. In: Proceedings of the Second Biennial Conference on Innovative Data Systems Research (2005)
15. Nierman, A., Jagadish, H.V.: ProTDB: Probabilistic Data in XML. In: Proceedings of International Conference on Very Large Databases (2002)
16. Hung, E., Getoor, L., Subrahmanian, V.S.: PXML: A probabilistic semistructured data model and algebra. In: Proceedings of IEEE International Conference on Data Engineering (2003)
17. Ljosa, V., Singh, A.: APLA: Indexing arbitrary probability distributions. In: Proceedings of IEEE International Conference on Data Engineering (2007)
18. Cheng, R., Xia, Y., Prabhakar, S., Shah, R., Vitter, J.: Efficient indexing methods for probabilistic threshold queries over uncertain data. In: Proceedings of International Conference on Very Large Databases (2004)
19. Lakshmanan, L., Leone, N., Ross, R., Subrahmanina, V.: Probview: A flexible probabilistic database system. ACM Transactions on Database Systems 22(3), 419–469 (1997)

Disclosure Risks of Distance Preserving Data Transformations

E. Onur Turgay, Thomas B. Pedersen, Yücel Saygın,
Erkay Savaş, and Albert Levi

Sabanci University
Istanbul, 34956, Turkey
onurturgay@su.sabanciuniv.edu,
{pedersen,ysaygin,erkays,levi}@sabanciuniv.edu

Abstract. One of the fundamental challenges that the data mining community faces today is privacy. The question "How are we going to do data mining without violating the privacy of individuals?" is still on the table, and research is being conducted to find efficient methods to do that. Data transformation was previously proposed as one efficient method for privacy preserving data mining when a party needs to outsource the data mining task, or when distributed data mining needs to be performed among multiple parties without each party disclosing its actual data. In this paper we study the safety of distance preserving data transformations proposed for privacy preserving data mining. We show that an adversary can recover the original data values with very high confidence via knowledge of mutual distances between data objects together with the probability distribution from which they are drawn. Experiments conducted on real and synthetic data sets demonstrate the effectiveness of the theoretical results.

1 Introduction

Data mining technology proved its success in many areas such as health, life-sciences, and security. On the other hand, the popularity of data mining ignited heated debates on the privacy aspects especially after the launch of large scale projects related to homeland security. In fact, some projects were stopped since they failed to meet the privacy concerns. According to a very recent article in Computer World by Jaikumar Vijayan "The chairman of the House Committee on Homeland Security, has asked Department of Homeland Security Secretary Michael Chertoff to provide a detailed listing of all IT programs that have been canceled, discontinued or modified because of privacy concerns" [15]. In addition to that, the Chairman also asked for information about the measures being taken to address privacy issues [15].

Measures to address privacy issues can be as simple as not collecting privacy sensitive information at all. Unfortunately, in many applications this is not possible. Therefore, advanced protocols based on statistics and cryptography are proposed to ensure privacy. Privacy preserving data management in general, is

B. Ludäscher and Nikos Mamoulis (Eds.): SSDBM 2008, LNCS 5069, pp. 79–94, 2008.

still an ongoing research topic, and efficient as well as secure methods without strong assumptions are yet to be proposed. In fact, recent results showed that the data sets transformed with perturbation based techniques can be recovered by a principle component analysis based attack[7]. In this paper, we present a general attack which is applicable in cases where only the pairwise distances among objects are known.

The main contribution of this paper is to demonstrate that *any* data transformation, from which an attacker can learn the mutual distances between data objects may disclose private information: (1) If the attacker knows a few of the data objects in the database, he can recover *all data perfectly*, (2) If the attacker has *a-priory knowledge* of the probability distribution from which the data is drawn, he can recover *all data* with *high precision*. Our attack is based on an attack of Liu *et al.*[7], but is improved so that it is applicable even to the privacy preserving method which was proposed in [8] to prevent the attack from [7]. Our attack is also improved in the sense that it is applicable to a wider range of scenarios than the attack of Liu *et al.*. We demonstrate the attack with known probability distribution on the Adult Census dataset from the UCI Machine Learning Repository [13], which is the same dataset used in [8], and show that the original data can be recovered with an error as low as 2%.

2   Related Work

Data perturbation is a widely used technique for privacy preserving data mining. Additive perturbation techniques proposed by Agrawal and Srikant are based on adding random noise to the original data which can then be filtered to recover the distribution of the original data[3]. Another scenario is where a group of organisations would like to perform collective data mining but would not like to share their data. An encryption based protocol for privacy preserving association rule mining in distributed environments is proposed in[6]. Similarly, secure multi-party computation based methods are applied to privacy preserving clustering in distributed environments[14]. In [2], authors propose an approach for privacy preserving data mining which maps the original data set into a new anonymised data set preserving the correlations among the different dimensions.

Security of random perturbation methods against partial disclosure through successive querying of the database by snoopers is studied in [10]. The effect of high dimensionality in randomisation was studied by Aggarwal in [1].

Many techniques for classification such as clustering only relies on the mutual distances between the objects in the database. In consequence several privacy preservation techniques which preserves mutual distances have been proposed. The authors of [4,5,11] have proposed perturbation techniques based on geometric transformations such as translation, rotation, and re-scaling of the dataset. With the exception of rescaling, these operations preserve distances. Even rescaling, while it does not preserve the exact distances, preserves the relative distances. Oliveira and Zaïane, propose techniques for securely computing the distances between each pair of data objects, and only reveal the resulting *dissimilarity matrix* to the third party, who can then perform clustering[12]. Oliveira and Zaïane prove that the dissimilarity matrix alone does not violate privacy *with the assumption that the attacker does not have domain knowledge.* However, in cryptography the well established Kerckhoffs' principle states that the security of a system must never rely on keeping the algorithm and/or the data secret — the only secret should be an easily exchangeable cryptographic key. The role of this principle in the case of privacy preserving data mining is well understood in the words of Bruce Schneier [9]:

> "Kerckhoffs' principle applies beyond codes and ciphers to security systems in general: every secret creates a potential failure point. Secrecy, in other words, is a prime cause of brittleness and therefore something likely to make a system prone to catastrophic collapse. Conversely, openness provides ductility."

While privacy is preserved in [12] when the adversary has no domain knowledge at all, it is unclear what happens if the adversary gains partial knowledge of the domain. A party involved in data mining, for instance, is likely to know the layout of the tables in the database, and anyone can easily gain access to national statistics about age, sex, income, e.t.c. Relying on this information to be kept secret from the adversary is unrealistic, and clearly violates Kerckhoffs' principle. Notice that knowing the distribution of the data is not the same as knowing the data. Even though anyone can see the distribution of patients with cancer according to e.g. age and income, we do not want anyone to learn the identity of a specific individual with cancer. In this paper we demonstrate that in a worst case scenario a secret database can be reconstructed very accurately if the adversary knows the table layout and knows the distribution of the data.

Our attack is based on the work by Liu *et al.*, where the authors point out that perturbation techniques which preserve distance between data objects can be attacked if the attacker knows a small set of data selected according to the same probability distribution as the original data set[7,8]. The attack applies principal component analysis to the perturbed data, and tries to fit it to the known data set. Liu *et al.* also propose an alternative transformation where the objects in the original data set are projected onto a subspace in a way that distance is preserved with high probability. They point out that the alternative approach is secure against the identified attack, but may not be secure against other attacks. Our attack is applicable to a wider range of scenarios than the attack of [7], since the attacker does not need the entire perturbed dataset: only the mutual distances and information about the probability distribution from which objects are chosen. Our attack is more general since: (1) In many cases the information about the probability distribution can be obtained from alternative sources (i.e. national statistical agencies), and (2) only the mutual distances from the original dataset are needed (not a perturbation of every object). Our attack also have some improvements in the computational cost: Our attack is polynomial in the number of attributes, whereas the attack in [7] is exponential in the number of attributes.

3   Problem Formulation

Throughout this paper we let n be the number of objects (i.e. rows) in the *target database* which is attacked. Each object in the database has d attributes. In other words, each object can be thought of as a vector in a d-dimensional vector space. For simplicity we assume that all attributes are from an alphabet Ω. We model the objects as random variables $X_1, \ldots, X_n$. We assume that these random variables are independent and identically distributed according to a *global probability distribution*, and let $P(x)$ denote the probability that $X_i = x$, for all $i \in \{1, \ldots, n\}$.

3.1   Distance Preserving Transformations and Dissimilarity Matrices

In this paper we show how to attack any distance preserving data transformation. The only thing we need for our attack are the pairwise distances between the objects in the database. We represent this information by a dissimilarity matrix as described below. We are not concerned with the actual transformation, or whether the data is centralised or distributed.

The dissimilarity matrix is an $n \times n$ matrix which contains the distances between each pair of data objects. We can describe the dissimilarity matrix as random variable D, which depends on the random variables $X_1, \ldots, X_n$ in that $D_{ij} = |X_i - X_j|$, for all $i, j \in \{1, \ldots, n\}$.

In our experiments we use databases containing numerical and boolean attributes, since they have well-defined distance measures. We use the Euclidean distance, and assign the values 1 and 0 to boolean values *true* and *false*, respectively. Textual and nominal data requires extra work, and are not addressed in this paper.

In data mining applications it is common to normalise the attributes before analysing the data. Normalisation prevents attributes of large magnitudes to dominate the small scaled attributes. In our work we assume that all attributes are normalised.

3.2   Motivating Scenario

Dissimilarity matrices of objects having only one attribute is a simple special case, where the distances between objects are equal to the differences in their attributes. In this section we will briefly study this special case to get some intuition.

Suppose we have a database containing the ages of randomly selected individuals within a country. For samples of sufficient sizes, it is acceptable to assume that the database has the same probability distribution of ages as nationwide.

Suppose we have a database of 5 individuals, x_1, x_2, x_3, x_4, x_5, with discrete ages 25, 95, 4, 60, 32, respectively. The corresponding dissimilarity matrix can be seen in Table 1.

If we know the age of two individuals, x_1 and x_2, say, we can easily find the age of all the other individuals: namely the unique age at the given distance

Table 1. Dissimilarity matrix of the age of 5 individuals

	x_1	x_2	x_3	x_4	x_5
x_1	0	70	21	35	7
x_2		0	91	35	63
x_3			0	56	28
x_4				0	28
x_5					0

from the two known ages. This corresponds to the "Hyper-lateration" attack described in Sec. 4 below.

To recover the ages from the dissimilarity matrix only, we start by finding the biggest distance in the matrix — in this case 91. The two individuals with biggest distance (x_2 and x_3) defines the boundaries of the database: one of them is the youngest, while the other is the oldest. We assign the age zero to any of the two points, x_2, say. Now the age of all the other individuals is their distance to x_2 (since all ages are known to be positive). The resulting ages can be seen in Table 2.

Table 2. Ages, if x_2 is assumed to be 0 years

x_2	x_4	x_5	x_1	x_3
0	35	63	70	91

Suppose that we know that more than half of the population is younger than 40 years. In that case the ages in Table 2 do not fit the probability distribution of the population — we have most likely chosen the wrong person as the youngest. When flipping the ages of x_2 and x_3 we get the ages seen in Table 3, which fits the global probability distribution better.

Table 3. Ages, if x_3 is assumed to be 0 years

x_3	x_1	x_5	x_4	x_2
0	21	28	56	91

Now that we have a good candidate dataset, the histogram of the candidate dataset is compared with the global probability distribution and the *statistical distance* between the two is computed (in our example, the dataset is too small to plot the histogram). By shifting the candidate dataset by small amounts (in this case by 1 year), and computing the statistical distance of the resulting probability distributions to the global probability distribution, we can find the best fitting candidate dataset. The shift trials are conducted until the oldest individual in the candidate dataset reaches the maximum possible age the global probability distribution (at 120 years, say). The candidate dataset that has the minimum statistical distance to the global probability distribution is chosen as the winning set.

In Sec. 5 we give an attack which can rotate a multidimensional candidate dataset to the true dataset with low error.

3.3   Attack Scenarios

Our underlying model is that a secret data base of n objects is drawn according to a global probability distribution of d dimensional vectors. An attacker is given the dissimilarity matrix of the data base. Besides the dissimilarity matrix he has some extra information available. This information can be:

Known sample. An attacker might be able to learn a few of the objects in the data base. If he knows at least $d + 1$ objects (and knows the corresponding entries in the dissimilarity matrix) he will be able to reconstruct the data base with high probability, as described in Sec. 4.

Known probability distribution. The probability distribution from which the objects are drawn may be known to an attacker. In Sec. 5 we show an attack using this information.

There are several ways in which an attacker might recover the necessary information. To get a known sample of the database, an attacker might get insider information from a person within the organisation which owns the database, or he might be able to inject information. In some cases it may even be realistic to assume that the attacker already knows some entries in the database (he had an operation in the hospital which has the target database of medical data). Knowledge of the global probability distribution can, in some cases, be obtained from national statistical societies. In other cases the attacker could be in possession of his own database with objects drawn from the same global probability distribution (a competing hospital).

Finally we assume that an attacker knows the schema of the database. The schema of the database will often depend on the software which is used by the organisation, and may be readily available. It may also follow public standards.

3.4   Principal Component Analysis

In the scenario where the attacker does not have a known sample from the database, but has knowledge about the global probability distribution of the data, we apply principal component analysis (PCA). PCA is a statistical method which identifies correlations in a dataset. It takes a dataset of random variables drawn from the d-dimensional vector space, and creates a vector basis which is best suited to represent the dataset. The basis is such that when data is projected onto the subspace spanned by "the most significant" basis vectors, only little information is lost.

More precisely, PCA computes the covariance matrix of the dataset. On entry $(i, j) \in \{1, \ldots, d\}^2$ the covariance matrix has the covariance

$$\mathrm{Cov}(X_i, X_j) = \mathrm{E}((X_i - \mu_i)(X_j - \mu_j)), \tag{1}$$

where μ_i and μ_j are the expected values of X_i and X_j, respectively. The eigenvalues and the corresponding eigenvectors of the covariance matrix are then computed. The eigenvectors (i.e. principal components) form the new vector basis, which we refer to as the eigen-basis. The eigenvectors with the largest eigenvalues are the most significant components, and point in the direction of the highest correlation.

The central observation is that the eigenvalues do not change when the dataset is rotated and/or mirrored. In particular, a dataset which is obtained by hyperlateration will have the same eigenvalues as the original dataset, and we expect that rotating the corresponding eigen-basis to the original eigen-basis will recover the data.

4 Attacking with Known Sample

In this section we assume that the attacker has a known sample of at least $d + 1$ objects from the data base, and knows the corresponding entries in the dissimilarity matrix.

Given three points in a two dimensional vector space, which do not lie on a line, a fourth point with known distances to these three points can be placed by triangulation or trilateration. Trilateration generalises to points in a d-dimensional vector-space, so that we can uniquely place a point with known distances to $d+1$ distinct points, which span the vector-space. We call this procedure "hyperlateration". Applying hyper-lateration to our case; if we have a database with d attributes and n objects, and we are only given the dissimilarity matrix, we can find the original data if we can correctly place $d + 1$ distinct points (which span the full vector-space). In most databases there are considerably more data objects than attributes. We thus reduce the complexity of guessing all n objects in the database to guessing or obtaining $d + 1 \ll n$ objects.

4.1 Hyper-lateration

We now give the algorithm for hyper-lateration. Given $d + 1$ reference points, $p_0, \ldots, p_d$, in $\mathbb{R}^d$ the following theorem gives us a point x at distance δ_i to point p_i, for $i = 0, \ldots, d$.

Theorem 1. *Let $p_0, \ldots, p_d$ be $d+1$ distinct points which span $\mathbb{R}^d$. Any point x is uniquely determined by the set of distances $\{\delta_i\}_{i=0}^d$, where δ_i is the distance from x to point p_i, for $i \in \{0, \ldots, d\}$.*

Proof. Hyperlateration is the task of solving the equations

$$\delta_i^2 = \sum_{j=1}^{d} (\overline{x}_j - \overline{p}_{ij})^2 = \sum_{j=1}^{d} \overline{x}_j^2 - 2\overline{x}_j \overline{p}_{ij} + \overline{p}_{ij}^2, \tag{2}$$

for all $i \in \{0, \ldots, d\}$. These equations are simplified by subtracting the equaiton $\delta_0^2 = \sum_{j=1}^{d} \bar{x}_j^2 - 2\bar{x}_j \bar{p}_{0j} + \bar{p}_{0j}^2$ from all the other equaitons:

$$\delta_i^2 - \delta_0^2 = \sum_{j=1}^{d} 2\bar{x}_j (\bar{p}_{0j} - \bar{p}_{ij}) + \bar{p}_{ij}^2 - \bar{p}_{0j}^2, \tag{3}$$

for all $i \in \{1, \ldots, d\}$. The result is a system of d linear equations which we write as

$$M\bar{x} = \bar{d} + \bar{\delta}, \tag{4}$$

where $M_{ij} = 2(\bar{p}_{0j} - \bar{p}_{ij})$, and $\bar{d}_i = \sum_{j=1}^{d} (\bar{p}_{0j}^2 - \bar{p}_{ij}^2)$ only depend on the known points, and $\bar{\delta}_i = \delta_i^2 - \delta_0^2$ depends on the distances. This system of equations has a unique sollution exactly when M is non-singular, which is the case if and only if $\bar{p}_1, \ldots, \bar{p}_d$ are linearly independent.

4.2   The Hyper-lateration Attack

If an attacker knows a sample of $d+1$ objects from the database, he may be able to recover the entire database if he sees the dissimilarity matrix. The success of this attack depends on two things: (1) the known objects should be represented by distinct points which span the full vector-space, and (2) the attacker must know the corresponding entries in the dissimilarity matrix (i.e. he must know the distances between the known objects and any other object). If these two conditions are met, the attacker will be able to fully recover the database without any error.

The attack consists of the following steps:

1. Pre-compute the matrix M, it's inverse, and the vector $\bar{d}$ from Eq. 4.
2. For each row in the dissimilarity matrix, which corresponds to an unknown object, compute the vector $\bar{\delta}$ from Eq. 4, and solve the system of linear equations.

The first step can be done in time $O(d^3)$, while the second step (assuming that $n > d$) requires time $O((n-d)d^2)$. The overall time is $O(d^2 n)$.

5   Attacking without Known Sample

If the attacker does not have a known sample of the target database, he can still attack the dissimilarity matrix if he knows the "shape" of the data. In this section we show how the attacker can map a candidate dataset obtained from the dissimilarity matrix into the real data. The attacker obtains a candidate dataset by randomly fixing $d+1$ points so that they are consistent with the dissimilarity matrix. This gives a candidate dataset where the relative position of all points is true (up to mirroring in any axis) — in other words: a perturbation of the original data in the target database. By applying a principal component analysis attack, similar to the one presented by Liu et al.[7], to the candidate dataset, we show how the candidate dataset can be mapped to a dataset which fits well to the real data of the target database.

5.1   The PCA Attack

In the PCA attack the attacker is given only two pieces of information:

- The dissimilarity matrix of the database.
- A representation of the global probability distribution from which the data is drawn.

Given this data, the goal of the attacker is to reconstruct the secret database.

Generating a candidate dataset, which is consistent with the dissimilarity matrix, is straightforward by hyper-lateration as described in Sec 4.1. Unfortunately the candidate dataset found by hyper-lateration can be an arbitrarily rotated and mirrored version of the real data. Such a rotation and mirroring of a dataset can be seen as a perturbation of the dataset. Liu et al. proposed using PCA to rotate and mirror a perturbed dataset back to its original position by comparing the principle components of a known sample from the same global probability distribution with the perturbed data's principle components and then generate a rotation matrix which rotates the perturbed data to the actual position by a simple matrix multiplication[7]. The principle components generated by PCA are very good representatives of the general shape of the probability distribution, as they directly depend on the variances and covariance values of individual attributes. As the distance matrix we are attacking is assumed to be of a data mining application, the correlations between variables will most likely generate strong principle components. The results of [7] can directly be applied to our case, since the result of our hyper-lateration process is special case of data perturbation.

Although one of the main uses of PCA is to project data to lower dimensions without loosing statistical information, we only use it to find principle components of both the candidate dataset and the known probability distribution, and try to construct a rotation matrix to match the principle components of these two probability distributions.

One limitation of PCA is that it is invariant under mirroring of the data. In other words: when using PCA to rotate the candidate dataset, we may end up with a dataset which is mirrored along any of the principal components. There are 2^d possible mirror images of which we have to find the one that matches the global probability distribution best. To test the quality of a candidate we compute the statistical distance between the probability distribution which can be computed from the candidate dataset to the real probability distribution. For a candidate dataset C and global probability distribution P we compute

$$\delta(C, P) = \sum_{v \in \Omega^d} \left| \frac{\#v}{\|C\|} - P(v) \right|, \tag{5}$$

where $\#v$ is the number of occurrences of v in C (this can be computed efficiently by only summing over v which occur in C).

Our attack can be described in the following steps:

1. Perform hyper-lateration on the dissimilarity matrix, to get a candidate dataset.

2. Compute the covariance matrices of the global probability distribution, and the candidate dataset.
3. Find eigenvalues and eigenvectors of the two matrices.
4. Match the eigenvectors of the two covariance matrices pairwise, and find a rotation which will rotate the eigen-basis of the candidate dataset to the eigen-basis of the global probability distribution.
5. For each eigenvector, measure the statistical distance between the known probability distribution and the candidate dataset obtained from both directions of the eigenvector — choose the one that is closest to the known distribution, and continue to next eigenvector.

Steps 2 and 3 comprises the PCA. Notice that since the target database is drawn from the global probability distribution, the covariance matrix made from target database should match the covariance matrix of the global probability distribution fairly accurately.

Our attack differs from the attack of Liu *et al.* on two points:

– Liu *et al.* do an exhaustive search amongst all 2^n possible mirroring of the eigenvectors, whereas we find the direction of eigenvectors one at the time.
– Liu *et al.* use multivariate two-sample hypothesis test to find the best candidate dataset. We compute the statistical distance defined in Eq. 5. The approach by Liu *et al.* requires time $O((n+d)^2)$, whereas our approach can be done in time $O(n \log n)$ (sort the vectors in the candidate dataset and count their frequencies).

Step 1 has time complexity $O(d^2 n)$. Step 2 also takes time $O(d^2 n)$ (we assume that the covariance matrix of the global probability distribution has been pre-computed) and Step 3 takes time $O(d^3)$. Finding the rotation in Step 4 takes $O(d^2 n)$. The final step has time complexity $O(dn \log n)$. In total our attack has time complexity $O(d^3 + d^2 n + dn \log n)$ (which is $O(n^3)$ when $d < n$).

5.2   Characteristics of Vulnerable Datasets

While our attack is based on a statistical method for mapping the hyper-latereted data to the global probability distribution, the characteristics of data can change the accuracy of the output considerably. For example, a dataset with a circular shape (no correlation between attributes) in 2-dimensional space cannot be mapped to its original position by using PCA.

The covariance matrix is the main identifier of the success of our attack, as it is the only input to PCA. Its eigenvalues and eigenvectors define the alignment of data and are the basis for finding the rotation which maps a candidate dataset to the real data of the target database. We therefore study the connection between the properties of the covariance matrix and the success of the attack. The covariance matrix contains the covariance values between each pair of attributes and the variances of single variables in its diagonal.

In order to see the effect of the covariance values, we implemented an algorithm that constructs a multivariate Gaussian distribution which has a given covariance matrix (see Sec. 5.2 below). By using this tool, we can observe error rates with different configurations of the covariance matrix.

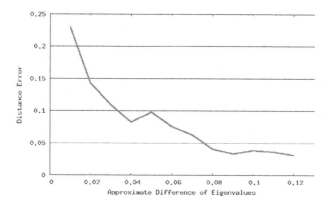

Fig. 1. Effect of eigenvalue differences (synthetic data)

Recall that the PCA-based attack works by finding the eigen-basis of the candidate dataset, and comparing it to the eigen-basis of the known properties distribution. In order to match the eigenvectors pairwise, all the corresponding eigenvalues must be different. We thus expect that the attack works best, when there is a large average difference between the eigenvalues of the covariance matrix (so that they are easily paired with the eigenvalues of the known probability distribution). To test this, we constructed data with difference covariance matrices, and plotted the average distance between eigenvalue versus the error percentage of the output of the attack. As can be seen from the result of the tests given in Fig. 1, there is a clear relationship between the eigenvalue difference and the error percentage. When the difference between the eigenvalues grows, the error drops. On the other hand, when the eigenvalues are very similar, the error percentage increases dramatically.

A dataset has a high average difference between eigenvalues of the covariance matrix when the correlations between different pairs of attributes differ. In other words: datasets where some attributes are highly correlated, while others are only weakly correlated are more vulnerable to the PCA attack.

In some scenarios it may not be realistic to assume that an attacker has access to statistical data which contains the correlations between attributes. If, for instance, the attacker can only obtain statistics for each attribute independently, he will not be able to apply the PCA attack, since his description of the global probability distribution does not contain the correlations necessary for the PCA attack to work. In this case, however, other methods may be applies. Recall that PCA is only used to recognise "geometrical characteristics" of the dataset which can be used to rotate the candidate dataset to the real data. Since we know that our hyper-latereted candidate dataset has *the same shape* as the original data — up to rotation, displacement, and mirroring, any technique which can give a simple representation of the shape of a dataset, can be used to find a rotation from the candidate dataset to the real data. Instead of using PCA an attacker may try to recognise each attribute independently in the candidate dataset. We leave it to future work to find alternatives to PCA.

Generating Synthetic Datasets. To characterise the vulnerable datasets, we generated synthetic datasets with a given covariance matrix, V. We do this as described in this section.

A d-dimensional data set, described by a d-dimensional random variable $X = (X_1, \ldots, X_d)$ has the covariance matrix:

$$
\begin{bmatrix}
E[(X_1 - \mu_1)(X_1 - \mu_1)] & \cdots & E[(X_1 - \mu_1)(X_d - \mu_d)] \\
E[(X_2 - \mu_2)(X_1 - \mu_1)] & \cdots & E[(X_2 - \mu_2)(X_d - \mu_d)] \\
\vdots & & \vdots \\
E[(X_d - \mu_d)(X_1 - \mu_1)] & \cdots & E[(X_d - \mu_d)(X_d - \mu_d)]
\end{bmatrix},
$$

where $\mu_i = E[X_i]$ is the expected value of the ith attribute. This matrix can be rewritten as

$$
Cov_d(X) = E\left[(X - E[X])(X - E[X])^T\right]. \tag{6}
$$

We now see, that for an d-dimensional random variable X, and orthogonal matrix U:

$$
\begin{aligned}
Cov_d(UX) &= E\left[(UX - E[UX])(UX - E[UX])^T\right] \\
&= E\left[U(X - E[X])(X - E[X])^T U^T\right] \\
&= U E\left[(X - E[X])(X - E[X])^T\right] U^T \\
&= U Cov_d(X) U^T.
\end{aligned}
$$

We can now create a data set with the given covariance matrix V. Since V is self-adjoint it can be diagonalised. In other words, we can find orthogonal matrix U, and diagonal matrix D such that:

$$
V = UDU^T. \tag{7}
$$

The matrix U will have the ith eigenvector of V as the ith row, and the matrix D will have the ith eigenvalue in D_{ii}.

We now see that if $Cov_d(X) = D$ then

$$
Cov(UX) = U Cov(X) U^T = UDU^T = V. \tag{8}
$$

In other words — by generating X with independently distributed variables, where X_I has variance D_{ii}, then the random variable UX has the desired covariance matrix.

6   Experimental Results

To demonstrate the potential power of our attack we apply it to both real and synthetic datasets. We use the "Adult Census" and "Auto-MPG" datasets from the UCI Machine Learning Repository [13]. For our experiments we use a 1.6 GHz Pentium Notebook with 2 MB cache and 512 MB RAM running the Windows XP operating system. The implementation of our attack is programmed using ruby 1.8.6 with rb-gsl (gnu scientific library) bindings for mathematical operations.

The Adult Census dataset contains 48842 objects. The objects have 14 attributes, of which we are only using the attributes "age", "education-num", "sex", and "hours-per-week" (Liu *et al.* apply their attack to the attributes "age", "education-num", "hours-per-week").

The Auto-MPG dataset contains information on car brands, their engines, and their gas consumption. The dataset contains 398 objects with 8 attributes. We apply our attack to the attributes "mpg", "displacement" (engine volume), "horsepower", "weight", and "acceleration".

Before the attack is applied, we compute the global probability distributions, which is to be known to the attacker. To this end each test first selects a subset from the given dataset, and computes the statistics on that subset. The attack is subsequently applied to the dissimilarity matrix of another subset of objects (the target database). When the target database is overlapping with the data used for computing the probability distribution, the computed probability distribution fits closely to the target database. This may make the attack seem better than it is. We have selected non-overlapping sets, where possible. Unfortunately the Auto-MPG dataset only contains 398 objects, so some overlap is unavoidable.

As measure of success, we compute the average distance between the recovered data and their real values. The tests are repeated 30 times, and their average is taken. In the following graphs the distances are reported in percentage of the maximum distance[1], which is referred to as *distance error*, to make comparison easy.

In Fig. 2 the results of the attack on the Auto-MPG dataset is shown. Since the dataset is small, we use 200 objects for computing the known probability distribution. We attack target datasets of sizes from 50 to 400 in steps of 50. In the tests where the target database has less than 200 objects, we use non-overlapping sets. However, in the tests with more than 200 objects there is an overlap between the data used for computing the probability distribution and

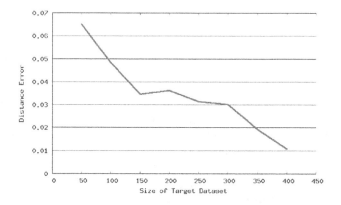

Fig. 2. Error percentage in Auto-MPG dataset with 5 attributes

---

[1] Since we normalise all data, the maximum distance of d-attribute objects is $\sqrt{d}$.

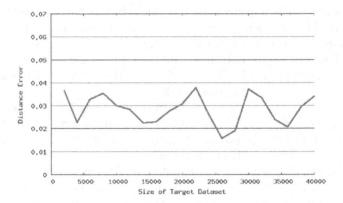

Fig. 3. Error percentage in adult census dataset with 4 attributes

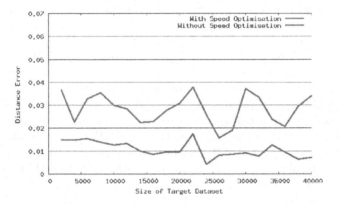

Fig. 4. Error percentage in adult census dataset with and without our speed optimisation

the target dataset. Even for a dataset as small as 50 objects, we can recover the data with an error of only 6.5%. When the size of the dataset grows, the error drops to approximately 2%, and therefore our attack becomes more effective.

In Fig. 3 the results of the attack applied to the Adult Census dataset is illustrated. The global probability distribution is computed from a set of 5000 objects, and the target datasets contain between 2000 and 40000 objects in steps of 2000. Since the Adult Census dataset contains 48842 objects the dataset used for computing the global probability distribution and the target dataset are non-overlapping. The tests show that in this very realistic scenario, we are able to recover the secret data with an error of only 3%, and on some cases as low as 2%. In terms of privacy this means that we can recover the age of individuals to a precision of 3 years; this is clearly a violation of privacy.

Since our attack does not do exhaustive search for the best mirroring of the dataset, but iteratively try to mirror in one direction at the time it will clearly not be as precise as when exhaustive search is used. To see how much the precision

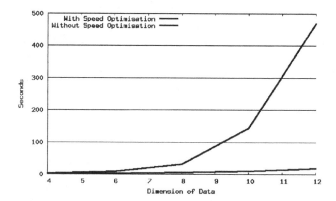

Fig. 5. Time of the attack on 1000 objects with and without our speed optimisation

suffers when using our optimisation we also make the above test on the Adult Census dataset. In Fig. 4 it is seen that the speed improvement reduces the precision of the attack, as expected.

To demonstrate the effect of our improvement on the speed of the attack, we perform a series of timings both with and without our improvement. The tests are performed on synthetic datasets of 1000 objects with 4 – 12 attributes. As can be seen from Fig. 5 our approach greatly reduces the time of the attacks.

For databases with many attributes our improved attack offers a realistic attack, where the original attack becomes infeasible.

7   Conclusions

Privacy preserving data mining is still an ongoing research topic where off-the-shelf software solutions are yet to be developed. Two of the main reasons for the lack of software solutions are the strong assumptions made by the existing methods, and possible privacy breaches. In this work we showed that distance preserving data transformation techniques proposed for privacy preserving data mining (1) make too strong assumptions for real life scenarios, and (2) compromise the privacy of individuals. Current distance preserving transformations assume that the adversary does not have background information about the released data. To prove that this is not a realistic assumption, we showed how an adversary can utilise public data sets to obtain statistics about the transformed data. We further demonstrated how this background information can be used in conjunction with the distance values to obtain the original data set. We conducted experiments on US census and Auto MPG data obtained from UCI [13] to show that the actual data can be recovered with very high accuracy. Our attack is an improvement of the attack of Liu *et al.* since it is applicable to *any* distance preserving map (including the projection map proposed in [8] which is secure against the attack from [7]).

It is still an open problem to quantify how much information can be disclosed from the dissimilarity matrix of a given dataset. We argued that, when the PCA attack is applied, the amount of disclosed information is related to the characteristics of the eigenvalues of the correlation matrix, but for other techniques other properties may govern the amount of leakage. It is an interesting problem to find alternatives to PCA.

References

1. Aggarwal, C.C.: On randomization, public information and the curse of dimensionality. In: ICDE, pp. 136–145 (2007)
2. Aggarwal, C.C., Yu, P.S.: A condensation approach to privacy preserving data mining. In: EDBT, pp. 183–199 (2004)
3. Agrawal, R., Srikant, R.: Privacy-preserving data mining. In: Proceedings of the 2000 ACM SIGMOD International Conference on Management of Data, Dallas, Texas, USA, May 16-18, 2000, pp. 439–450. ACM, New York (2000)
4. Chen, K.: Geometric Methods for Mining Large and Possibly Private Datasets. PhD thesis, Georgia Institute of Technology (2006)
5. Chen, K., Sun, G., Liu, L.: Towards attack-resilient geometric data perturbation. In: Proceedings of the 2007 SIAM International Conference on Data Mining, pp. 78–89 (2007)
6. Kantarcioglu, M., Clifton, C.: Privacy-preserving distributed mining of association rules on horizontally partitioned data. IEEE Trans. Knowl. Data Eng. 16(9), 1026–1037 (2004)
7. Liu, K., Giannella, C., Kargupta, H.: An attacker's view of distance preserving maps for privacy preserving data mining. In: Fürnkranz, J., Scheffer, T., Spiliopoulou, M. (eds.) PKDD 2006. LNCS (LNAI), vol. 4213, pp. 297–308. Springer, Heidelberg (2006)
8. Liu, K., Kargupta, H., Ryan, J.: Random projection-based multiplicative data perturbation for privacy preserving distributed data mining. IEEE Trans. Knowl. Data Eng. 18(1), 92–106 (2006)
9. Mann, C.C.: Homeland insecurity. The Atlantic Monthly 290(2) (2002)
10. Muralidhar, K., Sarathy, R.: Security of random data perturbation methods. ACM Trans. Database Syst. 24(4), 487–493 (1999)
11. Oliveira, S.R.M., Zaïane, O.R.: Privacy preserving clustering by data transformation. In: Proceedings of the 18th Brazilian Symposium on Databases, pp. 304–318 (2003)
12. Oliveira, S.R.M., Zaïane, O.R.: Privacy-preserving clustering by object similarity-based representation and dimensionality reduction transformation. In: ICDM 2004, pp. 21–30. IEEE Computer Society, Los Alamitos (2004)
13. UCI Machine Learning Repository, http://mlearn.ics.uci.edu/MLsummary.html
14. Vaidya, J., Clifton, C.: Privacy-preserving k-means clustering over vertically partitioned data. In: KDD 2003: Proceedings of the ninth ACM SIGKDD international conference on Knowledge discovery and data mining, New York, NY, USA, pp. 206–215. ACM Press, New York (2003)
15. Vijayan, J.: House committee chair wants info on cancelled dhs data-mining programs. Computer World (September 18, 2007)

Privacy-Preserving Publication of User Locations in the Proximity of Sensitive Sites

Bharath Krishnamachari, Gabriel Ghinita, and Panos Kalnis

Department of Computer Science
National University of Singapore
{kbharath,ghinitag,kalnis}@comp.nus.edu.sg

Abstract. Location-based services, such as on-line maps, obtain the exact location of numerous mobile users. This information can be published for research or commercial purposes. However, privacy may be compromised if a user is in the proximity of a sensitive site (e.g., hospital). To preserve privacy, existing methods employ the K-anonymity paradigm to hide each affected user in a group that contains at least $K - 1$ other users. Nevertheless, current solutions have the following drawbacks: *(i)* they may fail to achieve anonymity, *(ii)* they may cause excessive distortion of location data and *(iii)* they incur high computational cost.

In this paper, we define formally the attack model and discuss the conditions that guarantee privacy. Then, we propose two algorithms which employ 2-D to 1-D transformations to anonymize the locations of users in the proximity of sensitive sites. The first algorithm, called *MK*, creates anonymous groups based on the set of user locations only, and exhibits very low computational cost. The second algorithm, called *BK*, performs bichromatic clustering of both user locations and sensitive sites; BK is slower but more accurate than MK. We show experimentally that our algorithms outperform the existing methods in terms of computational cost and data distortion.

1 Introduction

The recent years have witnessed the widespread availability of positioning capabilities (e.g., GPS) in automobiles, handheld devices, etc. The emergence of novel applications based on user locations has created the potential for gathering large amounts of location data from mobile clients. Location data can also be collected from a variety of other sources. For instance, the Octopus system in Hong Kong, which employs a smart card for transportation and low-value purchases, can monitor the location where the card was used.

Location data can benefit a broad range of applications, such as alleviation of traffic congestion, or optimization of operations in a public transportation network. Nevertheless, Hu et al. [9] observed that publishing such data for research or planning purposes introduces serious privacy concerns. The location data can be joined with external information, such as schedules of hospital appointments, in order to reveal sensitive information about individuals. Consider the example in Figure 1a (adapted from [9]). Assume that the published location data for a specific day at $2pm$ consists of users $u_1 \ldots u_4$. Furthermore, hospital h publishes the appointment schedule of Figure 1b; note that the

B. Ludäscher and Nikos Mamoulis (Eds.): SSDBM 2008, LNCS 5069, pp. 95–113, 2008.

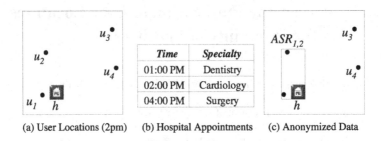

(a) User Locations (2pm)   (b) Hospital Appointments   (c) Anonymized Data

Fig. 1. Privacy violation in location data publishing

schedule is anonymous. Taken in isolation, neither of the two published datasets represents a privacy threat. However, by combining the two datasets, an attacker can infer that user u_1, who was the only one near the hospital at $2pm$, is consulting a cardiologist.

Previous work employed Spatial K-Anonymity (*SKA*) [10] to preserve privacy in Location-Based Services (*LBS*). SKA replaces the exact location of u with an Anonymizing Spatial Region (*ASR*), which encloses u as well as $K-1$ other users; therefore the identification probability of u does not exceed $1/K$ (K is a user-defined parameter called anonymity degree). The process of replacing exact locations with ASRs is called *cloaking*. Several algorithms for spatial cloaking have been proposed [5,6,8,10,13]. Most of the previous work focused on hiding the association between a query sent to an LBS and the actual querying user; therefore, each time a user u sends a query, a single ASR is built around u. Our problem, on the other hand, has two characteristics that make it more difficult: *(i)* the privacy of *all* users must be preserved, as opposed to anonymizing only a single querying user, and *(ii)* an additional set of sensitive locations must be taken into account; therefore, anonymization is performed with respect to the external data that an attacker may have access to (e.g., the hospital schedule).

Continuing the earlier example, let the anonymity degree be $K = 2$. Figure 1c shows $ASR_{1,2}$ that encloses users u_1 and u_2, as well as the sensitive site h. In the published data, the exact location of u_1 and u_2 is replaced by the ASR. From the attacker's point of view, u_1 or u_2 can be anywhere inside the ASR with equal probability. Moreover, the ASR encloses h, so it is closest to h than any other user. Consequently, the attacker can only assume that either u_1 or u_2 is having a cardiologist appointment with probability at most $1/K = 1/2$. Observe that the locations of u_3 and u_4 do not need to be cloaked, because these users are further away from the sensitive site (compared to the ASR), so they would not be associated with h. Note that, while cloaking preserves privacy, it also reduces the accuracy of the published data: a researcher that studies Figure 1c cannot know exactly where the users are located. A tradeoff emerges between privacy and the amount of information that is lost in the process of cloaking. The data distortion is the sum of areas of all ASRs; this metric must be minimized.

The previous example assumed that the attacker knows the identity of u_1, therefore the association between u_1 and "cardiologist" can be performed. In practice, when publishing location data, the identity of the mobile users is removed. Nevertheless, as shown in [4], a number of methods can be employed to infer the identity of a user based on his location (e.g., through trajectory reconstruction [15]). For the rest of the paper

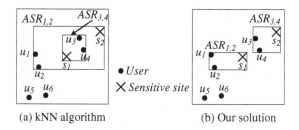

(a) kNN algorithm   (b) Our solution

Fig. 2. The kNN approach may result to excessive data distortion

we assume that an attacker is able, in the worst case, to acquire the identity of the user associated to each location in the published data.

A naïve solution to generate ASRs in the proximity of sensitive locations is the following: For each sensitive site s_i, use a k-Nearest-Neighbor (*kNN*) algorithm to assign the K nearest users to s_i. However, the data distortion depends on the order of processing the sites. In the example of Figure 2a, s_1 is processed first resulting to $ASR_{3,4}$ since u_3 and u_4 are the nearest users; then s_2 is assigned to $ASR_{1,2}$. On the other hand, our solution (Figure 2b) assigns s_1 to $ASR_{1,2}$ and s_2 to $ASR_{3,4}$; clearly the resulting data distortion is lower. Moreover, we show in our experiments that the kNN solution is slow. Recently, Hu et al. [9] formulated the problem as a version of the *set cover* problem and proposed heuristic algorithms. However, their solution suffers from the following drawbacks: *(i)* the approach for generating and publishing ASRs does not guarantee anonymity, *(ii)* the data distortion is high and *(iii)* the computational cost is very high.

In this paper we propose efficient solutions that do not suffer from the above-mentioned drawbacks. Our methods map the 2-D user locations to 1-D space. Dimensionality reduction has been acknowledged as a suitable method to achieve privacy for both relational data [7] and in Location-based services [10]. Figure 3 shows an outline of our approach: the locations of users and sensitive sites are mapped to the 1-D domain using the Hilbert [14] space-filling curve. The Hilbert curve has good locality properties: if two points are close to each other in the 2-D space, with high probability they will also be close in the 1-D transformation. We devise two methods to generate ASRs: *(i) Monochromatic K-anonymity (MK)* is a multi-stage method: first the set of users is partitioned into groups containing K to $2K - 1$ users; the partitioning is optimal

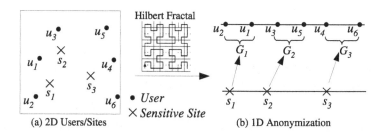

(a) 2D Users/Sites   (b) 1D Anonymization

Fig. 3. Using 2-D to 1-D mapping, s_1 is assigned to $ASR_{2,1}$, s_2 to $ASR_{3,5}$, etc.

(i.e., lowest possible group extents) in the 1-D space. Then, a greedy approach is used to assign sensitive sites to user groups. MK is fast because the partitioning phase is linear to the number of users. However, since the initial partitioning is independent of the sensitive sites, the resulting data distortion is not optimal. *(ii) Bichromatic K-anonymity (BK)* is a one-stage algorithm that performs an optimal assignment (in the 1-D space) of users to sites by simultaneously clustering both users and sensitive sites (hence "bichromatic"). Although the assignment is not optimal in the 2-D space, the resulting data distortion is lower compared to MK. The tradeoff is that BK is computationally more expensive, since the search space of the solution is considerably higher than for MK. Nevertheless, we show experimentally that both MK and BK are much faster and achieve lower data distortion compared to existing methods.

The rest of the paper is organized as follows: Section 2 defines formally the problem and surveys the related work. Section 3 describes the Monochromatic K-anonymity technique, whereas Section 4 introduces Bichromatic K-anonymity. The experimental evaluation is presented in Section 5. Finally, Section 6 concludes the paper with directions for future work.

2   Background and Related Work

This section formalizes the attack on the published location data and defines the anonymi-zation problem. It also presents the related work on relational databases and Location-based services.

2.1   Problem Definition

Let U be the set of user locations and S the set of sensitive sites. Both users and sites may have arbitrary shapes and are represented by their Minimum Bounding Rectangle (*MBR*). Consider that U is published in its original form. Then, an attacker can compromise privacy by joining U and S. Formally:

Definition 1 (Attack on Location Privacy). *Given U, S and the anonymity requirement K, an attack is defined as the result of the following spatial join:*

```
SELECT user.id, site.id
FROM U as user, S as site
WHERE distance(user.mbr, site.mbr) =
      SELECT MIN(distance(U.mbr, S.mbr))
      FROM U, S
      WHERE S.id = site.id
```

An attack is successful iff the probability of distinguishing a particular user u in any of the resulting tuples of the above query is larger than $1/K$.

In the example of Figure 1a, the spatial join will output the tuple $\langle u_1, h \rangle$, since u_1 is closest to h than any other user. Therefore the attacker can infer that u_1 has a cardiology appointment.

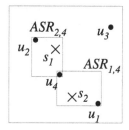

Fig. 4. User sharing violates privacy

To achieve anonymity, each sensitive site s is associated with an *anonymizing set*, denoted by $\mathcal{M}(s)$, of at least K users who are indistinguishable from each other. Instead of publishing the exact user locations, we publish the ASR, which is the MBR that encloses s and the users in $\mathcal{M}(s)$. Formally:

Definition 2 (Privacy-Preserving Location Publishing Format). *A privacy-preserving publication of U with respect to S is a mapping $\mathcal{M} : S \rightarrow 2^U$ (2^U is the set of all possible anonymizing sets). The published format consists of a collection of ASRs, one for each $s \in S$, where $ASR(s) = MBR(\{s\} \cup \mathcal{M}(s))$.*

Based on the attack model and publication format, we define below the K-anonymity condition for our problem:

Definition 3 (K-anonymous Location Publishing). *A privacy-preserving mapping $\mathcal{M} : S \rightarrow 2^U$ is K-anonymous iff* (i) $\forall s \in S$, $|\mathcal{M}(s)| \geq K$ *and* (ii) $\forall s_1, s_2 \in S$, $\mathcal{M}(s_1) \cap \mathcal{M}(s_2) = \emptyset$.

Condition *(i)* is imposed by the indistinguishability requirement, whereas condition *(ii)* specifies that the anonymizing sets of different sites should be disjoint. To demonstrate the need for the second condition, consider the example in Figure 4, where $K = 2$. S consists of sites s_1 and s_2, but the anonymizing sets of these sites overlap in user u_4. Therefore, only three distinct users are included in $\mathcal{M}(s_1) \cup \mathcal{M}(s_2)$. An attacker can infer that any of u_1, u_2 or u_4 was present at a sensitive site (either s_1 or s_2) with probability $2/3 = 0.66 > 1/K = 0.5$; hence, privacy is compromised.

Returning to the example of Figure 1c, the K-anonymity conditions are satisfied, since $|\mathcal{M}(h)| = |\{u_1, u_2\}| = 2$ and $\mathcal{M}(h)$ does not share users with any other site. This can be verified by executing the query of Definition 1. The exact locations of u_1 and u_2 have been replaced by the ASR, whose distance to h is 0. Therefore the query returns two tuples: $\langle u_1, h \rangle$ and $\langle u_2, h \rangle$; hence the probability of associating u_1 or u_2 with the cardiology appointment is at most $1/2$. Observe that u_3 and u_4 are not included in the query results, since their distance to h is larger than that of the ASR. Therefore, we can publish the exact locations of u_3 and u_4 without compromising privacy.

Besides providing privacy, the distortion (also called *generalization cost*) of the published data must be minimized. Similar to the related work in Location-based services [5,6,8,10,13], we measure the generalization cost by the sum of the areas of the published ASRs. Formally:

Definition 4 (Generalization Cost). *Given a set U of user locations, a set S of sensitive sites, and a privacy-preserving mapping $\mathcal{M} : S \rightarrow 2^U$, the generalization cost for site s is:*

$$GC(s) = Area(MBR(\mathcal{M}(s) \cup \{s\})) \qquad (1)$$

The overall (global) generalization cost of the entire mapping is:

$$GGC(\mathcal{M}) = \sum_{s \in S} GC(s) \qquad (2)$$

Recall that for some users we publish their exact locations (e.g., u_3 and u_4 in Figure 1c); no generalization cost is incurred for such users. Our problem is formally defined as:

Problem 1 (Optimal K-Anonymous Location Data Publication). *Given U, S and K, determine a K-anonymous mapping $\mathcal{M} : S \rightarrow 2^U$ such that the generalization cost $GGC(\mathcal{M})$ is minimized.*

2.2 K-Anonymity in Relational Databases

K-anonymity [16,17] was initially proposed in relational databases for privacy preserving publishing of detailed data (or *microdata*), such as hospital records. Although identifying attributes (e.g., name) are removed, microdata contains *quasi-identifying attributes (QID)* (e.g., $\langle Age, Zipcode \rangle$) that can be joined with external information, such as voting registration lists, to expose the identity of individual records. To address this threat, K-anonymity requires that each record must be indistinguishable from at least $K - 1$ other records, with respect to the QID. Two techniques are commonly used to achieve K-anonymity: *suppression*, where some of the attributes or tuples are removed, and *generalization*, which involves replacing specific values (e.g., phone number) with more general ones (e.g., only area code). Both methods lead to information loss. Algorithms for anonymizing an entire relation are discussed in [2,11]. Xiao and Tao [18] consider the case where each individual requires a different degree of anonymity, whereas Aggarwal [1] shows that anonymizing a high-dimensional relation leads to unacceptable loss of information due to the dimensionality curse. Machanavajjhala et al. [12] propose ℓ-diversity, an anonymization method that provides diversity among the sensitive attribute values of each anonymized group. Ghinita et al. [7] employ multidimensional to 1-D transformations to solve efficiently the K-anonymity and ℓ-diversity problems.

2.3 K-Anonymity for Location Data

Most related work in the area of location K-anonymity focuses on *query* privacy in Location-based Services (*LBS*). Users issue queries such as "find the closest hospital to my current location". Typically, there is a trusted Anonymizer Service (*AN*) between the users and the LBS. The users constantly update their location with AN. Queries are also sent through AN, which removes the user id and constructs an ASR that contains the querying user as well as at least $K - 1$ additional users. The AN forwards the ASR to the LBS, which computes the answer based on the ASR, instead of the exact user location. The result is also routed back to the querying user through the AN. In [8],

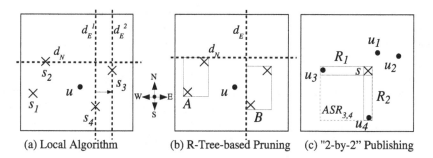

(a) Local Algorithm   (b) R-Tree-based Pruning   (c) "2-by-2" Publishing

Fig. 5. *Local* Algorithm

the anonymizer employs a quad-tree to index user locations. Given a query from u, the corresponding ASR is the lowest-level quadrant that contains u as well as $K - 1$ other users. In [13] a similar structure is used, but two neighboring quadrants are allowed to form an ASR, before ascending one level up in the quad-tree. In [6] queries from multiple users form a graph. The graph is searched for cliques (i.e., queries from near-by users), which are used to form the ASR. Kalnis et al. [10] identify the *reciprocity* property, a sufficient condition to guarantee anonymity. To enforce reciprocity, the users are split into disjoint buckets based on their 1-D Hilbert ordering; the same transformation is used in our work. The previous algorithms generate a single ASR independently for each query. This approach is not applicable to our problem, since we must publish an anonymized version of the entire dataset U; furthermore, anonymization depends on the set of sensitive sites S.

Location publishing in the proximity of sensitive sites was first discussed by Hu et al. [9]. They formulated the problem as a version of the *set cover* problem and proposed a heuristic algorithm called *Local* (see Figure 5). *Local* is a user-centric method: for each user $u \in U$ (for simplicity the example shows only one user) the location of u is incrementally enlarged to include sensitive sites in its bounding box. *Local* consists of four nested loops, corresponding to four directions originating at u (North, East, South, West), and each loop advances a plane-sweep line in its direction. In Figure 5a, the North sweeping line d_N is fixed, and the East line is advanced from d_E^1 to d_E^2 to cover sites s_4 and s_3, respectively. Out of all combinations along the four directions, the bounding box with the optimal *coverage* (measured as the area of the bounding box divided by the number of enclosed sites) is retained as the *candidate box* $\Omega(u)$. The candidate boxes are determined for all $u \in U$, and user u_0 with the lowest coverage is output, at which point a *coverage counter* of all sites enclosed by $\Omega(u_0)$ is increased. If the counter of a site s reaches K, s is removed from S, the candidate boxes of all other users that enclose s are updated, and the algorithm continues for the remaining sites, until all sites are covered at least K times. The complexity of *Local* is $O(|S|^4 \cdot (|S| + |U|))$, which is very high. [9] proposes an optimization based on the R-Tree spatial index (see Figure 5b). Instead of performing the plane-sweep with respect to individual sites, the algorithm considers the nodes of the R-Tree. Each node represents a "super-site", which is considered to be situated at the point inside the node that is closest to u, and has a weight equal to the number of sites rooted in the subtree of that node. It is shown that by using the super-site concept, a lower bound of the actual coverage is obtained, and the search space of the solution is reduced.

We show in Section 5 that the execution time (even with the R-tree optimization) and generalization cost of *Local* are very high. More importantly, *Local* allows the anonymizing sets of distinct sensitive sites to share users. Therefore, it does not guarantee privacy (recall condition *(ii)* in Definition 3). Furthermore, the authors of [9] propose a publication format, further referred to as *"2-by-2"*, that discloses a collection of MBRs for each site s: each MBR encloses s and a *single* user in $\mathcal{M}(s)$. However, this format discloses exact user locations, since each published MBR only contains two locations, and one of them (i.e., s) is known to the attacker. Figure 5c shows an example: Local chooses users u_3 and u_4 as part of $\mathcal{M}(s)$, because the two rectangles R_1 and R_2, which are very skewed, have small areas (hence, low generalization cost). An attacker can infer that the users are situated at the opposite extremities of R_1, respectively R_2, from s. This is similar to publishing the exact locations of all users, therefore privacy is compromised. Should we choose a secure publishing format like the one in Definition 2, i.e. $ASR_{3,4}$ in the example, the resulting area is very large. We will investigate this issue further in Section 5.

3   Monochromatic K-Anonymity (MK)

In this section, we present Monochromatic K-Anonymity (*MK*). MK is a multi-stage algorithm: In the first stage, it partitions the set U into groups with K to $2K - 1$ users each. In the subsequent stages, it uses a greedy approach to assign user groups to each site in S. The first stage of MK employs the *1DAnon* algorithm, which was used in [7] to partition relational data with 1-D quasi-identifiers. Below, we briefly explain *1DAnon*.

1DAnon takes as input the set U of user locations sorted according to their 1-D Hilbert values. We use u to denote a user, as well as his coordinate in the 1-D space. Furthermore, we denote by $|u_i - u_j|$ the 1-D distance between users u_i and u_j. Given a group of users $G = \{u_{begin}, \ldots, u_{end}\}$, where u_{begin} and u_{end} represent respectively the user with the minimum and maximum 1-D coordinate in G, we denote the *extent* of G in the 1-D space as: $1D\_Ext(G) = |u_{end} - u_{begin}|$. We refer to *begin* and *end* as the *boundaries* of G. *1DAnon* finds the optimal K-anonymous partitioning $\mathcal{U} = \{G_1, \ldots, G_{|\mathcal{U}|}\}$ of U, such that the

$$1D\_Cost(\mathcal{U}) = \sum_{G \in \mathcal{U}} 1D\_Ext(G) \qquad (3)$$

is minimized. Note that Eq. (3) is the one-dimensional equivalent of the GGC metric from Eq. (2). To find the optimal anonymous partitioning of U, *1DAnon* applies a dynamic programming recursive formulation which determines the best grouping for each prefix $\{u_1, \ldots, u_i\}$ (where $K \leq i \leq |U|$) of the user sequence. *1DAnon* returns a set of K-anonymous groups, each with size bounded between K and $2K - 1$. The computation cost of *1DAnon* is $O(K \cdot |U|)$, hence linear to the number of users. In our MK algorithm, we vary *1DAnon* slightly: Instead of $1D\_Cost$, we use the GGC metric. The user partitioning is not optimal in the 2-D space, but due to the good locality properties of the Hilbert ordering, the results are adequate in practice.

Monochromatic K-anonymity (MK)

Input: sets $U[1 \ldots n]$ and $S[1 \ldots m]$ sorted in ascending order of 1D Hilbert values

1. $\mathcal{U} = 1DAnon(U, K)$
2. **while** $|S| > 0$
3.   **foreach** $s \in S$
         // assign s to a user group s.t. the resulting area is minimized
4. $AS(s) = G_i$ s.t. $\forall j \neq i, Area(MBR(G_i \cup \{s\})) < Area(MBR(G_j \cup \{s\}))$
5.   **foreach** $G \in \mathcal{U}$
6.   **if** $\exists s$ s.t. $AS(s) = G$ **then**
         // choose the site that minimizes the resulting area
7.   choose s_0 s.t. $AS(s_0) = G$ and $\forall s \in S | s \neq s_0 \wedge AS(s) = G,$
              $Area(MBR(G \cup \{s_0\})) < Area(MBR(G \cup \{s\})))\}$
8.   **output** $MBR(G \cup \{s_0\})$
9. $U = U \backslash G$
10. $S = S \backslash \{s_0\}$

Fig. 6. Monochromatic K-Anonymity Pseudocode

Figure 6 shows the pseudocode of MK: the input consists of sets U and S, sorted in the 1-D Hilbert order of the locations of users and sensitive sites, respectively. The cardinalities of the two sets are denoted as $n = |U|$ and $m = |S|$. Initially, MK invokes *1DAnon* (line 1) and obtains $\mathcal{U}$, which is the partitioning of U into K-anonymous groups. Subsequently, MK assigns the sensitive sites to groups of $\mathcal{U}$. At each stage, each site $s \in S$ is assigned to the user group G that minimizes the area of $MBR(G \cup \{s\})$ (lines 3-4). We say that G is the anonymizing set of s, i.e. $AS(s) = G$. Note that, multiple sites can be assigned to the same group, whereas some groups may not be assigned any site. Since sites cannot share users (condition *(ii)* in Definition 3), collisions are solved (line 7) by choosing for each group the site that minimizes the area of $MBR(G \cup \{s\})$ (in case of ties, a random site is chosen). For each assigned site, we output (line 8) the MBR that encloses s_0 and its corresponding anonymizing set $\mathcal{M}(s_0) \equiv G$. U and S are updated by eliminating the users and sites that have been output (lines 9-10). If there still exist unassigned sites (line 2), the algorithm starts a new stage with the remaining users and sites. Since at most $2K - 1$ users belong to each user group, the algorithm is guaranteed to terminate if the inputs satisfy the condition $n \geq (2K - 1) \cdot m$.

In the worst case, at each stage all sites are assigned to the same user group, and the number of required stages is m. Each stage takes $O(m \cdot n)$ to find the closest group for each site. Hence, the complexity of MK is $O(K \cdot n + m^2 \cdot n)$; the first term corresponds to *1DAnon*. In practice, the number of stages is significantly smaller than m; in Section 5 we show that MK is very fast.

Figure 7 illustrates an example of applying MK for $U = \{u_1 \ldots u_6\}$ and $S = \{s_1 \ldots s_3\}$. Initially (Figure 7b), the *1DAnon* algorithm is executed, resulting in anonymous groups $G_1 \ldots G_3$. Then, sites s_1 and s_2 are assigned to G_1, whereas s_3 is assigned to G_3. Since the enclosing area of s_2 and G_1 is larger than that of s_1 and G_1, MK outputs s_1 with anonymizing set G_1, and s_3 with G_3. In the next stage (Figure 7c), the remaining users and sites are $\{u_3, u_5\}$ and s_2, which are output together. Note that,

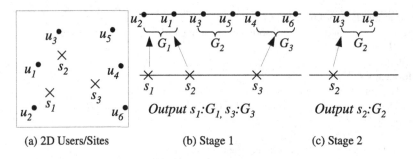

(a) 2D Users/Sites   (b) Stage 1   (c) Stage 2

Fig. 7. Example of MK

it is possible for MK to terminate, even though some users do not have any site assigned to them. For those users we publish their exact location, since their privacy is not threatened (refer to Definition 3).

4   Bichromatic K-Anonymity (BK)

This section introduces the Bichromatic K-anonymity (BK) algorithm. BK also uses the 1-D Hilbert transformation for U and S. However, the process of creating anonymizing sets considers simultaneously U and S (as opposed to MK, which partitions U independently). As a result, BK achieves lower generalization cost.

Before presenting BK, we will study a restriction of the problem to the 1-D space. We seek to find an optimal K-anonymous mapping $\mathcal{M}$ that assigns sites to user groups, such that the 1-D cost is minimized. In the 1-D domain the generalization cost of a mapping is:

$$1D\_Cost(\mathcal{M}) = \sum_{s \in S} 1D\_Ext(\{s\} \cup \mathcal{M}(s)) \qquad (4)$$

In Section 4.1 we identify three properties of an optimal 1-D mapping: *(i)* each anonymizing set contains *exactly* K users, *(ii)* each anonymizing set consists of users that are consecutive in the 1-D domain, and *(iii)* the extents of any two anonymizing sets do not overlap in the 1-D domain. Based on these properties, in Section 4.2 we present the BK algorithm, which employs dynamic programming to solve the problem in the 2-D space. Although the 2-D solution is not optimal, due to the good locality properties of the Hilbert ordering, BK achieves very low generalization cost in practice.

4.1   Properties of an Optimal 1-D Mapping

The following theorem states that there exists an optimal 1-D mapping where the anonymizing set of each site contains exactly K users.

Theorem 1. *Consider a set of user locations U and a set of sensitive sites S. Then, there exists an optimal mapping $\mathcal{M} : S \to 2^U$ such that, $\forall s \in S$, $|\mathcal{M}(s)| = K$.*

Proof. Let $\mathcal{M}'$ be the optimal mapping for U and S, and assume that $\exists s_0 \in S$ such that $G = \mathcal{M}'(s_0)$ and $|G| > K$. Let G' be the anonymizing set obtained by retaining only

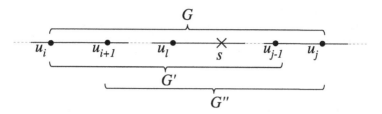

Fig. 8. Anonymizing sets consist of consecutive users in the 1-D sequence

K users from G. Define mapping $\mathcal{M}$ such that $\mathcal{M}(s_0) = G'$, and $\forall s \neq s_0, \mathcal{M}(s) = \mathcal{M}'(s)$. Then the K-anonymity condition for $\mathcal{M}$ is satisfied, according to Definition 3. Furthermore, since $G' \subset G$, we have that $1D\_Ext(G') \leq 1D\_Ext(G)$, hence the generalization cost of $\mathcal{M}$ does not exceed that of $\mathcal{M}'$. By applying the same reasoning for all groups with size larger than K, we obtain an optimal mapping $\mathcal{M}$ such that $\forall s \in S, |\mathcal{M}(s)| = K$. □

We further show that the anonymizing set of every site s consists of users that are consecutive in the user sequence. Formally:

Theorem 2. *There exists an optimal mapping $\mathcal{M}$ such that $\forall s \in S$, if $u_i, u_j \in \mathcal{M}(s)$ and $i < j$, then $\forall u_l$ with $i < l < j$, $u_l \in \mathcal{M}(s)$.*

Proof. Assume optimal mapping $\mathcal{M}'$ and that $\exists s \in S$ such that $u_i, u_j \in \mathcal{M}'(s), i < j$, and $\exists u_l, i < l < j$, such that $u_l \notin \mathcal{M}'(s)$. This situation is depicted in Figure 8. Then, we can replace $\mathcal{M}'(s)$ with either $G' = \mathcal{M}'(s) \backslash \{u_i\} \cup \{u_l\}$ or $G'' = \mathcal{M}'(s) \backslash \{u_j\} \cup \{u_l\}$. The privacy condition is still satisfied, since $|G'| = |G''| = K$; furthermore, $1D\_Ext(G') \leq 1D\_Ext(G)$ and $1D\_Ext(G'') \leq 1D\_Ext(G)$. Therefore, we obtain a new K-anonymous mapping $\mathcal{M}$ with generalization cost not exceeding that of $\mathcal{M}'$, hence optimal. □

We also prove that there exists an optimal 1-D mapping, where the extents of the anonymizing sets do not overlap. Formally:

Theorem 3. *There exists an optimal mapping $\mathcal{M}$ such that, $\forall s_1, s_2 \in S$, and $G_1 = \mathcal{M}(s_1)$, $G_2 = \mathcal{M}(s_2)$, the 1-D extents of G_1 and G_2 do not overlap.*

Proof. Denote by $\mathcal{M}'$ the optimal mapping for U and S, and assume that $\exists s_1, s_2 \in S$, $G_1 = \mathcal{M}'(s_1)$ and $G_2 = \mathcal{M}'(s_2)$, such that G_1 and G_2 overlap in their 1-D extents. Let u_{i_1}, u_{i_2} be the start boundaries and u_{j_1}, u_{j_2} be the end boundaries of groups G_1 and G_2, respectively. Without loss of generality, consider that $u_{i_2} < u_{j_1}$, a situation depicted in Figure 9. We build anonymizing groups $G_1' = G_1 \cup \{u_{i_2}\} \backslash \{u_{j_1}\}$ and $G_2' = G_2 \cup \{u_{j_1}\} \backslash \{u_{i_2}\}$, that is, we swap the end user of G_1 with the start user of G_2. Since $|u_{i_2} - u_{i_1}| + |u_{j_2} - u_{j_1}| \leq |u_{j_1} - u_{i_1}| + |u_{j_2} - u_{i_2}|$, the generalization cost of the mapping is not enlarged. Furthermore, sets G_1' and G_2' have the same cardinality as G_1 and G_2, hence the privacy requirement is satisfied. Therefore, the new mapping $\mathcal{M}$ obtained by replacing G_1, G_2 with G_1', G_2' is optimal. By applying the same reasoning for every pair of overlapping groups, the theorem is proved. □

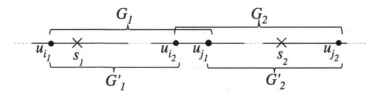

Fig. 9. Anonymizing sets do not overlap in their 1-D extent

4.2 The BK Algorithm in the 2-D Domain

BK is a dynamic programming algorithm, which is based on the properties of the 1-D ordering. BK finds the best mapping $\mathcal{M}$ by minimizing the 2-D GGC metric from Eq. (2). Recall that BK is not optimal in the 2-D domain.

Let $AS(s)$ be the anonymizing set of site s. Each AS has cardinality K (Theorem 1) and consists of consecutive users in the 1-D order (Theorem 2). Therefore, we can uniquely identify a particular AS by its start boundary i (i.e., the group starting at i consists of users $u_i \ldots u_{i+K-1}$).

BK determines recursively the optimal mapping for each sub-problem corresponding to prefixes $U_i = \{u_1 \ldots u_{i+K-1}\}$ of U and $S_j = \{s_1 \ldots s_j\}$ of S. Intuitively, U_i contains all users that may be part of anonymizing sets starting at boundary *at most* i. BK tabulates the values of a cost matrix $Cost[1 \ldots n][1 \ldots m]$, where element $Cost[i][j]$ contains the optimal solution to the sub-problem with inputs U_i and S_j.

According to Theorem 3, the users in $AS(s_j)$ must be after those in $AS(s_{j-1})$; therefore, not all start boundaries are acceptable for a given j. Let $a(j)$ be the minimum and $b(j)$ the maximum allowable start boundary for $AS(s_j)$. There must be enough users before $AS(s_j)$ to build AS for sites $s_1 \ldots s_{j-1}$. Similarly, sufficient users must remain after $AS(s_j)$, to form AS for $s_{j+1} \ldots s_m$. Formally:

$$a(j) = (j-1) \cdot K + 1, \ b(j) = n + 1 - (m - j + 1) \cdot K \tag{5}$$

Figure 10 illustrates the $Cost$ matrix and the possible choices of the start boundary for $AS(j)$.

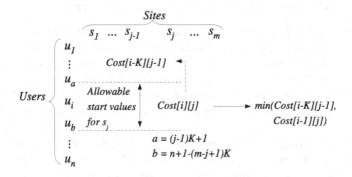

Fig. 10. BK: $Cost$ matrix tabulation

Bichromatic K-anonymity (BK)

Input: sets $U[1 \ldots n]$ and $S[1 \ldots m]$ sorted in ascending order of 1D Hilbert values

0. $min\_val = \infty, best\_start\_value = -1$

    /* populate first column */

1.   **for** $i = 1$ **to** $n - m \cdot K + 1$ **do** /* for every allowed start boundary of $AS(s_1)$ */

2.   **if** $GC(u_i, \ldots, u_{i+K-1}, s_1) < min\_val$ **then**

3. $min\_val = GC(u_i, \ldots, u_{i+K-1}, s_1)$

4. $best\_start\_value = i$

5. $solution[i][1] = best\_start\_value$

6. $Cost[i][1] = min\_val$

    /* populate remaining columns */

7.   **for** $j = 2$ **to** m **do**

8. $min\_val = \infty, best\_start\_value = -1$

9.   **for** $i = (j-1)K + 1$ **to** $n + 1 - (m - j + 1)K$ **do** /*for every start boundary of $AS(s_j)$*/

10.   **if** $(Cost[i - K][j - 1] + GC(u_i, \ldots, u_{i+K-1}, s_j)) < min\_val$ **then**

11. $min\_val = Cost[i - K][j - 1] + GC(u_i, \ldots, u_{i+K-1}, s_j)$

12. $best\_start\_value = i$

13. $solution[i][j] = best\_start\_value$

14. $Cost[i][j] = min\_value$

    /* output solution */

15. $group\_start = solution[n - K + 1][m]$

16. **output** $\mathcal{M}(s_m) \equiv AS(s_m) = (u_{group\_start} \ldots u_{group\_start+K-1})$

17. **for** $j = m - 1$ **downto** 1 **do**

18. $group\_start = solution[group\_start - K][j]$

19.   **output** $\mathcal{M}(s_j) \equiv AS(s_j) = (u_{group\_start} \ldots u_{group\_start+K-1})$

Fig. 11. Bichromatic K-Anonymity Pseudocode

Note that some users may not be included in any AS, hence there may be "gaps" left in the user sequence when forming an AS. As mentioned earlier, $Cost[i][j]$ stores the best cost of the solution for sub-problem U_i, S_j. According to Eq. (5), the AS of the last site in S_j can start at any value between $a(j)$ and i. Hence, $Cost[i][j]$ contains the minimum cost over all choices of start boundary i', $a(j) \leq i' \leq i$. The $Cost$ value is recursively determined as:

$$Cost[i][j] = \min\{Cost[i-1][j],\ Cost[i - K][j-1] + GC(u_i, \ldots, u_{i+K-1}, s_j)\} \quad (6)$$

If the first element of the *min* function is smaller, it signifies that choosing to start $AS(j)$ at i is more costly than if we start it at $i - 1$, or earlier. Hence, i should not be the start of $AS(s_j)$. Otherwise, $AS(s_j)$ should begin at i, and the value of $Cost[i][j]$ is updated as the sum of the immediate generalization cost associated to the area enclosing $\{s_j\} \cup \{u_i \ldots u_{i+K-1}\}$, and the recursive component $Cost[i - K][j - 1]$. The latter corresponds to the best cost obtained for the sub-problem U_{i-K}, S_{j-1} (the $i - K$ is dictated by the requirement that $AS(s_{j-1})$ must end before i, hence must have start boundary at most $i - K$).

Figure 11 shows the BK pseudocode. The values in the first column of the matrix (i.e., $Cost[*][1]$) are determined directly (lines 1-6) by computing all possible anonymizing sets associated to sensitive site s_1. Formally:

$$Cost[i][1] = \min_{1 \leq i' \leq i} GC(u_{i'}, \ldots, u_{i'+K-1}, s_1), \ a(1) \leq i \leq b(1) \quad (7)$$

In addition to the minimum cost value, we also need to retain the i' value that minimizes the above cost, to reconstruct the solution once the tabulation is completed. For this purpose we use an additional table $solution[1 \ldots n][1 \ldots m]$, which contains at element $solution[i][j]$ the start boundary of $AS(s_j)$ of the best solution to sub-problem U_i, S_j.

The main loop (lines 7-14) tabulates the contents of $Cost$ and $solution$ in increasing value of j (i.e., by columns), and in increasing value of i within each column, based on the best solution obtained previously for column $j-1$. Finally, the mapping $\mathcal{M}(s_j), 1 \leq j \leq m$, is obtained in lines 15-19 with the help of the $solution$ table. The cost of the best mapping corresponds to the minimum value in the final column m (recall that each entry in column j of $Cost$ stores the $accumulated$ cost of the solution to subproblem U_i, S_j). Formally:

$$BestCost = \min_{(m-1) \cdot K < i \leq n-K+1} Cost[i][m] \tag{8}$$

From Eq. (5), it results that the number of actual entries in each column j (i.e., the number of allowable i values) is $b(j) - a(j) + 1 = (n + 1 - m \cdot K)$. The total number of tabulated entries becomes $(n + 1 - m \cdot K) \cdot m \equiv O(m \cdot n)$. However, the tabulation proceeds column-by-column, and only the last column of $Cost$ needs to be retained at any time. Hence, the space complexity of storing $Cost$ is $O(n)$. Still, we need to store the entire $solution$ table. Nevertheless, only a constant $O(n)$ fraction (i.e., the current column) must be stored in main memory, while the rest can be saved to secondary memory and read one more time when the output is performed.

In terms of computational cost, BK needs to tabulate $O(n \cdot m)$ entries of $Cost$, and each entry requires $O(K)$ computation for determining the base-case cost of GC in Eq. (6) (line 10). The total cost is $O(m \cdot n \cdot K)$. Although this is asymptotically lower than MK, we show in Section 5 that BK is more expensive in practice, since the actual number of stages in MK is much smaller than the worst case analysis. Nevertheless, the generalization cost incurred by BK is considerably lower.

5   Experimental Evaluation

We implemented C++ prototypes of the proposed MK and BK algorithms, as well as the $Local$ technique from [9]. We also implemented a benchmark method based on nearest neighbor search, referred to as KNN. KNN picks sensitive sites in random order, and for each $s \in S$ and a given K, it includes in $\mathcal{M}(s)$ the K nearest users of s. Those users are then eliminated from U, to ensure that users are not shared among sites. For efficiency, in the KNN method we index the users with an in-memory R*-Tree [3].

Our experiments were run on a P4 3.0 Ghz machine with 1GB of RAM and Linux OS. We measured the execution time and the generalization cost GGC. GGC is expressed as the percentage of the sum of areas of all generalized locations, over the area of the entire dataspace (intuitively, this measures how much of the dataspace area is covered by the published locations). Formally:

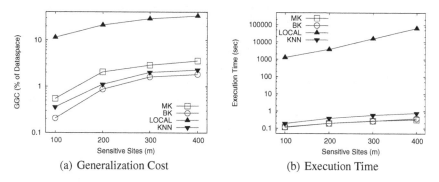

(a) Generalization Cost

(b) Execution Time

Fig. 12. Variable m, $K = 5$, 10000 Users

$$GGC(\mathcal{M}) = 100 \cdot \frac{\sum_{s \in S} Area(MBR(\mathcal{M}(s) \cup \{s\}))}{DomainArea}\% \qquad (9)$$

We used the NA[1] real dataset, consisting of $569,120$ locations on the North-American continent. We generated U and S sets of various sizes through random sampling from NA. In Section 5.1 we compare all algorithms for small input sizes, because of the very high overhead of *Local*, whereas in Section 5.2 we evaluate MK, BK and KNN for large inputs.

5.1 Comparison Against *Local*

In this experiment we set the number of users to $10,000$, $K = 5$, and vary the number of sensitive sites m. Figure 12a shows that even for such a small value of K, the GGC incurred by *Local* is one order of magnitude worse than that of other methods (roughly 10% of the dataspace). As discussed in Section 2.3, *Local* tends to include in $\mathcal{M}(s)$ users that are very close to site s in one of the x or y coordinates, but they may be far away in actual distance. Therefore, the resulting MBR is very large. We also measured GGC using the publication method proposed in [9] (recall from Section 2 that this format has serious privacy drawbacks). For $m = 200$, for instance, *Local* achieves a GGC value of 0.25, compared to 0.78 for KNN. However, *Local* performs poorly when a secure publishing format is used. Among the other methods, BK obtains the best GGC. In terms of execution time, Figure 12b shows that *Local* is several orders of magnitude slower than the other techniques. For 400 sites the absolute value is 18 hours. The results do no utilize the R*-Tree-based optimization described in [9]. However, a preliminary implementation that included that improvement did not show significant gains. Among the other algorithms BK and MK are very fast, outperforming KNN.

Figure 13 presents the results for variable K; $n = 10,000$ users and $m = 200$ sites. The only value for which *Local* achieves low GGC is $K = 2$. For this value, it is likely that a site s includes in its $\mathcal{M}(s)$ two users with close-by x or y coordinates, resulting

---

[1] http://www.rtreeportal.org

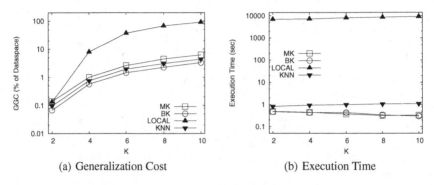

(a) Generalization Cost   (b) Execution Time

Fig. 13. Variable K, 200 Sensitive Sites, 10000 Users

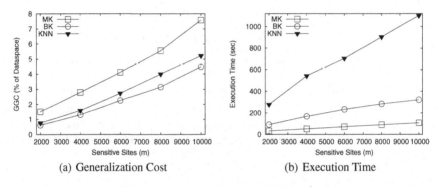

(a) Generalization Cost   (b) Execution Time

Fig. 14. Variable m, $K = 20$, 569120 Users

in an MBR with small area. As K increases, the resulting MBR becomes less skewed, and its area grows considerably.

5.2   Comparison of MK and BK Versus KNN

Below, we compare MK, BK and KNN for large input sizes that are relevant for practical applications (we exclude Local due to its excessive running time). Unless otherwise specified, U consists of the entire NA dataset (i.e., 569K users). Figure 14 compares the three algorithms for variable m and $K = 20$. There is a clear tradeoff between MK and BK: GGC is up to 2 times lower for BK compared to MK, but MK is up to 8 times faster. The execution time of MK is 42 seconds for the largest input. KNN is worse than BK in terms of GGC and it is also much slower.

In Figure 15, we vary K for $m = 4,000$ sensitive sites. BK maintains its advantage over KNN in terms of GGC, while being up to ten times faster. MK is the fastest method. Observe that the execution time of BK decreases with K, because the number of tabulated entries in the dynamic programming formulation is $(n + 1 - m \cdot K) \cdot m$. Intuitively, less candidate start boundaries need to be considered as K increases. For MK, there are two contrary effects as K increases: the initial cost of *1DAnon* is linear

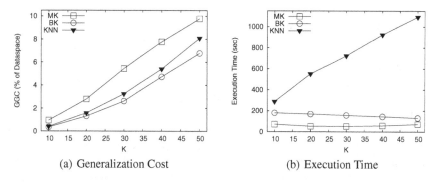

(a) Generalization Cost   (b) Execution Time

Fig. 15. Variable K, 4000 Sensitive Sites, 569120 Users

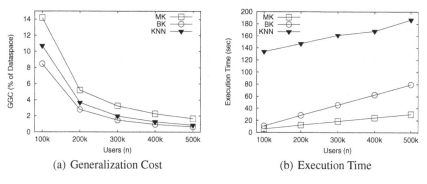

(a) Generalization Cost   (b) Execution Time

Fig. 16. Variable n, 2000 Sensitive Sites, $K = 20$

to K. However, fewer groups are generated and this reduces the user-to-site assignment phase of MK. As a result, the execution time remains almost constant.

Finally, in Figure 16 we fix $m = 2,000$, $K = 20$ and vary the number of users n. As n increases, the density of the users in the dataspace also increases, and more compact anonymizing sets can be formed. Therefore, GGC decreases with larger n for all methods. The execution time of KNN grows considerably with n, as more users need to be considered in the nearest-neighbor search. The execution time of both BK and MK is linear to n.

5.3   Discussion

Our two proposed methods, BK and MK, provide a clear tradeoff between generalization cost and execution time: BK is the best in terms of GGC out of all considered algorithms. It is also much faster than KNN and *Local*, but it is slower than MK. MK is faster at the expense of higher GGC (roughly 2 times worse than BK). Nevertheless, MK remains a good choice for applications where speed is essential; for instance, publishing real-time traffic updates.

Local cannot be used for any input size of practical value, due to its extremely high computational overhead. Furthermore, *Local* incurs very high generalization cost, if a secure location publishing format is used. Finally, the KNN method is outperformed by BK in terms of both GGC and execution time.

6 Conclusions

The collection of location data from mobile users has received considerable attention recently. To enable users with low-end communication devices (i.e., even without GPS) to access location-based services, certain LBS providers (e.g., GoogleMaps) have devised systems that calculate the user location from the identifiers of cellular network towers. As huge amounts of location data are becoming available, their privacy-preserving publication emerges as an important concern. In this paper we proposed two methods for the anonymous publishing of location data, which are fast and achieve low data distortion. Our methods are significantly better, compared to existing work.

In the future, we plan to study more complex attacks, based on traces of locations. By correlating information published at consecutive timestamps, an attacker may be able to gain additional knowledge and compromise the privacy of certain users. We also plan to address the scenario where the input location data is not entirely available before-hand, but instead it is generated in a streaming manner. This setting is more difficult, since data must be output before their expiration deadline; therefore, computational efficiency becomes a primary concern.

References

1. Aggarwal, C.C.: On k-Anonymity and the Curse of Dimensionality. In: Proc. of VLDB, pp. 901–909 (2005)
2. Bayardo, R., Agrawal, R.: Data Privacy through Optimal k-Anonymization. In: Proc. of ICDE, pp. 217–228 (2005)
3. Beckmann, N., Kriegel, H.-P., Schneider, R., Seeger, B.: The R*-Tree: An Efficient and Robust Access Method for Points and Rectangles. In: Proc. of ACM SIGMOD, pp. 322–331 (1990)
4. Bettini, C., SeanWang, X., Jajodia, S.: Protecting Privacy Against Location-Based Personal Identification. In: Jonker, W., Petković, M. (eds.) SDM 2005. LNCS, vol. 3674, pp. 185–199. Springer, Heidelberg (2005)
5. Chow, C.-Y., Mokbel, M.F.: Enabling Private Continuous Queries for Revealed User Locations. In: Papadias, D., Zhang, D., Kollios, G. (eds.) SSTD 2007. LNCS, vol. 4605, pp. 258–275. Springer, Heidelberg (2007)
6. Gedik, B., Liu, L.: Location Privacy in Mobile Systems: A Personalized Anonymization Model. In: Proc. of ICDCS, pp. 620–629 (2005)
7. Ghinita, G., Karras, P., Kalnis, P., Mamoulis, N.: Fast Data Anonymization with Low Information Loss. In: Proc. of VLDB, pp. 758–769 (2007)
8. Gruteser, M., Grunwald, D.: Anonymous Usage of Location-Based Services Through Spatial and Temporal Cloaking. In: Proc. of USENIX MobiSys, pp. 31–42 (2003)
9. Hu, H., Xu, J., Du, J., Ng, J.K.-Y.: Privacy-Aware Location Publishing for Moving Clients. Technical report, Hong Kong Baptist University (2007), http://www.comp.hkbu.edu.hk/~haibo/privacy_join.pdf
10. Kalnis, P., Ghinita, G., Mouratidis, K., Papadias, D.: Preventing Location-Based Identity Inference in Anonymous Spatial Queries. IEEE TKDE 19(12), 1719–1733 (2007)

11. LeFevre, K., DeWitt, D.J., Ramakrishnan, R.: Incognito: Efficient Full-Domain K-Anonymity. In: Proc. of ACM SIGMOD, pp. 49–60 (2005)
12. Machanavajjhala, A., Gehrke, J., Kifer, D., Venkitasubramaniam, M.: l-Diversity: Privacy Beyond k-Anonymity. In: Proc. of ICDE (2006)
13. Mokbel, M.F., Chow, C.Y., Aref, W.G.: The New Casper: Query Processing for Location Services without Compromising Privacy. In: Proc. of VLDB, pp. 763–774 (2006)
14. Moon, B., Jagadish, H., Faloutsos, C.: Analysis of the Clustering Properties of the Hilbert Space-Filling Curve. IEEE TKDE 13(1), 124–141 (2001)
15. Reid, D.: An algorithm for tracking multiple targets. IEEE Transactions on Automatic Control 24, 843–854 (1979)
16. Samarati, P.: Protecting Respondents' Identities in Microdata Release. IEEE TKDE 13(6), 1010–1027 (2001)
17. Sweeney, L.: k-Anonymity: A Model for Protecting Privacy. Int. J. of Uncertainty, Fuzziness and Knowledge-Based Systems 10(5), 557–570 (2002)
18. Tao, Y., Xiao, X.: Personalized Privacy Preservation. In: Proc. of ACM SIGMOD, pp. 229–240 (2006)

A Probabilistic Framework for Building Privacy-Preserving Synopses of Multi-dimensional Data[*]

Filippo Furfaro[1], Giuseppe M. Mazzeo[1,2], and Domenico Saccà[1,2]

[1] University of Calabria, Rende (CS) 87036, Italy
[2] ICAR-CNR, Rende (CS) 87036, Italy
{furfaro,mazzeo,sacca}@si.deis.unical.it

Abstract. The problem of summarizing multi-dimensional data into lossy synopses supporting the estimation of aggregate range queries has been deeply investigated in the last three decades. Several summarization techniques have been proposed, based on different approaches, such as histograms, wavelets and sampling. The aim of most of the works in this area was to devise techniques for constructing effective synopses, enabling range queries to be estimated, trading off the efficiency of query evaluation with the accuracy of query estimates. In this paper, the use of summarization is investigated in a more specific context, where privacy issues are taken into account. In particular, we study the problem of constructing *privacy-preserving synopses*, that is synopses preventing sensitive information from being extracted while supporting 'safe' analysis tasks. In this regard, we introduce a probabilistic framework enabling the evaluation of the quality of the estimates which can be obtained by a user owning the summary data. Based on this framework, we devise a technique for constructing histogram-based synopses of multi-dimensional data which provide as much accurate as possible answers for a given workload of 'safe' queries, while preventing high-quality estimates of sensitive information from being extracted.

1 Introduction

In the last three decades, a great deal of attention has been devoted to the problem of summarizing multi-dimensional data into synopses supporting the estimation of aggregate range queries. Several lossy compression techniques have been proposed, based on different approaches (such as histograms [11], wavelets [3], and sampling [7]). These techniques can be profitably applied in several application contexts (e.g., On-line Analytical Processing [7], query optimization [15], statistical and scientific databases [12]), where a high precision of query estimates is not mandatory, and fast query answers (affected by reasonable error rates) suffice to effectively support the tasks to be accomplished.

Intuitively enough, the experience acquired by the research community in designing effective lossy compression techniques could be applied in a new emerging scenario, where data should be published to support different analysis tasks,

---

[*] This work was supported by a grant from the Italian Research Project FIRB "TO-CAI", funded by MUR.

B. Ludäscher and Nikos Mamoulis (Eds.): SSDBM 2008, LNCS 5069, pp. 114–130, 2008.

with no risk for privacy issues. That is, the compression process could be driven so that the loss of information is exploited to hide sensitive information, while 'safe' information is enabled to be accurately extracted from the synopses. Indeed, most of summarization techniques proposed in the previously mentioned scenarios provide no warranty on the privacy preservation of sensitive information. In fact, the compression process accomplished by these techniques aims at reducing as much as possible the loss of information resulting from summarizing data in a limited amount of storage space, paying no attention to the risk that sensitive information could be extracted from the summarized data with a high degree of accuracy. This makes the problem of refining traditional compression techniques to deal with privacy-preserving issues intriguing, also due to its practical impact in many application contexts.

In this paper we focus our attention on histogram-based summarization techniques, which are widely used in the context of data compression. A histogram is a synopsis obtained by suitably partitioning the data domain into a set of blocks and then replacing the set of individual data inside each block with some aggregate data. First, we introduce a probabilistic framework for evaluating the quality of the estimates of sensitive information which can be obtained by accessing a histogram. Specifically, the quality of estimates is measured by evaluating the probability associated with confidence intervals of individual-data estimates. This framework can be used to assign a 'safety certificate' to histograms, as it provides a measure of the privacy threat owing to the summary data published through a histogram. Thus, we exploit the proposed probabilistic framework to devise a technique for constructing *privacy-preserving histograms*. Our technique is based on a greedy strategy for constructing a partition of data which aims at two objectives: on the one hand, the resulting histogram should provide as much accurate as possible estimates for a workload of queries considered 'safe'; on the other hand, the resulting histogram should provide low-quality estimates of individual data. Finally, we address future directions towards which our work could be extended.

2 Preliminaries

In this work, we focus our attention on multi-dimensional data defined on a domain whose dimensions are discrete, and the values associated with the points of the domain are non-negative real numbers. Specifically, a d-dimensional data set D is a set of tuples of the form $\langle p_1, \ldots, p_d, m \rangle$, where $p_1, \ldots, p_d$ identify a point in a multi-dimensional space of size $n_1 \times \cdots \times n_d$ and m is a measure associated with the point. Thus, D can be viewed as a d-dimensional array of size $n_1 \times \cdots \times n_d$, where n_i is the cardinality of the i-th dimension. Given a point $\mathbf{p} = \langle p_1, \ldots, p_d \rangle$ of the domain of D, where $p_i \in [1..n_i]$ ($\forall i \in [1..d]$), the value m associated with $\mathbf{p}$ will be denoted as $D[\mathbf{p}]$ (if D contains no tuple associated with point $\mathbf{p}$, then $D[\mathbf{p}] = 0$). A range over D is a d-tuple $\varrho = \langle \varrho_1, \ldots, \varrho_d \rangle$, where ϱ_i ($\forall i \in [1..d]$) is a pair of the form $\langle \varrho_i^l, \varrho_i^u \rangle$ such that $1 \leq \varrho_i^l \leq \varrho_i^u \leq n_i$. Basically, a range is a hyper-rectangular subset of the domain of D. We define the volume of a range ϱ as $\Pi_{i \in [1..d]}(\varrho_i^u - \varrho_i^l + 1)$ and denote it as $vol(\varrho)$.

A histogram over D is a synopsis of aggregate values which is constructed by first partitioning the domain of D into a number of non-overlapping ranges, called *buckets*, and then storing, for each bucket, some aggregate data summarizing the data set underlying it. Histograms can be used to support the estimation of aggregate range queries, which are evaluated by exploiting the summary data stored in its buckets. Specifically, we study the case that, for each bucket, the sum of the values of the points inside it is stored (as it will be clearer in the following, this allows sum-range queries to be estimated). In this scenario, a bucket of a histogram over D can be viewed as a pair $\langle \varrho, s \rangle$, where ϱ is a range over D and $s = \sum_{\mathbf{p} \in \varrho} D[\mathbf{p}]$. Given a bucket $\beta = \langle \varrho, s \rangle$, the terms ϱ and s will be referred to as the *range* and the *sum* of β, respectively, and will be denoted as $range(\beta)$ and $sum(\beta)$. Moreover, we will denote the volume of the range of β as $vol(\beta)$ and the average value of β (i.e., $\frac{sum(\beta)}{vol(\beta)}$), as $\mu(\beta)$. This kind of histogram (where buckets are associated with sums) supports the evaluation of sum-range queries, that is, queries asking for the sum of the values of the points of D inside a specified range. A range-sum query over D is an expression of the form $q = sum(\varrho_q)$, where ϱ is a range over D. The actual answer of q is the value $\sum_{\mathbf{p} \in \varrho_q} D[\mathbf{p}]$. Given a histogram $H = \{\beta_1, \dots, \beta_n\}$ over D, the estimated answer of q over H is $\tilde{q} = \sum_{i=1}^{n} vol(range(\beta_i) \cap \varrho_q) \cdot \frac{sum(\beta_i)}{vol(\beta_i)}$. Hence, the estimation is performed adopting linear interpolation, that is, assuming that the points inside a bucket β_i have the same value, namely the average value $\mu(\beta_i)$.

A point query on a data set D is a pair $q = \langle D, \mathbf{p} \rangle$ asking for the value $D[\mathbf{p}]$. Thus, q can be viewed as a range query where the specified range has volume 1, then the estimate of its answer obtained from histogram H is given by $\frac{sum(\beta)}{vol(\beta)} = \mu(\beta)$, where β is the bucket of H whose range contains $\mathbf{p}$. In the following, given a point query $q = \langle D, \mathbf{p} \rangle$ and a histogram H over D, we denote the bucket of H whose range contains $\mathbf{p}$ as $\beta(q)$ (observe that $\beta(q)$ is unique, as the buckets of H do not overlap).

Example 1. A two-dimensional data set D is shown in Fig. 1(a). A histogram on D is shown in Fig. 1(c). It has been obtained by first partitioning the domain of D (as in Fig. 1(b)) and then storing the boundaries and the sum of values of each block (bucket) of the partition. Consider the point query $q = \langle D, \langle 3, 4 \rangle \rangle$. In this case, the bucket involved in the query is $\beta(q) = \beta_1$, being the point $\langle 3, 4 \rangle$

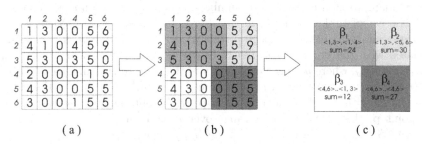

(a)   (b)   (c)

Fig. 1. A two-dimensional data set D (a), a partition of the domain of D (b) and a histogram H summarizing D (c)

inside the range of the bucket β_1. The estimated answer of q is $24/12 = 2$, since the sum and the volume of β_1 are 24 and 12, respectively.   □

3   A Probabilistic Framework for Estimating Individual Values from a Histogram

In this section we present a probabilistic framework supporting the estimation of individual values based on the summary data stored in a histogram. Specifically, this framework provides a measure of the quality of the estimates of individual data which can be obtained by exploiting the aggregate data stored in a histogram. The quality measure is given in terms of probability that the estimation of an individual value is within a confidence interval.

Given a histogram H on a data set D and a point query $q = \langle D, \mathbf{p} \rangle$, we model the answer of q estimated on H as a random variable $\tilde{q}_{s,b}$ defined over the sample space $\Omega(q) = [0, s]$, where s and b are the sum and the volume of $\beta(q)$. Basically, $\tilde{q}_{s,b}$ can assume all the values inside the interval $[0, s]$ as the actual value associated with $\mathbf{p}$ is non-negative and cannot exceed the overall sum of the bucket of H whose range contains $\mathbf{p}$. It is worth noting that this random variable does not depend on parameters other than s and b, as histogram buckets do not overlap and we assume independence among the values summarized into different buckets, thus the sum values and the volumes of the buckets different from $\beta(q)$ do not affect the estimation of q.

We now characterize the above-introduced random variable $\tilde{q}_{s,b}$.

Theorem 1. *Let D be a data set, H a histogram over D, $q = \langle D, \mathbf{p} \rangle$ a point query, and s and b be the sum and the volume of bucket $\beta(q)$ of H, respectively. The probability density function of the random variable $\tilde{q}_{s,b}$ is:*

$$f(x) = \begin{cases} \delta(0) & \text{if } s = 0; \\ \delta(s) & \text{if } b = 1; \\ \frac{b-1}{s} \cdot \left(1 - \frac{x}{s}\right)^{b-2} & \text{if } b > 1, \ s > 0, \ \text{and } x \in [0, s]; \\ 0 & \text{if } b > 1, \ s > 0, \ \text{and } x \notin [0, s]; \end{cases} \quad (1)$$

where $\delta(x)$ denotes the Dirac function, its cumulative distribution function, is:

$$F(x) = \begin{cases} H(0) & \text{if } s = 0; \\ H(s) & \text{if } b = 1; \\ 1 - \left(1 - \frac{x}{s}\right)^{b-1} & \text{if } b > 1, \ s > 0, \ \text{and } x \in [0, s]; \\ 0 & \text{if } b > 1, \ s > 0, \ \text{and } x < 0; \\ 1 & \text{if } b > 1, \ s > 0, \ \text{and } x > s; \end{cases} \quad (2)$$

where $H(x)$ denotes the Heaviside step function. The expected value and the variance of $\tilde{q}_{s,b}$ are

$$E(\tilde{q}_{s,b}) = \frac{s}{b} \tag{3}$$

and

$$\sigma^2(\tilde{q}_{s,b}) = \frac{b-1}{b+1}\left(\frac{s}{b}\right)^2, \tag{4}$$

respectively.

Proof. We first focus on the expressions for $f(x)$ and $F(x)$. In the case that $s = 0$, as the elements of D are non-negative, the actual value associated with each point inside β is 0. Hence, $\tilde{q}_{s,b}$ takes value 0 with probability 1.

In the case that $b = 1$, β contains a unique element, thus the definition of $f(x)$ derives from the fact that the value associated with $\mathbf{p}$ is exactly s (the sum associated with β).

We now consider the case that $b > 1$ and $s > 0$. In this case, clearly $f(x)$ is null for $x \notin [0,s]$, as individual values are assumed to be non-negative and their sum is s (thus, no individual value can be larger than s). For the same reason, $F(x) = 0$ for $x < 0$ (it is impossible that any individual value is less than 0) and $F(x) = 1$ for $x > s$ (it is certain that any individual value is less than or equal to s). Now we derive $f(x)$ and $F(x)$ for the most interesting case, that is $b > 1$, $s > 0$, and $x \in [0,s]$. We first characterize a discrete random variable V_j different from $\tilde{q}_{s,b}$, whose probability distribution will be exploited to derive $f(x)$ and which is defined as follows. Given a real number $\gamma > 0$ and a set $S = \{k_1 \cdot \gamma, \ldots, k_b \cdot \gamma\}$ of cardinality b, where, for each $i \in [1..b]$, $k_i \in N$, and $\sum_{i=1}^{b} k_i \cdot \gamma = s$, $Pr(V_j = x)$ denotes the probability that the value of $k_j \cdot \gamma$ is equal to x. Intuitively enough, V_j can be viewed as the translation of $\tilde{q}_{s,b}$ to the case that the domain of the values of D is discrete (i.e., the points D can be assigned only multiples of γ). Thus, V_j is a discrete random variable defined over the sample space $\Omega(V_j) = \{x | 0 \le x \le s \text{ and } x \text{ is a multiple of } \gamma\}$. We now show that

$$Pr(V_j = x) = \frac{\dbinom{\frac{s-x}{\gamma}+b-2}{\frac{s-x}{\gamma}}}{\dbinom{\frac{s}{\gamma}+b-1}{\frac{s}{\gamma}}}. \tag{5}$$

This formula can be explained as follows. If a value in S is equal to x, then the sum of the remaining $b-1$ elements is $s-x$. Therefore, $Pr(V_j = x)$ is equal to the ratio between all the possible value assignments to $b-1$ elements such that their sum is $s-x$ and all the possible assignments to b elements such that their sum is s. The formula derives from the facts that each element can be assigned a multiple of γ, and that all the possible value assignments to n elements such that their sum is y is equal to number of combinations with repetitions of n objects from which y have to be chosen, that is $\binom{y+n-1}{y}$.

We denote the cumulative distribution function of V_j as $F_V(x)$, and derive a formula for $F_V(x)$:

$$F_V(x) = Pr(V_j \leq x) = \sum_{k=0}^{\lfloor \frac{x}{\gamma} \rfloor} Pr(V_j = k \cdot \gamma) =$$

$$= 1 - \sum_{k=\lfloor \frac{x}{\gamma} \rfloor + 1}^{\frac{s}{\gamma}} Pr(V_j = k \cdot \gamma) = 1 - \frac{1}{\binom{b+\frac{s}{\gamma}-1}{\frac{s}{\gamma}}} \sum_{k=\lfloor \frac{x}{\gamma} \rfloor + 1}^{\frac{s}{\gamma}} \binom{b + \frac{s}{\gamma} - k - 2}{\frac{s}{\gamma} - k}$$

Let $i = \frac{s}{\gamma} - k$. We obtain:

$$\sum_{k=\lfloor \frac{x}{\gamma} \rfloor + 1}^{\frac{s}{\gamma}} \binom{b + \frac{s}{\gamma} - k - 2}{\frac{s}{\gamma} - k} = \sum_{i=0}^{\frac{s}{\gamma} - \lfloor \frac{x}{\gamma} \rfloor - 1} \binom{b - 2 + i}{i}$$

and by adopting the identity

$$\sum_{j=0}^{k} \binom{n+j}{j} = \binom{n+k+1}{k}$$

we obtain:

$$F_V(x) = 1 - \frac{\binom{b - 2 + \frac{s}{\gamma} - \lfloor \frac{x}{\gamma} \rfloor}{\frac{s}{\gamma} - \lfloor \frac{x}{\gamma} \rfloor - 1}}{\binom{b + \frac{s}{\gamma} - 1}{\frac{s}{\gamma}}}. \tag{6}$$

The cumulative distribution function $F(x) = Pr(\tilde{q}_{s,b} < x)$ of $\tilde{q}_{s,b}$ can be obtained as $F(x) = \lim_{\gamma \to 0} F_V(x)$. In fact, as γ tends to 0, the elements of set S can be assigned any real value in $[0, s]$ (under the constraint that their sum is s), thus at the limit the distribution functions F and F_V coincide. Then, we obtain:

$$F(x) = \lim_{\gamma \to 0} F_V(x) = 1 - \lim_{\gamma \to 0} \frac{\left(b - 2 + \frac{s}{\gamma} - \lfloor \frac{x}{\gamma} \rfloor\right)! \cdot \left(\frac{s}{\gamma}\right)! \cdot (b-1)!}{\left(\frac{s}{\gamma} - \lfloor \frac{x}{\gamma} \rfloor - 1\right)! \cdot \left(b + \frac{s}{\gamma} - 1\right)! \cdot (b-1)!} =$$

$$= 1 - \lim_{\gamma \to 0} \frac{\overbrace{\dfrac{(s - \lfloor \frac{x}{\gamma} \rfloor \cdot \gamma) + (b-2) \cdot \gamma}{\gamma} \times \cdots \times \dfrac{(s - \lfloor \frac{x}{\gamma} \rfloor \cdot \gamma)}{\gamma}}^{b-1 \text{ factors}}}{\underbrace{\dfrac{s + (b-1) \cdot \gamma}{\gamma} \times \cdots \times \dfrac{s + 1 \cdot \gamma}{\gamma}}_{b-1 \text{ factors}}} =$$

$$= 1 - \lim_{\gamma \to 0} \frac{(s - \lfloor \frac{x}{\gamma} \rfloor \cdot \gamma)^{b-1} + o(\gamma)}{s^{b-1} + o(\gamma)} = 1 - \left(1 - \frac{x}{s}\right)^{b-1}$$

From definition of probability density function of a continuous random variable, we have that the probability density function $f(x)$ and the cumulative distribution function $F(x)$ are related as follows:

$$F(x) = \int_0^x f(u)du.$$

By resolving the latter and exploiting the boundary condition $F(s) = 1$, we obtain the expression for $f(x)$ reported in the statement.

We now derive the expected value of $\tilde{q}_{s,b}$. From definition of expected value, we obtain:

$$E(\tilde{q}_{s,b}) = \int_{x=0}^s f(x) \cdot x \cdot dx = \frac{b-1}{s} \int_{x=0}^s \left(1 - \frac{x}{s}\right)^{b-2} \cdot x \cdot dx =$$

$$= \frac{b-1}{s} \left[-\left(1 - \frac{x}{s}\right)^{b-1} \frac{s}{b-1} \cdot x \right]_{x=0}^s + \frac{b-1}{s} \int_0^s \left(1 - \frac{x}{s}\right)^{b-1} \cdot \frac{s}{b-1} dx =$$

$$= \frac{b-1}{s} \left[\frac{s}{b-1} \left(1 - \frac{x}{s}\right)^b \frac{s}{b} \right]_{x=0}^s = \frac{s}{b}.$$

Similarly, from the definition of variance, we obtain:

$$\sigma^2(\tilde{q}_{s,b}) \int_{x=0}^s f(x) \cdot \left(x - \frac{s}{b}\right)^2 dx = \frac{b-1}{s} \int_{x=0}^s s\left(1 - \frac{x}{s}\right)^{b-2} \cdot \left(x - \frac{s}{b}\right)^2 dx =$$

$$= \frac{b-1}{s} \left[-\left(1 - \frac{x}{s}\right)^{b-1} \frac{s}{b-1} \left(x - \frac{s}{b}\right)^2 + \frac{2s}{b-1} \int \left(1 - \frac{x}{s}\right)^{b-1} \left(x - \frac{s}{b}\right) dx \right]_{x=0}^s =$$

$$= \left(\frac{s}{b}\right)^2 + 2 \left[-\left(1 - \frac{x}{s}\right)^b \frac{s}{b} \left(x - \frac{s}{b}\right) + \frac{s}{b} \int \left(1 - \frac{x}{s}\right)^b dx \right]_{x=0}^s =$$

$$= \left(\frac{s}{b}\right)^2 - 2\left(\frac{s}{b}\right)^2 + 2\frac{2s}{b} \left[-\left(1 - \frac{x}{s}\right)^{b+1} \frac{s}{b+1} \right]_{x=0}^s =$$

$$= -\left(\frac{s}{b}\right)^2 + \frac{2s}{b} \frac{s}{b+1} = \frac{b-1}{b+1} \left(\frac{s}{b}\right)^2. \qquad \square$$

The characterization of random variable $\tilde{q}_{s,b}$ can be exploited to determine the quality of the point-query estimates which can be obtained by accessing the summary data stored in H. In fact, a user owning the histogram can estimate the answer of a point query q as the expected value of $\tilde{q}_{s,b}$ (which corresponds to performing linear interpolation), and evaluate the quality of this estimate as the probability that the actual answer of q lies inside an interval containing $E(\tilde{q}_{s,b})$ as wide as desired. For instance, consider the data set D and the histogram H shown in Fig. 1, as well as the point query $q = \langle D, \mathbf{p} \rangle$, with $\mathbf{p} = \langle 3, 4 \rangle$. If only the aggregate data stored in the histogram is available, the answer of q is estimated as $E(\tilde{q}_{24,12}) = 2$, since the sum and the volume of the bucket β_1 of H containing the point $\langle 3, 4 \rangle$ are 24 and 12, respectively, as seen in Example 1. A user owning the histogram cannot infer the actual value associated with $\mathbf{p}$, but she can evaluate the probability associated to any confidence interval. For instance, a user could be interested in evaluating the probability that the value associated with $\mathbf{p}$ is inside $[1.8, 2.2]$, that is a 'narrow' range centered at the expected value. Using the results provided in Theorem 1, the user obtains that the probability that the actual answer is in $[1.8, 2.2]$ is $F(2.2) - F(1.8) = (1 - \frac{1.8}{24})^{11} - (1 - \frac{2.2}{24})^{11}$.

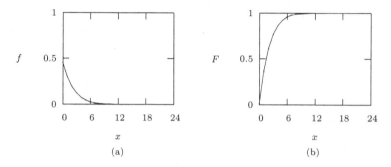

Fig. 2. Probability density (a) and distribution (b) functions of $\tilde{q}_{24,12}$

Fig. 2 depicts the probability density function (a) and the distribution function (b) of the random variable $\tilde{q}_{24,11}$.

Intuitively enough, as our framework can be used to measure the quality of estimates of queries asking for sensitive information, it can be exploited to determine whether a histogram can be considered safe or not w.r.t. a privacy standpoint. This matter is investigated in the following section.

3.1   Privacy and Histograms

Given a histogram H over a data set D, a *privacy breach* occurs if an adversary can retrieve from H "high"-quality estimates of individual data, that is she can reveal sensitive information by establishing with a high confidence level that an individual value is within a certain interval.

In the following we will devise a histogram construction technique which aims at preventing any user owning a histogram from establishing that the actual value associated with a point is "close" to its estimate with a probability higher than a certain threshold. This is tantamount to requiring that the estimated value of every individual data must be affected by a certain error with a probability at least equal to a certain threshold. For instance, a company publishing summary data about the incomes of its employees would like to impose that the estimate of the the income of a single employee evaluated by accessing the summary data is affected by at least 50% error with a probability greater than 70%.

In this example, we used the relative error to define the threshold guaranteeing the safeness of the summary data. However, different metrics could be used, such as the absolute error. In the following, we will consider the relative error as it is quite intuitive and it has been largely stressed in literature [6,8] that it represents a significant measure of the quality of the estimates. On the basis of this idea, we introduce the notion of *privacy preserving bucket* and *privacy preserving histogram*.

Definition 1. *Given a data set D, a histogram H on D, and two real numbers $\epsilon, \mathcal{P} \in (0, 1)$, a bucket β of H is said to be $\langle \epsilon, \mathcal{P} \rangle$-privacy-preserving if, for every point query $q = \langle D, \boldsymbol{p} \rangle$, where $\boldsymbol{p}$ is a point laid inside the range of β, it holds that:*

$$Pr\big(|\tilde{q}_{s,b} - E(\tilde{q}_{s,b})| \leq \epsilon \cdot E(\tilde{q}_{s,b})\big) < \mathcal{P}, \tag{7}$$

where s and b are the sum and the volume of β. □

Definition 2. *Given a data set D, a histogram H on D, and two real numbers $\epsilon, \mathcal{P} \in (0, 1)$, H is said to be $\langle \epsilon, \mathcal{P} \rangle$-privacy-preserving if every bucket of H is $\langle \epsilon, \mathcal{P} \rangle$-privacy-preserving.* □

According to Definition 2, a histogram H on a data set D is not privacy preserving (w.r.t. a pair $\langle \epsilon, \mathcal{P} \rangle$) if it does not protect the privacy of at least one point $\boldsymbol{p}$, that is the value associated with $\boldsymbol{p}$ is summarized in a bucket with sum s and volume b such that

$$Pr\big(|\tilde{q}_{s,b} - E(\tilde{q}_{s,b})| \leq \epsilon \cdot E(\tilde{q}_{s,b})\big) \geq \mathcal{P},$$

where q is the point query asking for the value of $\boldsymbol{p}$.

Hence, a pair $\langle \epsilon, \mathcal{P} \rangle$ defines a *privacy constraint*, and the values assigned to ϵ and $\mathcal{P}$ must be chosen according to the specific context where privacy must be guaranteed.

As $Pr\big(|\tilde{q}_{s,b} - E(\tilde{q}_{s,b})| \leq \epsilon \cdot E(\tilde{q}_{s,b})\big) = F\big(\frac{s}{b} \cdot (1 + \epsilon)\big) - F\big(\frac{s}{b} \cdot (1 - \epsilon)\big)$, where $F(\cdot)$ is the cumulative distribution function of $\tilde{q}_{s,b}$ derived in Theorem 1 (see Equation 2), under the assumption that $s > 0$, we find that[1]:

$$Pr\big(|\tilde{q}_{s,b} - E(\tilde{q}_{s,b})| \leq \epsilon \cdot E(\tilde{q}_{s,b})\big) = \left(1 - \frac{1-\epsilon}{b}\right)^{b-1} - \left(1 - \frac{1+\epsilon}{b}\right)^{b-1}. \tag{8}$$

Interestingly, from Equation 8, it turns out that the probability associated with a confidence interval of an estimate does not depend on the sum of the bucket summarizing the value to be estimated. Thus, in the following, we will refer to $Pr\big(|\tilde{q}_{s,b} - E(\tilde{q}_{s,b})| \leq \epsilon \cdot E(\tilde{q}_{s,b})\big)$ simply as $P(b, \epsilon)$.

In Fig. 3, the diagrams of $P(b, \epsilon)$ against ϵ and b are shown. It is worth observing that $P(b, \epsilon)$ is monotone increasing w.r.t. ϵ and monotone decreasing w.r.t. b. This means that a privacy constraint $\langle \epsilon, \mathcal{P} \rangle$ implies a lower bound on

---

[1] The case $s = 0$ can be disregarded, as it implies that every individual value inside the bucket is 0 with probability 1.

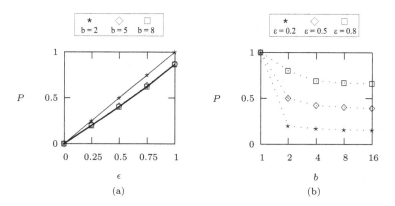

Fig. 3. $P(b, \epsilon)$ vs. ϵ, for different values of b (a), and vs. b for different values of ϵ

the volume of the buckets of a histogram: in order to satisfy the constraint $P(b, \epsilon) < \mathcal{P}$, a histogram must consist of only buckets having at least volume $b_{min} = \lfloor b^\star + 1 \rfloor$, where $b^\star$ is the solution of the equation

$$P(b^\star, \epsilon) = \mathcal{P}. \tag{9}$$

Solving Equation 9 is not possible through analytical methods, but the value of b_{min} can be efficiently computed[2] starting from $b = 1$ and iteratively incrementing b by one and computing $P(b, \epsilon)$, by adopting Equation 8, until $P(b, \epsilon) < \mathcal{P}$ holds.

The monotonicity of $P(\epsilon, b)$ w.r.t. b is at the basis of the property stated in the following proposition.

Proposition 1. *If a histogram H summarizing a data set D is not $\langle \epsilon, \mathcal{P} \rangle$-privacy-preserving, then there is no split-sequence of its buckets that can yield a $\langle \epsilon, \mathcal{P} \rangle$-privacy-preserving histogram.*

Proof. If a histogram is not $\langle \epsilon, \mathcal{P} \rangle$-privacy-preserving, then it has at least a bucket β of volume b with $P(\epsilon, b) \geq \mathcal{P}$. Any split of β yields new buckets with volume $b' < b$. It is easy to see that $\frac{\partial P}{\partial b} < 0$, thus $P(\epsilon, b') > P(\epsilon, b) \geq \mathcal{P}$ holds too. This implies that any histogram obtained from H by splitting β is not $\langle \epsilon, \mathcal{P} \rangle$-privacy-preserving. $\qquad \square$

The result is quite intuitive after considering that a privacy constraint $\langle \epsilon, \mathcal{P} \rangle$, as discussed above, is satisfied only if each bucket of the histogram has volume not less than b_{min}, where b_{min} can be computed by solving Equation 9. Thus, if a bucket β has volume $b < b_{min}$, any sub-bucket of β will have volume less than b_{min}.

In the following section we will introduce a greedy algorithm for constructing privacy preserving histograms, which exploits the property stated in Proposition 1.

---

[2] The time needed to compute b_{min} for one million different combinations of ϵ and $\mathcal{P}$ is only 2.7 seconds with a Pentium IV 3.2 GHz.

4   A Greedy Algorithm for Constructing Privacy-Preserving Histograms

In many practical cases, summarized data can be effectively exploited for performing statistical analysis. Several summarization techniques have been devised to support an efficient query evaluation, aiming at providing query answers affected by the least possible error. When privacy-constraints are defined over published data, the summarization has to take them into account. That is, on the one hand, answers of queries should be accurate enough to enable statistical analysis to be performed. On the other hand, published data must prevent sensitive information from being inferred.

In this work, we focus our attention on constructing privacy-preserving histograms which can be profitably exploited for statistical data analysis. Specifically, we consider the problem of constructing a privacy preserving histogram which aims at providing as much accurate as possible estimates of sum range queries considered 'safe'. More formally, given a privacy constraint $\langle \epsilon, \mathcal{P} \rangle$ and a query workload W consisting of m sum range queries defined over a multidimensional data set D, we consider the problem of constructing a $\langle \epsilon, \mathcal{P} \rangle$-privacy-preserving histogram H summarizing D which minimizes the error over the queries in W. In order to measure the error over a workload we consider the sum of squared errors of the range queries in the workload. That is, if $W = \{sum(\rho_1), \ldots, sum(\rho_m)\}$, being q_i the exact answer of the sum range query in W over the range ρ_i and $\tilde{q}_i$ the approximate one, the overall estimation error w.r.t. W is defined as:

$$SSE(D, H, W) = \sum_{i=1}^{m} (q_i - \tilde{q}_i)^2 .$$

Constructing the optimal histogram for a query workload over a multidimensional data set has been proved to be an NP-hard problem (in [14], Muthukrishnan et al. showed that constructing the optimal histogram of a two-dimensional data set is NP-hard when the query workload consists of all the possible point queries). Thus, we show how our probabilistic framework can be exploited in a greedy algorithm for building (possibly non-optimal) histograms for a given query workload.

Our algorithm (see Fig. 4) works as follows. It takes as input a data set D, a query workload W, and a privacy constraint $\langle \epsilon, \mathcal{P} \rangle$, and returns a $\langle \epsilon, \mathcal{P} \rangle$-privacy-preserving histogram summarizing D. It starts from a histogram consisting of a unique bucket (corresponding to the whole data domain), and it iteratively refines the current histogram by taking a bucket and splitting it into two smaller buckets. Being H' and H'' the histogram at the beginning and at the end of the current iteration, respectively, the choice of the most suitable split for a bucket β of H' is accomplished by function bestSafeSplit, which returns, among all the splits yielding two privacy preserving sub-buckets of β, the split which maximizes the difference $SSE(D, H', W) - SSE(D, H'', W)$, where H'' is the histogram obtained from H' by replacing β with the pair of buckets resulting

---

Input: A data set D, a query workload W, and a privacy constraint $\langle\epsilon,\mathcal{P}\rangle$
Output: An $\langle\epsilon,\mathcal{P}\rangle$-privacy-preserving histogram summarizing D

```
begin
    Histogram definitive=new Histogram();
    Histogram refinable=new Histogram();
    Bucket β=new Bucket(range(D), sum(D));
    refinable.add(β);
    while (!refinable.isEmpty()) do begin
        β=refinable.remove();
        ⟨β',β''⟩=β.bestSafeSplit(D,W,ϵ,𝒫);
        if (⟨β',β''⟩==null) then
            definitive.add(β);
        else begin
            refinable.add(β');
            refinable.add(β'');
        end;
    end;
    return definitive;
end;
```

Fig. 4. A greedy algorithm for constructing privacy preserving histograms

from the split. If no split exists for β yielding two privacy preserving buckets, then β is considered as a *definitive* bucket, and will be not considered for further splits in the subsequent iterations. In fact, from Proposition 1 we have that, if a non-privacy-preserving bucket were created by splitting β, at least one non-privacy-preserving bucket would exist at every subsequent iteration, and then the final histogram would not be privacy-preserving. The algorithm ends when there is no bucket of the current histogram which can be safely split. In the pseudo-code implementation shown in Fig. 4, buckets which can be still considered for being split are maintained in the histogram refinable, while definitive buckets are put in the histogram definitive. Thus, at each iteration, the current histogram is the union between the sets of buckets stored in refinable and definitive.

Observe that any bucket of refinable can be chosen to be split at each iteration[3]. In fact, the split of a bucket at a given iteration does not influence the possibility to split the other buckets in refinable.

We now analyze the complexity of the algorithm. We assume that D contains N points distributed across a multidimensional domain of size n^d (i.e., d dimensions each of size n) and that W contains m queries. The number of iterations of

---

[3] Indeed, in the case that the histogram size were bounded by a maximum amount of storage space, the choice of the bucket to be split at each iteration could be performed according to some greedy criterion (e.g., the bucket giving the largest contribution to the overall error could be chosen).

the algorithm is $O(N)$, as each iteration increases by one the number of buckets, and the final number of buckets cannot be larger than N (a bucket must contain at least one non-null value). Thus, the complexity of the algorithm depends on the cost of the bestSafeSplit function, which is called $O(N)$ times. At each call of bestSafeSplit, all the $O(d \cdot n)$ possible splits must be tried, and for each split a range query of cost $O(N)$ must be performed for each query of the workload, in order to evaluate the SSE reduction provided by the split. The cost of checking if each split is safe is constant. In fact, according to Definition 2, in order to verify if a split is safe, the value b_{min} could be computed before starting the iterations, and then the function bestSafeSplit simply checks if the two buckets resulting from the split have volume greater than b_{min} and sum greater than 0. Therefore, the time complexity of the algorithm is $O(N^2 \cdot m \cdot n \cdot d)$.

Remark. Our probabilistic framework is suitable for being embedded in summarization techniques constructing histograms whose bucket do not overlap. This is due to the fact that, in this case, a point query can be estimated by accessing one bucket only. Several well-known techniques have this characteristic, such as *MHIST* [15] and *MinSkew* [1]). However, some techniques constructing histograms whose bucket overlap could exploit our probabilistic framework as well. For instance, when a bucket is nested inside another bucket, representing a 'hole', the estimate of a point query still depends on a unique bucket. Two techniques belonging to this class are *CHIST* [5] and *STHoles* [2].

5   Extending the Basic Results in Further Directions

In this section we trace further directions towards which our work could be extended:

- managing privacy when additional information is known about original data in buckets;
- managing other forms of privacy constraints;
- managing privacy when buckets overlap.

5.1   Managing Privacy When Additional Information Is Known about Original Data in Buckets

The results derived in Section 3 are based on the assumption that nothing is known about the original data inside each bucket, except that their sum is s and they are distributed in the bucket range of volume b. In many real cases, further information could be available due to the specific application context. For instance, if the measure associated with points is represented by integers, the probability distribution associated with the random variable representing the estimate of individual values would not be that derived in Theorem 1. In this case, the random variable would be discrete, thus its sample space would be $\{0, 1, 2, \ldots, s\}$ instead of $[0, s]$. However, the corresponding random variable could be characterized even easier than the continuous case previously studied.

In fact, the new random variable probability distribution would be represented by Equation 5, with $\gamma = 1$. Then, its cumulative distribution function would be represented by Equation 6, again with $\gamma = 1$. That is,

$$Pr(\tilde{q}_{s,b} \leq x) = 1 - \frac{\binom{b - 2 + s - x}{s - x - 1}}{\binom{b + s - 1}{s}}.$$

Another issue which is worth investigating is the case that other aggregate data (such as the count of non-null values, the minimum or the maximum value) inside each bucket is known. This may happen if either this summary information is explicitly represented in the histogram along with the sums of the buckets (to enhance the estimation process) or it is retrieved from different sources.

5.2   Managing Other Forms of Privacy Constraints

The definition of privacy provided in Section 3.1 can cover a large number of practical cases, in which exact individual values have to be protected. However, some other forms of privacy are worth investigating, due to their practical impact. According to our approach, guaranteeing the privacy of an individual value means limiting the confidence level associated with a confidence interval whose width is proportional to the expected value. It would be interesting to study the case that the width of the confidence interval is defined by an absolute value (rather than a relative one), that is that the confidence interval is expressed in the form $[E(\tilde{q}) - \Delta, E(\tilde{q}) + \Delta]$, where $\tilde{q}$ is the estimate of an individual value which must be protected, and Δ is a real number. In particular, it would be interesting to devise an algorithm managing mixed forms of constraints, where the width of confidence intervals can be expressed by either relative or absolute values. In fact, using an absolute value is more suitable for buckets summarizing "small" values, whereas a relative value is more suitable for buckets summarizing "large" values (where the meaning of "small" and "large" depends on the specific application context). This is due to the fact that adopting a relative value for describing intervals centered at "small" values would result in defining "narrow" intervals, for which guaranteeing low confidence levels would not suffice to preserve privacy.

5.3   Managing Privacy When Buckets Can Overlap

Even though classical histograms are based on partitions of the multi-dimensional data domain (thus, their buckets do not overlap) some of the most performing techniques, such as *GENHIST* [9], exploit bucket overlapping in order to summarize the data set more accurately. To this aim, our framework should be extended to enable taking into account the possibility that the estimation of a single point depends on the aggregate data stored in a number of buckets. For instance, in the case that a point **p** is within the ranges of two overlapping buckets, the random variable representing the value associated with $\mathbf{p}$ would depend on the random variables $\tilde{q}_{s',b'}$ and $\tilde{q}_{s'',b''}$, each representing the value of $\mathbf{p}$ given by one of the two buckets, independently. The random variable representing the expected value associated with $\mathbf{p}$ would be represented by the sum of the two random variables $\tilde{q}_{s',b'}$ and $\tilde{q}_{s'',b''}$. Thus, its probability density function could be obtained by computing the convolution of the probability density functions of $\tilde{q}_{s',b'}$ and $\tilde{q}_{s'',b''}$. Computing the convolution of many probability density functions could be practically infeasible. However, for a large number of random variables, that is, when the value associated to a point inside the intersection of a large number of overlapping buckets must be estimated, for the central limit theorem, the random variable could be very well approximated by a normal distribution that could be completely characterized by knowing the expected values and the variances of the random variables which have to be summed.

6   Related Work

The problem of managing privacy in statistical databases has received a lot of attention in the last few years, and several works dealing with data summarization and privacy issues have been proposed. However, few works providing formal frameworks for checking the privacy preservation of summarized data have been developed.

Some works provide techniques for summarizing data with quality guarantees [6,10]. However, in these works, the quality is intended as a measure of the "distance" between a synopsis and the optimal synopsis consuming the same amount of storage space. Thus, no guarantee is provided on the error rates of query estimates which could be exploited to measure the safeness of a synopsis. Our probabilistic framework, instead, does not aim at providing a technique for building optimal histograms, but provides a tool for evaluating the quality of individual value estimates, intended as confidence levels related to confidence intervals.

A work which deals with the privacy guaranteed by histograms is [4]. In this paper Chawla et al. consider points of a multidimensional space as individuals, which are not associated with any label. A privacy violation occurs when a user can isolate less than t points inside a spherical region of radius proportional to a value c (c and t are parameters which have to be chosen according to the practical context). This work, analogously to others based on the preservation of anonymity of individuals [16], is different from ours as it aims at masking the identity of individuals, that is the coordinates of the points inside the multidimensional domain (which are not associated with any measure). Our work, instead, deals with labelled points, more specifically, with points which are associated with an additive measure. Thus, our approach to privacy preservation is orthogonal w.r.t. [4]: we aim at protecting the measure associated with individuals, rather than their identity.

A thread of works where the attention is focused on the possibility to infer sensitive information by means of range queries on multidimensional data is that leaded by Malvestuto et al. [13]. They study the possibility to infer confidential information exploiting the answers of multiple range queries which, separately, could be considered safe. They design a query engine providing safe answers, which keeps track of past queries, and checks that the answer of each new query cannot be combined combined with the answers previously published in order to enable sensitive information to be inferred. Our approach is different since we assume that to release the whole data set is summarized and published. A very interesting point of contact between the issues studied in [13] and this paper could be the study of the possibility to release multiple safe histograms, each optimized for a different query workload. In fact, when different histograms are released, the fact that each of them is privacy preserving does not suffice to guarantee that confidential information cannot be disclosed, as a user owning different histograms on the same data set could exploit them jointly.

7   Conclusions

In this work we provided a novel approach for constructing effective histograms in the presence of privacy constraints. We introduced the notion of privacy-preserving histograms, that is histograms preventing a user owning them to obtain high quality estimates of individual values which must be kept confident. We defined a probabilistic framework for estimating individual values summarized in a histogram and, on the basis of our probabilistic framework, we proposed a greedy approach for constructing privacy-preserving histograms with high data utility, that is privacy-preserving histograms minimizing the estimation error for range queries belonging to a given query workload supporting statistical analysis tasks. Finally, we outlined the directions towards which our work could be extended.

To the best of our knowledge, this is the first work presenting a mechanism enabling the quality of the estimates of individual values which can be retrieved from a histogram to be measured. Our approach to the problem of preserving the privacy of data can be viewed as orthogonal to other ones, which aim at masking the identity of points belonging to a multi-dimensional domain. Our approach, in fact, aims at protecting a measure associated with the individuals, rather than protecting the identity of individuals.

References

1. Acharya, S., Poosala, V., Ramaswamy, S.: Selectivity estimation in spatial databases. In: Proc. of 1999 ACM SIGMOD Int. Conf. on Management of Data (SIGMOD 1999), Philadelphia (PA), USA, June 1-3, 1999, pp. 13–24 (1999)
2. Bruno, N., Chaudhuri, S., Gravano, L.: STHoles: a multi-dimensional workload aware histogram. In: Proc. of 2001 ACM SIGMOD Int. Conf. on Management of Data (SIGMOD 2001), Santa Barbara (CA), USA, May 21-24, 2001, pp. 211–222 (2001)
3. Chakrabarti, K., Garofalakis, M.N., Rastogi, R., Shim, K.: Approximate query processing using wavelets. The VLDB Journal 10(2-3), 199–223 (2001)

4. Chawla, S., Dwork, C., McSherry, F., Smith, A., Wee, H.: Toward Privacy in Public Databases. In: Kilian, J. (ed.) TCC 2005. LNCS, vol. 3378, pp. 363–385. Springer, Heidelberg (2005)
5. Furfaro, F., Mazzeo, G.M., Sirangelo, C.: Exploiting cluster analysis for constructing multi-dimensional histograms on both static and dynamic data. In: Ioannidis, Y., Scholl, M.H., Schmidt, J.W., Matthes, F., Hatzopoulos, M., Böhm, K., Kemper, A., Grust, T., Böhm, C. (eds.) EDBT 2006. LNCS, vol. 3896, pp. 442–459. Springer, Heidelberg (2006)
6. Garofalakis, M.N., Gibbons, P.B.: Wavelet synopses with error guarantees. In: Proc. of 2002 ACM SIGMOD Int. Conf. on Managment of Data (SIGMOD 2002), Madison (WI), USA, June 3-6, 2002, pp. 476–487 (2002)
7. Gibbons, P.B., Matias, Y.: New sampling-based summary statistics for improving approximate query answers. In: Proc. of 1998 ACM SIGMOD Int. Conf. on Managment of Data (SIGMOD 1998), Seattle (WA), USA, June 2-4, pp. 331–342 (1998)
8. Guha, S., Shim, K., Woo, J.: REHIST: Relative Error Histogram Construction Algorithms. In: Proc. of 30th Int. Conf. on Very Large Data Bases (VLDB 2004), Toronto, Canada, August 29-September 30, pp. 300–311 (2004)
9. Gunopulos, D., Kollios, G., Tsotras, V.J., Domeniconi, C.: Selectivity estimators for multidimensional range queries over real attributes. The VLDB Journal 14(2), 137–154 (2005)
10. Jagadish, H.V., Koudas, N., Muthukrishnan, S., Poosala, V., Sevcik, K., Suel, T.: Optimal histograms with quality guarantees. In: Proc. of 24th Int. Conf. on Very Large Data Bases (VLDB 2004), New York (NY), USA, August 24-27, pp. 275–286 (2004)
11. Ioannidis, Y.E.: The History of Histograms (abridged). In: Proc. of 29th Int. Conf. on Very Large Data Bases (VLDB 2003), Berlin, Germany, September 9-12, pp. 19–30 (2003)
12. Malvestuto, F.M.: A Universal-Scheme Approach to Statistical Databases Containing Homogeneous Summary Tables. ACM Transactions on Database Systems 18(4), 678–708 (1993)
13. Malvestuto, F.M., Mezzini, M., Moscarini, M.: Auditing sum-queries to make a statistical database secure. ACM Transactions on Information and Systems Security 9(1), 31–60 (2006)
14. Muthukrishnan, S., Poosala, V., Suel, T.: On Rectangular Partitioning in Two Dimensions: Algorithms, Complexity and Applications. In: Proc. 7th Int. Conf. on Database Theory (ICDT), Jerusalem, Israel, January 10-12 (1999)
15. Poosala, V., Ioannidis, Y.E.: Selectivity estimation without the attribute value independence assumption. In: Proc. of 23rd Int. Conf. on Very Large Data Bases (VLDB 1997), Athens, Greece, August 25-29, pp. 486–495 (1997)
16. Sweeney, L.: k-Anomity: A model for protecting privacy. Int. Journal on Uncertainty, Fuzziness and Knowledge-based Systems 10(5), 557–570 (2002)

Efficient Similarity Search for Tree-Structured Data

Guoliang Li, Xuhui Liu, Jianhua Feng, and Lizhu Zhou

Department of Computer Science and Technology,
Tsinghua University, Beijing 100084, China
{liguoliang,fengjh,dcszlz}@tsinghua.edu.cn,
liushine9@gmail.com

Abstract. Tree-structured data are becoming ubiquitous nowadays and manipulating them based on similarity is essential for many applications. Although similarity search on textual data has been extensively studied, searching for similar trees is still an open problem due to the high complexity of computing the similarity between trees, especially for large numbers of tress. In this paper, we propose to transform tree-structured data into strings with a one-to-one mapping. We prove that the edit distance of the corresponding strings forms a bound for the similarity measures between trees, including tree edit distance, largest common subtrees and smallest common super-trees. Based on the theoretical analysis, we can employ any existing algorithm of approximate string search for effective similarity search on trees. Moreover, we embed the bound into a filter-and-refine framework for facilitating similarity search on tree-structured data. The experimental results show that our algorithm achieves high performance and outperforms state-of-the-art methods significantly. Our method is especially suitable for accelerating similarity query processing on large numbers of trees in massive datasets.

1 Introduction

The use of tree-structured data in modern database applications is attracting the attention of the research community. Typical examples of huge repositories of rooted, ordered and labeled tree-structured data include the secondary structure of RNA in biology or the XML data on the web. In this paper, we study the similarity measure and similarity search on large trees in huge datasets. These problems form the core operation for many database manipulations, such as, approximate join, clustering, k-NN classification, data cleaning, data integration etc. Data cleaning deals with the identification and correction of data inconsistencies. Frequently, due to such inconsistencies (e.g., misspellings in strings), multiple representations of real-world objects appear in data collections. Among other inconveniences, such redundant information may lead to wrong evaluation results, confuse consistency maintenance, and, when integrated from various data sources, may artificially inflate data files. Therefore, besides data cleaning, detection of duplicates is a long-term research goal in the relational world, often denoted as the fuzzy duplicate problem.

B. Ludäscher and Nikos Mamoulis (Eds.): SSDBM 2008, LNCS 5069, pp. 131–149, 2008.

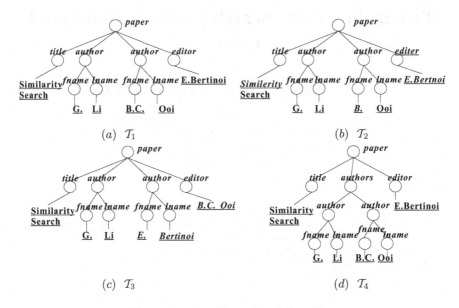

Fig. 1. Running Examples of Four Trees

Our findings are useful in numerous applications including XML data searching under the presence of spelling errors, efficient prediction of the functions of RNA molecules, version management for documents, etc. Trees provide an interesting compromise between graphs and the linear representation of data. They allow the expression of hierarchical dependencies where the semantics are specified implicitly by the relationship between their components; thus, the structure of the tree plays an important role in differentiating the data. More importantly, with growing importance of XML, this problem becomes even more urgent and, due to the structure of XML documents and their increased modeling flexibility, more challenging.

As a classical solution, relational DBMSs have correlated matching records of textual data by using similarity joins. A similarity join finds the pairs of tuples from two relations whose specified attributes are similar. The similarity of these attributes is expressed using a similarity function and a pair of tuples is qualified if the similarity function returns a value greater than a given threshold. However, extending this approximate operation to the tree structure brings a new quality to the correlation problem. As similar or even the same information could be embodied by quite different structures. For example, refer to the sample trees in Figure 1. Consider a situation where tree T_1 (a) has to be correlated to tree T_2 (b). Although they are obviously identical from the human observer, they would not be classified as equal because of textual variations. The use of an appropriate textual similarity measure would easily classify them as duplicate candidates. On the other hand, T_4 (d) refers to another paper. Although T_4 is similar to T_1 if considering textual similarity, the comparison of T_1 with T_4 would probably lead to non-duplicate detection when structural similarity is considered.

Hence, evaluation of structural similarity could help to classify the considered document fragments as non-duplicates. To address the related difficulties, we need accurate and efficient mechanisms to correlate trees thereby coping with additional complexities induced by the tree structures.

Although similarity search on numerical multidimensional data [21] and textual data [1,8,13,16] have been extensively studied, similarity search on tree-structured data has only recently attracted the attention of the research community [2,12,24]. The most commonly used distance measure on tree-structured data is the tree edit distance [4]. However, computing the tree edit distance can be very expensive both in terms of CPU cost and disc I/Os, rendering it impractical for huge datasets. To guarantee the filtration efficiency, the lower bound function should be a relatively precise approximation of the tree edit distance. At the same time, it should be computationally much less expensive than the real distance. Although some studies [2,12,24] are proposed to address this problem, they consider either the textual similarity or the structural similarity and cannot integrate them together for effective similarity search on trees.

To address above-mentioned problems, in this paper, we propose a filter-and-refine framework for effective similarity search on rooted, ordered, labeled trees by considering both the structure and the content. We transform the trees into strings with a one-to-one mapping. We use string edit distance to approximate tree similarity measures, including tree edit distance, largest common subtree distance and smallest common super-tree distance, and give a theoretical bound. We adopt the existing approximate string join techniques to filter out dissimilar strings based on the bound so as to filter the corresponding dissimilar trees.

To summarize, we make the following contributions:

- We propose to transform tree-structured data into strings with a one-to-one mapping and employ string edit distance to approximate the similarity between trees. Moreover, we prove that the edit distance of the corresponding transformed strings forms a bound for the similarity measures between trees, including tree edit distance, largest common subtree distance and smallest common super-tree distance.
- We devise a novel filter-and-refine framework for effective similarity search over tree-structured data by adopting effective filtration techniques borrowed from q-gram based methods in the literature of approximate string join.
- We have implemented our approach and the experimental results show that our approach achieves high performance and outperforms the existing state-of-the-art methods significantly.

The rest of the paper is organized as follows: In Section 2 we provide the background and an overview of the measures to evaluate the similarity of trees and strings. Section 3 presents a sequencing method, which transforms the trees to strings with a one-to-one mapping; while in Section 4 we demonstrate how to use string edit distance to approximate the similarity between trees. A thorough experimental study of our algorithms is conducted in Section 5. Section 6 reviews some related work. Finally, Section 7 concludes our paper.

2   Preliminaries and Related Work

2.1   Preliminaries and Notations

In this paper, we focus on the huge dataset $\mathcal{D}$ of rooted, ordered, labeled trees. Here, a tree is defined as a data structure $\mathcal{T}=(\mathcal{V}, \mathcal{E}, \text{root}(\mathcal{T}))$. $\mathcal{V}$ is a finite set of vertices. $\mathcal{E}$ is a relation on $\mathcal{V}$ where each pair $(u, v) \in \mathcal{E}$ represents the parent-child relationship between two nodes u and v. Node u is the parent of node v and v is one of the child nodes of u. There exists only one root note, denoted as $\text{root}(\mathcal{T})$, which has no parent. Every other node of the tree has exactly one parent and it can be reached through a path of edges from the root. The nodes which have a common parent u (i.e., all the children of u) are siblings. $|\mathcal{T}|$ is the size of $\mathcal{T}$, i.e., the number of nodes in tree $\mathcal{T}$. We call $\mathcal{T}$ a labeled tree if each node is a assigned a symbol from a fixed finite alphabet Σ. We call $\mathcal{T}$ an ordered tree if a left-to-right order among siblings in $\mathcal{T}$ is given. In our paper, we focus on rooted, ordered, and labeled trees.

2.2   Tree Similarity Measures

The measure of similarity between two trees $\mathcal{T}$ and $\mathcal{T}'$ has been well studied in combinatorial pattern matching. Various distance functions, such as edit distance, largest common subtree and smallest common super-tree, are proposed to measure the similarity between trees. In this paper, we consider matching problems based on these measures on top of rooted, ordered, and labeled trees.

For ease of the following discussions, we begin by introducing some notations. Given a tree $\mathcal{T}$, $\mathcal{V}(\mathcal{T})$ denotes the set of vertices in $\mathcal{T}$ and $\mathcal{E}(\mathcal{T})$ denotes the set of edges in $\mathcal{T}$. $\mathcal{L}$ denotes the set of leaves in $\mathcal{T}$.

Definition 1. ISOMORPHIC: $\mathcal{T}$ *is an isomorphic subtree of* $\mathcal{T}'$ *if there exists an injective mapping* $f \colon \mathcal{V}(\mathcal{T}) \to \mathcal{V}(\mathcal{T}')$ *satisfying the following conditions: i) if* $(u, v) \in \mathcal{E}(\mathcal{T})$, *then* $(f(u), f(v)) \in \mathcal{E}(\mathcal{T}')$ *and ii) if u is a preceding sibling of v,* $f(u)$ *is a preceding sibling of* $f(v)$. *Such a mapping* f *is called an isomorphic embedding* $f : \mathcal{T} \to \mathcal{T}'$.

If there is an isomorphic from $\mathcal{T}$ to $\mathcal{T}'$, we call that $\mathcal{T}$ is included in $\mathcal{T}'$, and $\mathcal{T}$ is called a subtree of $\mathcal{T}'$ ($\mathcal{T}'$ is also called a super-tree of $\mathcal{T}$). If $\mathcal{T} \neq \mathcal{T}'$, $\mathcal{T}$ is called a proper subtree of $\mathcal{T}'$.

Definition 2. LARGEST COMMON SUBTREE(LCST): $\mathcal{T}_l$ *is the largest common subtree of* $\mathcal{T}$ *and* $\mathcal{T}'$ *if* $\mathcal{T}_l$ *is a common subtree of* $\mathcal{T}$ *and* $\mathcal{T}'$ *and there does not exist another common subtree of* $\mathcal{T}$ *and* $\mathcal{T}'$, *which is a proper super-tree of* $\mathcal{T}_l$.

We can use the largest common subtree to measure the similarity between trees. As the more alike that $\mathcal{T}$ and $\mathcal{T}'$ are, the larger is their largest common subtree. We propose LCST distance to evaluate the similarity between trees.

Definition 3. LCST DISTANCE: *Given two trees* $\mathcal{T}$ *and* $\mathcal{T}'$, LCST *distance of* $\mathcal{T}$ *and* $\mathcal{T}'$, *denoted as* $\text{lcstd}(\mathcal{T}, \mathcal{T}')$, *is* $|\mathcal{T}| + |\mathcal{T}'| - 2 * |\text{LCST}(\mathcal{T}, \mathcal{T}')|$, *where* $\text{LCST}(\mathcal{T}, \mathcal{T}')$ *denotes the largest common subtree of* $\mathcal{T}$ *and* $\mathcal{T}'$.

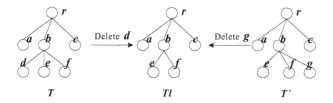

Fig. 2. Largest Common Subtree

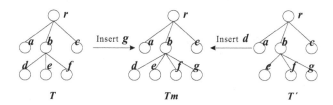

Fig. 3. Smallest Common Super-tree

Obviously, if T and T' are the same, their LCST distance is 0; on the contrary, if they do not share any common node, they are very dissimilar and their LCST distance is large. Note that the more alike that they are, the smaller is their LCST distance. For example, in Figure 2, T_l is the largest common subtree of T and T'. lcstd(T, T')=2.

Definition 4. SMALLEST COMMON SUPER-TREE(SCST): T_m *is the smallest common super-tree of* T *and* T' *if* T_m *is a common super-tree of* T *and* T' *and there does not exist a common super-tree of* T *and* T' *that is a proper subtree of* T_m.

We can use the smallest common super-tree to measure the similarity between trees. As the smaller the smallest common super-tree, the more relevant between the trees. We propose SCST distance to evaluate the similarity between trees.

Definition 5. SCST DISTANCE: *Given two trees* T *and* T', SCST *distance of* T *and* T', *denoted as* scstd(T, T'), *is* $2 * |SCST(T, T')| - |T| - |T'|$, *where* SCST$(T, T')$ *denotes the smallest common super-tree of* T *and* T'.

Obviously, if T and T' are the same, their SCST distance is 0; on the contrary, if they do not share any common node, they are dissimilar and their SCST distance is large. Note that the more alike that they are, the smaller is their SCST distance. For example, in Figure 3, T_m is the smallest common super-tree of T and T'. scstd(T, T')=2.

Another most commonly used distance measure on tree-structured data is the tree edit distance [4]. To introduce the tree edit distance, we begin by introducing some tree edit operations as follows:

- SUBSTITUTION: Substitute a node v to node u in T (i.e, change the label).

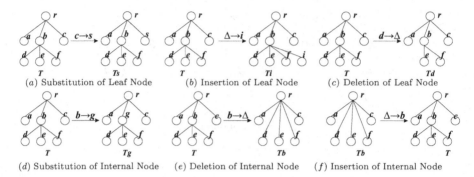

Fig. 4. Edit Operations on Trees

- DELETION: Delete a node v in $\mathcal{T}$ with parent v_p, making the children of v become the children of v_p. The children are inserted in the place of v as a subsequence in the left-to-right order of the children of v_p.
- INSERTION: Insert a node v as a child of node v_p, making v become the parent of a consecutive subsequence of the children of v_p.

Figure 4 illustrates tree edit operations. Note that $u \to v$ denotes substituting u with v. $u \to \Delta$ denotes deletion of u and $\Delta \to v$ denotes insertion of v.

Note that, insertion/deletion/substitution of an internal node is more expensive than insertion/deletion/substitution of a leaf node; insertion/deletion of a node will make two trees more dissimilar than the substitution of a node. Therefore, existing methods will assign different cost to different tree edit operations.

In this paper, we assign the cost of edit operations (denoted as λ_s for substitution, λ_d for deletion, λ_i for insertion) as follows:

$$\lambda_s(v) = \texttt{cSize}(v)$$

$$\lambda_d(v) = \begin{cases} 2 & \text{if } v \text{ is a leaf node} \\ 1 & \text{if } v \text{ is an internal node and } \texttt{parent}(v) = v \\ \texttt{cSize}(v) & \text{if } v \text{ is an internal node and } \texttt{parent}(v) \neq v \end{cases} \quad (1)$$

$$\lambda_i(v) = \begin{cases} 2 & \text{if } v \text{ is a leaf node} \\ 1 & \text{if } v \text{ is an internal node and } \texttt{parent}(v) = v \\ \texttt{cSize}(v) & \text{if } v \text{ is an internal node and } \texttt{parent}(v) \neq v \end{cases}$$

where $\lambda_s(v), \lambda_d(v)$ and $\lambda_i(v)$ denote the cost of substitution, insertion and deletion of node v respectively; $\texttt{cSize}$ denotes the number of children of v. $\texttt{parent}(v)$ denotes the parent of v and $\texttt{parent}(v) \neq v$ denotes that they have different labels. We note that if $\texttt{parent}(v) = v$, insertion/deletion of v will not make the tree much dissimilar as the children of v still share a parent with the same label as v; on the contrary if $\texttt{parent}(v) \neq v$, it will lead to the two trees much dissimilar, and thus we assign different cost for these two conditions.

We define the edit distance based on the edit operations. Assume that we are given a cost function defined on each edit operation. Let $\mathcal{T}$ and $\mathcal{T}'$ be two ordered

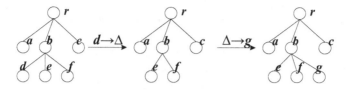

Fig. 5. An Edit Transformation between Trees

and labeled trees. An edit transformation between T and T' is a sequence of edit operations transforming T into T'. The cost of an edit transformation is the sum of the costs of the operations in the edit transformation. An optimal edit transformation between T and T' is an edit transformation between T and T' of minimum cost and this cost is the *tree edit distance*. The tree edit distance problem is to compute the edit distance and the corresponding edit transformation. An example of an edit transformation is shown in Figure 5.

A problem in the literature is the lack of an agreement on a definition of the edit operation/distance problem. The definition given here is by far the most well-studied and in our opinion the most natural [4].

The edit distance problem on ordered labeled trees was introduced by Tai [22] as a generalization of the well-known string edit distance problem. For the ordered version of the problems polynomial time algorithms exists. These are all based on the classic technique of dynamic programming and most of them are simple combinatorial algorithms. Tai presented an algorithm for the ordered version using $O(|T||T'||\mathcal{L}|^2|\mathcal{L}'|^2)$ time and space. Subsequently, this result has been improved by Zhang and Shasha [25] using $O(|T||T'|min(|\mathcal{L}|,|\mathcal{D}|)min(|\mathcal{L}'|,|\mathcal{D}'|))$ time and $O(|T||T'|)$ space, where $|\mathcal{D}|$ denotes the depth of T and $|\mathcal{L}|$ denotes the number of leaves in T. More recently, Klein [14] modified this algorithm to get a better worst case time bound of $O(min(|T|^2|T'|log(|T'|),|T'|^2|T|log(|T|)))$ under the same space bounds. We note that these algorithms are inefficient for similarity join on trees as formalized in DEFINITION 6, especially for large numbers of trees.

Definition 6. SIMILARITY JOIN ON TREES: *Given two sets of trees, treeSet and treeSet' and a distance threshold τ, let dist be a distance function on trees. The similarity join on the two sets of trees reports in the output all pairs of trees $< T \in \text{treeSet}, T' \in \text{treeSet'} >$, such that dist$(T,T')\leq\tau$.*

To address this problem, Guha et al. [9] proposed a pivot based approximate join algorithm on XML documents. However, the complexity of computing the proposed lower bounds is still $O(|T||T'|)$ (i.e., the complexity of tree edit distance computation), and it is not scalable to this problem. Kailing et al. [12] presented a set of filters grounded on structure and content-based information in trees. They proposed using the vectors of the height histogram, the degree histogram and the label histogram to represent the structure as well as content information of trees. The lower bound of the unordered-tree edit distance can be derived from the L_1 distance among the vectors. However, their filters are for unordered trees and cannot explore the structure information implicitly depicted by the order of siblings. Moreover, their lower bounds are obtained by considering structure and content information separately. In our approach, we suggest combining the two sources of information to provide accurate bounds. More recently, pq-Gram [2] transforms the trees into a set of pq-grams and uses pq-grams to approximately match trees. Binary tree based method [24] transforms tree-structured data into an approximate numerical multidimensional vector which encodes the original structure information, and employs the vectors to approximate the tree edit distance. However they emphasis on the structural similarity and neglect the textual similarity. That is, they only consider the exact match between the labels of nodes. For example, in Figure 1, their edit distance between T_1 and T_2 is 4 while that of T_1 and T_3 is 3. However, T_2 is much more similar to T_1 than T_3 from human observer. Alternatively, in this paper, we transform trees into strings and employ string edit distance to approximately match the trees by considering both the structural similarity and textual similarity, which is an effective and efficient approximation of the tree edit distance and LCST/SCST distance.

2.3 String Edit Distance

Most of existing methods usually use edit distance to evaluate the similarity between strings. The operations of string edit distance include substitution, insertion, deletion of a character and the cost of the three operations are always assigned to one. Similarity search and similarity join based on edit distance over textual data have been extensively studied [1,8,13,16]. Given a query set, retrieving all sets in a collection with similarity greater than some threshold is called similarity search. Given two input collections of sets, a set-similarity join identifies all pairs of sets, one from each collection, that have high similarity.

Gravano et al. [8] proposed q-grams to facilitate similarity match on textual data. Given a string S, a q-gram is a contiguous substring of S of length q. If S_1 and S_2 are within edit distance k, S_1 and S_2 must share at least $max(|S_1|,|S_2|)$-$(k$-$1)$*q-1 common q-grams. Recently, Li et al. [16] proposed a new technique called VGRAM to judiciously choose high-quality grams of variable lengths from a collection of strings for improving the performance. Indexing structures and merging algorithms were proposed to facilitate similarity searches in [10,11,15].

3 Sequencing

This section introduces a one-to-one sequencing method to transform the trees to strings. Prüfer (1918) proposed a method that constructed a one-to-one correspondence between a labeled tree and a sequence by removing nodes from the tree one at a time [18].

The construction of a sequence from tree T_n with n nodes labeled from 1 to n of the algorithm works as follows. From T_n, we remove a leaf with the smallest label to form a smaller tree T_{n-1}. Let a_1 denote the label of the node that was the parent of the removed node. Repeat this process on T_{n-1} to determine a_2 (the parent of the next node to be deleted), and continue until only two nodes joined by an edge are left. The sequence $(a_1, a_2, \cdots, a_{n-2})$ is called the Prüfer sequence of tree $\mathcal{T}_n$. From the sequence $(a_1, a_2, \cdots, a_{n-2})$, the original tree $\mathcal{T}_n$ can be reconstructed. The length of the Prüfer sequence of tree $\mathcal{T}_n$ is $n-2$. In fact, we can construct a Prüfer sequence of length $n-1$ for $\mathcal{T}_n$ by continuing the removal of nodes until only one node is left.

Any numbering scheme can be used in the above process to label a tree as long as it associates each node in the tree with a unique number between one and the total number of nodes. This guarantees a one-to-one mapping between the tree and the sequence. Without loss of generality, the post-order is used to uniquely number tree nodes. It helps a Prüfer sequence be constructed for a tree by using the node removal method. This sequence consists entirely of post-order numbers and is called NPS (Numbered Prüfer Sequence). When each number in an NPS is replaced by its corresponding label, a new sequence that consists of labels can be constructed, and this sequence is called LPS (Labeled Prüfer Sequence). On the basis of LPS, ELPS (Extended Labeled Prüfer Sequence) and ENPS (Extended Numbered Prüfer Sequence) can be constructed by extending leaf nodes of the document tree with dummy child nodes. Clearly the leaf node labels of the original tree are kept in ELPS. To facilitate similarity search, we introduce IPS (Inverted labeled Prüfer Sequence) and INPS (Inverted Numbered Prüfer Sequence), which invert the ELPS and ENPS respectively.

IPS embeds the structural relationships of trees, such as parent-child and ancestor-descendant relationships and the sibling order relationship. It is easy to figure out that *post-order* preserves the sibling order relationship, and IPS keeps parent-child and ancestor-descendant relationships as formalized in LEMMA 1.

Lemma 1. *Suppose $(e_1, e_2, \cdots, e_m)$ and $(n_1, n_2, ..., n_m)$ are respectively IPS and INPS of a tree $\mathcal{T}$. $\forall i,j, 1 \leq i < j \leq m$, we have,*

(1) *If $n_i > n_{i+1}$, e_i is the parent of e_{i+1}; and*
(2) *If $n_i < n_j$, e_j is an ancestor of e_i; and*
(3) *If $n_i > n_j$ and $\nexists t$, $i < t < j$, $n_i > n_t > n_j$, e_i is the parent of e_j.*

Proof. We first prove (1). As $n_i > n_{i+1}$, e_{i+1} must be removed before e_i according to the removal-based sequencing method. As e_{i+1} and e_i are neighbors, e_i is the parent of e_{i+1} according to the post-order encoding scheme.

We then prove (2). As $n_i < n_j$, e_i must be removed before e_j. According to the sequencing method, all the nodes, which have larger post-order than n_i, must be removed after e_i, thus those nodes must be before e_i in IPS based on the construction method of IPS. As $i < j$ and $n_i < n_j$, e_j must be an ancestor of e_i.

We finally prove (3). As $n_i > n_j$, e_i must be removed after e_j. e_i may be an ancestor of e_j or a following sibling of e_j. As $\nexists t$, $i < t < j$, $n_i > n_t > n_j$, e_i must be the parent of e_j, as there is no node between e_i and e_j which has larger post-order than n_j.

Example 1. In Figure 6, the circled numbers are the assigned (post-order) numbers to each node. We construct the strings of the corresponding trees according

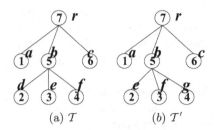

	$\mathcal{T}$	$\mathcal{T}'$
NPS	755577	755577
LPS	$rbbbrr$	$rbbbrr$
ENPS	17253545767	17253545767
EPS	$ardbebfbrcr$	$arebfbgbrcr$
INPS	76754535271	76754535271
IPS	$rcrbfbebdra$	$rcrbgbfbera$

(a) $\mathcal{T}$ (b) $\mathcal{T}'$

Fig. 6. Two Trees and Their Corresponding Strings

Table 1. Strings Transformed from Trees

Trees	Strings	Trees	Strings	Trees	Strings
$\mathcal{T}$	$S : rcrbfbebdra$	$\mathcal{T}'$	$S' : rcrbgbfbera$	$\mathcal{T}_l$	$S_l : rcrbfbera$
$\mathcal{T}_m$	$S_m : rcrbgbfbebdra$	$\mathcal{T}_s$	$S_s : rsrbfbebdra$	$\mathcal{T}_i$	$S_i : rcrbibfbebdra$
$\mathcal{T}_d$	$S_d : rcrbfbera$	$\mathcal{T}_g$	$S_g : rcrgfgegdra$	$\mathcal{T}_b$	$S_b : rcrfrerdra$

to our removal-based sequencing method. LPS of $\mathcal{T}$ is $rbbbrr$. ELPS can be constructed by inserting leaf nodes (i.e., a, d, e, f, c) into the corresponding positions of LPS. We note that the leaf node must be preceding and neighboring its parent. Accordingly, we get the ELPS of $\mathcal{T}$, $ardbebfbrcr$. We can get IPS of $\mathcal{T}$, $rcrbfbebdra$, by inverting its ELPS $ardbebfbrcr$.

Consider IPS and INPS of $\mathcal{T}$, as $n_3(7)>n_4(5)$, $e_3=r$ is a parent of $e_4=b$ according to LEMMA 1(1); as $n_7(3)<n_{10}(7)$, $e_7=e$ is a descendent of $e_{10}=r$ according to LEMMA 1(2); as $n_3>n_6$, and $n_4\not>n_6$ and $n_5\not>n_6$, $e_3=r$ is the parent of $e_6=b$ according to LEMMA 1(3). Accordingly, IPS captures the structural information of trees.

In addition, we give the IPS of $\mathcal{T}$, $\mathcal{T}'$ and $\mathcal{T}_l$ in Figure 2 and $\mathcal{T}_m$ in Figure 3, $\mathcal{T}_s,\mathcal{T}_i,\mathcal{T}_d,\mathcal{T}_g$ and $\mathcal{T}_b$ in Figure 4, as illustrated in Table 1.

4   Tree Similarity Distance Transformation

In this section, we employ string edit distance to approximate LCST distance, SCST distance and edit distance of trees. For ease of the following discussions, we introduce some notations. Given two trees $\mathcal{T}$ and $\mathcal{T}'$ and let S and S' denote the corresponding strings transformed from $\mathcal{T}$ and $\mathcal{T}'$ respectively as described in Section 3. Let $\text{ted}(\mathcal{T},\mathcal{T}')$ denotes the edit distance between $\mathcal{T}$ and $\mathcal{T}'$ in which the cost of each edit operation is assigned to one. $\text{ed}(\mathcal{T},\mathcal{T}')$ denotes the edit distance between $\mathcal{T}$ and $\mathcal{T}'$ where we assign the cost of edit operations as described in Equation 1. $\text{ed}(S,S')$ denotes the edit distance between S and S' in which the cost of each edit operation is assigned to one.

4.1   Edit Distance Transformation

Firstly, consider the substitution operation, suppose T' is constructed with a substitution operation $v \to u$ from T, obviously we have,

$$\text{ed}(S, S') = \text{cSize}(v) \tag{2}$$

where $\text{cSize}(v)$ denotes the number of children of v.

Secondly, consider the insertion operation, suppose T' is constructed with an insertion operation $\Delta \to v$ from T, obviously we have,

$$\text{ed}(S, S') = \begin{cases} 2 & \text{if } v \text{ is a leaf node} \\ 1 & \text{if } v \text{ is an internal node and } \text{parent}(v) = v \\ \text{cSize}(v) & \text{if } v \text{ is an internal node and } \text{parent}(v) \neq v \end{cases} \tag{3}$$

Thirdly, consider the deletion operation, suppose T' is constructed with a deletion operation $v \to \Delta$ from T, obviously we have,

$$\text{ed}(S, S') = \begin{cases} 2 & \text{if } v \text{ is a leaf node} \\ 1 & \text{if } v \text{ is an internal node and } \text{parent}(v) = v \\ \text{cSize}(v) & \text{if } v \text{ is an internal node and } \text{parent}(v) \neq v \end{cases} \tag{4}$$

Based on above analysis of the string edit distance for different tree edit operations, we give a bound of tree edit distance as formalized in THEOREM 1.

Theorem 1. *Given two trees T and T' and their transformed strings S and S', we have*

$$\text{ted}(T, T') \leq \text{ed}(S, S') = \text{ed}(T, T') \leq C_{max} * \text{ted}(T, T') \tag{5}$$

where $C_{max} = max(2, max_{u \in T \cup T'} \{\text{cSize}(u)\})$.

Example 2. For example, in Figure 4, consider T_s, $\text{ted}(T, T_s) = 1$ and $\text{cSize}(c) = 1$, thus $\text{ed}(S, S_s) = 1$. Consider T_g, $\text{ted}(T, T_g) = 1$ and $\text{cSize}(b) = 3$, thus $\text{ed}(S, S_g) = 3$ as illustrated in Table 1. Consider T_i, $\text{ted}(T, T_i) = 1$. As node i is a leaf node, $\text{ed}(S, S_i) = 2$. Consider T_b, $\text{ted}(T, T_b) = 1$. As b is an internal node and $\text{cSize}(b) = 3$, $\text{ed}(S, S_b) = 3$. Consider T_d, $\text{ted}(T, T_d) = 1$. As d is a leaf node, $\text{ed}(S, S_d) = 2$. Consider the construction of T from T_b, $\text{ted}(T_b, T) = 1$. As node b is an internal node and $\text{cSize}(b) = 3$, $\text{ed}(S_b, S) = 3$ as illustrated in Table 1.

Based on Theorem 1, we can use the string edit distance to approximate the tree similarity. More importantly, if we assign different cost for different operations as described in Equation 1, the tree edit distance is the same as the string edit distance. Obviously, insertion/deletion of an internal node lead to two trees more dissimilar than insertion/deletion of a leaf node; insertion/deletion of a node will make the two trees more dissimilar than the substitution of a node. Thus, our assignment is meaningful for tree edit distance.

Accordingly, we can give a filter-and-refine strategy for effective similarity search on trees in terms of tree edit distance based on string edit distance. Consider similarity join in DEFINITION 6, we transform the trees into strings offline and translate similarity join on trees to similarity join on strings. If we select ed as the distance function over trees, we first find the string pairs $<\mathcal{S},\mathcal{S}'>$, such that $\text{ed}(\mathcal{S},\mathcal{S}')\leq\tau$ and then return the corresponding trees.

If we select $\texttt{ted}$ as the distance function over trees, we propose a filter-and-refine framework according to THEOREM 1 as follows. For two strings $\mathcal{S}$ and $\mathcal{S}'$, if $\text{ed}(\mathcal{S},\mathcal{S}')>\tau*C_{max}$, then $\texttt{ted}(\mathcal{T},\mathcal{T}')\geq\frac{1}{C_{max}}*\text{ed}(\mathcal{S},\mathcal{S}')>\tau$, thus $\mathcal{T}$ and $\mathcal{T}'$ do not similarity match. Accordingly, we can filter out such trees. On the other hand, if $\text{ed}(\mathcal{S},\mathcal{S}')\leq\tau$, $\texttt{ted}(\mathcal{T},\mathcal{T}')\leq\text{ed}(\mathcal{S},\mathcal{S}')\leq\tau$, thus $\mathcal{T}$ and $\mathcal{T}'$ must similarity match. Accordingly, we can directly filter out many dissimilar trees. More importantly, we can employe q-gram based techniques [8,15,16] for processing string similarity search so as to effectively answer tree similarity queries.

4.2   LCST Distance Transformation

This section proposes to employ string edit distance to approximate the LCST distance. We give a bound of LCST distance based on tree edit distance as formalized in LEMMA 2. Based on THEOREM 1, we give a bound of LCST distance based on string edit distance as stated in THEOREM 2.

Lemma 2. *Given two trees $\mathcal{T}$ and $\mathcal{T}'$, we have* $\texttt{lcstd}(\mathcal{T},\mathcal{T}')\geq\texttt{ted}(\mathcal{T},\mathcal{T}')$.

Proof. Suppose $\mathcal{T}_l$ is the LCST of $\mathcal{T}$ and $\mathcal{T}'$. We can transform $\mathcal{T}$ to $\mathcal{T}'$ as follows. We first delete the nodes in $\mathcal{V}(\mathcal{T})$-$\mathcal{V}(\mathcal{T}_l)$ from $\mathcal{T}$ and generate a tree $\mathcal{T}_l$, and then insert the nodes in $\mathcal{V}(\mathcal{T}')$-$\mathcal{V}(\mathcal{T}_l)$ to $\mathcal{T}_l$ and get $\mathcal{T}'$. Thus, we can construct $\mathcal{T}'$ from $\mathcal{T}$ with $\texttt{lcstd}(\mathcal{T},\mathcal{T}')$ operations. Accordingly, $\texttt{lcstd}(\mathcal{T},\mathcal{T}')\geq\texttt{ted}(\mathcal{T},\mathcal{T}')$.

Theorem 2. *Given trees $\mathcal{T}$ and $\mathcal{T}'$,* $\texttt{lcstd}(\mathcal{T},\mathcal{T}')\geq\texttt{ted}(\mathcal{T},\mathcal{T}')\geq\frac{1}{C_{max}}*\text{ed}(\mathcal{S},\mathcal{S}')$.

Based on THEOREM 2, given two strings $\mathcal{S}$ and $\mathcal{S}'$, if $\text{ed}(\mathcal{S},\mathcal{S}')>\tau*C_{max}$, then $\texttt{lcstd}(\mathcal{T},\mathcal{T}')\geq\frac{1}{C_{max}}*\text{ed}(\mathcal{S},\mathcal{S}')>\tau$, thus $\mathcal{T}$ and $\mathcal{T}'$ do not similarity match. Accordingly, we can directly filter out such trees based on THEOREM 2.

4.3   SCST Distance Transformation

This section proposes to employ string edit distance to approximate the SCST distance. We give a bound of SCST distance based on tree edit distance as formalized in LEMMA 3. Based on THEOREM 1, we give a bound of SCST distance based on string edit distance as stated in THEOREM 3.

Lemma 3. *Given two trees $\mathcal{T}$ and $\mathcal{T}'$, we have* $\texttt{scstd}(\mathcal{T},\mathcal{T}')\geq\texttt{ted}(\mathcal{T},\mathcal{T}')$.

Proof. Suppose $\mathcal{T}_m$ is the SCST of $\mathcal{T}$ and $\mathcal{T}'$. We can transform $\mathcal{T}$ to $\mathcal{T}'$ as follows. We first insert the nodes in $\mathcal{V}(\mathcal{T}_m)$-$\mathcal{V}(\mathcal{T})$ to $\mathcal{T}$ and generate a tree $\mathcal{T}_m$, and then delete the nodes in $\mathcal{V}(\mathcal{T}_m)$-$\mathcal{V}(\mathcal{T}')$ from $\mathcal{T}_m$ and get $\mathcal{T}'$. Thus, we can construct $\mathcal{T}'$ from $\mathcal{T}$ with $\texttt{scstd}(\mathcal{T},\mathcal{T}')$ operations. Accordingly, $\texttt{scstd}(\mathcal{T},\mathcal{T}')\geq\texttt{ted}(\mathcal{T},\mathcal{T}')$.

Theorem 3. *Given trees $\mathcal{T}$ and $\mathcal{T}'$,* $\texttt{scstd}(\mathcal{T},\mathcal{T}')\geq\texttt{ted}(\mathcal{T},\mathcal{T}')\geq\frac{1}{C_{max}}*\text{ed}(\mathcal{S},\mathcal{S}')$.

Based on THEOREM 3, given two strings $\mathcal{S}$ and $\mathcal{S}'$, if $\text{ed}(\mathcal{S},\mathcal{S}')>\tau*C_{max}$, then $\texttt{scstd}(\mathcal{T},\mathcal{T}')\geq\frac{1}{C_{max}}*\text{ed}(\mathcal{S},\mathcal{S}')>\tau$, thus $\mathcal{T}$ and $\mathcal{T}'$ do not similarity match. Accordingly, we can directly filter out such trees based on THEOREM 3.

4.4   Structural Similarity and Textual Similarity

In this section, we present how to seamlessly integrate the structural similarity and textual similarity into our framework.

Consider the four trees in Figure 1, the traditional methods [2] only consider the structural similarity. In their methods, $ed(T_1,T_2)=4$. $ed(T_1,T_3)=3$. Obviously, T_2 is much more relevant to T_1 than T_3 from human observer.

In our method, we transform the trees into strings. If we only consider the structural similarity, we take the label of the node as an element in the string; otherwise, if we take both the structural similarity and textual similarity into account, we take each character in the label as an element in the string.

For example, consider the trees in Figure 1, we have,

S_1=*paper · editor · E. Bertinoi · paper · author · lname · Ooi · author · fname · B. C.·paper·author·lname·Li·author·fname·G.·paper·title·Similarity Search.*

S_2=*paper·*<u>*editer*</u>*· E. Bertnoi·paper · author · lname · Ooi · author · fname ·* <u>*B.*</u>*· paper · author · lname · Li · author · fname · G. · paper · title ·* <u>*Similerity*</u> *Search.*

S_3=*paper · editor ·* <u>*B.C. Ooi*</u>*·paper · author · lname ·* <u>*Bertinoi*</u>*·author · fname · E.·paper · author · lname · Li · author · fname · G.·paper·title·Similarity Search.*

S_4=*paper · editor · E. Bertinoi · paper ·* <u>*authors*</u> *· author · lname · Ooi · author · fname · B. C. ·* <u>*authors*</u> *· author · lname · Li · author · fname · G. · paper · title · Similarity Search.*

If we only consider the structural similarity, we have $ed(T_1,T_2)=4$. $ed(T_1, T_3)=3$. $ed(T_1,T_4)=2$. We find that $ed(S_1,S_2)=4$. $ed(S_1,S_3)=3$. $ed(S_1,S_4)=2$. Note that the string edit distances are exactly the same as the tree edit distances. On the contrary, if we consider both the structural similarity and textual similarity, we have $ed(S_1,S_2)=5$. $ed(S_1,S_3)=16$. $ed(S_1,S_4)=13$. Thus, T_2 is much more relevant to T_1 than T_4, which in turn is much more relevant to T_1 than T_3. This is consistent with human observer. Accordingly, we can seamlessly integrate the structural similarity and textual similarity for facilitating similarity join on trees.

5   Experimental Study

In this section, we compare the performance of our filter-and-refine similarity search algorithm, which employs string edit distance to approximate tree distance against the approximate XML join method [9] (denoted as AppJoin in the Figures). In our methods, we employed q-gram based methods [16] for similarity search on strings. We used the inverted index to maintain the q-grams [11,13] and adopted the merge based algorithm[15] for processing string similarity search so as to effectively answer tree similarity queries.

We employed the synthetic and real datasets: DBLP [1], TreeBank [2], SIGMOD Record [3] and XMark [4] for our experiments. (1) XMark is synthetic and generated

---

[1] http://dblp.uni-trier.de/xml/
[2] http://www.cs.washington.edu/research/xmldatasets/
[3] http://www.sigmod.org/record/xml/
[4] http://www.xml-benchmark.org/

Table 2. Characteristics of datasets

Datasets	Average # of elements	Maximal depth	Maximal fan-out
DBLP	13.6	4	8
SIGMOD Record	12.2	4	8
XMark	10.4	6	6
Treebank	11.8	8	6

by an XML data generator; (2) DBLP is a collection of papers and articles, which consists of bibliographic information on major computer science journals and proceedings; (3) SIGMOD Record, which is a collection of papers in database area and obtained from SIGMOD homepage; (4) Treebank is obtained from the University of Washington XML repository. The DTD of Treebank is very deep recursive. In the experiment, we generated 10000 XML documents for each dataset and we randomly selected 100 queries for similarity searches.

Different characteristics of selected trees are shown in Table 2. The results shown in this paper were all averaged on the queries. We adopted the CPU time consumption as performance measures. Through the experiments on real datasets, we show our algorithm's sensitivity to different features of the data. We also present our experiments on real dataset to show the algorithm's performance on different query characteristics.

We conducted all the experiments on a computer with AMD 5600 2.8GHz CPU and 2GB of RAM. We implemented the algorithms in C++.

5.1   Pruning Power

This section presents the pruning power of our proposed methods. We selected 10000 trees as the data collection, and randomly selected 100 queries and submitted them to the data collection. We varied different values of τ to evaluate the average number of pruned trees, where the pruned trees denote the dissimilar trees that are directly filtered out by the algorithms. Figure 7 gives the experimental results. In the Figures, SSTD denotes our proposed method for Similarity Search over Tree-structured Data. SSTD(ED), SSTD(TED), SSTD(LCSTD), and SSTD(SCSTD) denote our methods which respectively adopt ed, ted, lcstd and scstd similarity functions.

We observe that our proposed methods can prune many dissimilar trees and thus improve the efficiency of similarity search on trees. Thus, we can embed the bounds based on the similarity functions into our filter-and-refine framework for effective approximate tree join. Moreover, with the increase of τ, SSTD(LCSTD) and SSTD(SCSTD) drop down while SSTD(TED) and SSTD(ED) vary slightly. This is because the ed function can help to filter out many more irrelevant trees than lcstd and scstd functions as described in Section 4. This reflects the effectiveness of edit distance based methods which can accurately evaluate the similarity between trees.

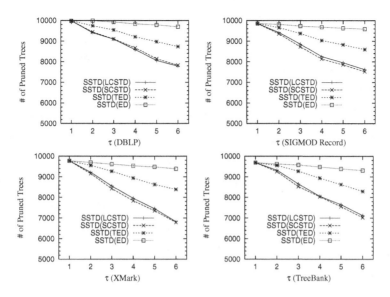

Fig. 7. Average # of Pruned Trees vs. Different Values of τ

5.2   Similarity Search Performance

The experiments described in this part were conducted to compare the performance of the filtration algorithms for the queries with different parameters.

Firstly, we selected 10000 trees as the data collection, and randomly selected 100 queries and submitted them to the data collection. We varied different values of τ to evaluate the average elapsed time of processing the one hundred queries. Figure 8 illustrates the experimental results.

We see that our algorithms achieve high performance and outperforms AppJoin significantly. This is because our method can prune many dissimilar trees based on string edit distance by employing effective filtration techniques, such as, q-grams filtration and distance pruning techniques, which can improve the efficiency of string similarity search and thus accelerate tree similarity search. Moreover, with the increase of τ, the performance of AppJoin drops down sharply while our methods vary slightly. This further reflects the benefits of our proposed techniques. In addition, we see that the edit distance based method achieves higher performance than those based on LCST distance and SCST distance. This is because string edit distance is much better to approximate tree edit distance than LCST distance and SCST distance.

Secondly, we selected different numbers of trees as the data collections, and randomly selected 100 queries and submitted them to the data collections. We varied the sizes of data collections, i.e., the numbers of trees in the data collections, to evaluate the average elapsed time of processing the one hundred queries. Figure 9 illustrates the experimental results obtained.

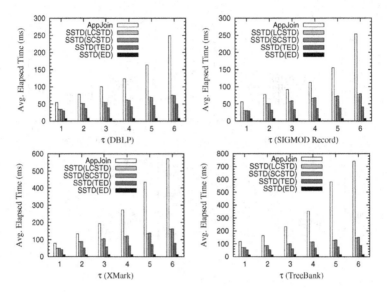

Fig. 8. Average Elapsed Time vs. Different Values of τ

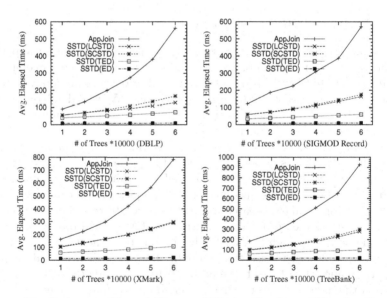

Fig. 9. Average Elapsed Time vs. Different Numbers of Trees($\tau=3$)

We see that our algorithms still achieve high performance and outperform AppJoin. Furthermore, with the increase of the numbers of trees, the elapsed time of our methods varies a little while that of AppJoin increases dramatically. This reflects the scalability of our methods.

5.3   Structural Similarity and Textual Similarity

This section evaluates the approximate tree search by combining both structural similarity and textual similarity as described in Section 4.4. We still selected 10000 trees on each dataset as the data collection, and randomly selected 100 queries and submitted them to the data collection. The total elapsed time of processing these 100 queries is illustrated in Table 3. We observe that our method achieves very high performance and the average elapsed time is even less than 4ms. This further reflects the benefits of our proposed methods.

Table 3. Elapsed Time of Combining Structural and Textual Similarity

Elapsed Time of Processing 100 Queries(ms)	$\tau=2$	$\tau=4$	$\tau=6$	$\tau=8$	$\tau=10$	$\tau=12$
DBLP	81	109	161	241	311	373
SIGMOD Record	90	118	173	245	323	381
XMark	82	101	155	234	318	365
Treebank	87	114	177	241	311	378

6   Related Work

In the literature, "approximate string matching" also refers to the problem of finding a pattern string approximately in a text. There have been many studies on this problem. Gonzalo Navarro [17] gives an excellent survey. The prefix filter [6] technique was proposed for evaluating joins using pure relational processing. It is designed for edit/hamming distance, Jaccard and some simple weighted variants. It can be modified to work for all weighted similarity measures for selection queries. Arasu et al. [1] designed a signature scheme that can be used as a filter for identifying candidate sets with hamming distance smaller than k from a query set. It was used for answering set similarity joins based on edit distance and Jaccard. It is not clear how to extend this work for weighted metrics and selection queries. Both exact and approximate algorithms for set similarity joins between sets with un-weighted elements have been proposed as well.

Specialized set similarity join algorithms using cosine similarity between sets have also been considered [3]. Kahveci et al. [11] proposed an index for substring matches within edit distance k from a query. Sahinalp et al. [19] proposed VP-trees for answering nearest neighbor queries for edit distance. An exhaustive comparison of methods based on edit distance and variants appears in [17]. Several algorithms (e.g., [5], [7], [8]) have been proposed for answering approximate string queries efficiently. Their main strategy is to use various filtering techniques to improve the performance. These filters can be adopted with slight modifications to be written as SQL queries inside a relational DBMS. Other related studies include [20], [23] on similarity joins. These algorithms find, given two collections of sets, those pairs of sets that share enough common elements. Similarity selections and similarity joins are in essence different. The former could be treated as a special case of the latter, but algorithms developed for the latter might not be efficient for the former.

More recently, Kim et al. [13] proposed a technique called "n-Gram/2L" to improve space and time efficiency for inverted index structures. Li et al. [16] proposed a new technique called VGRAM to judiciously choose high-quality grams of variable lengths from a collection of strings. Li et al. [15] studied the problem of how to efficiently find a collection of strings those similar to a given query string. They developed several merge algorithms that can greatly improve the performance. Hadjieleftheriou et al. [10] demonstrated a length bounding property. They proposed three new algorithms based on TA/NRA style processing on inverted lists. Their shortest-first algorithm achieved truly interactive responses.

7 Conclusion

We have studied the problem of effective similarity search over tree-structured data. We proposed a filter-and-refine framework to improve the efficiency of similarity search over trees by using approximate string search techniques. We transformed trees to strings with a one-to-one mapping strategy. We employed the string edit distance to approximate the similarity measures between trees, including tree edit distance, largest common subtree distance, and smallest common super-tree distance, and gave the theoretical bounds. We embedded the bound into our filter-and-refine framework for facilitating effective similarity search on tree-structured data. Moreover, we adopted the existing approximate string search techniques to improve the performance of similarity search on trees. The experimental results show that our methods achieve high performance and outperform the existing methods significantly.

Acknowledgement

This work is partly supported by the National Natural Science Foundation of China under Grant No.60573094, the National High Technology Development 863 Program of China under Grant No.2007AA01Z152 and 2006AA01A101, the National Grand Fundamental Research 973 Program of China under Grant No.2006CB303103.

References

1. Arasu, A., Ganti, V., Kaushik, R.: Efficient exact set-similarity joins. In: VLDB (2006)
2. Augsten, N., Bohlen, M., Gamper, J.: Approximate matching of hierarchical data using pq-grams. In: VLDB (2005)
3. Bayardo, R.J., Ma, Y., Srikant, R.: Scaling up all pairs similarity search. In: WWW (2007)
4. Bille, P.: A survey on tree edit distance and related problems. Theoretical Computer Science 337(1-3), 217–239 (2005)
5. Chaudhuri, S., Ganjam, K., Ganti, V., Motwani, R.: Robust and efficient fuzzy match for online data cleaning. In: SIGMOD (2003)

6. Chaudhuri, S., Ganti, V., Kaushik, R.: A primitive operator for similarity joins in data cleaning. In: ICDE (2006)
7. Gionis, A., Gunopulos, D., Koudas, N.: Efficient and tunable similar set retrieval. In: SIGMOD (2001)
8. Gravano, L., Ipeirotis, P.G., Jagadish, H.V., Koudas, N., Muthukrishnan, S., Srivastava, D.: Approximate string joins in a database (almost) for free. In: VLDB, pp. 491–500 (2001)
9. Guha, S., Jagadish, H.V., Koudas, N., Srivastava, D., Yu, T.: Approximate xml joins. In: SIGMOD (2002)
10. Hadjieleftheriou, M., Chandel, A., Koudas, N., Srivastava, D.: Fast indexes and algorithms for set similarity selection queries. In: ICDE (2008)
11. Kahveci, T., Singh, A.K.: Efficient index structures for string databases. In: VLDB (2001)
12. Kailing, K., Kriegel, H.-P., Schonauer, S., Seidl, T.: Efficient similarity search for hierarchical data in large databases. In: Bertino, E., Christodoulakis, S., Plexousakis, D., Christophides, V., Koubarakis, M., Böhm, K., Ferrari, E. (eds.) EDBT 2004. LNCS, vol. 2992, pp. 676–693. Springer, Heidelberg (2004)
13. Kim, M.-S., Whang, K.-Y., Lee, J.-G., Lee, M.-J.: n-gram/2l: A space and time efficient two-level n-gram inverted index structure. In: VLDB (2005)
14. Klein, P.: Computing the edit-distance between unrooted ordered trees. In: Bilardi, G., Pietracaprina, A., Italiano, G.F., Pucci, G. (eds.) ESA 1998. LNCS, vol. 1461. Springer, Heidelberg (1998)
15. Li, C., Lu, J., Lu, Y.: Efficient merging and filtering algorithms for approximate string searches. In: ICDE (2008)
16. Li, C., Wang, B., Yang, X.: Vgram: Improving performance of approximate queries on string collections using variable-length grams. In: VLDB (2007)
17. Navarro, G.: A guided tour to approximate string matching. ACM Computing Surveys, 31–88 (2001)
18. Prufer, H.: Neuer beweis eines satzes uber permutationen. Archiv fur Mathematik und Physik 27, 142–144 (1918)
19. Sahinalp, S.C., Tasan, M., Macker, J., Ozsoyoglu, Z.M.: Distance based indexing for string proximity search. In: ICDE (2003)
20. Sarawagi, S., Kirpal, A.: Efficient set joins on similarity predicates. In: SIGMOD (2004)
21. Seidl, T., Kriegel, H.-P.: Optimal multi-step k-nearest neighbor search. In: SIGMOD (1998)
22. Tai, K.-C.: The tree-to-tree correction problem. Journal of the Association for Computing Machinery (JACM) 26, 422–433 (1979)
23. Ukkonen, E.: Approximate string matching with q-grams and maximal matches. Theor. Comput. Sci. 92(1), 191–211 (1992)
24. Yang, R., Kalnis, P., Tung, A.K.H.: Similarity evaluation on tree-structured data. In: SIGMOD (2005)
25. Zhang, K., Shasha, D.: Simple fast algorithms for the editing distance between trees and related problems. SIAM Journal of Computing 18, 1245–1262 (1989)

Hierarchical Graph Embedding for Efficient Query Processing in Very Large Traffic Networks

Hans-Peter Kriegel, Peer Kröger, Matthias Renz, and Tim Schmidt

Ludwig-Maximilians-Universität München, Oettingenstr. 67, 80538 Munich, Germany
{kriegel,kroegerp,renz,schmidtti}@dbs.ifi.lmu.de
http://www.dbs.ifi.lmu.de

Abstract. We present a novel graph embedding to speed-up distance-range and k-nearest neighbor queries on static and/or dynamic objects located on a (weighted) graph that is applicable also for very large networks. Our method extends an existing embedding called reference node embedding which can be used to compute accurate lower and upper bounding filters for the true shortest path distance. In order to solve the problem of high storage cost for the network embedding, we propose a novel concept called hierarchical embedding that scales well to very large traffic networks. Our experimental evaluation on several real-world data sets demonstrates the benefits of our proposed concepts, i.e. efficient query processing and reduced storage cost, over existing work.

1 Introduction

Similarity queries in large traffic networks are important database operations in applications such as location-based services, traffic network monitoring, traffic information systems, etc. Typically, traffic networks such as road networks are modeled by graphs. Nodes of the graph represent crossings such as road intersections or junctions, whereas edges represent connections such as roads or railways between nodes. The data objects representing points of interest such as cars, service stations, etc. are distributed over this road network, i.e. are located at nodes or on edges or may move along the graph. The distance between objects in the network is measured by means of the shortest path distance which can be computed by the Dijkstra algorithm.

In today's applications usually a high number of online queries on networks of hundreds of thousands or even millions of nodes have to be answered in real-time. Obviously, a more efficient solution than computing Dijkstra for all these query nodes is utterly necessary for such scenarios. A filter/refinement approach is envisioned, applying a cheaper filter step in order to efficiently partition the data objects into a set of true hits and/or true drops, and a set of candidates, that need to be further analyzed. In order to decide about true hits, we need an upper bounding distance approximation, whereas a lower bounding distance approximation is needed to decide about true drops. The remaining set of candidates that cannot be discarded from or included in the result set by means of the filter step, need to be refined, i.e. the true network distance needs to be computed.

Here, we propose a novel filter/refinement query processor for very large graph networks based on a hierarchical network graph embedding. Section 2 introduces preliminary definitions and discusses related work. In Section 3, we show how the so-called

B. Ludäscher and Nikos Mamoulis (Eds.): SSDBM 2008, LNCS 5069, pp. 150–167, 2008.

reference node embedding can be extended to reduce the storage cost for very large networks. We show how this novel embedding can be computed efficiently for static and dynamic objects located on graph networks and derive efficient lower- and upper bounds for the network distance from the hierarchical embedding. Section 4 sketches our multi-step query processor. Section 5 presents an experimental evaluation of the proposed concepts and Section 6 concludes the paper.

2 Preliminaries and Related Work

2.1 Preliminaries

Let $\mathcal{D}$ be a database of objects that are located in a traffic network, e.g. cars or pedestrians in a network of streets. The traffic network is represented by an undirected weighted graph $\mathcal{G} = (N, E, W)$ called *network graph*, where N denotes the set of nodes, $E \subseteq N \times N$ denotes the set of edges and the function $W : E \rightarrow \mathbb{R}^+$ associates a *weight* $w(n_i, n_j)$ to each edge $(n_i, n_j) \in E$. The *network distance* between two nodes $n_i, n_j \in N$, denoted by $d_{net}(n_i, n_j)$, equals $w(n_i, n_j)$ if n_i, n_j are adjacent, i.e. $(n_i, n_j) \in E$, else it equals the length of the shortest path from n_i to n_j. The length of a path is defined as the sum of the weights of all participating edges.

If an object o is located on an edge $(n_i, n_j) \in E$, $d_i(o)$ and $d_j(o)$ denote the distance of o to the adjacent nodes n_i and n_j, respectively. The network distance between two objects $o_i, o_j \in \mathcal{D}$, $d_{net}(o_i, o_j)$, is the length of the shortest path between o_i and o_j. Thereby, we assume that o_i and o_j are additional "virtual" nodes of the graph. Thus, if o_i is located on edge (n_{i_1}, n_{i_2}) we introduce additional "virtual" edges (o_i, n_{i_1}) and (o_i, n_{i_2}) with weights $w(o_i, n_{i_1}) = d_{i_1}(o_i)$ and $w(o_i, n_{i_2}) = d_{i_2}(o_i)$, respectively. If o_i is located on a node n, we do not need to introduce additional edges or nodes but can work with n instead of o_i. Note, that by introducing the additional "virtual" nodes for objects, the network distance is still a function $N \times N \rightarrow \mathbb{R}$. Whenever we use d_{net} as a function on $\mathcal{D} \times \mathcal{D}$ in the following, we assume the introduction of virtual nodes for the according objects if necessary.

Based on the network distance, proximity queries are given as follows. Given a query object q located on $\mathcal{G}$ and a distance threshold $\varepsilon \in \mathbb{R}^+$, a *distance range query* (DRQ) returns the set $DRQ(q, \varepsilon) = \{o \in \mathcal{D} \,|\, d_{net}(q, o) \leq \varepsilon\}$. Given a query object q located on $\mathcal{G}$ and a number $k \in \mathbb{N}^+$, a *k-nearest neighbor query* (kNNQ) returns the set $NNQ(q, k)$ containing k objects such that $\forall o \in NNQ(q, k), \hat{o} \in \mathcal{D} \setminus NNQ(q, k) : d_{net}(q, o) \leq d_{net}(q, \hat{o})$.

2.2 Related Work

Proximity queries in traffic networks are based on network distances defined by the shortest path between two objects, e.g. computed by the Dijkstra algorithm [1] and its variants [2]. These algorithms expand the path from the starting node towards the target node using a priority queue of visited nodes sorted by ascending distance from the starting node. The A* algorithm [3] applies heuristics to prune the search space and direct the graph expansion. Materialization techniques [4,5,6] suffer from increasing storage cost. In [7] the authors divide the graph into regions and gather information whether an edge is on a shortest path leading to a specific region. All these approaches provide only a speed-up for the exact distance computation but cannot be used as a filter step.

In [8] the Euclidean distance between graph nodes/objects is used as a lower bounding filter in order to guide an incremental network expansion for refinement. This approach works well only for high-proximity queries (i.e. small query range ε or small nearest-neighbor coefficient k) and dense object distributions, otherwise a large portion of the network for distance computation need to be retrieved. Furthermore, this approach does not provide an upper bounding distance function to filter out true hits, resulting in a larger amount of refinements.

In [9] one of the graph embedding technique from [10] is applied in order to estimate the network distance between two nodes. An extended dynamic embedding for moving objects is presented. In addition, it is shown how the graph embedding can be used to compute an approximate shortest path between two objects. The accuracy of the approximation depends on the density and distribution of the objects in space. A severe drawback of the approach is that the embedded space involves 40 to 256 dimensions. In addition, it does not offer any solution for the computation of the exact distances of the candidates in the refinement step.

In [11] *distance signatures* are computed and managed for each data object o in the network graph containing a vector of distance approximations between o and all other data objects in the network graph. These distance approximations are then used to efficiently determine the candidates of a proximity query in a filter step. Subsequently, the exact distances of the candidates are computed online in the refinement step. The obvious drawback of this proposal is that the storage and query cost directly depend on the number of objects. Furthermore, this approach does not support an efficient re-embedding necessary to answer proximity queries on moving objects that frequently change their positions.

The work of this paper is based on the network graph embedding originally proposed independently by two research groups [12,13] and [14]. While the work in [12,13] only explores a lower bound, the authors in [14] also derive an upper bound for the network distance. In addition, the authors in [12,13] focus only on speeding up the shortest path computations whereas in [14], the authors propose a multi-step query processing framework for supporting proximity queries in traffic networks. The details of the embedding is reviewed in Section 3.1.

In [15] a Voronoi diagram on the network space is computed and each Voronoi cell that represents the region of the nearest neighbor in the network is represented by a 2D polygon. These Voronoi-cell polygons are indexed to support kNN queries. The performance of this approach mainly depends on the density and distribution of the objects in the network. Dense network graphs on which the data objects are sparsely distributed lead to large Voronoi cells with a lot of adjacent neighbor cells. In this case, the computation of the kNN would have a poor performance.

In this paper, we do not focus on another class of proximity queries in road networks called continuous proximity queries (as studied e.g. in [16,17]).

3 Network Graph Embedding

3.1 Basics

Our approach is based on a special form of a Lipschitz embedding of the traffic network using singleton reference sets which we call *reference nodes* according to [14] (in [12,13], these reference nodes are called *landmarks*). The embedding transforms the nodes of a given network graph and the objects located on that graph into a k-dimensional vector space. Let $\mathcal{G} = (N, E, W)$ be a network graph and $N' = \langle n_{r_1}, \ldots, n_{r_k} \rangle \subseteq N$ be a subsequence of $k \geq 1$ reference nodes. The embedding, or transformation, of the native space N into a k-dimensional vector space $\mathbb{R}^k$ is a mapping $F^{N'} : N \cup \mathcal{D} \rightarrow \mathbb{R}^k$, where $|N'| = k$ is the dimensionality of the vector space. A *reference node embedding* of $\mathcal{G}$ based on $N' \subset N$ defines the function $F^{N'}$ as follows. For each $n \in N$, $F^{N'}(n) = (F_1^{N'}(n), \ldots, F_k^{N'}(n))^{\mathbf{T}}$, where $F_i^{N'}(n) = d_{net}(n, n_{r_i})$ for $1 \leq i \leq k$. Objects can be embedded analogously. For each $o \in \mathcal{D}$ located on a node n, $F^{N'}(o) = F^{N'}(n)$. For each $o \in \mathcal{D}$ located on an edge $(n_1, n_2) \in E$, $F^{N'}(o) = (\hat{F}_1^{N'}(o), \ldots, \hat{F}_k^{N'}(o))^{\mathbf{T}}$, where $\hat{F}_i^{N'}(o) = \min\{d_1(o) + F_i^{N'}(n_1), d_2(o) + F_i^{N'}(n_2)\}$.

In Figure 1 a reference node embedding of some objects located on a sample network graph using reference nodes $N' = \langle n_8, n_7 \rangle$ is illustrated.

The reference node embedding has two major advantages. First, if the graph structure remains fixed (which is obviously a realistic assumption) and the embedding of the graph nodes (that do not change) is performed offline in a preprocesing step and is then stored, a re-embedding of moving objects can be done very efficiently. Second, the reference node embedding can be used to compute upper and lower bounds for the network distance. In [14] it is shown that the distance $D(x, y) = \max_{i=1..k} |x_i - y_i|$ in the embedded space lower bounds the distance d_{net} in the native space. In addition, it is shown that the distance function $D^*(x, y) = \min_{i=1...k} (x_i + y_i)$ is an upper bound of d_{net}. In summary, the reference node embedding approach is very suitable to efficiently support similarity queries over both static and dynamic objects in traffic networks.

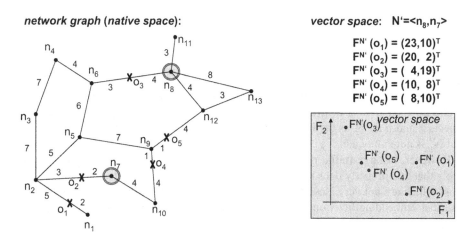

Fig. 1. Network graph embedding

3.2   The Idea of Hierarchical Network Embedding

Beside these two significant advantages, the reference node embedding proposed in [12,13] and [14] — in the following called *flat* embedding — has one major short-coming. The performance gain of the embedding heavily depends on the number of reference nodes used. Though it is shown in [14] that even a low number of reference nodes is sufficient in order to achieve significant performance boosts on small and medium-sized networks, it is also indicated that on large-scale networks, the number of reference nodes necessary to approximate the network distance sufficiently well and to speed-up similarity query processing is considerably large. However, a large set of reference nodes leads to high storage cost because we have to store $O(|N| \cdot |N'|)$ distances for the embedding. In addition, also the computational cost of the embedding and re-embedding process and of the query processor increases with increasing $|N'|$. Especially the increase of query processing cost (due to higher CPU cost to determine the distance between $|N'|$-dimensional points and due to higher I/O cost caused by the fact that higher dimensional points can be indexed less efficiently) is a severe handicap of the flat embedding approach.

Obviously, the reason for this bad scalability of the flat reference node embedding on large networks is the increasing dimensionality of the resulting embedding vectors in the vector space $\mathbb{R}^{|N'|}$. This is somewhat arbitrary because finally only one reference node is taken into account for a distance estimation as D and D^* aggregate over the distances to all reference nodes such that only the "best" reference node is taken. Usually a small subset of the reference nodes suffices for the distance estimation between an object o and any other object in the graph. It is easy to see that the smaller is the distance of a reference node to a particular object o, the better is this reference node for all distance approximations w.r.t. o.

In this paper, we propose a solution to the limited scalability of the flat reference node embedding that is inspired by these considerations. Given an object o there are reference nodes that are more relevant and less relevant for o in N'. So why not use only the relevant reference nodes in N' for the embedding of o? This should decrease the dimensionality of the resulting embedded vectors without downgrading the distance approximations considerably.

3.3   Two-Level Network Embedding

A first approach is to use for each object o only the K nearest reference nodes $N'_o \subseteq N'$, where $K \ll N'$. Obviously, the lower and upper bounding distance approximations D and D^* can still be used to approximate the network distance $d_{net}(x, y)$ between two objects x and y as far as the intersection of the corresponding reference node sets N'_x and N'_y is not empty, i.e. $N'_x \cap N'_y \neq \emptyset$.

However, in large traffic networks with a large reference node set N', it is more likely that this property does not hold for most of the pairs of objects, in particular for those which are not very close to each other. To overcome this problem, we introduce a further embedding level on top of the current embedding. A comprehensive graph $\mathcal{G}' = (N', E', W')$ is built using all reference nodes N' as nodes and all shortest paths between these nodes in the original graph $\mathcal{G}$ as edges E'. The weights W' are determined analogously. The idea is illustrated in Figure 2.

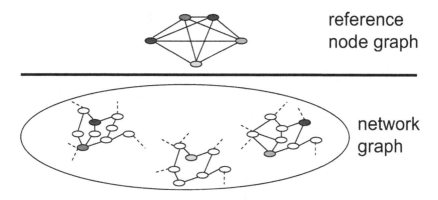

reference
node graph

network
graph

Fig. 2. Schema of a 2-level reference node embedding

Formally, let $\mathcal{G} = (N, E, W)$ be the network graph and $N' \subseteq N$ a set of reference nodes (landmarks) with $|N'| \geq K$. For each node or object $o \in N \cup \mathcal{D}$ let $N'_o = \langle r^o_1, \ldots, r^o_K \rangle$, $r^o_j \in N'$ be the set of K local reference nodes relevant for o. The 2-level embedding, or transformation, of the native space $N \cup \mathcal{D}$ into a K-dimensional vector space $\mathbb{R}^k$ is a mapping $\tilde{F}^{N'} : N \cup \mathcal{D} \to \mathbb{R}^K$ together with a reference node graph $\mathcal{G}' = (N', E', W')$. A *2-level reference node embedding* of $\mathcal{G}$ and $\mathcal{D}$ based on $N' \subset N$ is a pair $(\tilde{F}^{N'}, \mathcal{M}')$ consisting of the mapping function $\tilde{F}^{N'}$ and the reference node matrix $\mathcal{M}'$ that is the weighted adjacent matrix of $\mathcal{G}'$.

The function $\tilde{F}^{N'}$ is defined as follows.

$$\tilde{F}^{N'}_o(o) = \begin{cases} (d_{net}(r^o_1, o), \ldots, d_{net}(r^o_K, o))^{\mathbf{T}} & \text{if } o \in N \text{ is a node} \\ \tilde{F}^{N_n}(n) & \text{if object } o \in \mathcal{D} \text{ is located on } n \in N \\ (S^{N'}_1(o), \ldots, S^{N'}_k(o))^{\mathbf{T}} & \text{if } o \in \mathcal{D} \text{ is located on } (n_i, n_j) \in E \end{cases}$$

where $S^{N'}_i(o) = \min\{d_1(o) + \tilde{F}^{N'}_i(n_1), d_2(o) + \tilde{F}^{N'}_i(n_2)\}$.

Let us note that a re-embedding of moving objects using $\tilde{F}$ is still very efficient as long as we assume that the graph structure remains fixed because then the embedding of the graph nodes performed in a preprocessing step can be stored.

The reference node graph $\mathcal{G}' = (N', E', W')$ is a graph over all reference nodes N', where $E' = \{(n_i, n_j)|n_i, n_j \in N'\}$ is the set of all pairwise connections between the reference nodes in N' and $W'(n_i, n_j) = d_{net}(n_i, n_j)$ is the shortest path between the corresponding reference nodes $n_i, n_j \in N'$ in the original graph $\mathcal{G}$.

Because the set of edges is implicitly defined, we can store and represent this reference node graph by its weighted adjacency matrix which we call the *reference node matrix*. This matrix has the following general form.

$$\mathcal{M}' = \begin{bmatrix} 0 & d_{net}(r_1, r_2) & \ldots & d_{net}(r_1, r_k) \\ d_{net}(r_2, r_1) & 0 & \ldots & d_{net}(r_2, r_k) \\ \vdots & \vdots & \ddots & \vdots \\ d_{net}(r_k, r_1) & d_{net}(r_k, r_2) & \ldots & 0 \end{bmatrix}$$

In summary, the pair $(\tilde{F}^{N'}, \mathcal{M}')$ defines a 2-level reference node embedding of a graph $\mathcal{G}$.

3.4 Distance Approximations

Based on $\tilde{F}^{N'}$ and $\mathcal{M}'$, we can now define a distance function $\tilde{D}$ for objects in $x, y \in \mathcal{D}$ in the embedded space that lower bounds d_{net} as follows.

$$\tilde{D}(\tilde{F}^{N_x}(x), \tilde{F}^{N_y}(y)) = \max_{k \in N_x, l \in N_y} \begin{cases} M_{i_k, i_l} - \tilde{F}_k^{N_x}(x) - \tilde{F}_l^{N_y}(y) & \text{(case A)} \\ \tilde{F}_k^{N_x}(x) - M_{i_k, i_l} - \tilde{F}_l^{N_y}(y) & \text{(case B)} \\ \tilde{F}_l^{N_y}(y) - M_{i_k, i_l} - \tilde{F}_k^{N_x}(x) & \text{(case C)} \\ 0 & \text{(case D)} \end{cases}$$

where i_p represents the index of the $r_p^n \in N_n$ in $\mathcal{M}'$ and where the following cases appear: case A: $d_{net}(r_i, r_j) > d_{net}(n_a, r_i) + d_{net}(n_b, r_j)$, case B: $d_{net}(n_a, r_i) > d_{net}(r_i, r_j) + d_{net}(n_b, r_j)$, case C: $d_{net}(n_b, r_j) > d_{net}(n_a, r_i) + d_{net}(r_i, r_j)$ and case D otherwise.

Figure 3 illustrates the definition of $\tilde{D}$. On the left hand side, case A ($k = 1$ and $l = 4$, i.e. r_1 and r_4 determine the distance approximation) is visualized. On the right hand side, case B and case C (symmetric) are depicted.

Lemma 1 (Lower bounding property). *Let $(\tilde{F}, \mathcal{M}')$ be a 2-level reference node embedding of nodes and objects of a network $\mathcal{G} = (N, E, W)$ w.r.t. a set of reference nodes N' and local reference node sets N_o for all nodes or objects $o \in N \cup \mathcal{D}$. For each $x, y \in N \cup \mathcal{D}$ the following property holds.*

$$\tilde{D}(\tilde{F}^{N_x}(x), F^{N_y}(y)) \leq d_{net}(x, y).$$

Proof. Without loss of generality, let $r_i \in N_x$ und $r_j \in N_y$ be the reference nodes that determine $\tilde{D}$. Since d_{net} is a metric, the following considerations hold.

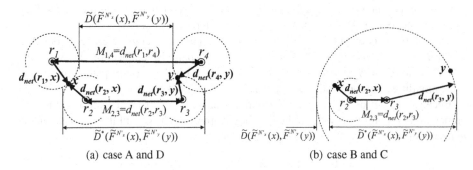

(a) case A and D   (b) case B and C

Fig. 3. Illustration of the distance approximation derived from a reference node embedding

Case A occurs if $d_{net}(r_i, r_j) > d_{net}(n_a, r_i) + d_{net}(n_b, r_j)$. Then,

$$\begin{aligned}
\tilde{D}(\tilde{F}^{N_x}(x), \tilde{F}^{N_y}(y)) &= M_{i_i, i_j} - \tilde{F}_i^{N_x}(x) - \tilde{F}_j^{N_y}(y) \\
&= d_{net}(r_i, r_j) - d_{net}(x, r_i) - d_{net}(y, r_j) \\
&\leq d_{net}(r_i, n_a) - d_{net}(n_b, r_j) \\
&= d_{net}(x, r_j) - d_{net}(y, r_j) \\
&\leq d_{net}(x, y)
\end{aligned}$$

Case B occurs if $d_{net}(n_a, r_i) > d_{net}(r_i, r_j) + d_{net}(n_b, r_j)$. Then,

$$\begin{aligned}
\tilde{D}(\tilde{F}^{N_x}(x), \tilde{F}^{N_y}(y)) &= \tilde{F}_i^{N_x}(x) - M_{i_i, i_j} - \tilde{F}_j^{N_y}(y) \\
&= d_{net}(x, r_i) - d_{net}(r_i, r_j) - d_{net}(y, r_j) \\
&= d_{net}(x, r_i) - d_{net}(r_j, r_i) - d_{net}(y, r_j) \\
&\leq d_{net}(x, r_j) - d_{net}(y, r_j) \\
&\leq d_{net}(x, y)
\end{aligned}$$

Case C occurs if $d_{net}(n_b, r_j) > d_{net}(n_a, r_i) + d_{net}(r_i, r_j)$. Then,

$$\begin{aligned}
\tilde{D}(\tilde{F}^{N_x}(x), \tilde{F}^{N_y}(y)) &= \tilde{F}_j^{N_y}(y) - M_{i_i, i_j} - \tilde{F}_j^{N_y}(y) \\
&= d_{net}(y, r_j) - d_{net}(r_i, r_j) - d_{net}(x, r_i) \\
&\leq d_{net}(y, r_i) - d_{net}(x, r_i) \\
&\leq d_{net}(x, y)
\end{aligned}$$

Otherwise, in case D, we have

$$\tilde{D}(\tilde{F}^{N_x}(x), \tilde{F}^{N_y}(y)) = 0 \leq d_{net}(x, y)$$

Analogously, we can define a distance function $\tilde{D}^*$ for objects in $x, y \in \mathcal{D}$ in the embedded space that upper bounds d_{net} as follwos.

$$\tilde{D}^*(\tilde{F}^{N_x}(x), \tilde{F}^{N_y}(y)) = \min_{k \in N_x, l \in N_y} \{M_{i_k, i_l} + \tilde{F}_k^{N_x}(x) + \tilde{F}_l^{N_y}(y)\}$$

where i_p is defined as above. Figure 3 illustrates the definition of $\tilde{D}^*$.

Lemma 2 (Upper bounding property). *Let $(\tilde{F}, \mathcal{M}')$ be a 2-level reference node embedding of nodes and objects of a network $\mathcal{G} = (N, E, W)$ w.r.t. a set of reference nodes N' and local reference node sets N_o for all nodes or objects $o \in N \cup \mathcal{D}$. For each $x, y \in N \cup \mathcal{D}$ the following property holds.*

$$\tilde{D}^*(\tilde{F}^{N_x}(x), \tilde{F}^{N_y}(y)) \geq d_{net}(x, y).$$

Proof. Let $x, y \in N \cup \mathcal{D}$. Since d_{net} is a metric, for each pair of reference nodes $r_i \in N_x$ und $r_j \in N_y$ the following holds:

$$
\begin{aligned}
\tilde{D}^*(\tilde{F}^{N_x}(x), \tilde{F}^{N_y}(y)) &= M_{i_i, i_j} + \tilde{F}_i^{N_x}(x) + \tilde{F}_j^{N_y}(y) \\
&= d_{net}(r_i, r_j) + d_{net}(x, r_i) + d_{net}(y, r_j) \\
&= d_{net}(x, r_i) + d_{net}(r_i, r_j) + d_{net}(r_j, y) \\
&\geq d_{net}(x, r_j) + d_{net}(r_j, y) \\
&\geq d_{net}(x, y)
\end{aligned}
$$

Let us note that for directed graphs, d_{net} is no metric distance function (it is not symmetric). However, lower and upper bounds for the network distance on the 2-level embedding can be defined analogously also for directed graphs. Since in this case $d_{net}(x, y)$ is not symmetric we have to distinguish between the two traversal directions ($x \rightarrow y$ and $y \rightarrow x$) for which we have to take the corresponding directed edge weights into account.

3.5 From Two-Level to Multi-level Network Embeddings

The proposed 2-level reference node embedding scales very well even for very large graphs as far as the number K of relevant reference nodes for each object is considerably small. We will see this in our experiments (cf. Section 5). However, for very large graph networks that require a high reference node density, K can again be large. In addition, the storage cost for the reference node matrix $\mathcal{M}'$ obviously scale quadratic with the number of global reference nodes N'. In such scenarios, $\mathcal{M}'$ will no longer fit into main memory. This will increase the query processing time dramatically since for determining the distance approximations, we steadily need random access to the elements in $\mathcal{M}'$.

To solve this problem, we propose to introduce further embedding levels, i.e. to generalize the 2-level reference node embedding to a multi-level reference node embedding. Such a multi-level embedding can be constructed bottom-up starting with a 2-level embedding. The reference node set is partitioned at each level. Each of these partitions is assigned to one of the objects/nodes in the network as the corresponding relevant reference node set. The reference node partitions may overlap and neighboring nodes/objects should get nearly the same reference node partition assigned. For each partition, a complete reference node graph is constructed on the second embedding level. Thereby, the size of each partition should be chosen such that the reference nodes on each level are completely connected (i.e. each reference node is reachable from each other) when combining all reference node graphs on a level. Furthermore, the reference node matrices of the corresponding reference node graphs should fit into a memory page. From the resulting reference node graph on embedding level i a predefined number of nodes is selected to form an embedding level $i + 1$ analogously. This procedure is iterated until only one "graph" remains that is complete and fits into main memory. The idea is illustrated in Figure 4.

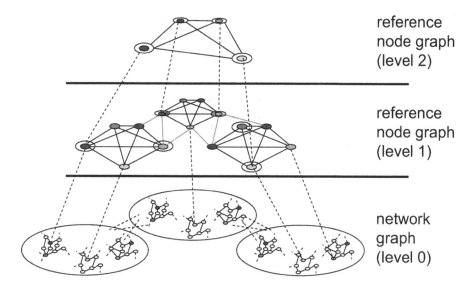

Fig. 4. Schema of a multi-level reference node embedding

3.6   Choosing the Reference Node Set

It is easy to see that the choice of the reference node set affects the quality of the distance approximations. The problem of how to choose the reference nodes is two-fold. First, the global set N' has to be chosen adequately, and second, for each node or object $o \in N \cup \mathcal{D}$, the local set of relevant reference nodes N'_o needs to be selected, too. Obviously, the choice of the global set N' affects the possibilities for the selection of the local sets N'_o.

For a flat embedding, $D(x, y) = d_{net}(x, y)$ for two objects $x, y \in \mathcal{D}$, i.e. the approximation error is zero, if there is at least one $r_i \in N'$ such that either $x \in \mathcal{P}_{best}(r_i, y)$ or $y \in \mathcal{P}_{best}(r_i, x)$, where $\mathcal{P}_{best}(a, b)$ denotes the shortest path between nodes/objects a and b. On the other hand, the approximation obtained from reference node r_i is very coarse if r_i is located such that $d_{net}(x, r_i) \approx d_{net}(y, r_i)$ and $r_i \in \mathcal{P}_{best}(x, y)$. On the other hand, $D^*(x, y) = d_{net}(x, y)$ for two objects $x, y \in \mathcal{D}$, i.e. the approximation error is zero, if there is at least one $r_i \in N'$ such that $r_i \in \mathcal{P}_{best}(x, y)$. The more disconnected a reference node r_i is from $\mathcal{P}_{best}(x, y)$, the coarser is the approximation obtained from this reference node. The same considerations hold true for a 2-level or even for a multi-level reference node embedding.

Intuitively, the probability that these conditions for accurate distance approximations are fullfilled is higher if the reference nodes are close to the objects. Thus, if the distribution of the objects in the network is unknown, the set of global reference nodes N' should be evenly distributed over the network because then, the probability that all objects have at least one reference node in their local vincinity is maximized. In addition, for each node or object $o \in N \cup \mathcal{D}$, the set of relevant reference nodes N'_o should be selected as the K-nearest reference nodes of o from N'. This further ensures that the reference nodes N_o' are in proximity of o.

On the other hand, if information about the object distribution, the characteristics of object movement, and/or the distribution of query locations is known, the set of local relevant reference nodes N_o' for all nodes or objects $o \in N \cup \mathcal{D}$ could be selected individually. Ideally, hot spots, i.e. nodes that are often part of shortest paths during query execution, should be chosen as reference nodes. Since the set N_o' of reference nodes relevant for node/object o can be dynamically adjusted rather easily, we can even learn the location of hot spots by monitoring for each node how often it is visited during a shortest path computation.

3.7 Efficient Shortest-Path Computation

Analogously to the flat embedding, our hierarchical reference node embedding can be successfully applied as heuristics for the A*-search algorithm to compute the true network distance. The A*-search method is a special case of a best-first search algorithm using heuristics. In contrast to the Dijkstra algorithm whose search is only backward-oriented (blind search), the A*-search method is an informed search method, i.e. it also looks in the forward direction using a lower bounding network distance approximation, e.g. the Euclidean distance. Here, we propose to use the distance function $\tilde{D}$ of the vector space resulting from our multi-level K-closest reference node embedding as estimator function. In addition, we can use the upper bounding distance estimation $\tilde{D}^*$ in order to identify the branches of the search tree that do not need to be expanded. Since these branches do not need to be considered throughout the remaining search steps, we do not need to maintain them which reduces the memory cost.

4   Multi-Step Query Processing

The upper and lower bounding distance estimations introduced above can be used in a filter step as well as for speeding-up the refinement step using the modified A* algorithm. In the following, we present the multi-step DRQ and kNNQ using our embedding function $F^{N'}$ implementing a multi-level K-closest reference node embedding. As mentioned above, for static objects, the graph embedding has to be performed only once in a preprocessing step before any query is launched. The re-embedding for dynamic objects can be computed rather efficiently on the fly (cf. Section 3).

The DRQ over the embedded objects and nodes can directly prune all objects for which the distance approximation $\tilde{D}$ is greater than ε as true drops without refining them. All objects are added to the result list if the distance estimation $\tilde{D}^*$ is lower or equal to ε. Only the remaining candidates need to be refined.

For the kNNQ we use the algorithm proposed in [18] which is shown to be optimal w.r.t. the number of candidates that are refined. The algorithm is illustrated in Figure 5. It uses a ranking of the objects in ascending order of their lower bounding filter distance $\tilde{D}$ and performs an iterative refinement as long as the lower bound of the next object in the ranking is smaller or equal to the current K-th nearest neighbor distance.

---

kNNQ$(q,k,\mathcal{G})$

SortedList *results,candidates*;
initialize ranking := $RQ(q,\mathcal{D})$;
candidates←first k objects from *ranking*;
$d_{min} = k^{th}$ smallest $D(F^{N'}(q), F^{N'}(o))$ of $o \in$ *candidates*;
$d_{max} = k^{th}$ smallest $D^*(F^{N'}(q), F^{N'}(o))$ of $o \in$ *candidates*;
$d_{f\_next} = D(F^{N'}(q), F^{N'}(o))$ of o=*ranking*.top_element;
do {
    update d_{min}, d_{max}, and d_{f_next};
    if $d_{min} \geq d_{f\_next}$ **then**
      candidates.add(*ranking*.top_element);
      update d_{min}, d_{max}, and d_{f_next};
    for all $c \in$ *candidates* **do**
      if $D^*(F^{N'}(q), F^{N'}(c)) < d_{min}$ **then** add c to *result*;
      if $D(F^{N'}(q), F^{N'}(c)) > d_{max}$ **then** prune c;
    if $|results|+|candidates| > k \vee d_{f\_next} \leq d_{max}$ **then**
      for all $c \in$ *candidates* with $D(F^{N'}(q), F^{N'}(c)) \leq d_{min}$
                   $\wedge \ d_{max} \leq D^*(F^{N'}(q), F^{N'}(c))$ **do**
      if $d_{net}(q, c) \leq d_{k-nn}(q, result)$ **then** add c to *result*;
    else add all remaining $c \in$ *candidates* to *result*;
} **while** $(d_{f\_next} \leq d_{max} \vee |candidates| > 0)$
return *result*;

Fig. 5. The kNNQ algorithm

5   Experimental Evaluation

Due to space limitations, we focus on a two-level embedding in our experiments. We used real road networks of San Joaquin County ("TG", 18,300 nodes) and San Francisco ("SA", 175,000 nodes). The network objects were simulated through randomized samples of the graph nodes. The graph was stored on disk implementing the approach proposed in [8] using R*-trees with a block size of 8 KB and an average storage load of 70% each. The R*-trees are used to manage the nodes, the edges and the street segments in form of polylines. An embedding vector is a further attribute of a node. The reference nodes were chosen by spatially ordering all graph nodes along a Hilbert curve. We then uniformly distributed the reference nodes along this curve. Datasets without an embedding are denoted by REF, flat embeddings by 1RNE and two-level embeddings by 2RNE. All experiments were performed on a workstation featuring a 1.8 GHz CPU, 2GB RAM, a random disk with page access time of 6 ms, and a transfer rate of 86MB/s. The cache size was set to 5% of the dataset size. In all experiments, we performed 1,000 random queries and averaged the results.

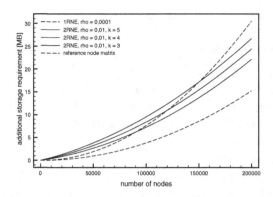

Fig. 6. Size of the embedding, w.r.t. size of the network graph

5.1 Storage Requirements

Figure 6 shows the storage requirements of different embeddings. We compared a flat embedding (1RNE) with an *object density* of $rho = 0,0001$ (*rho* = # objects / # graph nodes) with several two-level embeddings (2RNE) using different numbers K of *relevant reference nodes*[1] per node and object. For each 2RNE we assumed a considerably higher object density of $rho = 0,01$. In addition, the size of the reference node matrix $\mathcal{M}'$ is depicted. In the following, M denotes the size of the reference node matrix $\mathcal{M}'$, i.e. about $\sqrt{M}$ (1st-level) reference nodes are used for $\mathcal{M}'$. It can be observed, that using a 2RNE we can use approximately two orders of magnitude more reference nodes compared to a 1RNE with quite similar storage cost. Obviously, using more global reference nodes increases the quality of the distance approximations, and, thus boosts the overall performance.

5.2 Multi-step Query Processing

In this experiment, we assumed a capacity of 80 byte per embedded node and object of the network graph. We used $K = 3$ relevant reference nodes per node and object resulting in an overall number of 400 and 700 reference nodes for TG and SF, respectively. The reference node distance matrix $\mathcal{M}'$ thus required 0.61 MB (TG) and 1,88 MB (SF) RAM, respectively. Because of its small size, the distance matrix $\mathcal{M}'$ was kept in main memory in all experiments. The results of distance range query processing on the SF dataset are depicted in Figure 7. The most important advantage of the 2RNE over the 1RNE approach is that it can use significantly more reference nodes which increases the quality of the filter distances. Especially for less selective queries, the filter selectivity is significantly better than using a 1RNE with comparable storage requirements. The scalability of the 2RNE approach w.r.t. the object density is linear similar to the 1RNE approach. The results on the TG dataset are similar (not shown due to space limitations). In summary, the results show that the 2RNE approach is superior to 1RNE especially

---

[1] The number K of relevant reference nodes corresponds to the number of reference nodes assigned to each graph node on each embedding level.

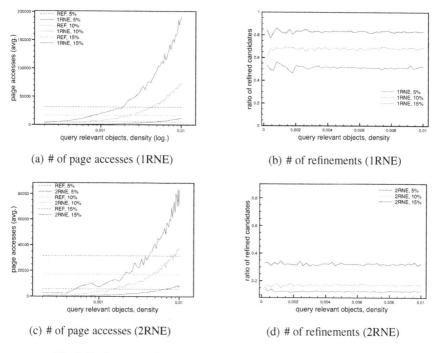

(a) # of page accesses (1RNE)

(b) # of refinements (1RNE)

(c) # of page accesses (2RNE)

(d) # of refinements (2RNE)

Fig. 7. Performance of DRQ w.r.t. the object density *rho* on SF dataset

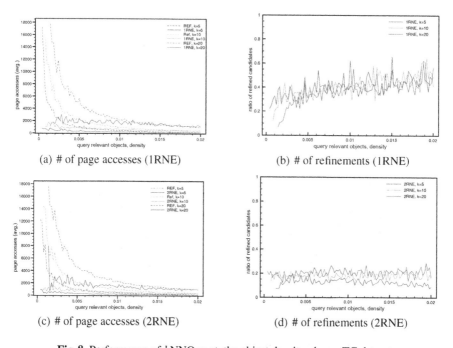

(a) # of page accesses (1RNE)

(b) # of refinements (1RNE)

(c) # of page accesses (2RNE)

(d) # of refinements (2RNE)

Fig. 8. Performance of *k*NNQ w.r.t. the object density *rho* on TG dataset

on large graphs with comparable storage requirements. The results of kNNQ processing on the TG dataset are depicted in Figure 8. On the SF dataset we made similar observations (not shown due to space limitations). Here, especially for high object densities, the 2RNE approach outperforms the 1RNE approach on large datasets. Again, the higher number of reference nodes that can be used in the 2RNE approach yields a significantly better distance approximation.

5.3 Shortest Path Algorithm

In this experiment, we concentrate on the cost required for the refinement step, i.e. the cost of the exact distance computation. The benefits of our novel distance approximations for computing the shortest path can be observed in Figure 9. A sample shortest path (marked in blue) is computed by four different methods. For each method, the corresponding search space containing all visited edges is marked in orange. While the Dijkstra algorithm (cf. Figure 9(a)) needs to access nearly the complete displayed part of the graph, the A* algorithm using the Euclidean distance as lower bounding distance estimation (cf. Figure 9(b)) requires a considerably reduced search space. The novel

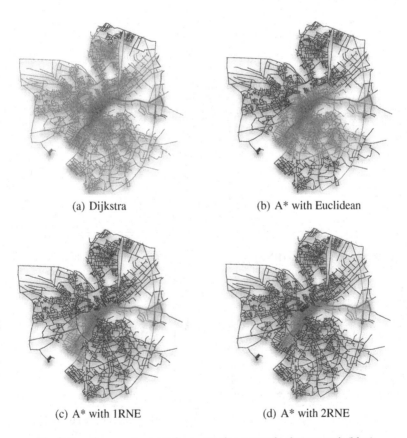

(a) Dijkstra   (b) A* with Euclidean

(c) A* with 1RNE   (d) A* with 2RNE

Fig. 9. Search space (orange) for computing a sample shortest path (blue)

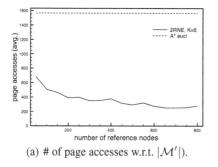
(a) # of page accesses w.r.t. $|\mathcal{M}'|$.

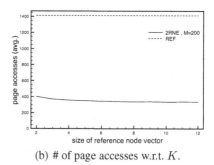
(b) # of page accesses w.r.t. K.

Fig. 10. Performance evaluation of the shortest path algorithm

A* algorithm with upper and lower bounding distance estimations derived from a flat reference node embedding with $|N'| = 50$ reference nodes (cf. Figure 9(c)) further reduces the search space. Finally, our 2-level reference node embedding with $|N'| = 100$ global reference nodes and $K = 5$ relevant reference nodes per object (cf. Figure 9(d)) requires the smallest search space of all competitors.

Figures 10(a) and 10(b) show the average number of disk page accesses required for one distance computation between two objects (nodes) w.r.t. the size of the reference node matrix $\mathcal{M}'$ and the number K of reference nodes used for the two-level embedding. As can be observed, our novel shortest path algorithm again significantly outperforms the A* search using the Euclidean distance and Dijkstra (not shown for clarity reasons). We also observed, that increasing the number K of relevant reference nodes per object does not significantly increase the quality of the distance approximation (not shown due to space limitations). Our experiments suggest that $K = 5$ is a reasonable choice despite it is a rather small value. As a consequence, the storage requirements for each object are rather low. In turn, this allows us to use a higher number of global reference nodes.

5.4 Comparison with Other Approaches

Finally, we compare the performance of our approach to that of state-of-the art approaches. We chose the distance signature (DS) approach [11] as comparison partner because it outperforms other methods such as the network voronoi diagram [15]. The DS method was parameterized as described in [11]. For the comparison, we computed a two-level (2RNE) embedding for $M = 256$ and $K = 5$. For an object density $rho = 0.01$, the 2RNE embedding occupies half of the space required by DS. Please note that the object density linearly influences the memory footprint of our technique, in contrast to the DS approach where the relationship between object density and memory consumption is quadratic. The DRQ experiments in Figure 11 show that the signature approach is significantly outperformed by our approach. The two-level embedding is able to outmatch DS although it occupies significant less memory, i.e. needs far less precomputed distance information.

In summary, our experimental evaluation empirically showed the following facts: First, the integration of our novel upper and lower bounding distance approximations

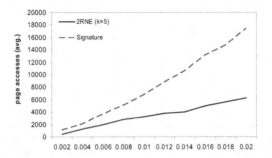

Fig. 11. RNE vs. Signature, w.r.t. the object density *rho*

into the A* algorithm is superior to state-of-the-art methods for shortest path computation. Second, our novel two-level (or even multi-level) embedding outperforms the flat embedding on large graphs because it allows an even more accurate lower and upper bounding distance approximation.

6   Conclusions

We proposed a hierarchical graph embedding of very large networks that is suitable for static and dynamic objects. From the embedding, we derived accurate upper and lower bounds for the network distance that can be used to implement a filter/refinement architecture for similarity search in large traffic networks. In addition, our embedding allows an acceleration of the refinement step by applying an informed A*-search using our novel distance approximations. Our experiments show that our novel approach outperforms a simple flat embedding and other existing competitors in terms of pruning power in the filter step and overall performance. Furthermore, it turned out that our informed search in the refinement step is much more efficient than comparable approaches due to a dramatically reduced search space.

References

1. Dijkstra, E.W.: A Note on Two Problems in Connection with Graphs. Numerische Mathematik 1, 269–271 (1959)
2. Corman, T.H., Leiserson, C.E., Riverst, R.L.: Introduction to Algorithms. MIT Press, Cambridge (1990)
3. Kung, R., Hanson, E., Ioannidis, Y., Sellis, T., Shapiro, L., Stonebraker, M.: Heuristic Search in Data Base Systems. Expert Database Systems (1986)
4. Agrawal, R., Dar, S., Jagadish, H.: Direct Transitive Closure Algorithms: Design and Performance Evaluation. TODS 15(3) (1990)
5. Ioannidis, Y., Ramakrishnan, R., Winger, L.: Transitive Closure Algorithms Based on Graph Traversal. TODS 18(3) (1993)
6. Jung, S., Pramanik, S.: HiTi Graph Model of Topographical Roadmaps in Navigation Systems. In: Proc. Int. Conf. on Data Engineering (ICDE 1996) (1996)

7. Köhler, E., Möhring, R.H., Schilling, H.: Acceleration of Shortest Path and Constrained Shortest Path Computation. In: Nikoletseas, S.E. (ed.) WEA 2005. LNCS, vol. 3503, pp. 126–138. Springer, Heidelberg (2005)
8. Papadias, D., Zhang, J., Mamoulis, N., Tao, Y.: Query Processing in Spatial Network Databases. In: Proc. Int. Conf. on Very Large Databases (VLDB 2003) (2003)
9. Shahabi, C., Kolahdouzan, M., Sharifzadeh, M.: A Road Network Embedding Technique for k-Nearest Neighbor Search in Moving Object Databases. Geoinformatica 7(3), 255–273 (2003)
10. Linial, N., London, E., Rabinovich, Y.: The geometry of graphs and some of its algorithmic applications. In: Proc. IEEE Symp. Foundations of Computer Science (1994)
11. Hu, H., Lee, D.L., Lee, V.C.S.: Distance Indexing on Road Networks. In: Proc. Int. Conf. on Very Large Databases (VLDB 2006) (2006)
12. Goldberg, A.V., Werneck, R.F.: Computing Point-to-Point Shortest Paths from External Memory. In: Proc. of the 7th WS on Algorithm Engineering and Experiments (ALENEX). SIAM, Philadelphia (2005)
13. Goldberg, A.V., Kaplan, H., Werneck, R.F.: Reach for A*: Efficient point-to-point shortest path algorithms'. In: Proc. of the 8th WS on Algorithm Engineering and Experiments (ALENEX). SIAM, Philadelphia (2006)
14. Kriegel, H.P., Kröger, P., Kunath, P., Renz, M., Schmidt, T.: Proximity Queries in Large Traffic Networks. In: Proc. 15th Int. Symposium on Advances in Geographic Information Systeme (ACM GIS 2007), Seattle, WA (2007)
15. Kolahdouzan, M., Shahabi, C.: Voronoi-Based K Nearest Neighbor Search for Spatial Network Databases. In: Proc. Int. Conf. on Very Large Databases (VLDB 2004) (2004)
16. Kolahdouzan, M., Shahabi, C.: Continuous K-Nearest Neighbor Queries in Spatial Network Databases. In: Proc. of STDBM 2004 (2004)
17. Cho, H.J., Chung, C.W.: An efficient and scalable approach to cnn queries in a road network. In: Proc. Int. Conf. on Very Large Databases (VLDB 2005) (2005)
18. Kriegel, H.P., Kröger, P., Kunath, P., Renz, M.: Generalizing the Optimality of Multi-Step k-Nearest Neighbor Query Processing. In: Papadias, D., Zhang, D., Kollios, G. (eds.) SSTD 2007. LNCS, vol. 4605, pp. 75–92. Springer, Heidelberg (2007)

Monitoring Aggregate k-NN Objects in Road Networks

Lu Qin[1], Jeffrey Xu Yu[1], Bolin Ding[1], and Yoshiharu Ishikawa[2]

[1] The Chinese University of Hong Kong, China
{lqin,yu,blding}@se.cuhk.edu.hk
[2] Nagoya University, Japan
ishikawa@itc.nagoya-u.ac.jp

Abstract. In recent years, there is an increasing need to monitor k nearest neighbor (k-NN) in a road network. There are existing solutions on either monitoring k-NN objects from a single query point over a road network, or computing the snapshot k-NN objects over a road network to minimize an aggregate distance function with respect to multiple query points. In this paper, we study a new problem that is to monitor k-NN objects over a road network from multiple query points to minimize an aggregate distance function with respect to the multiple query points. We call it a continuous aggregate k-NN (CANN) query. We propose a new approach that can significantly reduce the cost of computing network distances when monitoring aggregate k-NN objects on road networks. We conducted extensive experimental studies and confirmed the efficiency of our algorithms.

1 Introduction

With the development of positioning technologies such as the Global Positioning System (GPS), many applications are developed in transportation domains by taking advantages of monitoring object movements in road networks where the position and distance of objects are constrained by spatial networks. An important type of these queries is a k nearest neighbor (k-NN) query, which is widely used in location-based services, traffic monitoring, emergency management. Existing solutions focused on either monitoring k-NN objects over a road network from a single query point (observation point) [1], or computing the snapshot k-NN objects over a road network to minimize an aggregate distance function with respect to the multiple query points [2]. In this paper, we study a new problem that is to monitor k-NN objects over a road network from multiple query points to minimize an aggregate distance function with respect to multiple query points. We call it a continuous aggregate k-NN (CANN) query. In brief, it deals with the network distance instead of Euclidean distance, and it monitors the top-k objects, where an object is ranked based on an aggregate function value of the distances between the object and multiple query points. As an example, consider people in n companies/organizations need to schedule meetings in downtown frequently. The room availabilities in hotels and restaurants is monitored, and the best place is selected to reduce the total travel time for people to meet.

The main difficulties for processing CANN query are as follows. First, when there are a large number of objects in the road network or there are a large number of CANN

B. Ludäscher and Nikos Mamoulis (Eds.): SSDBM 2008, LNCS 5069, pp. 168–186, 2008.

queries, the cost of computing network distances becomes the bottleneck. Second, an object is ranked in the road network based on an aggregate function value in terms of the network distances to a set of query points. Unlike computing a CANN query for a single query point in the road network where the order of visiting edges can be determined using an expansion tree from the query point, computing the CANN query from multiple query points makes it difficult to find an order of visiting edges.

The main contributions of this paper are summarized below. (1) We study a new problem of processing the continuous aggregate nearest neighbor queries (CANN) over large road network. To the best of our knowledge, this is the first attempt to study this problem. (2) We propose new approaches that do not need to expand tree to compute CANN queries. Our approach can reduce the cost of computing network distances significantly. (3) We conducted extensive performance studies, and confirmed the efficiency of our new approaches.

The rest of the paper is organized as follows. Section 2 gives the problem statement. Section 3 introduces two existing solutions. Section 4.2 discusses our new approaches followed by discussions on implementations in Section 5. Section 6 shows our experimental results. The related work is given in Section 7. Finally, Section 8 concludes this paper.

2   Problem Definition

Road Network is an undirected weighted connected graph, $G(V, E)$, where V is a set of nodes (road intersections), and E is a set of edges (roads). An edge, $e \in E$, connects two nodes n_i and n_j. A positive number, $len(e)$, denotes the length of the edge e. *(Data or Query) points* lie on edges of road network G. We use $pos_e(p)$ to denote the position of a point p on $e = (n_i, n_j)$ by the distance from point p to node n_i on edge e, provided $i < j$.

Network Distance: For two nodes $n_i, n_j \in V$, the network distance $d(n_i, n_j)$ is the length of the shortest path between n_i and n_j in the road network. The network distance between a point, p that lies on the edge $e = (n_i, n_j)$, and a node, n_k, is computed as $d(p, n_k) = min\{pos_e(p) + d(n_i, n_k), (len(e) - pos_e(p)) + d(n_j, n_k)\}$. For any two data points p and p', if p and p' are on different edges, their network distance is computed as $d(p, p') = min\{pos_e(p) + d(p', n_i), (len(e) - pos_e(p)) + d(p', n_j)\}$. Otherwise, $d(p, p')$ is $min \{|pos_e(p) - pos_e(p')|, pos_e(p) + d(p', n_i), (len(e) - pos_e(p)) + d(p', n_j)\}$.

Figure 1 shows a simple road network. There are 6 nodes and 6 edges. The number in the brackets under each edge e_i denotes its length $(len(e_i))$. For instance, e_4 is the edge that connects nodes n_3 and n_4, and the length of e_4 is $len(e_4) = 80$. In Figure 1, a data point is indicated by a cross. The position of a data point is marked in the brackets above it. For instance, p_3 lies on edge e_3, and its position is $pos_{e_3}(p_3) = 70$. The network distance between two nodes, n_1 and n_6, is $d(n_1, n_6) = 30 + 80 + 30 = 140$, along the shortest path $e_1 \rightarrow e_4 \rightarrow e_6$, the network distance between data point p_3 and node n_4 is $d(p_3, n_4) = min\{pos_{e_3}(p_3) + d(n_2, n_4), (len(e_3) - pos_{e_3}(p_3)) + d(n_6, n_4)\} = min\{70 + 110, 50 + 30\} = 80$, and the network distance between two data points, p_3

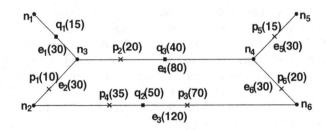

Fig. 1. Road Network

and p_2, that are on two different edges e_3 and e_4, is $d(p_3, p_2) = min\{pos_{e_3}(p_3) + d(n_2, p_2), (len(e_3) - pos_{e_3}(p_3)) + d(n_6, p_2)\} = min\{70 + 50, 50 + 90\} = 120$.

Problem Statement (CANN **Query**): Given a road network $G(V, E)$ and the set of data points (moving objects) $P = \{p_1, p_2, \cdots\}$ over $G(V, E)$. A continuous aggregate nearest neighbor query is denoted as CANN(Q, k, h), where $Q = \{q_1, q_2, \cdots\}$ is a set of fixed query points over $G(V, E)$, k is a positive number (> 0), and h is an aggregate function (sum, min, max). Here, for a data point, $p_i \in P$, $h(p_i) = h\{d(p_i, q_1), d(p_i, q_2), \cdots, d(p_i, q_{|Q|})\}$, regarding the query points Q. The CANN (Q, k, h) query is to monitor the top-k data points in P that has the smallest h function values while all data points are moving.

Consider a CANN(Q, k, sum) where $Q = \{q_1, q_2, q_3\}$, $k = 3$ against $G(V, E)$ (Figure 1). Here, $sum(p_1) = sum\{d(p_1, q_1), d(p_1, q_2), d(p_1, q_3)\} = 35 + 60 + 60 = 155$, $sum(p_2) = 155$, $sum(p_3) = 255$, $sum(p_4) = 200$, $sum(p_5) = 280$, and $sum(p_6) = 255$. The top-3 result is $\{p_1, p_2, p_4\}$.

3   Existing Solutions

While many recent researches have focused on continuous monitoring of nearest neighbors over dynamic objects, we first propose the solution for CANN query in road networks. Mouratidis et al.'s work in [1] is the one closest to ours. They gave two algorithms, IMA and GMA, to process continuous nearest neighbor queries over a road network, when there is a single query point, i.e., CANN(Q, k, h) where $|Q| = 1$ (a special case of CANN).

The incremental monitoring algorithm (IMA) retrieves the initial top-k data points using the shortest path expansion tree of the query point for a single CANN query. The group monitoring algorithm (GMA) groups multiple CANN queries that lie on the same edge, as a group, to process them together, based on IMA. IMA keeps expanding the tree and updating the top-k result until the next edge to be expanded has minimal distance that is no less than the kth distance in the current result. When data points move, the result for the query is maintained by incrementally expanding or shrinking the expansion tree.

Figure 2 shows an example to explain the expansion tree for CANN$(\{q_3\}, k, sum)$, where $k = 3$. Assume the current top-3 result is $\{p_1, p_2, p_5\}$, and the edges (called partial edges) that may partially affect the new top-3 results when data points move are

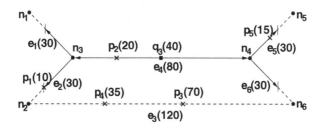

Fig. 2. Expansion tree and partial Edges

$P = \{e_1, e_2, e_5, e_6\}$. Suppose the data point p_1 moves out of the partial edges. IMA needs to expand the expansion tree from nodes n_3 and n_4 and retrieve all the data points on the edges in P to obtain the new top-k result $\{p_2, p_4, p_5\}$. In summary, when data points move, IMA does not need to recompute the result from scratch, for a CANN query, but it needs to retrieve the data points, that lie on the partial edges, which is time consuming. In this paper, we propose a new approach that does not need to use an expansion tree for CANN, and allows multiple query points in one CANN query.

4   A New Non-tree-expanding Approach

The high online processing cost for CANN queries dues to the frequent update of the expansion tree. In this paper, we propose a new approach that does not need an expansion tree. In brief, for a new CANN(Q, k, h) query registered, we construct a query graph, $G_Q(V_Q, E_Q)$, based on CANN and the road network G. The query graph, $G_Q(V_Q, E_Q)$, is static when processing CANN(Q, k, h) (no update is needed). It facilitates computing the value of aggregate function $h(p)$, for a data point $p \in P$. With the assistance of query graph G_Q, we can efficiently monitor the top-k results, when the data points move.

4.1   Query Graph Construction

The query graph $G_Q(V_Q, E_Q)$ facilitates computing the value of aggregate function $h(p)$ in CANN(Q, k, h), for a data point, $p \in P$. We require that, given the position of p on edge e, $pos_e(p)$, the value of aggregate function $h(p)$ can be computed efficiently. We first discuss the relationship between the distance function $d(q, p)$ / the aggregate function $h(p)$ and the position of p.

Distance function w.r.t. $pos_e(p)$**:** Consider a data point p on an edge $e = (n_i, n_j)$ in G_Q, and a query point $q \in Q$. The distance $d(p, q)$ between q and p can be specified as a function of $pos_e(p)$, denoted as $f_{e,q}$: $f_{e,q}(pos_e(p)) = d(p, q)$. We note that function $f_{e,q}(\cdot)$ is a continuous piecewise-linear function in the domain $[0, len(e)]$. We discuss the main idea behind $f_{e,q}(pos_e(p))$ followed by the discussion on how to compute it.

An example is illustrated in Figure 3(a) over the road network G (Figure 1). Take the edge $e_4 = (n_3, n_4)$ in G as an example. Its three functions, $f_{e_4,q_1}(pos_{e_4}(p))$, $f_{e_4,q_2}(pos_{e_4}(p))$, and $f_{e_4,q_3}(pos_{e_4}(p))$, for three different query points, q_1, q_2, and q_3,

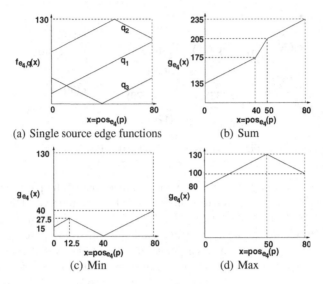

Fig. 3. Edge functions on e_4

are shown in Figure 3(a). Note: on x-axis, $[0, len(e_4)]$, is $pos_{e_4}(p)$, the distance from n_3. The curve of $f_{e_4,q_1}(pos_{e_4}(p))$ suggests that the shortest distance between q_1 and any data points p on e_4 should first go to the end node n_3 of e_4, and then go to p. The curve of $f_{e_4,q_2}(pos_{e_4}(p))$ suggests that the shortest distance between q_2 and any data points p on e_4 may come from two different ends of e_4 (from either n_3 or n_4). When the data point p is on the left side of $[0, len(e_4)]$ before the peak value of $f_{e_4,q_2}(pos_{e_4}(p))$, the shortest distance between q_2 and p should come from the end of n_3; when the data point p is on the right side of $[0, len(e_4)]$ after the peak value of $f_{e_4,q_2}(pos_{e_4}(p))$, the shortest distance between q_2 and p should come from the end of n_4.

Function $f_{e,q}(pos_e(p))$ can be computed as follows. Assume $e = (n_i, n_j)$, where $i < j$. With Dijkstra's single-source shortest-path algorithm, we obtain the shortest distance from q to every node in G. There are two cases.

i) q is not on edge e: If $|d(q, n_i) - d(q, n_j)| = len(e)$, $f_{e,q}(pos_e(p))$ is a 1-piece linear function of $pos_e(p) \in [0, len(e)]$. In this case, its 1-piece segment is $(0, f_{e,q}(0))$-$(len(e), f_{e,q}(len(e)))$, where $f_{e,q}(0) = d(q, n_i)$ and $f_{e,q}(len(e)) = d(q, n_j)$. Otherwise, $f_{e,q}(pos_e(p))$ is a 2-piece linear function, and its two linear segments are $(0, d(q, n_i))$-(x, y), and (x, y)-$(len(e), d(q, n_j))$, where x and y are computed as follows.

$$\begin{cases} x = \frac{d(q,n_j) - d(q,n_i) + len(e)}{2} \\ y = \frac{d(q,n_j) + d(q,n_i) + len(e)}{2} \end{cases} \quad (1)$$

ii) q is on edge e: Query point q split e into two parts, from n_i to q and from q to n_j respectively. Consider q as a node, function $f_{e,q}(pos_e(p))$ on each part shares high similarity to case i), thus we omit further explanation. The curve of $f_{e_4,q_3}(pos_{e_4}(p))$

shows such an example. But notice that function $f_{e,q}(pos_e(p))$ of $pos_e(p) \in [0, len(e)]$ may be a 3-piece linear function here. The 3-piece case happens only if q is on e.

From above discussions, we have the following lemma.

Lemma 1. $f_{e,q}(\cdot)$ is a continuous piecewise-linear function with at most 3 linear pieces on domain $[0, len(e)]$.

Aggregate function w.r.t. $pos_e(p)$: Since distance function $f_{e,q}(pos_e(p))$ is a continuous piecewise-linear function of $pos_e(p)$, given $\mathsf{CANN}(Q, k, h)$ and $Q = \{q_1, q_2, \cdots\}$, the aggregate function value for any data point p on edge e, regarding all query points, can also be specified as a continuous piecewise-linear function of $pos_e(p)$, denoted by

$$g_e(pos_e(p)) = h\{f_{e,q_1}(pos_e(p)), ..., f_{e,q_{|Q|}}(pos_e(p))\} \tag{2}$$

for $pos_e(p) \in [0, len(e)]$. Since $f_{e,q}(\cdot)$ has at most 3 linear pieces, $g_e(\cdot)$ has at most $O(|Q|)$ linear pieces.

Lemma 2. $g_e(\cdot)$ is a continuous piecewise-linear function with at most $O(|Q|)$ linear pieces on domain $[0, len(e)]$.

Reconsider the example in Figure 3(a) for the three query points, q_1, q_2, and q_3. The aggregate function on edge e_4 for $h = sum, min,$ and max, are shown in Figure 3(b), Figure 3(c), and Figure 3(d), respectively.

Constructing the query graph $G_Q(V_Q, E_Q)$: Given a $\mathsf{CANN}(Q, k, h)$ query over a road network $G(V, E)$, we define a query graph, $G_Q(V_Q, E_Q)$, to efficiently compute the value of $h(p)$ given $pos_e(p)$, the position of a data point p on edge e. The idea to construct G_Q is to segment edges in G, such that aggregate function $g_e(\cdot)$ w.r.t. $pos_e(p)$ is a 1-piece linear function within each segment.

Formally, suppose on an edge, $e = (n_i, n_j)$ in E, $g_e(\cdot)$ is a z-piece linear function, then e needs to be segmented into a sequence of edges, $(n_{k_0}, n_{k_1}), (n_{k_1}, n_{k_2}), \cdots,$ $(n_{k_{z-1}}, n_{k_z})$, where $n_i = n_{k_0}$ and $n_{k_z} = n_j$, such that $g_e(\cdot)$ is a 1-piece linear function on each segment $[pos_e(n_{k_{l-1}}), pos_e(n_{k_l})]$ $(1 \leq l \leq z)$. All such nodes n_{k_l}, for $0 \leq l \leq z$, will be included in V_Q, and all the segmented edges $(n_{k_{l-1}}, n_{k_l})$, for $1 \leq l \leq z$, will be included in E_Q. If $g_e(\cdot)$ is a 1-piece linear function, then there is no segmentation needed over an edge $e = (n_i, n_j)$ $(n_i, n_j$ are included in V_Q, and e is included in E_Q).

We explain how to segment an edge using an example (Figure 3(b)), for a $\mathsf{CANN}(Q, k, h)$ where $Q = \{q_1, q_2, q_3\}$, and $h = sum$. Consider edge $e_4 = (n_3, n_4)$, as shown in Figure 3(b), its aggregate edge function is a continuous 3-piece-segment linear function. Therefore, we add two new nodes into query graph G_Q, denoted, n_{k_1} and n_{k_2} at position 40 and 50 on the x-axis as shown in Figure 3(b). Note: 40 and 50 are the distance from n_3. $e_4 = (n_3, n_4)$ will be segmented into three edges, $(n_3, n_{k_1}), (n_{k_1}, n_{k_2}),$ and (n_{k_2}, n_4), in G_Q. Each of the three edges is associated with a 1-piece linear aggregate function.

It is important to note that in $G_Q(V_Q, E_Q)$, every edge is associated with a 1-piece linear function (a piece of $g_e(\cdot)$). We can compute the value of the aggregate function for any data point in any edge with the help of G_Q efficiently. Consider an edge $(n_{k_{l-1}}, n_{k_l})$

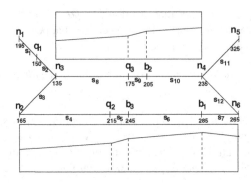

Fig. 4. Query graph

in G_Q, which is an edge segment of an edge e in G. Let $x_{l-1} = pos_e(n_{k_{l-1}})$ and $x_l = pos_e(n_{k_l})$. Let $y_{l-1} = g_e(x_{l-1})$ and $y_l = g_e(x_l)$ be the aggregate function values at nodes $n_{k_{l-1}}$ and n_{k_l}. When the position of a data point p, $pos_e(p)$, is within $[x_{l-1}, x_l]$, since $g_e(\cdot)$ is a 1-piece linear function on $[x_{l-1}, x_l]$, the aggregate function value at point p can be computed as:

$$g_e(pos_e(p)) = y_{l-1} + \frac{(y_l - y_{l-1}) \cdot (pos_e(p) - x_{l-1})}{(x_l - x_{l-1})}. \tag{3}$$

Figure 4 shows a query graph, $G_Q(V_Q, E_Q)$ over the road network G (Figure 1), for a CANN($\{q_1, q_2, q_3\}, k, sum$) query. There are totally 12 edges in G_Q, and each of them is marked as s_i for $1 \leq i \leq 12$. In addition to the original 6 nodes in $G(V, E)$, n_j, for $1 \leq j \leq 6$, there are 6 nodes $q_1 - q_3$ (for the three query points), and $b_1 - b_3$, which segment edges in E into linear pieces. The number below each node denotes the g_e value. The relationship between the the aggregate edge functions and the two horizontal edges are illustrated in Figure 4.

Lemma 3. *The time complexity for the construction of query graph $G_Q(V_Q, E_Q)$ is $O((n \cdot \log n + m \cdot \log |Q|) \cdot |Q|)$, where $n = |V|$ and $m = |E|$, given graph $G(V, E)$.*

Proof. For each query point q in Q, the complexity to find the distances from source q to every other node in G is $O(n \cdot \log n + m)$. In sum, we need $O((n \cdot \log(n) + m) \cdot |Q|)$ time. Moreover, since $g_e(\cdot)$ has at most $O(|Q|)$ linear pieces (Lemma 2), $|V_Q|$ and $|E_Q|$ are both bounded by $O(|Q| \cdot m)$. To segment an edge $e \in E$ into a sequence of edges in E_Q, we need $O(|Q| \log |Q|)$ time (sort all the linear pieces and scan them). Therefore, the total time complexity is $O((n \cdot \log n + m \cdot \log |Q|) \cdot |Q|)$.   □

4.2   Basic Top-k Monitoring Algorithm

In this subsection, we introduce our basic algorithm to monitor the top-k result for a set of CANN queries, $\{C_1, C_2, \cdots\}$, where $C_i = $ CANN(Q_i, k_i, h_i), over a road network G with data point set P.

For each query, C_i, the query graph is denoted as $G_{Q_i}(V_{Q_i}, E_{Q_i})$. Because of the property of query graphs we discussed in the previous subsection (recall Lemma 2), in

Algorithm 1. $IRC(C_i)$

1: $C_i.top \leftarrow \emptyset$; $C_i.k \leftarrow +\infty$;
2: $e \leftarrow head(C_i.E)$;
3: **while** $e \neq \emptyset$ **and** $low(e) \leq C_i.k$ **do**
4:     update $C_i.top$ and $C_i.k$ with data points on e;
5:     $e \leftarrow next(C_i.E)$;

---

the following part, we can assume the aggregate function value at a given data point p w.r.t. query C_i can be computed in constant time (according to Equation (3)).

All edges in E_{Q_i} are sorted in the ascending order of the aggregate function lower bounds within the edges. The sorted edge list is denoted by $C_i.E$. A pointer is associated with the ordered list $C_i.E$, and four operations are defined: i) $head(C_i.E)$ – set the pointer to the first edge in $C_i.E$ and return this edge; ii) $current(C_i.E)$ – return the edge pointed by the pointer currently; iii) $next(C_i.E)$ – move the pointer to the next edge and return this edge (or return $emptyset$ if the pointer points to the end of $C_i.E$); iv) $prev(C_i.E)$ – move the pointer to the previous edge and return this edge.

Initial Top-k Result Computation: The algorithm IRC (Algorithm 1) computes the top-k_i data points for a query C_i. In line 1, $C_i.top$, used to keep the set of the top-k_i data points for C_i, is initialized as empty; $C_i.k$, used to record the k_i-th smallest aggregate value of the data points kept in $C_i.top$, is initialized as $+\infty$. In line 2, $head(C_i.E)$ returns the first edge in $C_i.E$. In the while statement (line 3-6), it computes the top-k_i data points for C_i by scanning the ordered list $C_i.E$. In line 3, $e \neq \emptyset$ means $C_i.E$ has not been scanned to the end yet, and $low(e)$ denotes the aggregate function's lower bound within the edge e.

The case $low(e) \leq C_i.k$, called *edge e is influenced*, indicates that there may be some data points on e, which can be included in $C_i.top$. In this case, the top-k list ($C_i.top$) and the k_i-th smallest aggregate value in $C_i.top$ are updated using all the data points on the edge e (line 4).

Figure 5 shows an example over the road network G (Figure 1), for the query $C_i = $ CANN($\{q_1, q_2, q_3\}, 3, sum$). The label for each segment, s_l, for $1 \leq l \leq 12$ is illustrated in Figure 4. The x-axis shows the aggregate function values and the y-axis shows the list of edges $C_i.E$. All edges are listed in ascending order of the aggregate function lower bound on them, and each data point is marked as a cross in edges. Suppose the current set of data points is $P = \{p_1, p_2, \cdots, p_7\}$($p_7$ that lies on s_9 is not drawn on Figure 1). After visiting edges from s_2 to s_4, the data points, p_1, p_2 and p_4, are added to $C_i.top$. In the next iteration, the edge s_9 is visited. Note: s_9 is on edge $e_4 = (n_3, n_4)$ over the road network G from the position 40 to 50. On position 40 and 50, its aggregate function values are 175 and 205, respectively. This information is recorded in $C_i.E$. Here, p_7 is over s_9, and therefore on e_4 in the road network G. Note: $pos_{e_4}(p_7) = 45$. IRC computes its value, for p_7, $175 + \frac{(205-175) \times (45-40)}{50-40} = 190$, which is smaller than the current $C_i.k = 200$ for the data point p_4. Therefore, p_4 is removed from $C_i.top$ and p_7 is added. The value $C_i.k$ is updated to be 190. Then, when visiting the next edge s_{10}, the smallest value is 205 which is larger than $C_i.k = 190$, and it stops. The top-3 for C_i is then $C_i.top = \{p_1, p_2, p_7\}$.

Algorithm 2. MTR

1: let P_{del} be the set of removed data points;
2: let P_{ins} be the set of added data points;
3: **for** every data point p in P_{del} **do**
4:     suppose p lies on edge e;
5:     delete p from e (using an object index);
6:     **for** every C_i in that is influenced by e **do**
7:         **if** p in $C_i.top$ **then**
8:             delete p from $C_i.top$;
9:             $C_i.k \leftarrow +\infty$;
10: **for** every data point p in P_{ins} **do**
11:     suppose p lies on edge e;
12:     insert p into e (using object index);
13:     **for** every C_i that is influenced by e **do**
14:         update $C_i.top$ and $C_i.k$ using p;
15: **for** every C_i **do**
16:     **if** $C_i.k$ is greater than its previous value **then**
17:         $IRC(C_i)$;

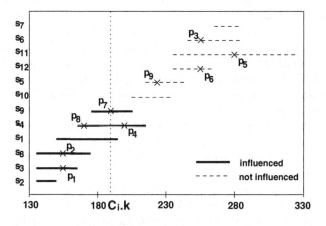

Fig. 5. Example for IRC and MTR

Monitor Top-k Result: Algorithm 2 shows top-k monitoring for a list of CANN queries. Here, the movement of a data point is considered as: first to delete it from P; then to insert a new data point into P. Let the set of deleted data points and the set of newly inserted data points be P_{del} and P_{ins}, respectively. (line 1-2). In Algorithm 2, in the first for statement (line 3-9), it updates $C_i.top$ if the deleted data points affects the top-k_i results. In the second for statement (line 10-14), it updates $C_i.top$ if the inserted data points affects the top-k_i results. In the first two for-statement, there is no need to scan $C_i.E$. In the third for-statement (line 15-17), if $C_i.k$ is changed and is greater than its previous $C_i.k$ value, it calls $IRC(C_i)$ to recompute the top-k_i results.

Reconsider the example (Figure 5) over the road network G (Figure 1), for the query $C_i = \mathsf{CANN}(\{q_1, q_2, q_3\}, 3, sum)$. First, suppose p_9 that lies on s_5 is inserted. The

Algorithm 3. $ForwardUpdating(C_i)$

1: $e \leftarrow current(C_i.E)$;
2: **while** $e \neq \emptyset$ **and** $low(e) \leq C_i.k$ **do**
3:     update $C_i.top$, $C_i.k$, $C_i.can$ with data points on e;
4:     $e \leftarrow next(C_i.E)$;

insertion of p_9 does not change the current top-3 results for C_i, as shown in Figure 5. Second, suppose p_7 is deleted which is in $C_i.top$. It leads to invoke $IRC(C_i)$ to recompute the top-k result. The new result is $C_i.top = \{p_1, p_2, p_4\}$. Then, suppose p_8 (lies on s_4) is inserted, which lies on the influenced edges (solid lines). It does not request recomputation. The new result is $C_i.top = \{p_1, p_2, p_8\}$.

4.3 Bidirectional Top-k Monitoring Algorithm

There are two drawbacks in the MTR algorithm. First, it needs to recompute top-k, for C_i, when $C_i.k$ increases (line 16-17) in MTR, which is time consuming. Second, it may scan some edges in $C_i.E$ which is unnecessary. In this section, we introduce a new incremental monitoring algorithm, to avoid the two drawbacks. The new algorithm keeps an additional structure called *candidate list*, denoted as $C_i.can$, for query C_i, which always stores the points lies on the influenced edges, but not in $C_i.top$. These points may be included in $C_i.top$, when some points in $C_i.top$ are deleted. As an example, consider Figure 6, for $P = \{p_1, p_2, \cdots, p_{10}\}$ (here p_7 to p_{10} is different from those in Figure 5) for a query $C_i = $ CANN ($\{q_1, q_2, q_3\}$, 4, *sum*). Suppose $C_i.top = \{p_1, p_2, p_9, p_4\}$ and $C_i.can = \{p_7, p_{10}\}$. Below, we give two procedures, namely forward updating and backward updating, followed by the introduction to the new monitoring algorithm.

Forward Updating: As shown in Algorithm 3, this procedure is similar to that of IRC (Algorithm 1). The differences are as follows. First, it does not need the initialization

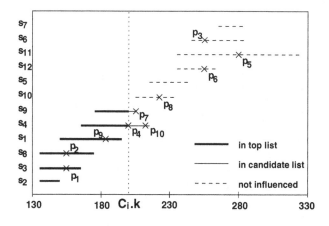

Fig. 6. Example for BUA

Algorithm 4. $BackwardUpdating(C_i)$

1: $e \leftarrow prev(C_i.E)$;
2: **while** $low(e) > C_i.k$ **do**
3:    delete data points on e from $C_i.can$;
4:    $e \leftarrow prev(C_i.E)$;
5: $next(C_i.E)$;

Algorithm 5. BUA

1: let P_{del} be the set of removed data points;
2: let P_{ins} be the set of added data points;
3: **for** every point p in P_{del} **do**
4:    suppose p lies on edge e;
5:    delete p from e (using an object index);
6:    **for** every C_i where e is influenced **do**
7:       **if** p in $C_i.top$ **or** p in $C_i.can$ **then**
8:          update $C_i.top$, $C_i.k$, $C_i.can$ by deleting p;
9: **for** every point p in P_{ins} **do**
10:    suppose p lies on edge e;
11:    insert p into e (using the object index);
12:    **for** every C_i where e is influenced **do**
13:       update $C_i.top$, $C_i.k$, $C_i.can$ by inserting p;
14: **for** every C_i **do**
15:    **if** $low(current(C_i.E)) \leq C_i.k$ **then**
16:       $ForwardUpdating(C_i)$;
17:    **else**
18:       $BackwardUpdating(C_i)$;

---

step. Second, the candidate list is updated in line 3. The forward updating procedure repeat updating $C_i.top$, $C_i.k$, and $C_i.can$ when not all the influenced edges are visited.

Backward Updating: This procedure, as shown in Algorithm 4, removes from $C_i.can$ the data points on every edge e in $C_i.E$, if e is not influenced any more.

The BUA Algorithm: Our new incremental bidirectional updating algorithm (BUA) is shown in Algorithm 5. We explain it using the example in Figure 6. Suppose initially, the set of data points is $P = \{p_1, p_2, \cdots, p_8\}$, for a query $C_i = $ CANN $(\{q_1, q_2, q_3\}, 3, sum)$. After $ForwardUpdating(C_i)$ for initialization, we can get the initial result $C_i.top = \{p_1, p_2, p_4\}$ and $C_i.can = \{p_7\}$. Then, suppose $P_{ins} = \{p_{10}\}$ and $P_{del} = \{p_4\}$. When inserting p_{10}, it lies on the influenced edge s_4 but has an aggregate function value that is less than $C_i.k$. So p_{10} is inserted into the candidate list of C_i, $C_i.can$. When deleting p_4, it is in the $C_i.top$. So it is removed from the $C_i.top$ and p_7 will be moved from $C_i.can$ to the $C_i.top$. At this time, the lower bound of the current edge $low(s_{10}) \leq C_i.k$ (the aggregate function value of p_7). So the forward updating is invoked, s_{10} becomes influenced in C_i. The data point p_8 that lies on s_{10} is also added to $C_i.can$. The current result becomes $C_i.top = \{p_1, p_2, p_7\}$ and $C_i.can = \{p_8, p_{10}\}$. Note that in case of the MTR algorithm, the result of C_i have to be recomputed because

$C_i.k$ increases. In the next time stamp, suppose $P_{ins} = \{p_9\}$ and $P_{del} = \phi$. After p_9 is used to update the result of C_i, it is added into $C_i.top$ and p_7 is moved from $C_i.top$ to $C_i.can$. At this time, we have the lower bound of current edge $low(s_5) > C_i.k$ (the aggregate function value of p_9). So the backward updating is invoked, and s_{10} is not influenced any more. The data point p_8 that lies on s_{10} is also removed from $C_i.can$. The result becomes $C_i.top = \{p_1, p_2, p_9\}$ and $C_i.can = \{p_7, p_{10}\}$.

4.4 Analysis

Suppose there are n nodes and m edges in the network, for each query $CANN(Q, k, h)$, there are s segments in the query graph on average, and the average number of objects on each segment is o, the buffer size for each query is b. The average number of segments that influence the result of a query is r, we have $o \cdot r \geq k$. We assume that the objects are uniformly distributed on all edges and the portion of objects that changes at each timestamp is $\lambda (0 \leq \lambda \leq 1)$. For convenience, we ignore the cost for operations on the object index, which is not the dominate cost.

Lemma 4. *In the IRC algorithm, for each query, the time complexity to compute the initial results is $O(o \cdot r \cdot \log k)$, the memory used is $O(k + b)$ and the I/O cost is $O(\frac{r}{b})$.*

Proof. To compute the initial top-k result of a query, we need to retrieve all the objects that lie on the influence segments(i.e., the first r segments in the segment list of the query). The number of objects to be retrieved is $O(o \cdot r)$. Each object is used to update the top-k results, which can be implemented as a heap of size k. Each update can be done in $O(\log(k))$ time, so the total time complexity is $O(o \cdot r \cdot \log(k))$. For the memory cost, we need $O(b)$ to buffer the segment list, and $O(k)$ to store the results, so the total memory used is $O(k + b)$. We visit the first r segments in the segment list sequentially, so the I/O cost is $O(\frac{r}{b})$. $\square$

Lemma 5. *In the MTR algorithm, with a probability of 0.5, the result of a query is needed to be recomputed at each timestamp. For the query that does not need to be recomputed, the time complexity for updating at each time stamp is $O(\lambda \cdot o \cdot r \cdot \log k)$ and no I/O operation is needed. The memory used for each query is the same as in IRC.*

Proof. The result of a query needs to be recomputed iff after the deletion and insertion steps, the new top-k result expires, or $C_i.k$ value for the query C_i increases. This case happens when, for the two sets P_{del} and P_{ins}, P_{del} contains more objects with cost smaller than the former $C_i.k$. The probability of this situation is 0.5 for the uniformly distributed objects. For each query that does not need re-computation, the time cost is the updates of $\lambda \cdot o \cdot r$ objects that lie on the influence segments, each update cost $\log(k)$ time as the same in the IRC algorithm, so the total time complexity for updating at each timestamp is $O(\lambda \cdot o \cdot r \cdot \log k)$. The influence segments keeps the same after the updating steps, so no I/O operation is needed on the segment list. The memory cost is also the same as the IRC algorithm. $\square$

Lemma 6. *For the BUA algorithm, no re-computation is needed to update the result of a query at each timestamp, the time complexity for each query is $O(\lambda \cdot o \cdot r \cdot \log(o \cdot r))$. The memory used for each query is $O(o \cdot r + b)$. The I/O cost is $O(\frac{\lambda \cdot r}{b})$ in the worst case.*

Proof. For the BUA algorithm, it uses an extra candidate list for each query to record the candidate objects that lie on the influence segments but not in the top-k result of a query. For the $\lambda \cdot o \cdot r$ changed objects that lie on the influence segments, the cost for updating each object is $O(\log o \cdot r)$ by using a heap to record all the objects that lie on the influence segments(i.e., all the objects in the top-k result and candidate list). The total time complexity is $O(\lambda \cdot o \cdot r \cdot \log (o \cdot r))$. For the memory cost, in addition to the $O(b)$ buffer size, we need $O(o \cdot r)$ cost to record all the objects that lie on the influence segments. The total memory cost is $O(o \cdot r + b)$. For the I/O cost, consider the worst case, when $\lambda \cdot o \cdot r$ objects move out of the influence segments and no object moves in, or $\lambda \cdot o \cdot r$ objects move into the influence segments and no object moves out. In the first case, we need to visit $O(\frac{\lambda \cdot o \cdot r}{o}) = O(\lambda \cdot r)$ segments which cost $O(\frac{\lambda \cdot r}{b})$ I/O operations for the forward updating. In the second case, we also need to visit $O(\lambda \cdot r)$ that cost $O(\frac{\lambda \cdot r}{b})$ I/O operations for the backward updating. So the I/O cost is $O(\frac{\lambda \cdot r}{b})$ in the worst case.   □

For the I/O cost of the BUA algorithm, in the average case, the number of objects that move into the influence segments is almost the same to the number of objects that move out, so the average I/O cost is very small in practice.

5   Implementation Details

In this section, we introduce the details for implementation including the data structures used and the storage model.We introduce three types of data structures that are constructed over the road network, data objects and queries respectively.

Edge Table. For every edge e in the road network, we store in the edge table two part of information. The first part is about the network structure, i.e., the edge $e.id$, the two nodes n_i and n_j it connects, the length of the edge $len(e)$, and the lists of edges to n_i and n_j, this part can be used to construct the query graph G_Q of a CANN query. The second part is the influence list of e maintaining a set of queries that e influence along with the set of influence edges in G_Q. Using this part of information, we can fast retrieve all queries that is influenced by e.

Object Index. Each object point p in the network can be represented as $(e.id, pos)$, where $e.id$ is the id of the edge it lies on, and pos is its position on e, i.e., $pos = pos_e(p)$. We use a index of a balanced tree to store all the object points in the network. It allows to retrieve all the objects that lies in a certain interval on a given edge e, or retrieve all the objects that over a certain edge s in a query graph of a CANN query. When the size of objects are large, the index can be stored external and a B+ tree can be used for storage.

Query Table. The query table stores the set of queries. For every query C_i in the query table, tree parts of information are stored. The first part is the query descriptor, i.e., $C_i.id$, Q, k and h. The second part is the list of top-k objects $C_i.top$ along with $C_i.k$ and the candidate list $C_i.can$. The third part is the sorted edge list in the query graph $C_i.E$, which is a external data structure on which only sequential access and read operation is allowed. Each edge in the list is represented as $s = (e.id, x_1, y_1, x_2, y_2)$, where e is the

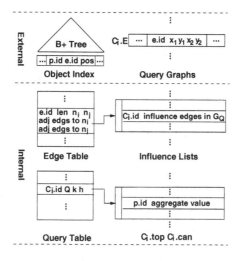

Fig. 7. Internal and External Structures

edge on the original graph G, x_1 and x_2 are the start and end positions of s on e, y_1 and y_2 are the aggregate values on x_1 and x_2 respectively. Based on the sequential property, a buffer can be used for each query when processing.

The main internal and external data structures used for processing are illustrated in Figure 7.

6 Experimental Studies

We conducted extensive experimental studies to test the performance of our algorithms. All algorithms are implemented using C++. We use the road-map in the Maryland State

Table 1. Parameters

Parameter	Default	Range
Number of edges	25K	10, 15, 20, 25, 30 (K)
Number of nodes	20K	5, 10, 15, 20, 25 (K)
Number of queries	5K	1, 3, 5, 7, 10 (K)
Number of query points	20	1, 10, 20, 30, 40
Number of objects	100K	10, 50, 100, 150, 200 (K)
Query distribution	Uniform	Gaussian, Uniform
Object distribution	Uniform	Gaussian, Uniform
Top-k	50	1, 25, 50, 100, 200
Object agility	10%	5, 10, 15, 20, 25 (%)
Buffer size	2K	1, 2, 3, 4, 5 (K)
Function	SUM	MIN, MAX, SUM

Table 2. Time to construct query graph

$|E|(K)$/T(ms)	10/114	15/214	20/319	25/408	30/505
$|N|(K)$/T(ms)	5/64	10/155	15/254	20/336	25/437
$|Q|$/T(ms)	1/26	10/178	20/343	30/506	40/675

in US extracted from US Census Bureau 2005 TIGER/Line.[1]. All the parameters including default values and ranges are listed in Table 1. Here, number of query points means the number of points in Q (i.e., $|Q|$) for each query, the query distribution is distribution of all query points, and the object agility is the percentage of objects that is changed per time stamp. The default graph is a subgraph of the above network with $20K$ nodes and $25K$ edges. When number of nodes varies, we use a subgraph of the network with the provided node number. When number of edges varies, we fix the node number to be 10K and generate a graph with the provided edge number. For each test that is to monitor the k-NN result, we process for 100 time stamps by generating the moving objects using the generator proposed in [3].We record the average performance for every time stamp. For the IRC algorithm, we mean to recompute the top-k result from scratch for every time stamp. Unless specified, we will use the default value for testing. All tests are conducted on a 2.8GHz CPU/1G memory PC running XP.

Query Graph Construction: We first test the time to construct the query graph for each query. We vary the number of edges, number of nodes and number of query points, and record the time to construct the query graph in each test. The result is shown in Table 2, the time to construct query graph is small (less than 0.7 second) for all tests. As each of the three parameters increases, the response time will increase steadily.

IMA,GMA vs BUA: With $|Q|$ fixed to be 1, we test the efficiency for IMA, GMA, and BUA algorithms. For each algorithm, we combine the different distributions(i.e., Gaussian and Uniform distribution) for queries and objects (e.g., U/G means the queries are uniformly distributed and the objects are in Gaussian distribution) with all other parameters setting to be the default values. As illustrated in Figure 8(a), our BUA algorithm always performs best and changes for distribution of both the queries and objects will not influence the efficiency of BUA algorithm much.

Network: We vary the number of edges and number of nodes for the network with an average of 4 objects on each edge and test the average processing time for IRC, MTR and BUA algorithms in each time stamp. We report our result in Figure 8(b) and Figure 8(c). For each test, the MTR algorithm is about 2-3 times faster than the IRC algorithm, and the the BUA algorithm is 2-4 times faster than the MTR algorithm. When the number of edges increases, the processing time for all three algorithms will increase, because as the network becomes denser, the number of influence edges will increase. When the number of nodes increases, the processing time for all three algorithms do not change much, because both the density of network and density of objects will not change as the network increases.

---

[1] Topologically Integrated Geographic Encoding and Referencing system:
http://www.census.gov/geo/www/tiger/

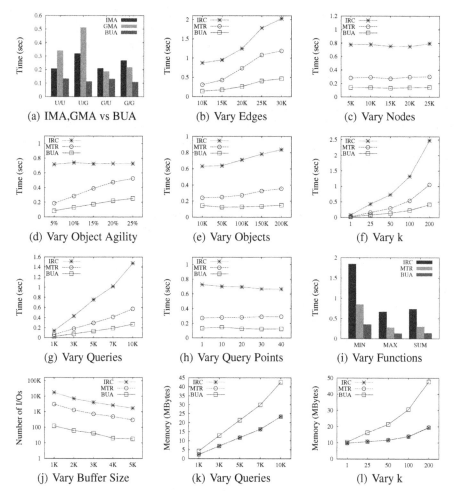

Fig. 8. Experimental Results

Objects: Figure 8(d) and Figure 8(e) shows the average processing time per time stamp for IRC, MTR and BUA algorithms when the object agility or the number of objects varies. When the object agility increases, the processing time for both MTR and BUA will increase steadily while IRC is not influenced because it will always computes each query from scratch at every time stamp. When the number of objects increases, the density of objects becomes larger, which increases the processing time. But the number of influenced edges will decrease, which decreases the processing time. We can see from Figure 8(e) that when the object number is larger than $50K$, the processing time for all the three algorithms all increase slowly.

Queries: There are mainly 4 parameters for the query: the top-k value, number of queries, number of query points in each query(i.e., $|Q|$), and the type of aggregate function for the query. In Figure 8(f) and Figure 8(g), when k increases or the number of queries increases, the processing time for IRC, MTR and BUA algorithms will increase steadily. In Figure 8(h), when the number of query points in each query increases, the processing time for all three algorithms will not influence much, because at one hand, the number of edges in each query's query graph in will increase which raise the complexity of algorithm; at the other hand, the length of edges in each query's query graph becomes shorter, and the objects lies on the influence edges become less, which lower the complexity. In Figure 8(i), we see that the MIN function consumes more for all three algorithms. It is because for MAX and SUM function, the best objects retrieved is more centralized for each query, while in the MIN function, each query point in a query can be considered as a center for the distribution of the top objects.

Total I/Os: We vary the buffer size used for every query C in the corresponding edge list $C.E$, and study the number of I/Os for IRC, MTR and BUA algorithms for each time stamp. As shown in Figure 8(j), as the buffer size increases, the number of I/Os will decrease steadily. The MTR costs about $\frac{1}{5}$ I/Os of IRC while BUA costs about $\frac{1}{20}$ of MTR, which is rather small, because the pointer for each query only moves forward of backward incrementally.

Memory: We finally test the memory used for algorithms of IRC, MTR and BUA. When the number of queries and top-k value vary, the result is shown in Figure 8(k) and Figure 8(l). As the query number or k increases, the memory used will increase steadily, for all three algorithms. The memory cost of IRC and MTR is the same as analyzed in Section 4.4. The memory cost for BUA is about 1.1 to 2.4 times of IRC.

7   Related Work

In this section, we survey k-NN search over road networks in two categories, namely, snapshot approaches and continuous monitoring approaches.

Snapshot approaches: Shahabi et al. in [4] applied an embedding technique to transform a road network to a high dimensional space, and used the Minkowski metrics for distance measurement in the embedded space. Jensen et al. in [5] proposed a foundation data model and a system prototype for k-NN queries in road networks. Shekhar et al. in [6] addressed the problem of finding the in-route nearest neighbor (IRNN). Papadias et al. in [7] proposed an architecture that integrates network and Euclidean information for query processing in spatial network databases. Tao et al. in [8] studied the time-parameterized k-NN queries when query points and objects change in certain speed and directions. Kolahdouzan et al. in [9] proposed to find the nearest points of interest to all the points on a path over road networks. They also performed k-NN over spatial networks in [10] based on the pre-computed first order Voronoi diagram. Yiu et al. in [17] first studied the aggregate nearest neighbor query in road networks, which explored the network around the query points until the aggregate nearest neighbors are discovered. UNICONS [11] developed a search algorithm which answers NN queries at any point of a given path. Huang et al. in [12] presented a versatile approach to k-NN computation in spatial networks using the island which is a sub-network in a certain area. Hu et al. in [13] proposed an approach that indexes the network topology based on a set of interconnected tree-based structures. Huang et al. in [14] focused on caching the query results in main memory and subsequently reusing these for query processing when there are multiple k-NN queries over a road network. Almeida et al. in [15] proposed a storage schema with a set of index structures to support Dijkstra based algorithms for k-NN queries in road networks. Deng et al. in [16] considered the problem of efficient multi-source skyline query processing in road networks.

Continuous monitoring approaches: In recent years, more works focused on continuous monitoring of NN queries over road networks. Ku et al. in [17] studied the adaptive NN queries in travel time networks. It developed a local-based greedy nearest neighbor algorithm and a global-based adaptive nearest neighbor algorithm that both utilize real-time traffic information to maintain the search results. Mouratidis et al. in [1] focused on monitoring nearest neighbors in highly dynamic scenarios.

8   Conclusion

In this paper, we studied a new problem (CANN query) that is to monitor k-NN objects over a road network from multiple query points to minimize an aggregate distance function with respect to the multiple query points. In order to reduce the cost of network distance computing, we proposed a new approach that computes a query graph offline for a CANN query. With the help of the query graph, the cost of computing aggregate function values for any possible data points on the road network is significantly reduced. In addition, we proposed two algorithms to monitor CANN queries. We conducted extensive experimental studies over large road networks and confirmed the efficiency of our algorithms.

Acknowledgment. This work was supported by a grant of RGC, Hong Kong SAR, China (No. 418206).

References

1. Mouratidis, K., Yiu, M.L., Papadias, D., Mamoulis, N.: Continuous nearest neighbor monitoring in road networks. In: VLDB, pp. 43–54 (2006)
2. Yiu, M.L., Mamoulis, N., Papadias, D.: Aggregate nearest neighbor queries in road networks. IEEE Trans. Knowl. Data Eng. 17(6), 820–833 (2005)
3. Brinkhoff, T.: A framework for generating network-based moving objects. GeoInformatica 6(2), 153–180 (2002)
4. Shahabi, C., Kolahdouzan, M.R., Sharifzadeh, M.: A road network embedding technique for k-nearest neighbor search in moving object databases. In: ACM-GIS, pp. 94–100 (2002)
5. Jensen, C.S., Kolárvr, J., Pedersen, T.B., Timko, I.: Nearest neighbor queries in road networks. In: GIS, pp. 1–8 (2003)
6. Shekhar, S., Yoo, J.S.: Processing in-route nearest neighbor queries: a comparison of alternative approaches. In: GIS, pp. 9–16 (2003)
7. Papadias, D., Zhang, J., Mamoulis, N., Tao, Y.: Query processing in spatial network databases. In: VLDB, pp. 802–813 (2003)
8. Tao, Y., Papadias, D.: Spatial queries in dynamic environments. ACM Trans. Database Syst. 28(2), 101–139 (2003)

9. Kolahdouzan, M.R., Shahabi, C.: Continuous k-nearest neighbor queries in spatial network databases. In: STDBM, pp. 33–40 (2004)
10. Kolahdouzan, M.R., Shahabi, C.: Voronoi-based k nearest neighbor search for spatial network databases. In: VLDB, pp. 840–851 (2004)
11. Cho, H.J., Chung, C.W.: An efficient and scalable approach to cnn queries in a road network. In: VLDB, pp. 865–876 (2005)
12. Huang, X., Jensen, C.S., Saltenis, S.: The islands approach to nearest neighbor querying in spatial networks. In: Bauzer Medeiros, C., Egenhofer, M.J., Bertino, E. (eds.) SSTD 2005. LNCS, vol. 3633, pp. 73–90. Springer, Heidelberg (2005)
13. Hu, H., Lee, D.L., Xu, J.: Fast nearest neighbor search on road networks. In: Ioannidis, Y., Scholl, M.H., Schmidt, J.W., Matthes, F., Hatzopoulos, M., Böhm, K., Kemper, A., Grust, T., Böhm, C. (eds.) EDBT 2006. LNCS, vol. 3896, pp. 186–203. Springer, Heidelberg (2006)
14. Huang, X., Jensen, C.S., Saltenis, S.: Multiple k nearest neighbor query processing in spatial network databases. In: Manolopoulos, Y., Pokorný, J., Sellis, T.K. (eds.) ADBIS 2006. LNCS, vol. 4152, pp. 266–281. Springer, Heidelberg (2006)
15. de Almeida, V.T., Güting, R.H.: Using dijkstra's algorithm to incrementally find the k-nearest neighbors in spatial network databases. In: SAC, pp. 58–62 (2006)
16. Deng, K., Zhou, X., Shen, H.T.: Multi-source skyline query processing in road networks. In: ICDE (2007)
17. Ku, W.S., Zimmermann, R., Wang, H., Wan, C.N.: Adaptive nearest neighbor queries in travel time networks. In: GIS, pp. 210–219 (2005)

RAM: Randomized Approximate Graph Mining

Shijie Zhang and Jiong Yang

EECS Department
Case Western Reserve Univ.,
10900 Euclid Ave, Cleveland OH 44106, USA
shijie.zhang@cwru.edu, jiong.yang@cwru.edu

Abstract. We propose a definition for frequent approximate patterns in order to model important subgraphs in a graph database with incomplete or inaccurate information. By our definition, frequent approximate patterns possess three main properties: possible absence of exact match, maximal representation, and the Apriori Property. Since approximation increases the number of frequent patterns, we present a novel randomized algorithm (called RAM) using feature retrieval. A large number of real and synthetic data sets are used to demonstrate the effectiveness and efficiency of the frequent approximate graph pattern model and the RAM algorithm.

1 Introduction

A large number of algorithms have been developed for mining exact graph patterns [7] [8] [9] [11] [18] [3], [12], [17], [5], [16]. However, in many important applications, the relationships modeled by edges may be inaccurate or incomplete. This is especially true in bioinformatics and social network analysis. Many types of biological data are known to be inaccurate, e.g., gene expression profiles, protein interaction networks, metabolic pathways. Besides, when describing networks of human social interactions, some relationship can be dual, e.g., two people can be both friends and enemies. Last but not least, approximate graph mining can be applied in procedure dependency graphs to discover neglected conditions, e.g., missing paths, conditions, and cases in the field of software engineering [2].

One example of such data is protein interaction networks. A challenging technical problem described in a recent review [1] is the prevalence of spurious interactions due to self-activators, abundant protein contaminants and weak, nonspecific integrations. Analysis based on the agreement of the interaction and expression data shows that less than half of these interactions are biologically relevant. Therefore one might observe an interaction among two proteins that are not functionally related in the cell at all. Due to the fact that various experiments use different confidence thresholds, there also exist a large number of false negative interactions.

Another important application of approximate graph mining lies in the study of cross-market customer segmentations [14]. In a specific market, the similarity among customers in market behavior can be modeled as a similarity graph. Each customer is a vertex in the graph, and two customers are connected by an edge if their behaviors in the market are similar. While we can identify customer types easily, the similarity level

B. Ludäscher and Nikos Mamoulis (Eds.): SSDBM 2008, LNCS 5069, pp. 187–203, 2008.

between customers in different markets is more difficult to evaluate. Thus, there may exist both false positive and false negative edges.

Due to edge distortion, exact graph mining algorithms may not find important patterns even when the support threshold is low. Therefore, in this paper, we first present a formal definition of frequent approximate patterns, which can be applied in graph databases when the edge relationships are less reliable. Then, we introduce our approximate graph mining algorithm.

One contribution of the paper is the formal definition of frequent approximate patterns. We propose two constraints on such a pattern: (1) the pattern should approximately be embedded in at least a certain number of graphs in the database; (2) the occurrence of each edge in the pattern should be higher than a minimum threshold. The second constraint gives users the ability to keep unrelated edges from affecting the frequency of patterns due to approximation.

Compared to exact graph pattern mining, approximate patterns tend to be larger, i.e., with more vertices and more edges. This adds two additional challenges for mining approximate graphs on top of the exact graph mining problems. First, with larger patterns, the total number of patterns increases. This can greatly impact the efficiency of any pattern mining algorithm. Secondly, canonical forms are commonly used for graph isomorphism tests. When the graph patterns grow large, it becomes extremely inefficient to compute the canonical forms and storing all the canonical forms as strings may lead the program to run out of memory.

With these two challenges in mind, we introduce a randomized algorithm, called RAM, based on feature retrieval. RAM mines frequent graph patterns in a depth first fashion. Instead of using canonical forms for isomorphism tests, we construct a vector of hash values for features. It greatly improves the efficiency and scalability of the mining process. However, it may produce false negatives, missing some patterns. To overcome this problem, we adopt a randomized algorithm. During each run, the insertion of edges and addition of vertices are ordered randomly. As a result, if during a run a pattern could be discovered with x probability, then by p runs, the probability to find that given pattern would be $1 - (1 - x)^p$. For example, even if 20% of the patterns may be missed during one run, then less than 1% of patterns would be missed after 3 runs. (Looking ahead, the accuracy of one run is in fact higher than 80% for the real data sets in our experiments.) Therefore, the RAM algorithm can achieve a high degree of accuracy with relatively short execution time.

The remainder of the paper is organized as follows. Section 2 is the related work. Section 3 defines the preliminary concepts. Some important properties of frequent approximate patterns are explained in Section 4. Section 5 presents a basic algorithm for the problem based on depth first search. Then, in section 6 we provide an optimized algorithm Monkey. Section 7 presents experiment results. In Section 8, some improvements are discussed, and Section 9 concludes our study.

2   Related Work

In recent years, a large number of algorithms have been designed to find exact frequent patterns. Inokuchi et al. proposed an Apriori-based algorithm [9], called AGM, to discover all frequent substructures. Kuramochi and Karypis further developed FSG [11], using a more efficient graph representation structure and edge-growth instead of vertex-growth. Yan et al. developed gSpan and designed DFS lexicographic order to support the mining algorithm. Huan et al. proposed FFSM [7] with a graph canonical form, called CAM, and a set of CAM tree operations. While AGM and FSG took advantage of apriori-like level-wise approaches, gSpan and FFSM adopted depth first search, which was more efficient for graph mining problems. Meanwhile, Huan et al. developed SPIN [8] and Nijssen et al. presented Gaston [13], both of which mined frequent patterns by first mining frequent trees or more basic patterns.

One important application of the graph mining algorithms is to find frequent patterns, motifs, and modules in biological networks. Koyuturk et al. [10] introduced an efficient algorithm specially designed for biological data. However, many edges in the dataset were unreliable, which indicated efficient approximate pattern mining algorithms should be designed for biological datasets with edge distortions.

In [19], Yan et al. presented Grafil to perform approximate graph indexing with edge relaxation. However, there are very few approximate graph mining algorithms in the current state of graph-based data mining. Holder et al. proposed SUBDUE [6] to discover approximate substructure pattern based on the minimum description length principle and optional background knowledge. Since SUBDUE used a computationally-constrained beam search, it cannot discover the complete set of frequent patterns. Furthermore, SUBDUE is not designed to find patterns in datasets with edge distortions. In [20], approximate pattern mining in the itemset setting has been proposed.

To the best of our knowledge, hashing has not been used in graph mining, but it is used in frequent itemset mining. In [15], the authors proposed a hashing method to find frequent 2-itemsets. Multiple 2-itemsets were hashed into a single entry. If any of these 2-itemset occurred in a transaction, the count of the hashing entry incremented. This can be used to efficiently discover frequent 2-itemsets. On the other hand, in [4], randomized algorithm has been first used in pattern mining.

3   Preliminaries

In this section, some terminologies are introduced, and then the problem statement is given.

Definition 1. *A **labeled graph** G is a five element tuple $G = \{V, E, \Sigma_V, \Sigma_E, L_G\}$ where V is a set of vertices and $E \subseteq V \times V$ is a set of edges. Σ_V and Σ_E are the sets of vertex and edge labels respectively. The labeling function L_G defines the mappings $V \rightarrow \Sigma_V$ and $E \rightarrow \Sigma_E$. If there is no edge between vertex x and y in G, we assume a virtual edge e_{xy} between x and y and define $L_G(e_{xy}) = \varnothing$. $V(G)$ and $E(G)$ represent the vertex and edge set of G.*

Definition 2. *To model the frequent approximate patterns with edge distortion, we introduce three parameters, **support** γ, **variability** β, and **tolerance** α. We will illustrate the details of these parameters in the following definitions.*

1. Support γ confines the size of support set.
2. Variability β indicates the maximum number of missing edges from an embedding.
3. Tolerance α regulates the maximum number of missing edge from the support set.

Definition 3. *We define a labeled graph p is β edge isomorphic to q, denoted by $p \approx_\beta q$ iff*

1. $\exists$ bijection $f : V(p) \leftrightarrow V(q)$, $L_p(x) = L_q(f(x))$, $x \in V(p)$, $f(x) \in V(q)$
2. At most β different pairs of vertices (x,y) in $V(p)$, such that $L_p(e_{xy}) \neq L_q(e_{f(x)f(y)})$. Moreover, if $L_p(e_{xy}) \neq L_q(e_{f(x)f(y)})$, $L_p(e_{xy}) = \varnothing$.

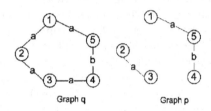

Graph q   Graph p

Fig. 1. β edge isomorphism ($\beta \geq 2$)

Figure 1 shows an example of β edge isomorphism. In the example, graph p is at least 2 edge isomorphic to q; however, graph q is not any edge isomorphic to p. The definition of β edge isomorphic is not symmetrical, i.e., $p \approx_\beta q$ cannot always lead to $q \approx_\beta p$, and p is not necessarily connected.

Graph G is β **edge subgraph isomorphic** to graph G', denoted by $G \subseteq_\beta G'$, iff there exists a subgraph G'' of G' such that $G'' \approx_\beta G$. G'' is defined as an **embedding** of G in G'. In figure 1, when q is a given pattern, then p can be considered as an embedding of q.

Definition 4. *Given a graph database $D = \{g_1, g_2, ..., g_n\}$, the β edge different support set of subgraph g, denoted by*
$sup(D, g, \beta)$, is defined as the subset of D to which g is β edge subgraph isomorphic.

$$sup(D, g, \beta) = \{g_i | g \subseteq_\beta g_i, g_i \in D\}$$

When $\beta = 0$, $sup(D, g, 0)$ is the subset of D to which g is exactly subgraph isomorphic.

Parameter **support** γ is used to indicate the minimum frequency threshold. However, if graph g contains some edge which rarely appears in the database, it is not beneficial to include g in the results. Therefore, we introduce another parameter **tolerance** α. If any edge of graph g appears less than $\gamma - \alpha$ times in $sup(D, g, \beta)$, we will not define g as frequent.

Definition 5. *Let $|E_g(e)|$ be the number of edges in graph g with the same edge and vertex label, e.g., in figure 2, $|E_g(e)| = 4$.*
*Graph g is β **edge** γ **frequent with tolerance** α (i.e., **frequent approximate pattern**) iff*

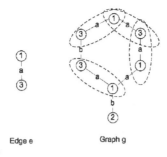

Edge e   Graph g

Fig. 2. Definition of $|E_g(e)|$

1. $|sup(D, g, \beta)| \geq \gamma$.
2. *For each edge e in g, $|S| \geq \gamma - \alpha$, where S is a subset of $sup(D, g, \beta)$ and each database graph g_i in S satisfies the following condition: g_i contains at least one embedding s, s.t. $s \approx_\beta g$, $|E_s(e)| = |E_g(e)|$.*

In figure 2, if graph g is a frequent approximate pattern, it indicates that there are at least $\gamma - \alpha$ embeddings of g in different graphs. Each of these embedding contains all the 4 marked edges. A further example will be presented in section 4.2.

Problem Statement: Given a graph database D and parameters support γ, variability β and tolerance α, we are to find all the connected graph g which is β edge γ frequent with tolerance α of edge missing in the support set.

4 Properties

In this section, we illustrate some important properties of approximate patterns.

4.1 Possible Absence of Exact Match

In the traditional graph mining models, a frequent pattern is exactly embedded in a minimum number of database graphs. However, after adding approximation, it is possible for a pattern to be a frequent approximate pattern even though it is not exactly embedded in the database. For example, assume that the parameters are $\alpha = 2$, $\beta = 1$, and $\gamma = 3$, the graph pattern g (as in figure 3) does not have any exact embedding in figure 4. However, g is β edge subgraph isomorphic to all the graphs in the database. Also, every type of edges occurs more than $1 (= 3 - 2)$ times. Therefore, g is a frequent approximate pattern in the database.

In our model, having exact embeddings is not a necessity to be defined as frequent. In many chemical and biological applications, important patterns may have several variations. As a result, traditional exact graph mining methods may miss these patterns when all the corresponding mutations are below the threshold. However, approximate graph mining may find the patterns, given the range of edge distortion.

Fig. 3. Frequent Approximate Pattern without Exact Match

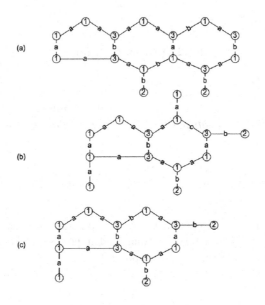

Fig. 4. Graph Database

4.2   Maximal Representation

There are two kinds of constraints. The first one is on the whole approximate frequent pattern, the other is on each type of embedding edges.

The second constraint guarantees the maximal representation property of approximate patterns. Any subgraph of pattern g, except g itself, cannot represent all the embeddings of g, i.e., for any non-trivial subgraph t of g, there exist at least $\gamma - \alpha$ embeddings of g which are not β edge isomorphic to t.

For example, assume a pattern p occurs in γ graphs, without the second property, i.e., the tolerance parameter, we can add any β edges into p, the resulting is still a frequent approximate pattern. Thus, we employ the tolerance parameter to prevent this.

Figure 5 further explains the property. Suppose the graph database and the parameters are the same as those in figure 4. For any non-trivial subgraph t of g, there exists at least one edge e (as marked) embedded in g but not in t. We also marked all the edges of the same type as e by ellipses. Suppose g is a frequent approximate graph pattern; then there are at least $\gamma - \alpha$ embeddings of g, which contain all the edges we have marked. Therefore, any of these embeddings contains e. According to definition 2, these

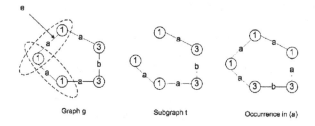

Fig. 5. Maximal Representation

embeddings are not β edge isomorphic to t. In this example, one embedding in graph (a) is not 1-edge isomorphic to t, as shown in figure 5.

However, if we do not include the second constraint in the definition, the maximal representation property can no longer hold. Many redundant patterns may be generated, whose embeddings can be represented by one of its subgraphs.

4.3 Apriori Property

The Apriori property claims that if a pattern satisfies the minimum support threshold, then any sub-pattern of it also satisfies the minimum support threshold. The property is very useful in the field of data mining.

By our definition of frequent approximate patterns, we have the following theorem.

Theorem 1. For a frequent approximate graph g, any subgraph of g is also a frequent approximate pattern.

Proof: For any subgraph s of g, there is a sequence of subgraphs of g: $a_0 = s$, $a_1 = s + e_1$, $a_2 = s + e_1 + e_2$, ..., $a_n = g$. a_i be a subgraph of a_{i+1} with one edge missing. By definition, when $k = n$, $a_n = g$ is a frequent approximate pattern. Suppose when $k = i$, a_i is a frequent approximate pattern; we therefore assume $a_i - a_{i-1} = e$. We want to prove that the approximate pattern a_{i-1} is frequent.

For every embedding t of a_i, if e is embedded in t as it is in a_i, we define $t' = t - e$. Otherwise, $t' = t$. From our definition, t' is also β edge isomorphic to a_{i-1}. Thus, $sup(D, a_{i-1}, \beta) \geq sup(D, a_i, \beta) \geq \gamma$. Also, for any edge e' in a_{i-1}, if $|E_t(e')| = |E_{a_i}(e')|$, we have $|E_{t'}(e')| = |E_{a_{i-1}}(e')|$. Thus, every type of edge exists no less than $\gamma - \alpha$ times in the graph database. Therefore, a_{i-1} is a frequent approximate pattern. By induction, we have $a_0 = s$ is also a frequent approximate pattern.

Theorem 1 guarantees that if graph h is not a frequent approximate pattern, we need not examine any supergraphs of h. This theorem is the foundation of our mining algorithm.

5 Basic Algorithm

In this subsection, we present a basic approximate graph mining algorithm based on depth first search, which is very similar to former exact graph mining algorithms.

Given parameters α, β, and γ, we first find frequent approximate edges by enumeration, and remove non-frequent edges in the database. Starting from these approximate frequent edges as a first set of patterns, we recursively grow edges in a depth first fashion. If all possible candidate patterns grown from pattern g have been checked, we backtrack to the pattern from which g is grown, and grow another edge to that pattern.

We first discuss the case when we grow an edge with a new vertex to an existing pattern g. Let e_{uv} be the growing edge, in which u exists in g but v is the new vertex to be introduced. Here we assume the label of vertex u is l_u and the label of v is l_v. If the pattern of $g + e_{uv}$ has not been reached before by the mining algorithm. We are to decide whether $g + e_{uv}$ could be a frequent approximate pattern. To achieve this, all the embeddings of g are examined. We test,

1. whether there are no less than γ database graphs which contain at least one embedding of g and a vertex v' with label l_v. These database graphs constitute the support set. Moreover, if each embedding in a database graph has exactly β edge difference from pattern g, and none of them are connected to v' with the same label as the growing edge e_{uv}, this database graph cannot be added to the support database graph set.
2. whether there are no less than $\gamma - \alpha$ database graphs which contain at least one embedding of g and vertex v'. Besides, if none of the embeddings are connected to v' with the same label as the growing edge e_{uv} and exactly contain all the edges in pattern g of the same type of e_{uv}, the database graph cannot be counted effectively.

If these two requirements are satisfied, the new pattern of $g + e_{uv}$ is also a frequent approximate pattern.

The procedure of adding a new edge to an existing pattern g without introducing any new vertex is similar. The search is also done in a depth first fashion. We do not need to identify the new vertex any more since the two endpoints of the new edge are already known. After discovering all frequent approximate patterns in which edge s is embedded, we remove s from the graph database to shrink the searching space.

In this basic algorithm, we adopt canonical adjacent matrix to determine whether the ongoing candidate pattern has been reached or not. The basic algorithm is a natural extension of exact graph mining algorithms to approximate graph mining. However, the calculations of canonical forms of graphs, especially for larger graphs, is extremely time-consuming. Additionally, frequent approximate graph models tend to generate many more patterns. We need to deal with these two difficulties in approximate graph mining. In the next section, we introduce a randomized algorithm, which is more efficient and flexible.

6   Algorithm RAM

A maximal frequent graph is a frequent graph all of whose supergraphs are infrequent. In approximate graph mining, taking edge relaxation into account, the average size of maximal frequent approximate pattern grows. Consequently, the number of non-maximal frequent approximate graphs is even larger. (This is due to the exponential growth of subgraphs, e.g., there are approximately $2^{\binom{n}{2}}/n$ subgraphs of a complete graph with n

nodes.) In the basic algorithm, we calculate the canonical form for every pattern. Here we present a Randomized Approximate Graph Mining algorithm (RAM) for better efficiency. RAM works in a depth first fashion just like the basic algorithm. Instead of using canonical forms, we retrieve a set of features to identify possible patterns. The set of features are carefully selected, and can be obtained from pattern graphs in polynomial time. To achieve better space management, we hash those feature sets into hash vectors. In the mining process, if a pattern with the same hash vector has been examined before, we will stop the searching from the pattern. The algorithm can also be applied to exactly mining large graphs. Next we will introduce the details of the algorithm.

6.1 Feature Retrieval

In the last section, we use canonical forms to distinguish graph patterns. However, the calculation of canonical forms of arbitrary graph patterns can be very expensive. Therefore, in this section, we use feature sets to identify patterns. The intuition is that more often than not a feature set is strong enough to distinguish graph patterns as long as it is well selected. Advantages of the substitution include: (1) calculating feature set is asymptotically easier than canonical forms; and (2) the space consumption of feature sets after hashing is very flexible.

The design of feature set is of crucial importance to the randomized algorithm. Although the design may vary for the graphs in different databases, there are some goals in common, which include: (1) to assure any two isomorphic graphs have the same feature set, which suggests that any kind of automorphism form of the same pattern should result in the same feature values; (2) to minimize the number of patterns which may share the same feature values with another pattern. When collisions are inevitable, patterns sharing the same feature values should at least have great topological similarity; (3) to minimize calculation - any feature selected can be calculated in polynomial time. In this section, we provide one possible design of a set of feature set, which takes advantage of a variety of topological information.

We map each vertex and edge label to a positive prime number. Suppose the total weight of the minimum spanning tree of graph g is $W(MST(g))$; the distance between two vertices v_1, v_2 is $dis(v_1, v_2)$; the degree of vertex v is $deg(v)$; the set of edges incident with vertex v is $e(g, v)$. Then, for a graph g, we can select the following set of six features $f_1, ..., f_6$.

1. f_1 relates to how big g is.
   $f_1(g) = |V||E|$.
2. f_2 relates to the type of edges.
   $f_2(g) = \Sigma_{e_{xy} \in E(g), Lg(e_{xy}) \neq \varnothing} L_g(x) L_g(y) L_g(e_{xy})$.
3. f_3 relates to the minimum spanning tree of g.
   $f_3(g) = W(MST(g))$.
4. f_4 relates to the connectivity of g.
   $f_4(g) = \Sigma_{v_1, v_2 \in V(g), v_1 \neq v_2} dis(v_1, v_2)$.
5. f_5 relates to the degree sequence of g and vertex labels.
   $f_5(g) = \Sigma_{v \in V(g)} deg^2(v) L_g(v)$.

6. f_6 relates to the degree sequence of g and edge labels.

$f_6(g) = \Sigma_{v \in V(g)} deg^2(v)(\Sigma_{e \in e(g,v)} L_g(e))$.

All the features can be obtained in polynomial time. In fact, since we adopt a depth first search as framework, all the feature values can be calculated in linear time by using the information of the predecessor. It is obvious that for a pair of graphs g_1, g_2, if $g_1 \approx g_2$, then $f_i(g_1) = f_i(g_2)$. For the benefit of space efficiency, we hash these features into a hash vector, i.e., $H_i = f_i \bmod P_i$, while P_i is a prime number. The selection of the prime numbers is based on the size of available space, i.e., P_i is proportional to the number of possible value of f_i and $\prod_i P_i$ is about the size of available space for the mining program.

6.2   A Randomized Algorithm for Approximate Graph Mining

In this subsection, we introduce the randomized algorithm RAM for approximate graph mining. The main differences between RAM and the basic algorithm are (1) RAM grows edges in a random order, and (2) RAM adopts hash vectors of feature sets instead of using canonical forms.

We use basic algorithm to mine the frequent approximate patterns with less than L edges first. Those small patterns tend to be missed by the randomized algorithm but can be mined efficiently. Expanding from those relatively small patterns, we reach candidate graph patterns by growing edges in a random order. Moreover, for any candidate graph g, instead of calculating the canonical form, we use the hash vector of the selected feature set of g. If a pattern with the same hash vector has already been found, graph g is considered to be reached (even if it is not). Otherwise, if g is frequent, we continue to grow edges from g.

After discovering all frequent approximate patterns in which edge s is embedded, we remove s from the graph database to shrink the search space. Also, we reset all sets of hash values. Because none of the patterns discovered afterwards contain s, it is safe to reset used hash vectors to avoid collisions. Algorithm RAM is shown in algorithm 1.

As we did not adopt canonical forms for efficiency, the algorithm may lose some patterns in a single run, i.e., if pattern A and B have exactly the same hash vector value, we may lose B if A has been reached first. However, since we grow edges in a random order, low miss rate can be achieved by multiple runs; i.e., in another run, B is recognized as frequent if it appears earlier than A. Furthermore, missing most of the non-maximal patterns is harmless to if their supergraph patterns can be discovered by the algorithm, i.e., if B and A are subgraphs of pattern C, we may find C by growing edges from A even if we failed to find B. Thus we recover B as a subgraph of C. Further analysis is presented in the following subsection.

6.3   Algorithm Analysis

As shown in the previous subsection, the randomized algorithm can miss patterns. We miss a pattern g when (1) a different graph with the same hash vector has been discovered already, and (2) all the connected subgraphs of g with one edge absent are missing.

When the feature set is well selected, we assume that their corresponding hash vectors are uniformly distributed. Then, the first factor is determined by the number of possible values of hash vectors, i.e. $\prod_{i=1}^{6} P_i$ in this paper for the features and hash functions we selected. Assume there are at most M approximate patterns with a given graph database and parameters which contain a same edge, the average number of graphs which share a same hash vector is $M/\prod_{i=1}^{6} P_i$. We define $P = min\{1, \prod_{i=1}^{6} P_i/M\}$, P quantifies the average probability of pattern being found without considering the second factor.

A connected subgraph with one edge of g absent is either a connected spanning subgraph, when e is not a bridge edge in g; or a non-trivial component of graph $g - e$ with $|V(g) - 1|$ vertices, when e is a bridge edge in g. The second factor is largely affected by the number of connected subgraphs with one edge absent, which is reversely related to the number of bridge edges of the pattern graph. The number of connected subgraphs with one edge absent varies a lot with the pattern graph, e.g., it is 2 for a line graph of size n while it is m for a cycle graph of size m. To simplify the problem, we assume that the number of connected subgraphs of a graph of size n with one edge absent is $n/2$.

Algorithm 1. *RAM*

   Input: Graph database D, parameter β, γ, α
   Output: FAG (frequent approximate graph) set $G(D)$
 1: Find one edge FAG set $G^1(D)$
 2: Find small FAG set (below L edges) $G^s(D)$
 3: $G(D) \leftarrow G^s(D)$
 4: **for** each graph $s \in G^s(D)$ **do**
 5:    $G(D) \leftarrow G(D) + RandomGrow(s, G^1(D), D)$
 6:    $G^s(D) \leftarrow G^s(D) - s$
 7:    **if** $\exists$ edge $e \in s, \forall s' \in G^s(D), e \in s'$ **then**
 8:      $G^1(D) \leftarrow G^1(D) - e$
 9:      Delete saved hash vectors
10:   **end if**
11: **end for**

   Subprocedure name: RandomGrow
   Input: FAG a, edge set S, Graph database D
   Output: frequent supergraphs of a
 1: $S' \leftarrow S$
 2: **while** $|S'| > 0$ **do**
 3:    Randomly select edge s from S'
 4:    **for** each candidate pattern $p \in a + s$ **do**
 5:      **if** p is frequent and $H(p)$ has not been saved **then**
 6:        output p and save $H(p)$
 7:        $RandomGrow(p, S, D)$
 8:      **end if**
 9:    **end for**
10:    $S' \leftarrow S' - s$
11: **end while**

Assume the average probability of a pattern graph of size n being found is P_n; combining above two factors, $P_n = P - P(1 - P_{n-1})^{(n/2)}$. Since we start the randomized algorithm only from larger patterns, the second part of the formula is negligible. Thus, we have $P_n \approx P$.

Supposing we would find approximately P of the maximal patterns in a single run, we would find approximately $1 - (1 - P)^q$ portion of the maximal patterns if we run RAM q times.

The hash vector of any pattern occupies only one bit in the multidimensional hash table. Therefore, P can be quite large. If the set of features is well designed, we can find most of the patterns at a much faster speed, as we avoid the tedious calculation of canonical forms. Moreover, if the amount of main memory is limited, we can adopt a hash function with small hash table, at the cost of losing more patterns in a single run. However, we can improve the accuracy by employing more runs.

7   Experiments Results

In this section, we report on our experiments, which validate the efficiency and effectiveness of the approximate graph mining algorithms. Due to Property 1, some patterns may not have any exact embedding. As a result, existing exact subgraph mining methods cannot be directly applied because they aim to find exact patterns. The performance of algorithm RAM is compared with that of the basic algorithm which can be considered a modification of existing depth first graph mining methods, e.g., FFSM [7], gSpan [18], etc. We do not compare RAM with any breadth first search algorithm because breadth first search is too space-consuming for approximate graph mining problems.

We use two kinds of datasets in our experiments: one real dataset and several synthetic datasets. The real dataset consists of 394 graphs. Each graph corresponds to a metabolic network of glycan. The network is generated from KEGG [22] using the second-level categories defined in [23]. In each graph, vertices represent enzymes, labeled with protein family ID. If there is an edge between enzymes (vertices) A and B, it indicates that A and B interact with each other. The synthetic data was generated using a method similar to that in [11]. The generator allows the user to specify the number of graphs, their average size, the number of seed graphs, the average size of seed graphs, the number of distinct labels, and the approximate level (the random probability that we change the edge label) of each edge.

In algorithm RAM, we adopt the set of hash functions for the real dataset. We set $P_1 = 53$, $P_2 = 53$, $P_3 = 43$, $P_4 = 43$, $P_5 = 13$, and $P_6 = 13$. All our experiments are performed on a 2.8GHZ, 2GB memory, Intel PC. Both algorithm RAM and the basic algorithm are compiled with VS2005.

7.1   Metabolic Pathways Dataset

We employ a categorized metabolic pathway dataset for the experiments. In the pathways graph databases, there are 394 graphs in total. The randomized algorithm starts when the size of the pattern graph is bigger than the size threshold 5. In each experiment, we only run RAM once.

Table 1. Results of Pathways Dataset (1)

(γ, β, α)	Exe.Time of Algorithm RAM	Basic Algorithm
(25,1,2)	9 s	25 s
(25,1,4)	12 s	33 s
(35,1,2)	8 s	17 s
(35,1,4)	10 s	26 s
(γ, β, α)	No. of Max. Patterns of RAM	Basic Algorithm
(25,1,2)	123	123
(25,1,4)	131	131
(35,1,2)	107	107
(35,1,4)	112	112

Table 2. Results of Pathways Dataset (2)

(γ, β, α)	Exe.Time of Algorithm RAM	Basic Algorithm
(25,1,2)	9 s	25 s
(25,2,2)	54 s	126 s
(35,1,2)	8 s	17 s
(35,2,2)	45 s	104 s
(γ, β, α)	No. of Max. Patterns of RAM	Basic Algorithm
(25,1,2)	123	123
(25,2,2)	461	467
(35,1,2)	107	107
(35,2,2)	431	431

First we show the runtime of both algorithms and the number of maximal approximate patterns. We set parameters (γ, β, α) to range from $(25, 1, 2)$ to $(35, 2, 4)$. As there are three different parameters, we show the results of both algorithms in the following tables. In table 1, in each subgroup, we keep β and γ unchanged and vary α from 2 to 4. In table 2, we vary β from 1 to 2. Lastly, in table 3, we vary γ from 25 to 35.

We can observe from the tables that overall, algorithm RAM is about 60% more efficient than the basic algorithm. On average, RAM can find no less than 99% of the total maximal patterns at a single run. When parameter γ decreases, the optimization of algorithm RAM tends to be more effective, due to the increase of frequent approximate patterns; the basic algorithm must calculate the canonical forms of every graph. The number of maximal frequent patterns grows quickly with parameter β, and it is consistent with the runtime of the algorithms.

When $\alpha = 2$, $\beta = 1$, and $\gamma = 25$, RAM found 123 maximal patterns, and the largest approximate pattern contained 14 edges and 14 vertices. The exact graph mining method only found 24 maximal patterns, and the maximal edge and vertex count of the exact patterns were 10 and 9, respectively. This indicates that approximate graph mining methods can find more and larger frequent patterns.

Last, we tested if the approximate patterns are useful in real applications. We divided the graph database into two parts. There were 300 database graphs in the first part and 94 in the second. We generated the exact and approximate patterns from the first set of patterns. Graphs in the second part were mixed with 100 arbitrary graphs as a test set.

Table 3. Results of Pathways Dataset (3)

(γ, β, α)	Exe.Time of Algorithm RAM	Basic Algorithm
(25,2,2)	54 s	126 s
(35,2,2)	45 s	104 s
(25,2,4)	63 s	149 s
(35,2,4)	50 s	125 s
(γ, β, α)	No. of Max. Patterns of RAM	Basic Algorithm
(25,2,2)	461	467
(35,2,2)	431	431
(25,2,4)	497	505
(35,2,4)	448	450

The exact and approximate patterns from the first part were used to predict whether a graph in the mixed set belongs to the original family. We found that approximate patterns yield about 15% higher accuracy than exact patterns on average, which indicates that approximate patterns better capture characteristics of the original data set.

7.2 Synthetic Dataset

In this section, we analyze the performance of the RAM algorithm on synthetic datasets. The synthetic graph dataset is generated as follows. First, a set of seed fragments is generated randomly, whose size is determined by a Poisson distribution with mean I. The size of each graph is a Poisson random variable with mean T. Seed fragments are then randomly selected and inserted into a graph one by one until the graph reaches its assigned size. We add an approximate level parameter P, which is the random probability that we change the edge label. More details about the synthetic data generator are available in [11]. A typical dataset may have the following setting: it has 500 graphs and uses 300 seed fragments with 30 distinct labels. On average, each graph has 40 edges and each seed fragment has 20 edges, the probability of change for any edge label is 0.05. This dataset is denoted as D500I20T40S300L30A0.05.

We first compared the performance of both algorithms with respect to the user input parameters β, γ, and α on the synthetic dataset D500I20T40S300L30A0.05. In test 1, we set α to 4 and β to 2, and vary γ from 50 to 200. In test 2, we set γ to 100 and β to 2, and vary α from 1 to 4. In test 3, we set γ to 100 and α to 2, and vary β from 1 to 4. Figure 6 shows the running time of both algorithms in tests 1, 2 and 3, respectively. Compared with the real dataset, RAM has a similar performance gain as in the synthetic datasets. However, it can be observed that the runtime increases significantly as β grows. We then tested the performance of the approximate graph mining algorithms with respect to different types of input datasets. Two characteristics of the datasets were selected for analysis: (1) the average size of graphs in the dataset (T), and (2) the number of graphs in the database (D). During the experiments, we set $\alpha = 4$, $\beta = 2$, and $\gamma = D/5$. Other non-varying parameters were the same as in the last experiments. In these tests, RAM can discover almost all maximal frequent approximate graphs that the basic algorithm can find, since we set P_i to be fairly large. To test the

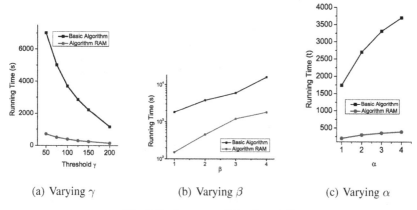

(a) Varying γ (b) Varying β (c) Varying α

Fig. 6. Performance of Varying α, β, γ

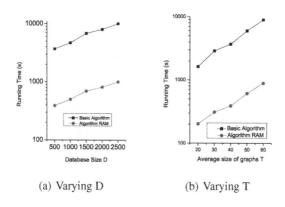

(a) Varying D   (b) Varying T

Fig. 7. Performance of Varying D, T

performance of RAM under a small space requirement, we set P_i to smaller values. We set $\alpha = 4$, $\beta = 2$, and $\gamma = 100$. Then, we varied the product of P_i, i.e., $\prod_{i=1}^{6} P_i$ from 10^4 to 10^8. Figure 8 shows the percentage of total maximal patterns algorithm RAM can find with different products of P_i. As we can see from the figure, the percentage of the total maximal patterns can be discovered increases greatly when products of P_i increases. Next, we tested the effects when we ran RAM multiple times with a small product of P_i. We set the product of P_i to 10^6, leaving the other parameters the same as those in the last experiment. Figure 8 indicates that the percentage of discovered maximal patterns rises significantly after we run RAM two or three times. However, when we ran it more times, increased amount of discovered patterns diminished. We will discuss the reason for this in a later section.

The empirical study clearly shows that RAM is much more time efficient than the basic algorithm under different sets of parameters , though it tends to lose a very small

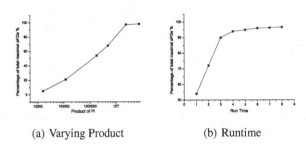

(a) Varying Product   (b) Runtime

Fig. 8. Percentage of total maximal FAGs varying product of P_i, and run time

fraction of maximal pattern with a small product of P_i and run times. However, when we execute RAM multiple times with a larger product of P_i, most frequent subgraphs can be discovered.

8 Discussion

In the experimental results section, we found that when there are more frequent approximate patterns than the available memory allows for, some patterns are more difficult to reach than others. To discover these patterns, we may have to run algorithm RAM many times. In this section, we discuss the reason and a possible solution.

If the hash functions are well designed, we can make the distribution of hash vectors close to uniform. However, the second factor, i.e., the number of connected subgraphs, is determined by the nature of the graph database. Even if we restrict the randomized search to larger patterns, some patterns may still be more vulnerable to not being discovered. To compensate the loss of patterns which have fewer subgraphs, one possible solutions is to add a probability parameter ϕ for those patterns, indicating the likelihood for continuing to search from those patterns when there is a hit on their hash value.

9 Conclusion

Graphs play an important role in modeling complex structural data, e.g., protein interaction networks, social networks, etc. In many applications, the graphs in the database may contain a number of unreliable edges. However, none of the current graph mining algorithms are designed to find the complete set of frequent patterns in graph databases with edge distortions.

We first introduced a formal definition of frequent approximate patterns in a graph database with edge distortion. Then we proposed a basic algorithm based on depth first search. As there are much more frequent patterns and embeddings, we presented the hashing-based randomized algorithm RAM. As shown in the experimental results, the approximate graph mining algorithm can find more and larger frequent patterns. More significantly, by approximate pattern mining, we may find important patterns that cannot be discovered by exact mining algorithms.

References

1. Bader, J., Chaudhuri, A., Rothberg, J., Chant, J.: Gaining confidence in high-throughput protein interaction networks. Nature Biotechnology 22(1), 78–85 (2004)
2. Chang, R., Podgurski, A., Yang, J.: Finding What's not there: a new approach to revealing neglected conditions in software. In: Proc. of ISSTA (2007)
3. Cong, G., Yi, L., Liu, B., Wang, K.: Discovering frequent substructures from hierarchical semi-structured data. In: Proc of SDM (2002)
4. Gunopulos, D., Mannila, H., Saluja, S.: Discovering All Most Specific Sentences by Randomized Algorithms Source. LNCS 1997(1997)
5. Hasan, M., Chaoji, V., Salem, S., Besson, J., Zaki, M.: ORIGAMI: Mining Representative Orthogonal Graph Patterns. In: Perner, P. (ed.) ICDM 2007. LNCS (LNAI), vol. 4597. Springer, Heidelberg (2007)
6. Holder, L., Cook, D., Djoko, S.: Substructure discovery in the subdue system. In: Proc. AAAI (1994)
7. Huan, J., Wang, W., Prins, J.: Efficient mining of frequent subgraphs in the presence of isomorphism. In: Proc. of ICDM (2003)
8. Huan, J., Wang, W., Prins, J., Yang, J.: SPIN: mining maximal frequent subgraphs from graph databases. In: Proc. of KDD (2004)
9. Inokuchi, A., Washio, T., Motoda, H.: An apriori-based algorithm for mining frequent substructures from graph data. In: Proceedings of PDKK (2000)
10. Koyuturk, M., Grama, A., Szpankowski, W.: An efficient algorithm for detecting frequent subgraphs in bioloical networks. Bionformatics 20, 200–207 (2004)
11. Kuramochi, M., Karypis, G.: Frequent subgraph discovery. In: Proc. of ICDE (2001)
12. Kuramochi, M., Karypis, G.: Finding frequent patterns in a large sparse graph. Data Min. Knowl. Discov. (2005)
13. Nijssen, S., Kok, J.: A quickstart in frequent structure mining can make a difference. In: Proc of KDD (2004)
14. Pei, J., Jiang, D., Zhang, A.: On Mining Cross-Graph Quasi-Cliques. In: Proc. of KDD (2005)
15. Park, J., Chen, M., Yu, P.: An effective hash based algorithm for mining association rules. In: Proc. SIGMOD, pp. 175–186 (1995)
16. Thomas, L., Valluri, S., Karlapalem, K.: MARGIN:Maximal Frequent Subgraph Mining. In: Proc. of ICDM (2006)
17. Yan, X., Han, J.: CloseGraph: Mining closed frequent graph patterns. In: Proc. of SIGKDD (2003)
18. Yan, X., Han, J.: gSpan: graph-based substructure pattern mining. In: Proc. of ICDM (2002)
19. Yan, X., Yu, P., Han, J.: Substructure similarity search in graph databases. In: Proc. of SIGMOD (2005)
20. Liu, J., Paulsen, S., Xu, X., Wang, W., Nobel, A., Prins, J.: Mining approximate frequent itemset from noisy data. In: ICDM (2005)
21. Zaki, M.: Efficiently mining frequent trees in a forest: algorithms and applications. In: IEEE TKDE (2005)
22. Kyoto Encyclopedia of Genes and Genomes, http://www.genome.jp/kegg/
23. Metabolic pathway categories in KEGG, http://www.kegg.com/kegg/pathway/map/map01100.html

Finding Frequent Items over General Update Streams

Sumit Ganguly, Abhayendra N. Singh, and Satyam Shankar

IIT Kanpur

Abstract. We present novel space and time-efficient algorithms for finding frequent items over general update streams. Our algorithms are based on a novel adaptation of the popular dyadic intervals method for finding frequent items. The algorithms improve upon existing algorithms in both theory and practice.

1 Introduction

There is a growing class of applications in areas of business and scientific data processing that continuously monitor large volumes of rapidly arriving data for detecting user-programmed scenarios, some of which may encode anomaly and exception conditions or desirable conditions. Although a deep analysis of the data can be done, it is both space and time consuming. Data streaming systems are designed to give fast, but possibly approximate answers to a class of queries while processing the input data in an online fashion. For example, consider a satellite data processing system where continuous and voluminous weather data has to be rapidly processed to give a forewarning of an emerging climate phenomenon. While deep analysis is possible, often, an early warning capability is very desirable, which though approximate, could then be used to trigger a deeper analysis. As another example, consider a biological experiment scenario where there are sensors attached to many biological subjects whose data is being continuously transmitted to a central server. Monitoring extremal aggregate conditions over to sensor readings are often useful indicators in such scenarios.

Central to the success of data streaming systems are highly space and time-efficient algorithms that can summarize input data streams while processing them in an online fashion. In this paper, we present novel algorithms for data stream processing in the same vein, specifically considering general data streams. In the general stream model, each input record indicates arbitrary insertions or deletions of an item, where, an item may be an IP-address, stock ticker, sensor-id, etc.. In this model, the sum of aggregate insertions (positive) and deletions (negative) for each item over the course of the stream may be either positive or negative. We address the problem of finding frequent items over general data streams.

The problems of finding frequent items and estimating item frequencies over data streams are among the most popular primitive operations over data streams [2,3,4,5,8,10]. Much of the research in this basic problem has centered around

B. Ludäscher and Nikos Mamoulis (Eds.): SSDBM 2008, LNCS 5069, pp. 204–221, 2008.

the insert-only streaming model [2,5,8,10] and the strict update models [3,6] respectively. For general streams, there are two known approaches towards the problem of finding approximate frequent items, namely, the non-adaptive group testing approach [4] and the reversible sketches approach [11]. In this paper, we present the random dyadic approach towards finding frequent items over general streams. The proposed algorithm is novel, and extends the applicability of the popular dyadic intervals technique for strict streams to general streams.

Data Streaming Model. A data stream σ over the domain $[1, n] = \{1, 2, \ldots, n\}$ is modeled as an unbounded sequence of records of the form $(pos, i, \delta v)$, where, pos is the current sequence index, $i \in [1, n]$ and $\delta v \in \mathbb{Z}$. Here, $\delta v > 0$ signifies insertion(s) of instance(s) of i and $\delta v < 0$ signifies deletion(s) of instance(s) of i. For each data item $i \in [1, n]$, its frequency $f_i(\sigma)$ is defined as

$$f_i(\sigma) = \sum_{(pos, i, \delta v) \,\in\, \text{stream}} \delta v, \quad i \in [1, n] .$$

In this paper, we consider the *general model*, where, the n-dimensional frequency vector $f(\sigma) \in \mathbb{Z}^n$. The frequency moment F_1 of a general stream is defined as the sum of the absolute values of the frequencies, that is, $F_1(\sigma) = \sum_{i \in [1,n]} |f_i(\sigma)|$. The second moment of the frequency vector is defined as $F_2(\sigma) = \sum_{i \in [1,n]} (f_i(\sigma))^2$. The data stream model of processing permits online computations over the input sequence using sub-linear space.

Conventions. (a) We will assume that the domain size n is a power of two. (b) By a data stream, we always mean the current state of the stream and hence we drop the stream argument σ; for example, f_i abbreviates $f_i(\sigma)$.

Problem definitions. In this paper, we consider the following two problems. Let $0 < \phi < \epsilon < 1$.

1. *Finding F_1-based frequent items,* denoted by APPROXFREQ$_1(\epsilon, \phi)$ is: return all $i \in [1, n]$ such that $f_i(\sigma) \geq \epsilon F_1$ and do not return any i such that $f_i \leq (\epsilon - \phi)F_1$. A randomized algorithm for this problem satisfies the above property for all items returned with probability $1 - \delta$.
2. *Finding F_2-based frequent items,* denoted by APPROXFREQ$_2(\epsilon, \phi)$ is: return all items $i \in [1, n]$ such that $|f_i| \geq (\epsilon F_2)^{1/2}$, and no i such that $|f_i| < ((\epsilon - \phi)F_2)^{1/2}$. A randomized algorithm satisfies the above properties with a total success probability of at least $1 - \delta$.

In this paper, we design randomized algorithms for finding F_1 and F_2-based frequent items whose space requirement is nearly linear in ϕ^{-1}.

Contributions. We present novel, space and time-efficient algorithms to solve the problems stated above. For the problem of finding frequent items, our technique extends the applicability of the popular dyadic intervals technique for strict streams to general streams. We present two algorithms for the problem

APPROXFREQ$_2(\epsilon, \phi)$ which improve the space requirement of the existing algorithm [4] by a factor of $O(\frac{1}{\phi})$. The solution to the F_1-based frequent items problem is shown to have better properties of precision and recall. The algorithms perform well in experiments and have rigorous space versus accuracy guarantees.

2   Review

In this section, we review relevant algorithmic techniques for processing general data streams.

2.1   Review: Finding Approximate Frequent Items

We review two approaches for finding approximate frequent items over general streams, namely, non-adaptive group testing [4] and reversible sketches [11].

Non-adaptive group testing. A collection of s hash tables $T_1, \ldots, T_s$ is kept, each consisting of b buckets numbered 1 to b. Associated with each hash table T_j is a pair-wise independent random hash function $h_j : [1, n] \rightarrow [1, b]$. Each bucket of a table contains a two dimensional array $U[0 \ldots 1, 1 \ldots \log n]$ of integer counters[1]. We refer to a specific entry of a bucket as $T_j[r].U[v][k]$, where, j is the table index in $[1, s]$, r is the bucket index in $[1, b]$, v is a bit value that is either 0 or 1 and k is a bit position with value from $[1, \log n]$. Corresponding to each stream record of the form (pos, x, Δ), the data structure (initialized to all zeros) is updated as follows. Let $x = x_{\log n} x_{\log n - 1} \ldots x_2 x_1$ be the binary representation of x.

$$T_j[h_j(x)].U[x_k][k] = T_j[h_j(x)].U[x_k][k] + \Delta$$
$$j = 1, \ldots, s, k = 1, \ldots, \log n \ .$$

For the problem APPROXFREQ$_1(\epsilon, \phi)$, where, $\phi < \epsilon$, b is set to $\lceil \frac{2}{\phi} \rceil$ and s is set to $O(\log((\phi\delta)^{-1}(\log(1/\phi))))$ in order to ensure that the problem APPROXFREQ$_1(\epsilon, \phi)$ is solved with error probability at most δ. In addition, a data structure for estimating F_1 of the stream to within a constant factor (say, $(1 \pm \frac{1}{8})$) is also kept. The procedure for inference is the following. A bucket $T_j[r]$ contributes at most one element x towards a set of candidate frequent items as follows. Let $\hat{F}_1 = (\text{estimate of } F_1) / (1 + 1/8)$. For each $j \in [1, s]$ and $r \in [1, b]$, the procedure RETRFREQUENT(j, r) is invoked for each hash table T_j and each bucket $r \in [1, b]$ of T_j to obtain a candidate set of non-NIL elements returned from the invocation RetrFrequent (j, r). These are the candidate frequent items–their frequencies are estimated by treating the data structure as a COUNT-MIN sketch structure [3] and $(x, \hat{f}_x)$ is returned as a frequent item and its estimate provided, $\hat{f}_x \geq (\epsilon - \phi)\hat{F}_1$. The space requirement of this technique

---

[1] For $k \in [1 \ldots \log n]$, $r \in [1, b]$ and $j \in [1, s]$, we have $T_j[r].U[0][k] + T_j[r].U[1][k] = \sum_{h_j(x)=r} f_x$. The latter quantity is stored in another counter associated with the bucket $T_j[r]$ thus reducing the storage associated with each bucket from $2 \log n$ counters to $1 + \log n$ counters. This optimization is done in the experiments.

procedure RETRFREQUENT(j, r) // $j \in [1, s], r \in [1, b]$
Returns $x \in [1, n]$ or NIL in case of perceived ambiguity.
$x := 0$;
for $k = 1$ *to* $\log n$ {
    if $(T_j[r].U[1][k] \geq (\epsilon - \phi)\hat{F}_1)$ *and* $(T_j[r].U[0][k] \geq (\epsilon - \phi)\hat{F}_1)$
       return NIL
    else if $(T_j[r].U[1][k] \geq (\epsilon - \phi)\hat{F}_1)$ $x := x + 2^{k-1}$
    else if $(T_j[r].U[0][k] < (\epsilon - \phi)\hat{F}_1))$ *return* NIL
}

is $O(\phi^{-1}(\log n)(\log F_1)(\log((\phi\delta)^{-1} \log \phi^{-1})))$ bits. The time required to process each stream update is $O((\log((\phi\delta)^{-1} \log \phi^{-1})) \log n)$.

The group testing approach was used by [4] to present algorithms for the problem APPROXFREQ$_2(\epsilon, \phi)$, that is, retrieve all items i such that $f_i > (\epsilon F_2)^{1/2}$ and not retrieve any items i with $f_i < ((\epsilon - \phi)F_2)^{1/2}$. The data structure has the same structure as the one described above; in addition to the array U kept for each hash table bucket $T_j[r]$, this structure also keeps $\log n$ AMS sketches, that is, $T_j[r].U[v][k]$ is an AMS sketch of the sub-stream defined by the items that map to bucket r of table j and have value v in bit position k. The asymptotic space requirement is $O(\frac{1}{\phi^2}(\log n)(\log F_1)(\log((\phi\delta)^{-1} \log \phi^{-1})))$ bits [4].

Reversible sketches. The reversible sketches paper [11] keeps $s = O(\log \frac{n}{\delta})$ tables T_j, where, each table has b buckets and each bucket is simply a counter that stores the sum of the frequencies of all the items that map to that bucket. A bucket $T_j[r]$ is considered to contain a potential frequent item provided, $T_j[r] \geq (\epsilon - \phi)F_1$. The reversible sketches does not keep any additional bits in the data structure to retrieve the items. Instead, the hash function is constructed in a modular manner that allows the retrieval of the items. The main problem with the approach is that the retrieval method can be very time-consuming (as we found in our experiments), since, the number of candidate frequent items can be as large as n^α, for α ranging from 0.5 to 0.9.

2.2   Review: Use of Dyadic Intervals

The dyadic intervals technique is a simple building block for design of algorithms for insert-only and strict streams. We briefly review the technique and its applications. Recall that we have assumed n to be a power of 2.

A dyadic interval at level l is an interval of size 2^l from the family of intervals of the form $[i2^l + 1, (i + 1)2^l]$, for $0 \leq i \leq \frac{n}{2^l} - 1$ and $0 \leq l \leq \log n$. The set of dyadic intervals of levels 0 through $\log n$ form a complete binary tree as follows. The root of the tree is the single dyadic interval $[1, n]$ and the leaf nodes are the singleton intervals. Moreover, for $0 \leq l < \log n$, each dyadic interval at level l of the form $I = [i2^l + 1, (i + 1)2^l]$ has two children at level $l - 1$, namely, the left and the right halves of I_h. The left child of I is the interval $[i2^l + 1, (2i + 1) \cdot 2^{l-1}]$ and the right child is the interval $[(2i + 1) \cdot 2^{l-1} + 1, (i + 1)2^l]$. The frequency of a dyadic interval I is defined as the sum of the individual frequencies of items in I, and is denoted as f_I.

The following observations can be made for strict streams (i.e, $f_i \geq 0$, for all $i \in [1, n]$). Since each level 0 item belongs to one and only one dyadic interval at a given level l, the sum of the interval frequencies at level l is the same as the sum of the item frequencies at level 0, which is F_1. That is, $F_1 = \sum\{f_I \mid I$ is a dyadic interval at level $l\}$, for each $l = 0, 1, \ldots, \log n$. If an item i is frequent (i.e., $f_i \geq \epsilon F_1$), then the dyadic interval I that contains i at any level l has frequency $f_I \geq f_i \geq \epsilon F_1$ and is therefore also frequent at level l.

Frequent items algorithm using dyadic intervals. An algorithm for solving APPROXFREQ(ϵ, ϕ) is as follows. For each level $l = 0, \ldots, \lfloor \log \epsilon n \rfloor$, a data structure for estimating the frequency of a given dyadic interval (for e.g., a COUNT-MIN sketch sketch or COUNTSKETCH) is kept. The elements at level l are the set of dyadic intervals interval I at level l and the frequency of an interval I is defined as the sum of the frequencies of the items that belong to I, that is, the leaves of the sub-tree of the dyadic binary tree rooted at I: $f_I = \sum\{f_i \mid i \in I\}$. The set of dyadic intervals at level l are identified with their starting position modulo 2^l. Corresponding to a stream update (pos, x, Δ), we propagate the update $(pos, \lfloor \frac{x}{2^l} \rfloor, \Delta)$ to the data structure at level l, for $l = 0, 1, \ldots, \lfloor \log(\epsilon n) \rfloor$. The inference procedure for finding frequent items is as follows. Start from the structure at level $l_{\max} = \lfloor \log(\epsilon n) \rfloor$ and estimate the frequencies of each of the $2^{l_{\max}}$ dyadic intervals at level l_0 using the data structure. Select those intervals whose estimated frequency is at least $(\epsilon - \frac{\phi}{2})F_1$; consider its left and right child, estimate their frequencies using the structure at the next lower level, retain only those intervals whose estimated frequency is at least $(\epsilon - \frac{\phi}{2})F_1$; this process is continued until the ground level is reached and the structure at level 0 is processed.

The main problem in applying this technique to general streams is that, since, item frequencies can be negative, a frequent item or interval at level l may be contained in an interval at level $l + 1$ that is not frequent at its level.

3   Algorithm COUNTSKETCH DYADIC

In this section, we present the algorithm COUNTSKETCH DYADIC for finding frequent items over general streams with respect to the second moment. That is, the problem APPROXFREQ$_2(\epsilon, \phi)$ is to retrieve all items i such that $|f_i| \geq (\epsilon F_2)^{1/2}$ and not return any i such that $|f_i| < ((\epsilon - \phi)F_2)^{1/2}$. The solution presented improves the space requirement of the current best algorithm by a factor of $O(\frac{1}{\phi})$ while preserving time-efficiency of processing stream updates and of retrieving the frequent items.

The basic idea is to randomly re-distribute the items in the dyadic intervals using random permutations. Let π be a random permutation of $[1, n]$ that is very nearly t-wise independent ($t = 3$ will suffice). A typical way of generating π is by the use of Fiestel permutations using Luby and Rackoff's technique [9]. The advantage of using Fiestel permutations is that it is very efficiently computed and the inverse permutation is also very efficiently computed as follows. Given a number x expressed using $2m$ bits, let L denote the top-order m bits and R

denote the low order m bits; thus $x = (L, R)$. A single round Fiestel permutation is a map $\pi : (L, R) = (R, L \oplus f(R))$, where, f is a t-wise independent hash function $f : [0, 2^m - 1] \to [0, 2^m - 1]$ and $\oplus$ denotes the bit-wise exclusive or operation. The inverse of a single-round Fiestel permutation is the map $(L, R) \to (f(L) \oplus R, L)$ and is thus easily computed. Luby and Rackoff show that four rounds of Fiestel permutations suffice to generate very nearly t-wise independent permutations such that the distance between the uniform distribution over $2m$ bits and the distribution of the Luby-Rackoff permutations is at most $t^2 \cdot 2^{-m}$. We note that for $t = 3$, there are known constructions for exactly 3-wise independent permutation families. However, for $t > 3$, constructions for exact independent random permutations are not known [7].

Let $\pi_1, \ldots, \pi_s$ be very nearly 4-wise independent permutations that are obtained in the manner explained above. For each π_j and each level $l = 0, \ldots, l_{\max}$, a COUNTSKETCH structure of height ck' and width w, where, $k' = \lceil \frac{1}{\phi} \rceil$ and the parameters c, w and $l_{\max}$ will be fixed in the analysis. For each $j = 1, 2, \ldots, s$, let $\xi_{j,x} \in \{-1, +1\}$ denote a four-wise random mapping for each $x \in [1, n]$ (i.e., an AMS sketch [1]). This family is independent of the sketches used by the COUNTSKETCH structures themselves. The processing of each stream record (pos, x, v) is as follows, for each $j = 1, 2, \ldots, s$ and $l = 0, 1, \ldots, l_{\max}$, the update $(pos, \lfloor \pi_j(x)/2^l \rfloor, v \cdot \xi_{j,x})$ is propagated to the COUNTSKETCH structure at level l corresponding to permutation π_j.

The retrieval of the frequent items is done as described in Section 2.2 with minor differences. The following procedure is repeated for each permutation index $j = 1, 2, \ldots, s$. The retrieval procedure starts from level $l_{\max}$ and scans all the dyadic intervals at this level and keeps those intervals whose estimated frequency is at least the threshold $((\epsilon - \frac{\phi}{2}) F_2)^{1/2}$. The children of such intervals are considered in turn–these are the candidate intervals at level $l_{\max} - 1$. Among these intervals, those whose estimated frequency crosses the threshold $((\epsilon - \frac{\phi}{2}) F_2)^{1/2}$ are retained, and the rest are discarded. The process continues to the next lower level in this manner until level 0 has been processed. The candidate intervals or items at a level are are those whose absolute value of the estimated frequency crosses the threshold $((\epsilon - \frac{\phi}{2}) F_2)^{1/2}$. An estimate $\hat{F}_2$ of F_2 that is correct to within a relative accuracy of $1 \pm \frac{1}{4}$ and probability $1 - \frac{\delta}{2}$ is used and can obtained using the FAST-AMS algorithm of [12] that requires space $O((\log \frac{1}{\delta})(\log F_1))$ bits and time $O(\log \frac{1}{\delta})$ for processing a stream update.

3.1 Analysis

The residual second moment [2] denoted by $F_2^{res}(k)$ is the sum of the squares of the frequencies of all items in the stream, except for the top-k frequencies in terms of absolute value. More formally, if $rank$ is a permutation of the items such that $|f_{rank(j)}| \geq |f_{rank(j+1)}|$, for $1 \leq j \leq n - 1$, then, $F_2^{res}(k) = \sum_{j=k+1}^{n} f_{rank(j)}^2$, defined for $k \in [0, n - 1]$.

For a permutation π_j, $j \in [1, s]$, $i \in [1, n]$ and level $l \in [0, l_{\max}]$, let $g_{j,l,i}$ be the frequency of the unique dyadic interval I to which $\pi_j(i)$ maps at level l. Let $\hat{g}_{j,i,l}$ denote the estimate obtained from the COUNTSKETCH structure for the unique dyadic interval at level l containing $\pi_j(i)$ at level l. Define the event NoCollision$_l(i)$ if the dyadic interval to which $\pi_j(i)$ maps at level l does not contain any of the top-k frequencies (except perhaps itself). Define

$$\text{NoCollision}(i, l_{\max}) = \text{NoCollision}_1(i) \text{ and } \text{NoCollision}_2(i) \text{ and } \ldots$$
$$\ldots \text{ and } \text{NoCollision}_{l_{\max}}(i) .$$

Lemma 1. *For* $1 \leq j \leq s$ *and* $i \in [1, n]$,

$$\Pr\left\{ |\hat{g}_{j,i,l} - f_i\xi_{j,i}| \leq \left(\frac{32 F_2^{res}(k')}{k'}\right)^{1/2}, \forall l : 0 \leq l \leq l_{\max} \right\} \geq \frac{5}{8} .$$

Proof. Fix a permutation π_j and abbreviate it by π and the corresponding sketch family as $\{\xi_i\}_{i\in[1,n]}$. Similarly, abbreviate $g_{j,i,l}$ by $g_{i,l}$, etc.. Fix a top-k element j, $j \neq i$. Let $l \in [0, l_{\max}]$. Due to t-wise independence of π_j, $t \geq 2$, the probability that i and j map to the same dyadic interval at level l is

$$\frac{\binom{n-2}{2^l-2}}{\binom{n-1}{2^l-1}} = \frac{2^l - 1}{n - 1} < \frac{2^l}{n} .$$

Therefore, $\Pr\{\text{NoCollision}_l(i)\} \geq 1 - \frac{k2^l}{n}$, by union bound. Since, NoCollision$_l(i)$ implies NoCollision$_{l'}(i)$, for $l' < l$, $\Pr\{\text{NoCollision}(i, l_{\max})\} \geq 1 - \frac{k2^{l_{\max}}}{n}$. Let $k' = 8\lceil\frac{1}{\phi}\rceil$. Fix an item i. For $j \in [1, n]$ and $j \neq i$, the indicator variable $u_{l,j}$ is defined as follows: it is 1 if j maps to the same dyadic interval at level l as i and is 0 otherwise. Thus,

$$g_{l,i} = f_i\xi_i + \sum_{j\neq i} f_j\xi_j u_{l,j} .$$

Assuming NoCollision$_l(i)$, we have by direct calculation

$$\mathsf{E}\left[(g_{l,i} - f_i\xi_i)^2\right] < F_2^{res}(k')\frac{2^l}{n} .$$

This repeats the arguments of Alon, Matias and Szegedy [1]. By Markov's inequality,

$$\Pr\left\{(g_{l,i} - f_i\xi_i)^2 < t F_2^{res}(k')\frac{2^l}{n}\right\} \geq 1 - \frac{1}{t}$$

or, equivalently,

$$|g_{l,i} - f_i\xi_i| < \left(\frac{t F_2^{res}(k')2^l}{n}\right)^{1/2} \text{ with prob. } 1 - \frac{1}{t} .$$

The expression $\frac{2^l}{n}$ is largest for $l = l_{\max}$. Therefore, letting $l_{\max} = \lceil \log \frac{n}{4k't} \rceil$ ensures that $\left(\frac{t F_2^{res}(k')2^l}{n}\right)^{1/2} \leq \left(\frac{F_2^{res}(k')}{4k'}\right)^{1/2}$. Therefore, with this choice of $l_{\max}$, we have

$$|g_{l,i} - f_i \xi_i| < \left(\frac{F_2^{res}(k')}{4k'} \right)^{1/2} \quad \text{with prob. } 1 - \frac{1}{t} . \tag{1}$$

Define $F_{2,l}$ to be the sum of the squares of the frequencies of the dyadic intervals at level l. For $i \in [1, n]$ and $r \in [1, \frac{n}{2^l}]$, let $v_{l,i,r} = v_{i,r}$ denote the indicator variable that is 1 if i is mapped to the dyadic interval $[r2^l + 1, (r+1)2^l]$. Therefore,

$$F_{2,l} = \left(\sum_{r=0}^{n/2^l - 1} \sum_{i=1}^{n} f_i v_{i,r} \xi_i \right)^2 .$$

By direct calculation, $\mathsf{E}[F_{2,l}] = F_2$ and $\mathsf{Var}[F_{2,l}] \le 5F_2$. Repeating the argument of COUNTSKETCH algorithm [2], with height $32k'$ and width w at each level,

$$|\hat{g}_{l,i} - g_{l,i}| \le \left(\frac{F_2^{res}(32k')}{4k'} \right)^{1/2} \quad \text{with prob. } 1 - 2^{-\Omega(w)} .$$

Combining with (1), we have,

$$\Pr \left\{ \forall l : 0 \le l \le l_{max} \left(|\hat{g}_{i,l} - f_i \xi_i| \le \left(\frac{F_2^{res}(k')}{k'} \right)^{1/2} \right) \right\}$$

$$\ge 1 - l_{max} \left(2^{-\Omega(w)} + \frac{1}{t} \right) .$$

Choosing $l_{max} = \lfloor \log \frac{\phi n}{32 \log(\phi n)} \rfloor$, $t = 8l_{max}$ and $w = O(\log \log l_{max})$, the error probability in the above expression is $\frac{2}{8}$. Since, the probability of NoCollision(i, l_{max}) is $\frac{7}{8}$, combining, we obtain the lemma. □

Theorem 1 summarizes the space, accuracy and time properties.

Theorem 1. *The algorithm* COUNTSKETCH DYADIC *with height* $ck' = 32\lceil \frac{1}{\phi} \rceil$, *width* $w = O(\log \log(\phi n))$, *maximum dyadic level* $l_{max} = \lfloor \log \frac{\phi n}{32 \log(\phi n)} \rfloor$ *and number of permutations* $s = O(\log \frac{1}{\phi \delta})$ *solves the problem* APPROXFREQ$_2(\epsilon, \phi)$ *with probability* $1 - \delta$ *with the following characteristics.*

$$\text{Space } O \left(\frac{1}{\phi} \left(\log \frac{\phi n}{\log(\phi n)} \right) \left(\log \frac{1}{\phi \delta} \right) (\log \log(n\phi)) (\log F_1) \right)$$

$$\text{Update Time } O \left(\left(\log \frac{\phi n}{\log(\phi n)} \right) (\log \log n) (\log \frac{1}{\phi \delta}) \right)$$

$$\text{Retrieval Time } O \left(\frac{\log(\phi n)}{\phi} (\log \log(n\phi)) (\log \frac{1}{\phi \delta}) \right) . \quad \square$$

The proposed algorithm improves the space requirement for solving the APPROXFREQ$_2(\epsilon, \phi)$ problem as compared to the variational deltoids algorithm [4] by reducing the dominant term in the space complexity expression from $O(\frac{1}{\phi^2})$ to $O(\frac{1}{\phi})$.

4   Algorithm COUNTSKETCH LINEAR

An improvement of the variational deltoids algorithm of [4] for the problem APPROXFREQ$_2(\epsilon, \phi)$ that reduces the dominant term in the space complexity expression from $O(\frac{1}{\phi^2})$ to $O(\frac{1}{\phi})$ can be designed although it appears to have higher constant factors than the COUNTSKETCH DYADIC algorithm discussed above. We briefly present the design and analysis of such an algorithm which we term as COUNTSKETCH LINEAR.

The data structure consists of s tables $T_1, \ldots, T_{s_1}$, each consisting of ck' buckets, where, $k' = \lceil \frac{1}{\phi} \rceil$, where, $c = 8$ and $s_1 = O(\log \frac{k' \log(1/\delta)}{\delta})$. Each bucket $T_j[r]$ has an array of sketches $U[v][k][s_2][s_3]$, where, $v \in \{0,1\}$ denotes a bit value, $k \in [1, \log n]$ denotes a bit position, $s_2 = O(1)$ (to be fixed later) and $s_3 = O(\log \log n)$. Corresponding to each table T_j, we keep $s_2 \cdot s_3$ independent families of AMS sketches denoted by $\xi_{x,j,u,w}$, where, $x \in [1,n]$, $j \in [1,s_1]$, $u \in [1,s_2]$ and $w \in [1,s_3]$. Each stream update of the form (pos, x, Δ) is processed as follows. Let $x = x_{\log n} x_{\log n - 1} \ldots x_2 x_1$ denote the binary representation of x.

$$T_j[h_j(x)].U[x_k][k][u][w] = \Delta \cdot \xi_{x,j,u,w},$$
$$j \in [1,s_1], k \in [1, \log n], u \in [1,s_2], v \in [1,s_3] \ .$$

The time taken to process each stream update is therefore $O(s_1 s_2 s_3 \log n) = O((\log \frac{\log(1/\delta)}{\phi \delta})(\log n)(\log \log n))$. A set of candidate frequent items is obtained by calling procedure $Retrieve(j,r)$, for $j \in [1,s_1]$ and $r \in [1,h]$ as presented in Figure 1. A second verification step is then performed wherein the frequency of each candidate frequent item x is estimated as $\hat{f}_x$ by treating the structure

---

procedure $Retrieve(j,r)$
Retrieves a potential candidate frequent item from $T_j[r]$
$x := 0$;
for $k := 1$ to $\log n$
    $c_0 := 0$; $c_1 := 0$;
    for $w = 1$ to s_3 do
        $\bar{U}[0][k][w] := \mathrm{avg}_{u=1}^{s_2}(T_j[r].U[0][k][u][w])^2$;

        $\bar{U}[1][k][w] := \mathrm{avg}_{u=1}^{s_2}(T_j[r].U[1][k][u][w])^2$;

        if $(\bar{U}[0][k][w] > \bar{U}[1][k][w])$ $c_0 := c_0 + 1$;

        else if $(\bar{U}[1][k][w] > \bar{U}[0][k][w])$ $c_1 := c_1 + 1$;

    endfor
    if $(c_1 > s_3/2)$ $x := x + 2^k$ elseif $(c_0 < s_3/2)$ return NIL ;
endfor
return x;

Fig. 1. Finding frequent items: Algorithm COUNTSKETCH LINEAR

as a standard COUNTSKETCH structure. The pair $(x, \hat{f}_x)$ is returned provided $|\hat{f}_x| \geq ((\epsilon - \frac{\phi}{2})\hat{F}_2)^{1/2}$. An estimate $\hat{F}_2$ such that $|\hat{F}_2 - F_2| \leq \frac{F_2}{4}$ is obtained using the FAST-AMS algorithm [12] using $O(\log \frac{1}{\delta})$ hash tables, each having $O(1)$ buckets.

Analysis of COUNTSKETCH LINEAR

Lemma 2. *Suppose* $s_2 \geq \frac{40\epsilon}{\epsilon - \phi/2}$, $h = ck' \geq 8\lceil\frac{1}{\phi}\rceil$. *If* $|f_x| > (\epsilon F_2^{res}(k'))^{1/2}$, *then, for any fixed* $j \in [1, s_1]$, *the probability that procedure* Retrieve$(j, h_j(x))$ *returns* x *is at least* $\frac{5}{8}$.

Proof. Fix a table index j. Let

$$X(v, k, w) = X_j(v, k, w) = \text{avg}_{u=1}^{s_2}(T_j[h_j(x)].U[v][k][u][w])^2,$$
$$G_{j,k}(x) = \sum\{f_y^2 \mid h_j(y) = h_j(x) \text{ and } y_k = x_k\} \text{ and}$$
$$H_{j,k}(x) = \sum\{f_y^2 \mid h_j(y) = h_j(x) \text{ and } y_k = \bar{x}_k\} .$$

By arguments of [1],

$$\mathsf{E}[X(x_k, k, w) - X(\bar{x}_k, k, w]] = G_{j,k}(x) - H_{j,k}(x),$$
$$\mathsf{Var}[X(x_k, k, w) - X(\bar{x}_k, k, w)] \leq \frac{5}{s_2}(G_{j,k}(x) + H_{j,k}(x))^2$$

By Chebychev's inequality,

$$\mathsf{Pr}\{X(x_k, k, w) - X(\bar{x}_k, k, w) \leq 0\} \leq \frac{\mathsf{Var}[X(x_k, k, w) - X(\bar{x}_k, k, w)]}{(\mathsf{E}[X(x_k, k, w) - X(\bar{x}_k, k, w)])^2}$$
$$\leq \frac{5}{s_2} \cdot \frac{G_{j,k}(x) + H_{j,k}(x)}{G_{j,k}(x) - H_{j,k}(x)} \quad (2)$$

Define the event NoCollision$_j(x)$ as: none of the top-k' items map to the same bucket as x in table T_j (except perhaps x itself). Therefore,

$$\mathsf{Pr}\{\text{NoCollision}_j(x)\} \geq 1 - \frac{k'}{ck'} = 1 - 1/c .$$

We have $G_{j,k}(x) \geq f_x^2 \geq \epsilon F_2^{res}(k')$. Assuming NoCollision$_j(x)$,

$$\mathsf{E}[H_{j,k}(x) \mid \text{NoCollision}_j(x)] \leq \frac{F_2^{res}(k')}{ck'}$$

and therefore by Markov's inequality,

$$\mathsf{Pr}\left\{H_{j,k}(x) \leq \frac{8F_2^{res}(k')}{ck'}\Big|\text{NoCollision}_j(x)\right\} \geq \frac{7}{8} .$$

Let $k' = \lceil \frac{1}{\phi} \rceil$ and $c = 16$. Then, $\frac{8F_2^{res}(k')}{ck'} \leq \frac{\phi F_2^{res}(k')}{2}$. Substituting in (2) and assuming $\text{NoCollision}_j(x)$,

$$\Pr\{X(x_k, k, w) - X(\bar{x}_k, k, w) \leq 0\} \leq \frac{5\epsilon}{s_2(\epsilon - \phi/2)} \leq \frac{1}{8}, \text{ if } s_2 \geq \frac{40\epsilon}{\epsilon - \phi/2} \quad . \quad (3)$$

Note that the probability in (3) depends on (a) $\text{NoCollision}_j(x)$, which holds for all k if it holds for any one, and, (b) is derived for any $G_{j,k}(x)$ and $H_{j,k}(x)$ satisfying $G_{j,k} \geq f_x^2$ and $H_{j,k}(x) \leq \frac{F_2^{res}(k')}{k'}$. Since, this is the worst case, the property holds for all k, as stated below. Suppose $s_2 \geq \frac{40(\epsilon+\phi)}{\epsilon-\phi}$. Then,

$$\Pr\{X(x_k, k, w) - X(\bar{x}_k, k, w) > 0, \ \forall k \in [1, \log n] \mid \text{NoCollision}_j(x)\} \geq \frac{7}{8} \quad (4)$$

Let $W(x, k)$ be the number of w's in $[1, s_3]$ for which $X(x_k, k, w) > X(\bar{x}_k, k, w)$. Then, $\mathsf{E}\big[W(x, k) \mid \text{NoCollision}_j(x)\big] \geq \frac{7s_3}{8}$ and by Chernoff's bounds,

$$\Pr\left\{W(x, k) < \frac{s_3}{2} \mid \text{NoCollision}_j(x)\right\} < e^{-9s_3/56} < \frac{1}{8\log n},$$

$$\text{if } s_3 \geq \frac{56}{9}\ln(8\log n) \ .$$

Combining using union bounds,

$$\Pr\{W(x, k) \geq 0.5s_3, \forall k \in [1, \log n]\} \geq 1 - \frac{\log n}{8\log n} = \frac{7}{8} \quad . \quad (5)$$

Combining the error probability using union bound, namely, $\frac{1}{8}$ for $\text{NoCollision}(x)$, the total error probability is at most $\frac{2}{8}$. Therefore, the probability that x is retrieved as a frequent item by procedure $Retrieve(j, r)$ is at least $\frac{6}{8}$.   □

Note that for $\phi < \epsilon$, $1 \leq \frac{\epsilon}{\epsilon-\phi/2} \leq 2$. We therefore have the following theorem.

Theorem 2. *Suppose* $|\hat{F}_2 - F_2| \leq \frac{F_2}{4}$ *with probability* $1 - \delta/2$, $s_1 = O$ $(\log \frac{\log(1/\phi\delta)}{\phi\delta})$, $s_2 = O(1)$, $s_3 = O(\log\log n)$ *and the height of the hash tables is* $ck' = O(\lceil \frac{1}{\phi} \rceil)$. *Then the algorithm* COUNTSKETCH LINEAR *solves the* APPROXFREQ$_2(\epsilon, \phi)$ *with probability* $1 - \delta$ *with the following characteristics.*

$$\text{Space} \quad O\left(\frac{1}{\phi} \cdot (\log n)(\log\log n)\left(\log \frac{\log(1/\phi\delta)}{\phi\delta}\right)(\log F_1)\right)$$

$$\text{Update Time} \quad O\left((\log n)(\log\log n)\log \frac{\log(1/\phi\delta)}{\phi\delta}\right)$$

$$\text{Retrieval Time} \quad O\left(\frac{Space}{\log F_1}\right) \ . \quad □$$

A comparison of Theorems 1 and 2 shows that the properties of COUNTSKETCH LINEAR and COUNTSKETCH DYADIC are similar although COUNTSKETCH LINEAR has slightly worse constants. Both algorithms improve over the space requirement of $O(\frac{1}{\phi^2} \cdot \text{poly-log}(n, F_1))$ of the variational deltoids algorithm [4].

5   Algorithm Count-Min Dyadic

In this section, we present an extension of the COUNT-MIN algorithm for finding F_1-based frequent items for general streams by using the dyadic intervals technique. We use s random permutations $\pi_1, \ldots, \pi_s$. Corresponding to π_j, we keep a dyadic intervals based data structure for levels 0 through $l_{\max}$ as described in Section 2.2. Corresponding to each permutation π_j and each dyadic level, we keep a COUNT-MIN sketch structure of height k' and width w, where, h and w are parameters that will be fixed later. Corresponding to a stream update (pos, x, Δ), the update $(pos, \pi_j(x), \Delta)$ is propagated to the jth dyadic intervals structure. Finally, during inference of frequent items, we use the jth dyadic based structure using the algorithm described in Section 2.2, to retrieve a set of candidate items S_j, then apply the inverse permutation π^{-1} to each candidate item to obtain $\pi^{-1}(S_j)$. This step is done for each $j = 1, 2, \ldots, s$. Finally, we return those items x that occur in at least two-thirds (or a majority) of the $\pi^{-1}(S_j)$'s and return the median estimate of its estimated frequency.

Analysis. Fix a permutation index j and abbreviate $\pi = \pi_j$. We will use the notation in the statement of Theorem 3. Let $k = \lceil \frac{1}{\epsilon} \rceil$. Here top-$k$ frequencies are determined in terms of the absolute value of f_j's. For a dyadic interval I at level l, define the random variable

$$g_I = \sum_{\pi(x) \in I} f_x \; .$$

Let $g_l(i)$ denote the frequency of the node I at level l to which the item i maps.

Lemma 3. *Let $t = 8\lceil \log(\phi n) \rceil$, $l_{\max} = \lfloor \log \frac{\phi n}{4t} \rfloor$ and $w = \log \log l_{\max}$. Then,*

$$\Pr\left\{ \forall l : 0 \le l \le l_{\max} \left(|\hat{g}_l(i) - f_i| \le \frac{\phi F_1}{2} \right) \right\} \ge \frac{5}{8} \; .$$

Proof. Let $g_l(i)$ denote the frequency of the dyadic interval I at level l to which the item i maps. Assume $\text{NoCollision}_l(i)$ holds. Then, $\mathsf{E}\big[|g_l(i) - f_i|\big] \le \frac{F_1(k)2^l}{n}$. By Markov's inequality,

$$\Pr\left\{ |g_l(i) - f_i| \le \frac{t F_1(k)2^l}{n} \right\} \le \frac{1}{t} \; .$$

Define $F_{l,1}$ as the sum of the absolute values of the frequencies of the family of dyadic intervals at level l. Then, $F_{l,1} \le F_1$. If $k' \ge 8\lceil \frac{1}{\phi} \rceil$, by COUNT-MIN structure guarantees, $|\hat{g}_l(i) - g_l(i)| \le \frac{\phi F_{l,1}}{4} \le \frac{\phi F_1}{4}$, with probability $1 - 2^{-\Omega(w)}$, for each l. By triangle inequality, and using union bound to add the error probabilities,

$$\Pr\left\{ \forall l : 0 \le l \le l_{\max} \big(|\hat{g}_l(i) - f_i| \le \frac{\phi F_1}{4} + \frac{t F_1 2^l}{n} \big) \right\}$$

$$\ge 1 - l_{\max}\left(2^{-\Omega(w)} + \frac{1}{t} \right) \; .$$

Substituting $t = 8\lceil \log(\phi n)\rceil$, $l_{\max} = \lfloor \log \frac{\phi n}{4t} \rfloor$ and $w = \log \log l_{\max}$, we have $\frac{l_{\max}}{t} \leq \frac{1}{8}$ and $\frac{t2^l}{n} \leq \frac{t2^{l_{\max}}}{n} \leq \frac{\phi}{4}$. $\qquad\qquad\square$

The property of the algorithm is summarized in the following theorem.

Theorem 3. *The algorithm* COUNT-MIN DYADIC *with height* $k' = 8\lceil \frac{1}{\phi} \rceil$, *width* $w = O(\log \log(\phi n))$, *maximum dyadic level* $l_{\max} = \lfloor \log \frac{\phi n}{32 \log(\phi n)} \rfloor$ *and number of permutations* $s = O(\log \frac{1}{\phi \delta})$ *solves the problem* APPROXFREQ(ϵ, ϕ) *with probability* $1 - \delta$ *with the following characteristics.*

$$\text{Space } O\left(\frac{1}{\phi} \left(\log \frac{\phi n}{\log(\phi n)} \right) \left(\log \frac{1}{\phi \delta} \right) (\log \log(n\phi))(\log F_1) \right)$$

$$\text{Update Time } \quad O\left(\left(\log \frac{\phi n}{\log(\phi n)} \right) (\log \log n)(\log \frac{1}{\phi \delta})) \right)$$

$$\text{Retrieval Time } \quad O\left(\frac{\log(\phi n)}{\phi} (\log \log(n\phi))(\log \frac{1}{\phi \delta}))) \right) \quad .$$

$\qquad\qquad\square$

6   Experimental Comparison

In this section, we present an experimental comparison of our algorithms with the relevant algorithms in the literature. For the problem of finding F_1-based frequent items, we compare our COUNT-MIN DYADIC algorithm with the reversible hash method of [11] and the absolute deltoids based group testing technique of [4]. For the problem of finding F_2-based frequent items, we compare our algorithms COUNTSKETCH DYADIC and COUNTSKETCH LINEAR with the variational deltoids group testing technique of [4].

Experimental testbed. Our experiments were run on Intel Pentium dual core 2.80 Ghz processor with 2Gb of main memory running Fedore Core version 6. We tested the algorithms against zipfian distributions. The algorithms under comparison were given the same space (in number of bytes) and run against the same input data. In fact, since our hash function code works for table sizes in powers of 2, we give additional advantage by rounding up the space to the nearest power of 2, for algorithms in the literature that we are comparing with.

The zipdiff(z_1, z_2) distribution. The input data was generated to simulate general streams, with positive and negative frequencies, as follows. Two random frequency vectors distributed as per normalized zipfian distribution *zipf* with parameters z_1 and z_2 are generated and their difference is taken. Varying z_1 and z_2 gives us the various test data. Such distributions are denoted as *zipfdiff(z_1, z_2)*. Such distributions typically have a set of relatively high positive values as the top frequencies of *zipf(z_1)* and a set of relatively high (in absolute value) negative values distributed as the top frequencies of *zipf(z_2)*. The item frequencies are chosen in a manner that the top frequencies in terms of absolute value of either distributions do not conflict [2].

We compare the algorithms on the standard measures of *precision* and *recall*. Recall is the percentage of the frequent items that are detected as frequent by the algorithm; thus $1-$ recall is the fraction of false negatives. Precision is the fraction of frequent items among the set of frequent items; thus $1-$ precision is the fraction of false positives.

The reversible hash algorithm [11] performs well only for a limited range of the input when there are very few frequent items in the data. Otherwise, we found that the reversible hashing algorithm generates a very large number of false positive frequent items to the tune of about two to three orders of magnitude (or more) larger than the actual number of frequent items and then attempts to eliminate them in a verification phase. In summary, for the range of tests that we performed and report below, the time required to find frequent items by the reversible hashing method was found to be higher than the other methods by at least factors of 1000 to 10000 (order of ms versus order of minutes). We therefore do not report specific experimental observations relating to the reversible hashing method.

Experiment 1: COUNT-MIN DYADIC *vs. Absolute deltoids.* Figure 2 presents the experimental evaluation of the COUNT-MIN DYADIC method and the absolute deltoids method of [4]. We consider frequency distribution over items with frequency distributed as the difference of zipfian distributions zipf(z) with parameters z_1 and z_2 respectively. We report results for the following three distributions. Distribution A: *zipfdiff*(0.1,0.9), distribution B: *zipfdiff*(0.4,0.5), distirbution C: *zipfdiff*(0.3,0.7). The number of distinct items was fixed at 2.1 million items (2^{21}). The total space used by the algorithms is given in the tables. For COUNT-MIN dyadic, either 6 or 7 tables were used for each permutation, the number of permutations was set to 1 (which was surprisingly sufficient), the height of the tables was varied from 2^{12} to 2^{14} (in powers of 2) and the number of levels was set to between 19 and 21 ($l_{max} = 32 - \log(height) + 1$). The parameters of the absolute deltoids algorithm was set so that the total space used is no less than the DYADIC algorithm–this translates to table height ranging from 2^{11} to 2^{13} (in powers of 2) and the number of tables being set to one more than that for the instance of COUNT-MIN DYADIC being compared with.

Results and Conclusions for Experiment 1. The precision of both algorithms is close to 100% in the sense that the items reported as frequent are truly frequent (almost always). We therefore do not report precision in the tables. The two algorithms are distinguishable by their recall; the COUNT-MIN dyadic method

---

[2] This can be done in multiple ways, namely, randomized, where, the ranking of the items in terms of each of *zipf*(z_1) and *zipf*(z_2) is randomized, leading to very low probability of conflict of the few top-k items in each distribution. We perform this in a deterministic manner, where the the ranking of the items in terms of frequencies for the first distribution *zipf*(z_1) is the standard order $1, 2, \ldots, n$ whereas, the ranking of the items for the second distribution is $s, s+1, \ldots, n, 1, 2 \ldots, s-1$, where, s is a shift parameter much larger than k.

Distribution	Space (in size of) (doubles)	Threshold αF_1 α	Actual No of frequent items	Recall Absolute Deltoids [4]	Recall COUNT-MIN DYADIC
zipfdiff (0.1, 0.9)	210540	2^{-9}	11	9	10
		2^{-10}	20	14	16
		2^{-11}	40	19	24
	409600	2^{-9}	11	10	11
		2^{-10}	20	17	17
		2^{-11}	40	24	29
		2^{-12}	86	37	52
	778240	2^{-9}	11	11	11
		2^{-10}	20	18	20
		2^{-11}	40	29	32
		2^{-12}	86	49	61
		2^{-13}	179	73	100
zipfdiff (0.4, 0.5)	210540	2^{-9}	0	0	0
		2^{-10}	0	0	0
		2^{-11}	0	0	0
	409600	2^{-9}	0	0	0
		2^{-10}	0	0	0
		2^{-11}	0	0	0
		2^{-12}	3	1	1
	778240	2^{-9}	0	0	0
		2^{-10}	0	0	0
		2^{-11}	0	0	0
		2^{-12}	3	1	2
		2^{-13}	8	6	11
zipfdiff (0.3, 0.7)	210540	2^{-9}	3	2	3
		2^{-10}	7	4	4
		2^{-11}	13	5	8
	409600	2^{-9}	3	3	3
		2^{-10}	7	4	4
		2^{-11}	13	8	9
		2^{-12}	26	11	16
	778240	2^{-9}	3	3	3
		2^{-10}	7	5	4
		2^{-11}	13	10	11
		2^{-12}	26	16	18
		2^{-13}	72	22	26

Fig. 2. F_1-based frequent items: Comparing absolute deltoids method [4] with COUNT-MIN DYADICmethod. Number of items = 2^{21}.

is consistently superior to the absolute deltoids algorithm. The results are presented in Figure 2.

Experiment 2. In this experiment, we evaluate the COUNTSKETCH DYADIC, COUNTSKETCH LINEAR and the variational deltoids algorithm. We consider data whose frequency is distributed as zipfian difference *zipfdiff*(z, z), for parameters

Distribution	Space (in size of) (doubles)	Threshold $(\alpha F_2)^{1/2}$ α	Actual No of frequent items	Recall, Precision Variational Deltoids [4]	Recall, Precision COUNTSKETCH DYADIC	Recall, Precision COUNTSKETCH LINEAR
zipfdiff (0.3, 0.3)	307240	2^{-9}	2	0	0, 0	1,0
		2^{-10}	8	0	3, 3	2,1
		2^{-11}	24	0	4, 4	3,1
		2^{-12}	76	0	10, 8	3,1
		2^{-13}	232	0	26, 19	3,1
	573440	2^{-9}	2	0	0, 0	0
		2^{-10}	8	0	4, 4	0
		2^{-11}	24	0	7, 7	0
		2^{-12}	76	0	18, 18	1,0
		2^{-13}	232	0	38, 37	1,0
	1064960	2^{-9}	2	0	0, 0	1,1
		2^{-10}	8	0	4, 4	1,1
		2^{-11}	24	0	10, 10	3,2
		2^{-12}	76	0	26, 26	3,2
		2^{-13}	232	0	54, 53	3,2
zipfdiff (0.4, 0.4)	307240	2^{-9}	17	0	8, 8	5,5
		2^{-10}	42	0	19, 19	7,7
		2^{-11}	99	0	39, 39	8,8
		2^{-12}	232	0	60, 59	10,9
		2^{-13}	540	0	115, 96	10,9
	573440	2^{-9}	17	2,2	11, 11	6, 6
		2^{-10}	42	3,3	24, 24	6, 6
		2^{-11}	99	0	44, 44	6, 6
		2^{-12}	232	0	91, 91	7,7
		2^{-13}	540	0	154, 149	7,7
	1064960	2^{-9}	17	6	12, 12	16, 14
		2^{-10}	42	8	28, 28	21, 19
		2^{-11}	99	2	56, 56	21, 20
		2^{-12}	232	0	109, 109	22, 22
		2^{-13}	540	0	184, 184	24, 24
zipfdiff (0.5, 0.5)	307240	2^{-9}	42	10, 10	27, 27	8, 7
		2^{-10}	84	4, 4	50, 50	9, 8
		2^{-11}	167	0	77, 77	9, 9
		2^{-12}	334	0	125, 122	9, 9
		2^{-13}	644	0	210, 183	10, 10
	573440	2^{-9}	42	14, 14	29, 29	25, 22
		2^{-10}	84	16, 16	56, 56	29, 28
		2^{-11}	167	3 , 3	95, 95	30, 30
		2^{-12}	334	0	162, 162	31,31
		2^{-13}	644	0	256, 256	31, 31
	1064960	2^{-10}	84	26,26	66, 66	41, 39
		2^{-11}	167	20,20	119, 119	44, 42
		2^{-12}	334	7, 7	208, 208	47, 44
		2^{-13}	644	1, 1	359, 359	48, 46

Fig. 3. Comparing COUNTSKETCH DYADIC/ LINEAR vs. variational deltoids

$z = 0.3, 0.4$ and 0.5. The number of distinct items was fixed at 4 million items. The total space used by the algorithms is given in Figure 3 and varies between 2.5— 10% of the space required to actually store the data. In comparison, in experiment 1, it was varied between $10 — 40\%$ of the size of the data. Thus, the experiments in this category use significantly less space (percentage wise) than the first experiment and significantly stresses the retrieval capabilities of the algorithms. The parameter choices are as follows. For COUNTSKETCH DYADIC, the settings are the same as those of COUNT-MIN DYADIC wherever possible. That is, the number of random permutations used is 1, the number of levels is kept between 19 and 21 and the number of tables is kept between 5 and 7. Recall that for the COUNTSKETCH LINEAR algorithm, s_2 is the number of sketches in each group whose average (of the squares) is taken, and s_3 is the number of such groups; for each bit value 0 or 1, for each bit position 1 through $\log n$ and each bucket of each table. In our experimentation, s_2 is set to 1 and s_3 to 5. These settings are significantly smaller than the theoretical bounds. For the variational deltoids algorithm, the number of tables were kept between 5 and 7. Since the space provided to the algorithms is the same, the main parameter that varies is the height of each of the tables, subject to the above settings.

Results of Experiment 2. The results of the experiments are summarized in Figure 3. Corresponding to each of the three algorithms tested, the precision and recall are shown in the same column (except when recall is 0). The nature of the results are both surprising and conclusive. It appears that COUNTSKETCH DYADIC is significantly superior in terms of both precision and recall to the COUNTSKETCH LINEAR algorithm, whereas the performance of the variational deltoids algorithm is quite poor. The recall is not 100%, given that the space provided to the algorithms is very small. Further, as expected, both precision and recall improve with increased space. It is an unexpected observation that COUNTSKETCH DYADIC is substantially superior to the other two algorithms.

7   Conclusions

We present novel and practical space and time-efficient algorithms for finding frequent items, absolute range sums and absolute quantiles over general streams.

Acknowledgements

We thank Tejas Gandhi and M. Ravibabu for implementing the reversible hashing algorithm of [11].

References

1. Alon, N., Matias, Y., Szegedy, M.: The space complexity of approximating frequency moments. J. Comp. Sys. and Sc. 58(1), 137–147 (1998)
2. Charikar, M., Chen, K., Farach-Colton, M.: Finding frequent items in data streams. In: Widmayer, P., Triguero, F., Morales, R., Hennessy, M., Eidenbenz, S., Conejo, R. (eds.) ICALP 2002. LNCS, vol. 2380, pp. 693–703. Springer, Heidelberg (2002)

3. Cormode, G., Muthukrishnan, S.: An Improved Data Stream Summary: The Count-Min Sketch and its Applications. J. Algorithms 55(1)
4. Cormode, G., Muthukrishnan, S.: What's New: Finding Significant Differences in Network Data Streams. In: Proc. IEEE INFOCOM (2004)
5. Demaine, E.D., López-Ortiz, A., Munro, J.I.: Frequency estimation of internet packet streams with limited space. In: Möhring, R.H., Raman, R. (eds.) ESA 2002. LNCS, vol. 2461, pp. 348–360. Springer, Heidelberg (2002)
6. Gilbert, A., Kotidis, Y., Muthukrishnan, S., Strauss, M.: How to Summarize the Universe: Dynamic Maintenance of Quantiles. In: Proc. VLDB, Hong Kong, August 2002, pp. 454–465 (2002)
7. Kaplan, E., Naor, M., Reingold, O.: Derandomized Constructions of k-Wise (Almost) Independent Permutations. In: Chekuri, C., Jansen, K., Rolim, J.D.P., Trevisan, L. (eds.) APPROX 2005 and RANDOM 2005. LNCS, vol. 3624, pp. 354–365. Springer, Heidelberg (2005)
8. Karp, R.M., Shenker, S., Papadimitriou, C.H.: A Simple Algorithm for Finding Frequent Elements in Streams and Bags. ACM TODS 28(1), 51–55 (2003)
9. Luby, M., Rackoff, C.: How to construct pseudorandom permutations and pseudorandom functions. SIAM J. Comp. 17(1), 373–386 (1988)
10. Misra, J., Gries, D.: Finding repeated elements. Sci. Comput. Programm. 2, 143–152 (1982)
11. Schweller, R., Li, Z., Chen, Y., Gao, Y., Gupta, A., Zhang, Y., Dinda, P., Kao, M.-Y., Memik, G.: Monitoring Flow-level High-speed Data Streams with Reversible Sketches. In: Proc. IEEE INFOCOM (2006)
12. Thorup, M., Zhang, Y.: Tabulation based 4-universal hashing with applications to second moment estimation. In: Proc. ACM SODA, New Orleans, Louisiana, USA, January 2004, pp. 615–624 (2004)

Efficiently Discovering Recent Frequent Items in Data Streams*

Ferry Irawan Tantono[1], Nishad Manerikar[1], and Themis Palpanas[1]

University of Trento

Abstract. The problem of frequent item discovery in streaming data has attracted a lot of attention lately. While the above problem has been studied extensively, and several techniques have been proposed for its solution, these approaches treat all the values of the data stream equally. Nevertheless, not all values are of equal importance. In several situations, we are interested more in the new values that have appeared in the stream, rather than in the older ones.

In this paper, we address the problem of finding *recent* frequent items in a data stream given a small bounded memory, and present novel algorithms to this direction. We propose a basic algorithm that extends the functionality of existing approaches by monitoring item frequencies in recent windows. Subsequently, we present an improved version of the algorithm with significantly improved performance (in terms of accuracy), at no extra memory cost. Finally, we perform an extensive experimental evaluation, and show that the proposed algorithms can efficiently identify the frequent items in ad hoc recent windows of a data stream.

1 Introduction

The problem of frequent item discovery in streaming data has attracted much attention, because it is relevant to many different applications across various domains [12,13,15]. A naive approach to deal with this problem is to keep a count of each distinct item. Yet, in general, we assume that our main memory is not large enough to hold counters for all the distinct items. Several techniques that can efficiently solve the problem have been proposed in the literature that also take into account the special characteristics and requirements of streaming data [10,17,23]. These techniques are approximate, but they can provide the correct answer with high probability and they have been empirically proven to produce accurate results.

The above approaches treat all the values of the data stream equally. Note though, that not all the values that have appeared in the data stream are of equal importance. In several situations, we are more interested in the values that have appeared in the stream in the recent past, rather than in the distant past.

---

* This work was partially supported by the FP7 EU Large-scale Integrating Project **OKKAM – Enabling a Web of Entities** (contract no. ICT-215032). For more details, visit http://www.okkam.org

B. Ludäscher and Nikos Mamoulis (Eds.): SSDBM 2008, LNCS 5069, pp. 222–239, 2008.

Similar observations have also been made in other works, where the problems of time-variant data summarization [5,25], clustering [3], and storage [8] have been studied.

The same is true for the problem of frequent item identification in data streams. A few indicative examples are described below.

– In the financial domain, we are interested in finding stocks that are traded the most in a stock exchange system. This knowledge is crucial for applications that deal with automatic trading, pre-trade analysis, post-trade execution, and market monitoring [26].
– In the communications and network operators industry several applications need to monitor the frequency of occurrence of packets traveling between specific nodes in the network [12]. This information is in many cases at the core of the business of companies in this area.
– Retail shops and online businesses are interested in identifying the products that sell the most. The results of this analysis can be used for launching special promotions, performing inventory management, and in other applications [19].

The applications in the above examples require estimates in the item frequencies for the recent past, rather than for the entire history of the data stream. Moreover, in certain cases the users would like to be able to query about the item frequencies in different windows in the recent past, and compare these values among themselves.

In this paper, we propose solutions for the discovery of *recent* frequent items in streaming data given a small bounded memory. These solutions are based on existing *sketching* techniques, which we extend in order to be able to effectively operate on the recent past. We describe the *TiTiCount*[1] algorithm that can be used to efficiently answer queries for frequent items in ad hoc recent windows. The algorithm uses a tilted timeframe for the representation of the past, which allows the algorithm to provide item frequency estimates for a number of different windows in the past, using a small amount of memory. At the same time, these estimates are more accurate for the most resent windows, and the accuracy of the estimates diminishes as we go further in the past. We also present a query answering method that takes into account the size of the window intervals used by our algorithm, and provides better frequency estimates than the straightforward approach.

Furthermore, we propose *TiTiCount+*, an enhanced algorithm for query answering. In this case, when a query for some item frequency in a particular window comes in, the query answering algorithm makes use of the information stored in the specified window of interest, but also uses the information stored in certain neighboring windows. Based on this extra information, the algorithm is able to refine the item frequency estimates, leading to more accurate results, with minimal additional processing. As we will describe in more detail later on, this scheme also leads to superior performance in the case where the distribution of the data stream is non-stationary.

---

[1] Tilted Timeframe Count.

In summary, in this work we make the following contributions.

– We describe algorithms that can estimate the frequency counts of hot items in the recent past of a data stream. Our approach efficiently supports queries on ad hoc recent windows, and can store information about arbitrary points in the past, depending on the user preferences and available memory budget.
– We propose a simple method that accounts for the size of our summary structures, and leads to more accurate item frequency estimates in query answering when compared to the straight-forward approach.
– We extend the above algorithm with a technique that combines the information stored in different parts of our data representation structures in order to improve the accuracy of the results. As we empirically demonstrate, the above technique results in a significant performance improvement at a negligible additional processing cost.
– Finally, we perform an extensive experimental evaluation using synthetic and real data. The results show the behavior of the algorithms in different conditions, and demonstrate the effectiveness of the proposed approach.

The rest of the paper is organized as follows. We start by giving some necessary background for the problem of mining data streams for frequent items in Section 2. In Section 3, we describe the problem of recent frequent items formally. Section 4 describes the development of our algorithm on the basis of two existing algorithms. Our experimental evaluation is presented in Section 5. Finally, we discuss related work in Section 6 and conclude in Section 7.

2   Background

We assume a data stream S that is composed of a stream of integer numbers, where each integer represents the occurrence of a data item in S.

Let N be the current length of the data stream S, i.e., N is the current number of transactions. Further assume that the data stream contains M distinct values. A *frequent item* is an item whose frequency is greater than ϕN, where the *support* parameter ϕ is a user-defined threshold in the interval $[0.0, 1.0]$.

Several algorithms have been proposed for efficiently mining frequent items in data streams. The Frequent (FREQ)[18] and the Lossy Counting (LC)[23] algorithms are based on maintaining approximate frequency counts, while Combinatorial Group Testing (CGT)[11], Count-Min (CM)[10], CCFC [7] and hCount (HC)[17] are based on *sketches*. We have conducted extensive experiments in order to compare the performance of these algorithms. Our implementation of the *FREQ, LC, CM, CGT*, and *CCFC* algorithms was based on the Massive Data Analysis Lab code-base [2]. The *hCount* algorithm was implemented from scratch, using the same optimizations as the other algorithms. We ran experiments on both synthetic and real datasets, and measured time and space usage for all the above six algorithms, averaged over several independent runs. In order to evaluate the quality of the results obtained, we used the two standard measures of *recall* (percentage of the true frequent items that are found by the algorithm) and *precision* (percentage of items identified by the algorithm, which are truly frequent).

In the interest of space, we only briefly summarize our results (details are in the full version of this paper). We ran experiments with varying the support threshold ϕ. The results indicate that the performance of *CCFC* is affected when support is low, but its recall improves when the support level is high. Regarding precision, *hCount* and *CCFC* are consistently the top performers, with the other algorithms improving their performance as the support threshold is increased.

We also measured the scalability and time requirements of the algorithms by running experiments with 10 to 100 million transactions. The results show that all algorithms scale linearly in time with respect to the number of transactions. *CCFC* requires the longest time, whereas *LC* and *FREQ* are the most time-efficient, with *hCount* performing very close to the fastest algorithms.

The qualitative results from all our experiments are summarized in Table 1 (a more detailed discussion of the experiments can be found elsewhere [22]). Based on these experiments (similar results have also appeared elsewhere [17]), we selected the *hCount* algorithm as the frequency estimation component of our approach, because it has several desirable characteristics. Namely, it exhibits a consistently good performance across various conditions, it has low time complexity, and is relatively easy to implement.

Note that this choice is not restrictive in any way, and in our techniques *hCount* could be replaced with any other suitable frequency estimation algorithm.

Table 1. Performance Summary

Algorithm	Characteristics
FREQ	Fast. Low precision.
CGT	Fastest of the sketch-based. Cannot handle lower support
CM	Less space than CGT but more time, cannot handle lower support
CCFC	Slow, fairly good accuracy
LC	Fast, good recall and precision
HC	Fast, good recall and precision

3   Recent Frequent Items

In this section, we formally define the problem of recent frequent item discovery, and we give a brief overview of our approach.

3.1   Problem Definition

Let the data stream S be represented by $\{T_1, T_2, \ldots, T_n\}$, where T_i denotes the ith item, and T_n is the latest (most recent) item in the stream. In this work, we assume that each item, T_i, is represented by a single integer, and corresponds to a transaction[2].

---

[2] For the remainder of this paper, we will use the terms *item* and *transaction* interchangeably.

Let $w = [w_{min}, w_{max}]$ define a window in the history of the stream, where w_{min} refers to the index of the least recent point in the window, and w_{max} to the index of the most recent one. The length, or size (in terms of number of transactions), of window w is $|w| = w_{max} - w_{min}$. Further, assume that ϕ, $0 < \phi \le 1$ is a user-defined parameter that determines which items are frequent, according to the following definition.

Definition 1. [Frequent Item] *An item is called* frequent *with respect to a window* w *if it appears in at least* $\phi|w|$ *transactions within* w.

We can now define the recent frequent item problem for a stream S.

Problem 1. [Recent Frequent Item (RFI)] Given a threshold ϕ, and a window w, where $n - w_{min} \le L$, we want to identify the frequent items in w, for a predetermined parameter $L \gg 1$.

We make two remarks regarding the above definition of the problem. First, both the window w and the threshold ϕ are part of the query, and can be different for each query. Also note that the query window of interest w, is ad hoc, and can refer to any interval in the recent history of the stream. The parameter L determines how far in the past the query window can refer to. Essentially, L defines the least recent transaction that can be part of the query window, and in practice can be very large. That is, for a window $w = [w_{min}, w_{max}]$, $n - L \le w_{min} < w_{max} \le n$.

Second, we define the window size in terms of the number of transactions, rather than time, because the data rates of streams are often times variable. Hence, windows defined in terms of number of transactions are more appropriate. Nevertheless, the techniques we propose can in principle work for both cases.

3.2 Proposed Approach

Previous works have studied the problem of identifying frequent items in the entire history of a data stream [7,10,11,17,18,23]. What is fundamentally different in our case is that we wish to identify frequent items in arbitrary (recent) window intervals of the stream. In order to solve the *RFI* problem, we have to store information about the item frequencies in various time-points in the past, which will allow us to answer queries for ad hoc windows.

A simple solution to the above problem is to divide the recent history of the stream, that is, the last L transactions, in fixed-size intervals, and estimate the item frequency counts for each one of these windows. This scheme allows us to answer queries even if they are not aligned to the interval boundaries; in this case, we provide an approximate answer.

However, the drawback of this approach is that the memory requirements are rather high. For $L \gg 1$, we need to keep information on a large number of intervals. Note that the number of intervals is also directly related to the accuracy of the query answers we can provide. Therefore, reducing the memory requirements comes at the cost of performance.

In order to overcome the above limitation and efficiently solve the *RFI* problem, we propose the use of *tilted-time* window intervals. (Similar approaches have been studied in other applications as well [3,5,8,25]. Though, as we describe later on, we propose novel operation schemes that allow our algorithms to offer significant performance improvements.) Under this scheme, we divide the history of the stream in increasingly larger intervals as we move in the past (resulting in more accurate item frequency estimations for the most recent window intervals, and increasingly less accurate for the window intervals further in the past). Therefore, we can significantly reduce the memory requirements, while still being able to answer queries from different time horizons. In the algorithms we propose, we assume *logarithmic* tilted-time windows, where each subsequent older window interval is twice the size of the previous interval. In this case, we can cover the entire space of L transactions with just $K = \log L$ windows.

In the following section, we describe algorithms that efficiently and effectively solve the *RFI* problem, using the tilted-time windows scheme. We also propose techniques that can significantly improve the accuracy of the algorithms using the same amount of memory. These improvements are more pronounced for queries involving the older, larger window intervals. Thus, we effectively alleviate the disadvantage that the tilted-time windows have on the intervals referring to stream values further in the past.

4   Algorithms for Recent Frequent Items

In this section, we present algorithms for the *RFI* problem. We start by briefly describing the main skeleton of the algorithms, which is the same for all of them. Subsequently, we discuss in more detail specific features of each algorithm, and the benefits it brings along.

As we mentioned earlier, we use the *hCount* sketch in order to estimate the frequency counts within a given window interval. The *hCount* algorithm [17] maintains an array of $M \times H$ counters, where M and H are parameters determined by the data characteristics and the allowed error. The algorithm uses H hash functions that map the occurrence of an item to H of the counters, which are subsequently incremented by one. The estimate of an item's frequency is computed as the minimum value of all the counters to which the item maps.

In our case, instead of a single window interval, the algorithm has to operate with K intervals. These windows follow a tilted time-frame as follows. The first window, w_0, covers the b most recent stream values, that is, transactions $T_{n-b+1}, \ldots, T_n$. The parameter b defines the size of w_0, and is called *batch size*. The second window, w_1 is also of size b, and covers the next b transactions. Then, the size of each subsequent window is double the size of the previous one. In general, the size of the i-th window is given by the formula $w_i = 2^{i-1}b$, $0 < i < K$.

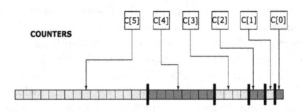

Fig. 1. Tilted-time windows

```
Let n :=current transaction number
    b := batch size

When new transaction T_n arrives:
1    use hCount to determine the set of counters C related to T_n
2    for each counter in C
3        update the counts of c
4    if (n mod b) == 0
5        call PerformShift()

When query for frequency of item i in window interval [t^q_{min}, t^q_{max}] arrives:
6    use hCount to determine the set of counters C corresponding to T_n
7        call GetFreqEst(C, [t^q_{min}, t^q_{max}])
```

Fig. 2. Main skeleton of the proposed algorithms

In order to account for all K window intervals, we extend the *hCount* sketch to an array of $M \times H \times K$ elements by replacing each one of the $M \times H$ counters $c_{m,h}$ in the original structure with an array $c_{m,h}[]$ of K counters, for $0 \le m < M$, $0 \le h < H$[3]. These arrays of counters correspond to the K windows, as shown in Figure 1. The first element (in some cases also the second, as we will explain later) of these arrays, $c[0]$, stores the counts for newly incoming stream values (according to the *hCount* algorithm). The subsequent elements store the historical values of the counts that refer to the corresponding window interval. In essence, they keep track of the history of the item frequencies.

There are two main operations that we need to have in place (outlined in Figure 2). First, the shifting of the counter values $c[]$ so that they correspond to the current window intervals. This operation is triggered every time the window that receives the new stream values gets full, that is, every b transactions. Second, the item frequency estimation mechanism, used to provide the estimate of an item frequency within a given window interval.

In the next sections, we describe in more detail different solutions that we propose for the above two operations.

---

[3] For the remainder of the text, we omit the indices m, h when we refer in general to the array of counters $c[]$.

4.1 Basic Algorithm

The straightforward approach to implement the shifting operation is to use intermediate windows (and corresponding counters). As shown in Figure 3, the counters corresponding to the first window, $c[0]$, are always receiving the new data (depicted in gray), and counter values shift sequentially every b transactions.

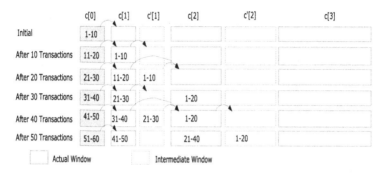

Fig. 3. Shifting with Intermediate Windows (batch size $b = 10$). Gray boxes denote windows receiving new stream values.

Answering item frequency questions in this model is simple as well. When a query for the frequency of an item in a specific window interval w_q comes in, we identify the counters that store information on time intervals overlapping with w_q, and we sum the estimates from these counters. Note that if the query interval w_q is not aligned with the counter time intervals, then we introduce errors in the estimation, since we are counting frequencies over intervals that do not belong in the w_q (this is true for the two ends of the query interval).

The advantage of this algorithm, which we call *NaiveCount*, is its simplicity. Though, this advantage comes at the expense of memory (for the intermediate windows). The required memory for *NaiveCount* is $(2K - 1)S$, where K is the number of windows and S is the memory required by *hCount* (or any other similar technique that can be used here), and the number of shift operations is in the worst case K. In the following sections, we show how we can reduce the memory requirements, while at the same time improving the accuracy of the results.

4.2 Reducing the Memory Requirements

We observe that we can reduce the memory requirements of the algorithm by discarding the intermediate windows. Under the new shifting scheme (see Figure 4), we keep track of which counters correspond to which window intervals, which allows us to directly move counter contents to the next window. (We also employ *lazy* shifting, by allowing also the second window to process new stream

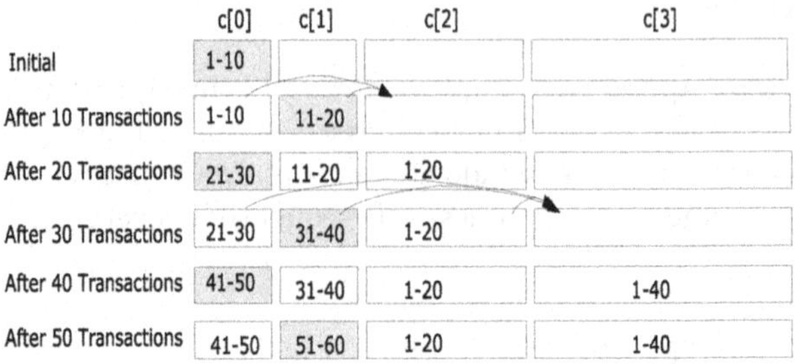

	c[0]	c[1]	c[2]	c[3]
Initial	1-10			
After 10 Transactions	1-10	11-20		
After 20 Transactions	21-30	11-20	1-20	
After 30 Transactions	21-30	31-40	1-20	
After 40 Transactions	41-50	31-40	1-20	1-40
After 50 Transactions	41-50	51-60	1-20	1-40

Fig. 4. Shifting without Intermediate Windows (batch size $b = 10$). Gray boxes denote windows receiving new stream values.

values, thus, only shifting contents when necessary and saving some shift operations.) In this case, the memory requirements are KS (K is the number of windows and S the size of the sketch), while the number of shift operations is in the worst case K.

Even though this algorithm, *TiTiCount*, needs almost half the amount of memory of *NaiveCount* by not using any intermediate windows, the accuracy of its results is not affected. This is because the intermediate windows are only used to facilitate the shifting operation.

What is more interesting is that with *TiTiCount* we can actually improve the accuracy of the results. In *NaiveCount*, we notice that whenever the edges of the query window w_q are not aligned with the edges of the window intervals corresponding to the counters $c[]$, we introduce an error in the results. Consider query q with $w_q = [100, 950]$ of Figure 6. The edges of w_q are not aligned with the edges of w_4 and w_0. Nevertheless, for the calculation of the result *NaiveCount* will consider the counts corresponding to the entire intervals w_4 and w_0, even though part of them falls outside w_q.

TiTiCount resolves this problem, and only takes into account the portions of the windows that are covered by the query. This is achieved by considering in the result the *weighted fraction* of the estimate provided by the counters $c[]$ that corresponds to the fraction of the counter window overlapping the query window. In this work, we apply a linear model in this computation (i.e., the fraction is directly computed as the amount of overlap), but other, more sophisticated techniques can be applied (e.g., even limited knowledge on the distribution of the frequencies within a window could lead to an even more accurate non-linear model). However, as we show in the experimental evaluation of the algorithms, this simple idea improves the quality of the results substantially.

4.3   Exploiting Redundant Information

Taking a close look at the shifting operation, we observe that during specific time intervals, information pertaining to the same data stream transactions is stored in more than one counters at the same time. For example, referring back to Figure 4 (example of shifting for *TiTiCount*), we observe that information

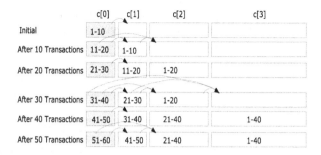

Fig. 5. Value added shifting (batch size $b = 10$). Gray boxes denote windows receiving new stream values.

regarding transactions $1 - 20$ is stored both in $c[2]$ and $c[3]$ (see bottom of the figure). Note that the counters do not store the same information, as $c[3]$ corresponds to a larger time interval than $c[2]$. Nevertheless, there is a certain amount of information redundancy, and in the following paragraphs we explain how the *TiTiCount+* algorithm uses it in order to further improve the accuracy of the results.

In order to exploit the above side-effect of shifting, we modify the shifting operation as follows. We no longer employ the lazy shifting scheme used by *TiTiCount*, but instead have the first window process all the new data stream transactions. This results in an increased number of shift operations, which in the worst case can be as many as $K(K - 1)/2$. However, the required shifts are on the average much less, and as we empirically demonstrate, the additional cost in the total running time is very small. The memory requirements are the same as before, namely, KS.

When we apply the above shifting mechanism, the way the various window intervals are placed with respect to each other is governed by the following properties[4].

Lemma 1. [Window Property 1] *If two window intervals overlap, then the smaller window interval is completely contained in the larger one.*

Lemma 2. [Window Property 2] *All window intervals that overlap have one common boundary, and this common boundary is the most recent edge of these intervals.*

The above properties are very important, because they constitute the base of the query answering algorithm. The main idea of the algorithm is to always use the counters corresponding to the *smallest* possible window interval in order to estimate some frequency count. When a query w_q comes in, it is split into subqueries $w_{sq_1}, \ldots, w_{sq_j}, \ldots, w_{sq_J}$ that align with the boundaries of the counter window intervals. Then, results for each subquery are derived as follows.

---

[4] In the interest of space, we omit the proofs, which can be found in the full version of this paper.

- *Smallest Interval:* If w_{sq_j} can be answered using multiple counters then use the counter that corresponds to the smallest window interval to compute the result.
- *Subtraction Operation:* If the window interval, w_l, of the counter that is to be used to answer w_{sq_j} overlaps with a smaller window interval, w_s, then subtract the values of the w_s counter from the w_l counter, and subsequently compute the result.

The above steps lead to correct results, because the properties stated in Lemmata 1 and 2 ensure the window intervals are aligned in such a way that the subtraction operation is feasible. The following example explains how this algorithm works.

Example 1. Assume we have five window intervals, $w_0, \ldots, w_4$, and that the current transaction number is 990, as shown in Figure 6. A query q comes in, asking for frequent items in interval $[100, 950]$. The algorithm splits q in five subqueries, according to the boundaries of the window intervals with which it overlaps. Then, the algorithm computes frequency estimates for each subquery as follows. The estimate for q_5 is derived from the w_0 counter by applying the weighted fraction model (i.e., the estimate will be $(950 - 901 + 1)/(990 - 901 + 1)$ times the result returned by the counter). Frequency estimates for q_4 and q_3 are derived directly from the counters of intervals w_1 and w_2, respectively. For q_2, the estimate is directly computed after subtracting the values of the w_2 counter from the w_3 counter. Finally, for q_1 the algorithm first subtracts the values of the w_3 counter from the w_4 counter, and then applied the weighted fraction model, since q_1 is not interested in the first 100 transactions of w_4.

We can now demonstrate the advantage that the subtraction operation provides to *TiTiCount+* for producing estimates with significantly improved accuracy, when compared to *TiTiCount*. Using the same example as above, assume that all the items in the interval $[1, 400]$ have the value x, and all the items in the interval $[401, 990]$ have the value y. Suppose, a query comes that asks for the frequency of value y in interval $[1, 400]$. In this case, *TiTiCount* will use just the w_4 counter with the weighted fraction model, returning an answer of 200. On the other hand, *TiTiCount+* will subtract the contents of the w_3 counter from those of the w_4 counter, and correctly return 0 as an answer. Evidently,

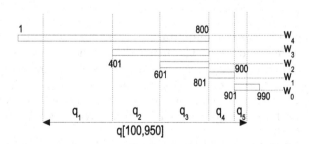

Fig. 6. Example of query answering for *TiTiCount+*

this advantage of *TiTiCount+* is magnified when the distribution of the values in the data stream change over time.

5   Experimental Evaluation

We implemented our proposal and conducted a series of experiments to evaluate the efficiency of our techniques in a variety of settings. Apart from the three algorithms we describe in this paper, we also implemented algorithm *Linear* to compare against our approach. *Linear* is similar to *TiTiCount*, except that instead of tilted time window intervals, it uses window intervals of fixed size.

In our experiments we used both synthetic and real datasets. The synthetic datasets we used were generated according to a Zipfian distribution with Zipf parameter 1.1, unless noted otherwise. We generated datasets with up to 100 million items, with both stationary and non-stationary distributions. The real datasets we used were as follows.

- kosarak [1]: It consists of anonymized click-stream data of a Hungarian online news portal, expressed as a sets of integers. It has about 8 million individual items.
- retail [4]: It contains retail market basket data from an anonymous Belgian store. This dataset has about 0.9 million individual items.

We implemented all our algorithms in C using the gcc compiler under Linux Fedora Core 5. The experiments were run on a dual Intel Xeon 2.8Ghz machine.

5.1   Evaluating the Accuracy

In the first experiment, we compare the algorithms *NaiveCount*, *TiTiCount*, and *TiTiCount+* in terms of the accuracy of the results they provide. We measure recall, defined as the percentage of the true frequent items that are found by the algorithm, and precision, defined as the percentage of items identified by the algorithm that are truly frequent. We ran experiments using several query window intervals, where in each interval we were looking for the frequent items ($\phi = 0.005$). In Figure 7, we report the results for nine of these queries (the results for the rest of the queries we tried were similar). The queries we used as test cases are listed in Table 2 (we report the boundaries of the query window intervals). All experiments used a batch size $b = 1,000$, they were repeated 15 times, and results were averaged.

Figures 7(a) and 7(b) show the recall and precision for the three algorithms, when run over a dataset with a stationary distribution. We observe that all three algorithms have virtually perfect recall rates. However, precision varies. *TiTi-Count* and *TiTiCount+* average precision rates close to 90%, with *TiTiCount+* performing slightly better. The performance of *NaiveCount* is notably worse, averaging a mere 45%.

In Figures 7(c) and 7(d), we show the results of the same experiment, when run over a dataset with time-variant distribution. In this case, the stream was

Table 2. Query window intervals used as test cases

n=50000			n=60000			n=70000		
No.	t_{min}^q	t_{max}^q	No.	t_{min}^q	t_{max}^q	No.	t_{min}^q	t_{max}^q
1	5000	45000	4	5000	55000	7	20000	45000
2	35000	45000	5	35000	55000	8	40000	55000
3	25000	40000	6	5000	50000	9	40000	65000

generated by concatenating several small datasets. These datasets were all generated by sampling a Zipfian distribution, but each one of them had a different set of frequent items.

These experiments represent a more challenging setting for our algorithms, and the results demonstrate the qualitative difference among them. *TiTiCount+* is consistently the best performer among the three, with significantly better performance than *TiTiCount* in several cases. The *NaiveCount* algorithm performs very poorly in terms of precision, which explains its high recall rates.

The reason *TiTiCount+* produces even more accurate results than *TiTiCount* for the time-varying dataset is because the *TiTiCount* algorithm relies solely on the weighted fraction mechanism to arrive at frequency estimates. Evn though this is an improvement over the *NaiveCount* algorithm, this mechanism works well only for stationary distributions, where the item frequencies remain relatively stable across different window intervals. In contrast, *TiTiCount+* using the subtraction mechanism can effectively alleviate this problem and produce better estimates. This explains the large difference in performance observed in test cases $7 - 9$.

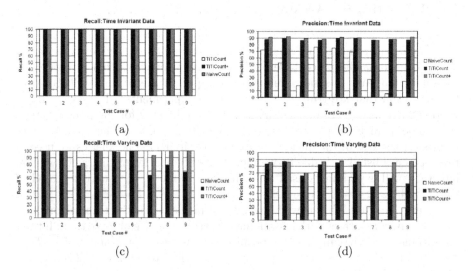

Fig. 7. Performance on time varying and non-varying data distributions

We also performed tests by varying the skew parameter of the Zipfian distribution. The trends in these experiments are similar, and we omit them for brevity. For the remainder of the discussion, we do not consider the *NaiveCount* algorithm.

In the following experiment, we tested the performance of the algorithms as a function of the size of the query window interval, and we also compare them to *Linear*. We use *Linear* only as an indication of how good the performance of our algorithms would be if they had enough memory to use fixed- instead of tilted-time window intervals. For our experiment, batch size $b = 1,000$, and number of windows $K = 11$. This means that our algorithms can answer queries about item frequencies for the past $1,000,000$ transactions. In order for *Linear* to be able to answer the same class of queries, we have to use 1000 windows (for window size equal to b), which requires two orders of magnitude more space than our algorithms. We also compared against *LinearConst*, which the *Linear* algorithm that is given the same amount of space as our algorithms (resulting in a window size of 100,000).

The experiment was run on the *kosarak* dataset, using 120 randomly generated queries following a Gaussian distribution (mean $9N/10$, stddev $N/8$). Figures 8(a) and 8(b) depict the results of the experiment for recall and precision, respectively. The graphs show that *TiTiCount+* outperforms *TiTiCount* across the entire range of query sizes. It is interesting to note that while *TiTiCount* exhibits a steady recall rate across the experiment, *TiTiCount+* improves its performance as the size of the queries increase. This happens because larger queries are more effectively managed by the subtraction mechanism of *TiTiCount+*.

As expected, *LinearConst* performs the worst (its somewhat high precision numbers are explained by the low performance in recall), and *Linear* is almost always the winner in both metrics. Note though, that the performance of *TiTi-Count+* is very close to *Linear*, which demonstrates the effectiveness of the subtraction mechanism.

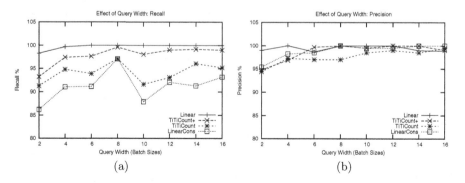

Fig. 8. Performance with respect to query width (dataset: kosarak, batch size $b = 1,000$)

5.2   Finding Top-k Items

In some situations, it is desirable to know the top-k most frequent items in a stream, or their cumulative frequency. Our algorithms can be adapted to determine those values. In this experiment, we tested $TiTiCount+$ for the accuracy of the estimated frequencies of the top-k items, and compared its results to the exact answers.

Similar to the previous experiment, we ran random queries of different sizes, asking for the cumulative frequencies of the top-k items, for several values of k. The results are illustrated in Figure 9, for both real datasets. The top-k items were correctly identified in all cases. The graphs show that the cumulative frequencies reported by $TiTiCount+$ were consistently very accurate (less than 0.05% error for our experiments).

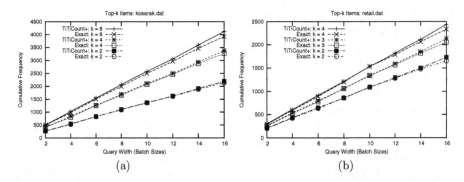

Fig. 9. Top-k items: Estimated and actual cumulative frequencies

5.3   Scalability

In order to evaluate the scalability of the proposed algorithms, we ran experiments to measure the update times of $TiTiCount$ and $TiTiCount+$. The update time is the time required to update the internal data structures every time a new transaction arrives, including shifting operations We tested the algorithms with data streams of 100 million transactions, and we report the cumulative update time in Figure 10. The reported times are averages over five independent runs. The results show that both algorithms scale linearly with the number of transactions, with $TiTiCount+$ being slightly less efficient, because of the higher worst case cost of the shift operation that it implements.

6   Related Work

In the recent years, numerous studies have focused on problems related to streaming data, ranging from practical applications to theoretical questions [16,24]. There is a wealth of work on the problem of identifying frequent items in streaming data. The Frequent (FREQ)[18] and the Lossy Counting (LC)[23]

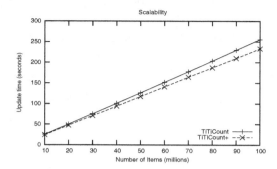

Fig. 10. Scalability: Variation in update time with increasing number of data items

algorithms maintain a number of counts, which are pruned as new items arrive in the data stream. Other algorithms, such as Combinatorial Group Testing (CGT) [11], Count-Min (CM)[10], CCFC [7] and hCount (HC) [17] are based on *sketches*. The sketches are designed so that they provide accurate results for the frequent item discovery problem, while requiring limited memory resources.

The important difference between these works and our approach is that we want to be more flexible in identifying frequent items, placing more importance on recent frequent items. For mining recent frequent items, an intuitive approach is to use time-decaying approximations. This technique has been used in several diverse areas, such as online time series summarization [25,5], streaming data clustering [3], and data warehousing [8].

Other works have studied the problem of efficiently identifying and maintaining frequent itemsets over streaming data [6,9,20]. In this case, we are interested in *sets* of items that appear frequently together. Specialized techniques and algorithms have been developed for the solution of this problem. Some of these works are also based on sliding windows [21], or tilted time windows [14], in order to focus on the transactions in the recent past of the data stream. The *FP-Stream* approach [14] uses a tilted timeframe similar to our work. However it makes use of the *FP-Tree* structure, which has been specifically designed for itemsets, rather than items. An efficient implementation of the above approach for the problem we solve in this study is not straightforward. Moreover, in our work we describe novel shifting schemes for the tilted timeframe, which are used by *TiTiCount+* in order to deliver significant performance improvements.

7   Conclusions

The problem of frequent item identification has attracted lots of attention in the past years, and has found many interesting applications across diverse domains. This work is motivated by the need of many real-world applications to identify frequent items in the *recent* past of a data stream, rather than over the entire history.

In this paper, we propose novel algorithms for the discovery of *recent* frequent items in a data stream. The proposed algorithms are based on the *sketching* technique, and are very flexible in that they are designed to answer queries for frequent items in ad hoc window intervals in the recent past of the data stream. Based on our observations, we also describe extensions of the basic algorithm that can significantly improve the accuracy of the query results, while maintaining the same memory usage and at negligible additional processing cost.

We have evaluated the performance of the proposed techniques on real and synthetic data streams. The results show that the algorithms can efficiently operate using few space and time resources, while maintaining a high quality approximation in query answering.

Acknowledgments. The authors would like to thank the anonymous referees for their constructive criticism and comments.

References

1. Frequent itemset mining dataset repository, university of helsinki (2008), `http://fimi.cs.helsinki.fi/data/`
2. Massive data analysis lab, rutgers university (2008), `http://www.cs.rutgers.edu/~muthu/massdal.html`
3. Aggarwal, C.C., Han, J., Wang, J., Yu, P.S.: A framework for clustering evolving data streams. In: VLDB, pp. 81–92 (2003)
4. Brijs, T., Swinnen, G., Vanhoof, K., Wets, G.: Using association rules for product assortment decisions: A case study. In: Knowledge Discovery and Data Mining, pp. 254–260 (1999)
5. Bulut, A., Singh, A.K.: Swat: Hierarchical stream summarization in large networks. In: ICDE, pp. 303–314 (2003)
6. Chang, C.-H., Yang, S.-H.: Enhancing swf for incremental association mining by itemset maintenance. In: Whang, K.-Y., Jeon, J., Shim, K., Srivastava, J. (eds.) PAKDD 2003. LNCS (LNAI), vol. 2637, pp. 301–312. Springer, Heidelberg (2003)
7. Charikar, M., Chen, K., Farach-Colton, M.: Finding frequent items in data streams. In: Widmayer, P., Triguero, F., Morales, R., Hennessy, M., Eidenbenz, S., Conejo, R. (eds.) ICALP 2002. LNCS, vol. 2380, pp. 693–703. Springer, Heidelberg (2002)
8. Chen, Y., Dong, G., Han, J., Wah, B.W., Wang, J.: Multi-dimensional regression analysis of time-series data streams. In: VLDB, pp. 323–334 (2002)
9. Cheung, D.W.-L., Han, J., Ng, V.T.Y., Wong, C.Y.: Maintenance of discovered association rules in large databases: An incremental updating technique. In: ICDE, pp. 106–114 (1996)
10. Cormode, G., Muthukrishnan, S.: An improved data stream summary: the count-min sketch and its applications. J. Algorithms 55(1), 58–75 (2005)
11. Cormode, G., Muthukrishnan, S.: What's hot and what's not: tracking most frequent items dynamically. ACM Trans. Database Syst. 30(1), 249–278 (2005)
12. Estan, C., Varghese, G.: New directions in traffic measurement and accounting. In: SIGCOMM, pp. 323–336 (2002)
13. Fang, M., Shivakumar, N., Garcia-Molina, H., Motwani, R., Ullman, J.D.: Computing iceberg queries efficiently. In: VLDB, pp. 299–310 (1998)

14. Giannella, C., Han, J., Pei, J., Yan, X., Yu, P.: Mining frequent patterns in data streams at multiple time granularities. In: NSF Workshop on Next Generation Data Mining (2003)
15. Gibbons, P.B., Matias, Y.: Synopsis data structures for massive data sets. In: DIMACS Series in Discrete Mathematics and Theoretical Computer Science (1999)
16. Gilbert, A.C., Kotidis, Y., Muthukrishnan, S., Strauss, M.: Surfing wavelets on streams: One-pass summaries for approximate aggregate queries. In: VLDB, pp. 79–88 (2001)
17. Jin, C., Qian, W., Sha, C., Yu, J.X., Zhou, A.: Dynamically maintaining frequent items over a data stream. In: CIKM 2003: Proceedings of the twelfth international conference on Information and knowledge management, pp. 287–294. ACM Press, New York (2003)
18. Karp, R.M., Shenker, S., Papadimitriou, C.H.: A simple algorithm for finding frequent elements in streams and bags. ACM Trans. Database Syst. 28(1), 51–55 (2003)
19. Kohavi, R., Provost, F.J.: Applications of data mining to electronic commerce. Data Min. Knowl. Discov. 5(1/2), 5–10 (2001)
20. Lee, C.-H., Lin, C.-R., Chen, M.-S.: Sliding window filtering: an efficient method for incremental mining on a time-variant database. Inf. Syst. 30(3), 227–244 (2005)
21. Lin, C.-H., Chiu, D.-Y., Wu, Y.-H., Chen, A.L.P.: Mining frequent itemsets from data streams with a time-sensitive sliding window. In: SDM (2005)
22. Manerikar, N., Palpanas, T.: Frequent Items in Streaming Data: An Experimental Evaluation of the State-of-the-Art. Technical Report DISI-08-017, University of Trento (March 2008)
23. Manku, G.S., Motwani, R.: Approximate frequency counts over data streams (2002)
24. Muthukrishnan, S.: Data streams: algorithms and applications. Foundations and Trends in Theoretical Computer Science 1(2) (2005)
25. Palpanas, T., Vlachos, M., Keogh, E.J., Gunopulos, D., Truppel, W.: Online amnesic approximation of streaming time series. In: ICDE, pp. 338–349 (2004)
26. Whitney, A.T., Shasha, D.: Lots o' ticks: Real-time high performance time series queries on billions of trades and quotes. In: SIGMOD Conference, p. 617 (2001)

Prioritized Evaluation of Continuous Moving Queries over Streaming Locations

Kostas Patroumpas and Timos Sellis

School of Electrical and Computer Engineering
National Technical University of Athens, Hellas
{kpatro,timos}@dbnet.ece.ntua.gr

Abstract. Existing approaches to the management of streaming positional updates generally assume that all active user requests have equal importance, ignoring the possibility of any priorities concerning delivery of results in mission-critical mobile applications. Query prioritization could be assigned either explicitly after users' preferences or implicitly by the processing engine itself to better regulate system load. In this work, we specifically examine priority-based evaluation of ranked continuous range queries against locations of moving objects streaming into a central processor. We define a versatile model with alternative scoring functions for deciding evaluation strategies adaptable to the relative importance of queries and the current distribution of objects. We also propose a processing mechanism enhanced with ranked priorities, which exploits shared computation and enables critical requests to receive response more frequently than less demanding ones. A comprehensive experimental study with performance results offers concrete evidence that such a scheme is capable of efficiently handling numerous moving queries of varying priorities and spatial extents with minimal system overhead.

1 Introduction

Proliferation of location-enabled mobile devices (like phones, PDAs, or GPS) has given rise to many modern monitoring applications, such as location-based advertising, car navigation systems, smart tourist guides, wildlife protection systems etc. From a data management perspective, the main challenge is to cope with *streams* of massive positional updates that arrive to a central server from numerous moving sources (e.g., humans, animals, or machines) at high rates. This information cannot be easily managed in a typical spatiotemporal database, and not just because of the enormous bulk of data that keep steadily accumulating. It is mainly the necessity to provide real-time response to several long-running user requests that calls for immediate handling and online processing of the incoming data items. Several types of such *continuous queries* have recently attracted much research interest, particularly range [2,14] or k-nearest neighbor search [13,20], and skyline computation [12]. Most approaches adhere to a "push-based" model for processing streams of moving objects' locations. For instance, considering continuous range search, it is the arrival of positions that triggers

B. Ludäscher and Nikos Mamoulis (Eds.): SSDBM 2008, LNCS 5069, pp. 240–257, 2008.

reevaluation of queries covering them. In contrast, according to a "pull-based" policy employed by traditional DBMS's, queries should check incoming locations and compute their own answer.

In this paper, we turn our focus on evaluating *continuously moving range queries* that also indicate a user-defined preference to receiving results promptly or frequently. This degree of interest is expressed with a (possibly time-varying) *rank* value associated to each request. A high-ranked query indicates that the user needs urgently or frequently information about objects within her area of interest. In contrast, a small rank value signifies a lower priority for this query, i.e., the client is content to receive notification less often. It is plausible that emergency messages asking for ambulances, police cars, fire brigade automobiles etc., should be prioritized over other requests searching for restaurants or bars in a certain area. Of course, a user may submit multiple requests with diverse ranks, implying that some of them are urgent, while others may be answered with a short delay or even less frequently.

Besides, it is most likely that overlapping areas of interest among many range queries offer opportunities for common processing, while requests that are spatially isolated clearly call for separate handling. Hence, the central server may deliberately assign priorities among active queries; e.g., by giving precedence to areas with a high concentration of queries, multiple requests could be processed faster. In that case, queries would be scheduled for execution so as to better utilize available resources and not on the basis of their perceived significance.

Our objective is to investigate *priority-based* evaluation strategies that take into account a *user-specified importance* of queries. A naïve execution scheme would be to process each query separately in descending rank order, each time starting from the top-ranked request. Apart from wasting resources, it is also probable that low-ranked queries would hardly get any response as soon as the incoming rate of positional updates escalates or the system gets overwhelmed with too many pending requests. With a constant demand for timely results, the processing engine might be forced to drop tuples or ignore requests.

Thus, we opt for solutions that intend to share computation among queries, while still respecting their rankings. In order to exploit common spatial predicates, we organize a simple, yet flexible grid partitioning for indexing both queries and locations. Accordingly, we are able to organize greater groups of range queries with a collective rank value that can be used to guide prioritized examination of requests. Among the designing principles of the system is that it should answer queries reasonably often and always provide fresh results computed against recently recorded object locations and query ranges. Fairness is another concern, such that high-ranked requests are not extremely favored to the detriment of the rest. A wise policy could also take into account object density when deciding execution priorities. The major challenge, though, is robustness; this framework must successfully cope with an increased number of objects and queries that are both moving continuously.

Overall, this paradigm suggests that queries should not necessarily be given equal importance when processed in location-based monitoring applications. To the best of our knowledge, this is the first work that attempts evaluation of prioritized moving queries over streaming positional updates.

The contributions of this work can be summarized as follows:

- We introduce a ranking model with several *scoring functions* that offer collective assessment for the spatial distribution of query priorities, as a means of identifying regions that involve many important queries.
- We suggest alternative strategies that *prioritize evaluation* of queries in greater groups according to their aggregated rankings.
- We conduct extensive experiments with large datasets to validate rank-aware execution policies and demonstrate the capabilities of this framework.

The remainder of this paper is organized as follows: Section 2 briefly reviews related work. Section 3 outlines the general processing framework and the underlying spatial index. Section 4 discusses the salient characteristics of the ranking model and examines alternative scoring schemes. In Section 5, evaluation strategies are introduced, along with a measure for the quality of service achieved. Experimental results are reported in Section 6. Finally, Section 7 concludes the paper and offers directions for future research.

2   Related Work

Abundance of available data in real-world applications requires some type of ranking for its efficient retrieval [10]. Ranked search in databases refers mainly to top-k query answering [1], which intends to limit the cardinality of results according to a user-specified parameter k. This trend has been particularly investigated with respect to rank-aware query optimization, for proper selection of cost-effective query plans (e.g., for joins [7]).

A new paradigm for personalized queries in databases [8] takes advantage of users' profiles in order to provide most relative answers. A generalized model [9] for combining and selecting preferences is utilized for generation of ranked results. Although we make use of two ranking functions suggested with this model, our starting point and objectives are utterly different; in our approach, users specify priorities for queries and not preferences to particular pieces of information, so evaluation proceeds in quite different fashion.

In [15], the novel concept of quality contracts is used to combine users' preferences for both Quality of Service (QoS) and Quality of Data (QoD) for data-intensive web sites, where users require short response times for queries and freshness of information. Quality contracts are defined with step and linear functions that specify time deadlines and freshness constraints. The proposed two-level technique initially allocates processing resources between queries and data updates, but then it allows queries and updates to have their own priorities. Our spatiotemporal scenario differs a lot, as we consider that object locations are always kept up-to-date and have no preferences; furthermore, our interest is in maximizing QoS by sharing computation among queries, also making use of suitable ranking functions to combine priorities.

Recent works on data streams address effective load shedding in a distributed mobile setting [3] or consider prioritized transmission to a central portal of query results computed locally at numerous sensor nodes [21]. In the former approach, the objective is to shed differing amounts of data from distinct regions in order to minimize total error in results, so the notion of a user-assigned importance for the submitted requests is disregarded. In the latter framework, prioritization mainly signifies delivery order for results, but without sharing computation among queries running at the same priority level.

Generic frameworks [6] for continuous monitoring or moving objects or sophisticated algorithms for specific query types (e.g., [4,13]) do not consider priorities of requests over streaming locations. The novelty of this work is that we emphasize on prioritized query evaluation and not at all on delivering ranked query results. In our context, ranks and priorities[1] affect strictly to the execution order of queries in the central processor; we do not intend to return the top-k objects found within the range of a given query, but to provide response to queries as frequently as their rank prescribes. Nor are we interested in rating the "importance" of data, according to some kind of preferences, as needed in several decision-making applications [18]. Our approach also distinguishes itself from the usual notion of ranking in spatial databases, utilized as a means of sorting objects according to their increasing distance from a query point [5].

3   Processing Framework

3.1   System Model

We assume a central processor capable of monitoring numerous objects moving in a given area (e.g., vehicles circulating in the road network of a city). At regular intervals, each object oid sends a message to the server, informing about its current position. This message may be considered a tuple of the form $\langle oid, x_i, y_i, \tau \rangle$, where τ is a *timestamp* value that denotes the time when 2-d point coordinates (x_i, y_i) were actually measured for that object. There is no distinction of class or importance among objects, so all are treated in equal terms.

On the other hand, users are able to register queries that search for objects falling within a specified range. For simplicity, we assume that all query ranges are rectangles, but any other polygonal area could be defined as well. An important characteristic is that query ranges may be *moving* as well, exactly like objects. Queries are also specified as tuples $\langle q_j, a_j, \rho_j, \tau \rangle$, signifying that query q_j will investigate spatial area a_j for qualifying locations at time τ. Each query q_j is also assigned a *rank* value $\rho_j \in [0..1]$ that expresses the degree of user's interest to get a quick or frequent response (Fig. 1a). Ranks close to 1 denote urgent requests asking for immediate response, while values close to 0 mean that users do not really care too much about how soon they will receive an answer and how often this information gets refreshed. Query rankings are amenable to dynamic changes with time, so users can modify their preferences at will.

---

[1] In the sequel, terms *"priority"* and *"rank"* of a query are used interchangeably.

Multiple continuous queries may be active at any given time, each one with varying specifications concerning current range and rank. We assume that at a given time instant τ_i, the current system workload includes M queries and receives messages from N objects in total. Usually $M < N$, but in practice this workload may fluctuate across time. However, location updates exhibit a sequential pattern, hence delayed or out-of-order items never appear at incoming streams of object positions or query ranges. Commonly to a streaming scenario, the arrival rate of object locations and query specification updates could surpass the processing rate of the system. Hence, it may occur that some requests will be left unanswered for some time or locations will be dropped without processing them (load shedding). Alternative strategies are needed to provide reliable response to as many queries as possible and also respect their priorities.

We opt for a processing policy that runs periodically in *execution cycles* (as in [11,19]), each one lasting for a predefined *time granule* T, i.e., an interval that spans between successive clock ticks (like heartbeats in [17]). At the beginning of a new cycle i, we assume that fresh location and query updates are available for processing and they all refer to the same heartbeat. Therefore, updates get admitted in waves and all items in cycle i are associated to the same timestamp value τ_i. Although in practice it cannot be expected that all sensor readings have been simultaneously collected, we assume that data given for processing are always *synchronized*, after mapping detailed time indications to the coarser time granule utilized by the system. For instance, incoming items may have timestamps expressed in milliseconds, but they will be rounded to seconds, if execution cycles are scheduled each second ($T = 1$ sec). So, queries will be definitely examined against the same object locations and will provide valid answers referring to the most recent time instant.

3.2   Spatial Indexing

In presence of numerous objects and queries that continuously change their spatial features, an indexing method is apparently required to speed up range search. Traditional spatial access methods suffer from the increased rate of updates inherent in positional streams, so the execution model must be modified (e.g., [14]). As in other recent approaches (e.g., [13,20]), we preferred an index based on regular *grid partitioning* of the Euclidean plane. This static subdivision into $c \times c$ rectangular cells is common for object locations and query rectangles and it covers the entire area of interest E where they actually move.

At every timestamp value τ, all current object locations and query rectangles are hashed against grid cells. Each object can only appear in one cell, but a query may affect multiple cells; depending on grid granularity, a query rectangle may be covering a cell in total or overlapping it in part (Fig. 1b). Thus, each cell maintains a list of point locations currently found inside its area and a separate list of all queries it overlaps with. Cell contents concerning previous time instants need not be retained, since queries always ask for currently recorded locations.

To evaluate a query, those grid cells that overlap with its rectangle are firstly identified. For cells completely within query range, we can readily provide their

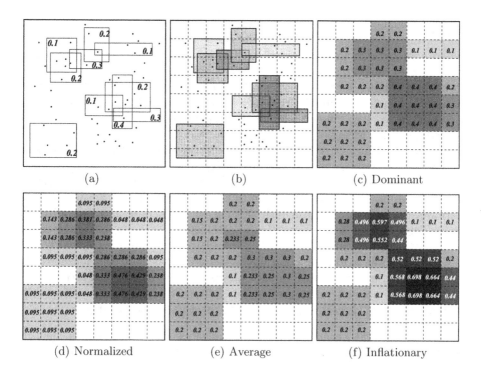

(a)   (b)   (c) Dominant

(d) Normalized   (e) Average   (f) Inflationary

Fig. 1. (a) Snapshot with $N = 50$ objects and $M = 10$ range queries with varying ranks. (b) Spatial indexing using a regular grid partitioning into 8×8 cells. (c) Dominant, (d) normalized, (e) average and (f) inflationary scoring functions for ranking cells.

associated object locations as a response. For partially covered cells, their related objects must be probed against the query rectangle to discover the qualifying ones. In line with the well-known technique of *"filter-and-refinement"* [16], this indexing structure identifies candidates just about the specified range and thus, prunes the search space considerably.

4   Ranking Model

Conforming rigorously to user-specified query rankings under limited system resources, would always favor processing of top-ranked queries at the expense of many starving requests of lower priority that would rarely be given the chance to execute. In this section, we develop a model that estimates a collective ranking of multiple queries in order to better organize their examination in groups of common interest and classified importance.

4.1   Rank Aggregation for Multiple Queries

To avoid penalizing the majority of user requests to the advantage of very few high-ranked ones, we consider several alternative *scoring functions* that examine a set of range queries overlapping with a given cell and return a *collective query rank*. This single value characterizes the entire cell area and offers a coarse estimation of the combined importance for all its associated requests. In essence, we perform a mapping from the set of rank values of numerous range queries to a much smaller set of *scores* assigned to spatially disjoint areas. When it comes to processing, instead of trivially evaluating each query individually, we may examine each cell in turn and identify qualifying object locations for all its related queries. Visiting order for cells is influenced by their collective rank, so evaluation at each cycle will most probably start from areas with higher density in urgent requests. Assuming that a mixture of queries with diverse ranks would appear in each cell, it is evident that even lower-priority queries could have a fair share in processing if they pertain to higher-priority areas.

However, queries are continuously moving and their rank may be subject to possible fluctuations across time. Hence, each cell may involve a different set of queries at every execution cycle, so its collective rank must be constantly updated. The chosen scoring function should be easy to calculate even for increased number of queries with varying ranks. Collective query rank of a cell is computed by taking into account range queries that fully cover or partially overlap with its area. Next, we suggest a family of representative scoring functions:

i) Dominant. This scheme dictates that collective rank of a cell c_k is equal to the highest priority among those currently assigned to its associated queries:

$$\sigma(c_k) = \max_{q_i \in c_k}(\rho_i) \qquad (1)$$

The rationale behind this function is that a very urgent request should prevail over others in the same cell and get precedence at any cost, no matter if the remaining queries are not so critical. Thus, a high-ranked query that covers several cells will give increased priority to all of them, even if some cells involve only few queries or have requests with rather low rank values. As a side-effect, such a choice could perhaps diminish the relative importance of other urgent queries except for the top-ranked one. However, by identifying the most pressing request in each cell, this function offers a clear view of the observed magnitude of rankings across the entire area of interest (Fig. 1c).

ii) Normalized. Towards an unbiased assessment on the distribution of query rankings in cells, this model emphasizes on the relative importance of each cell with respect to the cumulative ranking of entire query workload Q. Specifically:

$$\sigma(c_k) = \frac{\sum_{q_i \in c_k} \rho_i}{\sum_{q_j \in Q} \rho_j} \qquad (2)$$

Hence, cells are characterized by the regularized total intensity of rankings observed therein. This function conveys a proportional propensity, attempting to examine cells with impartiality and to weigh them up with respect to total demand. According to this formula, a cell with a single high-ranked query may be of similar priority to another cell overcrowded with queries of little interest

(Fig. 1d). Note that, although a single query rank value may contribute to many cell scores, it holds that the computed ratio $\sigma(c_k) \leq 1$ for any grid cell c_k.

iii) Average. This function suggests an egalitarian approach, by striking a balance in each cell between the amount of local query workload and the sum of their ranks. Such a policy can be simply expressed as the average value of query ranks per cell, assuming that m_k queries currently overlap with a given cell c_k:

$$\sigma(c_k) = \frac{1}{m_k} \cdot \sum_{q_i \in c_k} \rho_i \tag{3}$$

This objective ratio tends to smooth down the effect of extreme ranks (Fig. 1e). Yet, abundance of many low-ranked queries could prevent early execution of an urgent request that also appears in the same cell, due to its flattened score $\sigma(c_k)$.

iv) Inflationary. In contrast to previous ones, this function favors especially query "clusters" of greater interest. The more high-ranked requests present in a given cell, the more increased its inferred score should become. More formally:

$$\sigma(c_k) = 1 - \prod_{q_i \in c_k} (1 - \rho_i) \tag{4}$$

In essence, queries of higher priority are given superior influence on overall cell score. Presence of many such queries would probably boost aggregated ranks close to 1, potentially exceeding the maximum rank value assigned to any query overlapping this cell (Fig. 1f). Note that such "inflationary" [8] scores get more pronounced with higher rank values and considerable query workload.

Discussion. Overall, these functions aim at a hierarchical treatment of rankings, by capturing their actual trend and characterizing cells accordingly. With the exception of the dominant model, all other schemes also depend on the number of active queries, as well as on the distribution of their current rank values. Of course, many other schemes could have been devised, e.g., by taking the median of rank values in each cell or applying a "reserved" behavior [9]. There is no clear rule concerning which function is more suitable for assessing query rankings collectively, since they have similar computational complexity. As we experimentally verified (Section 6), inflationary and dominant models generally cope better with diverse query ranges compared to regularized functions (normalized, average). Proper choice of scoring function may be affected by the actual workload and the rank patterns of registered queries, but it chiefly depends on the operational goals of the application at hand.

4.2 Determining Cell Ranking Scores

Notwithstanding the crucial role of rankings, successful evaluation of queries also depends on the observed pattern of object locations. This sounds fair enough, since concentration of many objects may constitute a potential trend that requires notification before long, even for a few low-priority queries.

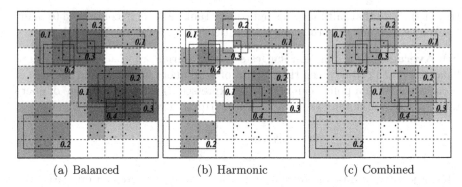

(a) Balanced   (b) Harmonic   (c) Combined

Fig. 2. Cell classification according to scores computed with an inflationary function (rankings as in Fig. 1), assuming that significance of queries is twice as much as that of objects. Graduated gray color reflects the magnitude of inferred cell ranking scores.

Therefore, prioritization should ideally strike a balance between query rankings and object distribution, so we adopt a composite model for deciding the final cell rankings. Let a grid cell c_k with m_k queries and n_k objects located there at timestamp τ. The estimated importance of this cell is expressed by its *ranking score* β_k, which is based on three factors:

1. A *collective query rank* $\sigma(c_k)$ of requests pertaining to this cell, as captured by the chosen scoring function. Intuitively, the more high-ranked queries currently present in a cell, the more increased its ranking should become.
2. The *percentage* $P_k = \frac{n_k}{N}$ of objects currently located within c_k. All cells have equal areas due to uniformity of grid partitioning, hence this fraction expresses the present *density* of moving objects in a cell. In most typical object distributions, some cells are expected more crowded than others, so their P_k value will be higher. Note that all objects are assumed equally important with no particular preference assigned to them, since ordering results in not needed for range queries.
3. A system-wide *regulation parameter* $\lambda \geq 0$ used to leverage the relative importance between the set of queries and the set of objects during processing.

We suggest three alternative options for determining cell ranking scores:

i) Balanced Score. The overall ranking score of cell c_k can be calculated as:

$$\beta_k = \lambda \cdot \sigma(c_k) + (1 - \lambda) \cdot P_k \qquad (5)$$

where parameter $\lambda \in [0..1]$. If $\lambda = 1$, the final cell score will take into account only query ranks, ignoring object distribution altogether. Smaller λ values indicate an increasing influence of objects into the final ranking scores. In case that $\lambda = 1/2$, objects and queries are treated in equal terms for computing the final score. If $\lambda > 1/2$, then query rankings are considered more important than object densities, hence their increased influence on final cell scores (Fig. 2a illustrates cell scores when $\lambda = 2/3$). In case that $\lambda = 0$, rankings are completely ignored, and query prioritization is based solely on current object distribution.

ii) Harmonic Score. Another scheme for assessing overall ranking score can be based on the *weighted harmonic mean* of queries and objects currently in the cell (reminiscent of a performance measure from Information Retrieval [10]):

$$\beta_k = \frac{(1+\lambda) \cdot \sigma(c_k) \cdot P_k}{\lambda \cdot \sigma(c_k) + P_k} \tag{6}$$

When $\lambda = 1$, queries and objects are considered equally important, but such an even weighing might not be always ideal. If $\lambda < 1$ the importance of queries gets accentuated, while values $\lambda > 1$ emphasize density of objects. In case that $\lambda = 1/2$, the collective rank of queries weighs twice as much as the total density of objects (Fig. 2b). As λ approaches zero, the impact of queries becomes more pronounced compared to the significance of objects. When P_k is very low (i.e., few objects in this cell), the overall score will drop, even if critical queries affect the same area. Whenever $\sigma(c_k) = 0$ or $P_k = 0$, we can safely avoid examination of that cell, since there are no pending queries or no objects at all, respectively.

iii) Combined Score. This method computes a ranking score based on queries and then it regularizes it by the mixed density of queries and objects alike:

$$\beta_k = \frac{m_k + n_k}{M + N} \cdot \sigma(c_k) \tag{7}$$

The fractional term expresses the "popularity" of that specific cell with the pending queries and current locations, not making any distinction between the two sets. Naturally, among cells that are equally congested, the one that has greater score σ will get higher priority (Fig. 2c).

Discussion. Note that $\sigma(c_k)$, P_k, $\beta_k \in [0..1]$, so the calculated score can be used as a *measure* for assessing the relative importance of cells. After sorting cells by their β_k, we can determine an advantageous order for visiting them and evaluating their queries. Anyway, all score formulae guarantee that $\max \beta_k = 1$. In fact, this can only occur at the highly improbable case that all queries and objects are found within the same cell c_k and all queries have rank 1.

The rationale behind suggestion of the aforementioned scoring schemes is illustrated in Fig. 2. Intuitively, balanced scores are meaningful for each cell referencing objects or queries, although probably weighing them unevenly. In contrast, harmonic scores only appear in cells that contain both queries and objects; thus, a cell without objects need not be visited at all, whereas objects need not be examined if no query covers their cell. As for combined scores, they are assigned to cells overlapping with queries, no matter if any objects are actually located in them.

5   Prioritized Query Evaluation

Based on ranking scores of all cells, an execution order can be determined for providing fresh results to continuous queries. In this section, we suggest alternative policies for prioritized examination of queries in cells and we discuss possible estimates for the quality of service achieved.

5.1   Utilizing Ranking Scores in Query Execution

As already pointed out, evaluation takes place in *execution cycles* that last for a fixed time T. During this period T, location and query updates of identical timestamp τ are first inserted into the data structures linked to the grid index, while rankings get computed for all affected cells. As long as T has not yet expired, cells are visited and results are generated for their associated queries, attempting to examine as many cells as possible. Apparently, an evaluation strategy should start from queries affecting the most important cells, which are exactly those with the highest ranking scores. When the deadline is reached (i.e., available T has expired), a new cycle starts with fresh stream items and query ranges, discarding previous state altogether and ignoring any unprocessed cells.

A strict time limit T incurs that there is no guarantee that a query q_j will receive exact results at any execution cycle. Unless all cells covered by its spatial extent a_j are examined, the response to a given query will be probably incomplete; unvisited cells may contain locations that fall inside range a_j, but these results will be missed. Therefore, answer could be emitted with an estimated *confidence margin* f_j for each query q_j. This factor should express how representative this answer actually is, showing the degree of processing for each query. For an approximate estimation of the accuracy of answers given to query q_j, we suggest to compare the number of cells C_j overlapping its extent a_j to the number of cells V_j that actually contributed to its current result (of course, $V_j \subseteq C_j$). Then, the answer to every single query q_j at time τ is returned with a confidence margin $f_j = \frac{|V_j|}{|C_j|}$. In case that a collective estimation is needed, we can get either the percentage of queries answered completely or the percentage of those that received no response at τ. Both estimates can be used as an empirical indication of efficiency, as we point out in Section 6. A detailed study concerning accuracy of query results is a challenging topic by itself and it is left for future work.

At the end of each execution cycle (i.e., every successive timestamp τ), the subset R of processed queries that received a response can be compared against the entire set Q of active queries. A global *success ratio* at each τ can be estimated collectively for all cells c_k of the grid:

$$\gamma(\tau) = \frac{\sum_{c_k} \sum_{q_i \in R} \rho_i}{\sum_{c_k} \sum_{q_j \in Q} \rho_j} \tag{8}$$

This ratio simply expresses how well the system coped with prioritized execution of active queries satisfying user demands, and can be considered as an indication for the Quality of Service (*QoS*) achieved by the processing mechanism. Note that success ratio is calculated on original user-specified query rankings and not on artificial scores computed during evaluation. In the optimal case that $\gamma(\tau) = 1$, all queries have been answered completely at this execution cycle.

5.2   Cell Examination Strategies

As soon as the ranking scores are computed at τ, the system can start visiting cells and probe locations against their associated queries, until the end of that cycle. Within each particular cell, queries are processed in descending rank order (i.e., the one assigned from users). However, several options are still available. Should we examine all queries within a cell before proceeding to the next, even for queries of negligible interest just because they happen to pertain to a prioritized cell? Or, is there any means of responding first to high-ranked requests throughout the entire area E of interest and afterwards, if time permits, to not-so-urgent remaining ones? Further, how can we cope with queries of low priority that only occasionally (or never) get a chance to execute? To tackle such issues, we devised three alternative evaluation strategies that concern the *degree of processing* that takes place within or among cells.

Exhaustive Evaluation. As its name suggests, this strategy provides response to all queries in a given cell, before visiting the successive one. Evidently, it starts processing cell c_k that currently has the highest ranking score β_k and examines all queries $q_j \in c_k$, even those of very low rank ρ_j. Then, processing continues with other cells in descending rank order, until deadline T is reached.

Stratified Evaluation. Obviously, cells can be classified into l *strata*, according to their calculated scores (e.g., $l = 4$ in Fig. 2a). Such a classification may resemble either to an *equi-sum histogram* where all cells in each class sum up to the same overall ranking or to a *quantile* by dividing cells into disjoint sets of equal range. This strategy follows a prioritized rotating scheme by examining earlier the cells of ith stratum every r^i execution cycles ($i = 0, 1, \ldots$). For instance, assuming that $r = 2$, cells in top class $i = 0$ still provide answers at every (2^0) cycle, but next class $i = 1$ takes precedence every 2^1 cycles, class $i = 2$ every 2^2 cycles, class $i = 3$ every 2^3 cycles and so on.

Essentially, this policy each time gives precedence for execution to a different class of queries (i.e., a stratum of similar rankings), while all the rest get processed in descending ranking order. Thus, cells with lower scores will sometimes be prioritized before top-ranked ones. The frequency of such an out-of-order prioritization depends on the number l of classes. Of course, cells belonging to the same class are still examined according to their ranking scores.

This scheme aims at preventing *starvation* of queries; while boosting execution of the most important ones, it still ensures that low-priority user requests eventually receive some results, even not so frequently.

Threshold-guided Evaluation. Consider a processing policy that iterates through cells in descending rank order, but for each cell c_k it only answers queries $q_j \in c_k$ that have $\rho_j \geq \theta$, where $\theta \in [0..1]$ is a flexible threshold value. After leaving cell c_k, its ranking score β_k gets properly adjusted by excluding all processed queries and their ranks. In case that all cells have been visited and there is still some time available until T expires, this technique can start complementary rounds of processing. In each round, cells are visited according to their modified scores, so the top cell may be different each time. Intuitively, this policy intends to deliver results primarily to an *élite* comprised of demanding clients. As soon as higher-priority requests are responded completely, then the system can take care of the rest. However, choice of θ is a delicate issue. If θ is too large, only high-ranked requests will be endorsed; a smaller value may retract the desired benefits by falling short to serve all targeted queries q_j with $\rho_j \geq \theta$ within the deadline.

Therefore, we propose a variation not requiring an explicit θ, but controlling itself the fraction of prioritized requests for each cycle τ. To this goal, global success ratio $\gamma(\tau - 1)$ obtained immediately before the current cycle (Eq.(8)) can be utilized as an expected target, along with a *local success ratio* $\gamma_k(\tau)$ computed dynamically for the examined cell c_k. This latter ratio characterizes the degree of progress for the set of queries Q_k pertaining to c_k as follows:

$$\gamma_k(\tau) = \frac{\sum_{q_i \in R_k} \rho_i}{\sum_{q_j \in Q_k} \rho_j} \qquad (9)$$

This quotient resembles to Eq.(8), but it is computed locally for currently examined cell c_k and it only involves the set R_k of already processed requests among those in Q_k. Hence, during evaluation at τ, this strategy should continue examining cell c_k as long as the local success ratio $\gamma_k(\tau)$ has not yet attained the most recent global ratio, i.e., $\gamma_k(\tau) < \gamma(\tau - 1)$.

Yet, it should be ruled out the risk that success ratio γ could steadily deteriorate cycle after cycle. To avoid this possibility, we have finally chosen a more *optimistic* variant, by purposely raising the expected target, say by a small $\delta = 10\%$. In this fashion, processing for a given c_k continues as long as $\gamma_k(\tau) < (1 + \delta) \cdot \gamma(\tau - 1)$. As soon as all cells have been examined and there is still time left for additional processing, this best-effort strategy will start another round on and on, eager to further improve the global success ratio.

6   Experimental Validation

In this section, we report performance and qualitative results from an experimental validation of the prioritized query evaluation strategies, and also discuss appropriate parameterization of the ranking model utilized.

6.1   Experimental Setup

We generated synthetic datasets for objects and queries moving at diverse speeds along the road network of greater Athens (an area of about 250 km^2). By calculating shortest paths between nodes chosen randomly across the network, we were able to create samples of 200 timestamps from each such route. Thus, we obtained a point set of locations for $N = 100000$ objects, and similarly, the centroids of $M = 10000$, 20000, and 50000 query rectangles. Since our focus is primarily on queries, N remains always fixed while M is scaling. Spatial range of queries is expressed as percentage (%) of the entire monitored area of interest E. The interarrival time of streaming messages from objects and queries was fixed, assuming that all of them reported their positional updates concurrently at regular time intervals (in this case, at every timestamp). This is actually the most

Table 1. Experiment parameters

Parameter	Values
Number N of objects	**100000**
Number M of range queries	10000, **20000**, 50000
Query range a (% of universe E)	1, **2**, 3, 4, 5
Execution cycle duration (T)	1, **2**, 3, 4, 5 seconds
Grid size ($c \times c$)	8×8, $\mathbf{16 \times 16}$, 24×24, 32×32
Regulation (λ)	0.01, 0.1, 0.33, 0.5, **0.66**, 0.75, 1
Number of classes (l)	4, 6, 8, 10

intensive situation, as *agility* of movement is set to 100%, so system must cope simultaneously with massive location updates and renewal of all query results.

With respect to query rankings, at each timestamp we assigned ranks according to a Zipfian distribution with parameter $s = 1$ for ten discrete values $\{0.1, 0.2, \ldots, 1\}$. This scheme imitates a real-world situation where "the more you pay, the sooner you get results", but few people can afford to pay expensively for better service. Thus, high-ranked queries ($\rho = 1$) represent a rather small fraction, while the majority consists of requests of negligible interest ($\rho = 0.1$).

Evaluation strategies were implemented in C++ and compiled with gcc on an Intel Core 2 Duo 3GHz CPU running GNU/Linux with 2GB of main memory. We ran simulations using different parameter settings for each experiment. Due to space limitations, we show results just from some representative ones. All results are calculated averages of the measured quantities for 200 time units. Table 1 summarizes experimentation parameters and their respective ranges; the default value is shown in bold.

6.2   Experimental Results

System Configuration. The first set of experiments aimed at fixing some basic parameters of the system. Primarily, we needed to specify granularity c of the partitioning, since this grid indexes all incoming stream items and also controls query execution. Figure 3a plots the per cycle cost of hashing a fixed number of objects and diverse query workloads into grid cells. This particular diagram refers to query rectangles that cover $a = 2\%$ of universe E, but similar trends occur for all ranges tested. As expected, cost depends on input size; however, as grid granularity gets finer, maintenance cost escalates, especially for larger query numbers. This is due not only to data adjustments in more cell lists, but also because scores must be calculated for many more cells. To take a decision, we examined global success ratio γ under different grid sizes. As indicatively shown in Fig. 3b for exhaustive evaluation of $M = 50000$ queries under a balanced inflationary scheme with a strict $T = 5$ sec, coarser subdivisions are more stable under all query ranges. Thus, we picked $c = 16$ for subsequent experiments, since this yields better quality γ of results for most spatial extents.

Another crucial system parameter concerns duration T of each execution cycle. A too short period will not be sufficient to respond to many queries, especially

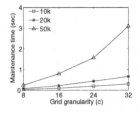
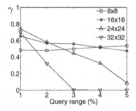
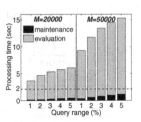

Fig. 3. Grid cell granularity   **Fig. 4.** Processing time

for greater workload M; on the other hand, a greater T will delay deliverance of fresh results to high-ranked queries. Figure 4 illustrates the processing cost for complete response to 20000 and 50000 queries of several ranges under a balanced inflationary scheme (similar graphs obtained for other scoring functions). Indeed, system overhead for index maintenance and rank aggregation at the cell level is minimal compared to actual evaluation of queries. Quite predictably, evaluation cost deteriorates –yet sublinearly– for larger query rectangles and greater number of requests. Besides, Fig. 7 depicts the success ratio achieved for $M = 20000$ queries, using the same scoring scheme but with several specified deadlines. A strict deadline T leaves many queries partially or completely unanswered, hence the drop in overall QoS. In effect, a proper choice of T is a trade-off between the number of queries and their actual ranges. In the sequel, unless otherwise specified, we examine $M = 20000$ queries with a strict $T = 2$ sec, in order to compare performance of evaluation strategies under the ranking model.

Ranking Schemes. As explained in Section 4.1, each of the scoring functions interprets distribution of query ranks in a different fashion. This is reflected in Fig. 5, which shows success ratio for $M = 20000$ queries at $T = 2$ sec, using a balanced scoring scheme with $\lambda = 0$. In that fashion, object distribution is purposely ignored, so as to get more insight into the role of rank aggregation. Not surprisingly, biased schemes (dominant, inflationary) seem to excel in quality compared to egalitarian ones (average, normalized). Similar conclusions can be drawn from Fig. 6 that depicts a breakdown of queries answered completely, only partially or not at all, irrespective of their rank. This effect can be attributed to cells that gain a high ranking score due to the presence of urgent requests, so this turns in favor of all their affected queries. In contrast, an impartial approach (especially a normalized scheme) tends to downgrade collective cell rankings and makes discrimination of regions with important queries much harder, thus reducing overall quality. It appears that an *inflationary function* copes slightly better than a dominant one with diverse spatial ranges and query numbers, hence it is utilized in all subsequent experiments.

Next comparison refers to alternative scoring schemes for cells (Fig. 8). Quality of results was derived from an exhaustive examination; for scoring cells, queries were considered twice as significant as objects. Interestingly enough, the combined score prevails over the other two, because it leans towards intensifying aggregated query ranks in proportion to cell density in queries and objects alike.

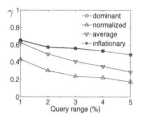

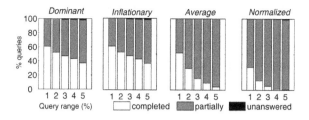

Fig. 5. Scoring functions

Fig. 6. Response to queries using scoring functions

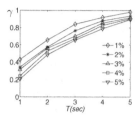

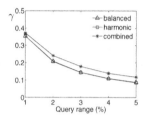

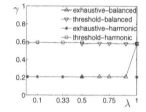

Fig. 7. Execution deadline

Fig. 8. Cell scoring

Fig. 9. Regulation effect

Cell ranking scores seem to get overly "spoiled" when considering object distribution P_k as an autonomous factor in Eq.(5) and Eq.(6). Despite suspicions that this might have been caused from inappropriate choice of regulation parameter, both balanced and harmonic schemes seem unaffected from moderate values of λ. As reflected in Fig. 9 for exhaustive and threshold evaluation of query ranges at $a = 2\%$ of E, quality remains almost stable with the exception of balanced score when objects are not taken into account at all ($\lambda = 1$). We believe that regulation between opposing demands for query responses and object freshness requires more investigation, and we plan to study dynamic determination of a flexible λ across execution, instead of fixing it beforehand.

Evaluation Strategies. As depicted in Fig. 9, threshold-guided evaluation consistently surpasses the exhaustive policy almost by a factor of 3 in overall quality. With respect to stratified evaluation (Fig. 10a), the combined scoring scheme is again the most beneficial to QoS. In addition, under all scoring schemes, quality slightly deteriorates as the number of classes increases (Fig. 10b). Indeed, with a strict deadline $T = 2$ sec, only top classes will be given complete response, while the remaining ones should await their turn in the rotating scheme to get priority. The greater the number of query classes, the less often each one gets prioritized. From Fig. 11, we also conclude that stratified execution is slightly better than exhaustive, but cannot really compete with the threshold-guided policy. Superiority of threshold-guided evaluation is justified considering that, at every cycle, this strategy at least tries not to fall short of the quality achieved at the preceding cycle, if not gaining a better QoS (in all simulations we set $\delta = 10\%$). In addition, if there is time left for additional processing, this self-regulating policy continues to evaluate more queries, further improving success ratio γ.

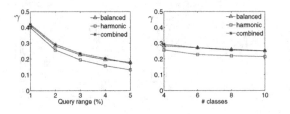

Fig. 10. Stratified evaluation

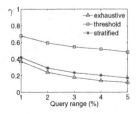

Fig. 11. Evaluation strategies

As already shown in Fig. 6, only rarely queries are left entirely unanswered, when time deadline is very short and the number of queries and objects is overwhelming. Typically, each strategy responds completely to a significant portion of the actual query workload, whereas most of the remaining requests receive at least some partial results. The returned locations are always accurate, so no false positives exist among those found for any query. However, some qualifying object locations may not be retrieved at all, because they happen to fall in a cell that was not visited during that cycle.

7   Conclusions and Future Work

In this work, we set forth a novel approach for evaluating moving range queries with user-specified priorities over streaming positions of moving objects. We developed a ranking model for identifying regions with higher concentration in urgent queries and object locations, in order to examine together groups of queries and not separately each one. We further proposed adaptive evaluation policies, that achieve varying degrees of quality in answering scaling query workloads.

In the future, we plan to extend the ranking model according to estimated cost and query selectivity in combination with users' preferences. Besides, by incorporating an aging-aware prioritization scheme that favors long-penalized queries, the processing mechanism can be fine-tuned even more, offering better accuracy in results. Finally, applying similar ranking schemes for other query types (k-nearest neighbor search, in particular) is a challenging research topic.

References

1. Carey, M.J., Kossmann, D.: On Saying "Enough Already!" in SQL. In: SIGMOD, pp. 219–230 (1997)
2. Cai, Y., Hua, K.A., Cao, G.: Processing Range-Monitoring Queries on Heterogeneous Mobile Objects. In: MDM, January 2004, pp. 27–38 (2004)
3. Gedik, B., Liu, L., Wu, K.L., Yu, P.S.: Lira: Lightweight, Region-aware Load Shedding in Mobile CQ Systems. In: ICDE, April 2007, pp. 286–295 (2007)
4. Gedik, B., Liu, L.: Mobieyes: A Distributed Location Monitoring Service using Moving Location Queries. IEEE Transactions on Mobile Computing 5(10), 1384–1402 (2006)

5. Hjaltason, G., Samet, H.: Ranking in Spatial Databases. In: Egenhofer, M.J., Herring, J.R. (eds.) SSD 1995. LNCS, vol. 951, pp. 83–95. Springer, Heidelberg (1995)
6. Hu, H., Xu, J., Lee, D.L.: A Generic Framework for Monitoring Continuous Spatial Queries over Moving Objects. In: SIGMOD, pp. 479–490 (2005)
7. Ilyas, I.F., Shah, R., Aref, W.G., Vitter, J.S., Elmagarmid, A.K.: Rank-aware Query Optimization. In: ACM SIGMOD, June 2004, pp. 203–214 (2004)
8. Koutrika, G., Ioannidis, Y.: Personalization of Queries in Database Systems. In: ICDE, April 2004, pp. 597–608 (2004)
9. Koutrika, G., Ioannidis, Y.: Personalized Queries under a Generalized Preference Model. In: ICDE, April 2005, pp. 841–852 (2005)
10. Manning, C.D., Raghavan, P., Schütze, H.: Introduction to Information Retrieval. Cambridge University Press, Cambridge (2008)
11. Mokbel, M., Xiong, X., Aref, W.G.: SINA: Scalable Incremental Processing of Continuous Queries in Spatiotemporal Databases. In: ACM SIGMOD, June 2004, pp. 623–634 (2004)
12. Morse, M., Patel, J.M., Grosky, W.I.: Efficient Continuous Skyline Computation. In: ICDE, April 2006, pp. 108–110 (2006)
13. Mouratidis, K., Hadjieleftheriou, M., Papadias, D.: Conceptual Partitioning: An Efficient Method for Continuous Nearest Neighbor Monitoring. In: ACM SIGMOD, June 2005, pp. 634–645 (2005)
14. Prabhakar, S., Xia, Y., Kalashnikov, D.V., Aref, W.G., Hambrusch, S.E.: Query Indexing and Velocity Constrained Indexing: Scalable Techniques for Continuous Queries on Moving Objects. IEEE Transactions on Computers 51(10), 1124–1140 (2002)
15. Qu, H., Labrinidis, A.: Preference-Aware Query and Update Scheduling in Web-databases. In: ICDE, April 2007, pp. 356–365 (2007)
16. Rigaux, P., Scholl, M., Voisard, A.: Spatial Databases with Application to GIS. Morgan Kaufmann, San Francisco (2001)
17. Srivastava, U., Widom, J.: Flexible Time Management in Data Stream Systems. In: PODS, June 2004, pp. 263–274 (2004)
18. Tao, Y., Hristidis, V., Papadias, D., Papakonstantinou, Y.: Information Systems 32, 424–445 (2007)
19. Wei, Y., Son, S., Stankovic, J.: RTSTREAM: Real-Time Query Processing for Data Streams. In: ISORC, April 2006, pp. 141–150 (2006)
20. Yu, X., Pu, K.Q., Koudas, N.: Monitoring k-Nearest Neighbor Queries Over Moving Objects. In: ICDE, April 2005, pp. 631–642 (2005)
21. Zhang, Y., Hull, B., Balakrishnan, H., Madden, S.: ICEDB: Intermittently Connected Continuous Query Processing. In: ICDE, April 2007, pp. 166–175 (2007)

A Comparative Evaluation of XML Difference Algorithms with Genomic Data

Cornelia Hedeler and Norman W. Paton

School of Computer Science, The University of Manchester, Oxford Road,
Manchester M13 9PL, UK
{chedeler,npaton}@cs.manchester.ac.uk

Abstract. Genome sequence data and annotations are subject to frequent changes resulting from re-assembly and re-annotation, or community feedback based on experimental evidence, giving rise to new data releases. These releases are rarely accompanied by a description of the changes, making it difficult for biologists working with the data to identify and work through the consequences of the changes that have taken place. This paper explores the extent to which existing XML difference algorithms, namely X-Diff, JXyDiff and 3DM, can be used to identify and document genome changes, in particular investigating: (i) their ability to detect typical changes in genome sequence documents; and (ii) the ease with which the difference report can be used to determine whether genes of interest are affected by changes to the genome. The evaluation compares the performance of the algorithms both with synthetic modifications and for detecting changes in a public genomic database. Typical behaviours of the algorithms are identified and a root cause analysis carried out.

1 Introduction

Genomic data, including the annotation of predicted genes and proteins, is available for an increasing number of genomes. The genomic data for each of those genomes undergoes regular re-annotation, and in many cases even re-assembly of the sequence data, resulting in new releases that replace previous versions. In particular in the early stages of a genome release, re-annotation generally involves automatic annotation of the (re-assembled) sequence from scratch using a computational analysis pipeline (e.g., the Ensembl analysis pipeline [1,2]). In the later stages, i.e., a few years after the initial genome release, this process might be complemented or increasingly replaced by manual changes to the annotation based on experimental evidence provided by the community. In contrast to the manual annotation, which explicitly introduces a change to the previous annotation and therefore, if captured appropriately, provides a delta description between two different releases of the same genome (e.g., the summary of chromosome sequence and annotation updates provided by the Saccharomyces Genome Database (SGD) [3]), the automatic re-annotation of a genome does not result in such a delta description. Therefore, very little or no information about the

B. Ludäscher and Nikos Mamoulis (Eds.): SSDBM 2008, LNCS 5069, pp. 258–275, 2008.

changes carried out is made available by providers of genomic data. For example, Ensembl [4] only provides very general information on the type of data updated as news for each new release, but no detailed information on gene level changes. In contrast, the Fungal Genome Initiative (FGI) at the Broad Institute[1] provides a list of the changes between two releases for some genomes. The information provided on the changes, however, is limited and only lists the identifiers of newly predicted, updated, split and deleted genes, but no further details. The limited information on changes between genome data releases presents significant challenges to biologists and bioinformaticians working with the genomic data and the corresponding annotation, in some cases even resulting in the continued use of an out-of-date genome release.

A number of data providers make previous genome releases available in archives (e.g., Ensembl, EMBL [5,6]), provide tools for converting data from one release to another release of a genome (Ensembl), or provide tools for comparison of two genome versions, the result of which can be inspected manually (EMBL). Those tools, however, are not generally applicable. The converter provided by Ensembl is currently only available for converting mouse assembly data between the current release and the previous one, and is not available for conversion between other combinations of releases or other genomes. The comparison tool provided by EMBL can be used to compare any two versions of a genome, but it compares the files line by line and highlights lines that have been removed or inserted between the two versions. Lines that contain changes are also presented as removed and inserted. The user has to inspect the highlighted file manually to identify the consequences for genes of interest, a labour-intensive task when carried out for a whole genome and/or a number of releases. Changes between two releases of the same genome can be of varied nature (see the following section). Without a history of changes, however, it is hard for users of the data to track genes of interest through time, i.e., various genome releases and determine, e.g., whether a missing gene has been removed, renamed, moved to a different location, or merged with a neighbouring gene.

Genomic data is represented in a variety of formats, amongst others XML (e.g., EMBL XML Schema), and converters are available to convert data between different formats and also to convert it into the EMBL XML format (e.g., converters are provided by EMBL [5] and BioJava[2]). A number of XML difference tools and algorithms have also been published previously (e.g., X-Diff [7], XyDiff [8], 3DM [9]). In the original publications these algorithms have mainly been applied to merging of XML documents that have been edited by different people (3DM), and for detecting changes of XML documents on the web (X-Diff, XyDiff). In this paper we evaluate the applicability of XML difference algorithms on XML documents representing genomic data and associated annotation. In so doing, we seek both to obtain insights into the algorithms and to identify an effective means of understanding the changes that have been made to genomes.

---

[1] http://www.broad.mit.edu/annotation/fgi/
[2] http://biojava.org

The paper is structured as follows. Section 2 provides a detailed description of the problem, including an introduction to the format in which genomic data is represented. Section 3 introduces the XML difference algorithms evaluated here. Section 4 describes the experimental setup and discusses the results. This is followed by an evaluation of the XML difference algorithms on real data in Section 5. Section 6 concludes by reviewing the lessons learned.

2   Problem Description

In this section the XML representation of genomic data is introduced. Furthermore, a number of typical changes to genomic data are presented.

Genomic data is represented in a number of flat file formats established by the major data providers, including Genbank [10] and EMBL [5]. For both Genbank and EMBL formats, XML representations of the data are available, namely Insd XML and EMBL XML, respectively. Due to its slightly more intuitive representation of genomic data, we have chosen to use the EMBL XML representation. The majority of genomic data is made available separately for each chromosome of a genome. For this reason, we have chosen a chromosome as the unit in which we analyse changes in genomic data. Therefore, we do not consider changes that affect multiple chromosomes, such as a move of a predicted gene between chromosomes; such changes would be detected as a deletion in one chromosome and an insertion in another.

A gene is a defined strand of DNA that contains regions that code for a protein (exons) and those that do not (introns). The complete sequence of exons for a gene is also called a coding sequence (CDS). The whole DNA sequence of a gene is transcribed from DNA to (pre-)mRNA, followed by a process called splicing during which the introns are removed and the exons spliced together to form the messenger RNA (mRNA). The mRNA is then translated into a protein.

In both the well established flat file formats and the corresponding XML representations a predicted gene is represented as follows: (i) an element capturing information on the gene; (ii) an element describing the corresponding mRNA; (iii) an element containing information on the coding sequence (CDS); and (iv) elements with information on the corresponding exons (elements `<feature name="gene">`,`<feature name="mRNA">`,`<feature name="CDS">`, and `<feature name="exon">`, respectively, in Figure 1). Usually, information on the predicted genes appear in the order of the genes on the chromosome, i.e., the order among siblings is important, and the genomic sequence is included as a whole for the chromosome and not for each gene separately. However, to ease the identification of changes in the sequence of a gene, we retrieve the sequence for each gene and include it alongside the corresponding gene information. An example of the resulting XML representation of the elements describing the gene, its mRNA, CDS, genomic sequence and exon is provided in Figure 1. Usually, the descriptions of all exons for all genes can be found at the end of the document (indicated by '...' in the example). As can be seen in the example, all the elements corresponding to a gene are located at the same level in the hierarchical structure

```
<EMBL>
    <entry accession="chromosome:SGD1.01:I:1:230308:1" lastUpdated="8-APR-2007" name="I">
        ...
        <feature name="gene">
            <qualifier name="gene">YAL003W</qualifier>
            <qualifier name="note"> Elongation factor 1-beta (EF-1-beta) </qualifier>
            <location complement="false" type="single">
                <locationElement complement="false" type="range">
                    <basePosition type="simple">142176</basePosition>
                    <basePosition type="simple">142255</basePosition>
                </locationElement> </location> </feature>
        <feature name="mRNA">
            <qualifier name="gene">YAL003W</qualifier>
            <qualifier name="note">transcript_id=YAL003W</qualifier>
            <location complement="false" type="single">
                <locationElement complement="false" type="range">
                    <basePosition type="simple">142176</basePosition>
                    <basePosition type="simple">142255</basePosition>
                </locationElement> </location> </feature>
        <feature name="CDS">
            <dbreference db="RefSeq_peptide" primary="NP_009398.1"/>
            <qualifier name="gene">YAL003W</qualifier>
            <qualifier name="protein_id">YAL003W</qualifier>
            <qualifier name="note">transcript_id=YAL003W</qualifier>
            <qualifier name="translation"> MASTDFSKIETLKQLNASLADKSYIEGTAVSQA...</qualifier>
            <location complement="false" type="single">
                <locationElement complement="false" type="range">
                    <basePosition type="simple">142176</basePosition>
                    <basePosition type="simple">142255</basePosition>
                </locationElement> </location> </feature>
        <sequence length="987" type="DNA" version="0.0"> agttgcgcatgaatttctcc...</sequence>
        ...
        <feature name="exon">
            <qualifier name="note">exon_id=YAL003W.1</qualifier>
            <location complement="false" type="single">
                <locationElement complement="false" type="range">
                    <basePosition type="simple">142176</basePosition>
                    <basePosition type="simple">142255</basePosition>
                </locationElement> </location> </feature>
        ...
    </entry>
</EMBL>
```

Fig. 1. Example of genomic data represented using a variant of EMBL XML [5,6]

of the XML representation. There are no parent elements for each gene that contain all the associated elements; instead the parent element of all elements describing all the genes is the element describing the chromosome on which the genes are located. It can also be seen that the elements contain partly redundant information, such as the location and the identifier of the gene.

With increasing knowledge of a genome, its genomic data and associated annotation can change in a number of different ways. Changes include modifications of the genomic sequence itself, the identification of new genes with their associated proteins, and the removal of a previously predicted gene that is no longer thought to be a gene. Further changes include the merging of two neighbouring genes into one gene or the splitting of one gene into two neighbouring genes. Examples of such changes can be found in the change history provided by SGD for the yeast genome.[3] In the remainder of the paper we focus on five types of

---

[3] http://www.yeastgenome.org/cache/genomeSnapshot.html#ChrSeqAnnotUpdates

changes, which are explained in more detail below. A number of other changes, such as the update of the name or identifier of a gene or the update of its location, are essentially updates to the value of a text node. As is shown later, an update to the value of a text node does not present a challenge to the XML difference algorithms.

Change to the genomic sequence. The genomic sequence can undergo changes, for example, to correct mistakes introduced in an earlier release. To change the sequence of a gene, the value of the corresponding text node with name **sequence** is updated.

Identification of a new gene. When a new gene that has been missed previously is identified, all elements capturing information on the gene, its mRNA, CDS, sequence and exons are inserted. As the genes appear in the document in the same order as on the chromosome, the first four elements (gene, mRNA, CDS, sequence) for the new gene are inserted after the element containing the sequence of the preceding gene and before the elements describing the following gene. As the exons are listed at the end of the document in the same order as on the chromosome, the elements with information on the exons of the new gene are inserted between the exons of the preceding gene and those of the following gene.

Removal of a previously predicted gene. Analysis of the genome and its sequence can reveal that a previously identified gene isn't actually a gene. Thus, all elements describing this gene, its mRNA, its CDS, its sequence and its exons need to be removed.

Both identification and removal of a gene result in a change of the positions of elements following the inserted or removed elements within the sequence of elements describing all the genes on a chromosome.

Merging of two neighbouring genes, gene1 and gene2, into one gene. Biological experiments can reveal that two separate genes actually correspond to a single gene. In such a case the annotation needs to be updated as follows (see also Figure 2):

- Update of the elements corresponding to *gene1*. Changes include (i) update of the identifiers of the elements describing the gene, its mRNA and CDS; (ii) update of the location of *gene1* and insert of additional location elements for the exons of *gene2*, which after the merge belong to *gene1*; and (iii) update of the sequence and translation of *gene1* by appending the corresponding sequence and translation of *gene2*.
- Update the identifiers of all the exons belonging to *gene1* and *gene2* before the merge to reflect that they belong to the merged gene.
- Delete the elements (with information on gene, mRNA, CDS, sequence) corresponding to *gene2*.

Splitting of one gene into two neighbouring genes, gene1 and gene2. As the complementary change to merging of two neighbouring genes, a single gene can

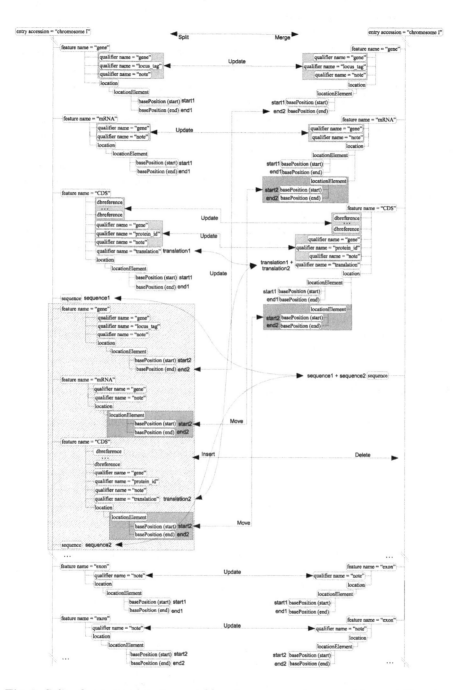

Fig. 2. Split of one gene into two neighbouring genes and merge of two neighbouring genes into one gene

also be split into two neighbouring genes. The annotation is updated as follows (see also Figure 2):

- Insert elements (gene, mRNA, CDS, sequence) corresponding to *gene2* with the appropriate part of the translation and sequence that belongs to *gene2* after the merge and the appropriate location elements. This information can be obtained from the single gene that is to be split.
- Update of the elements corresponding to the single gene to reflect that it represents *gene1* after the split. Changes include (i) update of the identifiers of the elements describing the gene, its mRNA and CDS; (ii) update of the location of the gene and delete of the additional location elements for the exons that belong to *gene2* after the split; and (iii) update of the sequence and translation of *gene1* by removing the part of the sequence and translation that belongs to *gene2* after the split.
- Update of the identifiers of all the exons belonging to the gene to be split to reflect that they belong to either *gene1* or *gene2*. If there is only one exon, this needs to be split by updating the existing exon appropriately to reflect that it belongs to *gene1* and inserting a new element with information on the exon belonging to *gene2*.

As shown above, changes to genomic sequence data and its annotation generally lead to a number of (different) changes to the document, not something the techniques evaluated here were designed to support.

3   XML Difference Algorithms

Three published XML difference algorithms for which working Java implementations were obtained are evaluated in this paper. The key properties of the algorithms are summarised in Table 1 and described below.

X-Diff [7]. The X-Diff algorithm involves the following steps:

- *Preprocessing:* In the preprocessing phase, both XML documents are parsed into tree representations and using XHash (a special hash function similar to DOMHash [13], but working on unordered trees), hash values for all nodes in both trees are calculated. The hash value of an element node a is calculated based on the hash values of its child nodes and the resulting value, therefore, represents the entire subtree rooted at the node a.

Table 1. Summary of key properties of XML difference algorithms

Algorithm	Source	Ordered/ Unordered Tree	Changes detected
X-Diff	[7]	Unordered	Insert, Delete of leaf nodes or subtrees; Update of values of text- or attribute nodes
JXyDiff	XyDiff [8,11]	Ordered	Insert, Delete, Move of leaf nodes or subtrees; Update of values of text- or attribute nodes
3DM	[9,12]	Ordered	Insert, Delete, Move of leaf nodes; Update of values of text- or attribute nodes

- *Matching:* The matching step consists of the following three steps.
  - To reduce the search space, subtrees with the same hash value are filtered out.
  - Starting from the leaf node pairs and moving upwards, nodes of the same type (i.e., text-, element-, or attribute node) and with matching ancestor names are matched, and their edit distance computed using a cost model with a uniform distance of 1 for update, insert and delete. To compute the edit distance between subtrees, the minimum-cost maximum flow algorithm [14,15] is used to find the minimum-cost bipartite matching. Both matching and edit distance are stored.
  - Starting from the root node and using the matchings and edit distances calculated in the previous step, create minimum-cost matchings between nodes of the two trees, allowing only one-to-one matchings and matching only child nodes of parents that are matched.
- *Edit script:* Starting from the root nodes and based on the minimum-cost matching and the edit distance, a minimum-cost edit script is generated. Nodes or subtrees found in the base document, but not found in the matching, are marked as deleted; nodes or subtrees found in the updated document, but not in the matching, are marked as inserted; and leaf nodes that are found in the matching but have different values are marked as updated.

JXyDiff (Java implementation of XyDiff [8,11]). The algorithm, called Bottom-Up, Lazy-Down (BULD) propagation, consists of the following steps:

- *Preprocessing:* Starting from the leaf nodes, hash values are calculated based on content of the node itself and the hash values of its children. Similar to X-Diff, the hash value of a node represents the entire subtree rooted in that node. In addition to the hash value, a weight is calculated for each node as follows: for a text node the weight is the size of the content and for an element node the weight is the sum of the weights of its child nodes. Subtrees represented by their root nodes are inserted into a priority queue where they are ranked by their weight.
- *Matching:* Starting with the heaviest subtree of the updated document (when there are several subtrees with the same weight, the first one in the queue is chosen), nodes with the same hash value (representing the entire subtree rooted at that node) are identified in the base document. If there is only one node in the base document with the same hash value, they are matched. If there are no nodes with the same hash value and the node is an element node, its children are inserted into the priority queue. If there are several nodes with the same hash value, the node whose parent matches the parent of the node from the updated document is chosen. The matching is followed by an optimisation phase in which already matched nodes are used to propagate matches further to nodes not matched in the previous matching step. During this phase, nodes are matched when their parents and/or children are matched and they have the same label. Bottom-up propagation of the matchings is controlled by the weight of the matching subtrees, i.e., the heavier/larger a subtree the further the matching is propagated.

– *Edit script:* Based on the matchings, the edit script is generated. Unmatched nodes in the old document are marked as deleted, and those unmatched in the new document are marked as inserted. Matched text nodes with changed content are marked as updated. Nodes that are matched, but without matching parents, are marked as moved. JXyDiff also detects moves of nodes within the same parent, i.e., changes to the order of matched siblings under matched parents. This is done by finding the largest order-preserving subsequence and adding move operations for the remaining pairs of nodes. As this step is expensive for large sequences of elements, in such cases the sequence is cut into smaller subsequences of a fixed maximum length and the same process applied to all the smaller subsequences. This improves the performance, but does not guarantee the optimal number of moves.

3-Way Merge and Diff (3DM) [9,12]. 3DM is a merge and diff tool that can merge 3 documents, the base document and two updated versions of the document for the purpose of reintegrating changes from two independently modified copies into a single document containing all the modifications. It can, however, also be used to find differences between two documents by providing two copies of the same document, e.g., the base document and the updated document as input. The algorithm consists of the following steps:

– *Matching:* For each node in the base document, find exact or close matching nodes in the updated document. The similarity of close matching nodes is based on the q-gram string distance measure [16]. Q-grams are substrings of length q and the q-gram distance is the number of q-grams that appear in only one of the two strings. For all pairs of matched nodes, match the subtrees by depth-first traversal starting with the two matched nodes. Continue as long as the child nodes are matched too. Select the best matching subtree.
– *Post-processing:* The postprocessing phase can be divided into the following steps:
  • Remove matches of small copies: All nodes in the updated document whose matching node in the base document has several matches in the updated document are checked, and matches to nodes that are part of a subtree containing only little information are removed to avoid copying small amounts of data.
  • Propagate matches: Using the structure of the document, nodes so far not matched are matched if their parents and left or right siblings are matched.
  • Set type of match: All matches are classified as structural, content or full (structural and content) matches.
  • Merge documents: Starting from the root node and based on the matchings between the base document and each of the two updated documents, the merged document and the edit script is created. This is done by pairing up children of matched nodes, determining the sequence of these pairs according to any moves made, and merging the contents of the matched pairs. The merged node is added to the merge tree.

– *Edit script:* During the merging step, the edit script is created. Nodes found in either of the updated copies but not the base version of the document are marked as inserted; matched nodes with different contents are marked as updated; nodes that are matched, but appear in a different order are marked as moved; and nodes not found in either of the two updated copies but in the base document are marked as deleted.

4   Experiments Involving Controlled Modifications

In this section the experimental setup is introduced and the results of the experiments are discussed. The experiment consists of two parts: (i) evaluation of the XML difference algorithms on synthetic modifications to enable controlled exploration of a number of different kinds of change; and (ii) exploration of the use of the algorithms with real modifications between different releases of genomic data obtained from Ensembl [4].

Experimental setup: Beginning with a base document of genomic data from chromosome 10 of yeast containing about 400 genes, changes are introduced in a systematic manner. For each type of change and each pairwise combination of changes mentioned in Section 2, new documents are produced with n (n = 4, 20, 40, 60, 80) of the genes subject to each change. The changes are introduced randomly, but conform to the constraints imposed by the well established representation of the data: for example, elements describing a gene, its mRNA, its CDS and its sequence are neighbouring siblings, genes appear in the order they are on the chromosome, and exons appear at the end of the document in the same order as they are on the chromosome. In addition to the updated documents, a change report for each updated document is produced, detailing the changes introduced, gathering corresponding gene-level changes, and presenting them in a manner meaningful to biologists, as illustrated in Figure 3.

Using each of the XML difference algorithms, the base document is compared with each of the updated documents. In a post-processing step, the edit scripts produced by each of the algorithms are processed to: (i) identify the changes reported; (ii) gather changes affecting the same gene; and (iii) reproduce as much as possible of the change report corresponding to each updated document.

A number of observations could be made in the post-processing phase for the majority of edit scripts produced by JxyDiff and 3DM, and for this reason, are summarised here: (i) Edit scripts produced by JXyDiff tend not to be minimal, in that they contain a number of moves of sibling elements within the same parent, an observation reported previously [7]. These move operations are a result of incorrect matching of subtrees (where, for example an updated subtree is matched incorrectly to a different subtree, that is then updated accordingly followed by move operations to restore the order of the siblings). To restore the order of the siblings, large sequences of siblings are split into smaller sequences to improve the performance of the analysis step that seeks to detect moves of nodes within the same parent (see Section 3). As no moves are introduced as changes, the reported move operations are not included in the change report

```
<changeReport>
   <insert_biological_concept concept="gene" name="newGene55">
      <insert> <feature name="gene">
         <qualifier name="gene">newGene55</qualifier>
         <qualifier name="locus_tag">newGene55_YEAST</qualifier>
         <qualifier name="note">Uncharacterized protein newGene55</qualifier>
         <location complement="false" type="single">
            <locationElement complement="false" type="range">
               <basePosition type="simple">95483</basePosition>
               <basePosition type="simple">95483</basePosition>
            </locationElement>
         </location> </feature> </insert>
      <insert> <feature name="mRNA">
         ... </feature> </insert>
      <insert> <feature name="CDS">
         ... </feature> </insert>
      <insert>
         <sequence length="0" type="DNA" version="0.0">agtgaataatttaa...</sequence>
      </insert>
      <insert> <feature name="exon">
         ... </feature> </insert>
   </insert_biological_concept>
   <delete_biological_concept concept="gene" name="YAL011W">
      <delete> <feature name="gene">
         ... </feature> </delete>
      <delete> <feature name="mRNA">
         ... </feature> </delete>
      <delete> <feature name="CDS">
         ... </feature> </delete>
      <delete>
         <sequence length="1878" type="DNA" version="0.0">agtttctgggttt...</sequence>
      </delete>
      <delete> <feature name="exon">
         ... </feature> </delete>
   </delete_biological_concept>
</changeReport>
```

Fig. 3. Example of a change report, developed to present changes in a manner meaningful to biologists

generated from the edit scripts. (ii) Edit scripts produced by 3DM contain the parent element as well as all the child elements that are affected by the change. In cases where the child element is reported with the same type of change as its parent element, i.e., the child element is subsumed by its parent element, the child element is not included in the change report generated from the edit script. If child and parent elements are reported with different types of changes, both are included in the change report. The generated change reports are then evaluated with respect to the quality of the results.

Experiment 1: *Change of genomic sequences.* In this experiment, which requires the fewest atomic element-level changes to the document, the value of the text node **sequence** is updated. This experiment analyses the ability of the algorithms to detect updates to values of text nodes. As the change introduced affects only a single node, this experiment is the least challenging. The number and types of changes reported by each difference algorithm are compared with those in the change report associated with the updated document.

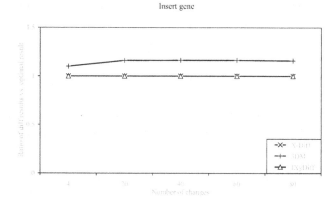

Fig. 4. Experiment 2, Introduction of new genes, relative performance

The following was observed: (i) Both X-Diff and 3DM detect all the introduced changes correctly and as the correct type of change. (ii) JXyDiff, however, detects all the introduced changes correctly, but as delete and insert of the text node `sequence` and not as an update, resulting in twice as many reported changes.

Experiment 2: *Identification of new genes.* In the second experiment, all the elements describing a gene, its mRNA, CDS, sequence and exons are inserted at appropriate positions in the document, following the constraints imposed by the representation of the data.

The following was observed: (i) Both X-Diff and JXyDiff detect all the introduced changes correctly. (ii) 3DM detects all the introduced changes correctly as inserts. In addition to the correctly identified inserts of elements, however, it detects a subset of the inserted sequence elements also as copies and updates of other `sequence` elements. The algorithm does not restrict the number of matches that can be identified between an element in the updated document and elements in the base document. As mentioned in Section 3, 3DM calculates the similarity of nodes using the q-gram string distance measure, and applies a threshold on the similarity to determine which nodes should be matched. As genomic sequence is basically a string of arbitrary length of the alphabet {a, c, t, g}, it is quite likely that two strings of sufficient length will have a sufficient number of q-grams in common to be regarded as similar enough to be copies of each other. As the q-gram string distance doesn't take into account the length of the string, sequences of very different lengths are matched based on their q-gram string distance, and one reported as an updated copy of the other. This results in a higher number of changes reported. The numbers of changes reported by each algorithm are compared with those in the change report associated with the updated document, and the ratios are plotted in Figure 4.

Experiment 3: *Removal of previously predicted genes.* In this experiment, all the elements representing a gene, its mRNA, CDS, sequence and exons are deleted.

The following was observed: all three XML difference algorithms detect the deletion of the subtrees correctly.

Experiment 4: *Change of genomic sequences and identification of new genes / Change of genomic sequences and removal of previously predicted genes.* In this experiment, the changes introduced in Experiment 1 are combined with those introduced in Experiment 2 or 3.

The following was observed: all three algorithms were able to identify the changes, exhibiting the same behaviour as in Experiments 1-3 with the corresponding single changes.

Experiment 5: *Identification of new genes and removal of previously predicted genes.* In this experiment, the changes introduced in Experiment 2 are combined with those introduced in Experiment 3.

The ratios of the numbers of changes reported by each algorithm compared with those in the corresponding change report are plotted in Figure 5. The following can be observed: (i) 3DM detects all the inserted elements and all the deleted elements correctly. As before in Experiment 2, however, a number of the inserted `sequence` elements are also detected as copies of other `sequence` elements combined with subsequent updates, resulting in a slightly higher number of changes reported. In contrast, JXyDiff and X-Diff report a far greater number of changes. (ii) JXyDiff detects inserts and deletes of elements describing the gene, mRNA, sequence and exons correctly, but fails to do so for a number of elements containing information on the CDS. In such cases, the child elements `dbreference`, `qualifier` and `basePosition` are reported as updates, deletes or inserts, resulting in the high number of changes reported. To restore the order of the elements, a large number of move operations of the siblings describing the elements of genes are reported. The incorrect matching of the inserted and deleted CDS elements principally results from the optimisation phase (see Section 3) in which matches of children, e.g., of the `location` elements, are

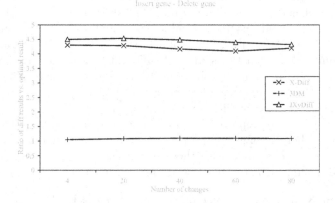

Fig. 5. Experiment 5, Introduction of new and removal of previously predicted genes, relative performance

propagated bottom-up to their parents. As the propagation is controlled by the weight of the matched subtree, this step can result in incorrect matchings of the CDS subtrees but not of the subtrees representing the other elements, as the CDS subtrees tend to have the larger number of children and therefore tend to be heavier. (iii) X-Diff reports the majority of deleted and inserted elements as updates, inserts or deletes of their corresponding leaf nodes, resulting in a high number of changes. As mentioned in Section 3 and listed in Table 1, X-Diff has the following properties: it regards the XML documents as unordered trees, uses a uniform cost model, creates a match starting with the leaf nodes, and matches nodes of the same type if their ancestors have matching names. These properties result in an almost arbitrary matching of the leaf nodes of the inserted and deleted subtrees representing the elements of the inserted and deleted genes. In the source documents though, order among siblings is significant.

Experiment 6: *Merging of two neighbouring genes.* In this experiment two neighbouring genes are merged. For detailed information on the changes introduced see Figure 2.

The ratios of the number of changes reported by each algorithm compared with those in the corresponding change report are plotted in Figure 6. The following can be observed: (i) X-Diff correctly identifies the deleted elements (representing one of the two neighbouring genes), but fails to match the updated elements correctly, resulting in a number of additional deletes, updates and inserts. For example, elements representing exons or CDS are matched with those representing mRNAs. These incorrect matchings are a result of handling the XML document as an unordered tree. However, no consistent pattern was observed for these incorrect matchings. (ii) JXyDiff identifies only a fraction of the deleted elements and matches very few of the updated elements correctly. This results in a large number of deletes, inserts, updates and moves to compensate for the incorrect matchings. In some cases, elements representing exons are matched with elements representing CDS or genes, however, no consistent

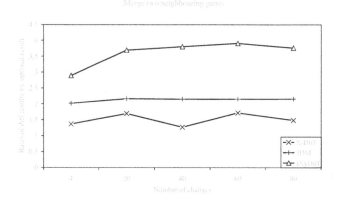

Fig. 6. Experiments 6, Merging of two neighbouring genes, relative performance

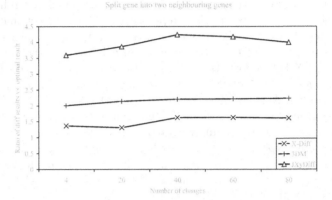

Fig. 7. Experiments 7, Splitting of one gene into two neighbouring genes, relative performance

pattern throughout all changes was observed. (iii) 3DM correctly identifies the majority of deleted elements and matches the updated elements. However, the majority of updates are not reported as such, but as deletes and inserts, resulting in about twice as many changes reported as in a minimal description. Furthermore, a number of elements are reported both as inserted and as updated, a result of the lack of a restriction on the number of matchings between an element in the updated document and elements in the source document.

Experiment 7: *Splitting of one gene into two neighbouring genes.* In this experiment one gene is split into two neighbouring genes. Detailed information on the changes introduced are shown in Figure 2. The ratios of the number of changes reported by each algorithm compared with those in the corresponding change report are plotted in Figure 7. The following was observed: (i) X-Diff identifies all inserted elements correctly and matches the majority of updated elements correctly. However, in some cases elements representing mRNAs are matched with elements representing genes or exons, resulting in detection of additional updates, deletes or inserts. (ii) JXyDiff identifies all the inserted elements correctly, but matches only a few of the updated elements correctly. This results in the identification of a large number of inserts, deletes, updates and additional moves to restore the order of the siblings and compensate for incorrect matchings. (iii) 3DM identifies all the inserted elements correctly but doesn't always match the updated elements correctly, resulting in a significantly increased number of updates, inserts and deletes. In cases where elements are matched correctly, the updates are sometimes reported as inserts and deletes.

5   Evaluation Involving Real Genomic Data

To explore the performance of the XML difference algorithms on detection of real modifications, we compared two different releases (41 and 42) of the genomic data

Table 2. Number of changes detected between different versions of genomic data

Chromosome	X-Diff	JXyDiff	3DM
3	1805	13223	1864
5	1130	18475	679

of yeast chromosomes 3 and 5 obtained from Ensembl. The numbers of changes reported by each of the tools for both chromosomes are shown in Table 2. The following observations were made:

X-Diff: While X-Diff performs reasonably well in the case of singular changes within an element (describing gene, mRNA, CDS, or exon) and detects the majority correctly, the fact that it represents XML documents as unordered trees leads to incorrect reporting of changes in the following cases amongst others: (i) Large number of changes within one element (e.g., inserts, updates, deletes of elements `dbreference` or `qualifier`). In such cases, instead of reporting the various updates, inserts and/or deletes within the element, they may be matched incorrectly with other elements in the document. (ii) Update of the start and end location of an element. As pairs of `basePosition` elements are affected and the order among siblings is not taken into account in X-Diff, updates to the start position can be confused with updates to the end position and vice versa. The order of the start and end positions indicates the direction in which the gene is transcribed.

JXyDiff: For the most part, JXyDiff seems unable to match elements correctly, resulting in the reporting of significantly more changes in comparison to the other two algorithms. The numbers shown for JXyDiff in Table 2 exclude the more than 8,000 and 10,000 moves reported for chromosomes 3 and 5, respectively. In particular CDS elements tend to be mismatched, resulting in a large number of updates of all the subelements and attributes. Mismatching of CDS elements was observed earlier in the synthetic experiments (see Experiment 5) and is the result of the optimisation phase in which matches of leaf nodes are propagated bottom-up to their ancestors. The extent of the propagation is controlled by the weight of the subtree.

3DM: 3DM performs reasonably well in detecting changes between the two versions of genomic data. The following cases, amongst others, though, lead to reporting of additional changes: (i) Insert or delete of `dbreference` or `qualifier` elements within one element describing a CDS. In such cases, a newly inserted element might be matched to an existing element, which is reported as updated to represent the new element, and then the original is reported as inserted, therefore increasing the number of changes reported. Similar incorrect matchings can be observed for deleted elements, which aren't reported as deletes, but as a number of updates followed by deletes of different similar elements. This occurs more frequently in cases of a larger number of changes within one element, but also for newly inserted elements describing a gene or mRNA. (ii) Update of sequence elements. In such cases, the updated sequence might be reported as both an update and a copy followed by an update of another sequence element. (iii) Moves of elements, i.e., changes to the order of elements on the chromosome. In this case, not only are the elements affected by the move reported as such, but also a number of neighbouring elements, again resulting in an increase in the number of changes reported. The majority of these mismatches are the result of the q-gram distance measure used to match elements, and a lack of restriction on the number of matches of elements in the updated document with elements in the base document. Both cases have been observed in the experiments with synthetic data.

6   Conclusions

This comparative evaluation has shown that all the algorithms tested contain features that are effective at detecting a subset of changes relevant within the context of genomic data. However, none are useable without significant post-processing of the edit scripts. We have identified the following features of the representation of genomic data and of the algorithms that affect performance.

The task of matching versions of genomic data correctly and detecting changes between them is hindered by the following properties of the fairly flat structure of the data representing genomic sequences and their annotation: (i) Large numbers of siblings. (ii) Related elements are not easily identifiable as such (e.g., elements describing gene, mRNA, CDS, exons and sequence of a gene are only identifiable as related by the values of their qualifier child elements). (iii) Very similar contents of nodes and subtrees. Elements or subtrees sometimes can only be distinguished by values of attribute or descendant text nodes. The following properties of the algorithms affected their performance: (i) Treating XML documents as unordered trees can result in inappropriate matchings. (ii) Fuzzy matching using q-gram string distance measure can result in incorrect matchings of elements, in particular sequence elements. (iii) Propagating matches bottom-up to nodes with the same labels. This, in combination with the similarity of the contents of nodes and subtrees can result in incorrect matchings.

This suggests that selection and use of genome difference algorithms requires: (i) good understanding of the data over which the algorithm is to be used during algorithm selection; and (ii) significant tailoring of results to compensate for the production of non-minimal and challenging-to-interpret edit scripts.

References

1. Curwen, V., Eyras, E., et al.: The Ensembl Automatic Gene Annotation System. Genome Res. 14, 942–950 (2004)
2. Potter, S.C., Clarke, L., et al.: The Ensembl Analysis Pipeline. Genome Res. 14, 934–941 (2004)
3. Dwight, S.S., Balakrishnan, R., et al.: Saccharomyces Genome Database: Underlying Principles and Organisation. Brief Bioinform. 5, 9–22 (2004)

4. Flicek, P., Aken, B.L., et al.: Ensembl 2008. Nucl. Acicds Res. 36, D707–D714 (2008)
5. Kulikova, T., Akhtar, R., et al.: Embl Nucleotide Sequence Database in 2006. Nucl. Acids Res. 35, D16–D20 (2007)
6. Cochrane, G., Akhtar, R., et al.: Priorities for Nucleotide Trace, Sequence and Annotation Data Capture at the Ensembl Trace Archive and the Embl Nucleotide Sequence Database. Nucl. Acids Res. 36, D5–D12 (2008)
7. Wang, Y., DeWitt, D.J., Cai, J.Y.: X-diff: An Effective Change Detection Algorithm for Xml Documents. In: 19th International Conference on Data Engineering (ICDE 2003), pp. 519–530 (2003)
8. Cobena, G., Abiteboul, S., Marian, A.: Detecting Changes in Xml Documents. In: 18th International Conference on Data Engineering (ICDE 2002), pp. 41–52 (2002)
9. Lindholm, T.: A Three-Way Merge for Xml Documents. In: DocEng 2004: 2004 ACM Symposium on Document Engineering, pp. 1–10 (2004)
10. Benson, D.A., Karsch-Mizrachi, I., Lipman, D.J., Ostell, J., Wheeler, D.L.: Genbank. Nucl. Acids Res. 35, D21–D25 (2007)
11. Marian, A., Abiteboul, S., Cobena, G., Mignet, L.: Change-Centric Management of Versions in an Xml Warehouse. In: 27th International Conference on Very Large Data Bases (VLDB 2001), pp. 581–590 (2001)
12. Lindholm, T.: A 3-way Merging Algorithm for Synchronizing Ordered Trees - the 3DM Merging and Differencing Tool for Xml. Master's thesis, Department of Computer Science, Helsinki University of Technology (2001)
13. Maruyama, H., Tamura, K., Uramoto, R.: Digest Values for DOM (Domhash). IBM Research, Tokyo Research Laboratory (2000),
http://www.research.ibm.com/trl/projects/xml/xss4j/docs/rfc2803.html
14. Tarjan, R.E.: Data Structures and Network Algorithms. CBMS-NSF Regional Conference Series in Applied Mathematics (1983)
15. Zhang, K.: A New Editing Based Distance Between Unordered Labeled Trees. In: Apostolico, A., Crochemore, M., Galil, Z., Manber, U. (eds.) CPM 1993. LNCS, vol. 684, pp. 254–265. Springer, Heidelberg (1993)
16. Ukkonen, E.: Approximate String-Matching with q-grams and Maximal Matches. Theoretical Computer Science 92, 191–211 (1992)

Adaptive Request Scheduling for Parallel Scientific Web Services

Heshan Lin[1], Xiaosong Ma[1,2], Jiangtian Li[1], Ting Yu[1], and Nagiza Samatova[1,2]

[1] Department of Computer Science, North Carolina State University
[2] Computer Science and Mathematic Division, Oak Ridge National Laboratory

Abstract. Scientific web services often possess data models and query workloads quite different from commercial ones and are much less studied. Individual queries have to be processed in parallel by multiple server nodes, due to the computation- and data-intensiveness of the processing. Meanwhile, each query is performed against portions of a large, common dataset. Existing scheduling policies from traditional environments (namely cluster web servers and supercomputers) consider only the data or the computation aspect alone and are therefore inadequate for this new type of workload.

In this paper, we systematically investigate adaptive scheduling for scientific web services, by taking into account parallel computation scalability, data locality, and load balancing. Our case study focuses on high-throughput query processing on biological sequence databases, a fundamental task performed daily by millions of scientists, who increasingly prefer to use web services powered by parallel servers. Our research indicates that intelligent resource allocation and scheduling are crucial in improving the overall performance of a parallel sequence database search server. Failure to consider either the parallel computation scalability or the data locality issues can significantly hurt the system throughput and query response time. Also, no single static strategy works best for all request workloads or all resources settings. In response, we present several dynamic scheduling techniques that automatically adapt to the request workload and system configuration in making scheduling decisions. Experiments on a cluster using 32 processors show the combination of these techniques delivers a several-fold improvement in average query response time across various workloads.

1 Introduction

There is a growing trend to provide parallel scientific computation services through the web interface, especially for computation- and data-intensive tasks such as scientific database queries, data mining, and visualization. Rather than having users download large volumes of shared data and run stand-alone applications, scientific web services allow them to perform common data processing/analysis tasks through intuitive web interfaces. For example, an online bio-sequence search service can be viewed as the equivalent of web search engine in the bioinformatics world.

B. Ludäscher and Nikos Mamoulis (Eds.): SSDBM 2008, LNCS 5069, pp. 276–294, 2008.

These services are very appealing to scientists due to several reasons. First, for many researchers the existence of web data processing services reduces or even eliminates the purchase and maintenance cost of owning local clusters. Second, many popular data processing tasks access shared public datasets (such as well-known sequence databases or satellite images) that are constantly updated. Having such datasets managed by a parallel web server enables individual scientists to access the latest data without worrying about downloading, storing, and updating large datasets. Last but not least, providing parallel scientific data processing through transparent and intuitive web interfaces hides the painful details of parallel computing. It lets domain scientists obtain the performance of powerful clusters without dealing with tedious and challenging tasks such as machine administration, batch job submission and monitoring, manual data staging, and after all, learning or even writing parallel software.

Given a computing platform (typically a cluster of back-end servers and one or more frontend servers), a collection of shared datasets, and a collection of applications to run on demand as services, efficient scheduling is crucial to the parallel web server's performance, in terms of the average request response time. However, existing scheduling strategies from two related application fields, namely commercial clustered web servers and space-shared parallel computers, are inadequate for this new type of workload. Below we briefly describe the reasons (more detailed discussion will be given in Section 5).

Scientific web services tend to be both computation- and data-intensive, performing non-trivial algorithms over large amounts of shared data. In contrast, commercial web servers typically stream contents or perform low-cost relational database queries. Hence their scheduling algorithms concentrate on data locality optimization and load balancing. Also, a back-end server node usually handles many client requests simultaneously with multiple open connections. With scientific data analyzing services, the CPU and the memory resources required to timely process a request are often far beyond those can be offered by a single node. Consequently, a group of these nodes is dedicated to every request in a tightly synchronized manner.

In this sense, request processing in scientific web services is closer to batch job processing on parallel computers, however with two major differences. First, on general-purpose parallel computers, batch jobs are mutually independent and rarely share data. Second, as shared computation platforms, parallel computers have no knowledge regarding each job's computation and I/O requirements, and the resources requested by each job (such as the number of processors and the maximum run time) are specified explicitly in job scripts. Therefore batch job scheduling usually pays no attention to data locality issues and has no control over the level of concurrency in each job. With scientific data services hosted by specialized data centers, data sharing is common and the parallel web server has much more knowledge about the services it provides.

Therefore, parallel scientific web services require careful examination of the intertwined computation and data management issues in making scheduling decisions. In this paper, we extended scheduling algorithms to work for parallel scientific web services, from those designed for the commercial cluster web servers and batch processing parallel computers. By adopting a novel combination of these extended algorithms, a parallel scientific web server will take into consideration both the data and the computation aspects: data locality, parallel execution efficiency, and load balancing. In addition, the combined strategies work fully adaptively, automatically adjusting scheduling strategies according to the server load levels and dynamic data access patterns.

We implemented our proposed scheduling algorithms, along with baseline strategies to compare with, in a parallel BLAST server prototype. BLAST [1] is a fundamental sequence database search task performed routinely by scientists. Given a query sequence, the BLAST family tools search through a database of known sequences and return sequences that are "similar" to it. Online parallel BLAST searches have become popular. In April 2005, the NCBI parallel BLAST web server received about 400,000 BLAST queries daily [19]. Such dedicated sequence search web servers often host multiple databases and provide different alignment algorithms. Meanwhile, the search workload is highly dynamic [2].

Our experiments on a cluster server performing parallel BLAST revealed that a careful choice in query concurrency and database-to-processor assignment may easily result in a dramatic difference in the average query response time. We confirmed that different query arrival rates and query composition ask for specialized strategies, and there are no "one-size-fits-all" solutions. The combination of the proposed adaptive strategies, however, achieves the best or close-to-best performance across a wide range of system load levels, with a several-fold improvement in average query response time in many cases.

Although our implementation and evaluation are based on parallel BLAST, this workload carries many common characteristics of scientific data processing applications, such as accessing large shared databases, map-reduce type of processing, and content-dependent execution time. We believe that the observations and experiences collected through this study can be utilized by many other applications.

2   Parallel BLAST Web Server Architecture

2.1   BLAST and Parallel BLAST

The BLAST [1] family algorithms search one or multiple input query sequences against a database of known nucleotide (DNA) or amino acid sequences. The input of BLAST is one or more query sequences and the name of the target database to search. For each query sequence, BLAST performs a heuristics based, two-phase search on all sequences in the database and returns those that are most similar to it. This requires a full scan of all the sequences in the database. For each of these sequences returned, BLAST reports its similarity score based on its alignment with the input query and highlights the regions with high similarity (called *hits*). Therefore, the BLAST process is essentially a top-k search, where k can be specified by the user, with a default value of 500.

Many approaches have been proposed to execute BLAST queries in parallel. Among them, the *database segmentation* [3,9,13,15] model has proved to be effective in processing the ever growing sequence data sets. With database segmentation, a sequence database is partitioned into multiple *fragments* and distributed to different cluster nodes, where the BLAST search tasks are performed concurrently on different database fragments. The local results generated by individual nodes for a common query sequence are merged centrally to produce the global results.

2.2   Parallel BLAST Web Server Architecture

Figure 1 illustrates the parallel BLAST web server architecture targeted in our study, with sample query and partial output. As in a typical cluster setting, each node has its own memory and local disk storage, as well as access to a shared file system. One of the cluster nodes serves as the front-end node, which accepts incoming query sequences submitted online, maintains a query waiting queue, schedules the queries, and returns the search results. The other nodes are back-end servers, often called "processors" in the rest of the paper for brevity.

For each query, the front-end node determines the number of processors to allocate, selects a subset of idle back-end nodes (called a *partition*) when they are available, and assigns these nodes to execute this query. After the parallel BLAST search, the results are merged by one of the nodes in the partition and returned to the client via the front-end node.

To save the database processing overhead, all the sequence databases supported by the parallel BLAST web server are pre-partitioned and stored in the shared storage. Figure 1 shows two sample databases, each partitioned into 4 fragments. The required database fragments will be copied to the appropriate back-end nodes' local disk before each query is processed, and are cached there using a cache management policy. Existing parallel BLAST implementations allow multiple database fragments to be "stitched" into a larger virtual fragment with little extra overhead. Therefore for the maximum flexibility in scheduling without creating physical fragments of many different sizes, we partition the database into the largest number of fragments allowed to be searched in

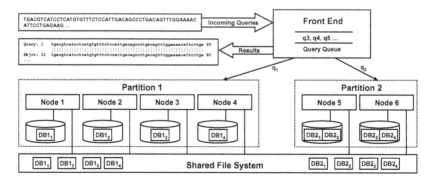

Fig. 1. Target parallel BLAST web server architecture

parallel. To simplify the scheduling and to achieve better load balance, both the database fragmentation and processor allocation are based on power-of-two numbers, which is natural considering the way clusters are purchased or built. Note that the fragments combined into a larger virtual fragment do not need to be in consecutive order. For example, when 16 processors are assigned to search a certain query against a database partitioned 64-way in a 64-processor cluster, one of them may be assigned to search fragments 0, 8, 45, and 57.

Assumptions: Before we move on to the scheduling strategies, we summarize assumptions made in this study: First, we assume a homogeneous environment, which is true for most clusters. Second, due to the space constraint, in this paper we discuss the scenario where the entire collection of databases can be accommodated at each cluster node's local disks.[1] This is likely the case for parallel BLAST servers, as the total size of formatted NCBI sequence databases is currently around 100GBs, while a cluster node can easily have hundreds of GBs of local disk space today. Finally, to simplify query workload generation, we assume that each query contains only one sequence. Although existing BLAST web servers may allow users to upload multiple query sequences, the standard NCBI BLAST engine processes input queries sequentially. The difference in search time between the shared and separate BLAST sessions for multiple query sequences is not significant and mainly lies in the initialization overhead. Our research results can be easily extended to handle multiple-sequence requests. In the rest of the paper, we use the terms "request" and "query" interchangeably.

3   Scheduling Strategies

In this section, we present scheduling strategies for parallel scientific web services, using parallel BLAST server as a case study. We extend two existing scheduling algorithms and integrate them to design adaptive algorithms that automatically adjust to various query workloads and cluster configurations. Like in many existing request scheduling studies, our major goal is to optimize the average query response time.

In Section 3.1 and Section 3.2, we discuss our extended scheduling algorithms respectively. The first one comes from the commercial cluster web server community and performs *data-oriented scheduling*. It determines which processors should be allocated for a specific query, considering existing data cached at these processors and their current load. The second one comes from the space-sharing parallel job scheduling community and performs *efficiency-oriented scheduling*. It determines the desired level of concurrency for processing a query, considering the specific query workload and the current system load. Both algorithms are extended substantially to fit the scenario of parallel scientific web services. Then in Section 3.3, we discuss our overall scheduling scheme and describe how we integrate the two scheduling algorithms.

---

[1] For systems equipped with insufficient local storage, we have developed additional optimizations, as described in our technical report [14].

3.1   Data-Oriented Scheduling

Like in other distributed or cluster web servers, data locality is a key performance
issue in parallel BLAST web servers. Figure 2 demonstrates the impact of going
down the storage hierarchy: main memory, local file system, and shared file
system. The experiments use sequential NCBI BLAST to search the `est-mouse`
and `nr` databases, which can fit into the memory of a single processor. For each
case, 10 sequences randomly sampled from the database itself are used as queries,
and the average search time is reported. In the "warm-cache" tests, we warm up
the file system buffer cache with the same query before taking measurements,
and in the "cold-cache" tests we flush the cache first. For "cold-cache-shared",
we force loading the database from the shared file system. The results indicate
that improving file caching performance and in particular, reducing remote disk
accesses can significantly improve the search performance.

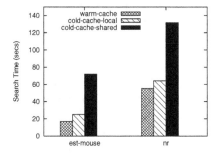

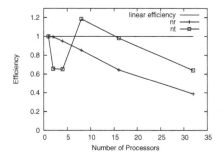

Fig. 2. Impact of data placement on the   **Fig. 3.** Parallel execution efficiency of
BLAST performance   BLAST

As mentioned earlier, in this paper we focus on the scenario where the entire
set of databases hosted by a parallel scientific web server can fit into the per-node
local disk space. Still, only a small fraction of those databases can be buffer-
cached in the main memory, and scheduling must be performed considering the
data locality issue. One intuitive locality-aware optimization is to assign queries
targeting different databases to disjoint pools of processors and let each processor
pool search the same database repeatedly. This way, the effective working set of
each processor is reduced. Creating static per-database processor pools, however,
is not flexible enough to handle the dynamic online query composition and will
likely cause serious system underutilization.

A similar problem has been addressed regarding general-purpose content-
serving cluster web servers. In this paper, we extend the LARD algorithm for
content request distribution proposed by Pai et al. [20] to the parallel scientific
web service context. Given a set of back-end servers, the LARD algorithm assigns
partitions of hosted targets to subsets of these servers. An incoming web request
will be routed to one of the servers assigned to its target, or the least loaded
server if it is the first request of the given target. Load balancing is performed

periodically to move requests from heavily loaded servers to lightly loaded ones. LARD exploits data locality to improve the server performance by assigning requests of the same target to the same set of processors.

Two major differences make our target system considerably more complex than a general-purpose cluster web server. First, multiple processors need to be co-scheduled to queries or co-transferred between pools. Second, a processor can handle only one query at any given time. Therefore queries cannot be piled to server nodes as they arrive, but need to wait for dispatch.

To handle these requirements, we extend LARD to a new algorithm called PLARD (Parallel LARD). To perform locality-aware assignment and load balancing, PLARD adopts a two-level scheduling mechanism. It establishes one global query queue (global_queue) and multiple per-database query queues (queue[DB_i]). Queries will be first appended to the global queue, and subsequently dispatched to one of the per-database queues. Similarly, because servers need to be assigned in groups, PLARD manages a global idle processor pool (global_pool), and multiple per-database processor pools (pool[DB_i]). Initially, all the processors are in the global pool. A scheduling operation will be triggered by either a query arrival or a query completion. Algorithm 1 gives the detail of the process of scheduling one query from the global queue.

Queries in the global queue will be scheduled in the first-come-first-serve (FCFS) order. When there are not enough resources for the next query, the scheduling attempt is aborted and the global scheduler waits until a query completes. This helps ensure fairness and prevents starvation. Also, this allows the recommended partition size to be recalculated as the system load changes.

Before moving a query from the global queue to a per-database pool, a recommended partition size will be calculated by the function get_recommended_size(). This function determines how many processors should be allocated to a target database, using algorithms such as the ones described in the next section. The target database-pool will be enlarged if the pool size is less than the recommended size. In case there are not enough processors to allocate from the global pool, the algorithm will seize processors from the most lightly loaded pool if there are fewer queries waiting in that pool's local queue than those waiting for the target database in the global queue.

After a query is assigned to a per-database processor pool, it goes to the local queue of that pool and is scheduled using an internal scheduling algorithm (such as a fixed partitioning policy or RMAP, as presented in the next section). This way, a relatively stable subset of server nodes are assigned to work on a certain database, maximizing the use of their collective buffer cache space.

Like in the original LARD, every time a query is scheduled the system performs load balancing. In PLARD, we move processors from the most lightly loaded pool (pool[DB_{min}]) to the most heavily loaded pool (pool[DB_{max}]), if one of the following conditions is satisfied:

1. queue[DB_{max}].length - queue[DB_{min}].length > T and
   queue[DB_{max}].length $\geq 2\times$ queue[DB_{min}].length, or
2. queue[DB_{min}].length = 0 and queue[DB_{max}].length > 1

Algorithm 1. PLARD

---

fetch the next query q from global_queue
$partition\_size \leftarrow$ get_recommended_size()
$m \leftarrow$ the number of queries waiting for $q.target\_db$ in global_queue
$candidate\_queues \leftarrow \bigcup$ queue$[DB_i]$, where DB_i not equal to $q.target\_db$
$increase\_size \leftarrow partition\_size$ - pool$[q.target\_db]$.size
if $increase\_size > 0$ then
    while global_pool.size $< increase\_size$ and $candidate\_queues$ not empty do
        $size\_needed \leftarrow increase\_size$ - global_pool.size
        find queue$[DB_j] \in candidate\_queues$ with smallest queue length
        if $m >$ queue$[DB_j]$.length then
            $num\_idle \leftarrow$ the number of idle nodes in pool$[DB_j]$
            $S \leftarrow$ release_idle_nodes(DB_j, $min(num\_idle, size\_needed)$)
            add S to global_pool
        end if
        remove queue$[DB_j]$ from $candidate\_queues$
    end while
    if $increase\_size \leq$ global_pool.size then
        $A \leftarrow$ allocate $increase\_size$ processors from global_pool
        add A to pool$[q.target\_db]$
    end if
end if
if pool$[q.target\_db]$ is not empty then
    append q to queue$[q.target\_db]$
end if
balance_load()

T in the above is a configurable threshold, which is set as 10 in our implementation. The number of processors moved during load balancing is set to be P_{min} of DB_{max}, where P_{min} is the minimum partition size allowed for a given database as described in Section 3.2. This helps reduce the internal fragmentation of a database pool during load balancing.

3.2 Efficiency-Oriented Scheduling

PLARD helps us optimize query processing performance by maximizing the use of cached data and improving load balance between server nodes. However, it does not consider the parallel processing scalability of the scientific applications that service the web requests. The latter turns out to be crucial in deciding how many processors should be allocated to each individual query, and can have a significant impact on the parallel web server's performance.

We illustrate the argument by examining parallel BLAST's performance scalability. Like most parallel applications, it is subject to the performance tradeoff between absolute performance and system efficiency when the level of concurrency increases. One obvious explanation is the higher parallel execution overhead associated with searching a single query using more processors. In addition, as BLAST performs top-k search, the task of processing and filtering of intermediate results grows with the number of processors. Figure 3 illustrates the performance trend of parallel BLAST from searching two widely used databases, the NCBI nr and nt, as benchmarked on our test cluster (to be described in Section 4.1). For each search workload, we plot the *efficiency*, which is defined as parallel speedup divided by the number of processors. Therefore a perfect linear efficiency is a flat line. For both nr and nt, the efficiency slides steadily as more processors are used for each query.

Systems such as the NCBI BLAST server reported periodic variances in the query arrival rate [2]. One intuitive heuristic is to control the number of processors allocated to each query based on the current system load: when the load is light, allocate more processors for smaller query response time; when the load is heavy and queries are piling up in the queue, allocate fewer processors for better system throughput (and consequently better average response time). This intuition is backed up by queuing theory and has been adopted in adaptive partitioning algorithms for parallel job scheduling [21]. In this work, we select the MAP algorithm [8], which improves upon the above work, as our base algorithm.

With MAP, both the waiting jobs and the jobs currently running are considered in determining the system load. It chooses large partitions when the load is light and small ones otherwise. More specifically, for each parallel job to be scheduled, a target partition size is calculated as

$$target\_size = Max(1, \lceil \frac{n}{q + 1 + f * s} \rceil),$$

where n is the total number of processors, q is the waiting job queue length, s is the number of jobs currently running in the system, and f $(0 \leq f \leq 1)$ is an adjustable parameter that controls the relative weight of q and s. In our experiments, we set the f value as 0.75, as recommended in the original MAP paper [8]. Once the target partition size is selected, the front-end node waits until these many processors become available to dispatch the query.

One may notice that in Figure 3 the nt curve does not monotonically decrease. Instead it peaks at 8 processors, with a super-linear speedup at that point. This is due to that the nt database cannot fit into the aggregate memory of 4 or fewer processors on our test platform. As BLAST makes multiple scans and random accesses to the sequence database, out-of-core processing causes disk thrashing and significantly limits the search performance. The nr database is much smaller and can be accommodated in a single compute node's memory, therefore does not show the same behavior.

This motivates us to propose Restricted MAP (RMAP), which augments the base MAP algorithm with a database-dependent and machine-dependent memory constraint. For a given database supported by a given cluster server, we select P_{min} and P_{max}, which define the range of partition sizes (in terms of the number of processors) allowed to schedule queries against this database. P_{min} is the smallest number of processors whose aggregate memory is large enough to hold the database. P_{max} is determined by looking up the saturation point in the speedup chart: it is the largest number of processors before the absolute search performance declines. In other words, after this point deploying more processors will not produce any performance gain. An initial benchmarking is needed to set P_{max} for each database, which is feasible considering the total number of different databases supported by a web server is often moderate[2].

For each query scheduled, when there are more idle processors available than p, the desired partition size calculated, RMAP adopts a simple node selection strategy called FA (First Available), where the first p idle processors by the processor rank will be assigned to work on the query. Database fragments will be assigned to these processors in a round-robin manner.

3.3 Combining PLARD and RMAP

We integrated PLARD and RMAP in our two-level query scheduler implementation for the parallel BLAST server prototype.

As shown in Algorithm 1, when dispatching a query from the global queue to a particular DB queue, the RMAP algorithm is first used to calculate a recommended partition size based on the global system state. More specifically, the queue length(q) is calculated by summing up all queries in the global queue and local DB queues, and the number of queries in the system(s) is the sum of queries being searched at all DB pools. If the number of idle processors in the processor pool of the target DB is smaller than the recommend partition size, the scheduling algorithm seeks to assign more processors to this pool by acquiring idle processors from the system idle processor pool and/or other relatively lightly-loaded DB pools.

When a partition with the recommended size can be provided, the query is moved into the local DB queue. There the RMAP algorithm will be called again to determine a proper partition size in local scheduling. At this point, each local RMAP scheduler uses the local system state, namely the local DB queue length as q and the number of queries being serviced in the local processor pool as s.

With this two-level scheduling approach, we adapt simultaneously to the intensiveness and the database access pattern of the dynamic query workload by leveraging strengths of both RMAP and PLARD. The two algorithms complement each other nicely under the new scheduling framework.

4   Performance Results

4.1   Experiment Configuration

In our experiments, we use five biological sequence databases downloaded from the NCBI public sequence repository. Table 1 summarizes several basic attributes of these databases. Among them, the first two are protein sequence databases (type "P") and the other three are nucleotide sequence databases (type "N"). The two types of the databases are searched using the `blastp` and `blastn`

---

[2] The number of all sequence databases offered by the NCBI web search is 21 at the time this paper is written.

Table 1. Database characteristics. Note the P_{min} values are multiples of 2, this is because our experiments are performed on a two-way SMP cluster, and we found using a compute node (2 processors) as the smallest scheduling unit yields better performance than does using an individual processor, as the former choice has better data locality.

Name	Type	Raw Size	Formatted Size	P_{min}	P_{max}
env_nr	P	1.7GB	2.5GB	2	32
nr	P	2.6GB	3.0GB	4	32
est_mouse	N	2.8GB	2.0GB	2	16
nt	N	21GB	6.5GB	8	32
gss	N	16GB	9.1GB	8	32

algorithms respectively. The size of each database shrinks after the database is formatted for search using the standard `formatdb` tool. For each of the databases, we also give the P_{min} and P_{max} pair, which defines the processor partition size range. As discussed in Section 3.2, P_{min} is determined by the memory constraint and P_{max} is determined by benchmarking the parallel execution scalability of the individual database search workload.

The parallel BLAST software we used is the popular mpiBLAST tool [9,13], available at http://mpiblast.org/. For queries, we sampled 1000 unique sequences from the five databases, with the number of samples from each database proportional to the formatted database size. Since sequence databases are constantly appended with newly discovered sequences, we hope this sampling method resembles the composition of real BLAST search workloads, which are driven by sequence discoveries. We compose online query traces by drawing queries randomly from this pool of unique sequences, setting the arrival interval with the Poisson distribution.

To create traces with the desired arrival rates, we benchmark the maximum throughput of the whole system. This maximum throughput is calculated in an aggressive manner: we measure the maximum throughput of each database' search workload by executing the corresponding subset from the 1000-query pool on the whole cluster using the smallest partition size (P_{min}). This way the system achieves best efficiency and data locality with the single-database workload and small partition size. We then derive the multi-database maximum throughput by taking a weighted average of the single-database peak throughput, according to the number of queries going to each database.

Unless noted otherwise, the experiments are performed using query traces that contain 600 query sequences sampled from the 1000-query pool above. Note that many of the charts use log2 scale on the y axis, due to the large distribution of performance numbers under different system load levels.

4.2   Test Platform

Our experiments were performed on the Orbitty Linux cluster located at North Carolina State University. Orbitty consists of 20 compute nodes, each equipped with dual Intel Xeon 2.40GHz processors sharing 2GB of memory. Due to its target workload, this cluster has 400GB per-node local storage space, which is large enough to host the entire collection of NCBI sequence databases. The interconnection is Gigabit Ethernet and a shared storage space of over 10TB is accessed through a Lustre server.

4.3   Data-Oriented Scheduling Results

First, we examine the effectiveness of improving data locality in query processing, by showing the impact of PLARD on three versions of fixed partitioning strategies. With fixed partitioning, the number of processors allocated to queries against the same database is fixed throughout the run. For each database, we choose three fixed partition sizes within its partition size range $[P_{min}, P_{max}]$: small (FIX-S), medium (FIX-M), and large (FIX-L).

Experiments are carried out using different levels of system load by adjusting the query arrival rate. A system load of 1 means the query arrival rate is equal to the maximum query throughput. All these strategies also use the default FA policy in selecting idle processors to schedule.

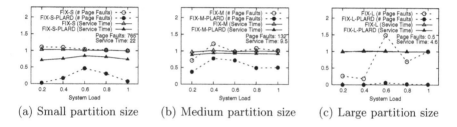

(a) Small partition size   (b) Medium partition size   (c) Large partition size

Fig. 4. Normalized average number of page faults and normalized average service time

Figure 4 portraits the impact of PLARD on the fixed partition size algorithms' file caching performance. Since BLAST uses memory mapped files, the number of page faults is a good indication of the amount of file I/O performed to retrieve the database fragments. For each of the fixed algorithms, we plot the average number of page faults per node (dashed lines) and the average query service time (solid lines), with and without PLARD. All the page fault numbers are normalized against the page fault number of the original algorithm (without PLARD) with the system load of 1. The same applies to the service times. The absolute values of these two pivot numbers are marked in the charts.

As expected, the PLARD algorithm does have a significant impact on the number of page faults. In particular, for FIX-S, the original page fault numbers of over 750 are reduced at least by half, and almost eliminated with the lightest and heaviest system loads. On average, the number of page faults is reduced by 79.87%. The original FIX-S page fault slightly declines as the system load intensifies since more processors will be actively used, and the chance of having cache hits increases due to the enlarged aggregate memory size, although there is no intentional, locality-aware query placement. With PLARD, however, the peak of page fault numbers appear in the medium load (0.6), where with the small partition size, the per-database processor pools are the most dynamic: processors are shifted between pools relatively more frequently, reducing the chances of cache hits within each database pool.

With FIX-M, the page fault reducing of PLARD is smaller but still considerable, with an average of 43.37% decrease. Here the peaks of the page fault numbers, both with and without PLARD, are different from those with FIX-S due to the larger partition sizes. For example, the lightest load achieves the best data locality since the query load is rather concentrated on a group of processors, facilitating in-memory data reuse, while the size of the group is large enough to spread the databases out and reduce the data access working set per node.

With FIX-L, the databases are so spread out so that all the fragments needed by a processor are almost always in the memory. Although the normalized curves look dramatic, the absolute numbers are very small. Even without PLARD, the cache misses are negligible, with an average page fault count of 0.37.

The improvement of service times using PLARD is a direct result of the improved data locality, as PLARD does not affect the computation efficiency of each query's processing with the fixed-partitioning algorithms. The degree of the improvement, however, declines as the partition size selected increases. This is because the number of page faults goes down faster than the service time does when larger partitions are used. Therefore the impact of page fault reduction plays a smaller and smaller role.

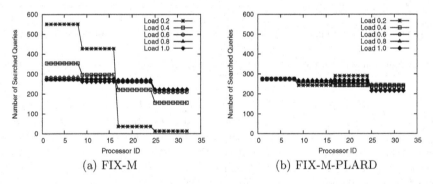

(a) FIX-M   (b) FIX-M-PLARD

Fig. 5. Query load distribution among processors with the medium partition size

Figure 5 provides additional information about the effect of PLARD, from its load balancing aspect. We illustrate this using FIX-M, the algorithm using the medium partition size. As discussed above, with the FA policy for processor assignment, the query processing workload distribution is skewed at the load level of 0.2. Most queries are assigned to the first 16 processors, with an additional 100+ queries assigned to the first 8 (please recall that the "medium partition size" varies from database to database). Heavier system loads force the queries to become more evenly distributed. With PLARD, the query processing assignments are well balanced among processors for all system load levels.

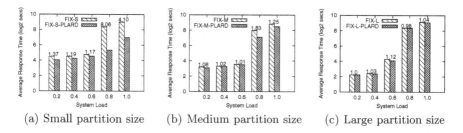

(a) Small partition size   (b) Medium partition size   (c) Large partition size

Fig. 6. Impact of PLARD on the average query response time. Note that the y axis uses the log2 scale, and the speedup factor brought by PLARD is shown at the top of each pair of bars.

Now we take a look at the overall impact of PLARD, by comparing the average query response time before and after. Because long waiting time with heavy system loads caused a wide distribution of response time, we show the numbers in log scale, with the speedup factor brought by PLARD labeled at the top of each pair of bars.

Figure 6 shows the comparison, again for each of the FIX algorithms using multiple system loads. As expected, the largest improvements are found with FIX-S, where the average response time is reduced by up to 4.1 times. As we have seen from Figure 4, the largest enhancement to data locality and the average query service time occurs with the small partition size. The changes in service time, in turn, has a varying impact on the query response time. With heavier loads, the reduced service time has a rather dramatic effect on decreasing the queue length and average query wait time. With light loads, the enhanced service time does not affect the per-query wait time much. For FIX-M, the best improvement is observed at the load of 0.8, with a speedup factor of 1.83. Not surprisingly, PLARD does not bring significant improvement to FIX-L.

4.4   Efficiency-Oriented Scheduling Results

Now we examine the impact of RMAP by enabling PLARD for all tests and compare the three FIX algorithms with RMAP.

Figure 7 portraits the results. As expected, no single fixed partitioning strategy performs consistently well. When the system load is light, the large partition size works best by using a large number of processors to reduce each query's response time. As the load increases, first the medium, then the small partition size becomes the winner. With heavier loads, smaller partition sizes help achieving better overall resource utilization by improving the parallel execution efficiency. The performance difference is significant: across the x axis, the difference between the best and worst average response time among the fixed partitioning strategies varies between 3.5 and 8 times. RMAP, on the other hand, closely matches the best performance from the three fixed partitioning strategies by automatically adapting to the system load.

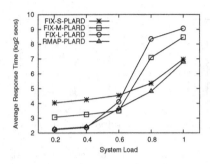
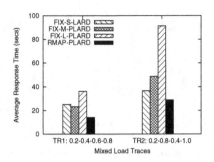

Fig. 7. Performance of combined RMAP and PLARD with fixed arrival rates (y axis uses log2 scale

Fig. 8. Performance of combining RMAP and PLARD on two 800-sequence traces with mixed arrival rates

The only point where the RMAP performance is visibly lower than the best fixed partition size algorithm is with the medium system load (0.6). Because the trace we used is not even-paced, the medium load is an unstable case for RMAP, where the scheduler adjusts the partition size (in both directions) most frequently. With frequent partition size changes, cache contents cannot be well utilized and more cold misses are introduced.

To verify this, we take a closer look at the behavior of FIX-M-PLARD and RAMP-PLARD. Table 2 summarizes a group of measurements taken from the experiments using the two algorithms, at the system load level 0.6 and 0.8. Because the partition sizes are power-of-two numbers, we calculate the average partition size by taking the arithmetic average after performing the log2 operation. The "total service time" is calculated as the total computation resource usage in an experiment. For each query, we calculate its resource usage as the product of its service time and the number of processors it used. We sum up the resource usage of all queries in a trace as the total service time.

From the page fault counts, we see that RMAP does hurt the data locality at load level 0.6. Consequently, RMAP adopts a slightly larger partition size than FIX-M does, but has a 8% higher average service time. The service time increase causes a similar increase in the waiting time and average response time.

Interestingly, RMAP caused a much larger increase in the number of page faults at the load of 0.8, yet the average response time of RMAP is 5 times better than that of FIX-M. This is caused by that RMAP has better parallel

Table 2. FIX-M-PLARD and RMAP-PLARD statistics at system load 0.6 and 0.8

System Load	0.6		0.8	
Policy	FIX-M-PLARD	RMAP-PLARD	FIX-M-PLARD	RMAP-PLARD
Average # Page Faults	93.00	132.74	63.64	224.88
Average Service Time (s)	8.81	9.53	8.76	15.47
Average Waiting Time (s)	2.50	2.95	128.25	12.51
Average Response Time (s)	11.31	12.48	137.02	27.98
Total Service Time (s)	69936	78700	69657	63977
Average Partition Size (log2)	3.71	3.88	3.71	2.78

computation efficiency there, which can be seen from the total service time: RMAP increased the total service time at 0.6 and decreased it at 0.8. Although the individual query's service time is longer than FIX-M, RMAP increases the whole-system throughput by automatically adopting a considerably smaller average partition size. With such a heavy system load, this had a dramatic effect on shortening the average query waiting time, and the average query response time consequently.

Finally, we evaluate the overall adaptivity of the combined RMAP-PLARD algorithm. Figure 8 shows two sets of experiments, each using a mixed load level trace containing 800 queries. In each trace, the average load level is adjusted several times, e.g., from 0.2 to 0.4, 0.6, and finally 0.8, for four equal-length intervals (in terms of the number of queries). Trace 1 adopts such an monotonically rising system load as in the above example, while trace 2 has a repeated up-down pattern. Again, for such mixed load traces, none of the fixed partition size algorithms consistently win, and each of them may suffer trace intervals where the selected partition size is undesirable. RMAP, on the other hand, successfully adapts to the varying query intensiveness and significantly outperforms all the fixed partition size algorithms, bringing an improvement factor of 1.63 and 1.26 in average response time over the best performing fixed algorithm for trace 1 and 2, respectively.

5   Related Work

There have been numerous studies on scalable distributed web-server systems, most of which were focused on efficient request routing and assignment for content serving, as surveyed by Cardellini et al. [6]. One closely related project is the LARD system [20], which performs content-based web request distribution to back-end servers considering both load balance and request locality for better memory cache performance. Research in this category, along with that on resource-intensive web request scheduling [24], often assumes that multiple requests can be served by the same back-end server simultaneously, or the request service time is known or can be predicted. In our target scenario, time-sharing the back-end servers is difficult given the closely-coupled message passing model used by a subset of servers performing parallel scientific applications, and the cost of each request could be quite unpredictable [11].

Regarding space-sharing of parallel computers, a wealth of job scheduling algorithms have been proposed and evaluated, as summarized by Feitelson [10]. However, with the prevailing use of message passing programming interfaces such as MPI [17] and contemporary batch parallel job execution environments, adaptive or dynamic allocation of resources is rarely used on parallel computers. Instead, jobs are given the exact number of processors as requested, using strategies such as FCFS plus backfilling [18]. Our work reveals a type of real-world workload that features so called "moldable parallel jobs" (those can be run on a flexible number of processors), where many existing scheduling strategies can be applied to. In this paper, we extend existing adaptive parallel job scheduling algorithms [8,21] to the high-performance scientific web service context.

Many projects have studied accelerating BLAST through parallel processing on SMP machines or clusters [3,4,5,7,9,12,13,15], with the current trend of enabling database segmentation [9,13]. Our case study of parallel BLAST server examines resource allocation and data placement issues related to handling online BLAST queries on a cluster web server, which can potentially work on top of any of the above underlying parallel BLAST implementations. Instead of making an individual parallel BLAST system more efficient, we focus on improving the overall resource utilization and exploiting data locality.

There have also been studies on high throughput BLAST online services. Wang and Mu described a distributed BLAST online service system [23], where the incoming query is assigned to the least-loaded SMP node and each node searches one entire target database. Wang et. al. introduced a service-oriented BLAST system built on peer-to-peer overlay networks [22]. This work assumes a heterogeneous environment with high communication cost. NCBI hosts a publicly accessible BLAST server on a farm of LINUX workstations [2,16]. For a given query, the system statically splits the search into 10-20 subtasks, each searching a different piece of the database. The subtasks are scheduled independently to the machines that have just searched the same piece of data when possible. A central machine tracks and merges results from subtasks for all queries. Due to the lack of design/implementation details about the NCBI BLAST server in the literature, we were not be able to do a direct comparison. However, we argue that the NCBI server is not able to factor in the parallel efficiency by using only static task partitioning. To the best of our knowledge, our paper presents the first systematic investigation of optimizing scientific web services by taking into account both parallel efficiency and data locality.

6   Conclusion and Future Work

Below we summarize the findings and contributions of this paper: (1)We identified the scheduling requirements of increasingly popular parallel scientific web services. (2) For our target workload, we extended and designed several adaptive scheduling strategies, namely PLARD for locality-enhancing resource partitioning, and RMAP for dynamic parallelism adjustments. These strategies automatically react to the query workload, both in terms of the request intensiveness and the data access pattern. (3) We integrated and implemented our proposed algorithms in a parallel BLAST sequence search prototype and performed extensive experiments using real-cluster tests. (4) Our results demonstrated that PLARD can significantly reduce the amount of file I/O. Meanwhile, RMAP outperforms its static counterparts across various query workloads. Combined together, our proposed strategies often deliver an several-fold performance gain.

This work can be extended in several directions. First, we would like to apply our scheduling algorithms to other parallel web services such as online scientific data mining. Another interesting topic will be investigating how to combine intelligent data prefetching and our adaptive scheduling to host service of huge datasets that cannot be fully cached in a server node's local storage.

Acknowledgment

This research has been supported by a DOE ECPI Award (DE-FG02-05ER25685), an NSF CAREER Award (CNS-0546301), and NSF grants CCF-0621470 and IIS-0430166. In addition, the work is supported by the joint appointment between NCSU and ORNL for Xiaosong Ma and Nagiza Samatova, and by the Scientific Data Management Center (http://sdmcenter.lbl.gov) under the Department of Energy's SCIDAC program. We thank Dr. John Blondin at North Carolina State University for facilitating our experiments on the Orbitty cluster, and we thank Andrew Brown for providing technical supports on the cluster.

References

1. Altschula, S., Gisha, W., Millerb, W., Meyersc, E., Lipmana, D.: Basic local alignment search tool. Journal of Molecular Biology 215(3) (1990)
2. Bealer, K., Coulouris, G., Dondoshansky, I., Madden, T., Merezhuk, Y., Raytselis, Y.: A fault-tolerant parallel scheduler for blast. In: SC 2004 (2004)
3. Bjornson, R., Sherman, A., Weston, S., Willard, N., Wing, J.: TurboBLAST(r): A parallel implementation of BLAST built on the TurboHub. In: IPDPS (2002)
4. Braun, R.C., Pedretti, K.T., Casavant, T.L., Scheetz, T.E., Birkett, C.L., Roberts, C.A.: Parallelization of local blast service on workstation clusters. Future Gener. Comput. Syst. 17(6), 745–754 (2001)
5. Camp, N., Cofer, H., Gomperts, R.: High-throughput BLAST, http://www.sgi.com/industries/sciences/chembio/resources/papers/HTBlast/HT_Whitepaper.html
6. Cardellini, V., Casalicchio, E., Colajanni, M., Yu, P.: The state of the art in locally distributed web-server systems. ACM Computing Surveys 34(2) (2002)
7. Chi, E., Shoop, E., Carlis, J., Retzel, E., Riedl, J.: Efficiency of shared-memory multiprocessors for a genetic sequence similarity search algorithm. Technical Report TR97-005, University of Minnesota, Computer Science Department (1997)
8. Dandamudi, S., Yu, H.: Performance of adaptive space sharing processor allocation policies for distributed-memory multicomputers. JPDC 58(1) (1999)
9. Darling, A., Carey, L., Feng, W.: The design, implementation, and evaluation of mpiBLAST. In: Proceedings of the ClusterWorld Conference and Expo, in conjunction with The HPC Revolution (2003)
10. Feitelson, D.: A survey of scheduling in multiprogrammed parallel systems. Technical Report IBM/RC 19790(87657) (1994)
11. Gardner, M., Feng, W., Archuleta, J., Lin, H., Ma, X.: Parallel genomic sequence-searching on an ad-hoc grid: Experiences, lessons learned, and implications. In: Löwe, W., Südholt, M. (eds.) SC 2006. LNCS, vol. 4089. Springer, Heidelberg (2006)
12. Grant, J., Dunbrack Jr., R., Manion, F., Ochs, M.: BeoBLAST: distributed BLAST and PSI-BLAST on a Beowulf cluster. Bioinformatics 18(5) (2002)

13. Lin, H., Ma, X., Chandramohan, P., Geist, A., Samatova, N.: Efficient data access for parallel BLAST. In: IPDPS, Washington, DC, USA (2005)
14. Lin, H., Ma, X., Li, J., T, Y., Samatova, N.: Processor and data scheduling for online parallel sequence database servers. Technical Report TR-2007-23. North Carolina State Univeristy (2007)
15. Mathog, D.: Parallel BLAST on split databases. Bioinformatics 19(14) (2003)
16. McGinnis, S., Madden, T.: BLAST: at the core of a powerful and diverse set of sequence analysis tools. In: Nucleic Acids Res. (2004)
17. Message Passing Interface Forum. MPI: Message-Passing Interface Standard (1995)
18. Mu'alem, A., Feitelson, D.: Utilization, predictability, workloads, and user runtime estimates in scheduling the ibm sp2 with backfilling. In: IEEE TPDS, vol. 12 (2001)
19. Ostell, J.: Databases of discovery. ACM Queue 3(3) (2005)
20. Pai, V., Aron, M., Banga, G., Svendsen, M., Druschel, P., Zwaenepoel, W., Nahum, E.: Locality-aware request distribution in cluster-based network servers. In: ASPLOS-VIII (1998)
21. Rosti, E., Smirni, E., Dowdy, L.W., Serazzi, G., Carlson, B.M.: Robust partitioning policies of multiprocessor systems. Perform. Eval. 19(2-3), 141–165 (1994)
22. Wang, C., Alqaralleh, B., Zhou, B., Till, M., Zomaya, A.: A BLAST service built on data indexed overlay network. e-science (2005)
23. Wang, J., Mu, Q.: Soap-HT-BLAST: high throughput BLAST based on Web services. BIOINFORMATICS -OXFORD- (2003)
24. Zhu, H., Smith, B., Yang, T.: Scheduling optimization for resource-intensive web requests on server clusters. In: SPAA (1999)

ViP: A User-Centric View-Based Annotation Framework for Scientific Data

Qinglan Li, Alexandros Labrinidis, and Panos K. Chrysanthis

Advanced Data Management Technologies Laboratory
Department of Computer Science, University of Pittsburgh
Pittsburgh, PA 15260, USA
{qinglan,labrinid,panos}@cs.pitt.edu

Abstract. Annotations play an increasingly crucial role in scientific exploration and discovery, as the amount of data and the level of collaboration among scientists increase. In this paper, we introduce ViP, a user-centric, view-based annotation framework that promotes annotations as first-class citizens. ViP introduces novel ways of propagating annotations, empowering users to express their preferences over the time and network semantics of annotations. To efficiently support such novel functionality, ViP utilizes database views and introduces new caching techniques. Through an extensive experimental study on a real system, we show that ViP can seamlessly introduce new annotation propagation semantics while significantly improving the performance over the current state of the art.

1 Introduction

Without a doubt, data management is playing a pivotal role in scientific exploration nowadays, constantly fueling the pace of discovery. In addition to efficiently managing the tsunami of experimental data generated, data management also facilitates effective collaboration among scientists, by recording *data provenance* [6] and *data lineage* [1,2,9,11,12], and by supporting *annotations* [3,4,5,15,23]. Data provenance and lineage essentially keep track of where the data is coming from (and what transformations it has been through), whereas annotations enable users to record additional information about the data stored (and propagate this information to all "related" data items).

In this paper, we present *ViP*, a novel annotation framework that introduces new annotation propagation methods, utilizes *views* both as a specification mechanism and as a user-interface mechanism, and employs caching techniques for improved performance compared to the state of the art [1].

Our interest in this research area came from our participation in the Center for Modeling Pulmonary Immunity (CMPI). CMPI is bringing together experimentalists and modelers to study pulmonary immunity in response to three bio-defense pathogens. Our group is responsible for the design and development of the *data sharing platform* (DataXS), where experimental data, analysis, and models will be shared among project participants. In such a diverse setting, the ability to record annotations and propagate

---

[1] This research was supported in part by NIH-NIAID grant NO1-AI50018.

B. Ludäscher and Nikos Mamoulis (Eds.): SSDBM 2008, LNCS 5069, pp. 295–312, 2008.
© Springer-Verlag Berlin Heidelberg 2008

them to all related data items and interested parties is crucial to the success of the project.

As part of the design process and during the implementation of our first prototype, we were able to identify two distinct *usage patterns* related to the specification and the propagation of annotations within a Database Management System (DBMS), which were not currently supported by the state of the art.

Support for user-centric time semantics for annotations. The first usage pattern that we observed was that *experimental data was almost always entered in the database in an order different than the one it was generated*. In fact, even data about the same experiment was entered at completely different times, since more than one labs were involved in generating the data (for example, one lab would generate the luminex data whereas a different lab would produce microarray data for the same tissues). Looking at annotations, this means that if one wanted to annotate data from a particular experiment with an observation about the tissues, it would **not** be enough to do this once, as additional experimental data may be added into the database later (which would not automatically "inherit" the annotation).

To address this, we propose the concept of *valid time* of annotations, where annotations should be propagated to all data items matching a certain description within the validity time interval specified by users. We refer to this feature as *"user-centric time semantics"*.

Support for propagation of annotations in user-defined ways. The second usage pattern that we observed was that *there exist many relationships, or paths, between data items that cannot be inferred by the existing database schema or their lineage*. Such paths materialize because, for example, tissues from multiple, independent experiments were processed together, in a single assay (for example, on a single plate that needed to be filled up to minimize costs). Annotations associated with this one assay would be propagated to all experiments that shared the plate, e.g., in the case of a contamination.

To address this, we propose to enable users to specify explicit *annotation paths*, thus allowing for more "interconnections" among data and knowledge. Annotations should be propagated along these paths, reaching "related" data items, as specified by users. Since these paths are essentially forming a network, we refer to this feature as *"user-centric network semantics"*.

User-friendly Implementation. First of all, we need to identify a way to formally define data items matching a certain criterion, to be used by our algorithms. The answer is somewhat obvious: use *database views*, which describe the results of a query. Similar approaches have been proposed in the past (e.g., [21]); ours significantly extends the use of views with additional semantics. Secondly, we need a realistic way for users to utilize the new features. Clearly, they are not to be expected to provide view definitions in SQL! In DataXS, a user can easily specify filtering conditions to locate certain data items, in other words, to specify views using a point and click interface (Figure 1); these views were initially implemented to provide an easy reference for frequently used queries (e.g., the In Vitro experiments tab in Figure 1), but can also be trivially used to implement all functionality of the ViP framework.

Fig. 1. DataXS User Interface

Contributions: This paper has both theoretical and practical contributions:

- based on our experience from a real system implementation, we introduce new annotation propagation methods, suitable for scientific data,
- we propose user-centric features (*user-centric time semantics, network semantics, and access control*) that enable users to personalize annotation propagation,
- we propose to use *views* as the formal mechanism to implement the new annotation propagation features and also as a user-interface paradigm,
- we utilize *caching* to significantly improve the performance over the state of the art,
- we experimentally evaluate the proposed ViP framework using a real system implementation and simulated workloads.

Roadmap: Section 2 presents the details of the proposed annotation framework, along with related work. Section 3 describes our implementation. Section 4 presents the results from our extensive experimental study using a real system.

2   Annotation Propagation Semantics in ViP

In this section, we present the details of our proposed semantics for annotation propagation. In each case, we also present the corresponding statements in ViP-SQL, our proposal for a simple extension to SQL that would handle the new semantics based on the concept of views.

2.1   User-Centric Time Semantics

We propose the concept of *valid time*, which is the validity time interval of an *annotation view* or *path*. It allows users to specify what time period to associate the annotations with corresponding data or to propagate the annotations via a certain path.

If we consider the time dimension of annotation propagation, we can easily distinguish four different cases:

- *now*, where an annotation is only propagated to data items currently in the database,
- *future*, where an annotation is only propagated to data items that are added in the database in the future,

- *now+future*, where an annotation is propagated to data items currently in the database, and also to those that are added in the database in the future. This is essentially the combination of the *now* and *future* approaches,
- *future interval*, where an annotation is propagated to data items that are added in the database in the time interval a user specifies.

These four cases represent all the possible alternatives if one considers the concept of future time semantics and also wants to give users the flexibility to specify the validity interval for their annotations.

Related Work. Most of the current annotation management frameworks utilize *now* time semantics, propagating annotations to only existing data items [2,3,4,13,14]. In contrast, our system supports now and future semantics (in all the four "flavors" described above), which we will assume for the rest of this paper. To the best of our knowledge, only the work in [21] functions similar to the *now+future* time semantics presented in this paper.

Motivating Example #1. To properly motivate the need for time semantics, let us assume a setup like that in the DataXS system, where experimental data are stored (in the Experiments table) and shared among project participants. Let us assume that a contamination happened in the ADMT Lab between Oct 1, 2007 and Oct 20, 2007, and we would like to annotate all experimental data accordingly, with *now+future* time semantics. Clearly, if we only attach an annotation to all the files matching the ADMT Lab and happened Oct 1 - 20, 2007 (at the time of annotating), we will miss all the files that are potentially added into the DataXS system at a later time, and still meet these conditions. As we discussed earlier, this is a typical usage pattern, making *now+future* time semantics a necessity. We can describe such an annotation in *ViP-SQL* as follows:

```
CREATE ANNOTATION V1 ON Experiments
    AS (SELECT ExpID FROM Experiments
        WHERE Lab = "ADMT" and Date >= "10/01/07"
              and Date <= "10/20/07")
    VALUE "ADMT Lab was contaminated between Oct. 1st
          & Oct. 20, 2007. Please use data with caution."
    VALIDTIME [now,  )
```

General Case. The main idea behind view-based annotation propagation is that we can attach an annotation to a **view**, i.e., a query definition that corresponds to a set of data items, instead of attaching it to individual data items. If we do not materialize the view, then the annotations will always be properly associated with the corresponding data items, according to *now+future* time semantics.

Given that annotations are associated with views instead of individual data items, the expected behavior in cases of modifications is straightforward (Figure 2). Assuming a view V_i with annotation a, the following actions can be defined:

- INSERT(data) into VIEW:
  if D_1 becomes a member of view V_i (either through insertion or an update or a creation of an annotation view), then it will also be associated with annotation a (attached to V_i) when it is queried.

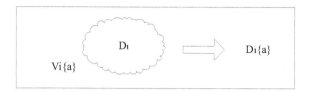

Fig. 2. View-based Annotation Propagation: User-centric Time Semantics. (Annotation a is associated with view V_i. Data item $D_1 \in V_i$ receives annotation a.)

- DELETE(data) from VIEW:
  if D_1 is no longer a member of view V_i (either through deletion or an update), then it will not be associated with annotation a when it is queried.
- DELETE(view):
  if V_i is deleted, then all the data items that were members of V_i and were associated with annotation a will no longer be associated with it.

2.2   User-Centric Network Semantics

Most annotation-enabled systems propagate annotations along data provenance paths. In other words, annotations are propagated over existing "schema" paths between source data and derived data. Although this happens over multiple derivation levels, it fails to capture relationships between data items that do not share a common source in the database. As we have witnessed from our involvement in the CMPI project, this can happen often in scientific databases.

Through the ViP framework, we propose to empower users to specify *explicit annotation paths* between data items, thus establishing additional annotation propagation paths. Such explicit paths are defined using views as follows:

- given a source view, V_s, and a destination view, V_d,
- an explicit annotation propagation path $V_s \rightarrow V_d$ is defined, such that any annotation that is added in a member of V_s must be propagated to all members of V_d.

Motivating Example #2. Continuing from Motivating Example #1, we have that the ADMT Lab and the Ross Lab are next to each other, and the ADMT Lab provides the Ross Lab with tissues for model analysis. As such, there is a need to propagate all annotations regarding ADMT Lab experiments to the Ross Lab (to properly record, for example, if there has been any contamination). We can describe such an annotation in *ViP-SQL* as follows:

```
CREATE ANNOTATION V2
ON Experiments
AS (select Date from Experiments
    where  Lab = "ADMT" and Treatment = "Influenza A")
TO Experiments
AS (select Date from Experiments
    where  Lab = "Ross" and Treatment = "Influenza A")
VALIDTIME [now, )
```

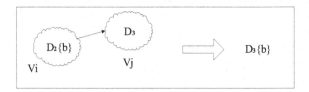

Fig. 3. View-based Annotation Propagation: User-centric Network Semantics (Disjoint source/destination). There exists an annotation propagation path from V_i to V_j. Data item $D_2 \in V_i$ has annotation b. Data item $D_3 \in V_j$ receives annotation b.

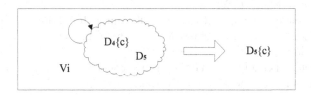

Fig. 4. View-based Annotation Propagation: User-centric Network Semantics (Identical source/destination). There exists an annotation propagation path from V_i to self. Data item $D_4 \in V_i$ has annotation c. Data item $D_5 \in V_i$ receives annotation c.

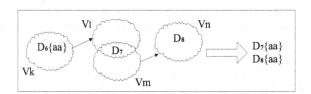

Fig. 5. View-based Annotation Propagation: User-centric Network Semantics (Overlapping source/dest.). There exists an annotation propagation path from V_k to V_l and another path from V_m to V_n. V_l and V_m overlap. Data item $D_6 \in V_k$ has annotation aa. Data item D_7 is a member of both V_l and V_m. Data item D_7 receives annotation aa. Data item D_8 receives annotation aa.

General Case. Considering the general case of using source/destination views to describe explicit paths for annotation propagation, we can see that such paths essentially form a **network**, hence the need for *network semantics*. With regards to view membership, the behavior is very similar to that in the case of *time semantics*, as presented in the previous section. With regards to the relation between the source and destination views, we consider the following cases:

- source and destination views are disjoint (Figure 3)
- source and destination views are identical (Figure 4)
- source and destination views are overlapping (Figure 5)

Related Work. In the context of metadata management, [21] considered implicit paths from queries to queries but they have not considered the explicitly-defined network paths as we do in this paper. In the context of schema mapping, there are multiple works that consider paths of "similar" tables [8,19].

2.3   User-Centric Access Control

We advocate that scientific annotation must have a strong access control component. First of all, much of the data is not public, so appropriate access controls need to be in place for the raw data, and the annotations on them. Secondly, even for public data, the annotations are often private, since they reflect additional analysis that is not ready to be made available to all. Thirdly, in many cases, even the way that raw data are associated with each other (i.e., by specifying explicit paths for annotation propagation) corresponds to private information that should not be made public. Given all these reasons, the ViP framework includes multiple user-centric access control features.

On the annotation level. First of all, we implement access control at the level of individual annotations. In other words, when an individual data item receives an annotation from a user, the user can specify who can access the annotation. We support arbitrary user hierarchies (i.e., specific users, groups of users, groups of groups of users, etc).

On the view level. We expect the majority of annotations to happen through views, to take advantage of time semantics. In this case, user access controls are also implemented, with the expected behavior.

On the path level. One important innovation of the ViP framework is the explicit path functionality (network semantics). We support three different access control features, as they apply to user-centric network semantics:

- **Access control:** Users would want to control who can take advantage of the explicit annotation propagation paths that they introduce. This is necessary for two reasons: (a) *confidentiality* of paths, i.e., not willing to make relationships between data public; and (b) *scalability* of paths from a information absorption point of view, i.e., not everybody is interested in everybody else's beliefs on which data is related. This of course means that certain paths will not be visible to some users.
- **Maximum HAP:** Given explicit annotation paths and the ensuing network semantics, an annotation can theoretically be propagated over an unreasonable number of paths, if left unconstrained. Towards this, the ViP framework includes a system variable, **MAX-HAP**, short for *maximum number of hops allowed to propagate*, which puts a system-wide upper bound over how many hops any annotation is allowed to propagate. The number of hops starts counting after we follow the first "direct" path (i.e., in Figure 5 the number of hops is 2). This was inspired by the TTL value of queries in unstructured peer-to-peer networks.
- **HAP on insert:** The ViP framework enables users to specify a variable, **HAP-i**, or *Hops Allowed to Propagate* at insertion, to indicate how far the newly-inserted annotation can be propagated. HAP-i = 0 means the annotator just wants to limit this annotation to data items specified in the view. HAP-i = 1 means the annotator allows this annotation to be propagated only to neighboring nodes. HAP-i = MAX-HAP means the annotator is not placing any restriction on the propagation of his/her annotation.

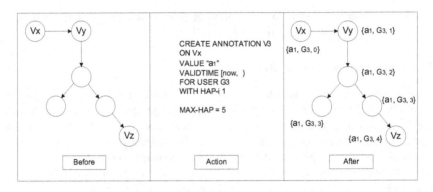

Fig. 6. User-Centric Annotation Propagation Example

• **HAP on query:** Although if $A \rightarrow B$ and $B \rightarrow C$ implies that $A \rightarrow C$, this may not be applicable for all cases (i.e., because of information "decay"). In cases of a network of paths (e.g., as in Figure 5), it may not be prudent to exhaustively follow all paths in the network to propagate annotations. Similarly with the *HAP on insert*, the ViP framework gives the option to specify a *maximum number of hops an annotation is allowed to propagate at query time*, or **HAP-q**. Given these three parameters (some of which are optional), the maximum number of hops followed is MIN(MAX-HAP, HAP-i, HAP-q). By setting HAP-i or MAX-HAP to 0, we effectively disable explicit annotation direct paths; by setting MAX-HAP to 1, we effectively disable cascading annotation propagation.

Motivating Example #3. We illustrate the user-centric semantics of the ViP framework using the example in Figure 6. Figure 6/Before has a network of paths; Figure 6/Action indicates that an annotation is added on node V_x; Figure 6/After shows how annotations would be propagated (the third number in the set corresponds to the number of hops required to reach each node). We see that the annotation a_1 is propagated to V_y within HAP 1 as (a_1, G_3, 1), and to V_z within HAP 4 (which is bigger than HAP-i). Clearly, users that neither belong to group G_3 nor specify a HAP-q high enough will not "see" annotation a_1. Besides, if HAP-i of a_1 is set to 0, even if users specify a high HAP-q will still not "see" annotation a_1. The queries and the results are shown in Table 1.

Related work. There is significant related work in personalization, especially in connection with information retrieval [16,18]. There is also additional work in user-centric data management, allowing users to express their preferences on the execution of their

Table 1. Queries and Results for Figure 6

HAP-i	Query	Result	User	HAP-q	Annotation
1	1	V_y	$U_1 \neg \subseteq G_3$	3	No a_1
1	2	V_z	$U_3 \subseteq G_3$	3	No a_1
0	3	V_z	$U_3 \subseteq G_3$	5	No a_1
MAX-HAP	4	V_z	$U_3 \subseteq G_3$	5	a_1

Table 2. Standard Annotation Management Features Comparison

Standard Features	DBNotes[3]	Mondrian[14]	ULDB[2]	bdbms[13]	MMS[21]	ViP
Annotation	Yes	Yes	Confidence	Yes	Yes	Yes
Provenance	Yes	Yes	Lineage	Yes	Yes	Yes
Time Semantics:						
· Implicitly-defined	No	No	No	No	Yes	Yes
· Explicitly-defined	No	No	No	No	No	Yes
Network Semantics:						
· Implicitly-defined	Limited	Limited	Limited	Limited	Yes	Yes
· Explicitly-defined	No	No	No	No	No	Yes
Propagation Type	Eager	On-demand	On-demand	Eager	On-demand	Hybrid
Annotation Storage	Naive	Naive	x-relations	Anno. table	q-type	A-table
Scalability	Small	Medium	Medium	Medium	Large	Large
Query	pSQL	Color algebra	TriQL	A-SQL	Predicate	ViP-SQL

queries, such as [17,20]. However, to the best of our knowledge, this is the first work to address in a unifying framework all the user-centric features that we proposed as part of ViP.

2.4 Discussion

There are many systems that support in isolation, some of the features that are part of ViP without any one single system incorporating all of them. Additionally, many of the semantics introduced by ViP are not found in other systems.

Most current systems, for example, do not support annotations that are also valid in the future (Table 2). Only MMS [21] supports future time semantics in an implicitly-defined way (i.e., without giving the user options to select as ViP is doing through the valid time concept).

One of ViP's novel ideas is the explicit paths for annotation propagation, which also have privacy controls. Although existing systems support implicit annotation propagation paths, none except for ViP supports explicit, user-defined annotation propagation paths. ViP supports large scale annotation management, thus employs a hybrid propagation scheme while [3,13] use eager propagation, whereas [2,14,21] use an on-demand scheme.

To the best of our knowledge, ViP brings user-centric features in many aspects that are not considered in most related work as shown in Table 3. ViP enables users to specify the propagation method. In DBNotes [3], users can specify *custom* propagation scheme to bind the source and target tuples while there is a join operation, so that the annotations that are associated to the source tuples will be propagated to the target tuples. ViP provides a stronger and more complex scheme, that is the *annotation path*.

Some systems consider *access control* on the data level, or even on the update authorization part [13]. Instead, we propose to fully support this feature in a broader domain, on annotations, annotation views, and annotation paths.

Table 3. User Centric Annotation Management Features Comparison

User-centric Features	DBNotes[3]	Mondrian[14]	ULDB[2]	bdbms[13]	MMS[21]	ViP
Time Semantics:						
· Valid Time	No	No	No	No	No	Yes
Network Semantics:						
· Propagation Method	Yes	No	No	Limited	No	Yes
Access Control:						
· Annotations	No	No	No	Limited	No	Yes
· Annotation Views	No	No	No	No	No	Yes
· Annotation Paths	No	No	No	No	No	Yes

3   Implementation

3.1   The ViP Framework

The ViP framework is illustrated in Figure 7. ViP-SQL queries are rewritten automatically into SQL queries evaluated by the annotation query processor, then registered with the annotation register and the path setup manager. They are sent to DBMS and the resulting annotation set is merged with the regular query result by the postprocessor for matching and presentation. Our DataXS application "fits" on top of this framework, providing a point-and-click user interface.

3.2   Annotations Registration

Explicit annotations could be a string or a file; while implicit annotations include annotation views and annotation paths. If it is an annotation view, the annotation register

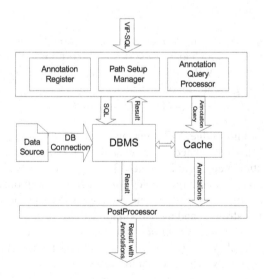

Fig. 7. ViP System Architecture

is responsible for insertion, deletion and updating. If it is an annotation path, the path setup manager will update the auxiliary table to record path source and target, with appropriate path query conditions. Obviously, sorting views or addressing the view containment problem [7,10,22] may bring significant computation and time complexity. To simplify the problem setting, we assume that the network formed by the annotation paths forms a directed acyclic graph, when ordered. All views are sorted by topological order to build a hierarchy/dependency tree, thus guarantee the correctness and completeness.

3.3 Implementing Auxiliary Tables

It is quite naturally to use auxiliary tables storing the attributes of the annotation views. Like MMS [21], ViP also uses auxiliary tables to store annotation view conditions, which will work as filters to drop unrelated annotation lookups. However, MMS uses Q-indexes (index on queries, which is similar to views in ViP) to maintain indexes on the Q-values (query values); as such, for every data change, all related index tables need to be updated. Unlike MMS, we use caching to improve the performance of computing annotations. The reason is that for the index to be useful, it would need to be efficiently updateable when data and annotations are inserted, deleted and updated; therefore, such index maintenance may require a high cost in space and time. In addition, from the usage pattern we observed in DataXS project, data updates happen more often than annotation views/paths updates, in which case an index approach would require a lot of Q-value updating. Thus, ViP relies on caching instead of indexing.

3.4 Querying Result with Annotations

We use ViP-SQL to allow users to retrieve regular results with annotations. A query with annotation is rewritten as standard SQL with preprocessing and postprocessing. Preprocessing checks the auxiliary table for possible early annotation filtering. If a query result is satisfied in an annotation view, then the annotation query processor will lookup the annotations associated with the query result. The cache is used to optimize system performance. We present the pseudocode for the corresponding algorithms accordingly.

Caching to Optimize Annotation Search. If a data tuple is not found in the cache, ViP will execute the annotation query and save its annotation result set into the cache. If a data tuple is found in the cache, we need to verify if it is still "fresh." Cache management will take no action if a *data item* is inserted, deleted, or updated in the database. Whenever an *annotation registration* is updated/inserted, our system will reset the cache appropriately. If an annotation registration is removed, our system will remove its related entries from the cache as well. The algorithm is shown in Figure 8.

Search Associated Annotations. To Search the annotations associated with of a data item, we need to search in both directions: its direct annotations (via annotation views) and its inherited annotations (via annotation paths) as shown in Figure 9.

```
hit_caching(Ti)
   Tj = search_in_cache_index(Ti.table, Ti.col, Ti.id)
   if Tj is found,
      compare(Ti.data, Tj.datasnapshot)
      if matches
         hit-counter++
         return Tj.CachedAnnotationQueryResult
   return false

insert_into_caching(Ti)
   if cache is full
      evict as LFU algo
   insert Ti to cache
   save a snapshot of data referred by Ti

after_annotation_delete(Ti)
   delete cached AnnotationQueryResult R
   where R.table = Ti.table and R.id = Ti.id
```

Fig. 8. Algorithm of Annotation Cache Management

```
search_associated_annotation(Ti)
   find_direct_associated_annotation(Ti)
   find_dependent_associated_annotation(Ti)
   return Ti.annotationQueryResult

find_direct_associated_annotation(Ti)
   A = search_in_Annotation_Attribute_table(Ti.table, Ti.col)
   for each annotation Aj in A
      compare_condition_parameter (Aj, Ti)
      if match, add Aj.id to Ti.annotationQueryResult

find_dependent_associated_annotation(Ti)
   H = search_in_Inhertance_Definition_table(Ti.table, Ti.col)
   for each Hj in H
      R = find_records_in_associated_table(Hj.inheritance_rule)
      R_column = Hj.inhertance_through_rule.attribute
      for each record Rm in R
         search_associated_annotation(Rm.R_column)
```

Fig. 9. Algorithm of Searching Associated Annotations

4   Experimental Results

We have implemented the ViP system as a Ruby on Rails application that interfaces to MySQL. We used simulated users, annotations, and query workloads to be able to scale our experiments to desired levels.

To the extent possible, we compare our system with MMS [21], the latest and the most related work. In [21], MMS was compared to other systems, specifically DBNotes [3] and MONDRIAN [14]. MMS showed significant benefits over those systems both in query times and storage space usage. That is because in DBNotes every relational table column is associated to one additional annotation column, and if a value in a tuple has more than one annotation, the tuple is recorded multiple times, one for each annotation. On the other hand, MONDRIAN associates one extra annotation column to each relation, plus one shadow column for each attribute to indicate whether the annotation refers to the respective attribute or not. In [21], the experimental results showed that MMS reduced the redundant space used in DBNotes and MONDRIAN; also, it decreased query

Table 4. Experiment Parameters

Parameter	Value	Parameter	Value
Data tuples	300,000	Queries	1,000
Annotation views	[1, 50,000]	Users	[1, 100]
Annotation paths	[1, 2,500]	Path Depth	[1, 10]

time even with the cost of updating the Q-index and querying additional metadata table. Our system works similar to MMS in the way that there are annotation tables instead of additional annotation columns. Thus, it is expected our system will perform similar to MMS when compared to DBNotes and MONDRIAN if the association between the data and the annotation is explicit and static. In this paper, we focused on implicit annotations, i.e., annotation propagation through annotation views and paths. Since both ViP and MMS can accommodate future tuples and use views to specify annotation registration, we compared our system with MMS mainly in terms of query time. For those features that ViP supports and MMS does not (such as the user-centric access control), we performed a sensitivity study of our framework.

Data. We gathered data from what has already been stored in our DataXS prototype. To test the scalability, we enlarged the dataset using uniform and Zipf distributions. The experiment parameters are shown in Table 4.

Annotation Traces. There are two types of annotations registered: annotation view and path. Annotation view is a query with static annotation(s) associated to it; annotation path is the establishment of an annotation(s) propagation from one annotation view to another annotation view. We generated annotation registrations using two different Zipf distributions: one to identify how many annotation views a data item should participate in, and another one to determine how many data items a particular annotation view should contain. Annotation traces include annotation insertion and update.

Query Traces. We generated queries with Zipf distribution on both (1) data tuples the query is associated with (2) query arrival sequence. Query conditions vary from 1 to 4 joins. All queries are read-only. Query time is measured in milliseconds unless otherwise indicated.

4.1 View-Based Annotation Propagation

We compared the query time of our system, ViP, with MMS. Both systems retrieved the same annotations associated with the same queries. In our first experiment, we varied the total number of annotation registrations (Figure 10). ViP always outperformed MMS due to its caching optimization. With more annotation views registered, ViP gained more benefit. In the case of 50,000 annotation views registered, ViP took about 25% less time, indicating that ViP works better for large numbers of annotation views.

We also measured the confidence interval of the result to make sure they are statistically significant. In the case of 1,000 queries with 50,000 annotation views, the 95% confidence interval for ViP mean query time (ms) is $(1468.06 \mp 7.36) = (1460.7,$

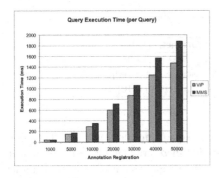
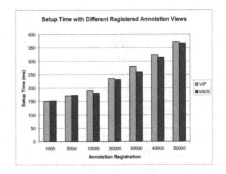

Fig. 10. Query Time   **Fig. 11.** Setup Time

1475.42); the 95% confidence interval for MMS mean query time (ms) is $(1878.91 \mp 4.05) = (1874.86, 1882.96)$. The results presented in the paper were acquired as the average value from 1000 repeated experiments with random parameter settings. Due to the limited space, not every confidence interval is listed in the paper; all results were similar to this experiment.

In all experiments, we started with 80% annotation views and paths insertions. When the query traces were executed, the remaining 20% of the annotation registrations were performed, with their arrival times uniformly distributed over the duration of the experiment. We assume each query or annotation registration operation is atomic. The query time includes (1) data query time, (2) annotation lookup time, (3) cache lookup time if cache is used, and (4) cache management time. The setup time includes (1) data insertion time, (2) annotation registration time, and (3) cache setup time. The setup time per query for both systems is shown in Figure 11. Although ViP took extra time to manage the cache, the overhead is negligible compared to the gain from the query time.

In the next set of experiments, we investigated the effect of various annotation densities, which is the percentage of data associated with annotation views. In Figure 12, 1000 queries were plotted in each subfigure to display the various query times. The density was changed from 50% to 200%, and the query time increased accordingly. In these figures, a vertical line corresponds to a cache hit (near 0 response time) on all annotations the query expects to return. We found in the extremely dense case, which is 200% in Figure 12(d), that ViP had so many cache hits, that the overall query time was reduced significantly. The detailed summary of average query time is presented in Table 5. Again, ViP works better in large scale of annotation views because of its optimized scheme. For fairness, we used only a 10% annotation density, which is the least beneficial setting for ViP, in all other experiments.

Table 5. Query Time with Different Annotation Densities

Anno. Density	40%	50%	60%	70%	80%	90%	100%	150%	200%
MMS Time (ms)	1804.81	1808.66	1812.50	1867.94	1878.02	1895.51	1928.80	1979.92	2178.67
ViP Time (ms)	1471.38	1419.73	1445.81	1499.44	1394.39	1386.27	1484.16	1483.84	1250.99

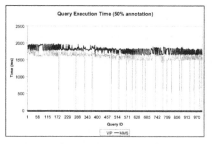

(a) Query Time of 50% Annotation Density

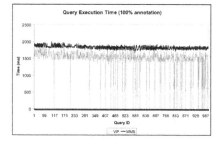

(b) Query Time of 100% Annotation Density

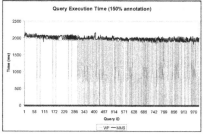

(c) Query Time of 150% Annotation Density

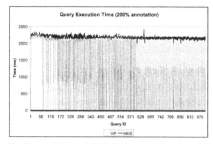

(d) Query Time of 200% Annotation Density

Fig. 12. Query Time with Different Annotation Densities

4.2   Annotation Propagation with Caching

In our optimization scheme, caching plays a major factor to improve system performance. However, the cache management time was insignificant compared to the query time, shown in Figure 13. Even with 50,000 annotation views, the cache management time is just about 3% in query time.

We performed a set of experiments to test the sensitivity of ViP to the cache size (Figure 14). We found that ViP worked best at 10% to 17.5% of the overall size. When

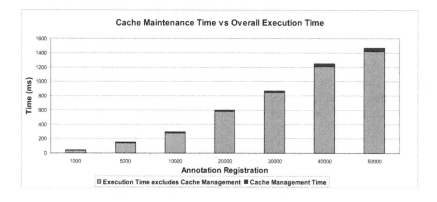

Fig. 13. Cache Management Time

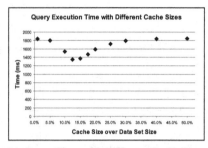 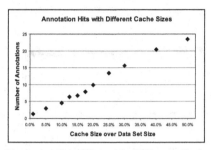

(a) Query Time with Different Cache Sizes   (b) Annotation Hits with Different Cache Sizes

Fig. 14. Various Cache Sizes

the cache size was larger than 30%, not much benefit was gained (i.e., query times did not improve much) from further cache size increases, although the cache hits may be increased. This is clearly because a larger size cache brings extra effort to lookup and manage the cache, so the overall query time will not be reduced.

4.3   View-Based Annotation Path Propagation

We conducted a set of experiments where varied the HAP variable (HAP-q) in annotation path propagation. HAP-i was set as MAX-HAP. We present the results in Table 6. It is obvious that with deeper hops search, more annotations got matched and more time it took to retrieve them. Nonetheless, ViP increased the query time gradually.

4.4   User-Centric Access Control

Another interesting feature of ViP is its user-centric access control features. Not only users may issue queries that include their search preference, but also users can specify public/private annotation views when they register the annotations. The first set of experiments, in Figure 15 and Figure 16 illustrate how the different search coverage affected the query times and the number of annotations found. The most restrictive user-specified condition decreased the query time as well as the associated annotations.

On the other hand, Figure 17 and Figure 18 present the query times with different percentages of public annotation views and annotation paths. In these cases, the remaining "private" annotation views and paths were uniformly distributed among all users. The query time almost decreased linearly as the public annotation views decreased; however, it decreased faster when the public annotation paths were decreased. Since

Table 6. Path Propagation in Network Semantics

HAP-q	1	2	3
Time (sec)	10.1445	11.1853	13.5833
Annotations Found	269	278	289

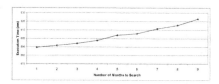

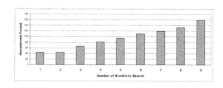

Fig. 15. Query Time for Different User Search Conditions

Fig. 16. Annotation Found for for Different User Search Conditions

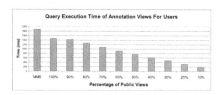

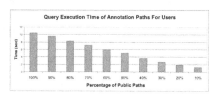

Fig. 17. Query Time with Different Public Annotation View Percentages

Fig. 18. Query Time with Different Public Annotation Path Percentages

annotation paths have the transitivity property, once the dependent views are not visible, it may speed up the query time exponentially. This essentially works like a first priority "filter" to reduce the query search time. In general, we expect such user-centric features to have a compound effect if used together, dramatically reducing query times.

5 Conclusions

In this paper we presented ViP, a view-based user-centric annotation framework. ViP introduced user-centric time semantics, network semantics, and access control for annotation propagation. Using database views as the underlying mechanism to implement these semantics enabled us to have a well-defined formal framework and also have a natural mapping to the existing user-interface, so that users of ViP do not have to learn SQL in order to specify their annotations. An other major advantage of ViP, compared to existing systems, is its use of caching techniques that significantly improve performance, as verified by our extensive experimental study on a real system.

References

1. Agrawal, P., Benjelloun, O., Sarma, A.D., Hayworth, C., Nabar, S., Sugihara, T., Widom, J.: Trio: A system for data, uncertainty, and lineage. In: Proc. of the VLDB conference (2006)
2. Benjelloun, O., Sarma, A.D., Halevy, A.Y., Widom, J.: ULDBs: Databases with uncertainty and lineage. In: Proc. of the VLDB conference, pp. 953–964 (2006)
3. Bhagwat, D., Chiticariu, L., Tan, W.-C., Vijayvargiya, G.: An annotation management system for relational databases. In: Proc. of the VLDB conference, pp. 900–911 (2004)
4. Buneman, P., Chapman, A., Cheney, J.: Provenance management in curated databases. In: Proc. of the ACM SIGMOD conference, pp. 539–550 (2006)

5. Buneman, P., Khanna, S., Tajima, K., Tan, W.-C.: Archiving scientific data. ACM Transaction Database Systems 29(1), 2–42 (2004)
6. Buneman, P., Khanna, S., Tan, W.-C.: Why and where: A characterization of data provenance. In: Van den Bussche, J., Vianu, V. (eds.) ICDT 2001. LNCS, vol. 1973, pp. 316–330. Springer, Heidelberg (2000)
7. Buneman, P., Khanna, S., Tan, W.-C.: On propagation of deletions and annotations through views. In: Proc. of the PODS conference (2002)
8. Chiticariu, L., Tan, W.-C.: Debugging schema mappings with routes. In: Proc. of the VLDB conference, pp. 79–90 (2006)
9. Chiticariu, L., Tan, W.-C., Vijayvargiya, G.: DBNotes: a post-it system for relational databases based on provenance. In: Proc. of the ACM SIGMOD conference (2005)
10. Cong, G., Fan, W., Geerts, F.: Annotation propagation revisited for key preserving views. In: Proc. of the CIKM, pp. 632–641 (2006)
11. Cui, Y., Widom, J.: Practical lineage tracing in data warehouses. In: Proc. of the ICDE, pp. 367–378 (2000)
12. Cui, Y., Widom, J.: Lineage tracing for general data warehouse transformations. The VLDB Journal, pp. 471–480 (2001)
13. Eltabakh, M.Y., Ouzzani, M., Aref, W.G.: bdbms – a database management system for biological data. In: Proc. of the CIDR (January 2007)
14. Geerts, F., Kementsietsidis, A., Milano, D.: Mondrian: Annotating and querying databases through colors and blocks. In: Proc. of the ICDE, p. 82 (2006)
15. Jagadish, H.V., Olken, F.: Database management for life sciences research. The SIGMOD Record 33(2), 15–20 (2004)
16. Koutrika, G., Ioannidis, Y.: Personalization of queries in database systems. In: Proc. of the ICDE, p. 597 (2004)
17. Labrinidis, A., Qu, H., Xu, J.: Quality contracts for real-time enterprises. In: Bussler, C.J., Castellanos, M., Dayal, U., Navathe, S. (eds.) BIRTE 2006. LNCS, vol. 4365, pp. 143–156. Springer, Heidelberg (2007)
18. Lauzac, S.W., Chrysanthis, P.K.: Personalizing information gathering for mobile database clients. In: Proc. of the ACM SAC, March 2002, pp. 49–56 (2002)
19. Melnik, S., Adya, A., Bernstein, P.A.: Compiling mappings to bridge applications and databases. In: Proc. of the ACM SIGMOD conference, pp. 461–472. ACM, New York (2007)
20. Qu, H., Labrinidis, A., Mosse, D.: Unit: User-centric transaction management in web-database systems. In: Proc. of the ICDE, April 2006, pp. 1–10 (2006)
21. Srivastava, D., Velegrakis, Y.: Intensional associations between data and metadata. In: Proc. of the ACM SIGMOD conference, pp. 401–412 (2007)
22. Tan, W.-C.: Containment of relational queries with annotation propagation. In: Proc. of the DBPL conference (2003)
23. Tan, W.-C.: Provenance in databases: Past, current, and future. Special Issue on Data Provenance, Bulletin of the Technical Commmittee on Data Engineering 32(4) (December 2007)

Ontology Database: A New Method for Semantic Modeling and an Application to Brainwave Data

Paea LePendu[1], Dejing Dou[1], Gwen A. Frishkoff[2], and Jiawei Rong[1]

[1] Computer and Information Science
University of Oregon, USA
{paea,dou,jrong}@cs.uoregon.edu
[2] Learning Research and Development Center
University of Pittsburgh, USA
gwenf@pitt.edu

Abstract. We propose an automatic method for modeling a relational database that uses SQL triggers and foreign-keys to efficiently answer positive semantic queries about ground instances for a Semantic Web ontology. In contrast with existing knowledge-based approaches, we expend additional space in the database to reduce reasoning at query time. This implementation significantly improves query response time by allowing the system to disregard integrity constraints and other kinds of inferences at run-time. The surprising result of our approach is that load-time appears unaffected, even for medium-sized ontologies. We applied our methodology to the study of brain electroencephalographic (EEG and ERP) data. This case study demonstrates how our methodology can be used to proactively drive the design, storage and exchange of knowledge based on EEG/ERP ontologies.

1 Introduction

With recent advances in data modeling and increased use of the Semantic Web, scientific communities are increasingly looking to ontologies to support web-based management and exchange of scientific data. Ontologies can be used to formally specify concepts and relationships between concepts within a domain. The resulting logic-based representations form a conceptual model that can help with storage, management and sharing of data among different research groups.

In addition to the representation of classes and properties, ontologies can store intensional knowledge in the form of general facts, often called rules, axioms or formulae, such as, "All Sisters are Siblings." Extensional data include specific facts, or ground terms, such as, "Mary and Jane are Sisters." Relational databases can effectively store and retrieve extensional data, but they lack obvious mechanisms to perform the inferences necessary to answer extensional queries over intensional data, as in, "Which individuals are Siblings?" Unlike a typical relational database, a knowledge base can support the deduction that Mary and Jane are siblings by using an inference engine.

B. Ludäscher and Nikos Mamoulis (Eds.): SSDBM 2008, LNCS 5069, pp. 313–330, 2008.

Intensional knowledge reduces the need to store large amounts of extensional data. For example, we do not need to store the fact, "Mary and Jane are Siblings," to know that it is true. The trade-off, however, is that inferences are required at run-time to generate this fact. What we have, therefore, is an example of the classical trade-off between time and space: the more extensional data we store, the less time it will take to answer queries about them. In this paper, we challenge traditional approaches for modeling knowledge-based or deductive database systems of this sort, which typically aim to find a balance between space and time requirements. Instead we propose that space is expendable and a great deal of inference (time) can be saved through the use of triggers and foreign-keys to forward-propagate inferences at load-time. Interestingly, when we compared our methods against existing benchmarks, we found we significantly improved query performance as expected, but load-time was remarkably unaffected.

In addition to these performance gains, we demonstrate that semantics can play an essential role in data management and query answering. In fact, both ontologies and database systems are important, leading us to propose a new methodology for database design, which we will call *ontology databases*.

To illustrate this idea, we describe the application of our methodology to brain electroencephalographic (EEG and ERP) data. In this application, we describe a database design that is ontology-driven. Moreover, we demonstrate how queries can be posed by domain experts at the ontology-level rather than using SQL directly. Database projects like ZFIN [8] and MGI [1], housing large central repositories for zebrafish and mouse genetic data, respectively, were later reinforced by the Gene Ontology [25] to help normalize knowledge across these kinds of repositories. By contrast, our Neural ElectroMagnetic Ontology (NEMO) project uses expert knowledge in the form of EEG/ERP ontologies to drive the data modeling and information storage and retrieval process.

The paper is organized as follows. We begin with related work (Section 2), followed by a description of our ontology-based modeling methodology and a performance analysis (Section 3). We then present a case study in which we applied our methodology to develop ontology databases for EEG/ERP query answering (Section 4). We conclude with a discussion and an outline of future work in Section 5.

2   Related Work

Ontologies can be regarded as a conceptual or semantic model for database design. Hull and King [19] provide a nice summary of semantic models of all kinds: Entity-Relational, Object-Oriented, Ontological and so on. While the notions in their survey make clear that there are firm connections between models, database implementations, and logics, we have been interested in exploring the question, "What is a semantic data model?" In particular, we wish to explore it from an ontology-based perspective that addresses practical issues in collaborative scientific research, especially, biomedical research. Increasingly, biomedical researchers are looking to develop ontologies to support cross-laboratory data sharing and integration. These ontologies can be found at ontology repositories around the world [34]. For example, more than 62 biomedical ontologies can be found at the National Center for Biomedical Ontology (NCBO) [6].

Pan and Heflin proposed a similar approach, which they call description logic databases (DLDB) [26]. DLDB is a storage and reasoning support mechanism for knowledge base facts (RDF triples), which has been compared to well-known systems such as Sesame [10]. Although we structure the database relations in a way that is similar to DLDB (i.e., unary and binary predicates become unary or binary relations), our implementation using triggers and foreign keys to support reasoning, as opposed to SQL views, allows for a significant performance gain by trading space for time by eagerly forward-propagating data at load-time. In this context, it is informative to consider the recent work by Paton and Díaz [27], which examines rules and triggers in active database systems.

Recent research on bridging the gap between OWL and relational databases by Motik, Horrocks and Sattler [24] provides unique insight into the expressiveness of description logics versus relational databases. The integrity constraints in databases can be described with extended OWL statements (axioms). An important contribution of this research is to show that the constraints can be disregarded while answering positive queries, if the constraints are satisfied by the database.

The idea of balancing space and time when we couple databases and reasoning mechanisms comes from seminal works by Reiter [28,30]. Reiter proposed a system that uses conventional databases for handling ground instances, and a deductive counterpart for general formulae. Since no reasoning is performed on ground terms, Reiter argues convincingly that in such a system queries can be answered efficiently while retaining correctness. OntoGrate [13] is precisely such a system for semantic query translation using ontologies. The key question that motivated our trigger-based approach was, "Since disk-space is rarely an issue these days, what would happen if we use even more space?"

The neuroscience community is a recognized leader in the development of biomedical ontologies. For example, the Human Brain Project has supported the development of a common data model and meta-description language [17] for neuroscience data exchange and interoperability. BrainMap [22] has designed a Talaraich coordinate-based archive for sharing and meta-analysis of brain mapping studies and literature, as well as a sharable schema for expression of cognitive-behavioral and experiment concepts. The fBIRN project [20] has pioneered several areas for neuroscience data sharing, including distributed storage resources and taxonomies of neuroscience terms (called BIRNlex). Our project will build on this prior work and extend it to incorporate ontology-based methods for reasoning. In addition to incorporating cognitive-behavioral and anatomy concepts represented in BrainMap and in fBIRN, NEMO will develop ontologies for temporal, spatial, and spectral concepts that are used to describe EEG and ERP patterns. In line with OBO "best practices," we will reuse ontology concepts from relevant domains. In fact, we are collaborating directly with ontology engineers and domain experts in the fMRI, as well as the EEG and ERP, communities.

The NEMO project brings some distinctive methods to bare on the problem of data sharing. Whereas most prior work on data sharing in the neurosciences has focused on the development of simple taxonomies or relational databases, NEMO uses ontologies to design databases that can support semantically based queries. What this means is that NEMO databases can be used to answer more complex queries, which cannot be handled by traditional (purely syntactic) database structures. For example, the popular Gene Ontology (GO) [25] provides a standard vocabulary and concept model for molecular functions, biological processes and cellular components in genetic research. The OWL [7] specification of GO is over 40 Megabytes in size [25] and terabytes of research data stored in model organism databases around the world such as ZFIN [8] and MGI [1] are all being marked-up according to the GO ontology. The NEMO working group is borrowing from this idea and taking it a step further [12,15]. More than a standard vocabulary of terms, the ontologies NEMO is developing will capture knowledge ranging from the experimental methods used to gather ERP data down to instrument calibration settings so that results can be shared and interpreted semantically during large-scale meta-analysis across laboratories.

3   Ontology-Based Data Modeling

We first present a new and general methodology, which takes a Semantic Web ontology as input and outputs a relational database schema. We call such a database an "ontology database," which is an ontology-based, semantic database model. As we will show in Section 4, after we load ERP data into the NEMO ontology database, we can answer queries based on the ontology while automatically accounting for subsumption hierarchies and other logical structures within each set of data. In other words, the database system is ontology-driven, completely hiding underlying data storage and retrieval details from domain experts, whose only interaction-interface happens at the ontology (conceptual) level.

3.1   The Procedural Extension

Although Description Logics (DL) [9] provide the formal logical foundation for OWL and Semantic Web ontologies, we do not require the full expressiveness of this logic for data modeling purposes in most scenarios we have encountered. It suffices to use rules of the form (reads "if C then D"):

$$C \Rightarrow D,$$

which exclude the analysis-by-cases and contrapositive reasoning provided by full DL inclusion axioms of the form (reads "C is subsumed by D"):

$$C \sqsubseteq D.$$

What this means is that we are drawing a line between databases and knowledge bases. For example, while it may be taken for granted in a knowledge-based system that, "X is either a Rock or it is not a Rock, no matter what X is," a database has no such reasoning capability. It can only say which is actually the case. As such, we technically only allow epistemic inclusion axioms with the **K** operator [9] which stands for "know" in the following rule (reads "Only when we *know* that C is true can we conclude D"):

$$\mathbf{K}C \sqsubseteq D.$$

The difference is evidenced by the fact that we can immediately conclude D (without any positive or negative witnesses of C) in:

$$(C \sqcup \neg C) \sqsubseteq D,$$

but not necessarily in:

$$(\mathbf{K}C \sqcup \mathbf{K}\neg C) \sqsubseteq D.$$

This restriction makes knowledge maintenance (reasoning) much easier: all we need to calculate is the *procedural extension* of a given set of facts and rules [9]. This can easily be done using database triggers and foreign keys with cascading deletes, the basic idea of which we outline below.

3.2   Triggers

Triggers are used for each rule to propagate data in a forward-chaining manner as facts are loaded into the ontology database. For example, suppose we have the following first-order rule (reads "all Sisters are Siblings"):

$$\forall x, y : \; Sisters(x, y) \rightarrow Siblings(x, y).$$

Whenever a new pair of sisters is inserted into the ontology database, such as $Sisters(Mary, Jane)$, a trigger fires, eagerly inserting $Siblings(Mary, Jane)$ as well. This process is depicted in Figure 1.

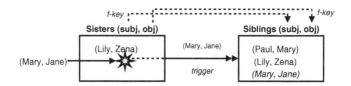

Fig. 1. This figure shows that upon asserting $Sisters(Mary, Jane)$ which means inserting $(Mary, Jane)$ into the Sisters-property table, the trigger causes $(Mary, Jane)$ to first be inserted into the Siblings-property table. Triggers generate knowledge in a forward-chaining manner for the Sisters-Siblings rule, $\forall x, y : \; Sisters(x, y) \rightarrow Siblings(x, y)$. Implicitly understood in this sub-property rule is also the contrapositive, $\forall x, y : \; \neg Siblings(x, y) \rightarrow \neg Sisters(x, y)$, an integrity check that foreign-keys can enforce, shown here as the dotted line.

Although the above is an example of a sub-property (Sisters is a sub-property of Siblings), triggers can be used for both sub-class and sub-property hierarchies. Each trigger is a straightforward encoding of the epistemic rule, in SQL:

```
CREATE TRIGGER subPropertyOf-Sisters-Siblings SUCH THAT
UPON DETECTING EVENT INSERT (x,y) INTO Sisters(subject,object)
FIRST EXECUTE INSERT (x,y) INTO Siblings(subject,object)
```

3.3   Foreign Keys with Cascading Delete

Foreign keys are used to check integrity constraints as usual, but by using the "on delete cascade" option, they also propagate deletions whenever facts are negated (which is not uncommon in scientific domains). For example, in the Sisters-Siblings sub-property rule of Figure 1 it is understood implicitly that if two people are *not* Siblings, then they cannot be Sisters either:

$$\forall x, y : \ \neg Siblings(x,y) \rightarrow \neg Sisters(x,y).$$

Semantically, we interpret the contrapositive to mean two things. First of all, it is an integrity constraint: if $Siblings(Mary, Jane)$ is not true, then it cannot be the case that $Sisters(Mary, Jane)$ is true, so an integrity check is performed to validate that $Siblings(Mary, Jane)$ is true before inserting $Sisters(Mary, Jane)$. Of course, care must be taken to ensure triggers and integrity checks happen in the correct order (note the "FIRST" keyword in the SQL trigger). Secondly, if deletions (negations) are performed, they must be propagated to ensure consistency is maintained, thus explaining the "on delete cascade" option. Indeed, this is the pattern for all sub-class and sub-property rules: they are both triggers (knowledge generating) and integrity constraints (knowledge checking), consistent with the semantics of inclusion axioms.

Integrity constraints also occur in domain and range restrictions on properties. In this case, we have foreign keys but no triggers. For example, when we assert $Sisters(x,y)$ we generally presume that x and y are People. That is, we mean:

$$\forall x, y : \ [\neg Person(x) \cup \neg Person(y)] \rightarrow \neg Sisters(x,y),$$

but not necessarily:

$$\forall x, y : \ Sisters(x,y) \rightarrow [Person(x) \cap Person(y)].$$

In other words, given the statement $Sisters(Mary, buddyTheFrog)$, we do not intend to automatically conclude that $buddyTheFrog$ is a Person but rather hope the assertion is rejected unless we know for sure that $buddyTheFrog$ is a Person (and not a Frog). This kind of reasoning is due in large part to the notion common in database systems that any fact not known to be true is presumed false, known as the *closed world assumption* [29].

Table 1. The *ontology database* methodology is summarized in this table. Here, respectively, *subj* and *obj* refer to the subject and object of a property, *MinCard* and *MaxCard* refer to cardinality, and *f-key* and *p-key* stand for foreign key (with an "on delete cascade" option) and primary key.

Logical Feature	FOL Formalism	Ontology DB Implementation
Structure		
$Class(A), Class(B)$ $Property(P)$	$A(x), B(y)$ $P(x, y)$	relation: $A(id), B(id)$ relation: $P(subj, obj)$
Restrictions		
$Domain(P, A)$ $Range(P, B)$	$\forall x, y : P(x, y) \rightarrow A(x)$ $\forall x, y : P(x, y) \rightarrow B(y)$	f-key: $P(subj)$ ref $A(id)$ f-key: $P(obj)$ ref $B(id)$
$MaxCard(P, 1)$	$\forall x, y, z : P(x, y) \wedge P(x, z)$ $\rightarrow y = z$	p-key: $P(subj)$
$MinCard(P, A, 1)$	$Domain(P, A)$ $\rightarrow (\forall x : A(x)$ $\rightarrow \exists y : P(x, y))$	f-key $P(subj)$ ref $A(id)$; trigger: on insert on A(id) insert ignore $P(id, null)$
Subsumption		
$subClassOf(B, A)$	$\forall x : B(x) \rightarrow A(x)$	trigger: **before** insert on B(id) insert ignore $A(id)$; f-key: $B(id)$ ref $A(id)$;
$subPropertyOf(Q, P)$	$\forall x, y : Q(x, y) \rightarrow P(x, y)$	trigger: **before** insert on Q(subj,obj) insert ignore $P(subj, obj)$; f-key: $Q(subj, obj)$ ref $P(subj, obj)$;
	Horn Rules & GMP	
	$\forall x_1, x_2 \dots x_m :$ $P_1(x_1, x_2) \wedge \dots$ $\wedge P_n(x_{m-1}, x_m) \rightarrow Q(x_i, x_j)$ $(1 \leq i, h \leq m, 1 \leq j, h \leq m)$	$\forall k \in [1..n]$ trigger(rule-premise-k): on insert on $P_k(x_{h-1}, x_h)$ update [rule-premise-table with P_k] trigger(rule-activate): on update on [rule-premise-table] if [all premises satisfied] then insert ignore $Q(x_i, x_j)$ $(1 \leq i, h \leq m, 1 \leq j, h \leq m)$

3.4   Modeling Summary

Table 1 summarizes the main logical features we implement in the ontology database methodology. These features can be categorized according to structures, restrictions and subsumptions which come from OWL, RDF [3] and general first-order logic. The database relational structure we have chosen (unary and binary predicates become unary and binary relations) is almost identical to the hybrid approach of DLDB [26], which combines approaches from prior works to effectively store RDF triples.

3.5   Logical Justification

Our ontologies are generally restricted to Horn Normal Form (HNF) [32], which is a disjunction with only one positive literal as in:

$$\neg p_1 \vee \neg p_2 \vee \ldots \vee \neg p_n \vee q.$$

These formulae can be written as implications without disjunctions on the right-hand side, like Datalog [33] rules, which we call implicative normal form (INF):

$$p_1 \wedge p_2 \wedge \ldots \wedge p_n \rightarrow q.$$

Generalized Modus Ponens (GMP) [32] is an inference rule based on the well-known modus ponens rule:

$$\frac{p_1' \wedge p_2' \wedge \ldots \wedge p_n' \quad p_1 \wedge p_2 \wedge \ldots \wedge p_n \rightarrow q}{SUBST(\theta, q)} \ GMP$$

GMP allows us to unify several antecedents simultaneously to prove a conclusion. It is well-known that GMP is sound and complete for knowledge bases in HNF (and therefore INF) [32]. A trigger is essentially a forward-chaining implementation of GMP, recursively calling other triggers as necessary. Because all definitions are acyclic, the procedure is guaranteed to terminate. Foreign-keys and null-valued triggers together provide the machinery for solemnization under existential constraints (such as, "All Employees have an SSN." [31]). According to this method, an ontology database therefore produces and maintains the procedural extension, guaranteeing that the database is a Herbrand Model for the given set of facts (see [32] for details on the Herbrand universe, interpretation and model).

3.6   General Performance Analysis

We tested our methodology using the Lehigh University Benchmark (LUBM) [18] ontology[1], and compared the load-time (see Figure 2) and query-answering (see Figure 3) performance against DLDB [26], an ontology data storage model not unlike our own.

The LUBM features an ontology for the university domain (e.g., faculty, courses, departments, etc.) together with a data generation tool for creating OWL datasets of arbitrary size and a set of queries for evaluating performance. The most significant difference between DLDB and our ontology database (OntoDB) is that DLDB uses SQL views instead of triggers to propagate subsumptions. In other words, our approach is like an *eager* evaluation strategy for subsumption inferences whereas DLDB is *lazy*. Because we propagate knowledge as data is loaded so as to increase query performance, we expected to incur a load-time hit. To our surprise, the load-time was largely unaffected even though query-time benefitted significantly. Our only explanation of this phenomenon is that the underlying database file system is optimized to perform several insertions (caused by triggers) in relatively constant-time – which might eventually be affected as the depth of the subsumption hierarchy grows. Naturally, our approach uses more disk-space (roughly 3-times the space), a trade-off we knew we had to make (space versus time has to give) [30]. Again, our results are summarized in Figures 2 and 3.

---

[1] All experiments were performed on an unremarkable personal laptop computer with a 1.8Ghz Centrino processor and 1Gb of RAM running MySQL 5.0 as the RDBMS.

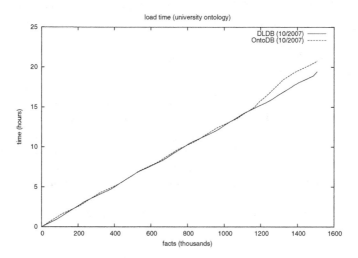

Fig. 2. The load-time results for the Lehigh University benchmark ontology data show that the load-time of our ontology database approach (OntoDB) is comparable to that of DLDB. The blips at around 1.2M and 1.4M are probably due to disk resizing or other background effects.

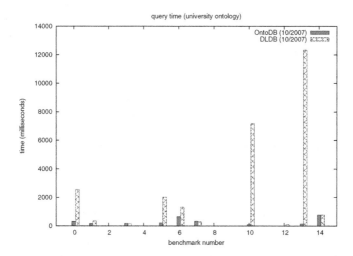

Fig. 3. The query-answering time results for the Lehigh University benchmark ontology data show that the query-time of our ontology database approach is often significantly faster than DLDB. Charted here are the running-times for 10 of the 14 benchmark queries published by Lehigh University. Queries 2, 4, 8, and 9 are not shown here due to scale, having extremely long running-times.

4   Case Study: Application of Ontology Databases to Brainwave Data

4.1   EEG and ERP Data Sharing in Clinical and Cognitive Neuroscience

The problem of data sharing in brain electromagnetic research, like that in other scientific fields, is challenged by data scale, multivariate parameterizations, and dimensionality. Neuroinformatics must address these challenges with robust data management and integration techniques. To this end, neuroinformatics researchers have developed a number of database and XML-based methods [21], providing effective solutions for annotation and storage of complex, large-scale datasets. Going beyond syntax and structure, the hard problems that remain are closely linked to the neuroscience community's requirements for rich semantic representation and integration of patterns across disparate experiment and laboratory procedures and paradigms.

The development of ontologies may be central to addressing these problems. Indeed, adoption of ontologies has already enabled major scientific progress in biomedical research [6,20,23,25] and is a rapidly growing area in bioinformatics and neuroinformatics research. The present work aims to extend and combine Semantic Web and database modeling technologies to address issues in ERP data representation and semantic query answering. The project is called "Neural ElectroMagnetic Ontology" (NEMO). Eventually, we hope that our ontology-based framework will support large-scale semantic data sharing, give rise to meta-analysis, and lead to major advances in brain functional mapping using ERP and related methods.

Electroencephalographic (EEG) data consist of changes in neuroelectrical activity measured over time (on a millisecond timescale), across two or more locations, using noninvasive sensors ("electrodes") that are placed on the scalp surface. A standard technique for analysis of EEG data involves averaging across segments of data ("trials"), time-locking to stimulus "events," to create *event-related brain potentials (ERPs)*. The resulting measures are characterized by a sequence of positive and negative deflections across time, at each sensor. For example, to examine brain activity related to language processing, the EEG may be recorded during presentation of words versus non-words, using 128 or more sensors (Figure 4). Averaging across trials within a given stimulus category accentuates brain activity that is related to processing the specific type of stimulus. In principle, activity that is not event-related will tend towards zero as the number of averaged trials increases. In this way, ERPs provide increased *signal-to-noise (SNR)*, and thus increased sensitivity to functional (e.g., task) manipulations.

The resulting datasets comprise rich sets of spatial, temporal, and functional (task-related) measurements. This case study describes the ontology that has been developed by domain experts and refined by data mining techniques to capture this knowledge. Furthermore, we demonstrate how the ontology database methodology can be used to automatically implement an effective storage and

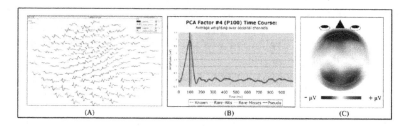

Fig. 4. (A) 128-channel EEG waveplot; positive voltage plotted up; responses to words versus non-words. (B) Time course of P100 factor for same dataset, extracted using Principal Components Analysis. (C) Topography of P100 factor (negative on top and positive at bottom). See [15] for details.

retrieval mechanism for ERP data that preserves the meaning and interpretation prescribed by domain experts.

4.2 ERP Ontology Development

In previous work, an ERP ontology for a limited domain (word recognition) was designed collaboratively with domain experts, using data collected in a series of visual word recognition experiments (see [12,16] for details). To support the development of an initial ERP ontology, based on automated data analysis and labeling, we applied data decomposition methods to help separate signal (brain activity) from noise (noncerebral artifacts) and to disentangle overlapping patterns [16]. More specifically, temporal Principal Components Analysis (PCA) was applied to ERP data consisting of 128 electrodes, 275 timepoints (sampling rate, 250Hz), 34 human subjects, and 4 experimental conditions (see [11] for details on PCA methods).

For each PCA factor, we extracted summary metrics representing spatial, temporal and functional dimensions of the ERP patterns of interest. Thus, the data represent the individual PCA factors, weighted across individual subjects and experiment conditions. These data were post-processed by ERP domain experts and represented as points in a 25 dimensional attribute space. In previous work, we characterized eight types of robust patterns, P100, N100, N2, N3, MFN, P1r/P2, N4 and P300 [16]. Rules for each pattern were based on results from prior literature. For example, the P100 rule was operationalized as follows:

$$\forall x, i, j : [\; PCA\_Factor(x) \land (80 < i) \land ti\_max(x, i) \land (i < 150) \land$$
$$factor Event(x, STIMON) \land factor Modality(x, VISUAL) \land$$
$$in\_mean\_roi(x, j) \land (0 < j) \land roi(P100v, OCC) \;]$$
$$\rightarrow occursIn(P100v, x)$$

where "ti_max" is the peak latency, "in_mean_roi" is the mean amplitude over a given region-of-interest (ROI), and ROI for "P100v" is specified as "occipital." In addition to "top-down" (expert-defined) pattern rules, we performed "bottom-up" (data mining) analysis using clustering-based classification to

discover class and property hierarchies and association rule mining to find axioms as a way to complement and refine the concepts and rules articulated by
domain experts [12,16]. Evaluation was performed against a "Gold Standard"
labeled dataset described in [16].

Our initial ERP ontology consists of classes, class taxonomy, properties and
their relationships. The ontology consists of roughly 29 classes, 40 properties,
27 sub-class relationships, and 3 super-properties. We show a partial view of
the ERP ontology in Figure 5. We would like to stress that this ontology has
undergone significant changes since the time of this writing. The latest version
of the NEMO ERP ontology is available online at http://aimlab.cs.uoregon.
edu/NEMO/NEMO_ERP.owl[2] and will soon be available on NCBO. Figure 5 shows

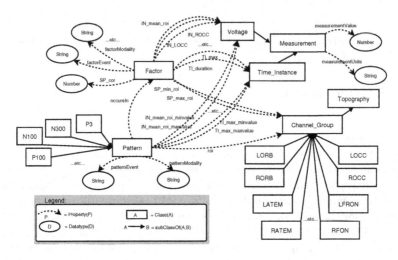

Fig. 5. A partial view of the ERP ontology

five basic classes, i.e., factor, pattern, channel group, topography and measurement. Factor objects have temporal, spatial and functional attributes (part of
which are listed in the graph, such as factorEvent, SP_cor and factorModality)
which are represented as properties of the factor class in the ERP ontology. TI-
max and IN-mean(ROI) are properties of factor which relates the measurements
(e.g., time-instance and voltage) which have both unit and value properties. The
pattern class has 8 sub-classes (P100, N100, etc.) which correspond to the 8
ERP patterns defined by domain experts [16,12]. The properties of the pattern
class are those used in expert rules or rules discovered by data mining. The
expert rules are represented as Horn rules whose body are conjunctions of predicates. The relationship between factor and pattern can be modeled using the
"occursIn" property. Each pattern has a region of interest, which is a channel
group belonging to the topography class. Each area on the scalp can be divided
into a left and right part. For instance, left occipital (LOCC) and right occipital

---

[2] Human readable OWLDoc: http://aimlab.cs.uoregon.edu/NEMO/OWLdoc_ERP

(ROCC) are sub-classes of channel group and the combination of them is called occipital (OCC) (not shown). The mean intensity (measured in microvolts) for each region of interest is calculated based on this relationship.

While the graph representation helps convey the general idea, we use a formal, first-order ontology language to represent the ontology internally. This internal language is (and has been) easily translated to and from standard ontology languages such as OWL [7] or OBO [2] for terminological knowledge and SWRL [5] for general Horn rules. We plan to contribute our ERP ontology to the National Center for Biomedical Ontology [6].

4.3 ERP Data Modeling Results

We applied our modeling methodology to the ERP ontology depicted in Figure 5 to investigate several properties: correctness, space, load-time, and query answering speed.

We worked with a visual word study data set in which there were 34 different human subjects, 25 different dimensions in the attribute space and a vector of 1152 different component factors after PCA decomposition. In essence, we were working with a relatively small matrix of data that was approximately 1152 rows by 30 columns in size.

For every class in the ERP ontology, we define a unary relation and for every property a binary relation. For every logical rule in the ontology specification, we generated the corresponding foreign keys, triggers, and primary keys in the database. Finally, for every data instance, we generated a unique internal object identifier. Altogether, the data essentially consists of 100,425 individual facts.

It took approximately 14 seconds to generate the database schema based on the ontology and load it into the MySQL RDBMS. It took 1.3 hours to load all of the individual facts. The entire ERP ontology database occupies roughly 10 MB of disk space, and contains over 145,000 facts (including new ones after all triggers). There are 29 tables for class concepts and 40 tables for properties. The class hierarchy has a depth of at most 5. The top-class in the hierarchy has 23,093 instances whereas the average-sized class has 1,152 instances. The ontology database generated 27 different triggers and 95 foreign-key constraints to maintain the procedural extension.

Figure 6 shows a visual representation of the entity-relation (ER) diagram for the ERP ontology database. Although too large and complex to show every detail in this paper, the diagram gives a rough idea of how many concepts (boxes) and dependencies (lines) are managed by the database (triggers are not shown).

As for query processing, we tested four different queries that exhibited the various properties of interest for our implementation: subsumption, data size (amount of data, joins, etc.), aggregate computations, and ease of formulation based on ontology concepts. A summary of the queries and properties they are meant to explore is shown in Table 2. Although we hoped to find at least some interesting and significant variations in query speed or formulation difficulty (according to domain experts), this was not the case. Each query proved extremely straightforward to formulate in SQL, and execution time was statistically unmea-

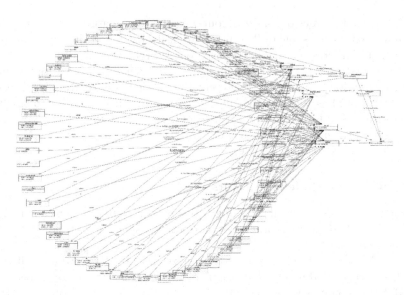

Fig. 6. Although too difficult to read in printed form, this visualization of the ER Diagram for the ERP ontology database gives a general sense of number and complexity of the concepts (boxes) and foreign-key relationships (lines). Central concepts such as "pattern," "factor," and "channel group" are toward the right-side of the image – they are the most densely connected nodes.

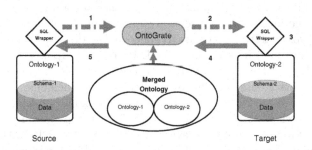

Fig. 7. A data integration scenario in which the user (1) issues a query using the semantics of the source ontology which (2) gets translated into the semantics of the target ontology using the inference engine in OntoGrate which is then (3) issued as a database query using a SQL syntax wrapper from which (4) target data is returned and finally (5) translated back into the source semantics for the user to interpret.

surable (somewhere between 0-40 ms) on our equipment. All answers returned by the database were 100% complete and sound (perfect recall and precision) as compared to answers expected by our domain experts. The answers and execution times for each query are also shown in Table 2. We would like to note that, although not the focus of this paper, the ontology database approach we describe adds the unique advantage that queries can be posed at the ontology-level

Table 2. This table lists the queries and answers verified by experts. Each query is meant to test various properties of interest.

Query / Answer	Property of Interest
(1) *Show the region of interest for all ERP patterns that occur between 0 and 300ms.* ``` Pattern ROI max_value min_value == N100 LOCC 229 151 N100 ROCC 229 151 N2 LPTEM 300 230 N2 LOCC 300 230 P100 ROCC 150 60 P100 LOCC 150 60 [Fetch MetaData: 0/ms] [Fetch Data: 10/ms] [Execution: 0/ms] ```	subsumption, data size
(2) *Which PCA factor do P100 patterns most often appear in?* ``` Pattern occurances Factor_Number =================================== P100 133 4 [Fetch MetaData: 0/ms] [Fetch Data: 0/ms] [Execution: 20/ms] ```	subsumption, aggregation, data size
(3) *What is the range of intensity mean for the region of interest for N100 patterns?* ``` Pattern in_mean_roi_min in_mean_roi_max == N100 -infinity -0.4 [Fetch MetaData: 0/ms] [Fetch Data: 0/ms] [Execution: 10/ms] ```	ease of formulation
(4) *Show the patterns whose region of interest is left occipital and occurs between 220 and 300ms.* ``` Pattern ROI max_value min_value ================================== N2 LOCC 300 230 [Fetch MetaData: 0/ms] [Fetch Data: 0/ms] [Execution: 10/ms] ```	subsumption, aggregation

by domain experts using languages such as SPARQL [4] or OWL-QL [14] and automatically translated to SQL using wrappers (see Figure 7).

5   Discussion and Future Work

In this paper, we have outlined a new framework for designing and implementing ontology databases. We have further presented a case study in which we applied our method to ERP data. This ontology-driven data modeling approach appears promising, working well for: (1) scientific application scenarios requiring rich semantics, and (2) "query-mostly" scenarios common to such domains in which large sizes of data must be queried and analyzed significantly more often than data are loaded. We have also argued that the ontology database methodology using triggers and foreign-keys is logically justified: that it correctly generates and maintains a logical model for a given ontology and set of data.

In terms of scalability, the LUBM is fairly complex but medium in size. GO, on the other hand, has over 36,000 different concepts arranged in a hierarchy roughly having depth 14. Although large in size, GO is mostly a class hierarchy with only one property ("part-of"). The main limiting factor for our approach will be the number of tables and triggers a database system can realistically support. MySQL, for example, is limited only by the number of files possible on the operating system. Unless other DBMSs have strict limitations, we do not see scalability to be a problem in general since ontologies do not typically grow to sizes on the order of millions of concepts. To be clear, we mean scalability in terms of the conceptual model, not the data instances which definitely pose scalability issues. We tested our system on a toy ontology up to size 40,000 and depth 20 and there was no visible difficulty. In future work, we will process GO itself and possibly incorporate data instances from ZFIN and MGI given our strong working relationship with those groups.

The next goal for the NEMO project is a comprehensive ontology-based modeling and integration system that will facilitate the representation and dissemination of ERP data across different EEG and ERP analysis methods, different experiment paradigms, and different laboratories. It is likely that the representation of EEG and ERP patterns that are associated with different analysis methods and different functional (experiment) paradigms will require multiple ontologies to be developed. Ontology-based integration in NEMO will study the mapping rules between these EEG and ERP ontologies. Given the mapping rules between different ontologies, once the user query comes in, various ERP databases with different ontologies can be searched for answers to the query. We reported an efficient ontology-based data integration system called *OntoGrate* that addresses this problem using an inference engine [13]. In general, we anticipate that this research can be generalized for integrating other types of neuroscience data (e.g., event-related fields (ERF) and functional magnetic resonance imaging (fMRI) data) and can support other biomedical ontology-based data sharing efforts (e.g., GO) in the future. Figure 7 highlights the main idea behind the query answering scenario under this model of integration.

Acknowledgements

We thank the other members in the NEMO working project group, and in particular Robert Frank, Allen Malony and Don Tucker, for their collaboration on related work. We also thank Jeff Z. Pan and Zena M. Ariola for valuable discussions on theoretical aspects of this work.

References

1. MGI: Mouse Genome Informatics, `http://www.informatics.jax.org/`
2. Open Biomedical Ontologies (OBO),
   `http://www.geneontology.org/GO.format.obo-1_2.shtml`
3. Resource Description Framework, `http://www.w3.org/RDF/`
4. SPARQL Query Language for RDF, `http://www.w3.org/TR/rdf-sparql-query/`
5. SWRL: A Semantic Web Rule Language Combining OWL and RuleML,
   `http://www.w3.org/Submission/SWRL/`
6. The National Center for Biomedical Ontology, `http://www.bioontology.org/`
7. Web Ontology Language (OWL), `http://www.w3.org/TR/owl-ref/`
8. ZFIN: The Zebrafish Information Network, `http://www.zfin.org`
9. Baader, F., Nutt, W.: Basic description logics. In: Description Logic Handbook, pp. 43–95 (2003)
10. Broekstra, J., Kampman, A., van Harmelen, F.: Sesame: A generic architecture for storing and querying rdf and rdf schema. In: International Semantic Web Conference, pp. 54–68 (2002)
11. Dien, J.: Addressing misallocation of variance in principal components analysis of event-related potentials. Brain Topography 11(1), 43–55 (1998)
12. Dou, D., Frishkoff, G., Rong, J., Frank, R., Malony, A., Tucker, D.: Development of NeuroElectroMagnetic Ontologies (NEMO): A Framework for Mining Brain-wave Ontologies. In: Proceedings of the 13th ACM International Conference on Knowledge Discovery and Data Mining (KDD), pp. 270–279 (2007)
13. Dou, D., LePendu, P.: Ontology-based integration for relational databases. In: ACM Symposium on Applied Computing (SAC), pp. 461–466 (2006)
14. Fikes, R., Hayes, P.J., Horrocks, I.: Owl-ql - a language for deductive query answering on the semantic web. J. Web Sem. 2(1), 19–29 (2004)
15. Frishkoff, G.A.: Hemispheric differences in strong versus weak semantic priming: Evidence from event-related brain potentials. Brain Lang. 100(1) (2007)
16. Frishkoff, G.A., Frank, R.M., Rong, J., Dou, D., Dien, J., Halderman, L.K.: A Framework to Support Automated Classification and Labeling of Brain Electromagnetic Patterns. In: Computational Intelligence and Neuroscience (CIN), Special Issue, EEG/MEG Analysis and Signal Processing (2007)
17. Gardner, D., Knuth, K.H., Abato, M., Erde, S.M., White, T., DeBellis, R.: Common data model for neuroscience data and data model exchange. J. Am. Med. Inform. Assoc. 8(1), 17–33 (2001)
18. Guo, Y., Pan, Z., Heflin, J.: Lubm: A benchmark for owl knowledge base systems. J. Web Sem. 3(2-3), 158–182 (2005)
19. Hull, R., King, R.: Semantic database modeling: survey, applications, and research issues. ACM Comput. Surv. 19(3), 201–260 (1987)
20. Keator, D.B., Gadde, S., Grethe, J.S., Taylor, D.V., Potkin, S.G.: A general xml schema and spm toolbox for storage of neuro-imaging results and anatomical labels. Neuroinformatics 4(2), 199–212 (2006)
21. Koslow, S.H., Subramaniam, S. (eds.): Databasing the Brain: From Data to Knowledge (Neuroinformatics). Wiley-Liss, Chichester (2005)
22. Laird, A.R., Lancaster, J.L., Fox, P.T.: Brainmap: The social evolution of a human brain mapping database. Neuroinformatics 3(1), 65–78 (2005)
23. Lindberg, D., Humphries, B., McCray, A.: The Unified Medical Language System. Methods of Information in Medicine 32(4), 281–291 (1993)

24. Motik, B., Horrocks, I., Sattler, U.: Bridging the gap between owl and relational databases. In: Proceedings of the 16th International Conference on World Wide Web (WWW), pp. 807–816 (2007)
25. G.Ontology Consortium. Creating the Gene Ontology Resource: Design and Implementation. Genome Research 11(8), 1425–1433 (2001)
26. Pan, Z., Heflin, J.: Dldb: Extending relational databases to support semantic web queries. In: Workshop on Practical and Scalable Semantic Systems (2003)
27. Paton, N.W., Díaz, O.: Active database systems. ACM Comput. Surv. 31(1), 63–103 (1999)
28. Reiter, R.: Deductive question-answering on relational data bases. In: Logic and Data Bases, pp. 149–177 (1977)
29. Reiter, R.: On closed world data bases. In: Logic and Data Bases, pp. 55–76 (1977)
30. Reiter, R.: On structuring a first order data base. In: Proceedings of the Canadian Society for Computational Studies of Intelligence, pp. 90–99 (1978)
31. Reiter, R.: What should a database know? J. Log. Program. 14(1&2), 127–153 (1992)
32. Russell, S., Norvig, P.: Artificial Intelligence: A Modern Approach, 2nd edn. Prentice-Hall, Englewood Cliffs (2003)
33. Ullman, J.D.: Principles of Database and Knowledge-Base Systems, vol. I. Computer Science Press (1988)
34. Yu, A.C.: Methods in biomedical ontology. J. of Biomedical Informatics 39(3), 252–266 (2006)

The hB-pi* Tree: An Optimized Comprehensive Access Method for Frequent-Update Multi-dimensional Point Data

Panfeng Zhou[1] and Betty Salzberg[2,*]

[1] Sybase, Dublin, CA 94568, USA
zhoup@sybase.com
[2] Northeastern University, Boston, MA 02115, USA
salzberg@ccs.neu.edu

Abstract. The R-tree [7] family is the most popular multi-dimensional index method. The R-tree, however, has overlaps among index entries and its
index page fanout decreases rapidly as data dimension increases. Furthermore, the R-tree has poor concurrency performance. For frequent-update
multi-dimensional point data sets, the hB-pi [5] tree is a better choice than
the R*-tree. But the hB-pi tree (and all other kd-tree based access methods) indexes the whole space no matter whether or not there is any data
in some sub-spaces. Indexing *empty space* (i.e., space without data inside)
leads to unnecessary data page accesses which increase with growing dimension. This paper addresses this problem by proposing the hB-pi* tree,
which efficiently indicates empty spaces and improves range query performances while preserving the hB-pi's high fan-out and good concurrency.
Our methods can be applied to any kd-tree based access methods, and our
claims are supported by extensive experimental evaluation.

Keywords: hB-pi* tree, Empty space, Multi-dimension access method.

1 Introduction

Indexing multi-dimensional point data sets has been extensively studied, and
numerous structures [6] have been developed. This paper focuses on indexing
low to medium (i.e., 2D to 8D) point data sets that undergo a significant load
of updates.

One of the most popular multi-dimensional access methods is the R-tree family, which can index both point data and non-point data. Spatially adjacent
objects are clustered into the same data page. Spatially adjacent data pages
are clustered into the same index page. This cluster process will repeat until
it reaches the root of the tree. Each entry in the data page is in the form of
<MBR, tid> where MBR is the minimal bounding box of the object O, and
the tid refers to the real position of O in the database. Each entry in the index
page is in the form of (MBR, ptr) where MBR is the minimal bounding box that

---

* This work was partially supported by the NSF under Grants IIS-0533625.

B. Ludäscher and Nikos Mamoulis (Eds.): SSDBM 2008, LNCS 5069, pp. 331–347, 2008.

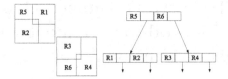

Fig. 1. Example of the R-tree

encloses all the MBRs in the child page (pointed to by ptr). Figure 1 illustrates a two level R-tree with the page capacity of 2. The R^*-tree [2], a variant of the R-tree [7], reduces the area, margin, and overlap of the index entries' MBRs, and it is robust against skewed data distributions. Furthermore, the R^*-tree introduces the concept of *Forced Reinsert*. That is, if an insertion causes a page P to overflow, a certain number of the entries in P (i.e., the entries whose centers are far away from the center of P) will be removed from P and reinserted into the tree. The *Forced Reinsert* can prevent the split, increase the storage utilization, and decrease the overlap and margins. The R^O-tree [14] further improves the performance of the R*-tree by storing outliers (i.e., an object which is located far from other objects) at index levels.

The R-tree is widely used because of its simplicity and efficiency. In spite of the R-tree's popularity, the R-tree has three problems. First, there are overlaps among index entries, and the overlaps increase rapidly with growing dimensionality of the data. According to [4], the overlap can reach 90% at the dimension of 6, where the *overlap* is defined as the percentage of data objects that fall in the overlapping portion of the space. Second, the R-tree's index entry contains the MBR whose storage cost linearly increases as the data dimension increases; thus, the fanout of the index pages decreases as the data dimension increases. Third, the R-tree has poor concurrency performance. For example, each data entry insertion might cause the update of MBRs to the root of the R-tree.

Although the R-tree has the problems listed above, some major database vendors [1] have used the R-tree to provide spatial database support for two reasons. First, most spatial data are 2D and 3D. Increasing overlaps and decreasing index page fanout are not serious problems for 2D and 3D data. Second, spatial databases tend to be static most of the time (e.g., some census data only change once per year). Concurrency is not a major concern for nearly static data sets.

However, for frequent-update multi-dimensional point data sets (especially those data sets whose dimensions are higher than 3D), the R-tree is not a good choice due to the reasons mentioned previously.

The R-tree is proposed as a multi-dimensional access method for point data and non-point data. In this paper, we consider only point data. For multi-dimensional point data, one set of popular access methods is the kd-tree based index methods. The kd-tree [3] is an in-memory binary search tree that recursively divides the k-dimensional space into non-overlap subspaces by using (k-1) dimensional hyper-planes. The kd-B-tree [13] is a variant of the kd-tree that pages secondary memory. The kd-B-tree is a balanced tree, but it cannot guarantee the minimum storage utilization because of the *force split*, which splits children of splitting parent index pages. The Bkd-tree [12] improves the storage utilization of the kd-B-tree by using a set of kd-B-trees. Henrich et al. propose the LSD-tree [9] (Local Split Decision Tree), which is another variant of the kd-tree. The LSD-tree is not a balanced tree because the heights of its external subtrees can differ at most by one.

Lomet and Salzberg propose the hB-tree [10] (holey Brick Tree), which can guarantee good storage utilization. The hB-tree is a balanced tree and adapts well to skewed data distributions. Later, Evangelidis et al. combine the hB-tree and the II-tree [11] into the hB-pi-tree [5]. The hB-pi-tree not only preserves the good properties from the hB-tree, but also inherits good concurrency and recovery control algorithms from the II-tree. The hB-pi tree divides the space into non-overlapping subspaces and is fairly insensitive to dimension increases because each kd-tree node only stores the split value of one split attribute. The advantages of the hB-pi tree exactly address the the impediments of the R-tree. That's why we argue the hB-pi tree is a good candidate to index multi-dimensional point data sets which update frequently and require high concurrency performance.

The hB-pi tree (and all other kd-tree based access methods) indexes the whole space no matter whether or not there is any data in some sub-spaces. Indexing the *empty space* (i.e., space without data inside) leads to empty data page accesses. The *empty data pages* are defined as the data pages that are accessed during a range query but do not contain any data in the query range. It has been observed that the real data in high-dimensional space are highly correlated and clustered, and that the data occupy only some subspaces of the whole space. So as the dimension increases, more and more empty data pages might be accessed.

In this paper, we attempt to overcome the problems listed above and combine the advantages of previous structures by proposing the hB-pi* tree, a comprehensive access method for frequent-update low to medium point data sets. The hB-pi* tree preserves the good properties of the original hB-pi tree such as non-overlap, insensitivity to dimension increases, and integration with good concurrency and recovery algorithms. The hB-pi* tree can also cluster the data as the R-tree, reinsert the data as the R*-tree, and detect the outliers as the R^O-tree.

A hB-pi* tree includes a modified hB-pi tree, which stores most data, a small hash table for data in sparse space (HTSS), which stores the data in *sparse space* (i.e., space with few or no objects inside), and a tiny auxiliary data page density queue (DPDQ), which maintains the densities of data pages in a priority queue (i.e., sorted by data page densities in ascending order). The hB-pi* tree involves several heuristics that take into account the density, area, and margin to improve query performance significantly. In the following, we abbreviate *sparse space* and (or) *empty space* as *sparse space* without causing any ambiguity.

The rest of the paper is organized as follows. Section 2 illustrates the access methods directly related to our work and analyzes their problems. Section 3 presents the structure of the hB-pi* tree and the corresponding construction and concurrency algorithms. Section 4 contains an extensive experimental evaluation, while section 5 summarizes the contributions and provides directions for future work.

2 Preliminaries

The hB-pi* tree is a kd-tree [3] based point access method. We will first illustrate the hB-pi tree, then explain empty space in the hB-pi tree.

2.1 hB-pi Tree

The hB-pi tree [5] is a comprehensive point access method different from other kd-tree variants [9,13]. First, the hB-pi tree can guarantee decent storage utilization, and it is a balanced tree. Second, the hB-pi tree adapts well to skewed data distributions [5]. Third, the hB-pi tree integrates the II-tree [11], which can efficiently support recovery and concurrency control.

Like all other hierarchical index methods, the hB-pi tree includes data pages and index pages. A data page contains one or more record lists (containing the real data) and zero or one kd-trees (storing the space decomposition information). A record list has records in the form $<a_1,a_2,..a_k,data>$ where a_1 to a_k are the attribute values and *data* is the real data value. The kd-tree node N is in the form of $<$*split-attribute, split-value, left-pointer, right-pointer*$>$ where *split-attribute* and *split-value* store the split dimension with the corresponding value. The *left-pointer* points to N's left child, which stores data whose value in *split-attribute* is smaller than or equal to the *split-value*, and the *right-pointer* points to N's right child, which stores data whose value in *split-attribute* is bigger than the *split-value*.

An index page only contains one kd-tree. A kd-tree node N is in the form of $<$*decoration, split-attribute, split-value, left-pointer, right-pointer*$>$ where *decoration* is the child page address where N was posted from, and it records the split order among data pages. The *decoration* will be used in the page consolidation (an example will be illustrated in the next section). All the other fields have the same meanings as those in data pages. An example is illustrated in Figure 2(a). The hB-pi tree starts with one data page A. After a certain number of data insertions, the original root data page A overflows. Part of the data in A will be moved to the newly created data page B. The data left in data page A is saved in two record lists (i.e., R1 and R2). A kd-tree is added to A with a side pointer pointing to page B. The index page P is created and it contains a kd-tree that stores the space decomposition information of the subspace. The *decoration* of both kd-tree nodes in P is A, because both nodes are posted from data page A. If a set of kd-tree nodes (e.g., nodes X10 and Y10) have the same decoration (e.g., A), only the kd-tree node at the highest level (e.g., node X10) will be decorated.

If the data page A further splits into data pages A and C (figure 2(b)), the *split path* (X10 right) is posted to the parent index page P. The *split path* is the path from the root of the kd-tree (e.g., X10 in A's kd-tree) in the splitting page (e.g., A) to the side pointer pointing to the new page (e.g., C). Starting from the kd-tree node (X10) in P (which is labeled with the decoration for A, the splitting page), the updating algorithm of the hB-pi tree will attempt to find where to place a copy of the path by matching the split path with the path starting from X10. It will find a full copy of this path already in P and only need to post the decoration for the new split on Y10 as illustrated.

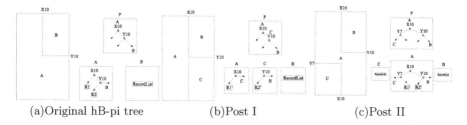

(a)Original hB-pi tree   (b)Post I   (c)Post II

Fig. 2. hB-pi tree example

If the data page A splits instead (not merely following the original split lines) into data pages A and C as in figure 2(c), the *split path* (X10 left -> Y7 left) is posted to P. In this case, the split path has not been posted yet. The split path itself must be posted as well as the new decoration.

2.2   Empty Spaces

The index entry of the hB-pi tree (and other kd-tree variants) is independent of the number of dimensions of the data set. As a consequence, the fanout of the index pages in the hB-pi tree is insensitive to dimension increases. The price we have to pay for fixed index entry size is that the whole data space is indexed in the hB-pi tree even if some sub-spaces do not contain a single data point. This can lead to unnecessary data page access in query operations. An example is illustrated in figure 3(a). Two data pages (D1 and D2) are divided by a split line (X8). Data points are represented by dots, and query range is represented by the shaded area. The query range does not contain any point in D2, but D2 will be visited because the space covered by D2 intersects with the query range. D2 is an *empty data page*, which is accessed during the range query, but does not contain any data in the query range.

As data dimension increases, more and more *empty data pages* might be accessed for two reasons. First, for the same size query range, more data pages might be intersected with the query range as the dimension increases. An example is illustrated in figure 4. The whole data space is divided into four parts (data pages) and query range always covers $\frac{1}{4}$ of the data space. For the 1-dimensional case, the query range intersects 2 data pages; for the 2-dimensional case, 4 data pages are intersected with the query range. Second, higher dimensional data is usually more skewed than lower dimensional data. As the dimension increases,

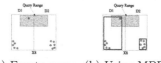

(a) Empty space (b) Using MBR

Fig. 3. Example of empty space

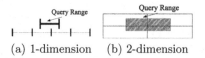

(a) 1-dimension   (b) 2-dimension

Fig. 4. Data pages intersect with query range

a higher percentage of the data pages that are visited during range query will be *empty data pages*.

A simple idea to reduce empty space is to introduce MBRs to the hB-pi tree. Each data page stores an MBR, which encloses all data points actually stored in the data page. Each index page stores an MBR, which encloses all data points actually stored in the corresponding sub-tree. The index entry in index pages stores the MBR of the lower level page together with the pointer that points to the corresponding lower level page. An example is illustrated in figure 3(b). Although the query range intersects with the data space of D2, the query range does not intersect with the MBR of D2. The range query will not access D2, and one empty data page access is saved. A similar idea is proposed in [8].

However, this simple idea of using MBRs has three disadvantages. First, the index entry size will increase linearly as the dimension increases because the index entry stores the MBRs, whose size increases as dimension increases. The index page fanout will decrease for high dimensional data as in the R-tree. Second, the outliers can make the MBR method (i.e., introducing MBRs to the hB-pi tree) quite inefficient. For example, data page D1 (figure 3(b)) will still be accessed because D1's MBR intersects with the query range. Third, each data entry insertion might cause the update of MBRs to the root of the hB-pi tree. This leads to the same poor concurrency performance as that of the R-tree. In our experiments, we found that the hB-pi tree with the MBRs hardly improves the query performance of the hB-pi tree, even without considering the concurrency issue. The performance of the hB-pi tree with MBRs is not included in this paper's experimental section. The hB-pi* tree, which will be illustrated in the next section, overcomes these problems while still efficiently reducing empty data page accesses.

3   hB-pi* Tree

The general idea of the hB-pi* tree is to use extra kd-tree nodes to indicate sparse spaces, and to store the outliers in an in-memory structure. At most one extra kd-tree node, which only stores the split dimension and the corresponding split value, is introduced during each data page split. Furthermore, the number of extra kd-tree nodes is also constrained by the bound on the size of the table storing the outliers to which they refer. Thus, although the number of index pages (not data pages) may increase, the experimental results show that, for a real data set, the hB-pi* tree has a similar number of index pages as that of the original hB-pi tree. Since the original hB-pi tree is independent of the

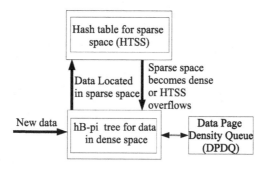

Fig. 5. Overview of the hB-pi* tree

number of dimensions of the data set, the hB-pi* tree is also not sensitive to data dimension.

In the following, section 3.1 first illustrates the structure of the hB-pi* tree. Thereafter, section 3.2 presents the insertion algorithm, section 3.3 demonstrates the page reentry algorithm. Deletion, query and concurrency algorithms are not provided in this paper due to space limitation. Detail algorithms can be found in [15].

3.1   Structure

The overview structure of the hB-pi* tree is illustrated in figure 5. An hB-pi* tree contains the following three parts: a durable tiny in-memory Data Page Density Queue (DPDQ), a durable small in-memory Hash Table for Sparse Space (HTSS), and a disk-based extended hB-pi tree. Durable in-memory structures (i.e., DPDP and HTSS) are copied to log checkpoints, and changes made to them are logged so that in case of system failure, they can be reconstructed (more details can be found in [15]). The extended hB-pi tree's recovery control algorithm is similar to that of the original hB-pi tree.

The DPDQ stores the densities of data pages in a priority queue in ascending order (i.e., the first data page in the DPDQ has the lowest density). Each entry has the form $<page_{id}, density>$, where $density$ is defined as the number of objects in the data page (indicated by $page_{id}$) divided by the space size of the data page. The DPDQ is used to detect the sparse space in data pages other than the currently overflowing data page.

The HTSS stores the data located in sparse space. Each entry has the form $<record\_list_{id}, data\_list>$. The $record\_list_{id}$ is the unique identifier assigned to each sparse space. Sparse spaces are detected by the hB-pi* tree insertion algorithm. The $data\_list$ is a record list that stores all the data in the sparse space indicated by the $record\_list_{id}$. If a sparse space becomes too dense with data, the corresponding record list might be re-inserted into the extended hB-pi tree.

The extended hB-pi tree stores all of the data in the dense space. However, the major difference between the original hB-pi tree and the extended hB-pi tree is that in the index pages of the original hB-pi tree, the pointers of the kd-tree nodes can only point to the other kd-tree nodes, the child pages at the lower level, or the sibling pages at the same level. However, in the index pages of the extended hB-pi tree, the pointers of the kd-tree node can also point to a record list in the HTSS. In the following section, the *extended hB-pi tree* is abbreviated as the *hB-pi tree* without causing any ambiguity.

Figure 6 illustrates the hB-pi* tree for the same data distribution of figure 3. The whole space first splits at Y3 (because of the data distribution). Point a (i.e., an outlier) is above the split line Y3 and it is stored in the record list E1 in the HTSS. The sub-space below Y3 further splits into two data pages at X5. The densities of D1 and D2 are inserted into the DPDQ. The index page (i.e., P) of the hB-pi tree is shown at the right-bottom corner. The kd-tree node Y3's right pointer points to the record list E1 in the HTSS.

If the query range is the same as that in figure 3, the query procedure can stop at the index page of the hB-pi tree when it travels the kd-tree in the index page. The query procedure retrieves the results (i.e., point a) directly from the record list E1 in the HTSS without accessing any data page.

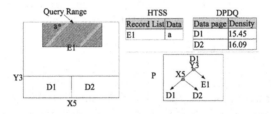

Fig. 6. Example of the hB-pi* tree

The insertion algorithm of the hB-pi* tree will be illustrated in the next section. This algorithm involves several heuristics to reduce the sparse space covered by the hB-pi tree, and to improve the query performance. Furthermore, the heuristics used by the algorithm are applicable to other kd-tree variants.

3.2   Insertion

Figure 7 shows the insertion algorithm of the hB-pi* tree. By using the kd-trees in the index pages of the hB-pi tree, the insertion algorithm first determines whether the new data will be stored in a data page (in the hB-pi tree) or a record list (in the HTSS). If the new data is located in a data page and entering the new data causes the data page to overflow, the page split algorithm will be invoked. If data is located in a record list, and the number of data records in the record list reaches the threshold (i.e., $\frac{1}{3}$ of the data page capacity) or the HTSS overflows, the re-insert algorithm will be executed. Finally, the updated information will be posted to parent pages.

A set of consecutive examples are used to demonstrate the evolving of the hB-pi* tree while new data is continuously inserted. In each example, the index page of the hB-pi* tree is used to illustrate the structure of the hB-pi* tree.

Algorithm Insert(dataRecord R)
1.   Find the data page (D) or the record list
     (L) that should contain data R
2.   If R is contained by D
3.   Add R to D
4.   If D overflows
5.   Invoke PageSplit on D
6.   Else if R is contained by L
7.   Insert R to L
8.   If the size of L is bigger than 1/3 page size
9.   Invoke PageReEntry on L
10.   Else
11.   If HTSS overflows
12.   Choose record list L' from HTSS
13.   Invoke PageReEntry on L'
14. If page split or page reentry occurs, post the update to parent pages

Fig. 7. The insertion algorithm

Figure 8 illustrates the initial status of the hB-pi* tree. The data page capacity of the hB-pi* tree is 12, and it starts with only one data page (D1). Most data is located below the line Y3, and only one outlier (i.e., a) is above the line Y3. The DPDQ and the HTSS are empty. There is no index page and the right-bottom corner of the figure 8 is empty.

In the following sections, the insertion method's two major sub-algorithms (i.e., the page split algorithm and the page reentry algorithm) will be illustrated in detail. Section 3.2 covers the major parts of the page split algorithm: the split detection algorithm and the procedure of splitting the sparse space from the currently overflowing data page. Section 3.3 illustrates the page reentry algorithm. In order to follow the examples according to the data insertion order, an example of how the page split algorithm splits the sparse space from the data page in the DPDQ is illustrated in Section 3.3.

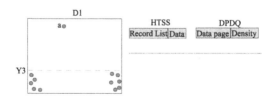

Fig. 8. The initial status of the hB-pi* tree

Sparse space in overflowing data pages. Because D1 already contains the maximum number of entries (e.g., 12 data entries), the data page D1 in figure 8 overflows when a new object is inserted. In this case, the original hB-pi tree splits D1 into two parts, as we have seen in Figure 3.1(a). However, much of the space in both parts is sparsely inhabited. The original hB-pi tree (and all the other kd-tree based access methods) indexes the whole space no matter whether or not there is any data in some parts of the space. Indexing sparse space will lead to empty data page access (i.e., visiting data pages that do not contain any data inside the query range) during range queries.

The hB-pi* tree is motivated by two facts. First, many real multi-dimensional data sets are highly correlated and clustered, and therefore most of the data occupies areas which are a small fraction of the whole space. For example, most data in figure 8 clusters at both bottom corners. Second, a few outliers may be isolated from the denser clusters of data. For example, there is only one data point above the Y3 line in figure 8. The main idea of the page split algorithm in the hB-pi* tree is to detect sparse spaces in data pages (not only in currently overflowing data pages) and to save outliers in an in-memory structure (i.e., HTSS).

Figure 9 formally describes the *PageSplit* algorithm. In the following, the definition of the sparse space is provided first; the example of the page split algorithm is illustrated afterwards; the explanation of the algorithm is presented at the end of this sub-section.

```
Algorithm PageSplit (Page D)
1.  Invoke DetectSplit on each dimesion of overflowing page D,
    and save the split results in cur_split_results
2.  Choose the split with largest dense space density from cur_split_results,
    and save the chosen split result as cur_split
3.  If sparse space exists in D
4.     Retrieve the first data page (Dₛ) in the DPDQ
5.     Invoke DetectSplit on each dimesion of page Dₛ
       and save the split results in sparse_split_results
6.     Invoke ChooseSplit on sparse_split_results,
       and save the chosen split result as sparse_split
7.     If cur_split's dense space density < sparse_split's dense space density
8.        Split the sparse space from D
9.        Split the dense space of D into two parts
10.    Else
11.       Split D into two data pages without splting sparse space
12.       Split the sparse space from Dₛ
13. Else
14.    Split D into two data pages without splting sparse space
15. Update the DPDQ
```

Fig. 9. The page split algorithm

The page split algorithm (figure 9) begins with detecting the sparse spaces in the currently overflowing data page (i.e., D1) along each dimension. The sparse space detection algorithm actually detects the *dense space* in data pages. A data page D contains *dense space* at dimension K, if and only if there exists a subspace (S) at one end of axis K that contains at least $\frac{2}{3}$ of D's data and S's density is at least $2^{\frac{dim}{2}}$ (i.e., *dim* is the dimension of data) times bigger than (or equal to) the density of D. The reason for using the metric $2^{\frac{dim}{2}}$ will be given at the end of this section. The subspace S is the *dense space* in D and the other space in D is the *sparse space*. If there is no *dense space* in D at dimension K, there is no *sparse space* in D at dimension K either. The sparse space detection algorithm will only detect the sparse space at both ends (i.e., low and high) of each dimension, and the reasons will be provided after the examples.

Let us continue using the data page (i.e., D1) split example in figure 8. First, the sparse space at X-dimension is detected by invoking the *DetectSplit* method. All data entries are sorted in ascending order by their X-values, and the results are saved in *sorted_result*. Then the sparse spaces at both ends of the

X-dimension are detected separately. The first $\frac{2}{3}$ entries of the *sorted_result* are stored in *low_data* and the last $\frac{1}{3}$ entries are saved in *high_data*. In this example, the density of *low_data* is smaller than twice (i.e., dim = 2 and $2^{\frac{dim}{2}} = 2$) of the density of D1; no dense space exists at the low end of the X-dimension, and no sparse space exists at the high end of the X-dimension. If the density of *low_data* is bigger than (or equal to) twice of the density of D1, the while loop (i.e., line 4-5 in figure 10) will be executed to put all points in dense space in the *low_data*. Next, the first $\frac{1}{3}$ entries of the *sorted_result* are stored in *low_data*, and the last $\frac{2}{3}$ entries are save in *high_data*. No sparse space exists at the low end of the X-dimension either. Therefore, there is no sparse space along the X-dimension.

Algorithm DetectSplit
1. page_density = data in the page/space size covered by the page
2. Sort the data ascendingly along the chosen dimension,
   and save the results in sorted_result
3. Assign the first 2/3 entries to low_data, the last 1/3 entries to high_data
4. While (density(low_data) > $2^{Dim/2}$* page_density) & (high_data is not empty)
5.   Move the first data in high_data to low data
6. If sparse space exists at high end, save the split result
7. Assign the first 1/3 entries to low_data, the last 2/3 entries to high_data
8. While (density(high_data) > $2^{Dim/2}$* page_density) & (low_data is not empty)
9.   Move the last data in low_data to high_data
10.If sparse space exists at low end, save the split result

Fig. 10. Detecting sparse spaces at each dimension

Next, the *DetectSplit* method is invoked to detect the sparse space along the Y-dimension. The sparse space exists at the high end of the Y-dimension (i.e., above Y3), and the space below Y3 is the *dense space* of D1. The split result is saved. In our example, there is only one split that contains sparse space. If there is more than one split that contains sparse space, the one with the largest dense space density will be chosen.

After the best split is found in the currently overflowing data page (i.e., D1), the sparse space from the first data page (i.e., the data page with the lowest density) in DPDQ also needs to be checked. Because the DPDQ is empty in our example, the density of dense space in *sparse_split* (i.e., the best split result from the DPDQ) is initially set to +∞. The density of dense space in D1 is smaller than that of *sparse_split*, so the sparse space (i.e., the space above Y3) is split from D1. The sparse record list (i.e., E1) is created in the HTSS, and the outlier is stored in E1. The dense space of D1 further splits into two parts (i.e., D1 and D2) at X5. The densities of D1 and D2 are inserted into DPDQ. The final status of the hB-pi* tree is illustrated in figure 11.

The sparse space detection algorithm only reports the sparse space at one end (i.e., low or high) of each dimension. The reason for this choice is twofold: (i) Indicating the sparse space at one end only needs one extra kd-tree node, and it can guarantee that the hB-pi* tree is still insensitive to the data dimension. Indicating the sparse space at multiple place in the data page would introduce multiple kd-tree nodes. The split algorithm would also be more complicated. (ii) Even if there is sparse space in the overflowing data page which is not at either end of the data page, it still might be detected in the future.

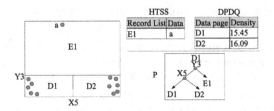

Fig. 11. Split sparse space from overflowing data pages

The DetectSplit algorithm (Figure 10) tests if density discovered in a sweep along one dimension is greater than $2^{dim/2}$ times the page density. because the formula $2^{\frac{dim}{2}}$ is the best criterion we found in experiments.

3.3   Page Reentry

The *PageReEntry* algorithm can re-distribute the data in the hB-pi* tree (i.e., move data from the HTSS to the hB-pi tree). The cost of the hB-pi* tree's *PageReEntry* algorithm is less than the re-insertion algorithm of the R*-tree. In the hB-pi* tree, the record list (e.g., E1) is either becomes a data page itself or consolidates with its decoration data page (e.g., D1) and no other data page is affected. In the R*-tree, the re-insertion algorithm will re-insert the data one by one. For a common experimental setting in the R*-tree (i.e., 4K bytes page size, 2-dimensional data set, 30% data in overflowing data page is re-inserted), one R*-tree re-insertion operation might include several dozens of data insertion operations and even more page accesses.

4   Experiments

To demonstrate the effectiveness of the hB-pi* tree, we performed an extensive experimental evaluation of the hB-pi* tree and compared it to the original hB-pi tree and the R*-tree. The experimental setup is discussed in section 4.1. The experimental results are illustrated in section 4.2.

4.1   Experimental Setup

All experiments were run on an Intel Pentium IV 2.66GHz CPU machine with 1G bytes main memory. In all experiments, the disk page size was 1K bytes. Each entry in a data page included its k-dimensional position values and an object ID (4 bytes). Values in all the dimensions were represented by 4-byte floats.

In our experiments, the range query and the exact match query were chosen to compare the three access methods because these two types of queries serve as basic operations for other queries such as nearest neighbor queries or partial range queries. Query performance was measured in terms of the number of page accesses. For each experiment, 100 random queries were executed and the average number of page accesses per query was reported.

All three access methods use the same amount of main memory for queries. The hB-pi* tree keeps all its index pages, the HTSS and the DPDQ in main memory. The hB-pi tree uses the same amount of memory to store all its index pages and data pages that contain most data entries. The R*-tree uses the same amount of memory to store its index pages. The index pages at higher levels (i.e., the root page is at the highest level) will be stored in main memory first. If index pages at a certain level cannot be all stored in main memory, the index pages that contain more index entries will be stored in main memory.

There are two important parameters for the hB-pi* tree, namely, the size of the HTSS and the size of the DPDQ. In all experiments, the size of the HTSS was 1% of the whole data set size. This is a practical assumption because the size of the main memory can easily reach 1% of the disk space in many real systems. In all experiments, the DPDQ stored the densities of the most sparse 20% of data pages. The bigger the DPDQ size, the more accurate the data page density information retrieved from the DPDQ can be. We have tested the bigger DPDQ size (e.g., stores the densities of all data pages) and found storing the densities of the sparsest 20% of data pages can have results similar to storing the densities of all data pages.

Each entry in the DPDQ stored a data page ID (4 bytes) and the corresponding density of the data page (4 bytes). For an 8-dimensional data set with 200,000 data (the raw data file occupies more than 10M bytes disk space), the corresponding DPDQ only required about 4K bytes space. For lower dimension data sets with the same amount of data, the size of the corresponding DPDQ would be even smaller. For low to medium (i.e., 2D to 8D) point data sets discussed in this paper, the size of the DPDQ is less than $\frac{1}{1000}$ of the data set size.

In all experiments, the criterion $2^{\frac{dim}{2}}$ (dim is the data dimension) was used in the sparse space detection algorithm. The criterion becomes stricter as the dimension increases.

4.2   Experimental Results

The house (HOUSE) data set is used in our experiments and it contains 84,362 eight dimensional records. Specifically, it contains house value, household income, latitude, and longitude, etc. We use PCA (Principle Component Analysis) method to generate the 2D to 8D data sets from the HOUSE data set. For example, the 2D data set contains the 2D values from the two dimensions with the biggest eigenvalues.

Figure 12(a) illustrates the number of index pages in the hB-pi tree, the hB-pi* tree and the R*-tree. As data dimension increases, the size of each data entry increases, and the number of data pages increases (for a fixed page size). More splits are posted to index pages and more index entries are inserted into index pages. So all three access methods have more index pages as data dimension increases. The R*-tree has far more index pages than the hB-pi tree and the hB-pi* tree because R*-tree's index entry contains an MBR whose size is increasing linearly with the data dimension. For each page split, the hB-pi tree and the hB-pi* tree only store the split dimension and the corresponding split values

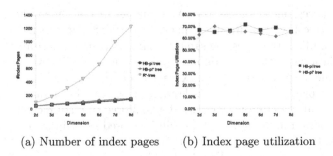

(a) Number of index pages   (b) Index page utilization

Fig. 12. Index pages

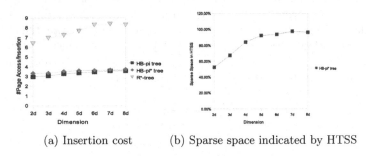

(a) Insertion cost   (b) Sparse space indicated by HTSS

Fig. 13. Tree construction and tree structure

that are insensitive to data dimensions. So the hB-pi tree and hB-pi* have much fewer index pages than the R*-tree.

If a data page overflows and it contains sparse space, the hB-pi* tree will use one extra kd-tree node to indicate the sparse space. Furthermore, for the same data set, the hB-pi tree and the hB-pi* tree have similar data pages utilization which designate two methods have similar number of data page splits. In the worst case (i.e., all overflowing data pages contain sparse spaces), for the same data set, the number of index pages of the hB-pi* tree will be about twice that of the hB-pi tree because at most one extra kd-tree node is introduced to the hB-pi* tree for each data page split. In the experiments, the hB-pi* tree has more index pages than the hB-pi tree (figure 12(a)), but much less than twice of the hB-pi tree's index pages. The reasons are: first, the hB-pi tree and hB-pi* tree have similar index page utilizations (figure 12(b)); second, sparse space only exists in a small number of data pages and only a few extra kd-tree nodes are added to the hB-pi* tree.

Figure 13(a) illustrates the insertion cost of the three access methods. The insertion cost is defined as the all pages accesses (from root page to leaf page) during the tree construction divided by the number of data entries being inserted. The insertion cost of the R*-tree is higher than the hB-pi tree and the hB-p* tree for two reasons. First the re-insertion operation of the R*-tree introduces extra page accesses. Second, the height of the R*-tree is higher than those of the hB-pi tree and the hB-pi* tree because of the R*-tree small index page fanout.

The insertion cost of the hB-pi tree and the hB-pi* tree is close to the final height of the tree as expected. The hB-pi* tree's insertion cost is higher than the hB-pi tree because the hB-pi* tree needs extra page access for page reentry and DPDQ retrieval operations. Since the page reentry method of the hB-pi* tree batch inserts data and the re-insertion of the R* tree re-inserts data one by one, the hB-pi* tree's page reentry cost is much lower than the re-insertion cost of the R*-tree.

Figure 13(b) illustrates what percentage of the whole space is taken up by the sparse space indicated by the HTSS entries. As the data dimension increases, higher and higher percentage of the space are indicated as sparse spaces and indexed by the HTSS. This experimental result is in line with the observation that the data tends to cluster in small sub-spaces in high dimension space. The sparse space indicated by the HTSS increases rapidly as data dimension increases when the data dimension is less than five. After the data dimension becomes larger than five, the sparse space indicated by the HTSS becomes stable. This observation indicates that detecting sparse space in even higher dimensional space (i.e., >8D) might not be as effective as in low-dimensional (e.g., <5D) space because most parts (i.e., higher than 90%) of the whole space have been indicated as sparse spaces and covered by the entries in the HTSS.

Figure 14 illustrates the query performance of the three access methods for different query sizes. Experimental results on 2D and 8D data sets are used to indicate the performace change of three access methods as data dimension increases (i.e., results on 3D-7D are skipped). For the 2-dimensional data set (figure 14(a)), the hB-pi* tree outperforms the R*-tree, and the R*-tree outperforms the hB-pi tree. For low dimensional (e.g., 2D) data sets, the index entry (i.e., MBR) size and overlap among index entries do not significantly affect the query performance of the R*-tree. On the other hand, the performance of the hB-pi tree is heavily impeded by the empty space. The average page access of the hB-pi* tree can be smaller than one page which indicates that some queries can be finished without visiting any data page.

Figure 14(b) illustrates the query performance for the 8-dimensional data set. The hB-pi* tree is better than the hB-pi tree, and the hB-pi tree is better than the R*-tree. For high-dimensional (e.g., 8D) data sets, the performance of the R*-tree deteriorates rapidly because of the increasing index page size and the overlaps among index entries. The hB-pi tree is overlap free and its index entry size is insensitive to data dimension, so the hB-pi tree can outperform the R*-tree. The hB-pi* tree inherits the advantages of the hB-pi tree. Furthermore, the hB-pi* tree can indicate sparse spaces and stores outliers in the HTSS. So the hB-pi* tree further outperforms the hB-pi tree.

Figure 15(a) illustrates the performance change of the three access methods as the data dimension increases (query range size is fixed at 0.01%). For low-dimensional (i.e., 2D-3D) data sets, the R*-tree outperforms the hB-pi tree. For high-dimensional (i.e., 5D-8D) data sets, the hB-pi tree outperforms the R*-tree. The hB-pi* tree outperforms the hB-pi tree and the R*-tree for all dimension (i.e., 2D-8D) data sets. Figure 15(b) further explains why the hB-pi*

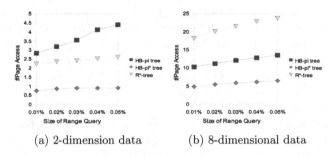

(a) 2-dimension data   (b) 8-dimensional data

Fig. 14. Range Query Performance

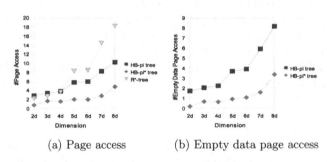

(a) Page access   (b) Empty data page access

Fig. 15. Range Query Performance

tree performs better than the hB-pi tree. As data dimension increases, data are clustered in sub-spaces and the empty page access increases in both the hB-pi tree and the hB-pi* tree. By indicating the sparse space and storing outliers in the HTSS, the hB-pi* tree can significantly reduce the empty data page access which is a serious problem in the original hB-pi tree (and all other kd-tree based methods).

5   Conclusions

In this paper, we addressed the problem of indexing frequent-update low to medium (i.e., 2D to 8D) point data sets. The original hB-pi tree (and all other kd-tree variants) indexes the whole space, even though some subspaces do not contain any data. Indexing empty space will lead range queries to access empty data pages. We proposed the hB-pi* tree to reduce the empty data page accesses. The hB-pi* tree uses an extended hB-pi tree to store most data in dense spaces and a HTSS to store the data in sparse spaces.

We illustrated the insertion algorithm and the concurrency algorithm of the hB-pi* tree in detail. The hB-pi* tree preserves the good properties of the hB-pi tree (non-overlap, insensitivity to data dimensions, guaranteed data page utilization, good concurrency algorithm, etc.) and combines the advantages of the R-tree family (clustering data, indicating outliers, etc.).

References

1. An, N., Kanth, K., Ravada, S.: Improving Performance with Bulk-Inserts in Oracle R-Trees. In: Proceedings of International Conference on Very Large Data Bases (VLDB), pp. 948–951 (2003)
2. Beckmann, N., Kriegel, H., Schneider, R., Seeger, B.: The R*-tree: An Efficient and Robust Access Method for Points and Rectangles. In: Proceedings of ACM/SIGMOD Annual Conference on Management of Data (SIGMOD), pp. 322–331 (1990)
3. Bentley, J.L.: Multidimensional Binary Search Trees in Database Applications. IEEE Transactions on Software Engineering 5(4), 333–340 (1979)
4. Berchtold, S., Keim, D.A., Kriegel, H.: The X-tree: An Index Structure for High-Dimensional Data. In: VLDB, pp. 28–39 (1996)
5. Evangelidis, G., Lomet, D.B., Salzberg, B.: The hB-Pi-Tree: A Multi-Attribute Index Supporting Concurrency, Recovery and Node Consolidation. In: Proceedings of International Conference on Very Large Data Bases (VLDB), pp. 1–25 (1997)
6. Gaede, V., Günther, O.: Multidimensional Access Methods. ACM Comput. Surv. 30(2), 170–231 (1998)
7. Guttman, A.: R-trees: a dynamic index structure for spatial searching. In: Proceedings of ACM/SIGMOD Annual Conference on Management of Data (SIGMOD), pp. 47–57 (1984)
8. Henrich, A.: The LSD^{h}Tree: An Access Structure for Feature Vectors. In: Proceedings of International Conference on Data Engineering (ICDE), pp. 362–369 (1998)
9. Henrich, A., Six, H.-W., Widmayer, P.: The LSD tree: Spatial access to multidimensional point and non-point objects. In: Proceedings of International Conference on Very Large Data Bases (VLDB), pp. 45–53 (1989)
10. Lomet, D.B., Salzberg, B.: The hBtree: A robust multiattribute search structure. In: Proceedings of International Conference on Data Engineering (ICDE), pp. 296–304 (1989)
11. Lomet, D.B., Salzberg, B.: Access Method Concurrency with Recovery. In: Proceedings of ACM/SIGMOD Annual Conference on Management of Data (SIGMOD), pp. 351–360 (1992)
12. Procopiuc, O., Agarwal, P.K., Arge, L., Vitter, J.S.: Bkd-Tree: A Dynamic Scalable kd-Tree. In: Hadzilacos, T., Manolopoulos, Y., Roddick, J.F., Theodoridis, Y. (eds.) SSTD 2003. LNCS, vol. 2750, pp. 46–65. Springer, Heidelberg (2003)
13. Robinson, J.T.: The K-D-B-Tree: A Search Structure For Large Multidimensional Dynamic Indexes. In: Proceedings of ACM/SIGMOD Annual Conference on Management of Data (SIGMOD), pp. 10–18 (1981)
14. Xia, T., Zhang, D.: Improving the R*-tree with Outlier Handling Techniques. In: GIS, pp. 125–134 (2005)
15. Zhou, P.: Querying Multi-dimensional Data and Spatio-temporal Data with Non-overlapping Access Methods. Northeastern University PhD Thesis (2006)

Breaking the Curse of Cardinality on Bitmap Indexes*

Kesheng Wu, Kurt Stockinger, and Arie Shoshani

Lawrence Berkeley National Lab, University of California, Berkeley, CA, USA
http://sdm.lbl.gov/

Abstract. Bitmap indexes are known to be efficient for ad-hoc range queries that are common in data warehousing and scientific applications. However, they suffer from the curse of cardinality, that is, their efficiency deteriorates as attribute cardinalities increase. A number of strategies have been proposed, but none of them addresses the problem adequately. In this paper, we propose a novel binned bitmap index that greatly reduces the cost to answer queries, and therefore breaks the curse of cardinality. The key idea is to augment the binned index with an Order-preserving Bin-based Clustering (OrBiC) structure. This data structure significantly reduces the I/O operations needed to resolve records that can not be resolved with the bitmaps. To further improve the proposed index structure, we also present a strategy to create single-valued bins for frequent values. This strategy reduces index sizes and improves query processing speed. Overall, the binned indexes with OrBiC great improves the query processing speed, and are 3 – 25 times faster than the best available indexes for high-cardinality data.

1 Introduction

A large data warehouse typically contains high-dimensional data with tens or even hundreds of attributes. Most popular indexing techniques are not effective for answering queries on these datasets; some use the term *the curse of dimensionality* to describe the poor performance [1]. The bitmap index is able to break this curse even when the dimensionality of the dataset is very high [2,3,4]. Hence, major commercial database systems, such as ORACLE, IBM DB2, and Sybase IQ, have implemented various bitmap indexes. In many cases, bitmap indexes not only take less disk space than the commonly used B-Tree indexes and their variations, but also answer queries much faster [2,3,5]. However, the current bitmap indexes also have serious limitations [6]. One of the most serious ones, which we call *the curse of cardinality*, is that both index sizes and query response time increase as the number of distinct values in an attribute increases. In this paper, we propose to break this curse with a novel binned index and demonstrate its effectiveness with both analyses and experimental measurements.

The number of distinct values of an attribute in a dataset is known as the *attribute cardinality*. A number of strategies have been proposed to improve the performance of bitmap indexes on high-cardinality attributes as discussed in the next section; among them we see binning as the most promising [7,8,9,10]. Instead of building one bitmap

---

* This work was supported by the Director, Office of Science, Office of Advanced Scientific Computing Research, of the U.S. Department of Energy under Contract No. DE-AC02-05CH11231.

B. Ludäscher and Nikos Mamoulis (Eds.): SSDBM 2008, LNCS 5069, pp. 348–365, 2008.

for each distinct value as in the basic bitmap index, a binned bitmap index builds one bitmap for a range (or bin) of values, which reduces the number of bitmaps used in the index. Typically a query range spans multiple *interior bins* and two *edge bins*. For example, an attribute "age" represented by integers between 0 and 100 might be divided into 10 bins uniformly with bin 1 for "age" between 0 and 9, bin 2 for "age" between 10 and 19, and so on. The query "age between 25 and 65" will have bins 4, 5, and 6 as interior bins, bins 3 and 7 as edge bins. The bitmaps can be used to identify rows in the interior bins and edge bins. Those in the interior bins are hits, but the ones in edge bins are only candidates and their base data has to be examined to determine whether they are hits. We call this process of examining the edge bins the *candidate check*. The candidate check is often slow and significantly diminishes the value of a binned index.

In this paper, we propose to solve this problem by augmenting the bitmap index with an Order-preserving Bin-based Clustering (OrBiC) structure. This structure clusters the values for each bin together in the same order as they appear in the bitmap for the bin. The most important reason that the candidate check takes a long time is that the values fall in a bin are scattered in data files. The OrBiC data structure stores the values of each bin together and reduces the time needed for the candidate check. Our analysis and measurements on both synthetic and application datasets show that the total size of the bitmap index with the OrBiC structure can be smaller than the size of the bitmap index without binning. Systematic timing measurements showed that our strategy significantly outperforms both unbinned bitmap indexes and conventional binned bitmap indexes without the OrBiC structure.

The second innovation in this paper is the use of a hybrid of single-valued bins and multi-valued bins. We give an algorithm for creating these single-valued bins for both integer values and floating-point values. This allows us to reduce the time for candidate checks and the size of OrBiC structures. On low-cardinality attributes, this hybrid binning strategy produces mostly single-valued bins. On high-cardinality attributes, it assigns the most frequent values to single-valued bins. It has the advantages of both binned and unbinned bitmap indexes. The combination of the OrBiC structure with hybrid-binning is especially effective for high-cardinality attributes, and results in three-fold to several dozen-fold improvement over well known indexing methods. In summary, our binned bitmap index with OrBiC essentially overcomes the curse of cardinality.

2 Background and Related Work

The strategies for improving the performance of bitmap indexes on high-cardinality attributes can be categorized into three broad categories: compression, encoding and binning. In this section, we briefly review the common strategies from each category and show how our method improves upon existing strategies.

Compression is typically used to reduce the size of each bitmap in a bitmap index. Different compression methods have been used, including general text compression and specialized bitmap compression [11]. General text compression methods are effective in reducing the sizes of bitmaps, but require a long time to decompress bitmaps to answer a query. To improve the query response time, a number of specialized compression methods have been designed, such as Byte-aligned Bitmap Code (BBC) [12] and Word-Aligned Hybrid code (WAH) [13]. In particular, the WAH compression index was shown to have the same theoretical optimality as the B+-Tree index, that is the time required to answer range queries is bounded by linear functions of the number of hits [13,14]. However, in timing measurements compressed bitmap indexes outperform the B+-tree index significantly [15]. We use WAH compression in this work because WAH compressed indexes were found to be 12 times faster than BBC compressed ones while using about 50% more space [5].

Compression alone does not fully address the difficulty of applying bitmap indexes to high-cardinality attributes. In the extreme case where every value is distinct, the compressed bitmap index can be larger than a typical B-Tree index. For a dataset with N tuples, a WAH compressed index containing N bitmaps requires $5N$ words to store the bitmaps (details in Sec. 4.1), which is larger than the size of a typical B-Tree implementation. The size of a BBC compressed index might be smaller than that of a WAH compressed index, but both indexes would take more time to answer a query than the projection index [3]. Therefore, encoding and binning technique are used in addition to compression especially for high-cardinality attributes.

Bitmap encoding methods are applied to reduce the number of bitmaps used in a bitmap index. Among the different encoding methods, the bit-sliced index [3] (also called the binary encoding [16]) produces the least number of bitmaps. One shortcoming of this encoding method is that it needs to access nearly every bitmap to answer any query. There is a number of other encoding methods that produce more bitmaps than the binary encoding, but only have to access a small number of bitmaps to answer a query [17,18]. Still, using encoding methods alone also does not fully address the performance issues of high-cardinality attributes. In the worst case, even the most compact binary encoding produces an index that is as large as the projection index. Furthermore, the projection index usually outperforms such a binary encoded index because the operations on the bitmap index are more complex than operations on the projection index.

For attributes with extremely high cardinality, combining compression and encoding methods does not produce indexes that are competitive with projection indexes. This is because bitmaps produced by compact encoding schemes such as the binary encoded bitmaps typically do not compress well. In this case, compression does little to reduce the index sizes, but could increase the query response time. Thus, binning may be the most promising technique for high cardinality attributes.

Binning places multiple distinct values into a single bitmap and therefore reduces the number of bitmaps required for the bitmap index [7,8,10]. It allows a user to control how many bitmaps to use in an index. As mentioned earlier, the disadvantage is that one needs to perform the candidate check to resolve the edge bins. Performing the candidate check usually ends up touching a majority of the disk pages storing the base data even though the number of false positives may be small. The reason is that reading data from disk is performed in pages (typically, 4 KB or 8 KB)[1] and the candidates are usually scattered throughout the base data. Therefore the time required to answer a query is usually longer than that of the projection index. Our challenge here is to reduce the time needed for candidate check, or eliminate such checks when possible.

---

[1] Most I/O system also performs read-ahead, which reads 128 KB or more in one operation.

There are a number of recent papers that address the issue of how to reduce the number of candidate checks [19,20]. They optimize the placement of bin boundaries to minimize the average query response time by taking into account of data distribution and query workloads. However, their strategy do not reduce the time required for each candidate check procedure. In this work, we present an auxiliary data structure that allows us to directly reduce the candidate check time and thus the overall query processing time. This complements the existing work. In practice, because the exact query workload is usually not available before indexes are built, our approach is likely to be more effective because it does not rely on knowing the query workload.

In this work, we assume the base data is not modified or infrequently modified. Such data are common in extremely large data warehouses, where the only updates are bulk loading of a large number of new records. Similarly, most scientific applications generate or collect data records never modify their data records either [21]. For this reason, a number of research database systems such as C-Store [22] and MonetDB [23] make similar assumptions. Usually these systems can efficiently append new records, but they also implement a number of strategies to accommodate a small number of updates. For example, a special mask for deleted entries can be maintained and an update to a row can be treated as a deletion followed by an append. Using these strategies, many datasets can be treated as read-only.

3 The New Binned Index Structure

In this section, we explain the new binned index structure. The two key elements of this index are: an Order-preserving Bin-based Clustering (OrBiC) data structure and a hybrid-binning strategy that uses single-valued bins together with multi-valued bins. Before describing them, we first briefly review the basic binned bitmap index.

To build a typical binned bitmap index, one first chooses the bin boundaries. Next, each value in a bin is represented in the corresponding bitmap by setting a bit to 1. Figure 1(a) shows an illustration of a bitmap index with two bins for an attribute whose values can be between 0 and 1. Bin 0 is for values between 0 and 0.5 (not including 0.5) and bin 1 is for values between 0.5 and 1 (including 0.5). For a query requesting the rows with values greater than 0.3, all rows in bin 1 satisfy the condition – an interior bin that we need to access the bitmap; some rows in bin 0 (an edge bin) also may satisfy the query condition. We need to examine the base data for bin 1 to determine if they are actually hits. Because the base data in bin 1 is usually scatter on disk, the *candidate check* process is typically expensive. The proposed new data structure is to cluster the values according to the bin number and reduce the I/O cost.

3.1 Order-Preserving Bin-Based Clustering

During a candidate check, all values that fall in the edge bin must be examined. The rows for a given edge bin are known as soon as the index is built. Therefore, it is possible to reorganize the data so that the values in an edge bin are stored consecutively. We call this additional data structure for storing the reorganized data as Order-preserving Bin-based Clustering (OrBiC).

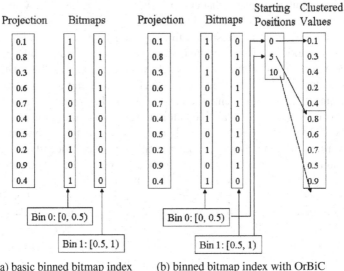

Fig. 1. An illustration of bitmap indexes with two bins

Typically, a bitmap index is built for a single attribute of a relation (a column of a table) at a time. In this setting, the OrBiC data structure can be thought of as a projection of the attribute, reordered according to the bin numbers, and preserving the relative order of values in each bin. To use these values efficiently, we also need to record the starting and ending positions of each bin. Because we preserve the relative order of values in each bin, their relative positions are the same as those bits that are 1 in the bitmap for the bin. This allows us to use the clustered values without their row identifiers.

An illustration of a bitmap index with two bins and the OrBiC structure is shown in Figure 1(b). In addition to the clustered values, we also store starting and ending positions of every bin. Since we assume the base data is read-only, the reordered values are of course also read-only. Their starting positions are never modified. This allows us to pack the clustered values in an array. We also pack the starting and ending positions together into one short array alongside the clustered values.

To perform a candidate check, we first read the starting positions to determine which values are needed. Since the values corresponding to each bin are packed consecutively on disk, they can be read sequentially. This significantly reduces the I/O cost associated with candidate checks as shown later in this paper. Additionally, since OrBiC data structure is part of a bitmap index, it does not affect the ordering of the base data and allow different attributes to have their own bitmap indexes with OrBiC.

3.2 Single-Valued Bins

Next we consider the issue of reducing the storage requirement for the OrBiC structure. Our approach is based on the observation that in real applications the distributions of data is hardly ever uniform, but skewed, that is some of the values appear much more frequently than the rest. For example, in a typical store, the sales records contain many more sales of lower priced items than higher priced ones. Similarly, in a climate simulation, there are fewer records with very high or very low temperature values. In these cases, removing the most frequent values from OrBiC significantly reduces the number of entries stored.

To facilitate the removal of frequent values, we store the actual minimum and maximum values in each bin. If a bin has the same minimum and maximum value, i.e., representing a single value, then no candidate check is ever needed for this bin. We call these bins the *single-valued bins*.

Knowing the minimum and the maximum also helps reduce the need for candidate check. In the example used in Figure 1(b), a query to find all records with values greater than 0.4 appear to have bin 0 as the edge bin. However, since the actual maximum of bin 0 is 0.4, we do not need to perform the candidate check.

3.3   Creating Single-Valued Bins

Now we discuss the mechanics of generating the single-valued bins. We first describe the procedure to specify bin boundaries to ensure that a single-valued bin actually holds only one value, and then describe a heuristic for assigning values to single-valued bins.

As mentioned before, our binning procedure starts with a set of bin boundaries. To create single-valued bins we need to specify the bin boundaries precisely so that the intended bins actually contain only one value each. Given a set of bin boundaries $\{b_0, b_1, \ldots, b_B\}$, we define a set of bins with closed left ends and open right ends. For example, the first bin contains values satisfying the following conditions $b_0 \leq x < b_1$ and the second bin contains values satisfying $b_1 \leq x < b_2$. Given this definition, to have single-valued bin for value b_i, we need to make sure that the next bin boundary is b_{i+1}, the smallest possible value that is larger than b_i. In digital computers, all numbers are discrete and it is possible to compute b_{i+1} quickly.

For integer attributes, the smallest possible value that is larger than b_i is $b_{i+1} = b_i + 1$. To compute the same for floating-point numbers, we rely on a parameter known as the *machine epsilon* (or the unit round off error) ϵ, which is defined to be the smallest number such that $1 + \epsilon > 1$ in floating-point arithmetic [24]. For a normal floating-point number b_i, we can compute b_{i+1} as $b_{i+1} = b_i(1 + \epsilon)$.

Now that we know how to specify the bin boundaries to make single-valued bins, the remaining challenge is to decide when to put a value in its own bin. In this work, we use a heuristic to produce approximate *equal-weight bins*. The overall goal is to make each bin have the same number of rows. Because there is no way to further divide a single value into multiple bins, a frequent value should be in its own bin. To make this decision, we need to know how many times each value appears in a dataset (i.e., the frequency). We may compute the exact frequencies or approximate them with sampling [25]. With the exact frequency counts, we can make more precise decisions, but it may take more time and space to collect the counts. We generate equal-weight bins by first identifying the most frequent value. If its frequency is no less than the average count for a bin, we place it in its own bin, otherwise we only have multi-valued bins. Once a single-valued bin is identified, the procedure is recursively applied to the left side and the right side of the single-valued bin. This heuristic requires the number of bins to be specified first, a topic we discuss in the next section.

4 Performance Analysis

In this section, we compute the worst case sizes and query processing costs of binned indexes with OrBiC. Our analyses use some earlier results on sizes of WAH compressed bitmaps [13]. In addition to understanding the performance characteristics of binning with OrBiC, we also use this study to explore options for deciding the number of bins to use.

4.1 Curse of Cardinality

To start with, we recall the main results about the most difficult case for compressed bitmap indexes, which is random data. We also use this opportunity to explain exactly what we mean by the curse of cardinality.

Let C denote the attribute cardinality of C and d_i denote the probability of value i in the dataset. Assuming each d_i is a constant independent of others, then the bitmaps generated for the basic bitmap index and the binned bitmap index are all *uniform random bitmaps* as defined in [13]. The key results we use for our analysis is the formula for the size of such a uniform random bitmap.

Following the definitions used in [13], we define w to be the number of bits in a word, N to be the number of rows in a dataset (also the number of bits in a bitmap of a bitmap index), and d to be the fraction of the bits that are 1 in a bitmap. The size of a uniform random bitmap is given by

$$m(d) = \left\lfloor \frac{N}{w-1} \right\rfloor + 2 - \left(\left\lfloor \frac{N}{w-1} \right\rfloor - 1 \right) \left((1-d)^{2w-2} + d^{2w-2} \right) \qquad (1)$$

$$\approx 3 + \frac{N}{w-1} \left(1 - (1-d)^{2w-2} - d^{2w-2} \right). \qquad (2)$$

The first part on the right-hand side of Equation (1), $\lfloor \frac{N}{w-1} \rfloor + 2$, is the maximum number of words that can be used by a WAH compressed bitmap. The remaining of the right-hand side is the expected number of words that can be removed by WAH compression [13]. Knowing the size of each bitmap, we can sum them up to give the total size of the bitmap index as $\sum m(d_i)$.

Note that Equation (1) is the exact formula from [13], while Equation (2) is a modification of the approximation given in [13]. This approximation is more accurate, particularly for very low bit densities. Since the bitmaps with only 0s are not stored in a bitmap index, the minimum bit density is $d = 1/N$. In this case, the two formulas from Equations (1) and (2) give the same value, 5, which is accurate in our experience. It is possible that every bitmap has a bit density of $1/N$ if every value of an attribute is distinct. In that case, there are N such bitmaps and the total size of bitmaps is $5N$

words[2]. This total size is larger than a typical B-Tree index which is observed to be 3 - $4N$ words, and it is also larger than the size of a projection index which uses exactly N words. This is one aspect of the curse of cardinality: even with an effective compression, the bitmap index size can be larger than commonly used indexes.

Associated with the increase in index sizes, the query processing time would also increase because of the increased time to perform I/O operations and to operate on the compressed bitmaps. For an attribute with typical high cardinality, say, $C < N/10$, the compressed index size is about $2N$ words, and WAH compressed indexes never take more time than scanning the vertical projections (also known as projection indexes). However, as the attribute cardinality further increases, the indexes would have more than $2N$ words and the WAH compressed indexes would take more time than the projection indexes for increasingly more queries. This is the second aspect of the curse of cardinality for bitmap indexes.

4.2   Sizes of Binned Indexes

Equation (1) gives us a way to compute the expected sizes of bitmaps used in an index. Since we assume the base data is read-only, these bitmaps will not change and therefore can be densely packed together one after another [26]. In an index file containing such a set of packed bitmaps, we also need to store the starting positions of bitmaps and bin boundaries. Since the bitmaps follow each other, we need to know the starting position of bitmap i and bitmap $i + 1$ in order to read the content of ith bitmap. To allow the last bitmap to be handled the same way as the rest, we store the position just after the last byte of the last bitmap as the starting position of a nonexistent bitmap. Altogether we store $B + 1$ starting positions for B bitmaps.

To define B bins for a variable that can take values between 0 and 1, we need to define $B - 1$ bin boundaries between 0 and 1. One may choose to store bin boundaries with or without the values 0 and 1. In our test software, we choose to store the value 1, but not the value 0. This allows us to easily count values less than b_i. This way, we store B bin boundaries for B bins. We also store the actual minimum and maximum value of each bin, which leads to an additional $2B$ values. Assuming that each value is stored in one word, the total size of the bin boundaries and the minimum and maximum values is $3B$ words.

Assuming no single-valued bins, the total number of the clustered values is exactly the number of rows, N. We need $B + 1$ values to record the starting position of each bin. Altogether, the total size (in number of words) of our binned index is

$$S = \sum m(d_i) + N + 5B + 2. \tag{3}$$

In later experiments, we use a series of synthetic data following Zipf distribution, where the value i, between 0 and $C - 1$, has the probability that is proportional to $(i + 1)^{-z}$.

---

[2] There are N words out of these $5N$ words that have the same value in each bitmap, therefore it is possible to replace these N words with one word [13]. However, doing so makes all the bitmaps depending on this single parameter which makes it more difficult to create the bitmap indexes. To simplify the testing software, we have chosen to keep a counter in each bitmap.

The constant z is known as the Zipf exponent ($z \geq 0$). When $z = 0$ and C is an integer multiple of B, we can simplify the formula for the total size of our binned index as follows.

$$S_{z=0} \approx \left(3 + \frac{N}{w-1}\left(1 - (1 - \frac{1}{B})^{2w-2} - (\frac{1}{B})^{2w-2}\right)\right) B + N + 5B + 2 \quad (4)$$

$$\approx 3N + 8B. \quad (5)$$

Note that the approximation in Equation (5) is accurate when $1 - (1 - 1/B)^{2w-2} \approx (2w - 2)/B$ is accurate, which is true for large B, say $B > 1000$. For smaller B, we need to go back to Equation (4) to compute the index size accurately. The expressions for the equal-weight bins and for non-zero Zipf exponents can be similarly computed even though they are not as easily simplified as in this special case.

4.3   Query Processing Cost

Next we compute the number of words accessed when answering an average query. To simplify the discussion, we only consider 1-sided range queries of the form $x > c$, where c is a constant we call the *query boundary*. The analysis we carry out here can be similarly applied to equality queries and 2-sided range queries. In this analysis, we only consider the amount of data to be read from disk in order to answer a query. To further simplify the discussion, we assume the query boundaries are a uniform sample of all distinct values that appears in the dataset. The main measure we use to judge the effectiveness of an indexing method is the average number of words needed to answer such a 1-sided range query.

To answer a range query with a binned bitmap index, two steps are needed. Step 1 operates on the bitmaps to identify which bins are fully contained in the query range and which are edge bins that require a candidate check. Step 2 performs the candidate check. For any 1-sided range query, there can be at most one edge bin. For simplicity, we assume a candidate check is always necessary. Before the evaluation can start, we always read the starting positions of the bitmaps, the bin boundaries, and the minimum values and the maximum values of each bin to memory. This process reads $4B + 1$ words. Clearly, this is the worst case scenario; one could cache these values to reduce the query response time.

In Step 1, the main cost is reading the bitmaps from disk. If we need to read more than half of the bitmaps (as measured by the number of words accessed), we can evaluate the complement of the query instead. This allows us to read no more than half of the words in the bitmap index. Given that the query boundaries are uniformly distributed in the domain of the attribute, the average number of words accessed in Step 1 is one quarter of the total size of the bitmaps ($0.25 \sum m(d_i)$).

In Step 2, the main cost is reading the values in the edge bin. We simply take this cost to be N/B words. The cost of reading the starting and ending positions of the bin could be taken as two words. However, because the underlying file system reads at least one page, we approximate it by assuming that all $B + 1$ starting positions are read. Overall, the total number of words read from disk is

$$R = 2 + 5B + N/B + 0.25 \sum m(d_i). \quad (6)$$

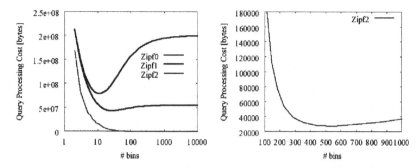

Fig. 2. The average number of words accessed to answer a 1-sided range query using bitmap indexes with equal-weight bins and OrBiC ($N = 10^8$, $C = 10^6$, $w = 32$)

With equal-weight bins and for nonuniform data, we can similarly compute the expected cost to process an average query. Because of the use of single-valued bins, the expressions for these cases are much longer than Equation (6). Instead of giving analytical expressions, we plotted them in Figure 2.

Figure 2 plots the average number of words accessed to answer a query. The figure shows the query processing costs for three synthetic attributes, Zipf0, Zipf1 and Zipf2 (named after their Zipf exponents). We identify the optimal number of bins to be about 13 for Zipf0, 25 for Zipf1 and 550 for Zipf2. We also note that the average query processing cost is not sensitive to the number of bins for Zipf1 and Zipf2. For example, in the case of Zipf2 any number of bins from 400 to 1000 leads to nearly the same query processing cost.

Next we compare the query processing cost of our new method against that of the unbinned bitmap index and the projection index. We first consider the unbinned bitmap index. Again we assume that one quarter of words has to be read on average. Thus, the query processing costs of an attribute where every value is distinct is $1.25N$ words. In Figure 2, we assumed $N = 10^8$, which means the query processing cost is at worst 5×10^8 bytes. The query processing cost of the projection index is always N words. In Figure 2, this corresponds to a query processing cost of 4×10^8 bytes. We clearly see that query processing of our binned index with OrBiC costs less than 4×10^8 bytes. With the optimal number of bins, the expected query processing cost for Zipf0 is about 8×10^7 bytes, which is about 1/5th of that of the projection index. In a more realistic case, where the unbinned index is close to $2N$ words, the average query processing cost is about 2×10^8 words, the binned index with OrBiC is about 3 ($\sim 2 \times 10^8/8 \times 10^7$) times as fast as the unbinned index. The query processing cost for nonuniform data is much less than that for Zipf0 as evidenced by experiments in the next section.

In Figure 3, we show a comparison between the query processing cost of a binned bitmap index with OrBiC and one without OrBiC. It is easy to see that without OrBiC we need to use more than 1000 bins in order for the query processing cost to be less than that of the projection index. However, with OrBiC, the query processing cost is always below that of the projection index.

Overall, the analyses here show that the new binned bitmap index with OrBiC and equal-weight binning outperforms well-known methods for answering range queries

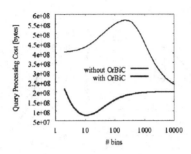

Fig. 3. The average number of words accessed to answer a 1-sided range query on Zipf0 using binned bitmap indexes with and without OrBiC

on read-only data. Since our analyses do not include the CPU time or I/O overhead such as disk seek time, the actual observed query response time could show different performance characteristics. Next, we conduct a number of tests on both synthetic and real application data to measure the actual performance.

5   Data Sets and Index Sizes

In this section we describe the synthetic and application datasets used for our performance evaluation. We also discuss the sizes of bitmap indexes with different numbers of bins and compare their sizes with unbinned bitmap indexes and the expected values computed in the previous section.

All experiments were conducted on a computer with dual 2.8 GHz Pentium 4 processors, 2 GB of main memory, and an IDE RAID storage system capable of sustaining 60 MB/sec for reads and writes. Our bitmap index software is implemented with C++ and compiled with gcc 4.1.0 using the compiler optimization flag -O5.

5.1   Zipf Data

The synthetic data set consists of three high-cardinality attributes following the Zipf distribution with Zipf exponents 0, 1, and 2. We refer to these three attributes as Zipf0, Zipf1 and Zipf2. The number of rows is 100 million. The total size of the data set is 1.2 GB. The three synthetic attributes have non-negative values less than 1 million. Their cardinalities are much higher than those used in the earlier tests [11,13].

5.2   Astrophysics Data

Our application data set is from an astrophysics application that studies supernova explosions. The data consists of 6 high-cardinality floating-point valued attributes with 110 million rows. The average attribute cardinality of these attributes is about 25 million. The total size of the data set is about 2.6 GB.

The distributions of two attributes are shown in Figure 4. Note the log scale on the y-axis. We show the distribution of x-velocity as the representative of the three velocity components. These attributes have some infrequent values, but the majority of the

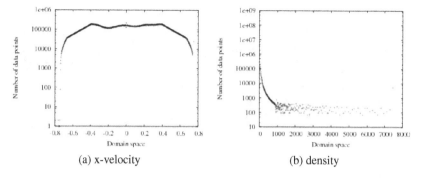

(a) x-velocity   (b) density

Fig. 4. Distribution of astrophysics data. Note the log scale on the y-axis

values have frequencies within the same order of magnitude. The other three attributes, density, entropy and pressure, have much higher skew in their distribution. We show the distribution of density as the representative. In this case, we see that the frequencies of values span eight orders of magnitudes.

5.3 Index Sizes

Next we examine the bitmap index sizes for both data sets. We start our discussion with the Zipf data.

For each of the three Zipf attributes Zipf0, Zipf1 and Zipf2, we generated bitmap indexes with 10, 20, 50, 100, 1000 and 10000 equal-weight bins, where each bin has about the same number of values. Each of these variants also include an instance with OrBiC and an instance without OrBiC. In addition, we also generated bitmap indexes with no binning. In total we generated 13 different bitmap indexes per attribute. The sizes of the bitmap indexes for the Zipf data are shown in Figure 5. We label the indexes with equal-weight binning (without OrBiC) as "binning", and the indexes with both equal-weight binning and OrBiC as "binning with OrBiC". For references, we also plotted the size of the base data as the solid horizontal line, the unbinned index size as the dashed line, and the expected sizes according to the analyses from the previous section as 'x'.

Note that the expected sizes of the bitmap indexes agree very well with the actual sizes in Figure 5. We see that the size of the bitmap index with binning (without Or-BiC) is always smaller than the bitmap index without binning. As the number of bins increases, the curves marked "binning" and "binning with OrBiC" become closer for Zipf2, because the OrBiC data structure stores less and less values as more and more bins become single-valued.

The sizes of the bitmap indexes for the astrophysics data are shown in Figure 6. Again we see that the bitmap index with binning is always smaller than the bitmap index without binning[3]. We also note that the bitmap index with binning and OrBiC

---

[3] Due to extensive resource requirements of bitmap indexes without binning we had to build the unbinned indexes on a server with more than 2GB of main memory.

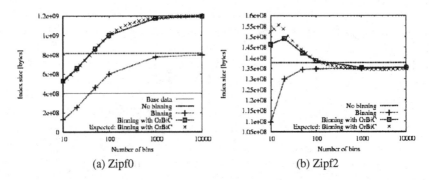

Fig. 5. Sizes of the bitmap indexes with different numbers of bins for Zipf data

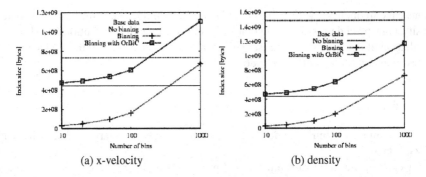

Fig. 6. Sizes of the bitmap indexes with different numbers of bins for astrophysics data

is only larger than the unbinned bitmap index in one case, namely for the attribute x-velocity with 1000 bins with OrBiC. These results clearly demonstrate that our novel binned bitmap index is able to take advantage of the non-uniformity present in the data to reduce the sizes of bitmap indexes.

6   Query Processing Time

In this section we report an experimental evaluation of our binned indexes on both synthetic and application data. The experiments are structured as follows. We first compare the performance of range queries for binned bitmap indexes with and without OrBiC. The results show that binned bitmap indexes with OrBiC are about a factor of 3 to 25 faster than those without OrBiC. Next, we run a set of tests to measure the relative performance of binned index with OrBiC against unbinned indexes and projection indexes. Because these indexes are known to significantly outperform the more popular B-Tree index [2,3,5], we do not compare with B-Tree indexes directly.

All experiments are based on so-called aggregation queries that are common in data warehousing and scientific applications. These types of queries provide statistical information on the result set rather than returning result records. A typical example of an aggregation query in an astrophysics application is as follows: "Count the number of cells where pressure > X". Note that before we executed each set of queries, we unmounted and remounted the file system containing both the data and the indexes in order to ensure cold cache behavior. The timing values reported here are elapsed time in seconds.

In our first set of experiments we evaluate the query performance of binned bitmap indexes with and without OrBiC. To simplify the following discussion, we fix the number of bins at 100. We start our performance evaluation with range queries over synthetic data. The cardinality of each synthetic attribute is 1,000,000. For each attribute we ran 10 range queries with ranges uniformly distributed over the entire attribute domain space. In particular, the query workload is as follows: $a \geq C/Q * q + \delta$ where a is the query attribute, C is the attribute cardinality, Q is the total number of queries, q is the query number and δ is a small value to make sure that the query range does not fall on a bin boundary. Note that if a query range falls on a bin boundary there is no need for a candidate check. Hence, the query performance for binned bitmap indexes with OrBiC is equal to those without OrBiC.

Figure 7(a) shows the query response time for binned bitmap indexes with and without OrBiC for Zipf0, i.e. uniformly distributed data. We see that binned bitmap indexes with OrBiC are about a factor of 3 faster than binned bitmap indexes without OrBiC. This agrees with Figure 3 for 100 bins. In Figure 7(a), we notice that both timing curves show a characteristic "A shape", because we evaluate the complement of the query if more than half of the bitmaps are involved.

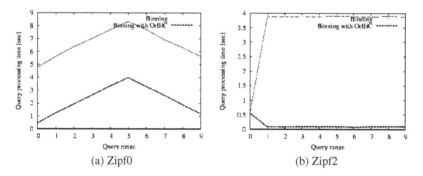

(a) Zipf0   (b) Zipf2

Fig. 7. Processing time of range queries over Zipf data. The figure shows a significant performance advantage of binned bitmap indexes with OrBiC over binned bitmap indexes without OrBiC.

Figure 7(b) show the query response time for binned bitmap indexes on Zipf2. Again we see that the binned bitmap indexes with OrBiC are significantly faster than those without OrBiC. In particular, the average performance improvement is a factor of 25.6 for Zipf2 (and 5.5 for Zipf1). These speedup values are larger than that for Zipf0 indicating that the advantage of using OrBiC increases as the skewness of data increases.

Figure 8 shows the query response time for range queries over two attributes of the astrophysics data set. Again we see a significant performance increase for binned bitmap indexes with OrBiC compared with bitmap indexes without OrBiC.

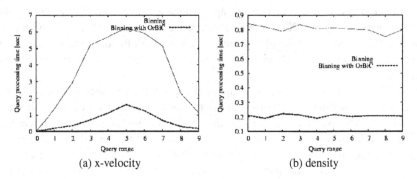

(a) x-velocity   (b) density

Fig. 8. Processing time of range queries over astrophysics data. The figure shows a significant performance advantage of binned bitmap indexes with OrBiC over binned bitmap indexes without OrBiC.

Table 1 summarizes the average performance improvements of binned bitmap indexes with OrBiC over binned bitmap indexes without OrBiC. The advantage of using OrBiC is the least for uniform data, which has a speedup value of 3. On our application data and synthetic data with skew, the speedup values are larger.

Table 1. Average speedup of binned bitmap indexes with OrBiC over binned bitmap indexes without OrBiC

Synthetic		Astrophysics			
Attribute	Speedup	Attribute	Speedup	Attribute	Speedup
Zipf0	2.94	density	3.91	x_velocity	5.65
Zipf1	5.50	entropy	12.61	y_velocity	4.82
Zipf2	25.62	pressure	4.40	z_velocity	4.28

The previous set of tests clearly confirms the advantage of using OrBiC. In the following tests, we compare binned indexes with OrBiC with two other types of indexes, the unbinned bitmap index and the projection index. In this set of tests, we use the average query response time over all queries to compare different indexing methods. We start with a comparison against the unbinned index in Figure 9.

In Figure 9, the vertical axis shows the average query response time to answer the same 10 range queries used in our previous tests. The horizontal axis shows the number of bins used by the binned index. Figure 9(a) is for the uniform data Zipf0 and Figure 9(b) is for the highly skewed data Zipf2. In the case of the uniform data, where the unbinned index require about 5.4 seconds to answer a query[4], the binned index with OrBiC is always better. In the best case, the new binned index with OrBiC is about three times faster than unbinned index, which agrees with the analysis given in Section 4.3.

---

[4] Assuming that 5.4 seconds were used to read a quarter of bitmaps, totaling about 8×10^8 bytes, the effective reading speed is 37 MB/s, which is about 2/3 of the maximum reading speed.

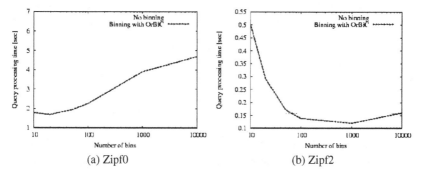

Fig. 9. The average query response time with the new binned index with OrBiC and the unbinned bitmap index

On highly skewed data, the unbinned index performs very well, using about 0.15 seconds instead of 5.4 seconds. However, even in this case, the new binned index with OrBiC can outperform the unbinned index with a wide range of choices as the number of bins.

Figure 10 shows the relative performance of the new binned index against the projection index. The vertical axis is the speedup of range queries using binned bitmap indexes over the projection index and the horizontal axis is the number of bins used by the binned index. Overall, we see that the speedup values are always greater than 1, indicating the new binned index with OrBiC is always faster than the projection index.

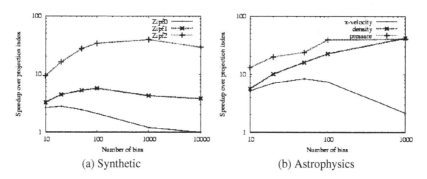

Fig. 10. Speedup of range queries using binned bitmap indexes over projection index

Figure 10(a) shows the results for synthetic data. For Zipf0 with 20 bins the speedup over the projection index is about a factor of 3. For Zipf1 and Zipf2, the highest speedup is about a factor of 6 and 40, respectively.

Figure 10(b) shows the average speedup for the astrophysics dataset. We can see that for attribute x-velocity bitmap indexes with about 50 bins have the highest speedup of about a factor of 8. Note that the attribute x-velocity is of moderate skewness. However, for the highly skewed attributes such as density and pressure, the optimal number of bins is much higher and is around 1000. Additional analyses are required to fully understand the dependency of the number of bins on the skewness of data. However, the benefit of using OrBiC is clear; the highest speedup over the projection index for the attributes density and pressure is about a factor of 40.

7   Conclusions

The basic unbinned bitmap indexes used in major commercial database systems are efficient for querying low-cardinality attributes and highly skewed data. However, they suffer from the *curse of cardinality*, i.e. their effectiveness decreases as attribute cardinality increases. We solve this important problem by using a novel binned bitmap index structure that performs efficiently even with extremely high-cardinality attributes. One key idea is to augment the bitmaps with an *Order-preserving Bin-based Clustering (OrBiC)* data structure. This data structure significantly reduces the cost of candidate checks. In addition, we use a hybrid-binning strategy that employs single-valued bins for frequent values to eliminate the need for candidate checks for these single-valued bins. This further enhances the search performance. We performed detailed analytical and experimental evaluations of our bitmap index structure and showed that our binned bitmap indexes are in most cases smaller in size and more efficient in answering queries than the unbinned bitmap indexes and the projection indexes.

In the worst case for a unbinned bitmap index, that is when the attribute has very high cardinality and uniform data distribution, our analysis provides definitive guidance on the number of bins to use. In addition, the predicted advantage has been verified experimentally as well. For more realistic data, where the single-valued bins enhance performance, a precise analysis of query processing cost (Equation (6)) is more complicated, and how to determine the optimal number of bins remains a challenge. Nevertheless, we experimentally demonstrated that our technique effectively broke the curse of cardinality.

The software that implements our binning technique has been released under LGPL and can be accessed at https://codeforge.lbl.gov/projects/fastbit.

References

1. Berchtold, S., Böhm, C., Kriegal, H.P.: The pyramid-technique: Towards breaking the curse of dimensionality. SIGMOD Record 27(2), 142–153 (1998)
2. O'Neil, P.: Model 204 architecture and performance. In: Second International Workshop in High Performance Transaction Systems. Springer, Heidelberg (1987)
3. O'Neil, P., Quass, D.: Improved query performance with variant indices. In: SIGMOD. ACM Press, New York (1997)
4. Wu, K., Otoo, E.J., Shoshani, A.: On the performance of bitmap indices for high cardinality attributes. In: VLDB, pp. 24–35. Morgan Kaufmann, San Francisco (2004)
5. Wu, K., Otoo, E., Shoshani, A.: A performance comparison of bitmap indices. In: CIKM. ACM Press, New York (2001)
6. Lewis, J.: Bitmap indexes - part 1: Understanding bitmap indexes (2006), http://www.dbazine.com/oracle/or-articles/jlewis3
7. Koudas, N.: Space efficient bitmap indexing. In: CIKM. ACM Press, New York (2000)

8. Shoshani, A., Bernardo, L.M., Nordberg, H., Rotem, D., Sim, A.: Multidimensional indexing and query coordination for tertiary storage management. In: SSDBM, pp. 214–225 (1999)
9. Stockinger, K., Duellmann, D., Hoschek, W., Schikuta, E.: Improving the performance of high-energy physics analysis through bitmap indices. In: DEXA. Springer, Heidelberg (2000)
10. Wu, K.L., Yu, P.: Range-based bitmap indexing for high cardinality attributes with skew. Technical Report RC 20449, IBM Watson Research, New York (1996)
11. Johnson, T.: Performance Measurements of Compressed Bitmap Indices. In: VLDB. Morgan Kaufmann, San Francisco (1999)
12. Antoshenkov, G.: Byte-aligned Bitmap Compression. Technical report, Oracle Corp. U.S. Patent number 5,363,098 (1994)
13. Wu, K., Otoo, E., Shoshani, A.: Optimizing bitmap indices with efficient compression. ACM Transactions on Database Systems 31, 1–38 (2006)
14. Comer, D.: The ubiquitous B-tree. Computing Surveys 11(2), 121–137 (1979)
15. Wu, K., Otoo, E.J., Shoshani, A.: Compressing bitmap indexes for faster search operations. In: SSDBM, pp. 99–108 (2002)
16. Wong, H.K.T., Liu, H.F., Olken, F., Rotem, D., Wong, L.: Bit transposed files. In: Proceedings of VLDB 1985, pp. 448–457. Stockholm (1985)
17. Chan, C.Y., Ioannidis, Y.E.: Bitmap Index Design and Evaluation. In: SIGMOD. ACM Press, New York (1998)
18. Chan, C.Y., Ioannidis, Y.E.: An Efficient Bitmap Encoding Scheme for Selection Queries. In: SIGMOD. ACM Press, New York (1999)
19. Rotem, D., Stockinger, K., Wu, K.: Minimizing I/O costs of multi-dimensional queries with bitmap indices. In: SSDBM. IEEE, Los Alamitos (2006)
20. Rotem, D., Stockinger, K., Wu, K.: Optimizing candidate check costs for bitmap indices. In: CIKM. ACM Press, New York (2005)
21. Gray, J., Liu, D.T., Nieto-Santisteban, M., Szalay, A., DeWitt, D., Heber, G.: Scientific data management in the coming decade. CTWatch Quarterly (2005)
22. Stonebraker, M., et al.: C-store: A column-oriented dbms. In: VLDB, pp. 553–564 (2005)
23. Boncz, P.A., Zukowski, M., Nes, N.: Monetdb/x100: Hyper-pipelining query execution. In: CIDR, pp. 225–237 (2005)
24. Golub, G.H., van Loan, C.F.: Matrix Computations, 3rd edn. The Johns Hopkins University Press (1996)
25. Thaper, N., Guha, S., Indyk, P., Koudas, N.: Dynamic multidimensional histograms. In: SIGMOD, pp. 428–439. ACM, New York (2002)
26. O'Neil, E., O'Neil, P., Wu, K.: Bitmap index design choices and their performance implications. In: IDEAS, pp. 72–84 (2007)

A New Approach for Optimization of Dynamic Metric Access Methods Using an Algorithm of Effective Deletion

Renato Bueno[1], Daniel dos Santos Kaster[2,*], Agma Juci Machado Traina[1],
and Caetano Traina Jr.[1]

[1] Department of Computer Science, University of São Paulo at São Carlos, SP, Brazil
{rbueno,agma,caetano}@icmc.usp.br
[2] Department of Computer Science, University of Londrina, Londrina, PR, Brazil
dskaster@uel.br

Abstract. The existing Metric Access Methods (MAM) assume the data elements represent immutable objects. However, many applications must handle complex data evolving over time. Health care, weather monitoring, and other applications require removing or updating elements. Most of the MAM presented in the literature either do not have the deletion operation described, or it is performed just marking the element as deleted without effectively removing it from the structure. In this paper we describe an algorithm that effectively removes any element from a metric tree. While maintaining the height-balancing of the structure, the proposed deletion algorithm uses mechanisms to enforce a reduced number of pages in the tree, improving the query performance. Based on the deletion algorithm, we propose a new way to optimize a MAM, which we call the *Push-pull* technique. It reduces the node overlap performing the deletion and reinsertion of elements close to the border of each node covering region. We also developed the *Smart Push-pull* algorithm, which uses statistical data about subtrees' overlapping to calculate how many elements should be removed from each node. The statistics are collected during the evaluation of the structure overlap, an operation employed to ascertain the need to trigger an optimization process. The experiments were run on the Slim-tree and showed a reduction of overlap and a query performance improvement over trees optimized by this technique as compared over trees optimized by the Slim-down method.

1 Introduction

Database Management Systems (DBMS) were initially developed to deal with numbers and small character strings. However, DBMS are being increasingly required to support more complex data types, such as images, video, audio, geo-referenced data and genetic sequences. Distinctly from traditional data, complex

---

* On leave at Department of Computer Science, University of São Paulo at São Carlos, SP, Brazil.

B. Ludäscher and Nikos Mamoulis (Eds.): SSDBM 2008, LNCS 5069, pp. 366–383, 2008.

data usually do not possess the total ordering property. Therefore, comparison operators ('<', '≥', '≤', '>') cannot be employed. Moreover, equality comparisons are almost useless, since it is very difficult to have two complex elements exactly equal. To compare data in those domains, the similarity among elements is the most relevant operation [1].

Although complex elements can be understood as points in a vector or a multidimensional space, there are domains that do not have a dimensionality, such as words and genetic sequences. However, similarity can be adequately expressed when data is represented in a metric space.

A metric space is formally defined as a pair $M = \langle S, d \rangle$, where S denotes the universe of valid elements and d is the function $d : S \times S \rightarrow \mathbb{R}^+$ that expresses the distance, or (dis)similarity, between elements in S. Thus, d must satisfy the following properties, $\forall s_1, s_2, s_3 \in S$: (1) symmetry: $d(s_1, s_2) = d(s_2, s_1)$; (2) non-negativity: $0 < d(s_1, s_2) < \infty$ if $s_1 \neq s_2$ and $d(s_1, s_1) = 0$; and (3) triangular inequality: $d(s_1, s_3) \leq d(s_1, s_2) + d(s_2, s_3)$.

Several Metric Access Methods (MAM) have been developed to speed up similarity query answering. However, most existing MAM consider that each data element represents an unchangeable object of the real world. Some of them are static and even those that are dynamic, allowing random insertions, hardly allow the deletion operation. In fact, it is common to perform the deletion operation by marking the element as deleted without effectively removing it from the structure. This alternative is enough when few deletions occur, but it is not acceptable when frequent deletions and/or updates occur.

For instance, consider an application for controlling epidemics and dangerous diseases, such as the recent cases of avian influenza. Patients suspect of infection by any of the monitored diseases are added to the database, storing also the results of their various medical exams. Storing the patients' exams enables searching for other people with similar health conditions and allows identifying associations in the population spread across the world. However, many false positives can be included that must be shortly removed upon notice.

Whereas deleting an element stored in a leaf node is straightforward (ignoring the minimum node occupancy rule), removing elements of an internal node that direct searches through the index is a tough task. When updates and deletions are frequent, just marking representatives as deleted becomes inappropriate, because they remain in the structure, increasing both the memory usage, the number of disk accesses and the number of distance calculations required.

In this work we propose an algorithm for effective deletion in metric access methods organized as hierarchical trees. Our approach enables deleting any element, including those stated as representatives. It employs a mechanism to reduce the structure reorganization, without requiring to rebuild all the subtree covered by a deleted representative. We show experiments confirming that subsequent queries over a tree that uses the effective deletion algorithm are faster than those over a tree with the marking as deleted technique.

Using the delete operation, we developed a new optimization technique for MAM already built. When the structure becomes "fat" (i.e. with many overlaps among nodes), the proposed technique removes elements near the border of each covering region and reinsert them in the structure. The technique, which we call *Push-pull*, is much cheaper than searching for a better location for each element in each level of the tree during the reinsertion, which can be prohibitively costly. We also developed the *Smart Push-pull* algorithm, which employs this technique and uses statistical data from the subtrees' overlapping to define how many elements should be removed from each node. The statistics are collected during the evaluation of the structure overlap, an operation employed to ascertain the need to trigger an optimization process. The experiments were run on the Slim-tree, but the presented algorithms can easily be applied to other hierarchical structures. As shown in the experiments, Push-pull reduces the overlap and speeds up executing queries, outperforming the current Slim-tree optimization technique Slim-down. Notice that this approach is feasible only when elements can be effective removed, reinforcing the importance of the proposed deletion algorithm.

In the next section we discuss basic concepts and related work to this research. Section 3 presents our algorithm for effective deletion and Section 4 details our new optimization technique for MAM. Section 5 shows the results of experiments performed with the proposed techniques. Finally, Section 6 concludes the paper.

2   Background and Survey

In this section we discuss the fundamental concepts involved in our work. Subsection 2.1 shows a review of interesting MAM and introduces the challenges of dealing with deletions in MAM. Subsection 2.2 presents the drawbacks related to node overlap in metric trees.

2.1   Deletion in Metric Access Methods

Many data structures were developed to index data in metric domains, where only the data elements and the distances (dissimilarity) among them are available. The methods proposed by Burkhard and Keller [2] were the starting point of MAM development. They introduced the techniques for recursive partitioning of metric datasets that led to the construction of MAM. Based on these ideas, various MAM were proposed, such as the *GH-tree* [3], *VP-tree* [4], *MVP-tree* [5], and *GNAT* [6]. However, all of those primordial structures are constructed in a single operation using the whole dataset, and neither insertions nor deletions are allowed.

The first dynamic MAM developed was the *M-tree* [7]. It is a height-balanced tree, where elements are stored in leaf nodes that correspond to regions of the metric space. Elements are grouped around special elements, named representatives, in order to cluster similar elements. Each representative is the center of a ball with a radius which covers all elements of the subtree rooted at it. Every element stored in the node or in any subtree rooted at the node has a distance to the node representative equal or smaller than its covering radius. The regions delimited by the nodes can intersect, generating the problem of node overlapping (see Subsection 2.2). The MAM *Slim-tree* [8] has a structure similar to the M-tree, however it improved M-tree presenting the first technique to measure and to reduce the amount of overlap between subtrees in metric spaces. Another dynamic MAM is the *DBM-tree* [9], which reduces the amount of overlap between nodes relaxing the height-balancing of the structure.

The development of dynamic MAM focused on supporting insertions and efficient query answering, but neglected the deletion of elements. Except by the DBM-tree, which allows the deletion by unbalancing the structure, none of them provide algorithms for the complete remotion of elements. However, removing the height-balancing mandate allow trees to, eventually, degenerate completely.

Deleting elements in metric trees can force large tree reorganizations. Although it is simple to remove elements that have not been used as representatives in index nodes, removing representatives require rebuilding the whole subtree centered on it. This operation can be very expensive, and can lead to the reconstruction of the whole tree when the removed element is in the root node. Therefore, the challenge of the removal algorithm is to reduce the required reorganization without degenerating the structure and without increasing node overlaps.

Variants of the *R-tree* [10] implement the deletion while maintaining the height-balancing property. However, the space division follows an approach totally different, which cannot be employed in metric spaces, as split rules cannot be directly used.

In almost every hierarchical MAM (and in all balanced MAM), it is suggested that the deletion of representative elements be performed just marking them as removed. This alternative, which we call here *m-delete* (mark-as-deleted), becomes inappropriate when applications perform a large number of deletions. In fact, maintaining deleted elements on the structure increases the consumption of memory and disk space, the number of disk accesses and also the number of distance calculations, as it forces comparisons with elements that do not exists any more.

In Section 3, we present the first algorithm for effective deletion of elements in height-balanced MAM, regardless where they are stored, in a leaf or index node.

2.2   Overlap in Metric Trees

The division of the metric space of almost every dynamic MAM does not generate disjunct regions, producing overlaps among nodes at a same level of the tree. This effect is undesirable, as it reduces the ability to prune subtrees. In fact, as opposed to search trees where the cost of a point query is always one disk access per level, the cost of a point query over a MAM also depends on the amount of overlaps.

The first approach to evaluate the amount of overlaps of the nodes of a tree was the *fat-factor*, defined over the Slim-tree [8]. The overlap between two nodes was defined as the number of elements in the corresponding subtrees covered by both regions, divided by the number of elements in both subtrees. Thus, the absolute fat-factor of a Slim-tree T with height H storing N elements on M disk pages is given by:

$$Fat(T) = \frac{I_C - H * N}{N} * \frac{1}{(M - H)} \qquad (1)$$

where I_C denotes the total number of node accesses required to answer a point query for each of the N elements stored in the metric tree.

The fat-factor is defined on the range $[0, 1]$, where large values denote trees with high degree of overlapping, and $Fat(T) = 0$ indicates an ideal tree (no overlap between nodes). The absolute fat-factor is a measure of the amount of elements that lie inside intersecting regions defined by nodes at same level of a MAM. However, if two trees storing the same dataset have a different number of nodes, the direct comparison of the corresponding absolute fat-factor will not give such an indication. To enable the comparison of two different trees that store the same dataset, they also proposed the *relative fat-factor*, which "penalizes" trees that use more than the minimum required number of nodes. The relative fat-factor considers not the height and number of nodes in the real tree, but that of the minimum tree:

$$rFat(T) = \frac{I_C - H_{min} * N}{N} * \frac{1}{(M_{min} - H_{min})} \qquad (2)$$

where $H_{min} = \lceil log_C N \rceil$, and the minimum number of nodes for a given dataset is $M_{min} = \sum_{i=1}^{H_{min}} \lceil N/C^i \rceil$, where C is the capacity of the nodes. The value of $rFat(T)$ varies from zero to a positive real number. The smaller the number, the fewer disk accesses will be needed to answer a query.

Supported by the fat-factor measure, Traina Jr. et.al [8] also proposed a technique to minimize the node overlap in metric trees, named *Slim-down*, which is intended to be executed after the tree construction. When sibling leaf nodes overlap themselves, the Slim-down performs the "migration" of the farthest element of a node into a sibling node that also covers the element. As the migration reduces the covering node radius without increasing the radius of any other node, the overlap is reduced. Nodes that become empty are removed, contributing to further overlap reduction. This procedure is repeated until no element migrates between siblings nodes. The authors reported results where Slim-down reached improvements from 10% to 40% in the number of disk accesses for range queries.

The Slim-down optimization algorithm restricts the covering radius shrinking to the leaf nodes of a subtree. Therefore, when two leaf nodes rooted at different index nodes overlap each other, no improvement is achieved. Skopal et. al [11] developed the *Generalized Slim-down* algorithm, which addresses this point traversing each level of the tree, starting on the leaf level. For each node on a given level, a better location for each element stored in the node is pursued executing a modified range query. The processing of a given level is repeated until no element is moved anymore. When a level is finished, the algorithm processes the next higher level. Although this algorithm can move elements between

Algorithm 1: Effective deletion algorithm

```
1    Delete(s_d)
2    INPUT: The element  s_d  to be deleted
3    OUTPUT: TRUE if  s_d  was deleted or FALSE otherwise

1        IF  PointQuery(s_d, root) is NULL
2            return FALSE
3        RecursiveDelete(s_d, root)
4        IF  ElementsToReinsert is not empty
5            reinsert elements in  ElementsToReinsert
6        return TRUE
```

any subtrees, it is very time-consuming. As shown by the authors' experiments, this optimization is up to two orders of magnitude costlier than building the whole structure from scratch, both regarding number of disk accesses and the wall clock time.

In Section 4, we present an optimization that also allows to migrate elements among subtrees and improves a tree better than the Slim-down does, but it is **much cheaper** than the Generalized Slim-down. Our technique employs deletion and reinsertion as occurs in the R^*-*Tree* [12], but whereas the elements reinserted in the R*-Tree are those stored in nodes split in the insertion path during the insertion operation, our approach is to find elements in overlapping regions after a tree was already built, performing a global tree optimization.

3   An Algorithm for Effective Deletion in Metric Access Methods

In this section we present the first deletion algorithm developed for dynamic height-balanced MAM. It is based on importing the sibling subtrees when the Minimum Node Occupation (MNO) rule is violated, taking into account that the node covering overlap must be maintained low. The pseudo-code for the deletion operation is presented in Algorithm 1.

To delete an element s_d, the *Delete* procedure (Algorithm 1) starts calling a *PointQuery*, to locate the path from the root to the leaf node where s_d is stored. If the element is not found, the procedure finishes (lines 1-2) returning NULL. Otherwise, the deletion is performed recursively, calling the *RecursiveDelete* procedure (line 3).

The *RecursiveDelete* (Algorithm 2) recursively traverses the tree until finding the leaf node where s_d is stored (lines 1-2) and deletes its entry (lines 33-34). If the leaf node violates the MNO property (underflow), its remaining entries are stored in the array *ElementsToReinsert*, and the leaf is deleted (lines 35-38). If only s_d is deleted and it was the node representative, then a new representative must be chosen and propagated to the upper levels (lines 39-41). If none of these situations occur, *RecursiveDelete* is completed (line 42).

Each recursive call to *RecursiveDelete* returns the action that must be executed in the node at the level above in response to the operations performed in

Algorithm 2: Recursive deletion algorithm

```
RecursiveDelete(s_d, Node)
INPUT: The element s_d to be deleted and the current Node in
       DeletePath
OUTPUT: NO_ACTION, CHANGED_REP or REMOVED_NODE
```

```
 1    IF Node is an index node
 2        SWITCH (RecursiveDelete(s_d, ChildNode))
 3            CASE NO_ACTION
 4                return NO_ACTION
 5            CASE CHANGED_REP
 6                update the representative of ChildNode entry with
                      the new one
 7                IF the representative of Node was changed
 8                    choose a new representative
 9                    return CHANGED_REP
10                return NO_ACTION
11            CASE REMOVED_NODE
12                set RemoveEntry to TRUE
13                IF ExportSubtrees is not empty //subtrees of child
                      node
14                    IF it was possible to export entries in
                          ExportSubtrees to other child of Node
15                        delete ChildNode
16                        set RemoveEntry to FALSE
17                IF RemoveEntry is TRUE
18                    delete the ChildNode entry from Node
19                    IF Node violates MNO //MNO of root is 2
20                        IF Node is the root
21                            set the node pointed by this entry as the
                                  new root
22                            delete Node
23                            return NO_ACTION
24                        Try to import an element
25                        IF it was not possible to import
26                            IF Node has less than MIN_CONCESSION
                                  entries OR Node is in the lower half
                                  of the tree
27                                store entries of Node on
                                      ExportSubtrees
28                                return REMOVED_NODE
29                    IF the representative of Node was changed
30                        choose a new representative
31                        return CHANGED_REP
32                return NO_ACTION
33    IF Node is a leaf node
34        delete s_d
35        IF Node violates MNO
36            store entries of Node on ElementsToReinsert
37            delete Node
38            return REMOVED_NODE
39        IF s_d was the representative of Node
40            choose a new representative
41            return CHANGED_REP
42        return NO_ACTION
```

the current $Node$. NO_ACTION means that nothing need to be changed in the upper levels nodes; CHANGED_REP means that the node representative was changed
and an update on its father node is required; and REMOVED_NODE indicates that

the node was deleted, thus the corresponding entry must be removed from its father.

After returning from a recursion that updated the representative (CHANGED_REP) (lines 5-7), the old representative is compared to the representative of the current *Node*. If both are the same, a new representative is chosen (lines 5-9) and is propagated to its parent, otherwise *RecursiveDelete* finishes (line 10).

When the node in the lower level was removed (REMOVED_NODE) and there are subtrees of the *ChildNode* to be exported (that is, the auxiliary array *ExportSubtrees* is not empty) *RecursiveDelete* exports them to another child of *Node*. If every subtree is exported, *ChildNode* is deleted (lines 13-16). Notice that it is possible that a subtree is not exported if *Node* has only one subtree and the value of MIN_CONCESSION = 1 (the minimal desired prunability of index nodes, explained later).

If a child node was deleted, its entry on *Node* is removed too (lines 17-18), what can, in turn, lead *Node* to violate the MNO. If this occurs at the root node, whose MNO is always 2, the node pointed by the remaining entry is promoted to be the new root, *Node* (the old root) is removed, reducing the height of the tree and *RecursiveDelete* finishes (lines 19-23). If the MNO violation occurs deeper, the algorithm attempts to import an entry from a sibling node that will not violate the MNO (line 24). The first attempt is to import an entry already covered by *Node*, so its covering radius does not increase. If none is found, the algorithm attempts to import the entry closest to the *Node*'s representative. If every sibling node will violate the MNO to export one element, then the siblings of *Node* are surely able to store the entries previously stored in the child node. Therefore, the subtrees of *Node* are stored in *ExportSubtrees* and the recursion returns one level setting REMOVED_NODE (lines 25-28). Finally, if the representative of *Node* and the *ChildNode* entry were the same, a new representative must be chosen and propagated to the upper level (lines 29-31).

After finishing *RecursiveDelete*, if a leaf was deleted, then its entries were stored in the array *ElementsToReinsert*. Hence, they are re-inserted again in the tree (Algorithm 1, lines 4-5), ending the deletion operation.

To reduce the required tree reorganization, a concession to the MNO rule is granted in the following case: if the node is in the upper half of the tree and all of the siblings are also with the minimum occupation, then the MNO is allowed to be violated, up to a limit defined by the user-defined parameter MIN_CONCESSION. Notice that accepting the MNO violation on nodes in the upper half of the tree is not critical, first because the root can naturally have a much lower occupation than other nodes, and second because the proportion of nodes in the upper levels of the tree is small. Thus the MNO violation causes small impact in the average occupation of the tree. Anyway, it is important to note that the MNO property can also be violated by the *m-delete* algorithm even at the leaf nodes, which is a much worse situation, as there are many more leaf nodes than index nodes at the upper half of the tree. Thus, the MIN_CONCESSION parameter is a way to allow the user to tune how much the MNO can be violated in the upper half of the tree.

To evaluate the proposed deletion algorithm, we implemented it over the Slim-tree MAM. Nevertheless, it can be applied to other height-balanced dynamic MAM having one representative per node. Section 5.1 presents the experiments performed.

4   An Overlap Reduction and Optimization Technique for Metric Access Methods

In this section we introduce a novel optimization technique based on the effective deletion algorithm. It searches for elements that are not close to the others on the node, thus increasing the covering radius. Nodes with large covering radius increase the overlap on the structure, thus worsening query execution. Our idea is to remove several elements in the periphery of leaf nodes and reinsert them at once.

The elements selected to be removed are the farthest from their representatives, so the covering radius of the leaf nodes tend to decrease significantly. As the insertion operation tries first to store new elements in the nodes that do not increase their covering radius, most elements will be reinserted either on nodes that already cover them or in subtrees requiring a small radius increasing, therefore achieving an overall overlap reduction. We call this the Push-pull technique.

Figure 1 illustrates the stepwise execution of Push-pull. Figure 1a shows a structure before applying any overlap reduction technique. It has two index nodes, each one with three leaf nodes. In this figure, it was considered the maximum node occupation as 8, the minimum occupation as 2 and the number of elements to be removed from each leaf node as 2. The elements to be removed are highlighted as stars in Figure 1b: those farthest from their respective representatives. Figure 1c shows the tree after the deletions. Notice that only two leaf nodes remained on the left index node, because one of its leaves was removed by violating the MNO (it became empty). Finally, Figure 1d presents the optimized structure, after the reinsertions.

For the sake of comparison, Figure 1e shows the same tree optimized by Slim-down. Notice that, as opposed to Slim-down, the Push-pull technique allows "migration" of elements between any leaf node, not being limited to siblings. Figure 1e, illustrates the overlap reduction between leaf nodes linked to the same index node, but not between leaf nodes linked to different index nodes. On the other hand, Push-pull, enabled overlap reduction even between leaf nodes linked to different index nodes (Figure 1d).

In the naïve implementation of the Push-pull technique, users need to provide the quantity of elements to be removed from each node as an input parameter. Through experimental evaluation, we found that the ideal percentage of elements to be removed vary from dataset to dataset, but it is limited by a saturation point (see Subsection 5.2). We also verified that the ideal number of elements to

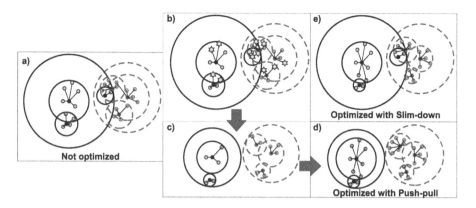

Fig. 1. Comparison of Push-pull and Slim-down optimizations. (a) the original structure; (b) and (c) intermediate steps during execution of Push-pull; (d) the final structure optimized with Push-pull; (e) the structure optimized with the Slim-down.

be removed from each node varies from node to node, according to their overlap.
Thus, we developed the Smart Push-pull algorithm to find automatically an
adequate value for the quantity of elements to be removed in a a given dataset,
based on statistics measured in the tree. This algorithm defines a near optimal
number of elements to be removed from each leaf node, and it can be different
for each node in a given tree.

Smart Push-pull uses the total number of disk accesses required to reach each
element through a point query, which is calculated during the computation of the
tree's fat-factor, and for each node it evaluates AVG_{Node} as the average number
of accesses it had to retrieve every entry stored in each leaf node. The maximum
value of AVG_{Node} among all the nodes is computed, in order to find the costlier
node (i.e. the node with the highest overlap), which is marked as the AVG_{Max}
node. Thereafter, the ideal number of elements to be removed ($\#Obj_{Del}$) from
each node is given as:

$$\#Obj_{Del} = \frac{AVG_{Node} - H}{AVG_{Max}} * Max\_Occup \tag{3}$$

where Max_Occup is the maximum node capacity and H is the height of the tree.
Notice that the optimal AVG_{Node} is the tree height, since it only occurs when
there is no overlap. Thus H is subtracted from AVG_{Node} in order to acquire
only the exceeding disk accesses. We limit the value of $\#Obj_{Del}$ to 40% of the
node capacity, because experimental evaluation indicated this value as very close
to the saturation point.

We present in Section 5.2 experiments that confirmed the effectiveness of the
Push-pull technique. The experiments also showed that the Smart Push-pull
algorithm outperforms the Slim-down method, the most known MAM optimization technique. We also obtained very good results that showed that Smart
Push-pull always performed close to the best case of naïve Push-pull using user-
defined settings, but relinquishing the user from this responsibility.

5   Experiments

We implemented the proposed algorithms over the Slim-tree MAM and evaluated them using several synthetic and real datasets, with varying properties such as dimension, number of elements and node capacity. In this paper we show results from the more representative datasets, presented in Table 1.

Table 1. Datasets used in the experiments

Name	Nr Elems.	Dim.	Node capacity	Description
Cities	5,507	2	26	Latitudes and longitudes of Brazilian cities (http://www.ibge.gov.br)
Letters	20,000	16	56	Attributes extracted from character images - UCI Machine Learning Archive (http://mlearn.ics.uci.edu/MLRepository.html)
ColorHisto	12,000	256	49	Color image histograms from Amsterdam Library of Object Images (http://www.science.uva.nl/ aloi)
SynthData	200,000	64	94	Synthetic vector with 100 clusters with Gaussian distribution in a 64D unit hypercube (generated by the tool DBGen [13])

The algorithms were implemented in C++ on the MAM Slim-tree using the library *Arboretum*[1]. The experiments were executed in a computer equipped with an AMD Athlon 2.6 GHz processor, 2Gb of RAM and a 250Gb SATA disk, running MS Windows XP Professional. The Slim-tree was set up with its recommended parameters, which includes *MST* for the split algorithm, *minDist* for the algorithm of choosing subtrees, and MNO of 25%. The value of MIN_CONCESSION was set with one. The next subsections show the results achieved for the deletion algorithm (Subsection 5.1) and the Push-pull optimization (Subsection 5.2).

5.1   Effective Deletion Algorithm

In this section we present results of the experiments comparing the proposed deletion algorithm with *m-delete*. As we pointed out in Subsection 2.1, the *m-delete* (mark-as-deleted) algorithm effectively deletes non-representative elements and just marks representative elements as deleted in non-leaf nodes.

The experiments of this section used the datasets Letters and Cities. Each dataset was initially divided in two parts, the first with 80% of the elements and the second with the remaining 20%. For each dataset, an initial tree was built with the first part of the dataset (16,000 elements for the dataset Letters and 4,500 for the dataset Cities). Next, approximately half of the indexed elements were randomly removed (8,000 for the dataset Letters and 2,000 for the dataset Cities).

First we show that the cost of effective deletion (*Effective delete*) is very close to the cost of *m-delete*. Table 2 displays statistics about the execution of the two

---

[1] An open source software library implementing several MAM (http://www.gbdi. icmc.usp.br/arboretum).

Table 2. Comparing the costs of the *m-delete* algorithm and the proposed deletion algorithm. These results were obtained when deleting respectively 8,000 and 2,000 elements from the datasets Letters and Cities.

Dataset/Algorithm	Number of pages	Total time (ms)	Disk accesses (avg.)	Distance calculations (avg.)
Letters				
m-delete	639	6,296	43	997.4
Effective delete	442	6,313	44.9	1,058.9
Cities				
m-delete	408	198	13.9	46.4
Effective delete	309	212	14.9	76.5

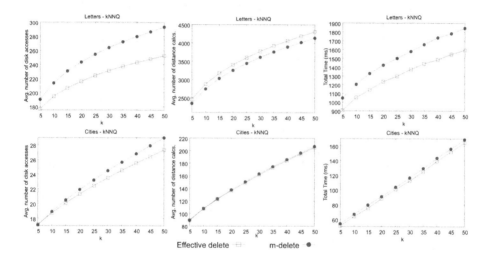

Fig. 2. Performance of *k*-NN queries (varying *k*), after deletions and insertions, over a Slim-tree employing the proposed delete algorithm (Effective delete) and over a Slim-tree using the *m-delete* algorithm. The results were obtained after the deletion of 8,000/2,000 elements (respectively for the datasets Letters and Cities) followed by the insertion of other 4,000/1,000 elements (Letters/Cities).

deletion methods, for the reorganization. As it can be seen, the average number of disk accesses required to perform the structure reorganizations were very similar. Regarding the number of distance calculations *m-delete* required 6% less distance calculations than ours for the Letters dataset and 40% less for the Cities dataset. The total time required to delete the stated amount of elements executing both algorithms were nearly the same for the Letters dataset, and the *m-delete* was only 7% faster for Cities dataset. However, after the execution of both algorithms, the number of pages left in the tree by the *m-delete* algorithm was 32% larger for Cities dataset and 44% larger for Letters dataset than the number of pages left by our algorithm, what improves query evaluation significantly.

In order to evaluate the query performance after deletions, we inserted the second part of the original datasets (4,000 elements for the dataset Letters and 1,000 for the dataset Cities) into the resulting trees. Figure 2 shows the results of running 500 k-NN queries over the resulting trees, with k ranging from 5 to 50.

As it can be seen in Figure 2, queries performed after the deletions and insertions executing the proposed deletion algorithm were always faster, with gains of up to 14%. Regarding disk accesses, the deletion algorithm also presented better results considering both datasets, with gains of up to 14% using the dataset Letters. The small increase in the number of distance calculations with the Letters dataset (maximum of 4.6%) is explained by the reduction of the number of pages in the tree, resulting from the use of the effective deletion algorithm: due to the increase of the average node occupation rate, the number of distance calculations to process each accessed node increases accordingly. However, this increase of distance calculations was compensated by the reduction of disk accesses, resulting in an overall faster query answering.

Analyzing the experimental results, we conclude that, for applications that perform many updates in the database, using the proposed algorithm for deletion is the better option, considering the overall performance and also the amount of disk usage.

5.2   Evaluating the Push-Pull Optimization

This section presents results of experiments comparing the Push-pull with the Slim-down techniques for overlap reduction. The first question these experiments aimed at answering is: how the Push-pull technique behaves varying the amount of elements removed per node? Considering our approach for automatically defining the ideal amount of elements to remove, the second question is: how Smart Push-pull compares with naïve Push-pull? The third question is: how Smart Push-pull compares to Slim-down?

In order to cope with these questions, in the first part of the experiments we built a Slim-tree for each dataset and optimized them using: (i) Slim-down, (ii) Smart Push-pull and (iii) naïve Push-pull with a varying number of elements removed per node (this number is indicated by a percentage of the node capacity). Then, we executed 500 k-NN queries over the trees, using $k = 10$.

Figure 3 shows the results obtained. Each column of graphics in this figure corresponds to a dataset (Letters, ColorHisto and Cities), and the rows correspond to the total time (in miliseconds), the average number of disk accesses, the average number of distance calculations and the relative fat-factor of each resulting tree.

Considering the total execution time of the queries (first row), the trees optimized with Push-pull obtained better performance with the augment of the amount of elements removed per node. However, it can be noticed a saturation point around 30% to 40% of entries removed. It is also visible that Smart Push-pull outperforms both Slim-down and naïve Push-pull having obtained better results in the majority of the cases. Another important analysis is that Smart Push-pull achieved results always near to the best case of naïve Push-pull, which is a very desirable result. The same behaviour can be noticed regarding the other measures, except for the average number of disk accesses

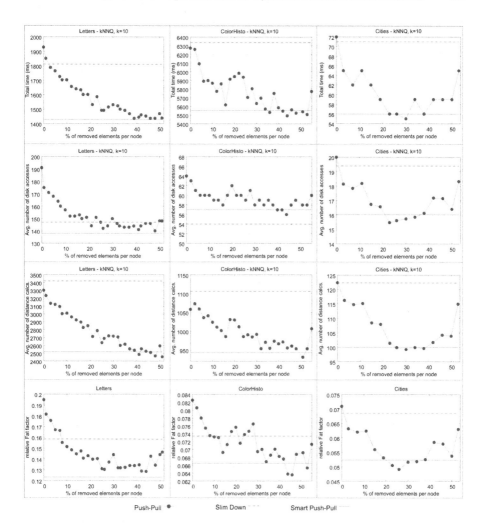

Fig. 3. Comparison of the total processing time (first row), average number of disk accesses (second row) and average number of distance calculations (third row) to process 500 10-NN queries, and relative fat-factor (fourth row): of Slim-trees optimized with naïve Push-pull varying the percentage of removed elements per node, a Slim-tree optimized with Slim-down and a Slim-tree optimized with Smart Push-pull.

for the ColorHisto, where Slim-down was 5% better than Smart Push-pull. In general, queries posed over the tree optimized by Smart Push-pull was up to 27% faster, required up to 18% less disk acesses, and performed up to 35% less distance calculations than the tree optimized by Slim-down.

Next, we evaluated the behavior of queries executed over trees not optimized, optimized by Slim-down and optimized by Smart Push-pull. Figure 4 shows measurements of executing 500 k-NN queries varying k between 5 and 50.

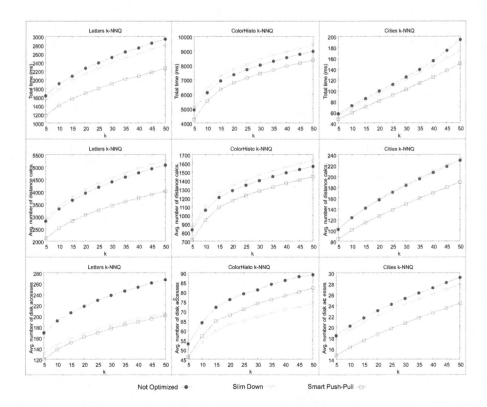

Fig. 4. Comparison of the total processing time (first row), average number of disk accesses (second row) and average number of distance calculations (third row), to perform 500 k-NN queries varying k between 5 and 50 over a Slim-tree not optimized, a Slim-tree optimized with Slim-down and a Slim-tree optimized with Smart Push-pull, varying k.

As it can be seen in Figure 4, the trees optimized by the Smart Push-pull algorithm have always evaluated k-NN queries faster, being up to 28% faster over the not optimized tree and 23% faster over the tree optimized by the Slim-down.

Another part of the experiments aimed to evaluate how Smart Push-pull behaves regarding the dataset size and how it compares to Slim-down in this aspect. In this experiment, we employed a 64D synthetic dataset, varying its size from 40 to 200 thousands elements. Figure 5 shows measurements of performing 500 *10*-NN queries over trees without optimization, over trees optimized by Slim-down, and over trees optimized by Smart Push-pull.

As it can be noticed in Figure 5, the tree optimized by Smart Push-pull have always answered k-NN queries faster. Queries posed over the non-optimized tree were up to 196% slower, required up to 140% more disk acesses and performed up to 125% more distance calculations than the tree optimized by Smart Push-pull. Moreover, k-NN queries posed over the tree optimized by our algorithm are up to 60% faster than when posed over tree optimized with Slim-down method, having

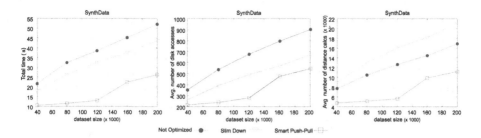

Fig. 5. Comparison of the total processing time (first graphic), average number of disk accesses (second graphic) and average number of distance calculations (third graphic), over Slim-trees not optimized, Slim-trees optimized with Slim-down and Slim-trees optimized with Smart Push-pull, with varying dataset size.

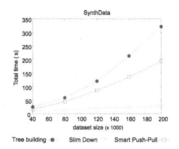

Fig. 6. Comparison of the time required to optimize a tree with Slim-down and Smart Push-pull, compared to the time required to construct the initial tree, with varying dataset size

achieved gains of up to 43% in number of disk accesses and 65% in number of distance calculations. In summary, Smart Push-pull have always produced trees that answered similarity queries faster than the trees optimized by Slim-down.

Finally, we developed experiments to evaluate the cost of the proposed algorithm. Figure 6 shows the total time required to optimize a tree using Slim-down and Smart Push-pull, compared with the time to build the tree, varying the dataset size. For this experiment we also employed 64D synthetic datasets varying its size from 40 to 200 thousands elements.

As it can be seen in Figure 6, the time required by Smart Push-pull was only between 60% and 77% of the time required to construct the original tree. As expected, the proposed technique has proven to be more expensive than the Slim-down method, as Push-pull allows reinserting the removed elements anywhere in the tree, distinctly from the local action of the Slim-down. In fact, the execution time of Smart Push-pull was from 2 to 6.5 times larger than the cost of Slim-down. However, the optimization is applied only few times over a tree if compared to the execution of queries, which our algorithm improved to run up to 150% faster than Slim-down did. Thus, depending on the query load, the Smart Push-pull can be a very good option.

6   Conclusions

The deletion operation is not described or implemented in most of the existing MAM. Deletion is usually performed just marking the data as removed, without really removing it from the tree. This solution is not satisfactory when the structure is frequently updated, because old data elements continue to be used internally during query operations, increasing the required number of disk access and distance calculations.

This work proposed an algorithm for the effective deletion of elements indexed in MAM, allowing to delete any element, including those used in the internal node structure without incurring in expensive tree reorganizations. The experiments performed using the Slim-tree MAM showed that the queries were always faster after the application of the proposed algorithm than marking the removed representatives.

This work also presented a new technique to optimize MAM, the Push-pull. It reduces the overlap between nodes, performing deletion and reinsertion of elements in the border of the leaf-nodes. We developed an algorithm to automatically define a number of elements to be removed near optimal for each leaf node, based on statistical data collected from subtrees overlap. This algorithm was called Smart Push-pull. The experiments showed that the trees optimized by the Smart Push-pull tend to answer queries more efficiently, having been up to 190% faster than the not optimized tree, and up to 150% faster than trees optimized by Slim-down.

Acknowledgments

This work has been supported by FAPESP (São Paulo State Research Foundation), CNPq (National Council for Scientific and Technological Development) and CAPES (Brazilian Federal Funding Agency for Graduate Education Improvement).

References

1. Faloutsos, C.: Indexing of multimedia data. In: Multimedia Databases in Perspective, pp. 219–245. Springer, Heidelberg (1997)
2. Burkhard, W.A., Keller, R.M.: Some approaches to best-match file searching. CACM 16(4), 230–236 (1973)
3. Uhlmann, J.K.: Satisfying general proximity/similarity queries with metric trees. Information Processing Letters 40(4), 175–179 (1991)
4. Yianilos, P.N.: Data structures and algorithms for nearest neighbor search in general metric spaces. In: ACM/SIAM/SODA, Austin, TX, USA, pp. 311–321. ACM, New York (1993)
5. Bozkaya, T., Özsoyoglu, Z.M.: Indexing large metric spaces for similarity search queries. ACM TODS 24(3), 361–404 (1999)
6. Brin, S.: Near neighbor search in large metric spaces. In: Dayal, U., Gray, P.M.D., Nishio, S. (eds.) VLDB, Zurich, Switzerland, pp. 574–584. Morgan Kaufmann, San Francisco (1995)

7. Ciaccia, P., Patella, M., Zezula, P.: M-tree: An efficient access method for similarity search in metric spaces. In: Jarke, M. (ed.) VLDB, Athens, Greece, pp. 426–435. Morgan Kaufmann, San Francisco (1997)
8. Traina Jr., C., Traina, A.J.M., Faloutsos, C., Seeger, B.: Fast indexing and visualization of metric datasets using Slim-trees. IEEE TKDE 14(2), 244–260 (2002)
9. Vieira, M.R., Traina Jr., C., Traina, A.J.M., Chino, F.J.T.: DBM-tree: A dynamic metric access method sensitive to local density data. In: Lifschitz, S. (ed.) SBBD, Brasília, DF, Brazil, pp. 33-47. SBC (2004)
10. Guttman, A.: R-tree: A dynamic index structure for spatial searching. In: ACM SIGMOD, Boston, MA, USA, pp. 47–57. ACM, New York (1984)
11. Skopal, T., Pokorný, J., Krátký, M., Snásel, V.: Revisiting M-tree building principles. In: Kalinichenko, L.A., Manthey, R., Thalheim, B., Wloka, U. (eds.) ADBIS 2003. LNCS, vol. 2798, pp. 148–162. Springer, Heidelberg (2003)
12. Beckmann, N., Kriegel, H.P., Schneider, R., Seeger, B.: The R*-tree: An efficient and robust access method for points and rectangles. In: Garcia-Molina, H., Jagadish, H.V. (eds.) ACM SIGMOD, Atlantic City, NJ, USA, pp. 322–331 (1990)
13. Ferreira, M.R.P., Bueno, R., Traina Jr., C.: DBGen - Gerador de dados sintéticos com distribuição fractal. In: Brayner, N., Dorneles, C.F. (eds.) Demo session - SBBD, Uberlândia, MG, Brazil, pp. 25-30 (2005)

An Ontology-Based Index to Retrieve Documents with Geographic Information*

Miguel R. Luaces, Jose R. Paramá, Oscar Pedreira, and Diego Seco

Database Laboratory, University of A Coruña
Campus de Elviña, 15071 A Coruña, Spain
{luaces,parama,opedreira,dseco}@udc.es

Abstract. Both *Geographic Information Systems* and *Information Retrieval* have been very active research fields in the last decades. Lately, a new research field called *Geographic Information Retrieval* has appeared from the intersection of these two fields. The main goal of this field is to define index structures and techniques to efficiently store and retrieve documents using both the text and the geographic references contained within the text.

We present in this paper a new index structure that combines an inverted index, a spatial index, and an ontology-based structure. This structure improves the query capabilities of other proposals. In addition, we describe the architecture of a system for geographic information retrieval that uses this new index structure. This architecture defines a workflow for the extraction of the geographic references in the document.

1   Introduction

Although the research field of Information Retrieval [1] has been active for the last decades, the growing importance of Internet and the World Wide Web have made it one of the most important research fields nowadays. Many different index structures, compression techniques and retrieval algorithms have been proposed in the last few years. More importantly, these proposals have been widely used in the implementation of document databases, digital libraries, and web search engines.

Another field that has received much attention during the last years is the field of Geographic Information Systems [2]. Recent improvements in hardware have made the implementation of this type of systems affordable for many organizations. Furthermore, a cooperative effort has been undertaken by two international organizations (ISO [3] and the Open Geospatial Consortium [4]) to

* This work has been partially supported by "Ministerio de Educación y Ciencia" (PGE y FEDER) ref. TIN2006-16071-C03-03, by "Xunta de Galicia" ref. PGIDIT05SIN10502PR and ref. 2006/4, by "Ministerio de Educación y Ciencia" ref. AP-2006-03214 (FPU Program) for Oscar Pedreira, and by "Dirección Xeral de Ordenación e Calidade do Sistema Universitario de Galicia, da Consellería de Educación e Ordenación Universitaria-Xunta de Galicia" for Diego Seco.

B. Ludäscher and Nikos Mamoulis (Eds.): SSDBM 2008, LNCS 5069, pp. 384–400, 2008.
© Springer-Verlag Berlin Heidelberg 2008

define standards and specifications for interoperable systems. This effort is making possible that many public organizations are working on the construction of spatial data infrastructures [5] that will enable them to share their geographic information.

During the last decades these two research fields have advanced independently. However, many of the documents stored in digital libraries and document databases include geographic references within their texts. For example, news documents reference the place where the event happened and often the place where the document has been written. Furthermore, the information in a spatial data infrastructure often includes documents with geographic information such as construction licences or urban planning information. Finally, geographic references can also be attached to web pages by using information from the text, the location of the web server, and many other information elements.

Even though it is very common that textual and geographic information occur together in information systems, the geographic references of documents are rarely used in information retrieval systems. Few index structures or retrieval algorithms take into account the spatial nature of geographic references embedded within documents. Pure textual techniques focus only on the language aspects of the documents and pure spatial techniques focus only on the geographic aspects of the documents. None of them are suitable for a combined approach to information retrieval because they completely neglect the other type of information. As a result, there is a lack of system architectures, index structures and query languages that combine both types of information.

Some proposals have appeared recently [6,7,8] that define new index structures that take into account both the textual and the geographic aspects of a document. However, there are some specific particularities of geographic space that are not taken into account by these approaches. Particularly, concepts such as the hierarchical nature of geographic space and the topological relationships between the geographic objects must be considered in order to fully represent the relationships between the documents and to allow new and interesting types of queries to be posed to the system.

In this paper, we present an index structure that takes these issues into account. We first describe some basic concepts and related work in Section 2. Then, in Section 3, we present the general architecture of the system and describe its components. The system architecture defines a workflow for constructing a document database where both the words and the geographic references in the documents are considered. The core of the system architecture is an index structure that enables the system to store and access efficiently the documents using both their textual references and their geographic ones. Finally, the system architecture includes two different user interfaces: one for final users that can be used to pose queries to the system and to display the results, and another for system administrators that can be used to manage the document collections. After that, in Section 4, we describe some types of queries that can be answered with this system and we sketch the algorithms that can be used to solve this queries.

Furthermore, Section 5 presents some experiments that we made to compare our structure with other ones that use a pure spatial index. Finally, Section 6 presents some conclusions and future lines of work.

2   Related Work

Inverted indexes are considered the classical text indexing technique. An inverted index associates to each word in the text (organized as a *vocabulary*) a list of pointers to the positions where the word appears in the documents. The set of all those lists is called the *occurrences* [1]. The main drawback of these indexes is that geographic references are mostly ignored because place names are considered words just like the others. If the user poses a query such as as *hotels in Spain*, the place name *Spain* is considered a word, and only those documents that contain that word are retrieved. A document containing only names of cities of Spain but not the exact word *Spain* is not retrieved by the system because it does not fulfil the textual query. Regarding indexing geographic information, many different spatial index structures have been proposed along the years. A good survey of these structures can be found in [9]. The main goal of spatial index structures is improving access time to collections of geographic data objects. One of the most popular spatial index structure and a paradigmatic example is the R-tree [10]. The R-tree is a balanced tree derived from the B-tree which splits space in hierarchically nested, possibly overlapping, minimum bounding rectangles. The number of children of each internal node varies between a minimum and a maximum. The tree is kept in balance by splitting overflowing nodes and merging underflowing nodes. Rectangles are associated with the leaf nodes, and each internal node stores the bounding box of all the rectangles in its subtree. The decomposition of space provided by an R-tree is adaptive (dependent on the rectangles stored) and overlapping (nodes in the tree may represent overlapping regions). A drawback of spatial index structures is that they do not take into consideration the hierarchy of space. Internal nodes in the structure are meaningless in the real world, they are just meaningful for the index structure. For example, imagine that we want to build an index for a collection of countries, provinces, and cities. These objects are structured in a topological relationship of containment, that is, a city is contained within a province that is itself contained within a country. If we build an R-Tree with these geographic objects, the containment hierarchy will not be maintained. The internal nodes of the R-Tree do not represent provinces or countries, and therefore, the hierarchy of space is not maintained in the index. It is not possible to associate some information to the node of a province and have the cities belonging to this province inherit this information because there is no relation at all between a province and its cities in the R-Tree index structure. Some work has been done to combine both types of indexes. The papers about the SPIRIT (Spatially-Aware Information Retrieval on the Internet) project [11,12,13,14,15] are a very good starting point. In [14], the authors conclude that keeping separate text and spatial indexes, instead of combining both in one, results in less storage costs but it could lead to higher response times. More recent works can be broadly classified into two categories depending on how they combine textual and spatial indexes. On the one hand, some proposal have appeared that combine textual and spatial aspects in an hybrid index [16,17]. On the other hand, some proposals define structures that keep separate indexes for spatial and text attributes [6,7,8]. Our index structure is part of this second group because this division has many advantages [8]. Furthermore, in [7,8], the authors survey the work in the SPIRIT project and propose improvements to the system and the algorithms defined. In their work they propose two naive algorithms: *Text-First* and *Geo-First*. Both algorithms use the same strategy: one index is first used to filter the documents (textual index in Text-First and spatial index in Geo-First), the resulting documents are sorted by their identifiers and then filtered using the other index (spatial index in Text-First and textual index in Geo-First). Nevertheless, none of these approaches take into account the relationships between the geographic objects that they are indexing.

A structure that can properly describe the specific characteristic of geographic space is an *ontology*, which is a formal explicit specification of a shared conceptualization [18]. An ontology provides a vocabulary of classes and relations to describe a given scope. In [19], a method is proposed for the efficient management of large spatial ontologies using a spatial index to improve the efficiency of the spatial queries. Furthermore, in [12,15] the authors describe how ontologies are used in query term expansion, relevance ranking, and web resource annotation in the SPIRIT project. However, as far as we know, nobody has ever tried to combine ontologies with other types of indexes to have a hybrid structure that captures both the topological and the spatial relationships between the geographic objects indexed.

3   System Architecture

Fig. 1 shows our proposal for the system architecture of a geographic information retrieval system. The architecture can be divided into three independent layers: the index construction workflow, the processing services and the user interfaces. The bottom part of the figure shows the index construction workflow, which, in turn, consists of three modules: the document abstraction module (described in Section 3.1), the index construction module (the textual part of this process is described in Section 3.2 and the spatial part of this process is described in Section 3.3), and the index structure itself (described also in Section 3.3).

The processing services are shown in the middle of the figure. On the left side, the *Geographic Space Ontology Service* used in the spatial index construction is shown. This service is used extensively in the index construction module, and therefore it is described in Section 3.3. On the right side, one can see the two services that are used to solve queries. The rightmost one is the *query evaluation service*, which receives queries and uses the index structure to solve them. Section 4 describes the types of queries that can be solved by this service, as well as the algorithms that are used to solve these queries. The other service is

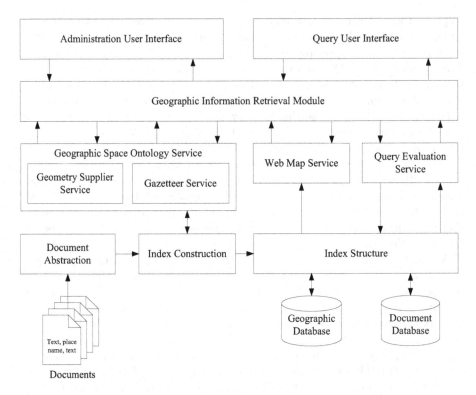

Fig. 1. System Architecture

a *Web Map Service* following the OGC specification [20] that is used to create cartographic representations of the query results. This service is not described in this paper. On top of these services a *Geographic Information Retrieval Module* is in charge of coordinating the task performed by each service to response the user requests. The topmost layer of the architecture shows the two user interfaces that exist in the architecture: the *Administration User Interface* and the *Query User Interface*. These user interfaces are described in Section 3.4.

3.1   Document Abstraction

Given that the system must be generic, it must support indexing several kinds of documents. These documents will be different not only because they may be stored using different file formats (plain text, XML, etc.), but also because their content schema may be different. A document collection may have a set of attributes that have to be stored in the index (such as *document id, author*, and *document text*), whereas other document collection may have a different set of attributes (such as *document id, summary, text, author*, and *source*).

To solve this problem, we have defined an abstraction that represents a *document* as a set of *fields*, each one with a value that is extracted from the document text. Each field can either be *stored, indexed*, or both. If a field is stored, its contents are stored in the index structure and they can be retrieved by a query. If a field is indexed, then this field is used to build the index structure. Furthermore, a field can be indexed textually, spatially, or in both indexes. The definition of a document as a set of fields is similar to the one used in the Lucene text search engine [21]. We have extended this idea adding the spatial indexing possibility.

In order to support different types of documents and different file formats, the document abstraction is exposed by the system as a programming interface that can be extended with particular implementations for different configurations of file formats and document schemas. In order to support a new configuration, a developer only has to implement the interface *DocumentFactory* that defines the operations that must be implemented in order to create *Documents*.

As an example for the validation of the system, we have indexed documents from the Financial Times collection [22]. The document collection is marked up in SGML (*Standard Generalized Markup Language*). Each document has a <DOCNO> tag including the TREC document identifier string and a <TEXT> tag including the main content of the document. Fig. 2 shows a partial example of a document in this collection.

```
<DOC>
     <DOCNO>FT941-6371</DOCNO>
     <TEXT>
          Senior European company executives are being invited to 'vote'
          for Europe's Most Respected Companies . . .
     </TEXT>
</DOC>
```

Fig. 2. Financial Times (TREC) example

To support this document collection, we defined a *TRECFTDocumentFactory* that builds documents with two fields. The first field contains the tag DOCNO content and it is stored but not indexed. The second field contains the tag TEXT content and it is not stored but indexed in both indexes.

3.2 Textual Indexing

As we said before, the index structure at the core of the system architecture contains both a textual index and a spatial index. We use Lucene [21] to implement a textual index. Lucene is a high-performance, full-featured text search engine library written entirely in Java. It is an open source project part of the Apache project. Lucene uses an object representation of the indexable documents. A *Document* in Lucene contains several *Fields*. A *Field* in Lucene is a pair (*name, value*) and information about whether it is stored and/or indexed. Field values are set using *Analyzers*. These analyzers implement several classical information retrieval techniques to reduce the number of indexed words and to improve the index performance such as *removing stopwords, stemmers*, etc. *StandardAnalyzer* is the most sophisticated analyzer built into Lucene's core. It is a parser with rules for email addresses, acronyms, hostnames, floating point numbers, as well as converting the value to lowercase and removing stop words.

In this stage of the workflow process, the system builds a Lucene index. Each of the documents built in the previous stage is inserted into the textual index. The document identifier is stored but not indexed in the textual index, and each field marked to be indexed in the textual index or in both indexes is indexed tokenized in the Lucene index but not stored.

3.3   Spatial Indexing

After building the textual index, the spatial index must be built. The spatial indexing is the most complex stage, and it comprises three steps. First, the system analyses the document fields that are marked as spatially indexable and extracts candidate location names from the text. In a second step, these candidate locations are processed in order to determine whether the candidates are real location names, and, in this case, to compute their geographic locations. There are some problems that can happen at this point. First, a location name can be ambiguous (*polysemy*). For instance, "*London*" is the capital of the United Kingdom and it is a city in Ontario, Canada too. Second, there can be multiple names for the same geographic location, such as "*Los Angeles*" and "*LA*". Finding geographical references in text is a very difficult problem and there have been some papers that deal with different aspects of this problem [6,23,24]. Web-a-where [23] uses "spatial containers" in order to identify locations in documents, MetaCarta (the commercial system described in [24]) uses a natural language processing method, and STEWARD [6] uses an hybrid approach. It is not the aim of this paper to deal with this problem but we describe how we obtain geographic references in order to complete the architecture description. Finally, the third step consists in building the spatial index with the geo-referenced locations computed in the previous step together with references to the documents containing them. We describe these three steps and the spatial index structure below.

Discovery of Location Names. For the discovery of candidate location names, all the spatially indexable fields are processed in order to discover the place names contained within. There are two *Linguistic Analysis* techniques that are widely used for this: *Part-Of-Speech* tagging and *Named-Entity Recognition*. On the one hand, Part-Of-Speech tagging is a process whereby tokens are sequentially labelled with syntactic labels, such as "verb" or "gerund". On the other hand, Named-Entity Recognition is the process of finding mentions of predefined categories such as the names of persons, organizations, locations, etc.

Our *Location Names Discovery* module uses the *Natural Language Tool Ling-Pipe* [25] to find locations. It is a suite of Java libraries for the linguistic analysis of human language free for research purposes that provides both Part-Of-Speech tagging and Named-Entity Recognition. LingPipe involves the supervised training of a statistical model to recognize entities. The training data must be labelled with all of the entities of interest and their types. In the system validation with the Financial Times collection, we use LingPipe trained with the MUC6 corpus (http://www.ldc.upenn.edu) labelled with locations, people and organizations. After the LingPipe processing, the module filters the resultant named entities selecting only the locations and discarding people and organization names.

Geo-referenciation of Location Names. After discovering a collection of candidate location names, the system must distinguish false candidates and geo-reference the real ones. In this context, geo-referencing a location name implies not only to obtain its coordinates in a particular coordinate system, but also to obtain all the data needed to include the place in a spatial index. We have developed a system based on an ontology of the geographic space that is built using a *Gazetteer* and a *Geometry Supplier*.

A Gazetteer is a geographical dictionary that contains, in addition to location names, alternative names, populations, location of places, and other information related to the location. In our test implementation we use *Geonames* [26] that provides a geographical database available under a creative commons attribution license. This database contains more than two million populated places over the world with their latitude/longitude coordinates in WGS84 (*World Geodetic System 1984*). All the populated places are categorized so that it is possible to classify them into different administrative division levels (continents, countries, regions, etc.).

However, Geonames (and Gazetteers in general) does not provide geometries for the location names other than a single representative point. But for our spatial index we need the real geometry of the location name (for example, the boundary of countries). We defined a *Geometry Supplier* service to obtain the geometries of those location names. As a base for this service we used the *Vector Map* (VMap) cartography [27]. VMap is an updated and improved version of the National Imagery and Mapping Agency's Digital Chart of the World. It supplies first and second level administrative division geometries in a proprietary format. However, there are free tools that can create *shapefiles* from that format, such as FWTools [28]. We have created a PostGIS [29] spatial database with these shapefiles and we have done several corrections and improvements over this database.

Even though our test implementation uses Geonames and VMAP, it has been designed so that these components are easily exchangeable. All accesses to these components are performed through generic interfaces that can be easily implemented for other components.

This step combines both services in order to geo-reference location names. First, an ontology of the geographic space is defined. In our test implementation, the geographic space is divided into three levels of administrative divisions (continents, countries and regions) and a level of populated places. These four levels are organized in a hierarchical structure where each level geographically contains all objects in the next level.

Then, for each candidate location name, an *ontology path* must be built. This path will be used in the construction of the spatial index structure. For this task, a hierarchical structure following the design pattern *Chain of Responsibility* [30]

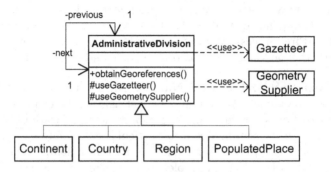

Fig. 3. Geo-references module

was implemented. Fig. 3 shows a brief class diagram of this component. The structure is composed of four levels (continent, country, region, and populated place), one for each level of the ontology, but it is easily extensible. Each level contains a connection to the gazetteer and to the geometry supplier in order to retrieve the data needed by the process. Then, an algorithm in two steps obtains all possible geo-references associated with a location name. In the first step, each level obtains from the gazetteer all the locations with the requested name. If the gazetteer does not return any location for a given candidate location name, the candidate is discarded. In the second step, the system builds the complete ontology path from bottom to top. For instance, if the requested location name was London, in the first step the system obtains two locations with this name. After that, it returns the paths *United Kingdom, England, London* and *Canada, Ontario, London.*

Spatial Index Construction. Fig. 4 shows a class diagram of the index structure. The main component of the index structure is a tree composed by nodes that represent location names. These nodes are connected by means of inclusion relationships (for instance, *Galicia* is included in *Spain*). The tree structure is built using the ontology paths computed by the process described in the previous section. In each node we store: (i) the keyword (a place name), (ii) the bounding box of the geometry representing this place, (iii) a list with the document identifiers of the documents that include geographic references to this place, and (iv) a list of child nodes that are geographically within this node. If the list of child nodes is very long, using sequential access is very inefficient. For this reason, if the number of children nodes exceeds a threshold, an R-Tree is used instead of a list.

Two auxiliary structures are used in the index. First, a *place name hash table* that stores for each location name its position in the index structure. This provides direct access to a single node by means of a keyword that is returned by the Gazetteer Service if the word processed is a location name. The second auxiliary structure is the textual index with all the words in the documents that is used to solve textual queries (this index is described in section 3.2).

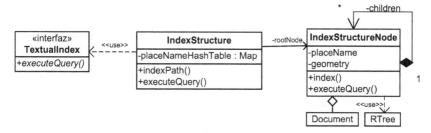

Fig. 4. Class diagram of the index structure

Keeping separate indexes for text and geographical scopes has many advantages. First of all, all textual queries can be efficiently processed by the text index, and all spatial queries can be efficiently processed by the spatial index. Moreover, queries combining textual and spatial aspects are supported. Furthermore updates in each index are handled independently, which makes easier the addition and removal of data. Finally, specific optimizations can be applied to each individual indexing structure.

However, this structure has two main drawbacks. First, the tree that supports the structure is possibly unbalanced penalizing the efficiency of the system. We present some experiments in Section 5. Our intention is to prove that it is not a very important problem. Second, ontological systems have a fixed structure and thus our structure is static and it must be constructed *ad-hoc*.

3.4   User Interfaces

The system has two different user interfaces: an administration user interface and a query user interface. The administration user interface was developed as a stand-alone application and it can be used to manage the document collection. The main functionalities are: creation of indexes, addition of documents to indexes, loading and storing indexes, etc. The main screen of this interface shows useful information about the loaded index such as the number of documents indexed, the fields of each of these documents, the number of location names in the index, etc.

Fig. 5 shows a screenshot of the query user interface. This interface was developed as a web application using the *Google Maps API* [31]. This API provides a number of utilities for manipulating maps and adding content to the map.

In the next section we sketch the types of queries that can be solved with this system. These queries have two different aspects: a textual aspect and a spatial aspect. The query user interface allows the user to indicate both aspects. The spatial context can be introduced in three ways that are mutually exclusive:

– *Typing the location name.* In this case, the user types the location name in a text box. This is the most inefficient way because the system has to obtain all geo-references associated with the typed place name and it is a time-expensive process.

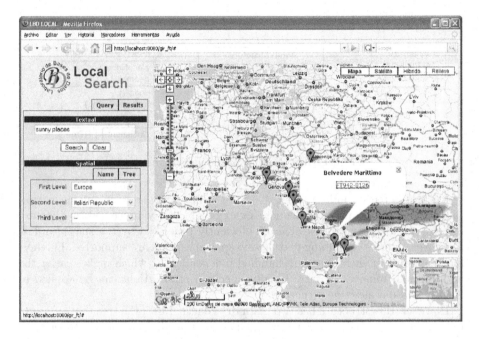

Fig. 5. Query User Interface

- *Selecting the location name in a tree.* In this case, the user sequentially selects a continent, a country within this continent, a region within the country, and a populated place within the region. If the user wants to specify a location name of a higher level than a populated place, it is not necessary to fill in all the levels. The operation is very easy and intuitive because the interface is implemented with a custom-developed component using the AJAX technology that retrieves in the background the location names for the next level. When the user selects a place in the component, the map on the right zooms in automatically to the selected place.
- *Visualizing the spatial context of interest in the map.* The user can navigate using the map on the right to select the spatial context of interest. The system will use the bounding box of this map as the query window if the user did not type a place name or did not select a location name.

4   Supported Query Types

The most important characteristic of an index structure is the type of queries that can be solved with it. The following types of queries are relevant in a geographic information retrieval system:

- *Pure textual queries.* These are queries such as *"retrieve all documents where the words* hotel *and* sea *appear"*.

- *Pure spatial queries.* An example of this type of queries is *"retrieve all documents that refer to the following geographic area"*. The geographic area in the query can be a point, a query window, or even a complex object such as a polygon.
- *Textual queries with place names.* In this type of queries, some of the words are place names. For instance, *"retrieve all documents with the word* hotel *that refer to Spain"*.
- *Textual queries over a geographic area.* In this case, a geographic area of interest is given in addition to the set of words. An example is *"retrieve all documents with the word* hotel *that refer to the following geographic area"*.

Inverted indexes can solve pure textual queries by retrieving from the inverted index the lists of documents associated to each word and then performing the intersection of the lists. Pure spatial queries can be solved by spatial indexes by descending the structure and taking into consideration only those nodes whose bounding box intersects with the geographic area of the query. This operation returns a set of candidate documents that has to be refined with the actual geographic reference in order to decide whether the document is part of the result or not.

Pure textual queries can be solved by our system because a textual index is part of the index structure. Similarly, pure spatial queries can also be solved because the index structure is built like a spatial index. Each node in the tree is associated with the bounding box of the geographic objects in its subtree. Therefore, the same algorithm that is used with spatial indexes can be used with our structure. However, the index structure that we propose can be used to solve the third and fourth types of queries, which cannot be easily solved using a textual index and a spatial index. For the case of the query with place names, our system can discover that *Spain* is a geographic reference by querying the Gazetteer service and then we can use the *place name hash table* in the structure to retrieve the index node that represents *Spain*. Thus, we save some time by avoiding a tree traversal. Fig. 6 shows how these type of queries can be solved by the index structure. The textual index of the structure can be seen on the right part of the figure, whereas the spatial index can be seen on the left part. When the user poses a query with the text *sunny places* and the place name *Spain*, the textual index is used to retrieve the list of documents that contain the words, and the index structure is used to compute the list of documents that reference the geographic area. These two lists can be seen at the bottom part of the figure. Then, the result to the query is computed as the intersection of both lists.

Regarding the fourth type of query, the textual index is used to retrieve the list of documents that contain the words and the ontology-based index structure is used to compute the list of documents that reference the geographic area. Then, the intersection of both lists is the result to the query. We analyze the performance of our structure to solve this type of queries in comparison with other proposals using a pure spatial index in Section 5. The conclusion of these experiments is that *the performance of our structure is acceptable in comparison with index structures using pure spatial indexes*.

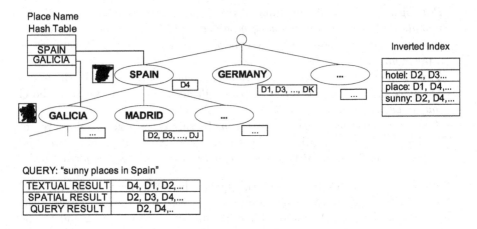

Fig. 6. Example of the index structure

Another improvement over text and spatial indexes is that our index structure can easily perform query expansion on geographic references because the index structure is built from an ontology of the geographic space. Consider the following query *"retrieve all documents that refer to Spain"*. The query evaluation service will discover that Spain is a geographic reference and the place name index will be used to quickly locate the internal node that represents the geographic object *Spain*. Then all the documents associated to this node are part of the result to the query. Moreover, all the children of this node are geographic objects that are contained within Spain (for instance, the city of Madrid). Therefore, all the documents referenced by the subtree are also part of the result of the query. The consequence is that the index structure has been used to expand the query because the result contains not only those documents that include the term *Spain*, but also all the documents that contain the name of a geographic object included in Spain (e.g., all the cities and regions of Spain).

5 Experiments

In the previous section we showed that our structure has a qualitative advantage over systems that combine a textual index with a pure spatial index because query expansion can be performed directly with our index structure (e.g. *retrieve all documents with the word* hotel *that refer to Spain*). Hence, our index structure supports a new type of query that cannot be implemented with a pure spatial index. However, unlike pure spatial index structures, our index structure is not balanced and therefore, the query performance can be worse.

In this section we describe the experiments that we performed to compare our structure with other ones based on a pure spatial index. We used the TREC FT-91 (Financial Times, year 1991) document collection [22], which consists of 5,368 news documents. Then, we built two indexes over this collection: one using our index structure as described in this paper, and another one using a textual index

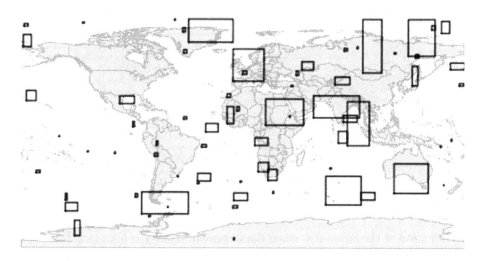

Fig. 7. Example of randomly generated query windows

and an R-Tree [10]. Furthermore, we developed an algorithm to generate random spatial query windows. This algorithm is based on the performance comparisons of the R*-Tree in [32] and it generates query windows where the ratio of the x-extension to the y-extension uniformly varies from 0.25 to 2.25 and the centres of the query rectangles are uniformly distributed all over the world. Fig. 7 shows several query windows generated using this algorithm.

We compared the structures with respect to four different query window areas, namely 0.001%, 0.01%, 0.1% and 1% of the world. We generated 100,000 random query windows for each area, and we averaged the computing time of each query execution. Table 1 shows the results of this experiment.

Table 1. Ontology-based index versus R-Tree

Query area	0.001%	0.01%	0.1%	1%
Our index	0.013	0.017	0.052	0.360
R-Tree	0.010	0.016	0.057	0.370

The first row of the table shows the results obtained with our structure (in milliseconds), and the second one shows the results obtained with the structure using an R-Tree. Both index structures have similar performance. The performance of our structure is a bit worse than the R-Tree when the query window is small but, surprisingly it is a bit better than the R-Tree when the query window is bigger. In order to explain this surprising result, we analyzed the performance in particular zones. We distinguished two relevant types of zones and we repeated the experiment generating random queries in both zones. First, we studied the

Table 2. Ontology-based index versus R-Tree (zones of high document density)

Query area	0.001%	0.01%	0.1%	1%
Our index	0.03	0.11	1.05	9.84
R-Tree	0.07	0.22	1.64	12.85

Table 3. Ontology-based index versus R-Tree (zones of low document density)

Query area	0.001%	0.01%	0.1%	1%
Our index	0.02	0.03	0.09	0.4
R-Tree	0.02	0.03	0.07	0.2

performance of the structures when the document density is high (see Table 2). In this case, the performance of our structure is higher than the R-Tree performance. We believe this is because our structure stores a list of documents for each location while the R-Tree uses a node for each one document.

Second, we studied the performance when the documents density is low (see Table 3). In this case, the R-Tree performance is better because the number of nodes in both structures is similar and the R-Tree is balanced whereas our structure may be unbalanced. For this reason, in the general case, when the query window is small the probability of that query window being in a high document density zone is small and, therefore, the R-Tree performance is better. However, when the query window is bigger that probability is higher and, therefore, the R-Tree performance is lower.

6   Conclusions and Future Work

We have presented in this paper a system architecture for an information retrieval system that takes into account not only the text in the documents but also the geographic references included in the documents and the ontology of the geographic space. This is achieved by a new index structure that combines a textual index, a spatial index and an ontology-based structure. We have also presented how traditional queries can be solved using the index structure. Finally, new types of queries that can be solved with the index structure are described and the algorithms that solve these queries are sketched.

Future improvements of this index structure are possible. We are currently working on the evaluation of the performance of the index structure, particularly we are performing experiments to determine the precision and recall. Moreover, *Toponym Resolution* techniques must be implemented to solve ambiguity problems when we geo-reference the documents. Another line of future work involves exploring the use of different ontologies and determining how each ontology affects the resulting index. Furthermore, we plan on including other types of spatial relationships in the index structure in addition to inclusion (e.g., adjacency). These relationships can be easily represented in the ontology-based structure, and the index structure can be extended to support them. Finally, it is necessary to define algorithms to rank the documents retrieved by the system. For this task, we must define a measure of spatial relevance and combine it with the relevance computed using the inverted index.

References

1. Baeza-Yates, R., Ribeiro-Neto, B.: Modern Information Retrieval. Addison Wesley, Reading (1999)
2. Worboys, M.F.: GIS: A Computing Perspective. CRC, Boca Raton (2004)
3. ISO/IEC: Geographic Information – Reference Model. International Standard 19101, ISO/IEC (2002)
4. Open GIS Consortium, Inc.: OpenGIS Reference Model. OpenGIS Project Document 03-040, Open GIS Consortium, Inc.(2003)
5. Global Spatial Data Infrastructure Association: Online documentation (Retrieved May 2007), http://www.gsdi.org/
6. Lieberman, M.D., Samet, H., Sankaranarayanan, J., Sperling, J.: STEWARD: Architecture of a Spatio-Textual Search Engine. In: Proceedings of the 15th ACM Int. Symp. on Advances in Geographic Information Systems (ACMGIS 2007), pp. 186–193. ACM Press, New York (2007)
7. Chen, Y.Y., Suel, T., Markowetz, A.: Efficient query processing in geographic web search engines. In: SIGMOD Conference, pp. 277–288 (2006)
8. Martins, B., Silva, M.J., Andrade, L.: Indexing and ranking in Geo-IR systems. In: GIR 2005: Proceedings of the 2005 workshop on Geographic information retrieval, pp. 31–34. ACM Press, New York (2005)
9. Gaede, V., Günther, O.: Multidimensional access methods. ACM Comput. Surv. 30(2), 170–231 (1998)
10. Guttman, A.: R-Trees: A Dynamic Index Structure for Spatial Searching. In: Yormark, B. (ed.) SIGMOD 1984, Proceedings of Annual Meeting, Boston, Massachusetts, June 18-21, 1984, pp. 47–57. ACM Press, New York (1984)
11. Jones, C.B., Purves, R., Ruas, A., Sanderson, M., Sester, M., van Kreveld, M., Weibel, R.: Spatial information retrieval and geographical ontologies an overview of the SPIRIT project. In: Proceedings of the 25th Annual International ACM SIGIR Conference on Research and Development in Information Retrieval, pp. 387–388 (2002)
12. Jones, C.B., Abdelmoty, A.I., Fu, G.: Maintaining ontologies for geographical information retrieval on the web. In: Meersman, R., Tari, Z., Schmidt, D.C. (eds.) CoopIS 2003, DOA 2003, and ODBASE 2003. LNCS, vol. 2888, pp. 934–951. Springer, Heidelberg (2003)
13. Jones, C.B., Abdelmoty, A.I., Fu, G., Vaid, S.: The SPIRIT Spatial Search Engine: Architecture, Ontologies and Spatial Indexing. In: Egenhofer, M.J., Freksa, C., Miller, H.J. (eds.) GIScience 2004. LNCS, vol. 3234, pp. 125–139. Springer, Heidelberg (2004)
14. Vaid, S., Jones, C.B., Joho, H., Sanderson, M.: Spatio-Textual Indexing for Geographical Search on the Web. In: Bauzer Medeiros, C., Egenhofer, M.J., Bertino, E. (eds.) SSTD 2005. LNCS, vol. 3633, pp. 218–235. Springer, Heidelberg (2005)
15. Fu, G., Jones, C.B., Abdelmoty, A.I.: Ontology-Based Spatial Query Expansion in Information Retrieval. In: Meersman, R., Tari, Z. (eds.) OTM 2005. LNCS, vol. 3761, pp. 1466–1482. Springer, Heidelberg (2005)

16. Zhou, Y., Xie, X., Wang, C., Gong, Y., Ma, W.Y.: Hybrid index structures for location-based web search. In: CIKM 2005: Proceedings of the 14th ACM international conference on Information and knowledge management, pp. 155–162. ACM, New York (2005)
17. Hariharan, R., Hore, B., Li, C., Mehrotra, S.: Processing Spatial-Keyword (SK) Queries in Geographic Information Retrieval (GIR) Systems. In: Proceedings of the 19th Int. Conf. on Scientific and Statistical Database Management (SSDBM 2007). IEEE Computer Society, Los Alamitos (2007)
18. Gruber, T.R.: A Translation Approach to Portable Ontology Specifications. Knowledge Acquisition 5(2), 199–220 (1993)
19. Dellis, E., Paliouras, G.: Management of Large Spatial Ontology Bases. In: Proceedings of the Workshop on Ontologies-based techniques for DataBases and Information Systems (ODBIS) of the 32nd International Conference on Very Large Data Bases (VLDB 2006) (September 2006)
20. Open GIS Consortium, Inc.: OpenGIS Web Map Service Implementation Specification. OpenGIS Project Document 01-068r3, Open GIS Consortium, Inc. (2002)
21. Apache: Lucene (Retrieved October 2007), http://lucene.apache.org
22. National Institute of Standards and Technology (NIST): TREC Special Database 22, TREC Document Database: Disk 4 (Retrieved November 2007), http://www.nist.gov/srd/nistsd22.htm
23. Amitay, E., Har'El, N., Sivan, R., Soffer, A.: Web-a-where: geotagging web content. In: SIGIR 2004: Proceedings of the 27th annual international ACM SIGIR conference on Research and development in information retrieval, pp. 273–280. ACM, New York (2004)
24. Rauch, E., Bukatin, M., Baker, K.: A confidence-based framework for disambiguating geographic terms. In: Proceedings of the HLT-NAACL 2003 workshop on Analysis of geographic references, Morristown, NJ, USA, pp. 50–54. Association for Computational Linguistics (2003)
25. Alias-i: LingPipe, Natural Language Tool (Retrieved October 2007), http://www.alias-i.com/lingpipe/
26. Geonames: Gazetteer (Retrieved September 2007), http://www.geonames.org
27. National Imagery and Mapping Agency (NIMA): Vector Map Level 0 (Retrieved September 2007), http://www.mapability.com
28. FWTools: Open Source GIS Binary Kit for Windows and Linux (Retrieved September 2007), http://fwtools.maptools.org
29. Refractions Research: PostGIS (Retrieved June 2007), http://postgis.refractions.net
30. Gamma, E., Helm, R., Johnson, R., Vlissides, J.: Design Patterns: Elements of Reusable Object-oriented Software. Addison-Wesley, Reading (1996)
31. Google: Google Maps API (Retrieved November 2007), http://www.google.es/apis/maps/
32. Beckmann, N., Kriegel, H.P., Schneider, R., Seeger, B.: The R*-tree: an efficient and robust access method for points and rectangles. SIGMOD Rec. 19(2), 322–331 (1990)

Mining Temporal Association Patterns under a Similarity Constraint

Jin Soung Yoo[1] and Shashi Shekhar[2]

[1] Department of Computer Science, Indiana University-Purdue University,
Fort Wayne, Indiana, USA
yooj@ipfw.edu
[2] Department of Computer Science and Engineering, University of Minnesota,
Minneapolis, Minnesota, USA
shekhar@cs.umn.edu

Abstract. We study the problem of mining all associated itemsets whose prevalence variations are similar to a given reference sequence from temporal databases. The discovered temporal association patterns can reveal interesting relationships of itemsets which co-occur with a particular event over time. A user-defined subset specification which consists of a reference sequence, a similarity function, and a dissimiliarty threshold is used for defining interesting temporal patterns and guiding the similarity search. We develop algorithms with exploring interesting properties for efficiently finding the similar temporal association patterns. Experimental results show that the proposed algorithms are efficient than a naive approach.

1 Introduction

We study the problem of mining all associated itemsets whose prevalence variations are similar to a given reference sequence from temporal databases. The reference sequence can be a user guided prevalence sequence pattern showing special shapes (e.g., seasonal, emerging and diminishing patterns over time) or a prevalence variation of item of interest (e.g., a scientific phenomenon like El Niño, a weather event such as a hurricane, a product sale in market basket data, and a stock exchange in the stock market). The mining results can reveal interesting relationships of data items which co-occur with a particular event over time, and also be used as filtered information for further analysis of trend, prediction, factors showing strong connections with a certain scientific event, and so on.

Application Examples
Earth scientists have been interested in the behavior of climates in a region which shows a strong connection with the El Niño phenomenon, an abnormal warming in the eastern tropical Pacific Ocean [17]. For example, the extreme climate variability in Australia is known to be connected with the El Niño phenomenon, which in turn can be associated with an El Niño index, e.g., the Southern Oscillation Index(SOI) [23]. Fig. 1 show an illustration of mining a similarity based temporal association patterns with Earth climate data. The monthly climate measurements in a region (e.g., Australia) can be represented to a temporal

B. Ludäscher and Nikos Mamoulis (Eds.): SSDBM 2008, LNCS 5069, pp. 401–417, 2008.
© Springer-Verlag Berlin Heidelberg 2008

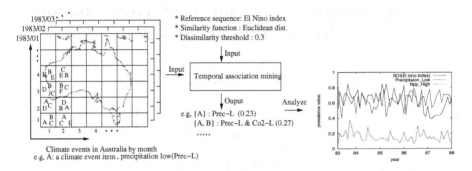

Fig. 1. An application example

database. The item types are climate events (e.g, low temperature, high precipitation) or other related events (e.g, high NPP(Net Primary Production)) in an ecosystem model. Given a similarity constraint which includes a reference sequence(e.g., SOI index), a similarity function(e.g, Euclidean distance) and a dissimilarity threshold to distinguish interest patterns, we are interested in discovering all subsets of items whose prevalence variances are similar with the specific reference sequence. In the example, we can notice that the temporal prevalence variation of a discovered itemset, precipitation_low, is similar with the SOI index sequence.

As another application example, for market data analysis, a retail analyst may be interested in such queries as finding all itemsets whose sales similarly change to that of an itemset of interest for a period of time. Consider sales of retail items during hurricane season in a region. While we can expect that sales of bottled water (one of survival kits) will increase with an increasing strength of hurricane threat, Wal-Mart recently discovered a surprising customer buying pattern. Not only did survival kits (e.g., flashlights, generators, tarps) show similar selling patterns with bottled water, but so did the sales of Strawberry Pop-Tarts (a snack item) [2]. The similarity based temporal association patterns can give important insight into many application domains such as Earth science, business and ecology.

Related Work

Although much work has been done on association patterns [3,14,19,20] and time series search [4,11,9], little attention has been paid to such a similarity based temporal association pattern. The closest related efforts have attempted to capture special temporal regulations of frequent association patterns such as cyclic association rule mining [18] and calendar-based association rule mining [15] in temporal association mining. Özden et al. [18] examined cyclic association rule mining, which detects periodically repetitive patterns of frequent itemsets over time. Cyclic associations can be considered as itemsets that often occur in every cycle with no exception. Li et al. [15] explored the problem of finding frequent itemsets along with calendar-based patterns. The calendar-based patterns are defined with a calendar schema, e.g., (year, month, day). For example, (*,10,31) represents the set of time points each corresponding to the 31st day of October.

However, real-life patterns are usually imperfect and may not demonstrate any regular periodicity. Other studies in temporal data mining have discussed the change of found association rules. Dong et al. [10] presented the problem of mining emerging patterns, which are the itemsets whose supports increase significantly from one dataset to another. Ganti et al. [12] presented a framework for measuring difference in two sets of association rules from two datasets. Liu et al. [16] studied the change of fundamental association rules between two time periods using support and confidence. When new transactions are added to the original dataset, the maintenance of discovered association rules with an incremental updating technique was proposed in [7]. In contrast, Agrawal et al. [6] addressed the problem of monitoring the support and confidence of association rules. First, all frequent rules satisfying a minimum threshold from different time periods are mined and collected into a rule base. Then interesting rules can be queried by specifying shape operators(e.g., ups and downs) in support or confidence over time. This problem model appears to be close to our problem. However, we use a user-defined numeric sequence for the interesting query pattern and search globally similar itemsets.

Our similarity constraint is formulated with a *similarity-profiled subset specification* which consists of a reference time sequence, a similarity function, and a dissimilarity threshold. The subset specification is used to define a user interest temporal pattern and guide the degree of approximate matching to prevalence changes of itemsets. For distinguishing interesting itemsets, we use a composite interest measure which consists of prevalence values of an itemset over time (i.e., a support sequence), and the dissimilarity degree of the support sequence to the reference sequence. We call the result itemsets to *similarity-profiled temporal associations*. Finding the similarity-profiled temporal association patterns is computationally challenging. A naive approach is to divide the mining process into two separate phrases. The first phrase computes the support values of all possible itemsets at each time point, and generates their support sequences. The second phrase compares the generated support time sequences with a given reference sequence, and finds the similar itemsets. However, the computational costs of generating the support time sequences of all combinatorial itemsets and then doing the similarity search become prohibitively expensive with increase of items. Thus we explore schemes which can conduct the support time sequence generation and the similarity search interactively, and develop two algorithms with differing database scan methods. The experimental results on synthetic and real data show that our algorithms are efficient and scalable rather than the naive approach.

The remainder of the paper is organized as follows. In Section 2, we formalize the problem of discovering temporal association patterns under a similarity constraint. Our algorithmic design concepts are discussed in Section 3. Section 4 presents the proposed algorithms. The experimental results are presented in Section 5. Section 6 summarizes our work and discusses future work.

2   A Formalization of the Problem

First, we formally define our problem, describe our subset specification in detail, and then introduce our interest measure.

2.1   Problem Definition

Given:

1) A set of items $\mathcal{I}$

2) A time period $\mathcal{T}=t_1 \cup \ldots \cup t_n,\ t_i \cap t_j = \emptyset, i \neq j$

3) A temporal database $\mathcal{D}=\mathcal{D}_1 \cup \ldots \cup \mathcal{D}_n,\ \mathcal{D}_i \cap \mathcal{D}_j = \emptyset, i \neq j$ Each transaction record $d \in \mathcal{D}$ is a tuple $< timestamp, itemset >$ where $timestamp$ is a time $\in \mathcal{T}$ that the record data is observed, and $itemset \subseteq \mathcal{I}$. $\mathcal{D}_i$ is a set of records included in time slot t_i.

4) A subset specification

   4a) A reference sequence $\boldsymbol{R} =< r_1,\ldots,r_n >$ over the time slots $t_1,\ldots,t_n$

   4b) A similarity function $f_{Similarity}(\boldsymbol{X},\boldsymbol{Y}) \mapsto \mathbb{R}^n$, where $\boldsymbol{X}$ and $\boldsymbol{Y}$ are numeric sequences.

   4c) A dissimilarity threshold θ

Find: A set of itemsets $I \subseteq \mathcal{I}$ which satisfies the given subset specification, i.e., $f_{Similarity}(\boldsymbol{S_I},\ \boldsymbol{R}) \leq \theta$, where $\boldsymbol{S_I} =< s_1,\ldots,s_n >$ is the sequence of support values of an itemset I over the time slots $t_1,\ldots,t_n$.

Objective: Reduce the computational cost.

We use the standard notion of *items* in traditional association rule mining. Items can be supermarket items purchased by a customer during a shopping visit, product pages viewed in a web session, climate events at a location, stocks exchanged within a hour, etc. Items can be grouped to form an *itemset*. We assume we are interested in a fixed time period. A time period can be a particular year or any arbitrary period of time. We model time as discrete, and thus, a total time period can be viewed as a sequence of time slots at certain time granularity. For example, one year period can be divided into monthly unit time slots. We denote an ith time slot by t_i. We assume our temporal database $\mathcal{D}$ is a set of timestamped transactions. Each transaction record consists of a set of items over a finite item domain $\mathcal{I}$ and a time point when the transaction is executed. The transaction dataset can be partitioned to disjoint groups of transaction records by time slots. We denote a part of $\mathcal{D}$ included in time slot t_i by $\mathcal{D}_i$. For example, Fig. 2 (a) shows an example dataset including three distinct items, A, B and C. It is partitioned into two groups of transactions related to time slots t_1 and t_2.

2.2   Subset Specification

A subset specification is a set of conditions that itemsets have to satisfy to become interesting patterns. Our subset specification for a similarity constraint consists of three components: a reference sequence, a similarity function, and a dissimilarity threshold. First, we assume that an arbitrary temporal pattern of interest can be defined as a reference sequence by user. A reference time sequence is a sequence of interesting values over time slots $t_1,\ldots,t_n$. We assume that the reference sequence values are in the same scale as the prevalence measure of itemsets or can be normalized to the same scale to allow for differences in level and scale. For example, in Fig. 2 (a), $<0.4, 0.6>$ is given as the reference sequence.

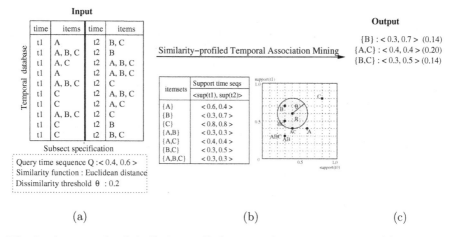

Fig. 2. An example of similarity-profiled temporal association mining (a) Input data (b) Generated prevalence time sequences, and the sequence search (c) Output itemsets

Second, we propose using a $\mathcal{L}_p$ norm ($p = 1, 2, \ldots, \infty$) based similarity function between the reference sequence and the prevalence time sequences of itemsets. Many similarity measures have been discussed in the time series database literature [8]. A $\mathcal{L}_p$ norm is a typical similarity measure and useful in many real application areas [4,9,11,22], and can be used as basic building blocks for more complex similarity models as in [5].

Definition 1. *For two time sequences* $\boldsymbol{X} =< x_1, \ldots, x_n >$ *and* $\boldsymbol{Y} =< y_1, \ldots, y_n >$, *the* $\mathcal{L}_p$ *norm between* $\boldsymbol{X}$ *and* $\boldsymbol{Y}$ *is defined as* $\mathcal{L}_p(\boldsymbol{X}, \boldsymbol{Y}) = (\sum_{i=1}^n |x_i - y_i|^p)^{\frac{1}{p}}$, *where* $p = 1, 2, \ldots, \infty$.

Especially, when $p=2$, $\mathcal{L}_2$ norm is called *Euclidean distance* and defined as $\mathcal{L}_2(\boldsymbol{X}, \boldsymbol{Y}) = (\sum_{i=1}^n |x_i - y_i|^2)^{\frac{1}{2}}$. It is known that the use of Euclidean distance is optimal(in the Maximum Likelihood sense) when measurement differences are independent, identically distributed Gaussian [22]. However, the disadvantage of $\mathcal{L}_p$ norm distance is that it has no obvious bound value of the maximum dissimilarity distance. It is hard to infer from its value whether the degree of similarity is small or large. Thus, we also use a *normalized Euclidean distance*, which gives the bound value for the maximum dissimilarity distance.

Definition 2. *For two time sequences* $\boldsymbol{X} =< x_1, \ldots, x_n >$ *and* $\boldsymbol{Y} =< y_1, \ldots, y_n >$, *the* **normalized Euclidean distance** *between* $\boldsymbol{X}$ *and* $\boldsymbol{Y}$ *is defined as*

$$Normalized\_\mathcal{L}_2(\boldsymbol{X}, \boldsymbol{Y}) = (\frac{1}{n})^{\frac{1}{2}} * \mathcal{L}_2(\boldsymbol{X}, \boldsymbol{Y}) = (\frac{\sum_{i=1}^n |x_i - y_i|^2}{n})^{\frac{1}{2}},$$

where n *is the number of time slots.*

The last component of the subset specification is a dissimilarity threshold. It indicates a maximum discrepancy to allow for interesting patterns.

2.3   Interest Measure

Prevalence measures such as support have been successfully used in traditional association rule mining and temporal association rule mining [21]. Support is the fraction of records that contain an itemset in a database.

Definition 3. *Let $\mathcal{D}$ be a transaction database. The* **support** *of an itemset I is defined as $s(I, \mathcal{D}) = |\{d \in \mathcal{D}|I \subseteq d\}|/|\mathcal{D}|$.*

Instead of support in an entire dataset for measuring the prevalence of an itemset, we measure the support value each time slot, and generate the n-dimensional support sequence from a n divided dataset.

Definition 4. *Let $\mathcal{D} = \mathcal{D}_1 \cup \ldots \cup \mathcal{D}_n$ be a temporal database of disjoint transaction sets. The* **support time sequence** *of an itemset I is defined as*

$$\boldsymbol{S_I} =< s(I, \mathcal{D}_1), \ldots, s(I, \mathcal{D}_n) >,$$

where $s(I, \mathcal{D}_t)$ is the support of an itemset I in a sub dataset $\mathcal{D}_t$.

As an interest measure for similarity-profiled temporal patterns, we use a $\mathcal{L}_p$ norm ($p = 1, 2, \ldots, \infty$) based distance between the support time sequence of an itemset I, $\boldsymbol{S_I}$, and the reference sequence $\boldsymbol{R}$. It is denoted by $D(\boldsymbol{R}, \boldsymbol{S_I})$. In the case of the normalized Euclidean distance, $D(\boldsymbol{R}, \boldsymbol{S_I})=Normalized\_\mathcal{L}_2$ $(\boldsymbol{R}, \boldsymbol{S_I}) = (\frac{1}{n})^{\frac{1}{2}} * \mathcal{L}_2(\boldsymbol{R}, \boldsymbol{S_I}) = (\frac{\sum_{i=1}^{n} |r_i - s_i|^2}{n})^{\frac{1}{2}} = \sigma(\boldsymbol{S_I})$. Statistically, the distance can be thought of as the deviation of the support time sequence $\boldsymbol{S_I}$ from the reference sequence $\boldsymbol{R}$. The range of value is the set of real numbers between 0 to 1 inclusive. A value of 1 reflects the most distant behavior between two sequences. If $D(\boldsymbol{R}, \boldsymbol{S_I})$ is less than a given dissimilarity threshold θ, the itemset I is called *similar itemset*. For example, Fig. 2 (b) shows the support time sequences of all possible itemsets from Fig. 2 (a) dataset. As seen in Fig. 2 (c), the result itemsets are {B}, {A, C} and {B, C} since their interest measure values do not exceed the dissimilarity threshold, 0.2.

3   Algorithmic Design

In this section, we discuss our algorithmic design concept for mining similarity-profiled temporal association patterns.

3.1   Upper Bound and Lower Bound of Support Time Sequence

Generating the support time sequences of itemsets is a core operation in discovering similarity-profiled temporal association patterns. The operation, however, is very data intensive and sometimes can produce the sequences of all combinations of items. We explore a way for estimating support time sequences without examining the input dataset. We define the upper bound and the lower bound of the support time sequence of an itemset using the support time sequences of its subsets.

Definition 5. *Let $\mathcal{D} = \mathcal{D}_1 \cup \ldots \cup \mathcal{D}_n$ be a temporal database of disjoint transaction sets. Let I be a size k itemset $\subseteq \mathcal{I}$, and $\mathcal{J} = \{J_1, \ldots, J_k\}$ be a set of all size $k-1$ subsets of I. The* **upper bound support time sequence** *of I, $U_I =<$ $u_1, \ldots, u_n >$ is defined to $< min\{s(J_1, \mathcal{D}_1), \ldots, s(J_k, \mathcal{D}_1)\}, \ldots, min\{s(J_1, \mathcal{D}_n), \ldots, s(J_k, \mathcal{D}_n)\} >$, where $s(J_h, \mathcal{D}_t)$ is the support of itemset $J_h \in \mathcal{J}$ in $\mathcal{D}_t$, $1 \le h \le k$, $1 \le t \le n$.*

Definition 6. *Let $\mathcal{D} = \mathcal{D}_1 \cup \ldots \cup \mathcal{D}_n$ be a temporal database of disjoint transaction sets. Let I be a size k itemset $\subseteq \mathcal{I}$, and $\mathcal{J} = \{J_1, \ldots, J_k\}$ be a set of all size $k-1$ subsets of I. The* **lower bound support time sequence** *of I, $L_I = < l_1, \ldots, l_n >$ is defined to $< max\{(s(J_1, \mathcal{D}_1) + s(I - J_1, \mathcal{D}_1) - 1), \ldots, (s(J_k, \mathcal{D}_1) + s(I - J_k, \mathcal{D}_1) - 1), 0\}, \ldots, max\{(s(J_1, \mathcal{D}_n) + s(I - J_1, \mathcal{D}_n) - 1), \ldots, (s(J_k, \mathcal{D}_n) + s(I - J_k, \mathcal{D}_n) - 1), 0\} >$, where $s(J_h, \mathcal{D}_t)$ is the support of itemset $J_h \in \mathcal{J}$ in $\mathcal{D}_t$, $1 \le h \le k$, $1 \le t \le n$, and $s(I - J_h, \mathcal{D}_t)$ is the support of an single item of $I - J_h$ in $\mathcal{D}_t$.*

For example, let suppose that in a sub dataset $\mathcal{D}_1$, $s(A, \mathcal{D}_1) = 0.6$, $s(B, \mathcal{D}_1) = 0.3$, $s(C, \mathcal{D}_1) = 0.8$, $s(AB, \mathcal{D}_1) = 0.3$, $s(AC, \mathcal{D}_1) = 0.4$ and $s(BC, \mathcal{D}_1) = 0.3$, and in a sub dataset $\mathcal{D}_2$, $s(A, \mathcal{D}_2) = 0.4$, $s(B, \mathcal{D}_2) = 0.7$, $s(C, \mathcal{D}_2) = 0.8$, $s(AB, \mathcal{D}_2) = 0.3$, $s(AC, \mathcal{D}_2) = 0.4$ and $s(BC, \mathcal{D}_2) = 0.5$. The upper bound support time sequence U_{ABC} of an itemset ABC is $< min\{0.3, 0.4, 0.3\}, min\{0.3, 0.4, 0.5\} >=< 0.3, 0.3 >$. The lower bound support time sequence L_{ABC} of ABC is $=< max\{(0.3 + 0.8 - 1), (0.4 + 0.3 - 1), (0.3 + 0.6 - 1), 0\}, max\{(0.3 + 0.8 - 1), (0.4 + 0.7 - 1), (0.5 + 0.4 - 1), 0\} > =< 0.1, 0.1 >$.

3.2   Lower Bounding Distance

We explore bounding distances to eliminate candidate itemsets which could not possibly be a best match with a reference sequence under a given dissimilarity threshold. In our concept of lower bounding distance, if the lower bounding distance of an itemset does not satisfy the dissimilarity threshold, its true dissimilarity distance also will not satisfy the threshold. Our lower bounding distance consists of two parts, *upper lower-bounding distance* and *lower lower-bounding distance*.

Definition 7. *For a reference sequence R and the upper bound support sequence U of an itemset, let $R^U =< r_1, \ldots, r_k >$ be a subsequence of R, and $U^L =< u_1, \ldots, u_k >$ be a subsequence of U where $r_i > u_i$, $1 \le i \le k$. The* **upper lower-bounding distance** *between R and U, $D_{Ulb}(R, U)$, is defined as $D(R^U, U^L)$.*

The upper lower-bounding distance between a reference sequence R and an upper bound support sequence U is a dissimilarity distance between subsequences of R, R^U, and subsequences of U, U^L, in which each element value r_i in R^U is greater than the corresponding element value u_i of U^L. For example, when Euclidean distance is the similarity function, the upper lower-bounding distance between R and U is $D_{Ulb}(R, U) = D(R^U, U^L) = (\sum_{i=1}^{n} f(r_i, u_i)))^{\frac{1}{2}}$, where if $r_i > u_i$, $f(r_i, u_i) = |r_i - u_i|^2$; otherwise, $f(r_i, u_i) = 0$. In the same way, the upper lower-bounding distance between a reference sequence R and the true support time sequence S of an itemset, $D_{Ulb}(R, S)$, is $D(R^U, S^L)$ where $S^L =< s_1, \ldots, s_k >$ be a subsequence of S, and $r_i > s_i$, $1 \leq i \leq k$.

Definition 8. *For a reference sequence R and a lower bound support time sequence L of an itemset, let $R^L =< r_1, \ldots, r_k >$ be a subsequence of R, and $L^U =< l_1, \ldots, l_k >$ be a subsequence of L where $r_i < l_i$, $1 \leq i \leq k$. The* **lower lower-bounding distance** *between R and L, $D_{Llb}(R, L)$, is defined as $D(R^L, L^U)$.*

The lower lower-bounding distance is a dissimilarity distance between subsequences of a reference sequence R, R^L, and subsequences of a lower bound support sequence L, L^U, in which each element value r_i in R^L are less than the corresponding element value l_i of L^U.

Definition 9. *For a reference sequence R, and the upper bound support time sequence U and lower bound support time sequence L of an itemset, the* **lower bounding distance**, *$D_{lb}(R, U, L)$ is defined as $D_{Ulb}(R, U) + D_{Llb}(R, L)$.*

Fig. 3 gives an example of lower bounding distances computed using the Fig. 2 (a) dataset. Fig. 3 (a) shows the upper lower-bounding distances of several true support time sequences. Fig. 3 (c) shows the upper lower-bounding distance of the upper bound support sequence of {A, B}, U_{AB} (AB_upper in the figure), and the lower lower-bounding distance of the lower bound support sequence of {A, B}, L_{AB} (AB_lower in the figure). The lower lower-bounding distance of L_{AB} is 0 since no value in L_{AB} is greater than the values of R. The lower bound distances of the two bound sequences of AB is $D_{Ulb}(U_{AB}, R) + D_{Llb}(L_{AB}, R) = 0.22 + 0 = 0.22$ as shown in Fig. 3 (b).

Next, we discuss an interesting property related to our upper lower-bounding distance. We first present a related lemma.

Lemma 1. *The support values of the support time sequence of an itemset are monotonically non-increasing with the size of itemset at each time slot.*

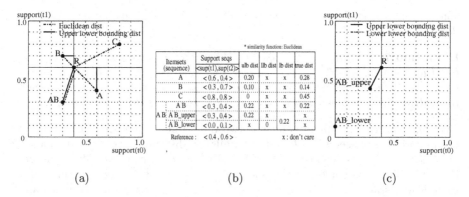

| | (a) | | (b) | | (c) |

Fig. 3. An example of lower bounding distances (a) Upper lower-bounding distances of true support sequences (b) Distance table (c) Upper lower-bounding distance and Lower lower-bounding distance of bounds of support sequence

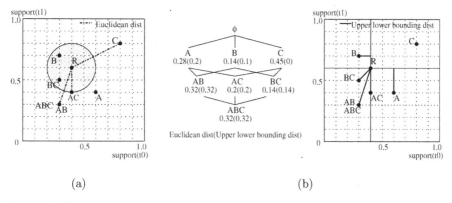

(a)   (b)

Fig. 4. (a) Non-monotonicity of the Euclidean distance with itemset size (b) Monotonically non-decreasing property of the upper lower-bounding distance with itemset size

Proof. Support has a monotonically non-increasing property with increasing size of the itemset [3]. All supersets of a given itemset have supports less than or equal to the support of that itemset, i.e., if I and J are itemsets and $J \subseteq I$, then $support(I) \leq support(J)$. The support time sequence of an itemset consists of supports of the itemset from a set of disjoint transactions in each time slot by Definition 4. Each support value of the prevalence time sequence follows the same monotonicity property.

Similarly, the monotonicity property of our interest measure can state that all supersets of a given itemset have dissimilarity distances greater than the dissimilarity distance of the itemset, i.e., if I and J are itemsets and $J \subseteq I$, then $D(\boldsymbol{R}, \boldsymbol{S}_I) \geq D(\boldsymbol{R}, \boldsymbol{S}_J)$. For example, Fig. 4 (a) shows the Euclidean distances between the support time sequences of $\{C\}$, $\{A,C\}$ and $\{A,B,C\}$, $\boldsymbol{S}_C$, $\boldsymbol{S}_{AC}$ and $\boldsymbol{S}_{ABC}$, and a reference sequence $\boldsymbol{R}$. As can be seen, $D(\boldsymbol{S}_C, \boldsymbol{R})$=0.45, $D(\boldsymbol{S}_{AC}, \boldsymbol{R})$=0.2 and $D(\boldsymbol{S}_{ABC}, \boldsymbol{R})$=0.32. Thus, $D(\boldsymbol{S}_{ABC}, \boldsymbol{R}) > D(\boldsymbol{S}_{AC}, \boldsymbol{R})$ but $D(\boldsymbol{S}_{AC}, \boldsymbol{R}) < D(\boldsymbol{S}_C, \boldsymbol{R})$. That means we observed the $\mathcal{L}_p$ norms-based interest measure does not show any monotonicity with the size of the itemset. However, we can notice that the upper lower-bounding distance is monotonic.

Lemma 2. *The upper lower-bounding distance between the (upper bound) support time sequence of an itemset and a reference time sequence is* **monotonically non-decreasing** *with the size of the itemset.*

Proof. We prove it using Euclidean distance. First, we prove the monotonicity of the upper lower-bounding distance to the true support time sequences. According to Definition 7, the upper lower-bounding distance between $\boldsymbol{S}_I =< s_1, \ldots, s_n >$ for a size k itemset I and $\boldsymbol{R} =< r_1, \ldots, r_n >$ is $D_{Ulb}(\boldsymbol{R}, \boldsymbol{S}_I) = (\sum_{i=1, r_i > s_i}^{n} (r_i - s_i)^2)^{\frac{1}{2}}$. For a size $k+1$ itemset $I'=I \cup \{i'\}$, where $i' \notin I$ and its support time sequence $\boldsymbol{S}_{I'} =< s_1', \ldots, s_n' >$, we need to prove that $D_{Ulb}(\boldsymbol{S}_I, \boldsymbol{R}) \leq D_{Ulb}(\boldsymbol{S}_{I'}, \boldsymbol{R})$. According to Lemma 1, the support is non-increasing with the size of itemset at each time slot, i.e., the support of I' is equal to or less than the support of I at each time slot such that $s_1 \geq s'_1, \ldots, s_n \geq s'_n$. If $s_i \geq s'_i$, $s_i < r_i$ and $s'_i < r_i$, then $r_i - s_i \leq r_i - s'_i$. Thus, we can get $(\sum_{i=1, r_i > s_i}^{n} (r_i - s_i)^2)^{\frac{1}{2}} \leq (\sum_{i=1, r_i > s'_i}^{n} (r_i - s'_i)^2)^{\frac{1}{2}}$, i.e., $D_{Ulb}(\boldsymbol{R}, \boldsymbol{S}_I) \leq D_{Ulb}(\boldsymbol{R}, \boldsymbol{S}_{I'})$. Second, the monotonicity of upper lower-bounding distance to the upper bound support time sequence can be similarly proved.

For example, in Fig. 4 (b), $D_{Ulb}(\boldsymbol{S}_A, \boldsymbol{R})=0.2$, $D_{Ulb}(\boldsymbol{S}_B, \boldsymbol{R})=0.1$ and $D_{Ulb}(\boldsymbol{S}_{AB}, \boldsymbol{R})=0.32$. Thus $D_{Ulb}(\boldsymbol{S}_A, \boldsymbol{R}) \leq D_{Ulb}(\boldsymbol{S}_{AB}, \boldsymbol{R})$ and $D_{Ulb}(\boldsymbol{S}_B, \boldsymbol{R}) \leq D_{Ulb}(\boldsymbol{S}_{AB}, \boldsymbol{R})$. We can also see that $D_{Ulb}(\boldsymbol{S}_{AB}, \boldsymbol{R}) \leq D_{Ulb}(\boldsymbol{S}_{ABC}, \boldsymbol{R})$, $D_{Ulb}(\boldsymbol{S}_{AC}, \boldsymbol{R}) \leq D_{Ulb}(\boldsymbol{S}_{ABC}, \boldsymbol{R})$, and $D_{Ulb}(\boldsymbol{S}_{BC}, \boldsymbol{R}) \leq D_{Ulb}(\boldsymbol{S}_{ABC}, \boldsymbol{R})$.

3.3   Database Scan Strategy

Support time sequences can be generated using different methods to scan the timestamped transaction database. We can consider two database scan methods: a *lattice-dominant scan* and a *snapshot-dominant scan*.

1) Lattice-dominant scan: The lattice-dominant scan method reads a whole database from time slot t_1 to time slot t_n for lattice itemsets of each depth, and generates the support time sequences of the itemsets over all time slots.

2) Snapshot-dominant scan: The snapshot-dominant scan method repeats the scanning of transaction records at each time slot, e.g, from the first time slot, by counting the supports of itemsets until it finds all candidate itemsets of different sizes. It then moves to the next time slot and repeats the process. This method incrementally generates support time sequences with the processed time slots.

4   Algorithms

We developed similarity-profiled association mining algorithms based on our algorithm design concept discussed in Section 3. We first discuss a naive approach for comparison, and then present the proposed algorithms.

4.1   Naive Approach

A naive approach for finding similarity-profiled temporal associations can be characterized using a two-phase paradigm. The first phase generates the history of supports for all possible itemsets at different time slots. The second phase compares the generated support time sequences with a reference sequence to find similar associated itemsets. In the second step, we can use advanced time series search algorithms using multi-dimensional index structures, e.g., R-tree [13]. However, exponentially increasing computational costs of generating the support time sequences of all combinatorial candidate itemsets become prohibitively expensive.

4.2   SPAMINEs

We propose a one-step approach to combine the generation of support time sequences and the sequence search. We propose two algorithms for the Similarity-Profiled temporal Association MINing mEthod(SPAMINE): a Lattice-dominant

SPAMINE(L-SPAMINE) and a Snapshot-dominant SPAMINE(S-SPAMINE). They are different in the database scan method.

4.2.1   L-SPAMINE

The L-SPAMINE algorithm uses the lattice-dominant database scan method to generate the support time sequences. Algorithm 1 shows the pseudocode. Fig. 5 provides an illustration of execution trace of L-SPAMINE using the example data in Fig. 2 (a).

[Steps 1 - 3] *Generate the support time sequences of single items and find similar items:* All singletons ($k = 1$) become candidate items(C_1). In the first scan of an entire time-stamped database, the supports of singletons are computed per each time slot and their support time sequences(S_1) are generated. If the distances between the support time sequences and a given reference sequence do not exceed a given dissimilarity threshold, the singletons are added to a result set(A_1). On the fly, if the upper lower-bounding distances of the support time sequences satisfy the threshold, the items are kept to B_1 for generating the next size candidate itemsets. In Fig. 5, only item B is a similar itemset but items A and C are also kept for generating the next size candidate itemsets.

[Step 6] *Generate candidate itemsets and their upper and lower bound support sequences:* All size k ($k > 1$) candidate itemsets(C_k) are generated using size $k - 1$ itemsets(B_{k-1}) whose upper lower-bounding distances satisfy the dissimilarity threshold. If any subset of size $k - 1$ of the generated itemset is not

Inputs:
E : A set of single items.
TD: A temporal transaction database
R : A reference sequence
D : A similarity function
θ : A dissimilarity threshold
Output: All itemsets whose support sequences are similar to R under D and θ
Variables :
k : Itemset size
C_k : A set of size k candidate itemsets
U_k : A set of upper bound support sequences of size k itemsets
L_k : A set of lower bound support sequences of size k itemsets
S_k : A set of true support sequences of size k itemsets
B_k : A set of size k itemsets whose upper lower-bounding distance $\leq \theta$
A_k : A result set of size k itemsets whose true distance $\leq \theta$
Main:
1) $C_1 = E$;
2) S_1= generate_support_sequences(C_1, TD);
3) (A_1, B_1)= find_similar_itemsets(C_1, S_1, R, D, θ);
4) $k = 2$;
5) **while** (not empty B_{k-1}) **do**
6)   (C_k, U_k, L_k) =generate_candidate_itemsets(B_{k-1}, S_{k-1});
7) C_k =prune_candidate_itemsets_by_lbd(C_k, U_k, L_k, R, D, θ);
8) S_k=generate_support_sequences(C_k, TD);
9)   (A_k, B_k) =find_similar_itemsets(C_k, S_k, R, D, θ);
10) $k = k + 1$;
11) **end**
12) **return** $\bigcup(A_1, \ldots, A_k)$;

Algorithm 1. L-SPAMINE algorithm

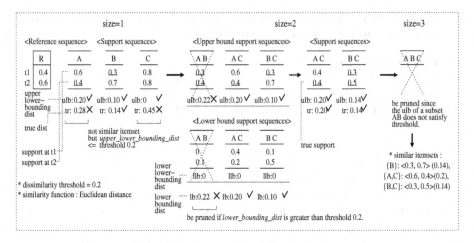

Fig. 5. An illustration of L-SPAMINE algorithm trace

in the B_{k-1}, the candidate itemset is eliminated using the monotonically non-decreasing property of the upper lower-bounding distance. Otherwise, the upper and lower bound support sequences of candidate itemsets are generated. Fig. 5 shows the generated size 2 candidate itemsets, and their upper bound and lower bound support time sequences.

[Step 7] *Prune candidate itemsets using the lower bounding distance:* The lower bounding distances of candidate itemsets are calculated. If the lower bounding distance of an itemset exceeds the dissimilarity threshold, the candidate itemset is eliminated from the candidate set. For example, in Fig. 5, the lower bounding distance of itemset {A, B} is 0.22 which is the sum of its upper lower-bounding distance and lower lower-bounding distance. Itemset {A, B} is removed from the set of candidate itemsets since the value is greater than the threshold 0.2.

[Step 8] *Scan database and generate the support time sequences:* The supports of candidate itemsets are computed during the scan of the temporal database, and their support time sequences(S_k) are generated.

[Step 9] *Find itemsets showing similar prevalence variations:* The true distance between the support time sequence of an itemset and the reference sequence is computed. If the value satisfies the threshold, the itemset is included in the result set(A_k). In Fig. 5, {A,C} and {B,C} are similar itemsets. On the fly, if the upper lower-bounding distances of itemsets satisfy the threshold, the itemsets are added to B_k for generating the next size candidate itemsets.

[Step 10] *Increase the examined itemset size:* The size of the examined itemsets is increased to $k = k + 1$. The above procedures(steps 6-10) are repeated until no itemset in B_k remains.

4.2.2   S-SPAMINE

The S-SPAMINE algorithm uses the snapshot-dominant scan method. It first repeats scanning transaction records at time t_1 with increasing itemset size until all candidate itemsets at time t_1 are found. The pruning strategy is the same as those of L-SPAMINE except S-SPAMINE uses the accumulated distances up to the current time slot. If the accumulated upper lower bounding distance is greater than a given threshold, the candidate itemset is pruned at the current time slot. After finding all size candidate itemsets, we move to the next time slot and repeat the procedures until the last time slot. Finally, itemsets whose accumulated true similarity value is less than the threshold are included in the result set.

5   Experimental Evaluation

We conducted a series of experiments to examine the effect of pruning by lower bounding distance, database scan method, number of time slots, number of items, and number of time slots. Our experiments were performed on synthetic and real datasets. Synthetic datasets were generated using the transaction generator designed by the IBM Quest project used in [3]. We modified it for generating a time-stamped transaction database, where each transaction has a time slot value. In the rest of paper, we use the following parameters to characterize the synthetic datasets we used. TD is the total number of transactions($\times$ 1,000), D is the number of transaction per time slot($\times$ 1,000), I is the number of distinct items, L is the average size of transaction, and T is the number of time slots. A reference time sequence was randomly generated in the scale of the support value at each time slot or generated by choosing a support value near a quartile e.g., a 25th, 50th, 75th percentile of the sorted supports at each time slot after calculating the supports of single items. For the experiment with a real dataset, we used a climate dataset. We compared the performances of L-SPAMINE, S-SPAMINE and the naive method. Since the naive method with no pruning scheme generates huge candidate itemsets with increase of itemset size, we used small datasets, especially in the number of items, in comparisons with the naive approach. Normalized Euclidean distance was used for the similarity function. All experiments were performed on a workstation with 2 Gbytes of memory running the Linux operating system.

5.1   Experiment Results

Effect of lower bounding pruning. In this experiment, we examined the effect of pruning by our lower bounding distance in L-SPAMINE. The TD100-D1-L10-I20-T100 dataset was used and a reference sequence was chosen near the second quartile. Fig. 6 (a) shows the pruning effect ratios per each level with different dissimilarity thresholds. The pruning effect ratio is the number of candidate itemsets which need a database scan over the total number of possible itemsets per level. L-SPAMINE generates dramatically fewer candidate itemsets compared with the naive approach. In the next experiment, we examined the pruning effect with different reference sequence types. The reference sequence values were chosen near different quartiles of sorted supports of single items. Fig. 6 (b) shows the results when the dissimilarity threshold was fixed to 0.2. As can be seen, our lower bounding pruning scheme is most effective when the reference sequence values are overall greater than the support sequence values.

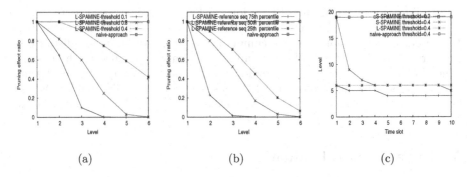

(a)   (b)   (c)

Fig. 6. (a) Effect of lower bounding pruning (b) Effect of reference sequence type (c) Effect of database scan method

Effect of database scanning method. We examined the effect of choice of database scan method with different thresholds. We used a synthetic dataset TD100-D1-L10-I20-T100 and choose a reference sequence near the second quartile. Fig. 6 (c) shows a max lattice level explored in each time slot. S-SPAMINE examined a little smaller lattice level with L-SPAMINE at threshold 0.2. However, with the increase of the dissimilarity threshold (e.g., 0.4), S-SPAMINE showed a dramatically large itemset search space in the beginning time slots, since the accumulated lower bounding distances were not enough for pruning. By contrast, L-SPAMINE showed a constant deployment of the itemset lattice independent with the time slot number.

Effect of number of items. We examined the effect of number of items with synthetic datasets of different number of items, TD10-D1-L10-I*-T10. Under the 0.2 threshold, which prunes most itemsets before around level(pass) 4 in these datasets, L-SPAMINE and S-SPAMINE showed a similar execution time and received little effect with the increase of itemset size. However, when the threshold value was increased to 0.3, S-SPAMINE showed dramatically increased execution time. The reason is that the pruning ability of S-SPAMINE was weak in the beginning time slots under this threshold and kept many lattice subsets. Fig. 7 (a) shows the results.

Effect of number of time slots. In this experiment, we examined the effect of time sequence length using synthetic datasets, TD*-D1-L6-I20-T* having different numbers of time slots. Query sequences were chosen near the second quartile in each dataset and the threshold was the same value, 0.2 in each dataset. Fig. 7 (b) shows that the execution time of both algorithms increased with increases in the number of time slots since the total number of transactions is increased with the number of time slots. However, L-SPAMINE receives less effect with number of time slots.

Experiment with a real dataset. Finally, we examined our algorithms with an Earth climate dataset available in UCI KDD Archive [1]. The data set contains

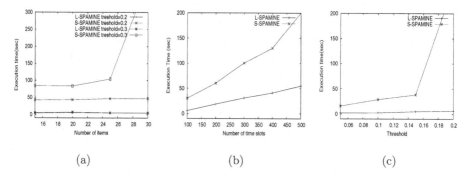

Fig. 7. (a) Effect of number of items (b) Effect of number of time slots (c) A real dataset

oceanographic and surface meteorological readings taken from a series of buoys positioned throughout the equatorial Pacific. Each buoy has five meteorological readings, e.g., humidity measure, almost every day from 1980 March to 1998 June. Since the reading value is a real number, we categorized the real value using its quartile values. Each reading item was categorized to four item types, e.g., humidity_high, humidity_mid1, humidity_mid2, humidity_low. Total 20 categorized items are generated. One transaction consists of five climate items of a day at a buoy. The total number of transactions was 178,080. When we chose a month as a time slot granularity, there were total 220 time slots. Fig. 7 (c) shows the experiment result with different thresholds. The S-SPAMINE shows a dramatic increase of execution time with the increase of the threshold. L-SPAMINE showed overall better performance than S-SPAMINE.

6   Conclusion

We formulated the problem of mining temporal association patterns under similarity constraints and proposed novel algorithms to discover them. The proposed algorithms substantially reduced the search space by pruning candidate itemsets using the lower bounding distance of the bounds of support sequences, and the monotonicity property of the upper lower bounding distance. Experimental results showed that L-SPAMINE algorithm is computationally efficient than S-SPAMINE and a naive method. We need further study for discovery accuracy with domain experts. On the other hand, time series literature proposes many different similarity measures for time sequence search [8]. We plan to explore various similarity functions for similarity-profiled association patterns. Another issue in similarity modeling is concerned about the components of sequence matching. We currently use a whole-sequence matching for discovering similar association patterns. We also plan to consider techniques for sub-sequence matching, e.g., for the case that the length of a reference sequence and of a support sequence is different.

References

1. Uci kdd archive, http://kdd.ics.uci.edu/
2. After Katrina: Crisis Management, The Only Lifeline Was the Wal-Mart. In: FORTUNE Magazine (October 3, 2005)
3. Agarwal, R., Srikant, R.: Fast Algorithms for Mining Association Rules. In: Proc. of International Conference on Very Large Data Bases(VLDB) (1994)
4. Agrawal, R., Faloutsos, C., Swami, A.: Efficient Similarity Search in Sequence Databases. In: Lomet, D.B. (ed.) FODO 1993. LNCS, vol. 730. Springer, Heidelberg (1993)
5. Agrawal, R., Lin, K.I., Sawhney, H.S., Shim, K.: Fast Similarity Search in the Presence of Noise, Scaling, and Translation in Time-series Database. In: Proc. of International Conference on Very Large Databases(VLDB) Conference (1995)
6. Agrawal, R., Srikant, R.: Mining Sequenetial Patterns. In: Proc. of International Conference on Data Engineering (ICDE) (1995)
7. Cheung, W., Han, J., Ng, V.T., Wong, C.Y.: Maintenance of Discovered Association Rules in Large Databases: An Incremental Updating Technique. In: Proc. of IEEE International Conference on Data Engineering(ICDE) (1996)
8. Das, G., Gunopulos, D.: Time Series Similarity Measures. In: Tutorial notes of ACM SIGKDD International Conference on Knowledge Discovery and Data Mining (2000)
9. Das, G., Gunopulos, D., Mannila, H.: Finding Similar Time Series. In: Proc. of Principles of Data Mining and Knowledge Discovery, European Symposium (1997)
10. Dong, G., Li, J.: Efficient Mining of Emerging Patterns: Discovering Trends and Differences. In: Proc. of ACM SIGKDD International Conference on Knowledge Discovery and Data Mining (1999)
11. Faloutsos, C., Ranganathan, M., Manolopoulos, Y.: Fast Subsequence Matching in Time-series Database. In: Proc. of ACM SIGMOD International Conference on Management of Data (1993)
12. Ganti, V., Gehrke, J., Ramakrishnan, R.: A Framework for Measuring Changes in Data Characteristics. In: Proc. of ACM SIGMOD International Conference on Principles of Database Systems (PODS) (1999)
13. Gunopulos, D., Das, G.: Time Series Similarity Measures and Time Series Indexing. SIGMOD Record 30(2) (2001)
14. Han, J., Pei, J., Yin, Y.: Mining frequent patterns without candidate generation. In: Proc. of ACM SIGMOD International Conference on Management of Data (2000)
15. Li, Y., Ning, P., Wang, X.S., Jajodia, S.: Discovering Calendar-Based Temporal Assocation Rules. In: Proc. of Internationl Symposium Temporal Representation and Reasoning (TIME) (2001)
16. Liu, B., Hsu, W., Ma, Y.: Discovering the Set of Fundamental Rule Change. In: Proc. of ACM SIGKDD International Conference on Knowledge Discovery and Data Mining (2001)
17. NOAA. El Nino Page, http://www.elnino.noaa.gov/
18. Ozden, B., Ramaswamy, S., Silberschatz, A.: Cyclic Association Rules. In: Proc. of IEEE International Conference on Data Engineering(ICDE) (1998)
19. Park, J.S., Chen, M., Yu, P.: An Effective Hashing-based Algorithm for Mining Association Rules. In: Proc. of ACM SIGMOD International Conference on Management of Data (1995)

20. Savasere, A., Omiecinski, E., Navathe, S.: An Effective Algorithm for Mining Association Rules in Large Databases. In: Proc. of International Conference on Very Large Databases(VLDB) (1995)
21. Tan, P., Kumar, V., Srivastava, J.: Selecting the Right Interestingness Measure for Association Patterns. In: Proc. of ACM SIGKDD International Conference on Knowledge Discovery and Data Mining (2002)
22. Yi, B.K., Faloutsos, C.: Fast Time Sequence Indexing for Arbitrary $\mathcal{L}_p$ norms. In: Proc. of International Conference on Very Large Data Bases(VLDB) (2000)
23. Zhang, P., Steinbach, M., Kumar, V., Shekhar, S., Tan, P., Klooster, S., Potter, C.: Discovery of Patterns of Earth Science Data Using Data Mining. In: Kantardzic, M.M., Zurada, J. (eds.) Next Generation of Data Mining Applications. IEEE Press, Los Alamitos (2004)

A General Framework for Increasing the Robustness of PCA-Based Correlation Clustering Algorithms

Hans-Peter Kriegel, Peer Kröger, Erich Schubert, and Arthur Zimek

Institute for Informatics, Ludwig-Maximilians-Universität München
{kriegel,kroegerp,schube,zimek}@dbs.ifi.lmu.de
http://www.dbs.ifi.lmu.de

Abstract. Most correlation clustering algorithms rely on principal component analysis (PCA) as a correlation analysis tool. The correlation of each cluster is learned by applying PCA to a set of sample points. Since PCA is rather sensitive to outliers, if a small fraction of these points does not correspond to the correct correlation of the cluster, the algorithms are usually misled or even fail to detect the correct results. In this paper, we evaluate the influence of outliers on PCA and propose a general framework for increasing the robustness of PCA in order to determine the correct correlation of each cluster. We further show how our framework can be applied to PCA-based correlation clustering algorithms. A thorough experimental evaluation shows the benefit of our framework on several synthetic and real-world data sets.

1   Introduction

Finding clusters in arbitrarily oriented subspaces is an important data mining task for many applications. The motivation behind this task is that in high dimensional data, one probably cannot find clusters due to several properties of high dimensional feature spaces. In contrast, clusters can usually be found in arbitrarily oriented subspaces of the original data space. The points of a subspace cluster are then located on a common lower dimensional hyperplane and exhibit a common correlation among a subset of the attributes. The task of finding clusters in arbitrarily oriented subspaces is also called correlation clustering.

The major challenge of correlation clustering is identifying the correct subspace of a cluster. Most correlation clustering algorithms [1,2,3,4,5,6] apply principal component analysis (PCA) to a subset of points in order to define the correct subspace in orientation and weighting of the transformed axes. PCA is a mature technique and allows the construction of a broad range of similarity measures grasping local correlation of attributes and, therefore, allows to find arbitrarily oriented subspace clusters. It is easy to see that the more points of this subset are cluster members that are located on the common hyperplane, the more accurate the procedure of determining the correct subspace (i.e. hyperplane) will be. However, a drawback common to all those approaches is the notorious *locality assumption*. Since cluster memberships of points are obviously not known

B. Ludäscher and Nikos Mamoulis (Eds.): SSDBM 2008, LNCS 5069, pp. 418–435, 2008.

beforehand, it is assumed that the local neighborhood, e.g. the ε-neighborhood or the k-nearest neighbors, of cluster points or cluster centers represents the correct subspace suitably well in its orientation and variance along axes. This assumption is widely accepted but it boldly contradicts the basic problem statement, i.e. "find clusters in a high-dimensional space", because high dimensional spaces are typically doomed by the *curse of dimensionality*. The term "curse of dimensionality" refers to a bundle of problems occurring in high dimensional spaces. The most important effect in the sight of clustering is that concepts like "proximity", "distance", or "local neighborhood" become less meaningful with increasing dimensionality of a data set (as elaborated e.g. in [7,8,9]). As a consequence of these findings, the discrimination between the nearest and the farthest neighbor becomes rather poor with increasing data dimensionality. This is by far a more fundamental problem than the mere performance degradation of algorithms on high dimensional data: The higher the dimensionality of a data set is, the more outliers will be placed inevitably in the set of neighboring objects.

As we will see in this paper, PCA is very sensitive to outliers. In other words, if the local neighborhood of cluster members or cluster centers to which PCA is applied in order to find the correct subspace of the corresponding cluster contains noise points that do not belong to the cluster, the subspace determination process will be misled. Thus, in view of the "curse of dimensionality", to successfully employ PCA in correlation clustering in high-dimensional data spaces may therefore require more sophisticated techniques of selecting a representative set of neighbors.

In this paper, after shortly reviewing existing approaches to correlation clustering (cf. Section 2), we evaluate the influence of outliers on PCA in general (cf. Section 3) and propose a general framework to determine the correct local subspace dimensionality and orientation for cluster members and cluster centers in a more robust way (cf. Section 4). In Section 5, we show how to apply the proposed framework for increasing the robustness the subspace determination process on existing correlation clustering approaches. Section 6 demonstrates the impact of the increased robustness of PCA on several data sets. The paper is concluded in Section 7.

2   Related Work

The first approach to *generalized projected clustering*, called ORCLUS [1], is a K-means like approach. It picks $K_c > K$ seeds at first and assigns the data base objects to these seeds according to a distance function that is based on an eigensystem of the corresponding cluster assessing the distance along the small eigenvectors only (i.e., the distance in the projected subspace where the cluster objects exhibit high density). The eigensystem is iteratively adapted to the current state of the updated cluster (i.e., based on the current neighborhood of the cluster center). The number K_c of clusters is reduced iteratively by merging closest pairs of clusters until the user-specified number K is reached. The method proposed in [10] is a slight variant of ORCLUS designed for enhancing multi-dimensional indexing. Initially, however, the eigensystems in both methods are based on the local neighborhood in the Euclidean space.

The algorithm 4C [2] is based on a density-based clustering paradigma [11]. Thus, the number of clusters is not decided beforehand but clusters grow from a seed as long as a density criterion is fulfilled. Otherwise, another seed is picked to start a new cluster. The density criterion is a required minimal number of points within the neighborhood of a point, where the neighborhood is ascertained based on distance matrices computed from the eigensystems of two points. The eigensystem of a point is based on the covariance matrix of the ε-neighborhood of the point in Euclidean space.

As a hierarchical approach, HiCO [4] defines the distance between points according to their local correlation dimensionality and subspace orientation – thus again based on a local neighborhood query – and uses hierarchical density-based clustering [12] to derive a hierarchy of correlation clusters.

COPAC [5] is based on similar ideas as 4C but disposes of some problems like meaningless similarity matrices due to sparse ε-neighborhoods instead taking a fixed number k of neighbors — which raises the question how to choose a good value for k but at least choosing $k > \lambda$ ensures a meaningful definition of a λ-dimensional hyperplane. Still, the Euclidean neighborhood critically influences the results.

The latest PCA-based correlation clustering algorithm is ERiC [6], also deriving a local eigensystem for a point based on the k nearest neighbors in Euclidean space. Here, the neighborhood criterion for two points in a DBSCAN-like procedure is an approximate linear dependency and the affine distance of the correlation hyperplanes as defined by the largest eigenvectors of each point. In finding and correctly assigning complex patterns of intersecting clusters, COPAC and ERiC improve considerably over ORCLUS and 4C.

Another approach based on PCA said to find even non-linear correlation clusters, CURLER [3], seems not restricted to correlations of attributes but, according to its restrictions, finds any narrow trajectory and does not provide a model describing its findings. However, even in this approach the PCA is applied to the local neighborhood of points in Euclidean space.

Note that the term "correlation clustering" relates to a different task in the machine learning community, where a partitioning of the data shall correlate as much as possible with a pairwise similarity function learned from past data [13].

3   Problem Analysis

To the best of our knowledge, all correlation clustering algorithms that use PCA as the method to determine the correct subspace of a cluster face the following problem. In order to determine the correct subspace of a cluster, a (considerably large) number of cluster members needs to be identified first such that PCA can be applied to them. On the other hand, in order to identify points of a particular cluster, the subspace of this cluster needs to be determined first. To escape from this vicious circle all algorithms rely on the locality assumption, i.e. it is assumed that the points in the local neighborhood of cluster members or cluster representatives sufficiently reflect the correct subspace of the corresponding cluster such that applying PCA to those neighboring points reports the cluster hyperplane.

As stated above, selecting a meaningful neighborhood becomes more and more difficult with increasing data dimensionality. A neighboring set of points will almost certainly contain outliers, i.e. points that do not belong to the cluster and, thus, are not located on the hyperplane of the cluster. Obviously, these outliers are not helpful to assign a meaningful local correlation dimensionality and orientation. On the other hand, all correlation clustering approaches available (cf. Section 2) rely on an arbitrarily chosen set of neighboring points. We therefore argue to choose a neighboring set of points in a more sophisticated way to enhance the robustness of local correlation analysis and, consequently, to enhance the robustness of correlation clustering algorithms.

3.1   Impact of Outliers on PCA

Correlation analysis using PCA is a *least squares fitting* of a linear function to the data. By minimizing the *mean square error*, outliers are emphasized in a way that is not always beneficial, as can bee seen in Figure 1. This data set consists of 5 points in a 2D space that are strictly positively correlated and, thus, are located on a common 1D hyperplane plus one additional outlier that is not located on that 1D hyperplane. When applying PCA on these six points and computing the strongest eigenvector of the corresponding covariance matrix, the resulting vector is directed towards the outlier (cf. Figure 1). This implies that in certain situations, adding only one single extra point to the correlation computation can cause the resulting strongest eigenvector(s) to flip into a completely different direction. Let us note that if the outlier point would have been closer to the other points it would, at a certain distance, not have made any difference on the vector orientation, but this distance threshold for the flip is rather small.

As a consequence, one needs to carefully select the points that are included into the computation of the cluster hyperplane. In addition, one can consider using a modified correlation analysis procedure which is less sensitive to the effect

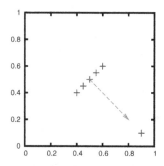

Fig. 1. Simple data set with 6 points and largest eigenvector after PCA

of outliers. In fact, there are obviously multiple strategies to handle these issues. The most obvious one – using outlier detection to remove outliers from the computation – can usually not be applied to this problem because we face the same vicious circle when searching for outliers as we face when detecting cluster points: in order to identify outliers that do not belong to any clusters, the subspaces of the clusters need to be determined first; in order to determine the correct subspace of a cluster, a (considerably large) number of cluster members needs to be identified first such that PCA can be applied to them; etc. Instead, we introduce two ideas to stabilize PCA for correlation clustering. First, we explore a local optimization strategy that handles the problem of picking appropriate neighboring points in a way that is easy to integrate in many correlation clustering algorithms. Second we will add a modified correlation analysis to further stabilize results which is based on the integration of a suitable weighting function into PCA.

3.2   Statistic Observations on Data Correlation

Without loss of generality, we assume that the points on which PCA is applied to find the correct subspace of a particular cluster are selected as the k-nearest neighbors (kNN) of cluster members or cluster representatives. Later, we will discuss the extension of our ideas to methods like ORCLUS that use neighborhood concepts other than kNN.

When comparing the relative strength of the normalized eigenvalues (i.e. the part of the total variance explained by them) computed for the kNN of a particular point w.r.t. increasing values of k (ranging from 0 to 50% of the data set), a behavior similar to that shown in Figure 3 can usually be observed. We used a 3D data set shown in Figure 2, with a set of 200 outlier points (noise), a correlation cluster of 150 points sharing a common 2D hyperplane (plane), and a correlation cluster of 150 points that are located on a common 1D hyperplane (line) that is embedded into the hyperplane of the 2D cluster. In Figure 3 there are three plots in this graph representing the behavior of the eigenvalues of a sample noise point, of a sample point on a 2D, and of a sample point on a 1D line in the data set (embedded within the 2D plane), respectively.

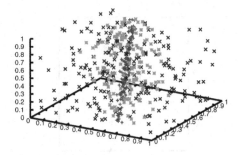

Fig. 2. Data set with a 2D plane and an embedded 1D line

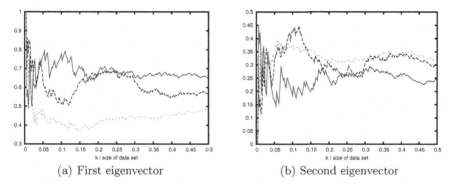

(a) First eigenvector   (b) Second eigenvector

Fig. 3. Relative strength of eigenvectors

Examining the noise point (green dotted lines in Figure 3) we observe a minimum relative strength of the first eigenvalue of about 0.4 for $k = 10\% - 15\%$ (cf. Figure 3(a)). Since the minimum possible value for the strongest eigenvector in a 3D data set is $1/3 = 0.33$, the noise point shows approximately no correlation when looking at its kNN with $k = 10\% - 15\%$ of the data set. The second eigenvector (cf. Figure 3(b)) shows similar behavior in that particular range of k confirming our conclusions.

For the point in the 1D cluster (red solid lines in Figure 3), the first eigenvector (cf. Figure 3(a)) explains 80% of the complete variance at around $k = 7\%$, i.e. using this value for k, the kNN of the particular point form the 1D line of the cluster. It is worth noting that the amount of variance explained for the 1D cluster case drops quickly when increasing k beyond this point. The reason for this is that – since the line is embedded in a plane – with increasing k more and more points of the kNN are points from the 2D cluster. As a consequence, the variance explained by the first eigenvector decreases, whereas the variance explained by the second eigenvector increases simultaneously (cf. Figure 3(b)). Then, at $k \approx 10\%$, we have again a very high strength of the first eigenvector (less points from the 2D cluster and more points from the 1D cluster are considered), etc. In other words, depending on the value of k, the kNN of the point form the 1D cluster line or the 2D cluster plane.

For evaluating the 2D cluster, the relevant graph (depicting the behavior of the second eigenvector) is shown in Figure 3(b). In a 3D data set, a value of around $1/3$ would be typical for uncorrelated data and is observable on noise points. For the sample point from the 2D cluster it peaks at almost 45% for about $k = 10\%$. Together with the first graph, this means that the first two eigenvectors explain almost the complete variance at that particular value for k. In other words, for $k = 10\%$, the kNN of this point reflect the 2D plane of the cluster sufficiently. Compared to this observation, the variance of the sample point from the 1D cluster embedded in the 2D cluster (red dotted line) along the first two eigenvectors is significantly below the expected value (which is not surprising, having seen that the first eigenvector reaches 80%).

These simple examples illustrate that it is essential to select a sufficient set of points by choosing a suitable value for k. A slight change in k can already make a large difference. Moreover, we have seen that it is rather meaningful to choose even significantly different values of k for different points.

4   A General Framework for Robust Correlation Analysis

The above presented considerations induce two important aspects. First, since PCA is a least square fitting and we cannot assume that there are no outliers in the kNN of a point, adjusting the weighting of the points during PCA should improve the results. Second, the selection of points to which PCA is applied can be improved by both micro-adjusting the value of k (to avoid sudden drops in the explained variance) as well as choosing significantly different k for different points in the data set. In the following, we will discuss both aspects in more detail. In fact, our framework for making PCA-based correlation analysis more robust uses both ideas.

4.1   Increasing the Robustness of PCA Using Weighted Covariance

As mentioned above, PCA is a common approach to handling correlated data. It is also commonly used for dimensionality reduction by projecting onto the λ strongest (i.e. highest) components. In correlation clustering, PCA is a key method to finding correlated attributes in data.

PCA operates in two steps. In the first step, for any two attributes, i.e. dimensions, d_1 and d_2 the covariance $\mathrm{Cov}(X_{d_1}, X_{d_2})$ of these two dimensions is computed. In the second step, the eigenvectors and eigenvalues of the resulting matrix (which by construction is positive, symmetric and semi-definite) are computed. The computation of eigenvectors and eigenvalues on a symmetric matrix is a standardized procedure which cannot be altered to make the overall process more robust. Instead, the stabilization has to be implemented during the first step.

Given an attribute X, we can model the values of k points in that particular attribute, denoted by x_i for the i-th point, as a random variable. Then, the covariance between two attributes X and Y is mathematically defined as

$$\mathrm{Cov}(X, Y) := E((X - E(X)) \cdot (Y - E(Y))), \tag{1}$$

where E is the expectation operator. Usually, one uses the mean of all values of the corresponding attribute as expectation operator, i.e.

$$E(X) = \frac{1}{k} \sum_{i=1}^{k} x_i =: \hat{x}, \tag{2}$$

so we have

$$\mathrm{Cov}(X, Y) := \frac{1}{k} \sum_{i=1}^{k} (x_i - \hat{x})(y_i - \hat{y}). \tag{3}$$

Obviously, all data points are treated equally in this computation. But given that we want to reduce the effect of outliers, it is more appropriate to use a different expectation operator. Given arbitrary weights w_i for all points i $(1 \leq i \leq k$ and $\Omega := \sum_{i=1}^{k} w_i)$, we can define a new expectation operator

$$E_w(X) := \frac{1}{\Omega} \sum_{i=1}^{k} w_i x_i =: \hat{x}_w. \tag{4}$$

With this new expectation operator, we can give each point in kNN a different weight. In particular, we can give potential outliers a smaller weight. Using $E_w(X)$, we can compute the covariance as given below.

$$\text{Cov}_w(X, Y) := \frac{1}{\Omega} \sum_{i=1}^{n} w_i (x_i - \hat{x}_w)(y_i - \hat{y}_w). \tag{5}$$

Steiner's translation still applies, which leads to the following slightly simpler equation.

$$\text{Cov}_w(X, Y) = \left(\frac{1}{\Omega} \sum_{i=1}^{n} w_i x_i y_i\right) - \left(\frac{1}{\Omega} \sum_{i=1}^{n} w_i x_i\right) \cdot \left(\frac{1}{\Omega} \sum_{i=1}^{n} w_i y_i\right). \tag{6}$$

This form is particularly nice for computation. It is also trivial to prove that if $w_i = 1$ for all i, we have $\text{Cov}(X, Y) = \text{Cov}_w(X, Y)$. If a point i is assigned the weight $w_i = 2$, the result would be the same as if we had two points with the same coordinates as i. If a point i is weighted by $w_i = 0$, the result is the same as if point i had not been included in the computation at all.

We can now use arbitrary weighting functions to calculate the weights to be used. Obviously, we again have the dilemma that we do not know which points are outliers and need to get assigned a lower weight. However, since all algorithms use the locality assumption, we can make the following considerations: On the one hand, it is usually very likely that taking the local neighborhood of points includes a lot of outliers. But on the other hand, the neighbors that are near to the query point will more likely be cluster members than the neighbors that are farther apart from the query point. So a distance-based weighting function will most likely weight cluster points higher and outliers lower.

Some examples of distance-based weighting functions are given in Figure 4. We have chosen parameters such that the value at $x = 0.0$ is about $f(0.0) \sim 1.0$ and at $x = 1.0$ it is about $f(1.0) \sim 0.1$. Weights too close to 0.0 are not very useful, because then, these points are not considered for the computation at all. The example weighting functions we have used in our experiments (cf. Figure 4) include a constant weighting of 1.0 (solid red line in Figure 4), a linearly decreasing function ranging from 1.0 to 0.1 (dashed blue line in Figure 4), an exponential fall-off (green dashed line in Figure 4), a sigmoid-curved fall-off (violet dotted line in Figure 4), a Gauss function (green dashed-dotted line in Figure 4), and the complementary Gauss Error Function *Erfc* (red dashed-dotted

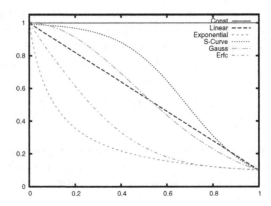

Fig. 4. Some weight functions

line in Figure 4). The last one is a function well-known from statistics related to normal distributions and, thus, probably the most sound choice.

In our experiments, all of the alternative weighting functions (except the constant weight) lead to similar improvements so there is no reliable measure or significance to establish a ranking between the different weighting functions. In fact, it is plausible that different functions are appropriate for different underlying causes in the data or assumptions in the clustering process (e.g. clustering algorithms assuming a Gauss distribution might benefit best from a Gaussian weighting function).

For distance-based weighting functions, several tasks arise. We have chosen to scale distances such that the outermost point has a distance of 1.0, i.e. a weight of 0.1, ensuring that this point has still some guaranteed influence on the result. This choice is somehow arbitrary, but it has at least the benefit of fairness. On the other hand, this fairness comes at the cost that all weights depend on the outermost point. When points are selected using a range query, the query range could offer a better normalization. When an incremental computation is desired, a completely different choice might be appropriate. Additionally, we are computing weights based on the distance to a query point. This is appropriate for situations where the data is obtained via kNN or ε-neighborhoods. When computing the correlation for an arbitrary set of points, the distance might need to be computed from the centroid or medoid of that set.

In the above described toy example of five cluster points plus one outlier (cf. Figure 1), the observed sensitivity to that outlier is significantly decreased, given that the outlier will only be weighted at around 0.1. Applying the weighting function to the 3D example data set of Figure 2 we also observe an increased robustness of the correlation analysis. Figures 5(a) and 5(b) depict the effect of a weighted covariance on the relative strength of the first eigenvector and the normalized sum of the first two eigenvectors, respectively, using the *Erfc* weighting. Compared to Figures 3(a) and 3(b) we can derive that many of the sudden drops have been erased, while the overall shape is well preserved. Especially

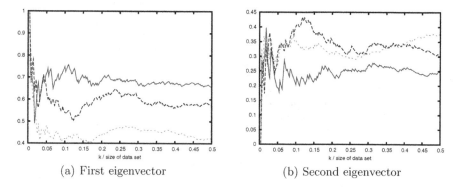

(a) First eigenvector   (b) Second eigenvector

Fig. 5. Relative strength of eigenvectors (with *Erfc* weight)

for higher values of k, sudden jumps have mostly disappeared. Therefore, this measure is useful to avoid choosing a particularly bad value of k, i.e. a k where the kNN of the particular point do not reflect the correct subspace of the corresponding cluster, by somewhat averaging with neighbors. Peaks usually are shifted towards a slightly higher value of k. This is natural since the added points are weighted low at first.

4.2 Auto-Tuning the Local Context of Correlation Analysis

Graphs such as Figure 3(a) show that even small differences in k can lead to significantly different results. Therefore, it is reasonable not to use a fixed value of k, i.e. a fixed number of neighboring points, but rather to adjust the value of k for each point separately. For example, one can use a globally fixed number of neighbors k_{max} and then individually select for each point the $k \leq k_{max}$ neighbors that are relevant for the particular point. As far as k_{max} is sufficiently large, we should in general be able to select a reasonable k, so that this strategy produces accurate results. Of course there are different strategies of selecting k. Since there are $O(2^{k_{max}})$ subsets of the given k_{max} points that could be used, simply trying all combinations of subsets of k points ($1 \leq k \leq k_{max}$) is not feasible. Probably the easiest strategy of $O(k_{max})$ complexity is to test for any k ($1 \leq k \leq k_{max}$) only the k nearest points, resulting in k_{max} tests. The next question that arises is how to evaluate the results of the k_{max} tests in order to report the best value for k. The obvious strategy of returning the result that maximizes the relative strength of eigenvalues has shown to be not very reliable because of jitter: one particular k value could result in a "perfect" hyperplane consisting mainly of points that form a subspace completely different to the subspace of the cluster. Figure 6 illustrates this effect: using only the three points in the red ellipsoid, we will hardly find the correct hyperplane of the cluster although all those three points are cluster members because they do not fit the subspace perfectly. Rather, the three points perfectly form a different line so the relative strength of the first eigenvalue will be very high (appr. 100%).

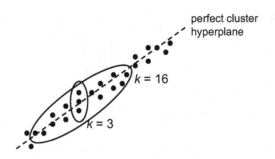

Fig. 6. Problems with jitter

In fact, we are more interested in a range of k values where we have a high and stable relative strength of eigenvalues, so we need a more elaborate filtering.

In our evaluations, we have chosen the strategy to use the k nearest points for correlation analysis, with $k_{min} \leq k \leq k_{max}$, where k_{min} is a minimum number of points such that the PCA is at least somewhat sensible at this data set dimension. The motivation behind the introduction of the lower bound k_{min} is that we need at least λ points to span a λ-dimensional hyperplane and $3 \cdot \lambda$ has been considered as a lower bound of points such that the detection of a λ-dimensional hyperplane by PCA is trustworthy rather than arbitrary. To avoid jitter and outlier effects, we use a sliding window to apply a dimensionality filter and average the variance explained by the largest eigenvalues.

Let

$$\text{ex}(E, \lambda) := \frac{\sum_{i=1}^{\lambda} e_i}{\sum_{i=1}^{d} e_i} \tag{7}$$

be the relative amount of variance explained by λ eigenvalues $E = \{e_i\}$ representing a hyperplane of dimensionality λ. Most correlation clustering algorithms rely on a level of significance $\alpha \leq 1$ to decide how many eigenvectors explain a significant variance and, thus, span the hyperplane of the cluster. Intuitively, the eigenvectors are chosen such that the corresponding eigenvalues explain more than α of the total variance. The number of those eigenvectors is called *local dimensionality* (of a cluster), denoted by λ_E, formally

$$\lambda_E = \min_{\lambda \in \{1...d\}} \{\lambda \mid \text{ex}(E, \lambda) \geq \alpha\}. \tag{8}$$

Let us note that almost all correlation clustering algorithms use this notion of local dimensionality. Typical values for α are 0.85, i.e. the eigenvectors that span the hyperplane explain 85% of the total variance along all eigenvectors.

As indicated above, for filtering out the best value of k, we are intuitively interested in a value where (i) the local dimensionality λ is stable, i.e. increasing or decreasing k by a small degree does not affect the value of λ, and (ii) $\text{ex}(E, \lambda)$ is maximal and stable, i.e. increasing or decreasing k by a small degree does not affect the value of $\text{ex}(E, \lambda)$. The motivation behind these considerations is that the value of k that fulfills both properties leads to the determination of a robust hyperplane, that maximizes the variance along its axis. In other words, using the neighbors determined by k, the hyperplane reflects all of these neighbors in a best possible way and there are most likely only very few neighbors that are outliers to this hyperplane. In addition, increasing or decreasing k, i.e. adding or deleting few neighbors, does not affect the correlation analysis.

To find the value of k that meets both properties, we determine $ex(E, \lambda)$ for all $k_{min} \leq k \leq k_{max}$. We then use a sliding window $W = [k_l, k_u]$ and choose $k = (k_l + k_u)/2$ such that for all k' in W (i.e. $k_l \leq k' \leq k_u$) the local dimensionality λ is the same and the average of $ex(E_{k'}, \lambda)$ is maximized. Additionally, if this maximum is at the very beginning or end of our search range (i.e. $k_l = k_{min}$ or $k_u = k_{max}$), we discard it. We can still obtain multiple maxima, one for each dimensionality λ. In this case we pick the lowest like all correlation clustering algorithms aiming at finding the lowest dimensional subspace clusters. Those are the most interesting ones since they involve the largest set of correlations among attributes.

5 Application to Existing Approaches

In the following, we discuss how our concepts can be integrated into existing correlation clustering algorithms in order to enhance the quality of their results. Exemplarily, we show this integration with two different types of algorithms, the latest density-based algorithm ERiC and the k-means-based algorithm OR-CLUS.

5.1 Application to Density-Based Correlation Clustering Algorithms

The integration of our concepts into ERiC is rather straightforward. ERiC determines for each data point p the subspace of the cluster to which p should be assigned (hereafter called the subspace of p). The subspace of p is computed by applying PCA to the kNN of p where k needs to be specified by the user.

Using our concepts, we can simply replace the parameter k by the global maximum k_{max} of neighbors that should be considered. Both the weighting and the auto-tuning can then be applied directly when computing the subspace of p. First, from the k_{max}NN of p, the optimal $k_p \leq k_{max}$ for detecting the subspace of p is determined as described in Section 4.2 based on a weighted covariance as described in Section 4.1. Second, the subspace of p is computed by applying PCA using a weighted covariance on the k_pNN of p (cf. Section 4.1).

The integration of our concepts into other density-based algorithms like CO-PAC, HiCO, and 4C can be done analogously.

5.2 Application to Partitioning Correlation Clustering Algorithms

ORCLUS determines the subspace of each cluster C by applying PCA to the local neighborhood of the center of C, denoted by r_C. The local neighborhood of r_C includes the set S_C of all points that have r_C as their nearest cluster representative.

Using our concepts, we can simply consider S_C as the maximum set of points that should be considered for PCA, i.e. $k_{max} = |S_C|$. Both the weighting and the auto-tuning can then be applied directly when computing the subspace of C. First, from the S_C, the optimal $k_C \leq k_{max}$ for detecting the subspace of C is determined as described in Section 4.2 based on a weighted covariance as described in Section 4.1. Second, the subspace of C is computed by applying PCA using a weighted covariance on the k_C points in S_C that are closest to r_C (cf. Section 4.1).

6   Experiments

6.1   Evaluation Methodology

In order to evaluate the results of our novel concepts integrated into ERiC and ORCLUS, we generated artificial data sets with a well-defined gold standard, i.e. we defined certain data distributions and all points in our data set are assigned to the distribution with the maximum density in that particular point. Since both ERiC and ORCLUS have different properties and, here, we are not interested in judging which algorithm is better for which data set, we generated different data sets for each algorithm.

To evaluate the quality of the clustering, we employ a pair-counting F-measure, considering the noise points to be a cluster on its own. This means that any two points in the data set form a pair if they belong to the same cluster (or noise). Let $C = \{C_i\}$ be a clustering (with C_i being the clusters in C, including the noise cluster). Then $P_C := \{(a, b) \mid \exists C_i : a \in C_i \wedge b \in C_i\}$ is the set of pairs in clustering C. The F-measure to evaluate how good a clustering C matches the gold standard D is then defined as

$$F(C, D) := \frac{2 \cdot |P_C \cap P_D|}{2 \cdot |P_C \cap P_D| + |P_C \setminus P_D| + |P_D \setminus P_C|}.$$

Obviously, $F(C, D) \in [0, 1]$, where $F(C, D) = 1.0$ means that the clustering C is identical to the gold standard D.

6.2   Synthetic Data

For evaluating the influence of our novel methods on both ORCLUS and ERiC, we used several synthetic data sets ranging from 3 to 100 dimensions. In the following discussion, we focus on some lower dimensional data sets for a clear presentation.

ERiC. We first focus on two 3D synthetic data sets that can be seen in Figure 7. Figure 8(a) gives the results for data set DS1 shown in Figure 7(a). We plotted the F-measure of the compared algorithms along the y-axis and varied the

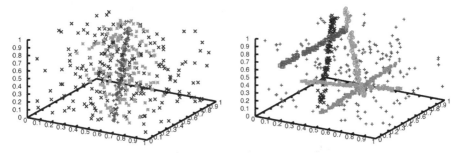

(a) DS1: a 1D lines (150 points) embed-   (b) DS2: five 1D lines (100 points each)
ded within a 2D plane (150 points) plus   plus 200 points noise
200 points noise

Fig. 7. 3D synthetic data sets used for evaluating ERiC

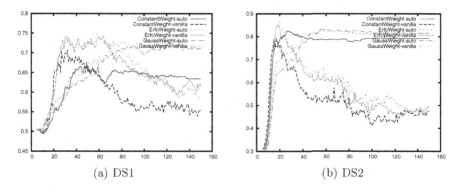

(a) DS1   (b) DS2

Fig. 8. Results of ERiC with different weight functions and auto-tuning

parameters k and k_{max} along the x-axis. The blue line represents the results of
the unmodified ERiC algorithm. Obviously the choice of k is nontrivial, a value
of about $k = 34$ gives the best results. The violet dotted line is the result when
using the Erfc weight in PCA. Obviously, the results are significantly better,
and any k in $35 < k < 65$ gives good results. Therefore choosing a good k has
become a lot easier using only the weighting approach. The green line with the
short dash-dot pattern depicts the result of ERiC using a Gauss weight. As it
can be seen, the results using a simple Gaussian weighting do not significantly
differ from the Erfc weighting results.

The remaining three lines show the results of ERiC when using the auto-
tuning of the parameter k, i.e. for each point the optimal $k \leq k_{max}$ is determined
separately (for these graphs, the x-axis represents the chosen k_{max} value). The
red line is using the traditional PCA without any weighting, while the dashed
green and the orange line with the long dash-dot pattern represent the results
using the Erfc and Gauss weights, respectively. The results show that k_{max} simply
needs to be chosen high enough in order to achieve reasonably good results. While

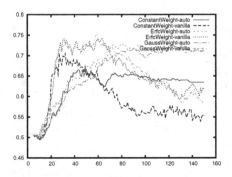

Fig. 9. Results of ERiC with different weight functions and auto-tuning on a sample 10D synthetic data set

these results do not reach the results of choosing the optimum k (which is not possible without knowing the gold standard), they approach the optimal value quite well. This observation dramatically simplifies the choice of the k/k_{max} parameter.

Figure 8(b) depicts the results on DS2 shown in Figure 7(b). Overall, the results on DS1 and DS2 are comparable. However, given that every point has just one "sensible" dimensionality – the other data set had points that had both a sensible 1D and 2D context – and the noise level is not as high, the effect of the weighted PCA on DS2 is not as high as on DS1. Since increasing the noise level will increase the difference between the non-weighted and weighted graphs, the weighting is especially interesting for noisy (e.g. higher dimensional) data.

All observations that could be made for the two 3D data sets could also be made for higher dimensional data sets. For example, Figure 9 shows the results of ERiC with different extensions for a sample 10D data set. Again, the version of ERiC using an Erfc weighted PCA in combination with the auto-tuned selection of k achieved the best overall F-measure. Also, as long as k_{max} is chosen sufficiently high, we get rather accurate results.

In summary, we observed that in all cases, the combination of the Erfc weighted PCA and the auto-tuned selection of k considerably increased the F-measure of the resulting clustering and significantly reduced the complexity of selecting sufficient input parameters compared to the original ERiC algorithm.

ORCLUS. The results of ORCLUS are harder to evaluate, because the results of ORCLUS depend on the order in which the data points are processed. Therefore, we generated 100 permutations of the original data, applied ORCLUS with optimal parameters to all of them, and averaged the results. The data set used in these computations was a 10-dimensional data set, containing 10 clusters of dimensionalities 2 to 5. The results are given in Table 1.

Each of these values was obtained by running ORCLUS and its variants on the same 100 permutations of the input data set and averaging the resulting F-measure values. The standard deviation over the 100 resulting F-measure values is given to show the dependence of ORCLUS on picking good seeds. It can

Table 1. Impact of the integration of our novel concepts into ORCLUS

Variant	Average F-measure	Standard Deviation
ORCLUS	0.667	0.046
ORCLUS + Gauss weight	0.684	0.055
ORCLUS + Exponential weight	0.676	0.054
ORCLUS + Erfc weight	0.683	0.061
ORCLUS + Linear weight	0.686	0.056
ORCLUS + Auto	0.751	0.070
ORCLUS + Auto + Gauss	0.763	0.069
ORCLUS + Auto + Exponential	0.754	0.075
ORCLUS + Auto + Erfc	0.754	0.075
ORCLUS + Auto + Linear	0.771	0.078

Table 2. Results on NBA data using ERiC with autotuning and Erfc weighting

cluster ID	dim	Description
1	4	"go-to-guys"
2	4	guards
3	4	reserves
4	5	small forwards

be observed that the benefits of using a weighted PCA are smaller (≈ 0.02) than those of using an auto-tuning PCA (≈ 0.09) and the combination of both actions further improves the results. Interestingly, in this experiment, a linear weighting function is slightly better (by up to 0.02) than a Gaussian or Erfc weighting. However, in general on different data sets, there is no significant difference observable comparing different weighting functions. In summary, using our novel concepts, the F-measure on this data set is improved by approximately 0.1 corresponding to a 10% quality boost.

6.3   Real-World Data

We applied the enhanced version of ERiC (using autotuning and Erfc weighting) on a data set containing average career statistics of current and former NBA players[1]. The data contains 15 features such as "games played" (G), "games started" (GS), "minutes played per game" (MPG), "points per game" (PPG), etc. for 413 former and current NBA players. We detected 4 interesting clusters each containing players of similar characteristics (cf. Table 2). In addition, several players were assigned to the noise set. Cluster 1 contains active and former superstars like Michael Jordan, Allen Iverson, Larry Bird, Dominique Wilkins, and LeBron James, etc. The second cluster features point and shooting guards. A third cluster contains only very few players that are not so well-known because

---

[1] Obtained from http://www.nba.com

Table 3. Results on Metabolome data using ERiC with autotuning and Erfc weighting

cluster ID	dim	Description
1	10	PKU
2	10	controll
3	11	PKU
4	12	PKU
5	13	PKU

they are usually reserves. The fourth cluster consists of small forwards. Let us note that we also applied the original ERiC algorithm (without the extensions) to the NBA data set but could not get any clear clusters. In summary, using our novel concepts, the algorithm ERiC is now able to detect some meaningful clusters on the NBA set.

In addition, we applied our novel concepts in combination with ERiC to the Metabolome data set of [14] consisting of the concentrations of 43 metabolites in 20,391 human newborns. The newborns were labelled according to some specific metabolic diseases. The data contains 19,730 healthy newborns ("control"), 306 newborns suffering from phenylketonuria ("PKU"), and 355 newborns suffering from any other diseases ("other"). The results are depicted in Table 3. As it can be seen, we could separate several of the newborns suffering from PKU from the other newborns. Again, the original version of ERiC could not find any comparatively good results.

7   Conclusion

Almost all correlation clustering algorithms suffer from an arbitrary selection of points in the local neighborhood of cluster members or cluster representatives from which the subspace of a cluster is determined by applying PCA. Choosing the local neighborhood, is a heuristics that is usually more meaningfull than random sampling. However, due to outliers in this selection, the process of finding the correct subspace is often misled because PCA is rather sensitive to outliers. In this paper, we discuss general concepts to enhance the robustness of PCA for finding the correct subspace in order to increase the effectiveness of any PCA-based correlation clustering algorithm. Thereby, we do not solve the problem of making a more suitable selection rather than the local neighborhood but try to ease the influence of outliers in this local neighborhood by a two-step approach: First, a weighting function is applied to the points when computing the covariance matrix for PCA in order to weight points that are potential outliers lower than points that are potential cluster members. Second, a method for selecting a suitable number of neighbors for each cluster member or cluster representative separately is presented. We further discuss how our general method can be integrated into existing correlation clustering algorithms based on different cluster paradigms. Our experiments show that the quality of the corresponding results can be significantly enhanced when using our new methods. In addition, the experiments show that our approach remarkably simplifies the selection of critical parameters. In summary, our method considerably enhances the robustness and usability of correlation clustering algorithms.

References

1. Aggarwal, C.C., Yu, P.S.: Finding generalized projected clusters in high dimensional space. In: Proc.SIGMOD (2000)
2. Böhm, C., Kailing, K., Kröger, P., Zimek, A.: Computing clusters of correlation connected objects. In: Proc.SIGMOD (2004)
3. Tung, A.K.H., Xu, X., Ooi, C.B.: CURLER: Finding and visualizing nonlinear correlated clusters. In: Proc.SIGMOD (2005)
4. Achtert, E., Böhm, C., Kröger, P., Zimek, A.: Mining hierarchies of correlation clusters. In: Proc.SSDBM (2006)
5. Achtert, E., Böhm, C., Kriegel, H.P., Kröger, P., Zimek, A.: Robust, complete, and efficient correlation clustering. In: Jonker, W., Petković, M. (eds.) SDM 2007. LNCS, vol. 4721. Springer, Heidelberg (2007)
6. Achtert, E., Böhm, C., Kriegel, H.P., Kröger, P., Zimek, A.: On exploring complex relationships of correlation clusters. In: Proc.SSDBM (2007)
7. Beyer, K., Goldstein, J., Ramakrishnan, R., Shaft, U.: When is nearest neighbor meaningful? In: Beeri, C., Bruneman, P. (eds.) ICDT 1999. LNCS, vol. 1540, pp. 217–235. Springer, Heidelberg (1998)
8. Hinneburg, A., Aggarwal, C.C., Keim, D.A.: What is the nearest neighbor in high dimensional spaces? In: Proc.VLDB (2000)
9. Aggarwal, C.C., Hinneburg, A., Keim, D.: On the surprising behavior of distance metrics in high dimensional space. In: Van den Bussche, J., Vianu, V. (eds.) ICDT 2001. LNCS, vol. 1973. Springer, Heidelberg (2000)
10. Chakrabarti, K., Mehrotra, S.: Local dimensionality reduction: A new approach to indexing high dimensional spaces. In: Proc.VLDB (2000)
11. Ester, M., Kriegel, H.P., Sander, J., Xu, X.: A density-based algorithm for discovering clusters in large spatial databases with noise. In: Proc.KDD (1996)
12. Ankerst, M., Breunig, M.M., Kriegel, H.P., Sander, J.: OPTICS: Ordering points to identify the clustering structure. In: Proc.SIGMOD (1999)
13. Bansal, N., Blum, A., Chawla, S.: Correlation clustering. Mach. Learn. 56, 89–113 (2004)
14. Liebl, B., Nennstiel-Ratzel, U., von Kries, R., Fingerhut, R., Olgemöller, B., Zapf, A., Roscher, A.A.: Very high compliance in an expanded MS-MS-based newborn screening program despite written parental consent. Preventive Medicine 34(2), 127–131 (2002)

Searching Correlated Objects in a Long Sequence

Ken C.K. Lee[1], Wang-Chien Lee[1], Donna Peuquet[1], and Baihua Zheng[2]

[1] Pennsylvania State University, University Park, PA16802, USA
cklee@cse.psu.edu, wlee@cse.psu.edu, djp11@psu.edu
[2] Singapore Management University, Singapore
bhzheng@smu.edu.sg

Abstract. *Sequence*, widely appearing in various applications (e.g. event logs, text documents, etc) is an ordered list of objects. Exploring correlated objects in a sequence can provide useful knowledge among the objects, e.g., event causality in event log and word phrases in documents. In this paper, we introduce *correlation query* that finds correlated pairs of objects often appearing closely to each other in a given sequence. A correlation query is specified by two control parameters, *distance bound*, the requirement of object closeness, and *correlation threshold*, the minimum requirement of correlation strength of result pairs. Instead of processing the query by scanning the sequence multiple times, that is called *Multi-Scan Algorithm (MSA)*, we propose *One-Scan Algorithm (OSA)* and *Index-Based Algorithm (IBA)*. OSA accesses a queried sequence once and IBA considers correlation threshold in the execution and effectively eliminates unneeded candidates from detail examination. An extensive set of experiments is conducted to evaluate all these algorithms. Among them, IBA, significantly outperforming the others, is the most efficient.

1 Introduction

Many datasets, such as event logs and textual documents, organize data objects in an ordered list, i.e., *sequence*. Both the data objects and their positions are captured by the sequence where the closeness of two objects in a sequence implies their relationships. We refer objects a and b as *correlated* if they often occur closely to each other. Efficiently identifying correlated objects has a large application base. For example, finding products likely to be selected by the same customers some time after their purchase of certain products is a key to the success of recommendations [4]. Detecting events usually happened some time after some others from an event log can provide hints to determine event causality in an event analysis [8]. Figuring out words frequently appearing together in documents will help identifying key phrases used and providing better understanding of documents [6].

Motivated by the importance of identifying correlated objects in a sequence, we introduce *correlation query* in this paper. Its definition is formalized in Section 3. In a sequence, objects can be classified into *object sets*, i.e., subsets of

B. Ludäscher and Nikos Mamoulis (Eds.): SSDBM 2008, LNCS 5069, pp. 436–454, 2008.

objects categorized by certain properties of interests. Two objects are said to be close if their distance along the sequence does not exceed a threshold, specified by a query parameter *distance bound*. A correlation query is to retrieve object set pairs that have a large portion of objects close to each other. Another query parameter *correlation threshold* is specified that two object sets (that we call them an *object set pair*) satisfy a correlation query when their correlation coefficient is greater than the specified threshold. A correlation coefficient (defined by cosine function in this paper) measures the strength of correlation in two object sets whose objects are closely located. A correlation query finds all the satisfied correlated object set pairs from a sequence.

Efficiently processing a correlation query is challenging because the number of close objects is subject to the specified distance bound. The most intuitive way is to scan the queried sequence to measure the numbers of close objects, and then determine the correlation coefficients. Following this idea, we propose a scan-based algorithm, namely *Multi-Scan Algorithm (MSA)*, to serve as the baseline algorithm. It examines a pair of candidate object sets in each scan. Suppose there are n objects sets. MSA scans the whole sequence $\binom{n}{2}$ times that is very time consuming. To overcome the shortcoming of MSA, we propose another scan-based algorithm, One-Scan Algorithm (OSA), which finishes the query within one sequence scan. Scan-based algorithms, however, have serious performance deterioration when the queried sequence is very long. Since only object set pairs with high correlation coefficients are needed and worth investigation, we propose *Index-Based Algorithm (IBA)*, which builds an index for every object set to capture the positions of mapped objects in the sequence. Given two indices, the number of close objects can be determined by merging the two indices and thus the correlation coefficient is calculated. Several effective optimization techniques, such as candidate screening, group matching, and early termination, are proposed to further boost up the search performance.

We conduct an extensive set of experiments on both synthetic and real datasets to evaluate the proposed search algorithms. MSA and OSA perform stably with various sequence properties and OSA significantly outperforms MSA. IBA runs even much faster than OSA due to effectiveness of optimization techniques, especially when search criteria is strict (i.e., a large correlation threshold and a small distance bound) and the cardinalities of object sets differ a lot. We also discuss some variants of correlation query including constrained correlation query, position correlation query and correlation spectrum query. Our contributions in this paper are summarized as below:

1. We introduce a new query type, called correlation query, which retrieves correlated object set pairs based on specified distance bound and correlation threshold.

2. We analyze the characteristics of correlation query and propose two scan-based algorithms, namely *Multi-Scan Algorithm* (MSA) and *One-Scan Algorithm* (OSA).

3. We also propose *Index-Based Algorithm* (IBA), that indexes objects in a sequence, and employs optimization techniques for better search performance.

4. We introduce variants of correlation query including constrained correlation query, position correlation query and correlation spectrum query.
5. We conduct an extensive set of experiments to evaluate the performance of the proposed algorithms. The results indicate that IBA performs better than the others and it is the most efficient algorithm for this correlation query.

The remainder of the paper is organized as follows. Section 2 reviews related work about correlation analysis in related domains. Section 3 formalizes the correlation query and discusses algorithm design criteria. Section 4 details our proposed algorithms. Section 5 discusses variants of correlation query. Section 6 evaluates the performance of proposed algorithms and presents our results. Section 7 concludes this paper.

2 Related Work

Subject to application needs and data characteristics, the definitions and measurements of object correlation are different [5,10]. In statistics, correlation measures the strength and direction of a linear relationship between two random variables (e.g. education and income). Two random variables are correlated when the values of both variables increase (or decrease) with similar amplitude simultaneously. In data mining where transaction databases are usually considered, finding association among objects is one of the most important search. Result objects are those frequently appearing in same transactions [3]. Association mining finds which pairs or groups of objects are often included in same transactions.

	y	$\bar{y}$	
x	f_{xy}	$f_{x\bar{y}}$	f_x
$\bar{x}$	$f_{\bar{x}y}$	$f_{\bar{x}\bar{y}}$	$f_{\bar{x}}$
	f_y	$f_{\bar{y}}$	N

Fig. 1. A 2×2 contingency table for x and y

Finding correlated objects is fundamentally different from association mining that correlated pairs of objects may not have high frequencies but strong correlations [13]. Currently, there are a number of correlation metrics (e.g., lift, cosine, χ^2 and Pearson's correlation coefficient) defined to quantify the strength of object correlation [10]. Most of the metrics are developed based on contingency table. Figure 1 shows a 2×2 contingency table for two objects, x and y where f_{xy} is the frequency (i.e., the counts) of baskets containing both x and y at the same time, and $f_{\bar{x}\bar{y}}$ is one containing neither x nor y. $f_{x\bar{y}}$ ($f_{\bar{x}y}$) represents the number of baskets containing x (or y) only. Based on these frequencies, x and y are highly correlated if f_{xy} is relatively large to f_x and f_y. To perform such correlation analysis, all the frequencies have to be collected in advance.

There are several related research studies exploring correlation in sequences, but they are different from what we focus in this paper. Existing studies concern the correlation between individual sequences from a pool of sequences [9,14], while our work is to explore a *single long* sequence and find out the correlation among objects according to *distance bound* a query parameter. Subject to the setting of distance bound, the frequency of close objects is not fixed. Thus, counting the number of close objects in prior is no longer feasible. Thus, new and efficient algorithms that can quickly identify correlated objects are demanded.

3   Problem Formulation

A sequence, S, is a list of objects $\langle o_1, o_2, \cdots o_{|S|} \rangle$, where o_i represents an object o located at position i in S and $|S|$ is the length of S. The distance between two objects o_i and o_j where o_i can be located either before or after o_j, denoted by $\delta_{i,j}$, is equal to $|j - i|$. Two objects o_i and o_j are said to be close if their distance is not greater than a distance bound w, i.e., $\delta_{i,j} \leq w$. Each object is classified to one of n object sets, i.e., $\mathcal{O} = \{O_i | i \in [1, n]\}$ according to application needs. The following is a running example.

Example 1 (Running Example). *Given a sequence $S = \langle a_1, b_2, a_3, a_4, b_5, b_6, a_7, c_8, c_9, d_{10}, d_{11}, c_{12} \rangle$ and four object sets, $\mathcal{O} = \{A, B, C, D\}$ with $A = \{a\}$, $B = \{b\}$, $C = \{c\}$ and $D = \{d\}$. The distance between a_7 and d_{10}, $\delta_{7,10}$, is 3, and that between a_7 and b_8, $\delta_{7,8}$, is 1. When w is set to 2, a_7 and b_8 are regarded to be close but a_7 and d_{10} are not.*   □

Our model considers one object in one sequence position for presentation clarity. It can be easily extended to have multiple objects located at a same position and use real number as positions [7,12]. Correspondingly, our proposed search algorithms are general enough to handle these variations. The correlation coefficient between two object sets is defined in Definition 1. We consider the *cosine* metric because of its wide acceptance. The coefficient $\phi_w(X, Y)$ ranges from 0 to 1. The larger the coefficient is, the stronger the correlation of two object sets exploits.

Definition 1 Object Set Correlation Coefficient. *The correlation coefficient between two object sets X and Y is defined in Equation (1).*

$$\phi_w(X, Y) = \frac{|XY|_w}{\sqrt{|X| \cdot |Y|}} \tag{1}$$

where $|X|$ and $|Y|$ are the numbers of objects in X and in Y, respectively and $|XY|_w$ is the number of close object pairs that depends on the setting of w. For convenience, we omit w from $\phi_w(X, Y)$ and $|XY|_w$ if the context is clear.   ■

To calculate $\phi(X, Y)$, $|X|$, $|Y|$ and $|XY|$ have to be determined. However, it is not that straightforward to measure $|XY|$ due to a *redundant count problem*. Let us consider the first 5 objects a_1, b_2, a_3, a_4, b_5 in S in the running example. If w is set to 2, b_2 is close to a_1, a_3 and a_4, and b_5 is close to a_3 and a_4. Based on this, while $|A|$ and $|B|$ are 3 and 2, respectively we would obtain 5 pairs of close objects (i.e., $|AB| = 5$), which is, however, incorrect. In fact, $|XY|$ represents the number of close object pairs that must be disjoint. In other words, once an object in set X is identified to be close to an object in set Y, it contributes only one to $|XY|$, no matter how many objects in set Y it is close to and vice versa. Back to the running example, we can only identify 2 disjoint close object pairs, e.g., $\langle a_1, b_2 \rangle$ and $\langle a_4, b_5 \rangle$ and $|AB|$ equals 2. Based on object set correlation coefficient, correlation query is formally defined in Definition 2 and exemplified in Example 2. Take the redundant count problem into consideration, our proposed algorithms to be discussed next guarantee the correctness of $|XY|$.

Definition 2 Correlation Query. *Given a sequence, a set of predefined object sets, $\mathcal{O}$, and two query parameters: distance bound, ω, and correlation threshold, t, a correlation query, $Q(S, \omega, t)$, returns all pairs of object sets $(X, Y) \in \mathcal{O} \times \mathcal{O}$ with $\phi_\omega(X, Y) > t$.* ∎

Example 2. *Given a correlation query $(S, 2, 0.5)$ using S and $\mathcal{O}$ specified in Example 1, the correlation coefficients of all object set pairs are derived according to Equation (1) and listed in Figure 2.*

XY	$|X|$	$|Y|$	$|XY|_\omega$	$\phi_\omega(X,Y)$
AB	4	3	3	0.87
AC	4	3	1	0.29
AD	4	2	0	0.00
BC	3	3	1	0.33
BD	3	2	0	0.00
CD	3	2	2	0.82

Fig. 2. Correlation coefficients

Given the four object sets, there are 6 object set pairs. As t is set to 0.5, only AB and CD are qualified and returned as the result set. □

4   Search Algorithms

In this section, we present three algorithms for correlation query, namely, *Multi-Scan Algorithm* (MSA), *One-Scan Algorithm* (OSA) and *Index-Based Algorithm* (IBA). MSA and OSA are scan-based while IBA is an index approach.

4.1   Multi-Scan Algorithm (MSA)

Multi-Scan Algorithm (MSA) is an iterative algorithm. In each turn, it examines one pair of object sets, say X and Y, and determines the corresponding $|X|$, $|Y|$ and $|XY|$ to compute $\phi(X, Y)$. It skips objects not belonging to candidate object sets. Given n sets of data objects, MSA iterates for $\binom{n}{2}$ object set pairs.

To tackle the redundant count problem that affects the correctness of $|XY|$, we allocate a sliding window W to buffer the ω recently examined objects. An object is only compared against those objects inside W to form close object pairs. If an object can be paired with multiple objects in W, the oldest object is matched so the recent ones are reserved to match with those later examined in order to maximize $|XY|$. Once an object is paired with a new object, it is deleted from the sliding window W to prevent double counting. A counter c_{XY} carries the number of close object pairs formed so far with zero as its initial value.

Figure 3(a) depicts the pseudo-code of MSA. It consists of a big loop (line 1-15). For each iteration, it examines one object set pair. It reads one object o from S each time (line 4). It compares o against a buffer W and updates counters (i.e., c_X, c_Y and c_{XY}) and W accordingly (line 6-11). By the end of each turn, it collects the examined object sets if the calculated correlation coefficient is greater than a correction threshold, t (line 14) and returns the result (line 16). Example 3 shows how MSA determines the correlation coefficient.

Example 3. *Suppose object sets A and B are examined and ω set to 2. First, three counters c_A, c_B and c_{AB} that are used to measure $|A|$, $|B|$ and $|AB|$, respectively, are all initialized to 0, and a sliding window, W, that buffers two recently accessed objects, is initialized with $(\bot, \bot)$, (where $\bot$ means no object). The trace of MSA examining A and B is shown in Figure 3(b) where each row presents a state right after an object is examined.*

Algorithm. MSA
input: a sequence S; a set of object sets $\mathcal{O}$,
        dist. bound ω; corr. threshold t;
output: a result set of object set pairs R;
Begin
1. **foreach** $(X, Y) \in \mathcal{O} \times \mathcal{O} \wedge X \neq Y$ **do**
2.   start at the head of S;
     $c_X \leftarrow 0; c_Y \leftarrow 0; c_{XY} \leftarrow 0;$
3.   **repeat**
4.   read o from S;
5.   **if** $o \in X \vee o \in Y$ **then**
6.   increase $c_X (c_Y)$ if $o \in X$ (Y) by 1;
7.   compare o against W;
8.   **if** o matches with o' **then**
9.   increase c_{XY} by 1;
10.   replace o' with o' in W; add o to W;
11.   **else** add o to W;
12.   **else** add $\bot$ to W;
13. **until** S end;
14. **if** $\frac{c_{XY}}{\sqrt{c_X \cdot c_Y}} > t$ **then** $R \leftarrow R \cup \{(X, Y)\}$;
15. **endforeach**
16. **return** R;
End.

object	W	matched	c_A	c_B	c_{AB}
$\langle$init$\rangle$	$(\bot, \bot)$	-	0	0	0
a_1	$(\bot, a_1)$	no	1	0	0
b_2	(a_1, b_2)	$\langle a_1, b_2 \rangle$	1	1	1
a_3	(b_2, a_3)	no	2	1	1
a_4	(a_3, a_4)	no	3	1	1
b_5	(a_4, b_5)	$\langle a_3, b_5 \rangle$	3	2	2
b_6	(b_5, b_6)	$\langle a_4, b_6 \rangle$	3	3	3
a_7	(b_6, a_7)	no	4	3	3
c_8	$(a_7, \bot)$	no	4	3	3
c_9	$(\bot, \bot)$	no	4	3	3
d_{10}	$(\bot, \bot)$	no	4	3	3
d_{11}	$(\bot, \bot)$	no	4	3	3
c_{12}	$(\bot, \bot)$	no	4	3	3

(a) The pseudo-code of MSA

(b) Trace of MSA for A and B

Fig. 3. Multi-Scan Algorithm

The search starts with examining a_1 ($\in A$) from S; c_A and W are updated to 1 and $(\bot, a_1)$, respectively. Next, b_2 is examined and it is close to a_1 in W. Both are marked as a_1 and b_2 so they are not available for other match and both c_B are c_{AB} are updated to 1. Next, a_3 is accessed and c_A is increased to 2. Since no buffered object available for matching, it is appended to W, while a_1 is shifted out. W becomes (b_2, a_3). Next, a_4 is scanned and W is replaced with (a_3,a_4) and c_A is increased to 3. Later, b_5 is examined and both a_3 and a_4 are close to it. To maximize c_{AB}, b_5 is matched with a_3, i.e, the older one in W and c_{AB} is updated to 2. This examination continues until S is completely scanned. At last, c_A, c_B and c_{AB} are 4, 3, and 3, respectively and hence the coefficient $\phi(A, B)$ is obtained as $c_{AB}/\sqrt{c_A \times c_B} = 3/\sqrt{4 \times 3} = 0.87$. □

MSA needs only a few counters and a ω-slot buffer. However, it is inefficient because of its blind scan of the sequence multiple times. As seen in Example 3, the last five objects scanned from S do not belong to either A or B and they do not affect $\phi(A, B)$ but MSA has to scan all of them. Similarly, when examining another pair of candidates, C and D, the head portion of the sequence that contains no related objects is also scanned. Finally, each scan incurs $O(\omega \cdot |S|)$ comparisons. Hence, the complexity of MSA is $O(n^2 \cdot \omega \cdot |S|)$.

4.2   One-Scan Algorithm (OSA)

One-Scan Algorithm (OSA) improves MSA by evaluating all object set pairs in one sequence scan. For each object set pair, it counts the numbers of close objects. During the sequence scan, it updates the respective counters. The pseudo-code of OSA is depicted in Figure 4(a). It compares each examined object o against a sliding window W and updates respective counters (line 2-12). After the scan, those with coefficient higher than the correlation threshold t are collected as a part of the query result (line 13-15) and finally the result is returned (line 16).

To address the redundant count problem, we associate objects in W with their matched partners if any. When an object $o \in O$ is examined against objects in W, it tries to match with an object available, i.e., not belonging to O and not being matched with any object belonging to O. In case multiple buffered objects are available to match, the oldest one is chosen. Example 4 illustrates OSA based on our running example.

Example 4. *Due to limited space, our discussion focuses only on object sets A, B and C and their counters c_{AB}, c_{AC} and c_{BC}. Assume that ω is set to 2. Figure 4(b) shows the trace. We use x:$\{y, z\}$ to denote a buffered object x and its paired objects, y and z. OSA first loads a_1 from S and buffers it in W, which becomes $(\bot, a_1:\{\})$. Next, b_2 is examined. It matches a_1 and contributes one to c_{AB}. Consequently, W becomes $(a_1:\{b_2\}, b_2:\{a_1\})$. Thereafter, a_3 and a_4 are studied and found that b_2 has already been matched with a_1. Now W becomes $(a_3:\{\}, a_4:\{\})$. Further, b_5 is matched with a_3 which is the oldest and available and c_{AB} is increased to 2. Next, b_6 is matched with a_4; thus, c_{AB} is updated to 3. For the next object a_7, no match is found. Next, c_8 is retrieved and it is matched with both b_6 and a_7. Consequently, both c_{AC} and c_{BC} are updated to 1.*

Algorithm. OSA
input: a sequence S; a set of object sets $\mathcal{O}$,
            dist. bound ω; corr. threshold t;
output: a result set of object set pairs R;
Begin
  1. start at the head of S;
     $c_X \leftarrow 0$; $c_Y \leftarrow 0$; $c_{XY} \leftarrow 0$;
  2. **repeat**
  3.   read o from S (assuming $o \in X$);
  4.   increase c_X by 1;
  5.   compare o with W;
  6.   **forall** o' in W matched with o
  7.   increase c_{XY} by 1 where $o' \in Y'$;
  8.   associate o' with o;
  9.   associate o with o';
 10.   **endforall**
 11.   add o and its associated objects to W;
 12. **until** S end;
 13. **foreach** $(X,Y) \in \mathcal{O} \times \mathcal{O} \wedge X \neq Y$
 14.   **if** $\frac{c_{XY}}{\sqrt{c_X \cdot c_Y}} > t$ **then** $R \leftarrow R \cup \{(X,Y)\}$
 15. **endforeach**
 16. **return** R;
End.

(a) The pseudo-code of OSA

exam	W	c_{AB}	c_{AC}	c_{BC}
$\langle$init$\rangle$	$(\bot, \bot)$	0	0	0
a_1	$(\bot, a_1{:}\{\})$	0	0	0
b_2	$(a_1{:}\{b_2\}, b_2{:}\{a_1\})$	1	0	0
a_3	$(b_2{:}\{a_1\}, a_3{:}\{\})$	1	0	0
a_4	$(a_3{:}\{\}, a_4{:}\{\})$	1	0	0
b_5	$(a_4{:}\{\}, b_5{:}\{a_3\})$	2	0	0
b_6	$(b_5{:}\{a_3\}, b_6{:}\{a_4\})$	3	0	0
a_7	$(b_6{:}\{a_4\}, a_7{:}\{\})$	3	0	0
c_8	$(a_7{:}\{c_8\}, c_8{:}\{a_7, b_6\})$	3	1	1
c_9	$(c_8{:}\{a_5, b_6\}, c_9{:}\{\})$	3	1	1
d_{10}	$(c_9{:}\{\}, d_{10})$	3	1	1
d_{11}	(d_{10}, d_{11})	3	1	1
c_{12}	$(d_{11}, c_{12}:\{\})$	3	1	1

(b) Trace of OSA

Fig. 4. One-Scan Algorithm

*The next object is c_9 which does not find any close object and hence is simply
inserted into W. The run continues until S is fully scanned. Finally, c_{AB}, c_{AC}
and c_{BC} are 3, 1, and 1, respectively, based on which, the correlation coefficients
of the object set pairs are calculated.*   □

For each object $o \in O$ retrieved from a sequence, OSA examines it against all
the objects in the sliding window W. Suppose there are n object sets, an object
in W can be associated with at most $n-1$ objects. The complexity of examining
an object is $O(\omega)$ and that of OSA is $O(\omega \cdot |S|)$ which is n^2 times faster than
MSA. However, OSA needs maintain $O(n^2)$ counters and a window with $O(n \cdot \omega)$
slots which incurs a higher space requirement.

4.3   Index-Based Algorithm (IBA)

Since correlation query retrieves object set pairs whose correlation coefficients
are higher than a given threshold based on Definition 2, evaluating all the object set pairs is unneeded especially when most of them do not provide higher
coefficients. Motivated by this observation, we propose Index-Based Algorithm
(IBA). IBA preserves multiple indices, each of which corresponds to one object
set. Each index maintains the positions of objects (in the sequence) belonging
to the corresponding object set in an ascending order. For instance, for object
set A in our running example, the index maintains $\langle 1, 3, 4, 7 \rangle$, i.e., a shorter

sequence. The index can be prepared off line and its small construction cost that involves only one sequence scan can be amortized by multiple correlation queries with different ω's. Also, statistics collected during index construction is useful to speed up the search.

Given two indices, the correlation coefficient of two corresponding object sets X and Y can be determined by a merge-like matching function. Initially, two pointers p_X and p_Y point to the head of both indices. Follow steps in a *comparison*, *match*, and *slide* strategy. In the comparison step, two positions pointed by p_X and p_Y are compared and the smaller one in the sequence is taken to compare against the buffer W, which keeps ω recently examined positions and corresponding object sets that contribute these position entries. If a match is found, the counter c_{XY} is increased by one, and both matched positions become unavailable for later match. Otherwise, the position is inserted into the buffer. Finally, the pointer located at the examined position slides to the next one and the same steps repeat. If one of indices reaches its end, another index is iteratively fetched. It continues until both indices are completely scanned. We use Example 5 to illustrate this matching.

Example 5. *The trace of IBA matching function (for object sets A and C, based on our running example) is depicted in Figure 5. An object with underline represents the one having smaller position, i.e., the examined object. In the indices, the positions of objects are stored. For illustration, we show the objects.*

A	C	W	c_{AC}
$\langle init \rangle$	$\langle init \rangle$	$(\perp,\perp)$	0
$\underline{a_1}$	c_8	$(\perp,a_1)$	0
$\underline{a_3}$	c_8	$(\perp,a_3)$	0
$\underline{a_4}$	c_8	(a_3,a_4)	0
$\underline{a_7}$	c_8	$(\perp,a_7)$	0
$-$	$\underline{c_8}$	$(\cancel{a_7},\cancel{c_8})$	1
$-$	$\underline{c_9}$	$(\cancel{c_8},c_9)$	1
$-$	$\underline{c_{12}}$	(c_9,c_{12})	1

Fig. 5. Trace of IBA for object sets A and C

First, all the four objects from A, i.e, a_1, a_3, a_4 and a_7, are retrieved as all of them are smaller than c_8, the head object of set C. Then, the index for A reaches its end and c_8, the head object of C is retrieved. It matches a_7 in W and c_{AC} is increased to 1. Thereafter, objects c_9 and c_{12} are examined and the end of set C is reached, indicating the completion of this matching function. Since c_{AC} (i.e., $|AC|$) equals 1 and $|A|$ and $|C|$ are 4 and 3, respectively, the correlation coefficient of sets A and C $\phi(A,C) = 1/\sqrt{4 \cdot 3} = 0.29$. $\square$

This matching function outperforms MSA because it only scans objects belonging to the targeted object sets but not the entire sequence as MSA does. It reduces the number of scanned objects from $O(|S|)$ to $O(|X| + |Y|)$, with X and Y indicating the examined object sets. However, it may still suffer from multiple scans of indices. Actually, the performance of IBA can be significantly improved when several optimization techniques are applied. In what follows, we first discuss three optimization techniques, namely, *candidate screening*, *group matching* and *early termination* and then explain how to integrate them into IBA to further boost up the search performance.

Candidate Screening. Candidate screening attempts to filter out object set pairs with their correlation coefficient definitely lower than a given correlation threshold, so the examination of those can be saved. Based on the cardinality and distribution of each object set, two coefficient values can be estimated respectively. In the following, we detail the two correlation coefficient estimations.

- **Estimation based on cardinalities.** As $|X|$ and $|Y|$ are the cardinalities of X and Y, respectively and they can be accounted during index building, the upper bound of the correlation coefficient between X and Y is $\frac{\min(|X|,|Y|)}{\sqrt{|X|\cdot|Y|}}$. For instance, the maximum correlation coefficient between A and D in our example is $\frac{\min(4,2)}{\sqrt{4\cdot2}} = 0.45$.
- **Estimation based on distributions.** The cardinality-based estimation is straightforward, but it is nothing related to ω. In fact, the number of close objects is highly dependent on ω and the distance between close objects. During the index construction for each object set, we account 1) the smallest and the largest positions of objects inside the object set to get the distance range; and 2) the distance between any two adjacent objects. For any two object sets, if their distance ranges are more than ω apart, they are guaranteed not correlated. Thus, the estimated coefficient should be zero. For instance, the ranges of A and D in our running example are $(1,7)$ and $(10,12)$. Consequently, the ranges of A and D are disjoint and their estimated coefficient is, of course, zero.

  If two object sets have their ranges overlap, their coefficient can be estimated based on the probability of finding close object pairs, as detailed in the following. Assuming distances between adjacent objects in an object set X follows normal distribution, we collect the mean (μ_X) and standard deviation (σ_X) of all the distances between adjacent objects during index construction. Other possible distributions will be studied in our future work. Consider A from our example. After building the index, $|A|$, μ_A and σ_A are collected as 4, 1.67 (i.e., $\frac{2+1+2}{3}$) and 0.58, respectively.

  We estimate the probability that the distance between objects of two object sets is not greater than ω, denoted by p. So, p is the probability that objects are close enough to match. Let $\delta_{X,Y}$ be the expected distance between objects in X and Y, and p can be estimated by $P(|\delta_{X,Y}| \leq \omega) = P(-\omega \leq \delta_{X,Y} \leq \omega)$, i.e., the probability that $\delta_{X,Y}$ lies within the range $[-\omega, \omega]$. To obtain p, we first obtain the standard normal variable Z based on Central Limit Theorem [11], i.e.,

$$Z = \frac{(\mu_X - \mu_Y) - \delta_{X,Y}}{\sqrt{\sigma_X^2/|X| + \sigma_Y^2/|Y|}}$$

where the value of Z follows normal distribution. We estimate p as $P(z_{lower} \leq Z \leq z_{upper})$ (i.e., $P(-\infty \leq Z \leq z_{upper}) - P(-\infty \leq Z \leq z_{lower})$), in which z_{lower} and z_{upper} are the lower and upper limits, respectively. To resolve this probability, z_{lower} and z_{upper} are computed as $z_{lower} = \frac{(\mu_X - \mu_Y) - \omega}{\sqrt{\sigma_X^2/|X| + \sigma_Y^2/|Y|}}$, and $z_{upper} = \frac{(\mu_X - \mu_Y) + \omega}{\sqrt{\sigma_X^2/|X| + \sigma_Y^2/|Y|}}$. Finally, the estimated maximum correlation coefficient is determined as $p \cdot \frac{\min(|X|,|Y|)}{\sqrt{|X| \cdot |Y|}}$.

Our approach first conducts cardinality-based estimation that is lightweight and discard those object set pairs with their estimations smaller than the given threshold. For those object set pairs passing the first estimation, distribution-based estimation is conducted and compared. Finally, the indices of those object set pairs passing both tests are examined with matching functions.

Group Matching. Instead of pairwise matching, matching among a group of object sets is preferred, thus avoiding the multiple index accesses if an object set is founded to be correlated to more than one object set simultaneously. The idea of group matching is pretty similar to OSA by maintaining several counters. The only difference is that multiple indices, rather than a single sequence, are traversed at the same time.

Early Termination. Early termination determines approximate the correlation coefficient of object set pairs without completely traversing the indices, thereby improving the response of the search. We maintain c_X, c_Y and c_{XY} to keep track of the numbers of examined objects in X, Y, and matched objects, respectively. In addition, we keep ω_X and ω_Y to bookkeep the number of buffered objects of X and Y that are still available (i.e., not yet matched). During matching, we estimate both the maximal correlation coefficient $max\phi(X,Y)$ and the minimal correlation coefficient $min\phi(X,Y)$.

The maximal coefficient $max\phi(X,Y)$ can be obtained if all remaining unexamined objects can be matched and calculated as $\frac{c_{XY} + \min(|X| - c_X + \omega_X, |Y| - c_Y + \omega_Y)}{\sqrt{|X||Y|}}$ at any point of time. Consider Example 5. Behind object c_8, there is no more object from A and two objects from C pending for the examination, with an empty buffer. Since the current c_{AC} is one, we can approximate the maximal correlation coefficient $max\phi(A,C)$ is $\frac{1 + \min(4 - 4 + 0, 3 - 1 + 0)}{\sqrt{4 \cdot 3}} = \frac{1}{\sqrt{4 \cdot 3}} = 0.29$. Since the maximum value of the coefficient is below the given threshold ($t = 0.5$), it is safe to skip the remaining objects (i.e., c_9 and c_{12}) from examination and assures that object set A and C are not correlated.

The minimal correlation coefficient, $min\phi(X,Y)$ can be determined if all the remaining unexamined objects do not match. It is expressed as $\frac{c_{XY}}{\sqrt{|X||Y|}}$. Once an object set pair with minimal coefficient larger than the given threshold, it is guaranteed to be one of the answer sets. Back to Example 5 and suppose $t = 0.2$. After the examination of object c_8, c_{AC} is one and there might not be any close object pair. Therefore, the minimal value of coefficient can be derived according

Algorithm IBA
input: a sequence S; a set of object sets $\mathcal{O}$,
             distance bound ω; correlation threshold t;
output: a result set of object set pairs R;
Begin
   1. **foreach** $(X, Y) \in \mathcal{O} \times \mathcal{O}$ **do**
   2. **if** (X,Y) pass *candidate screen* **then**
   3.   start from heads of I_X and I_Y;
   4.   **repeat**
   5.   read o with the smallest position from I_X and I_Y;
   6.   increase c_X if $o \in X$ (or c_Y if $o \in Y$) by 1;
   7.   compare o with W;
   8.   **if** match **then** increase c_{XY} by 1.
   9.   add o to W;
  10.   compute $max\phi$ and $min\phi$;
  11.   **if** $max\phi \le t$ **then goto** 14;
  12.   **if** $min\phi > t$ **then** $R \leftarrow R \cup \{(X, Y)\}$; **goto** 14;
  13.   **until** I_X and I_Y end;
  14.   **if** $\frac{c_{XY}}{\sqrt{c_X \cdot c_Y}} > t$ **then** $R \leftarrow R \cup \{(X, Y)\}$;
  15. **endforeach**
  16. **return** R;
End

Fig. 6. The pseudo-code of IBA

to $\frac{c_{AC}}{\sqrt{|A| \cdot |C|}}$, i.e., $min\phi(A, C) = \frac{1}{\sqrt{4 \cdot 3}} = 0.29$. Thus, it can be safely included as an answer set.

Putting all the techniques together, Figure 6 lists the pseudo-code of IBA. IBA first prepares a pool of candidate object set pairs. Then, it studies all the individuals with candidate screening and discards those uncorrelated based on the two estimated coefficients (line 2). The remainders are then examined through group matching. Here, the figure shows the matching function (line 5-9) for sake of simplicity and I_X and I_Y are the indices of X and Y, respectively. During the match, we validate if early termination applies to stop the matching without examining the rest of the indices (line 10-12). Finally IBA outputs the result object set pairs if their correlation coefficients (line 14) (or their minimal correlation coefficients obtained while the match is early terminated (line 12)) are greater than the correlation threshold of the query.

Let $1/f$ be a fraction of candidates passing the candidate screening. IBA examines n^2/f candidates with $f \in [1, n^2]$. As each matching function incurs $O(\omega \cdot |S|/n)$ comparisons, the complexity of IBA is $O(n \cdot \omega \cdot |S|/f)$. The performance of IBA depends on f that is affected by distance bound and correlation threshold. So, for a small ω or a large correlation coefficient, f will become large. When $f > n$, IBA will achieve better performance than OSA. To construct the index, a sequence needs to be scanned once and the cost of $O(|S|)$ is amortized by correlation queries.

5   Variants of Correlation Query

In this section, we discuss several variants of our correlation query, namely, constrained correlation query, position correlation query and correlation spectrum query, and discuss the extensions of our algorithms to support them.

Constrained Correlation Query. In our model, if multiple objects are available for matching, the farthest one within a window is picked to maximize the counts and thus the correlation coefficient. However, the matching in some cases is not arbitrary. For instance, in document analysis, a word is usually semantically related with closest one; in event causality analysis, one cause event must occur right before its consequence. Therefore, the presence order have to be considered in identifying close object pairs. Constrained correlation query takes additional matching constraints into consideration. Our proposed algorithms can be easily adjusted by incorporating matching rules, like matching the closest one. When an examined object from a sequence is compared with buffered objects, the matching rules are applied to find a right candidate to match.

Position Correlation Query. For some applications, it is also interesting to know the correlation of objects with respect to their positions in a sequence. For example, a company may be interested to explore the correlation of their products sold to certain days and event analysts want to identify what events are likely to happen at certain times. Specific to temporal data, this is also referred to as temporal autocorrelation. Putting the search into a generalized framework, position correlation query explores the correlation of objects to their positions in a sequence. This query can be extended to determine object periodicity in a sequence by specifying regular interval. To support this variant, our algorithms can be extended by buffering specific sequence positions rather than examined objects. The other parts of our algorithms remain the same to count the number of close objects and to determine correlation coefficients.

Correlation Spectrum Query. Correlation coefficients increase together with the number of close objects which is in turn controlled by ω. In some applications, we might suspect that two object sets are correlated but are not so certain about the setting of a distance bound which can produce a high correlation coefficient. A straightforward approach is to obtain the coefficient for each possible ω, which varies from 1 up to the length of the entire sequence. Correlation spectrum query returns the coefficients between two object sets according to a range of ω but not a single one. The proposed algorithms can be extended by keeping a large number of counters and a very large buffer. However, it may not be space and time efficient. We shall study this in our future direction.

6   Performance Evaluation

This section evaluates the performance of our three proposed algorithms, namely, Multi-Scan Algorithm (MSA), One-Scan Algorithm (OSA) and Index-Based Algorithm (IBA) for correlation query. We implemented them in GNU C++ and conducted experiments on Linux computers with Intel CPU 3.2GHz. We evaluate our algorithms based on synthetic and realistic data sequences with each sequence stored in one file. Synthetic data sequences are characterized by the sequence length (i.e., $|S|$), the number of object sets (i.e., n) and the variations of object set cardinalities (controlled by a factor s). The sequence length varies from 1M (2^{20} objects) to 5M with 2M as the default unless specified otherwise. The number of object sets (n) is ranged from 20 to 100 in step of 20 with 60 as the default. The cardinalities of object set are controlled by a skewness factor s. In generating synthetic sequence, the probability of objects in a sequence mapped to object sets follows Zipf distribution with s controlling the skewness of the distribution. The value of s varies from 1.5 to 3 in step of 0.5. This affects the cardinalities of object sets and the distributions of objects of an object set in a sequence. As a large s is set, both object set cardinalities and distributions vary a lot and only a few object sets would produce higher correlation coefficients.

We also use two realistic data sequences, i.e., EARTHQUAKE [2] and APRS [1]. EARTHQUAKE is an earthquake log. It remarks times, geographical coordinates and earthquake magnitudes. This log contains 446k records ordered according to time. We classify each entry based on coordinates into 100 equal-sized rectangular geographical regions. For EARTHQUAKE, $|S| = 446k$ and $n = 100$. Correlation query is evaluated on this earthquake log to search which pairs of geographical regions usually experienced earthquake at the same time (according to the setting of ω). APRS is a message log about radio base station broadcasting messages in United States. It includes times and names of base stations that broadcast. The log consists of 188k records related to 1000 base stations collected on Aug 23 2001, and it is ordered based on time. For APRS, $|S| = 188k$ and $n = 1000$. In this log, it only records base stations who broadcast messages but no information about their correspondents. Correlation query is used to find pairs of communicating base stations based on an observation that two communicating base stations would have multiple message exchanges within small time intervals, determined by ω.

Correlation query is evaluated based on two parameters, namely, distance bound (ω) and correlation threshold (t). The settings of ω is varied from 10, 100, to 1000 and t is varied among 0.4, 0.5, and 0.6. Two performance metrics are measured, namely, *elapsed time* and *I/O cost*. The elapsed time is the duration of time, in terms of seconds, from the time when an algorithm starts to the time when all the results are returned. The I/O cost measures the number of pages accessed from an underlying file storing the sequence. The page size is 4KB. The results to be present are the averages of 100 runs for each experiment setting.

6.1   Evaluation on Synthetic Data

The first set of experiments is based on synthetic data sequence. We evaluate all the factors, namely, ω, t, n, $|S|$ and s. We first evaluate the impact of ω on the search performance. The larger the ω is, the more the objects are considered to be close and hence the larger the resulted correlation coefficients are. Figure 7(a) and Figure 7(b) depict the results in terms of elapsed time and number of pages

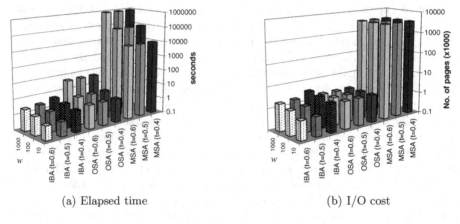

(a) Elapsed time   (b) I/O cost

Fig. 7. Impact of ω

accessed for various ω while $|S|$, n and s are fixed at $2M$, 60 and 2.0, respectively. From Figure 7(a), it can be observed that an increase of ω results in longer elapsed time. For both MSA and OSA, the size of the buffer is increased as ω grows thus increasing the lookup cost. Among all, MSA incurs the longest elapsed time, several orders of magnitude longer than OSA and IBA for same settings because of its multiple scans. On the other hand, IBA performs the best and at least 10 times faster than OSA. From the figure, we can see OSA and MSA are invariant to the correlation threshold setting (t from 0.4 to 0.6) but IBA performs better when a larger t is set. This is because the proposed optimization techniques become more effective when t is larger.

In Figure 7(b), observations similar to Figure 7(a) are made that MSA is the worst among all candidates. Both OSA and MSA incur constant I/O costs, due to a fixed number of scans. The performance of IBA varies depending on the number of object set pairs being investigated. When t is smaller (e.g., $t = 0.4$) or ω is larger (e.g., $\omega = 1000$), IBA becomes less competitive than OSA in terms of number of page accesses. This is because the optimization techniques proposed to speed up the performance of IBA do not take effect for a longer distance bound or a larger correlation threshold, without mentioning that IBA still suffers from multiple scans compared with OSA. However, the measurement of counts for correlation coefficient is CPU intensive. IBA, although accessing a little more pages, incurs less overheads in matching objects to measure the coefficient and hence its cost is payed off. As previously shown, IBA takes shorter elapsed time. Since MSA is identified as the weakest candidate, we omit it from the following discussion. Besides, we focus our remaining evaluation on the elapsed time.

Then, we evaluate the factor of n, the number of object sets. The immediate effect of n is on the size of a candidate pool and the number of candidates in matching for IBA. Figure 8(a) plots the results in terms of elapsed time against n. The other factors such as $|S|$, s and w are fixed at $2M$, 2.0 and 100, respectively. For IBA, the index construction time is 6.2 seconds for all n evaluated and the

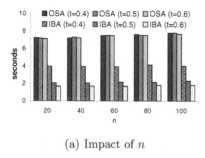

(a) Impact of n

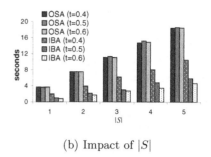

(b) Impact of $|S|$

Fig. 8. Impacts of n and $|S|$

indices are used for queries with various t. The performance of OSA is consistent to our analysis that it is invariant to n. On the other hand, IBA is more or less stable to n. Although the number of object set pairs grows as n does, the average sizes of indices are reduced due to fixed $|S|$. Further, most of the pairs are filtered out when the threshold t is set to be high. As a result, we can observe a significant difference between IBA and OSA especially when t is set to 0.6.

Next, we evaluate the impact of $|S|$, the length of sequence. Figure 8(b) shows the results in terms of elapsed time versus the length of sequence, $|S|$. The other factors such as n, s, and ω are set to 60, 2 and 100, respectively. Obviously, a longer sequence results in a longer elapsed time. OSA is invariant to t as explained before and its running time is linear proportional to $|S|$. IBA again runs much faster than OSA for all $|S|$ evaluated. From this, we can conclude that for a long sequence, IBA is superior to OSA, particularly when a larger t is specified. For $|S| = 1M, 2M, 3M, 4M$ and $5M$, the index construction times for IBA are 3.1, 6.2, 9.4, 12.4, 15.1 seconds, respectively.

Further, we examine the impact of s, the skewness parameter for object set cardinality variation. If the object cardinalities are very different, the correlation coefficients of object set pairs would not be high due to a number of unmatched objects. In this evaluation, we vary s among 1.5, 2, 2.5 and 3. When s is set to 3, the produced sequence has the most significant variation in the cardinalities of object sets, i.e., the most skewed sequence with regard to cardinalities.

The results are displayed in Figure 9(a). Here, the performance of OSA is improved together with the increase of s. This is because when s is large, certain object sets dominate the entire sequence and thus the buffer. As a results, the majority of the objects in the sequence belong to a small number of object sets, and the comparison between objects from the same set, which is expected to occur very frequently, can be saved. On the other hand, IBA performs well when s is set to 2 or above. For these settings, the object set cardinalities are skewed and most of the object set pairs that are identified not correlated will be eliminated at the beginning. However, when s is at 1.5, object sets are in similar sizes and hence the estimation based on cardinalities and distribution is not effective in candidate screening. Many object set pairs have to be examined

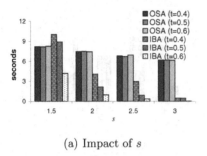
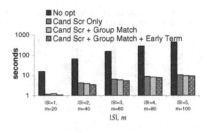

(a) Impact of s (b) Evaluation of IBA ($\omega = 100$)

Fig. 9. Elapsed time for various s and Evaluation of IBA optimization

in detail, causing a longer elapsed time. However, this cardinality variation can be detected during the IBA index construction. If s is small, OSA is preferred. Otherwise, IBA is more efficient especially when a larger threshold (t) is used.

Evaluated upon all the factors in synthetic data sequences, IBA is shown to perform the best. Now we investigate the effectiveness of proposed techniques to improve IBA. Recall that the three proposed techniques are candidate screening (labeled as Cand Scr), group matching (Group Matching) and Early Termination (Early Term). Instead of trying every possible combination of proposed techniques, we incrementally enable those techniques against IBA with no technique applied (No opt) and evaluate the performance in terms of elapsed time. In this experiment, we fix ω at 100 and t at 0.5. The results are shown in Figure 9(b), from which we can observe that candidate screening is the most effective approach that reduces the elapsed time by screening out irrelevant candidates. Group Matching and Early Termination can further slightly reduce the elapsed time.

6.2 Evaluation of Real Data

In this subsection, we evaluate the performance of OSA and IBA on real datasets. We vary both ω and t in our evaluation. This experiment tests the practicality of our algorithms in real situations. The results in terms of elapsed time for EARTHQUAKE and APRS sequences are shown in Figure 10(a) and 10(b), respectively. For EARTHQUAKE, ω is expressed as days, we evaluate 10 days, 100 days and 1000 days. For APRS, ω is expressed as 10 sec, 100 sec and 1000 sec. The results are consistent with those obtained from synthetic data. When ω is set to a small value (say, 10), both IBA and OSA can quickly determine the results since most of objects are not close and the buffer size is small. While ω is increased, IBA can save more elapsed time than OSA. As we explained above, this improvement is contributed by candidate screening technique which approximates the potential correlation coefficient to filter those unqualified candidates out of the detailed examination. From the result, IBA can be concluded as the best efficient search for correlation query.

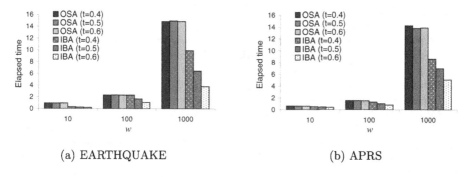

(a) EARTHQUAKE   (b) APRS

Fig. 10. Evaluation on real datasets

7 Conclusion

Sequence is widely used by various applications. In a sequence, objects that are often closely located are likely to be correlated to each other. In this paper, we identify a new query, namely *correlation query*, to search for object set pairs based on two parameters: 1) *distance bound* (ω) and 2) *correlation threshold* (t). The distance bound determines whether two objects are close in a sequence. Based on the number of close objects, we measure the strength of object correlation by cosine metric as the correlation coefficient. The larger the coefficient is, the stronger the correlation between corresponding object set pairs is interpreted. A correlation query then returns those object set pairs having corresponding correlation coefficient higher than the given correlation threshold. Three search algorithms, namely, Multi-Scan Algorithm (MSA), One-Scan Algorithm (OSA) and Index-Based Algorithm (IBA), are proposed in this paper to efficiently process correlation query. We conducted an extensive set of experiments to evaluate the performance of different algorithms. IBA, together with three optimization techniques, outperforming the other two for both real and synthetic sequences, is the most efficient algorithm to this correlation query.

Acknowledgement

This study was supported in part and monitored by the Advanced Research and Development Activity (ARDA) and the Department of Defense. The views and conclusions contained in this document are those of the author(s) and should not be interpreted as necessarily representing the official policies or endorsements, either expressed or implied, of the National Geospatial-Intelligence Agency or the U.S. Government. The work by Ken Lee and Wang-Chien Lee is also supported in part by the National Science Foundation under Grant no. IIS-0328881, IIS-0534343 and CNS-0626709.

References

1. APRS: Automatic Position Reporting System. [web], http://aprs.net/
2. U.S. Geological Survey Earthquake Hizards Program. [web],
   http://earthquake.usgs.gov/region/neic/
3. Agrawal, R., Imielinski, T., Swami, A.N.: Mining Association Rules between Sets of Items in Large Databases. In: Proceedings of the 1993 ACM SIGMOD International Conference on Management of Data, Washington, D.C, May 26-28, pp. 207–216 (1993)
4. Agrawal, R., Srikant, R.: Mining Sequential Patterns. In: Proceedings of the 11th International Conference on Data Engineering (ICDE), Taipei, Taiwan, March 6-10, pp. 3–14 (1995)
5. Han, J., Kamber, M.: Data Mining - Concepts and Techniques. Elsevier, Amsterdam (2006)
6. Li, Y., Chung, S.M.: Text Document Clustering Based on Frequent Word Sequences. In: Proceedings of the 2005 ACM International Conference on Information and Knowledge Management (CIKM), Bremen, Germany, October 31-November 5, pp. 293–294 (2005)
7. Mamoulis, N., Yiu, M.L.: Non-contiguous Sequence Pattern Queries. In: Bertino, E., Christodoulakis, S., Plexousakis, D., Christophides, V., Koubarakis, M., Böhm, K., Ferrari, E. (eds.) EDBT 2004. LNCS, vol. 2992, pp. 783–800. Springer, Heidelberg (2004)
8. Mannila, H., Toivonen, H., Verkamo, A.I.: Discovery of Frequent Episodes in Event Sequences. Data Mining and Knowledge Discovery 1(3), 259–289 (1997)
9. Papadimitriou, S., Sun, J., Yu, P.S.: Local Correlation Tracking in Time Series. In: Proceedings of the 6th IEEE International Conference on Data Mining (ICDM), Hong Kong, China, December 18-22, 2006, pp. 456–465 (2006)
10. Tan, P.-N., Kumar, V., Srivastava, J.: Selecting the Right Interestingness Measure for Association Patterns. In: Proceedings of the 2002 ACM SIGKDD International Conference on Knowledge Discovery, Alberta, Canada, July 23-26, 1994, pp. 32–41 (1994)
11. Walpole, R.E., Raymond H, M., Myers, S.L.: Probability and Statistics for Engineers and Scientists. Prentice Hall, Englewood Cliffs (1997)
12. Wang, H., Perng, C.-S., Fan, W., Park, S., Yu, P.S.: Indexing Weighted-Sequences in Large Databases. In: Proceedings of the 19th International Conference on Data Engineering (ICDE), Bangalore, India, March 5-8, 2003, pp. 63–74 (2003)
13. Xiong, H., Shekhar, S., Tan, P.-N., Kumar, V.: TAPER: A Two-Step Approach for All-Strong-Pairs Correlation Query in Large Databases. IEEE Transactions on Knowledge and Data Engineering (TKDE) 18(4), 493–508 (2006)
14. Zhang, P., Huang, Y., Shekhar, S., Kumar, V.: Correlation Analysis of Spatial Time Series Datasets: A Filter-and-Refine Approach. In: Whang, K.-Y., Jeon, J., Shim, K., Srivastava, J. (eds.) PAKDD 2003. LNCS (LNAI), vol. 2637, pp. 532–544. Springer, Heidelberg (2003)

Caching Dynamic Skyline Queries*

Dimitris Sacharidis[1], Panagiotis Bouros[1,**], and Timos Sellis[1,2]

[1] National Technical University of Athens
9 Iroon Polytechniou, Athens 157 80, Greece
{dsachar,pbour}@dblab.ntua.gr
[2] Institute for the Management of Information Systems — R.C. Athena
17 G. Mpakou, Athens 115 24, Greece
timos@imis.athena-innovation.gr

Abstract. Given a query tuple q, the dynamic skyline query retrieves the tuples that are not dynamically dominated by any other in the data set with respect to q. A tuple dynamically dominates another, w.r.t. q, if it has closer to q's values in all attributes, and has strictly closer to q's value in at least one. The dynamic skyline query can be treated as a standard skyline query, subject to the transformation of all tuples' values. In this work, we make the observation that results to past dynamic skyline queries can help reduce the computation cost for future queries. To this end, we propose a caching mechanism for dynamic skyline queries and devise a cache-aware algorithm. Our extensive experimental evaluation demonstrates the efficiency of this mechanism compared to standard techniques without caching.

Keywords: skyline, dynamic skyline query, caching.

1 Introduction

The skyline query has received considerable attention since its introduction in the database community [1]. Consider a data set P where each tuple is represented as a d-dimensional point. The *skyline query* returns all points in P not dominated by another point. A point p_i is said to *dominate* another point p_j if for all dimensions p_i has equal or smaller coordinate values than p_j and in at least one dimension p_i has strictly smaller value than p_j. Intuitively, assuming in all dimensions lower values are better, the skyline query retrieves the best tuples, irrespective of how a user assigns preference to each dimension. More formally, for any monotone preference function that assigns scores to tuples, the highest scored — most preferable — tuple is included in the skyline.

---

* This work has been funded by the project PENED 2003. The project is cofinanced 75% of public expenditure through EC - European Social Fund, 25% of public expenditure through Ministry of Development - General Secretariat of Research and Technology and through private sector, under measure 8.3 of OPERATIONAL PROGRAMME "COMPETITIVENESS" in the 3rd Community Support Programme.
** The author is partially supported by the Greek State Scholarships Foundation (IKY).

B. Ludäscher and Nikos Mamoulis (Eds.): SSDBM 2008, LNCS 5069, pp. 455–472, 2008.
© Springer-Verlag Berlin Heidelberg 2008

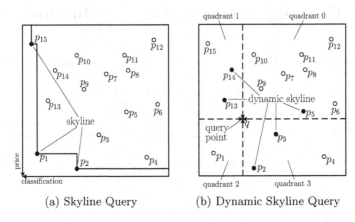

(a) Skyline Query   (b) Dynamic Skyline Query

Fig. 1. Skyline queries

Consider a table that contains entries about hotels, with attributes Name, Price and Classification. Naturally, considering only its price, a hotel is more preferable if it cheap. Similarly, regarding its classification, a high-starred hotel is more desirable. Figure 1(a) shows 15 hotels drawn as points in the two dimensional plane, where the x dimension is Classification, and the y is Price. The arrows in the axes indicate the direction of preference in each dimension, e.g., the more preferable high ranked (budget) hotels have lower x (y) values. Notice that there is no hotel that dominates p_1, p_2 and p_{15}, and, hence, they all belong in the skyline, as pointed in Figure 1(a). On the other hand, clearly, p_3 cannot be in the skyline as it is dominated by both p_1 and p_2. In fact, any hotel that resides in the right-hand side with respect to the line connecting the skyline hotels, is dominated.

A natural extension of the skyline query is its dynamic counterpart, introduced in [2]. Given a query point q, not necessarily in the data set P, the *dynamic skyline query* retrieves all points in P not dynamically dominated, with respect to q, by another point. A point dynamically dominates another, w.r.t. q, if it has closer to q's values in all dimensions, and has strictly closer to q's value in at least one. Returning to the hotel example, suppose that a user looks for hotels that match her budget and standards. For this reason, she specifies her "ideal" hotel q and wishes to retrieve all similar, in price and classification, hotels not dominated by others. Figure 1(b) illustrates the query point q and the dynamic skyline with respect to it. The dynamic skyline query w.r.t. q returns the hotels p_2, p_3, p_5, p_{13} and p_{14}, as shown in the figure. Notice that p_9 is not in the skyline, because p_{13} is closer to q in all dimensions, i.e., p_{13} matches the "ideal" hotel better than p_9 both in price and classification.

Dynamic skylines are useful in a variety of settings, where preferences are defined relatively to an exemplar, as in the ideal hotel scenario previously described. They also serve as the basic block for more complex queries. Seeing the skyline computation problem from the micro-economic perspective, other types of dominance related queries [3] also make sense. For example, hotel owners might be interested to find out for which ideal hotels specified by users their hotels belong in the skyline. The latter is known as the reverse skyline problem [4]. Further, a query can specify a set of exemplars, rather than one, and the dominance relationships are adjusted accordingly to consider the entire set. Examples of such queries include the multi-source skyline [5] and the spatial skyline [6].

A dynamic skyline query can be reduced to a static skyline query, subject to the transformation of all points' coordinates. In particular, given query q, a point p is transformed to p' such that the i-th coordinate of p' is computed as $p'^i = |p^i - q^i|$. Therefore, any method designed for the standard skyline query can be trivially applied to the dynamic case. Note that all such methods are designed to solve a single instance, given a single data set. Therefore, in the dynamic case, where multiple instances — one for each query point — need to be solved, the algorithm must run anew each time, examining all points for dominance. In this paper we show that results to past dynamic skylines queries can help reduce the cost of processing future queries. We present a caching mechanism that maintains the most useful past results and uses them to exclude from consideration certain points.

Figure 2 illustrates the intuition behind our caching mechanism. Assume that q_a, q_b and q_c are past dynamic skyline queries, as depicted in Figure 2(a). Observe that each query point partitions the space into 4 quadrants. Let q represent the dynamic skyline query under consideration, shown in Figure 2(b), and examine the upper-right quadrant, which contains the past query q_a. Assuming we have cached the result for query q_a, we know that p_7 is part of q_a's dynamic skyline and that it dominates points p_8, p_{11} and p_{12}, as seen in the upper right shaded area in Figure 2(b). Furthermore, since p_7 lies in the same quadrant (upper-right) with respect to q_a as q_a lies with respect to q, we conclude that these points are dynamically dominated by p_7 w.r.t. q, as well. Indeed, p_8 is farther, in all dimensions, from both q_a and q than p_7. With analogous reasoning and by examining the past query q_b (q_c) one can deduce that p_{15} (p_4) cannot be in the dynamic skyline of q since it is dominated by p_{14} (p_3).

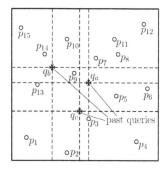

(a) Dynamic Skyline Queries

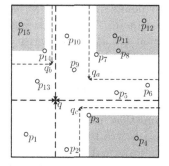

(b) Pruning Points

Fig. 2. Caching skylines

The contributions of this work can be summarized as follows:

1. We introduce the notion of *orthant skylines* and examine its relationship with dynamic skylines.
2. We extend the well-known Bitmap algorithm to compute the orthant skylines in parallel to the dynamic skyline, incurring small computation overhead.
3. We show how cached orthant skyline queries can help expedite the computation of future dynamic skyline queries. We propose three cache replacement policies for deciding which queries to expunge when the cache is full.
4. We perform an extensive experimental evaluation that demonstrates the efficacy of the caching mechanism, as portrayed by the significant reduction in query processing time in all settings.

Paper outline. First, in Section 2 we review literature on skyline related problems, focusing on the Bitmap algorithm, which is used throughout this work. Then, in Section 3 we present and formalize the basic notions discussed in this paper. The extension to the Bitmap algorithm for dynamic skylines and the caching mechanism are discussed in detail in Section 4. The most important experimental findings are presented in Section 5. Finally, Section 6 concludes the paper.

2   Related Work

Computing the points in the skyline, also known as finding the maxima in a set of vectors [7], has been thoroughly studied in the area of computational geometry where a large number of theoretical results exists. The first work to address the skyline computation problem in the context of databases was [1]. The authors discuss various techniques: they devise an algorithm that iterates over all points using block nested loops (BNL), propose a B-tree based approach, and also adapt the multidimensional divide and conquer algorithm [8] to handle external memory. An extension of the BNL algorithm that relies in presorting the points is introduced in [9]. The work in [10] introduces progressive algorithms that output points guaranteed to belong in the skyline without having to scan the entire data set. The Bitmap algorithm encodes all points using a bitmap representation and uses fast bitwise operations to extract the skyline points. We use Bitmap as the basis of our methodology for computing dynamic skylines in the presence of cache, and, thus, we present it in detail in Section 2.1. Another indexed method based on B-trees is also discussed in [10], where points are sorted according to their lowest valued coordinate.

Algorithms that use R-trees to index points have also been proposed. In [11] the authors observe that the nearest neighbor (NN) point to the beginning of the axes is always part of the skyline. This point segments the dataset into overlapping partitions according to its coordinates. Then, NN search is performed on each partition and the algorithm proceeds iteratively. Special care needs to be taken to remove duplicates resulting from the overlapping partitions. The branch and bound algorithm (BBS) introduced in [2] avoids the pitfalls of the nearest neighbor approach. BBS maintains the expanded R-tree entries into a heap in ascending order of their minimum distance to the beginning of the axes. The first point visited in this manner is the NN and belongs to the skyline. When an entry is de-heaped, only its children not dominated by the skyline points found so far are inserted into the heap. BBS is proved to examine only the nodes in the R-tree that can potentially contain skyline points, and, hence, is I/O optimal.

The notion of dynamic skyline was first introduced in [2], where a variant of the BBS algorithm was presented. Given a point p, the reverse skyline query [4] retrieves the points whose dynamic skyline includes p. The authors in [4] present algorithms that are based on finding the global skyline of p, a notion related to the orthant skylines defined in this work. Another related notion is the multi-source skyline query [5,6], in which a set of query points is specified and the result contains the points not dominated w.r.t. to the set.

When only a subset of the dimensions is considered, the skycube operator [12] returns the points that belong in the skyline. Some dominance related queries seen from the micro-economic perspective are presented in the data-warehouse framework of [3]. When the domain of a dimension is partially ordered, i.e., its values belong in a hierarchy, the skyline computation becomes more involved and the final result may require pruning as discussed in [13]. The notion of probabilistic skylines is defined [14] for the case where multiple tuples (or samples) correspond to randomly distributed objects in the data set.

2.1   The Bitmap Algorithm

The Bitmap algorithm was introduced in [10] for determining the skyline points efficiently when the domains of the defining dimensions are small and, most importantly, discrete. Briefly, Bitmap works as follows: (a) it pre-processes all points to obtain an appropriate bitmap representation, and (b) it checks each point for dominance against all points and outputs it if not dominated. The latter step is efficiently performed by fast bitwise AND/OR operations on the bitmap representations obtained in the former step.

Bitmap representation. For ease of presentation, we assume that all d dimensions have a domain of size n and its values belong in $\{0, 1, \ldots, n-1\}$, 0 being the most preferable value; the extension to dimensions with different domains is straightforward. A value u is represented by a bitmap of size n, where the u most significant bits are set to 0 and the remaining (the $n - u$ least significant bits) are set to 1. A d-dimensional point is, hence, represented as d bitmaps, one for each of its coordinates. We maintain the bitmap representation for all points in a bitmap table. For example, assume $n = 8$ and consider the point $p_1(0, 3, 7)$; its coordinates are represented as the bitmaps 11111111, 00011111 and 00000001, respectively. Figure 3(a) shows the bitmap table for 6 points, including that of p_1. The function of the bold and italicized bits will become apparent in the following.

Dominance check. The *dominance check* of point p identifies the points that dominate it. Clearly, a point belongs to the skyline if its dominance check

	D_1	D_2	D_3
$p_1(0,3,7)$	11*1*11111	00011111	00000001
$p_2(6,7,2)$	000*0*0011	00000001	0011*1*1111
$p_3(1,0,2)$	01*1*11111	11111111	0011*1*1111
$p_4(3,0,4)$	000*1*1111	11111111	00000*1*111
$p_5(2,2,6)$	00*1*11111	00111111	00000011
$p_6(6,4,7)$	000*0*0011	00001111	00000001

$A^1 = 101110 \quad B^1 = 101010$

$A^2 = 001100 \quad B^2 = 000000$

$A^3 = 011100 \quad B^3 = 011000$

$A = 001100 \quad B = 111010$

$C = A\&B = 001000$

(a) Bitmap table

(b) Dominance check for p_4

Fig. 3. Bitmap example

identifies no dominating point. Performing the check in the Bitmap algorithm involves extracting vertical *bitslices* and performing bitwise AND/OR operations. Let $p^i \in \{0, 1, \ldots, n-1\}$ denote the i-th coordinate of p, where $1 \le i \le d$. We extract two vertical bitslices, A^i and B^i, from the bitmap representation of the data set. We extract a single bit for each point; thus, each bitslice has length N, where N denotes the size of the data set. In particular, we obtain the bitslice A^i by juxtaposing the $(p^i + 1)$-th bit of the bitmap representation of the i-th coordinate for all points. Similarly, we obtain B^i by juxtaposing the preceding, i.e., the p^i-th, bit of the bitmap representation of the i-th coordinate for all points; note that when $p^i = 0$ we explicitly set B^i to all zeros. Figure 3(b) shows the A^i, B^i bitslices for the dominance check of $p_4(3, 0, 4)$. For the A^i bitmaps we extract the 4th, 1st and 5th bit, shown in bold in Figure 3(a), for the first, second and third dimension of each point. For the B^i bitmaps we extract the 3rd and 4th bit, shown in italics in Figure 3(a), for the first and third dimension of each point; B^2 is set to all zeros. Given point p, its bitslice A^i encodes which points (i.e., those whose corresponding bit in A^i is set to 1) are equally as good or better than p with respect to the i-th dimension. In other words if the k-th bit of A^i is set to 1, then the k-th point has equally good or better value than p in the i-th coordinate. On the other hand, the bitslice B^i encodes the points that are strictly better in the i-th dimension.

Let $A = A^1 \& A^2 \& \ldots \& A^d$ denote the bitwise AND operation of all A^i bitslices. A indicates the points that are equally as good or better than p in *all* dimensions. Consider a point p in the 2 dimensional space shown in Figure 4. All points, including p, that reside in the shaded area of Figure 4(a) have their bit in A set to 1; for all other points the bit is 0. Similarly, let $B = B^1 | B^2 | \ldots | B^d$ denote the bitwise OR operation of all B^i bitslices. Then, B indicates the points that are strictly better than p in *at least one* dimension. All points that reside in the shaded area, excluding those in the dashed line and p, shown in Figure 4(b) have their bit in B set to 1; for all other points the bit is 0. According to the definition of dominance, if a point has its corresponding bit set both in A and B, then it dominates p, and, hence, p is not in the skyline. On the other hand, if $C = A\&B$ has no bit set, then p is not dominated by any point, and thus belongs in the skyline. All points, excluding p, that reside in the shaded area

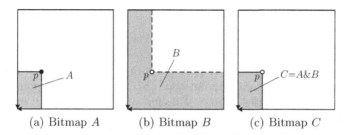

(a) Bitmap A (b) Bitmap B (c) Bitmap C

Fig. 4. Dominance check

shown in Figure 4(c) dominate p and thus have their bit in C set to 1; for all other points the bit is 0.

Returning to the dominance check of point $p_4(3,0,4)$ for the example illustrated in Figure 3(b), notice that $A = 001100$ and $B = 111010$; hence, $C = 001000$. Since the third bit in C is set to 1, it follows that $p_3(1,0,2)$ dominates $p_4(3,0,4)$. So, p_4 does not belong in the skyline.

3   Preliminaries

In this section, we formally define the notions of dominance, and skyline points. We assume a d-dimensional space, where D_i denotes the domain of the i-th dimension, $i \in \{1,\ldots,d\}$. Each domain D_i is totally ordered by $<$ assigning preference to the values of the domain. Consider values $u,v \in D_i$; u is more preferable than v iff $u < v$. The data set P contains $N = |P|$ d-dimensional points. Each point $p \in P$ belongs in the space $D = D_1 \times D_2 \times \cdots \times D_d$ and is represented by its coordinates, $p = (p^1, p^2, \ldots, p^d)$.

Definition 1 (Dominance). *Let $p_1, p_2 \in P$ and $i,j \in \{1,\ldots,d\}$. A point p_1 dominates another point p_2, denoted as $p_1 \prec p_2$, iff (i) for all dimensions, p_1^i is more, or equally preferable than p_1^i, i.e., $\forall i : p_1^i \leq p_2^i$, and (ii) in at least one dimension, let j, p_1^j is strictly more preferable than p_2^j, i.e., $\exists j : p_1^j < p_2^j$.*

A point not dominated by any other in the data set is called a skyline point. Intuitively, one cannot prefer a non-skyline point over a skyline point for any preference function that is monotonic in each dimension. In other words, the skyline contains the top-1 point for any preference function and, conversely, for a given skyline point there always exists a function under which this point is the top-1.

Definition 2 (Skyline). *The skyline of P, denoted as $SL(P)$, is the set of points in P that are not dominated by any other point of P, i.e., $SL(P) = \{p_1 \in P \mid \nexists p_2 \in P : p_2 \prec p_1\}$.*

The skyline query retrieves the points that belong in the skyline. For example, for the data set shown in Figure 1(a), the skyline query retrieves the points p_1, p_2 and p_{15}.

3.1 Dynamic Skyline

According to the definitions of dominance and skyline presented above, the most preferable point is the beginning of the axes $o = (0, \ldots, 0)$, assuming it exists in P, since it dominates all other points. As argued in Section 1, however, in many cases the most preferable point could be a user specified point q. In this case we need to express the notions of preference and dominance relative to q. Given a point $q = (q^1, \ldots, q^d) \in D$ (not necessarily in P), the value $u \in D_i$, for some i, is more preferable than the value $v \in D_i$ iff $|u - q^i| < |v - q^i|$. Based on this preference notion, we now provide the definitions of dynamic dominance and skyline.

Definition 3 (Dynamic Dominance). *Let $p_1, p_2 \in P$, $i, j \in \{1, \ldots, d\}$ and $q \in D$. Given a query point q, a point p_1 dynamically dominates, w.r.t. q, another point p_2, denoted as $p_1 \prec_q p_2$, iff (i) for all dimensions, p_1^i is more, or equally preferable, w.r.t. q, than p_2^i, i.e., $\forall i : |p_1^i - q^i| \leq |p_2^i - q^i|$, and (ii) in at least one dimension, let j, p_1^j is strictly more preferable, w.r.t. q, than p_2^j, i.e., $\exists j : |p_1^j - q^j| < |p_2^j - q^j|$.*

Definition 4 (Dynamic Skyline). *Given a query point $q \in D$, the dynamic skyline of P w.r.t. q, denoted as $DSL(P, q)$, is the set of points in P that are not dynamically dominated, w.r.t. q, by any other point of P, i.e., $SL(P) = \{p_1 \in P \mid \nexists p_2 \in P : p_2 \prec_q p_1\}$.*

Consider the example data set and query q shown in Figure 5. The dynamic skyline w.r.t. q contains the points p_2, p_3, p_5, p_{13} and p_{14} drawn with black solid circles in the figure.

The dynamic counterparts of the dominance and skyline notions, essentially, correspond to the standard notions applied to the transformed data set P' obtained by mapping each point $p = (p^1, \ldots, p^d) \in P$ to the point $p' = (|p^1 - q^1|, \ldots, |p^d - q^d|)$, given the query point q. Consider Figure 5; observe that any point p, where $p^j - q^j < 0$ for at least one dimension, has been mapped to a point p' in the upper right quadrant w.r.t. q. The mapping is shown with a dashed line and the mapped point is drawn as a dashed circle.

As illustrated in Figure 5, the query point q partitions space D into the 4 quadrants (2^d orthants in the d-dimensional case) defined by constraining the space to be higher or lower than q^j for each dimension j. An orthant can be identified by a number written in binary containing d bits where the j-th bit is 0 (1) if for the j-th dimension the orthant contains the values not smaller (smaller) than q^j. Figure 5 shows the 4 orthants and their ids assuming dimension order yx. We introduce the notion of orthant skylines, which is defined as the dynamic skyline when considering only the points in P inside an orthant.

Definition 5 (Orthant Skyline). *Given a query point $q \in D$, the o-th orthant skyline of P w.r.t. q, where $o \in \{0, \ldots, 2^d - 1\}$, denoted as $OSL(P, q, o)$, is the set of points in P that belong to the o-th orthant and are not dynamically dominated, w.r.t. q, by any other point of that orthant.*

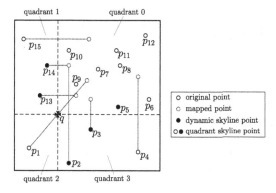

Fig. 5. Mapping points

Return to the example of Figure 5. The upper right quadrant (with id 0) skyline contains points p_5, p_9 and p_{10}, which are drawn with filled circles (black or grey). Note that an orthant skyline point can be dominated by the skyline point of another orthant and, hence, the former cannot belong in the dynamic skyline. For example, p_9 is dominated by the mapping of point p_{13} and, hence, is not part of the dynamic skyline. Similarly, p_{10} is dominated by the mappings of points p_2 and p_{14}, which happen to coincide. It is straightforward to see that the following lemma regarding the union of all orthant skylines w.r.t. q, termed global skyline in [4], holds.

Lemma 1. *The union of all orthant skylines w.r.t. q is a superset of the dynamic skyline w.r.t. q.*

4   Caching Dynamic Skylines

We first present the dynamic Bitmap algorithm, termed DBM, for obtaining the orthant skylines as well as the dynamic skyline in Section 4.1. Next, in Section 4.2, we demonstrate that caching queries and their orthant skylines can help reduce the execution time for future dynamic skyline queries. Finally we discuss cache replacement policies in Section 4.3.

4.1   Computing the Orthant Skylines

The DBM algorithm computes the orthant skylines and the dynamic skyline with respect to a query point q. Because the number of orthants is exponential to the dimensionality, it is crucial that our method finds the orthant skylines with little overhead compared to calculating only the dynamic skyline.

Initially, we construct the bitmap table for all points. In particular, we represent each coordinate of point $p \in P$ by converting the value $|p^i - q^i|$ (and not p^i) into the $|D_i|$ bits, as described in Section 2.1. We maintain a *global mask M* of length N to indicate the points that need to be considered. Initially all bits

---

Algorithm DBM

Input: data set P, query q, bitmap table T, masks M, $\{M_j\}$
Output: orthant skyline bitmaps $\{O_j\}$, dynamic skyline bitmap R

```
1  begin
2  |   foreach p ∈ P do
3  |   |   Let o be the orthant p resides in w.r.t. q
4  |   |   if M_o[p] = 0 then                              // pruned by cache
5  |   |   |   continue
6  |   |   else
7  |   |   |   GetBitmapC(T, p)
8  |   |   |   if C&M_o ≠ 0 then                           // case (I)
9  |   |   |   |   M_o[p] ← 0                              // not in the orthant
10 |   |   |   |   M[p] ← 0                  // and not in the dynamic skyline
11 |   |   |   else if C ≠ 0 then                          // case (II)
12 |   |   |   |   M[p] ← 0                   // not in the dynamic skyline

13 |   R ← M
14 |   O_j ← M_j for all j
15 |   return ⟨R, {O_j}⟩
16 end
```

Fig. 6. The Dynamic Bitmap algorithm (DBM)

are set to 1. In addition, DBM creates the *orthant masks* of length N, denoted M_j for $j \in \{0, \ldots, 2^d - 1\}$, to indicate which points belong to an orthant. Note that in the following we slightly abuse notation by referring to p's bit in M as $M[p]$. If p resides in the o-th orthant w.r.t. q, then $M_o[p]$ is set to 1. The orthant masks can be created in parallel to the bitmap table construction, as points are examined, by identifying the orthant a point resides in. Recall from Section 3.1 that the orthant id written in binary contains d bits, one for each dimension. Assume that point p is considered; the sign of $p^i - q^i$ designates the value of the i-th bit of p's orthant w.r.t. q, i.e., the bit is 0 if $p^i - q^i \geq 0$ and 1 if $p^i - q^i < 0$.

Figure 6 illustrates the DBM algorithm that examines each point in turn (Line 2). Given query point q, let p be the current point considered and let o denote the orthant p resides in w.r.t. q (Line 3). If p's bit in its orthant mask is set to 0, then we skip this point (Lines 4–5). Of course, in the case we examine here, this cannot happen, as the M_o mask is initialized to 1 for all its points; however, in Section 4.2, when the query cache is considered, this may no longer hold.

Next, DBM computes the C bitmap (Line 7) as discussed in Section 2.1 and performs two dominance checks, one for the orthant and one for the dynamic skyline. We distinguish three cases.

(I) $C\&M_o \neq 0$ denotes that p is dominated by some other point in its orthant (Line 8).

(II) $C \neq 0$ and $C\&M_o = 0$ denotes that p is not dominated by some other point in its orthant, but it is dominated by some point in another orthant (Line 11).

(III) $C = 0$ and $C\&M_o = 0$ denotes that p is not dominated by neither some other point in its orthant, nor by some point in another orthant.

In case (I), by Lemma 1, point p cannot belong to its orthant and the dynamic skyline. Hence, its bit in M and M_o is set to 0 (Lines 9–10). In case (II) p cannot be in the dynamic skyline but is part of it's orthant skyline. DBM set its bit to 0 only in M (Line 12); its M_o bit remains set to 1. Finally, in case (III) DBM retains p's bit in M and M_o to 1. After all points have been examined, the M mask identifies the dynamic skyline points, whereas the $\{M_j\}$ masks identify the orthant skyline points (Lines 13–14).

4.2 Dynamic Skylines Via Caching

In this section we show how caching of past queries and orthant skylines can help expedite dynamic skyline queries. We start by describing the cached Dynamic Bitmap algorithm, termed cDBM; we then discuss cache replacement strategies in Section 4.3. We assume that the cache Q stores for each past query q_i, the query itself and all its orthant skylines $OSL(P, q_i, j)$ for $j \in \{0, \ldots, 2^d - 1\}$. In particular, the orthant skyline $OSL(P, q_i, j)$ is represented as the N-length bitmap O_j^i in which p's bit is set to 1 if p is included in the j-th orthant skyline w.r.t q_i. Therefore, $Q = \{\langle q_i, \{O_j^i\}\rangle\}$, i.e., the cache needs to store 2^d bitmaps $\{O_j^i\}$ for each past query q_i; later, we provide a method to compress these bitmaps.

The intuition for using past orthant skylines lies in the fact that they can immediately and safely prune potentially large parts of the data set. Indeed, an orthant skyline contains precomputed information about the dominance checks in the particular orthant, as the next lemma suggests.

Lemma 2. *Consider queries $q_i, q \in D$ and a point $p \in P$, such that q_i belongs in the o-th orthant with respect to q, and p belongs in the o-th orthant w.r.t. q_i, and, thus, w.r.t. q as well. If p is not part of the o-th orthant skyline w.r.t. q_i, then, p is not part of the o-th orthant skyline w.r.t. q, and, hence, neither is part of the dynamic skyline w.r.t. q.*

Proof. Without loss of generality, assume $o = 0$. Since p is not part of the o-th orthant skyline w.r.t. q_i, it is dominated by (at least) one point, let p_a, i.e., $p_a \prec_{q_i} p$. Therefore, for $o = 0$ we have that $p_a^j - q_i^j \leq p^j - q_i^j$ for all j, and $p_a^k - q_i^k < p^k - q_i^k$ for at least one k, where $j, k \in \{0, \ldots, 2^d - 1\}$. Adding $q_i^j - q^j$ and $q_i^k - q^k$ to the previous inequalities, and since all quantities are non-negative, we obtain $p_a \prec_q p$. □

Figure 2(b) demonstrates Lemma 2 for the current query q and the past query q_a, which lies in the upper-right quadrant with respect to q. The skyline of the upper-right quadrant w.r.t. q_a contains a single point, p_7, which dynamically dominates p_8, p_{11} and p_{12} w.r.t. q_a. It is obvious that p_7 also dominates p_8, p_{11} and p_{12}, w.r.t. q, and hence, these, points cannot belong to the upper-right quadrant or dynamic skyline of q.

The cDBM algorithm for calculating the dynamic and orthant skylines is illustrated in Figure 8. For each query, cDBM first computes the masks (Line 4) and then calls the DBM algorithm (Line 5). Finally, the orthant skylines are inserted into the cache (Line 6).

Procedure ComputeMasks

Input: data set P, query cache Q, query q
Output: bitmap table T, masks M, $\{M_j\}$

```
1  begin
2  |    partition Q to the {Q_j} sets w.r.t. q
3  |    foreach p ∈ P do
4  |    |    Let o be the orthant p resides in w.r.t. q
5  |    |    M[p] ← 1                                    // initialize mask to 1
6  |    |    M_o[p] ← 1                                  // initialize o-th orthant mask to 1
7  |    |    M_j[p] ← 0, for all j ≠ o                   // initialize all other orthant masks to 0
8  |    |    update T with BitmapEncode(p, q)
9  |    |    foreach ⟨q_i, {O_j^i}⟩ ∈ Q_o do
10 |    |    |    Let o_i be the orthant p resides in w.r.t. q_i
11 |    |    |    if o_i = o then
12 |    |    |    |    Let O_o^i be the o-th orthant skyline of q_i from {O_j^i}
13 |    |    |    |    if O_o^i[p] = 0 then              // if p not in its orthant skyline w.r.t. q_i
14 |    |    |    |    |    M[p] ← 0                     // it cannot be in the dynamic skyline w.r.t. q
15 |    |    |    |    |    M_o[p] ← 0                   // neither in the j-th orthant skyline w.r.t. q
16 |    |    |    |    |    break

17 |    return ⟨T, M, {M_j}⟩
18 end
```

Fig. 7. Computing masks given the query cache

Algorithm cDBM

Input: data set P

```
1  begin
2  |    Q = ∅
3  |    foreach incoming q do
   |    |    // initialize masks and construct bitmap table
4  |    |    ⟨T, M, {M_j}⟩ ← ComputeMasks(P, Q, q)
   |    |    // calculate dynamic and orthant skylines
5  |    |    ⟨R, {O_j}⟩ ← DFB(P, q, T, M, {M_j})
6  |    |    update Q with ⟨q, {O_j}⟩             // run a cache replacement policy

7  end
```

Fig. 8. The cached Dynamic FastBitmap algorithm (cDBM)

The most important step of the cDBM algorithm is the ComputeMasks procedure, shown in Figure 7. Given query q, this procedure creates the masks M and $\{M_j\}$ applying Lemma 2 to determine which points need not be considered. Initially, the cache Q is partitioned into sets Q_j for $j \in \{0, \ldots, 2^d - 1\}$, such that Q_j contains the queries that reside in the j-th orthant w.r.t. q (Line 2). Then, each point p is examined in turn (Line 3). The bitmap representation of p with respect to q is computed and the bitmap table T is updated (Line 8), as discussed in Section 4.1. Let o denote the orthant p lies w.r.t. q (Line 4). Then, p's bit in the o-th orthant mask is set to 1, whereas in all other orthant masks it is set to 0 (Lines 6–7); of course, p's bit in M is set to 1 (Line 5). The algorithm continues by examining the past queries that reside in the o-th orthant, i.e., those in Q_o. Let q_i be such a query (Line 9) and let o_i be the orthant that p lies in with respect to q_i (Line 10). If p lies in the same orthant with respect to

q_i as q_i lies w.r.t. q, i.e., $o_i = o$ (Line 11), then Lemma 2 applies (Lines 12–16). Therefore, if p was not in o-th orthant skyline w.r.t q_i (Line 13), then it can be excluded from consideration in the orthant (Line 14) and the dynamic skyline (Line 15) w.r.t. q. If this was the case, then no other past query needs to be examined (Line 16).

Compressing Cached Queries. The caching mechanism discussed above has a large space overhead, as it requires storing 2^d bitmaps for each query in the cache. We address this issue making the following observation: a point can belong to only one orthant and, thus, can be part of only one orthant skyline per query. Given query q_i and its orthant skyline bitmaps O_j^i, for all $j \in \{0, \dots, 2^d - 1\}$, we construct a single orthant skyline bitmap O^i by disjuncting all O_j^is. Then, p's bit in O^i is set to 1, if point p belongs in the skyline of its orthant w.r.t. q. Note that the ComputeMasks procedure need not change; in Line 13 of Figure 7, the O^i mask can be used instead of the o-th orthant mask O_o^i.

4.3   Cache Replacement Policies

In this section we discuss replacement policies for our caching mechanism. The objective of these policies is the identification of the least *useful* query point among those in the cache that must be discarded together with its orthant mask. The first two policies we consider are the common Least Recently Used (LRU) and Least Frequently Used (LFU) policies, which keep track of the usage for each query in the cache. On the other hand, the Least Pruning Power (LPP) policy measures the pruning ability of each query and discards the least strong.

LRU and LFU policies. In Section 4.2 we used *all* queries in the cache to discover which points to exclude from consideration. However, given a query q, some of the queries in the cache are redundant, i.e., they identify points that can be pruned even if we don't consider these cached queries. This is exemplified in Figure 9(a), where q denotes the current query under consideration, and q_a, q_b, q_c and q_d are past queries in the cache that reside in the upper-right quadrant w.r.t. q. For each of these queries, their upper-right quadrant skyline can be used to prune some of the points that are contained in the dashed box ranging from the query to the upper right point in Figure 9(a). Observe that since q_c's box is entirely contained in q_a, the points the former can possibly prune can also be pruned by the latter; the same holds for q_d and q_b. It is easy to show that, given a query q, if a cached query q_c is dominated by another cached query q_a, then q_c can be safely disregarded when computing the masks for query q.

The previous observation suggests the following change to the ComputeMasks procedure in Figure 7. In Line 2, ComputeMasks partitions the cached queries according to the orthant they belong with respect to the query point q. Instead, we calculate the orthant skylines w.r.t. q of all queries in Q so that the set Q_j now contains the queries in Q that belong in the j-orthant skyline (and not all queries in the j-th orthant).

Consider the case when query point q is considered and that we compute the Q_j sets. In the Least Recently Used (LRU) policy, each time a cached query

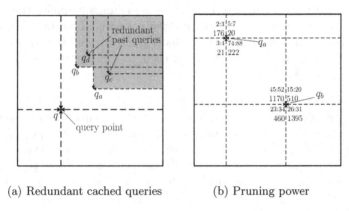

(a) Redundant cached queries   (b) Pruning power

Fig. 9. Cache Replacement Policies

belongs in the orthant skyline, i.e., it will be used to prune points, we annotate it with a timestamp to indicate that it was used for query q. Note that the timestamp can be a simple counter that increases with each query considered. When the cache has reached its capacity we choose to evict the cached query that was least recently used, i.e., has the smallest timestamp. To efficiently identify the cached query to be evicted, we maintain a priority queue with key the timestamp of each query. Therefore, updating the timestamp of an already stored query, inserting a new and evicting the least recently used one require time logarithmic to the size of the cache.

In the Least Frequently Used (**LFU**) policy, we maintain a usage counter for each query in the cache. Each time a cached query belongs in the orthant skyline w.r.t. the query point, i.e., it will be used to prune points, we increment its usage counter. We choose to always include in the cache the current query under consideration, and if the cache is full we evict the least frequently used query, i.e., that with the smallest usage counter. As before, a priority queue can be used to expedite the identification of the query to be evicted.

LPP policy. Intuitively, a useful cached query is one which has great pruning power, i.e., it can discard a large number of points for a lot of queries. Depending on the position of a cached query q_a relative to a query q, the pruning power of q_a can vary significantly. Let dp_a^j denote the number of points dominated by cached query q_a's j-th orthant skyline. Consider for example the two cached queries illustrated in Figure 9(b). If a query lies in the upper-left quadrant with respect to q_a, then q_a's skyline for the lower-right quadrant can prune 74 points, as indicated by the first of the three numbers per query and quadrant shown in Figure 9(b), i.e., $dp_a^3 = 74$. On the other hand, if a query lies in the lower-right quadrant w.r.t. q_a, then q_a's upper-left quadrant skyline can only prune 2 points, i.e., $dp_a^1 = 2$.

The pruning power of a query's orthant also depends on the probability of a query residing in the antisymmetric orthant. In the example of Figure 9(b), it is rather unlikely for a query to belong in the upper-left quadrant w.r.t. q_a, hence q_a's highly dominant quadrant will rarely be used. It is reasonable to assume that queries follow a similar distribution to the data set; therefore the probability of a query residing in any area is analogous to the number of data points this area includes. Given a cached query q_a, let np_a^j denote the number of points residing in the j-th orthant w.r.t. q_a. In Figure 9(b), np_a^j is shown as the second number in the triad of numbers for all queries and quadrants.

Given a query q_a, its pruning power for the j-th orthant, denoted as pp_a^j, is given by $pp_a^j = np_a^{\bar{j}} \cdot dp_a^j$, where $\bar{j}$ identifies j's antisymmetric orthant. In Figure 9(b), pp_a^j is shown as the third number typed in larger font for each query and quadrant. In the case of query q_a, for example, the upper-left quadrant skyline dominates 2 out of 3 points, and the lower-right quadrant contains a total of 88 points; hence, the pruning power of the q_a upper-left quadrant is $2 \cdot 88 = 176$. The pruning power pp_a of a query q_a is the sum of its pruning power for all orthants, i.e., $pp_a = \sum_j pp_a^j$. Intuitively, a low pruning power implies that the query is expected to prune only a few points. LPP always evicts from cache the query with the least pruning power. As before, a priority queue can be used.

Note that a query's pruning power for all orthants can be computed with little overhead. The number of points in an orthant np_a^j can be counted in the ComputeMasks procedure while examining where each point resides with respect to the query. Also, the number of points dominated by an orthant skyline dp_a^j can be calculated as the number of points in the orthant minus the orthant's skyline size; the latter can be found by simply using a counter in the DBM algorithm.

5   Experimental Evaluation

We present an extensive experimental evaluation of the DBM algorithm paired with the three cache replacement policies discussed in Section 4.3. In particular, we compare LRU, LFU, LPP with the Bitmap algorithm adapted to dynamic skyline query processing, denoted as NO-CACHE since it corresponds to the case where no cache is used. All algorithms are implemented in C++, compiled with gcc and executed on a 3 Ghz Intel Core 2 Duo CPU.

We use the generator from [15] to create data sets of three types of distribution:

- Independent: The attribute values are drawn from a uniformly random independent distribution.
- Correlated: Tuples whose attribute values are low, i.e., preferable, in one dimension have most likely low values in the other dimensions, as well.
- Anti-Correlated: Tuples whose values are low in one dimension have most likely high, i.e., not preferable, values in the other dimensions.

The dimensionality of the data set varies from $d = 2$ up to 6, where each dimension's domain contains a fixed number of discrete values, ranging from $|D| = 10$ up to 50. The size of the data set, N, is between 10 thousand and up

Table 1. Experimental parameters

Parameter	Values
N	10000, 20000, **50000**, 100000
d	2, 3, **4**, 5, 6
$|D|$	10, **20**, 50
$|Q|$	10, 20, **30**, 40, 50

to 100 thousand tuples. We test all cache replacement policies for different cache sizes that extend from $|Q| = 10$ to 50 queries. We perform $|Q| + 20$ dynamic skylines queries and we measure the average performance of all policies for the last 20 queries, so that the query cache is full in all cases. More specifically, we measure the average running time for each method and we count the average number of points pruned by the cache. In each experiment we vary a single parameter while we set the remaining ones to their default values. Table 1 summarizes the parameters involved and their ranges; the default values are shown in bold.

5.1   Experimental Results

In the first set of experiments we vary the cache size from $|Q| = 10$ to 50 while the data set is fixed to containing $N = 50000$ tuples with $d = 4$ attributes of cardinality $|D| = 20$. In this setting, the dataset is 1000 KB, whereas the cache size increases from 62.5 KB (6% of data size) for $|Q| = 10$, to 187.5 KB (19% of data size) for the default setting ($|Q| = 30$), and up to 312.5 KB (31% of data size) for the largest setting $|Q| = 50$. We measure the average running time and the average number of points pruned for 20 dynamic skyline queries when the cache is full.

Figure 10 presents the results of all cache replacement policies for the three data sets. The average number of pruned points are shown next to the time measurements for the LPP and LFU policies. We also draw the running time when the queries are processed witout cache, denoted as NO-CACHE, which is a straight line over $|Q|$. The expected behavior when the cache size increases is the

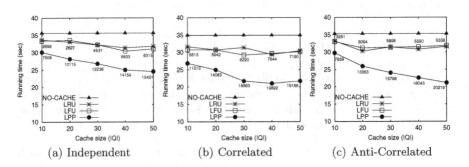

(a) Independent   (b) Correlated   (c) Anti-Correlated

Fig. 10. Varying the query cache size

number of pruned points to also decrease, resulting in shorter running times. This is clearly the case for the LPP policy which outperforms the usage-based policies. LRU and LFU, especially in the Anti-Correlated data set (Figure 10(c)), fail to take advantage of the larger cache. The reason for this behavior can be attributed to the fact that caching more queries recently (LRU) or frequently used (LFU) does not guarantee that future queries will benefit from them. In other words, the queries already seen are not representative of the queries to follow. On the other hand, LPP keeps in cache queries with great pruning power that can prove useful for *any* future query. For the maximum setting $|Q| = 50$, LPP immediately prunes 15421, 19166 and 20219 out of the 50000 points and decreases the processing time by 31%, 38% and 40% for the Independent, Correlated and Anti-Correlated data sets, respectively.

Figure 11 shows the effect of the distribution parameters on the caching mechanism. In this setting we draw the relative improvement in running time for the three cache replacement policies over the case of no cache, for the Correlated data set; similar results hold for the other distribution types. In Figure 11(a) we vary the data set size while keeping all parameters, including cache size Q, to their default values shown in Table 1. This implies that relative to the data size, the cache decreases as N grows. Still, Figure 11(a) shows that the policies can prune a rather significant part of the dataset (31% – 41% for LPP as shown by the labels in the figure), which is translated to an analogous running time improvement. Note that as the data set becomes denser, LFU and LRU's performance also increases.

In Figure 11(b) we vary the dimensionality of the data set, while the remaining parameters have their default values. The LPP policy is highly affected by the curse of dimensionality, i.e., as the space becomes sparser (since N is fixed) its pruning power rapidly decreases, i.e., from 78% down to 17%. The usage-based policies are also affected but to a lesser degree.

Finally, in Figure 11(c), we vary the domain cardinality for each dimension, when $N = 50000$, $d = 4$ and $|Q| = 50$. Larger $|D|$ values result in sparser data sets. However, unlike Figure 11(b), the cache replacement policies are not significantly affected. LPP in all cases improves running time by 38%.

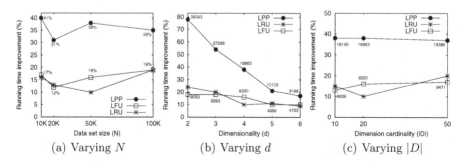

(a) Varying N (b) Varying d (c) Varying $|D|$

Fig. 11. Effect of distribution parameters

6   Conclusions and Future Work

In this paper we study the problem of dynamic skyline queries from a fresh perspective. We consider the case where we keep past queries and their results in a cache so as to expedite future query processing. For this end, we introduce the notion of orthant skylines and extend a well-known skyline algorithm to handle them. Then, we prove that results of orthant skyline queries can potentially exclude a large part of the data set from the costly dominance checks. We propose three cache replacement policies so that the cache always contains the most useful queries and their results. Through extensive experimental results on synthetically generated data set, we demonstrate the efficiency of the proposed caching mechanism: using less than 20% of the data size for the cache, we can reduce the processing time by 40%. In the future, we plan to apply the caching framework to other skyline processing algorithms, focusing on indexed approaches. Furthermore, we will investigate methods to increase the pruning power of the cached queries while at the same time reducing the space overhead.

References

1. Börzsönyi, S., Kossmann, D., Stocker, K.: The skyline operator. In: ICDE, pp. 421–430 (2001)
2. Papadias, D., Tao, Y., Fu, G., Seeger, B.: Progressive skyline computation in database systems. ACM Transactions on Database Systems 30(1), 41–82 (2005)
3. Li, C., Ooi, B.C., Tung, A.K.H., Wang, S.: Dada: a data cube for dominant relationship analysis. In: SIGMOD Conference, pp. 659–670 (2006)
4. Dellis, E., Seeger, B.: Efficient computation of reverse skyline queries. In: VLDB, pp. 291–302 (2007)
5. Deng, K., Zhou, X., Shen, H.T.: Multi-source skyline query processing in road networks. In: ICDE, pp. 796–805 (2007)
6. Sharifzadeh, M., Shahabi, C.: The spatial skyline queries. In: VLDB, pp. 751–762 (2006)
7. Kung, H.T., Luccio, F., Preparata, F.P.: On finding the maxima of a set of vectors. Journal of the ACM 22(4), 469–476 (1975)
8. Preparata, F.P., Shamos, M.I.: Computational Geometry: An Introduction. Springer, Heidelberg (1985)
9. Chomicki, J., Godfrey, P., Gryz, J., Liang, D.: Skyline with presorting. In: ICDE, pp. 717–816 (2003)
10. Tan, K.L., Eng, P.K., Ooi, B.C.: Efficient progressive skyline computation. In: VLDB, pp. 301–310 (2001)
11. Kossmann, D., Ramsak, F., Rost, S.: Shooting stars in the sky: An online algorithm for skyline queries. In: VLDB, pp. 275–286 (2002)
12. Yuan, Y., Lin, X., Liu, Q., Wang, W., Yu, J.X., Zhang, Q.: Efficient computation of the skyline cube. In: VLDB, pp. 241–252 (2005)
13. Chan, C.Y., Eng, P.K., Tan, K.L.: Stratified computation of skylines with partially-ordered domains. In: SIGMOD Conference, pp. 203–214 (2005)
14. Pei, J., Jiang, B., Lin, X., Yuan, Y.: Probabilistic skylines on uncertain data. In: VLDB, pp. 15–26 (2007)
15. Random dataset generator for SKYLINE operator evaluation, http://randdataset.projects.postgresql.org/

Plot Query Processing with Wavelets*

Mehrdad Jahangiri and Cyrus Shahabi

Computer Science Department
University of Southern California
Los Angeles, California 90089-0781
{jahangir,shahabi}@usc.edu

Abstract. Plots are among the most important and widely used tools for scientific data analysis and visualization. With a plot (a.k.a. range group-by query) data are divided into a number of groups, and at each group, they are summarized over one or more attributes for a given arbitrary range. Wavelets, on the other hand, allow efficient computation of (individual) exact and approximate aggregations. With the current practice, to generate a plot over a wavelet-transformed dataset, one aggregate query is executed per each plot point; hence, for large plots (containing numerous points) a large number of aggregate queries are submitted to the database. On the contrary, we redefine a plot as a range group-by query and propose a wavelet-based technique that exploits I/O sharing across plot points to evaluate the plot efficiently and progressively. The intuition behind our approach comes from the fact that we can decompose a plot query into two sets of 1) aggregate queries, and 2) reconstruction queries. Subsequently, we exploit and extend our earlier related studies to effectively compute both quires in the wavelet domain. We also show that our technique is not only efficient as an exact algorithm but also very effective as an approximation method where either the query time or the storage space is limited.

1 Introduction

Spreadsheets allow us to easily perform complex data analysis on scientific datasets. However, they cannot operate efficiently on very large multidimensional datasets generated by the current data acquisition methods. Current science practice is to store the original data in databases or ftp sites and then manually generate a smaller subset of the data (by sampling, aggregating, or categorizing) as a new "data product". Yet, this time-consuming process suffers from one major drawback. We lose the detailed information and end-up working with the secondhand dataset. Hence, this may result in a biased study of the

---

* This research has been funded in part by NSF grants EEC-9529152 (IMSC ERC) and IIS-0238560 (PECASE), unrestricted cash gifts from Google and Microsoft, and partly funded by JPL SURP program and the Center of Excellence for Research and Academic Training on Interactive Smart Oilfield Technologies (CiSoft); CiSoft is a joint University of Southern California - Chevron initiative.

B. Ludäscher and Nikos Mamoulis (Eds.): SSDBM 2008, LNCS 5069, pp. 473–490, 2008.

data by verifying our known hypothesis rather than being surprised with unknown facts. To address these shortcomings, we are investigating how to enable spreadsheet-type functionalities on the original large datasets in databases.

One of the mostly exercised functionalities of spreadsheets is to generate meaningful plots over the data. Here, we redefine a plot as a database query (range group-by query) and progressively process it. A plot query summarizes how a fact changes over a set of attributes and is visually represented in various forms of charts. These graphs are considered among the most effective visual aids for statistical analysis methods and are widely used to provide valuable insights over any dataset. For example, one can extract outliers, trends, clusters, or measurements such as the gradient or the area under the curve by quickly looking at a plot output.

Approximate plots with a limited number of I/Os are often acceptable enough to assist us to intuitively understand the general behavior of the data. The valuable insight provided by these queries comes from the easy-to-visualize relationship among the plot points. Thus it is essential to preserve this relationship in approximate or progressive answering rather than conserving the accuracy of each individual plot point.

In addition, scientists desire to have the plot output in various resolutions from time to time. In one scenario, the graphic software at the application side may be limited to only a small number of plot points. In another scenario, scientists may be only interested in large scale changes (e.g., annual climate change). Finally, the fine resolution data may carry some noise that its mining renders useless. With any of these scenarios, it is necessary to compute the plot query for a coarser resolution and avoid retrieving unwanted details of the data.

We decompose a plot query into two sets of fundamental queries, aggregate query and reconstruction query. Given these components: aggregation and reconstruction, and emphasizing on the plot drawing requirements: multiresolution and approximation/progreessiveness, we propose to utilize wavelet transform to provide efficient plot query processing. The intuition behind our proposal comes from the fact that we have observed that aggregate queries can be efficiently evaluated in the wavelet domain and the original data can be equitably reconstructed from wavelet-transformed data. Using the proposed method, we provide high-quality approximate plot results independent of the data distribution with very little I/O and computational overhead by using the most important query wavelet coefficients. We further extend our algorithm to progressive query processing by ordering the retrieved data. Our experimental results show that the approximate results produced by our progressive framework are very accurate long before the exact plot query is complete (below 10% of retrieval).

We begin our discussion with reviewing the related work in Section 2. Then, we define the plot query as a single database query and process it using the state of the art method in Section 3. Next in Section 4, we overview the wavelet preliminaries that we use throughout the rest of the paper. In Section 5 we present our efficient algorithm to process plot queries with wavelets. We extend our query framework by providing approximate and progressive query processing in Section 6. We extensively examine our technique with real-world multidimensional datasets in Section 7. Finally, we conclude our discussion in Section 8.

2   Related Work

The current practice for generating a plot over a transformed dataset is to compute the plot point-by-point by performing an aggregate query per plot point, which results in submitting a large number of aggregate queries to the database for large plots. Each item of the plot is an aggregate value over one or more measure attributes for a given dimension value.

Extensive studies have been done for evaluation of single aggregate queries in exact [5,10], approximate [2,4,6,7,8,16,22], or progressive [9,13,14,17,18,23] fashion. However, none of these techniques addressed I/O sharing among a set of queries as these techniques are essentially designed for individual query processing. More importantly in the case of approximation, these techniques minimize the approximation error per individual queries rather than minimizing the total approximation error of the entire set.

Simultaneous evaluation of multiple queries has been addressed in [3,24]. However, their primary focus is on resource sharing among the queries by either creating materialized views or computing partial datacubes. In addition, they are not designed for the case when relations are stored using pre-aggregation or transformation techniques. Similar to the single query methods, these techniques do not provide a comprehensive plan for approximation of the entire set of queries.

Toward addressing these shortcomings, we introduced a framework for progressive answering of multiple aggregate queries in [18]. The focus of this study was to minimize the structural error across a batch of queries and to share I/O among them. However, deploying this technique often requires submitting a large number of queries, which is neither memory efficient (both in the client side and the server side) nor communication efficient, as compared to the method we propose for plot query processing. In addition, this technique was efficient when we have no extra information about the relationships among the queries inside the batch. However for processing the plot queries, we have this extra information in advance when we form the batch of queries. Intuitively, our proposed plot processing method exploits this extra knowledge for more efficient processing of the plot.

In this paper, we redefine plots as database queries instead of a set of individual aggregate queries. To the best of our knowledge no other work has defined plots as database queries. Here, we decompose a plot query into two sets of 1) aggregate queries, and 2) reconstruction queries. Subsequently, we effectively compute both in the wavelet domain by extending our earlier studies [11,13,19].

In [19], we proposed a new wavelet technique for fast exact, approximate, or progressive polynomial aggregate query processing that data did not have to be compressed, unlike most of the prior studies in this area. The use of the wavelet decomposition was justified by the fact that we could reduce the query cost from range size to the logarithm of the data size, which is a major benefit especially for large range queries. We further enhanced our former method in [13] by providing practical solutions for real-world deployment in various scientific applications. In addition, our progressive query processing method was fundamentally different as we ordered the query coefficients, not the data (see [13,19,20]).

In [13,19,20], we have extensively studied the progressiveness in aggregate query processing by defining a significancy function for query coefficients. Here, we briefly overview our former work and extend it to plot processing. Progressive method for plot queries differs from aggregate queries in that the plot queries are combined of two sets of queries: aggregate queries and reconstruction queries. In this paper, we show that reconstruction query requires a different ordering function and study the near optimal ordering for plot processing in general.

In [11], we introduced two novel operations for wavelet transformed data to provide general purpose functionalities and work directly in the wavelet domain. Unlike other reconstruction methods [2] which are limited to only non-standard multidimensional wavelet, the work of [11] is general enough to support both forms of wavelet transformations; specifically we study the standard form of multidimensional transformation, which is the same way we perform multidimensional transformation in this paper. Here, we utilize these operations in performing our reconstruction query to efficiently reconstruct a subset of data from its wavelet coefficients.

3   Plot Query

We often wish to generate ad-hoc plots over large multidimensional datasets to understand the relationship between different parts of the data. Toward this end, we select a certain region of data, called *range*, and divide it to different groups based on a subset of its attributes, called *grouping attributes*. Subsequently, we compute a value, called a *plot value*, per group and draw the chart of plot values versus grouping attributes.

Consider a dataset with $a_1,...,a_d$ as its dimension attributes and D as its measure attribute. Let the range for each dimension i ($i \leq d$) be $[l_i, h_i]$ and let the first g dimensions ($g \leq d$) be the grouping dimensions, without loss of generality. For each combination of grouping dimensions, we compute the plot value by aggregating the measure values inside the range. The *sum* and *average* are the most widely used aggregation in this regard. In this paper, we focus on *sum* due to its simplicity. Extension to other aggregations is straightforward as discussed in [19].

We define a plot query as a query that prepares the data in the form of a set of tuples $(a_1, ..., a_g, G)$ which we articulate as a plot of G versus $(a_1, ..., a_g)$. We denote $(a_1, ..., a_g)$ and G by *grouping attributes* and *plot value*, respectively. In a chart, the grouping attributes appear on the category-axis as the independent variables, while the plot values appear on the value-axis as the dependent variable. Usually the number of the grouping dimensions is limited to 3 because of human limitation in visualizing more than 3-dimensional spaces. However, we can conceptually go beyond this limitation and prepare higher dimensional plots.

In relational databases, we can compute a plot query by executing the following SQL statement. In this paper, we propose an efficient method to process such queries progressively using Wavelets.

$$SELECT \quad a_1, ..., a_g, SUM(D)$$
$$FROM \quad Data$$
$$WHERE \quad l_1 \leq a_1 \leq h_1$$
$$...$$
$$AND \quad l_d \leq a_d \leq h_d$$
$$GROUP \ BY \quad a_1, ..., a_g;$$

Now we mathematically define a plot query as following:

Definition 1. *Given a d-dimensional datacube D with its first g dimensions as the grouping dimensions, and a range $[l_i, h_i]$ for each dimension i, the plot query is defined as:*

$$\{(a_1, ..., a_g, G)|\forall i \leq g, l_i \leq a_i \leq h_i,$$
$$G(a_1, ..., a_g) = \sum_{l_{g+1} \leq a_{g+1} \leq h_{g+1}} \cdots \sum_{l_d \leq a_d \leq h_d} D(a_1, ..., a_d)\} \quad (1)$$

To simplify our notation, we denote the grouping dimensions by x, $x = (a_1, ..., a_g)$, and the rest of the dimensions by y, $y = (a_{g+1}, ..., a_d)$. Similarly, we use (x, y) instead of $(a_1, ..., a_g, a_{g+1}, ..., a_d)$ throughout the paper. Therefore, we simplify our query definition as following:

$$\{(x, G)|l_x \leq x \leq h_x, G(x) = \sum_{l_y \leq y \leq h_y} D(x, y)\} \quad (2)$$

The equation above states that $G(x)$ is the sum of all values inside the range $[l_y, h_y]$ for each x point inside the grouping dimension range $[l_x, h_x]$. Let us continue our discussion with an illustrative example.

Example 1. Figure 1a demonstrates a 2-dimensional datacube with *product* and *time* as the dimension attributes and *sales* as the measure attribute. Let *time*

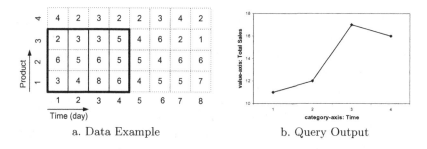

a. Data Example   b. Query Output

Fig. 1. Example of Plot Query

be the grouping dimension and *product* be the aggregating dimension. Now we would like to perform a plot query of sales versus time for $1 \leq time \leq 4$ and $1 \leq product \leq 3$.

To answer this query, we compute the daily total sales by performing a range aggregate query for each day. For example for $time = 2$, we have $sum(sales) = D(2,1) + D(2,2) + D(2,3) = 4 + 5 + 3 = 12$ (see Figure 1b). ❏

Lemma 1. *Given a d-dimensional data D with the domain size of N per dimension, a range size of M per dimension, and a set of g grouping dimensions, the I/O complexity of plot query processing is $O(M^d)$ if we naively perform a batch of aggregate queries over the original data.*

Proof. Our plot query consists of M^g range aggregate queries with the cost of $O(M^{d-g})$ for each when we perform the aggregate queries on the original data. Thus, the total complexity becomes $O(M^d)$.

Note that in the worst case where M becomes as large as N, the total complexity becomes $O(N^d)$ which is basically reading the entire database for a single plot query. Yet, this method suffers from two other major drawbacks in addition to its high I/O complexity. First, it does not provide any mechanism for approximate and/or progressive evaluation of the query. Second, it does not present a resolution-aware process to reduce the query complexity for coarser resolutions. It is straightforward to see that the complexity remains $O(M^d)$ for coarser resolutions since as the number of range aggregate queries reduces for coarser resolutions, the cost of each aggregate query increases proportionally.

Towards addressing these shortcoming, we utilize wavelets to process plot queries. Use of wavelets not only dramatically reduces the cost of aggregate queries but also addresses the multiresolution and approximation requirements. In Section 5, first we show how the use of wavelet transform reduces the complexity of each individual aggregate query. Then, we introduce our efficient algorithm in which we exploit the I/O sharing across the aggregations and show the excellent approximation of the plot in its entirety. But first we overview the preliminary concepts of Wavelet Transform that we use in the rest of the paper.

4   Discrete Wavelet Transform

Discrete Wavelet Transform (DWT) is defined as a series of pairwise operations on data by creating "rough" and "smooth" views of the data at different resolutions. In the case of Haar wavelets that we use throughout this paper, the "smooth" view consists of averages or summary coefficients, whereas the "rough" view consists of differences or detail coefficients. At each resolution, termed level of decomposition or scale, the summaries and details are constructed by pairwise averaging and differencing of the summaries of the previous level.

We denote by $u_{k,j}$ and $w_{k,j}$ the j-th summary and the j-th detail coefficient, respectively, for the k-th level of decomposition. Note that the j-th summary

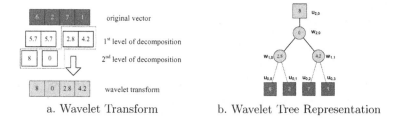

a. Wavelet Transform   b. Wavelet Tree Representation

Fig. 2. Discrete Wavelet Transform

coefficient of the 0-th level represents the untransformed data: $u_{0,j} = D(j)$, for $0 \leq j < N = 2^n$. Here, D represents the data of length N. We have:

$$u_{k,j} = (u_{k-1,2j} + u_{k-1,2j+1})/\sqrt{2} \qquad w_{k,j} = (u_{k-1,2j} - u_{k-1,2j+1})/\sqrt{2}$$

Example 2. Consider a vector of 4 values $D = \{6, 2, 7, 1\}$. To apply DWT on this vector, we first start by taking the pairwise summaries: $u_{1,0} = \frac{6+2}{\sqrt{2}} = 4\sqrt{2}$ and $u_{1,1} = \frac{7+1}{\sqrt{2}} = 4\sqrt{2}$, and the pairwise details $w_{1,0} = \frac{6-2}{\sqrt{2}} = 2\sqrt{2}$ and $w_{1,1} = \frac{7-1}{\sqrt{2}} = 3\sqrt{2}$. These coefficients form the first level of decomposition. We continue by constructing the summary and detail coefficients from the summaries. The final summary and all the details produced at all levels of decomposition form the Haar transform. Figure 2a illustrates the process. Notice that at each level of decomposition the summaries and details can be used to reconstruct the summaries of the previous level.   ❑

4.1   Wavelet Tree

We use the notion of *Wavelet Tree*, which exploits the relationships between wavelet coefficients, to simplify our presentation throughout the paper. Figure 2b shows a Haar wavelet tree for a vector of size 4; summary coefficients are shown with squares, whereas detail coefficients are shown in circles. The original data is drawn with dotted line as children of the leaf nodes of the tree. Haar wavelet tree is a binary tree where each node $w_{k,j}$ has exactly two children, $w_{k-1,2j}$ and $w_{k-1,2j+1}$. The summary coefficient $u_{n,0}$ is the root of the tree having only one child $w_{n,0}$. Wavelet tree portraits how a single point in time domain depends on those wavelet coefficients on the path to the root.

Lemma 2. *Let $\hat{D}$ be the wavelet transform of vector D of size N, $\hat{D} = DWT(D)$. Any value of D can be reconstructed using $O(\log N)$ coefficients from $\hat{D}$.*

Proof. Let $D[j]$ be the j-th value of D and n be $\log N$. At each level of decomposition k, there is exactly one wavelet coefficient $w_{k, \lfloor \frac{j}{2^k} \rfloor}$ that depends on $D[j]$. Thus, the total dependant coefficients are defined as $w_{k, \lfloor \frac{j}{2^k} \rfloor}$ for all $k \in [1, n]$ and $u_{n,0}$. This means that each data value can be reconstructed in time proportional to the tree height n.

Corollary 1. *Let $\hat{D}$ be the wavelet transform of vector D of size N. Any update on D results in updating $O(\log N)$ coefficients from $\hat{D}$.*

It is important to emphasize that we benefit from wavelet transform when we perform range aggregate queries.

Lemma 3. *Let $\hat{D}$ be the wavelet transform of vector D of size N. Any range-sum query on D can be computed by retrieving $O(\log N)$ coefficients from $\hat{D}$.*

Proof. For any range on D, we need to retrieve only the coefficients lying on the path from the boundaries of the range to the root of the wavelet tree. We refer the reader to [13,19] for the complete proof.

In this case for a range of size M, we retrieve $O(\log N)$ coefficients, independent of the range size, from $\hat{D}$. When $M \gg \log N$, the reduction in retrievals is significant.

Lemma 4. *Let $\hat{D}$ be the wavelet transform of vector D of size N. Any region of D with the size of M can be reconstructed using $O(M + \log \frac{N}{M})$ coefficients from $\hat{D}$.*

Proof. For any $D[j]$ in the range, we must retrieve the path from the leaf $D[j]$ to the root of the wavelet tree. By performing the union on all these paths, we have $O(M + \log \frac{N}{M})$ coefficients depend on the data values inside the range. We refer the reader to [11] for the complete proof.

4.2   Wavelet Matrix

Since the Wavelet Transform is a linear transformation, we can represent it by an $N \times N$ matrix W to transform the array D of length N as following. Here, λ represents the level of decomposition.

$$\hat{D}(i) = \sum_j W^\lambda(i,j)D(j)$$

For example, the table below shows the corresponding matrices for one level of transformation W^1 and two levels of transformation W^2. We omit the superscript λ when we refer to the maximum level of decomposition, that is, $\lambda = \log N$. We refer the reader to [15] for more details about creating these matrices.

Table 1. Wavelet Matrix

$$\begin{pmatrix} \frac{1}{\sqrt{2}} & \frac{1}{\sqrt{2}} & 0 & 0 \\ \frac{1}{\sqrt{2}} & -\frac{1}{\sqrt{2}} & 0 & 0 \\ 0 & 0 & \frac{1}{\sqrt{2}} & \frac{1}{\sqrt{2}} \\ 0 & 0 & \frac{1}{\sqrt{2}} & -\frac{1}{\sqrt{2}} \end{pmatrix} \qquad \begin{pmatrix} \frac{1}{2} & \frac{1}{2} & \frac{1}{2} & \frac{1}{2} \\ \frac{1}{\sqrt{2}} & -\frac{1}{\sqrt{2}} & 0 & 0 \\ \frac{1}{2} & \frac{1}{2} & -\frac{1}{2} & -\frac{1}{2} \\ 0 & 0 & \frac{1}{\sqrt{2}} & -\frac{1}{\sqrt{2}} \end{pmatrix}$$

$$W^1 \qquad\qquad\qquad W^2$$

Example 3. Consider the same vector of 4 values $D = \{6, 2, 7, 1\}$. We perform the first level of decomposition on this vector using the transformation matrix W^1 as : $\hat{D} = W^1 \cdot D = \{4\sqrt{2}, 2\sqrt{2}, 4\sqrt{2}, 3\sqrt{2}\}$. The second level of transformation is performed similarly by using W^2: $\hat{D} = W^2 \cdot D = \{8, 2\sqrt{2}, 0, 3\sqrt{2}\}$ ❏

4.3 Multidimensional Wavelets

The standard form of multidimensional wavelet transform is performed by applying a series of one-dimensional decompositions along each dimension. For example, to decompose a 2-dimensional array of size N^2, we first completely decompose one dimension and then the other, with the order not being important. This means that we first transform each of the N rows of the array to construct a new array and then take each of the N columns of the new array and again perform 1-d DWT on them. The final array is the 2-dimensional standard transform of the original array.

To represent a d-dimensional wavelet transformed data, we use d 1-dimensional wavelet trees. In fact, since each dimension is decomposed independently in multi-dimensional wavelet transform, therefore, there cannot be a single tree capturing the levels of decomposition. Every coefficient in the transformed data has d indices, one for each dimension. Each of these indices identifies a position in the 1-dimensional wavelet tree.

Theorem 1. *Given a d-dimensional wavelet-transformed datacube $\hat{D}$ with the domain size of N per dimension, the complexity of range aggregate query processing with wavelets is $O(\log^d N)$.*

Theorem 2. *Given a d-dimensional wavelet-transformed datacube $\hat{D}$ with the domain size of N and the range size of M per dimension, the complexity of data reconstruction from its wavelet transformed is $O((M + \log \frac{N}{M})^d)$.*

We refer the reader to [11,13] for the proof of these theorems.

Using the matrix notation, we represent the multidimensional transformation of a given multidimensional array D as following where W_x and W_y are the transformations along the dimensions x and y, respectively.

$$\hat{D}(x, y) = \sum_{s,t} W_x(x, s) W_y(y, t) D(s, t) \quad \Leftrightarrow \quad \hat{D} = W_x W_y D$$

As a rule of thumb throughout the paper, we use $\hat{D}$ when we refer to the full transformation of D along all its dimensions, whereas we denote the transformation of D only along y dimensions by $W_y D$.

5 Efficient Plot Query Processing with Wavelets

In this section we present our algorithm for efficient processing of plot queries. First, we employ the wavelet transform to reduce the cost of computing each plot value. Next, we propose our novel method in which we introduce a framework to share the coefficients across all the plot points.

Lemma 5. *Given a d-dimensional wavelet-transformed data $\hat{D}$ with the domain size of N per dimension, the range size of M per dimension, and the g grouping dimensions, the I/O complexity of plot query processing is $O(M^g \log^d N)$ if we perform a batch of aggregate queries using wavelets.*

Proof. The plot query consists of M^g range aggregate queries. Theorem 1 shows that the cost of each aggregate query decreases from M^{d-g} to $O(\log^d N)$ using wavelets. Thus, the total I/O complexity becomes $O(M^g \log^d N)$.

Despite the significant improvement compared to the naive method, this utilization of wavelet transform still suffers from the fact that it treats the plot as a set of aggregate queries. Thus, it does not share the common coefficients among the queries which results in several passes over data. In addition, it cannot approximate the plot in its entirety; instead, it approximates each aggregated value separately which may not necessarily lead to the best approximation of the plot.

To address these issues, we introduce an efficient algorithm in which we can process a plot query as a single query. We divide this process into two steps, aggregation and reconstruction, and describe each in turn. The aggregation phase deals with preparing the aggregated values for each plot point in the wavelet domain, whereas the reconstruction phase deals with converting these values back to the original domain.

5.1   Aggregation Phase

In this phase, we show how we recast a plot query as vector queries in the wavelet domain for its efficient processing. Let us simplify the plot equation by defining an aggregate query vector as following:

Definition 2. *The aggregate query vector consists of a set of 1's inside the range and 0's outside the range:*

$$Q(y) = \begin{cases} 1 \text{ if } l_y \leq y \leq h_y; \\ 0 \text{ otherwise.} \end{cases}$$

Now we rewrite the basic definition of the plot query (Eq. 2) as following:

$$\{(x,G)|l_x \leq x \leq h_x, G(x) = \sum_y D(x,y) \cdot Q(y)\}$$

This equation can be considered as a dot product of two vectors: the x column of D (noted as D_x) and Q, the data vector and the aggregate query vector, respectively. We denote the x column of D by D_x and rewrite the equation as following:

$$y(x) = \sum_z D_x(y) \cdot Q(y) \tag{3}$$

We wavelet transform both vectors of D_x and Q and utilize the following useful lemma:

Lemma 6. *Given a wavelet-transformed data vector $\hat{D}_x$ and a wavelet-transformed query vector $\hat{Q}$, we compute the plot values as following:*

$$G(x) = \sum_y \hat{D}_x(y) \cdot \hat{Q}(y) \tag{4}$$

Proof. It is proven that Discrete Wavelet Transform preserves the Euclidean norm. Thus, the generalized Parseval equality applies to DWT, that is, the dot product of two vectors equals to the dot product of the wavelet-transformation of the vectors (see [19] for more information).

Example 4. Consider the same data and query described in Example 1. Now, we would like to compute the plot values using Wavelets.

Figure 3 illustrates the process of computing $G(2)$ both in the original domain and in the wavelet domain. We select the D_x for $x = 2$ as the data vector and wavelet transform it to have $\hat{D}_x$. Then, we form the aggregate query vector Q with 1's inside the range R_y and 0's outside the range and wavelet transform Q to have $\hat{Q}$. Subsequently, we perform a dot product between $\hat{D}_x$ and $\hat{Q}$ to compute $G(x)$. ❑

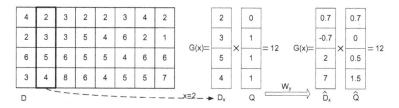

Fig. 3. Aggregation in the Wavelet Domain

Toward computing the plot values using Wavelets, we must be able to efficiently wavelet transform the query vector and the data vector. For the transformed query, we employ our efficient wavelet transformation algorithm (see [13]) to transform the vector by computing the coefficients only on its boundaries. For the transformed data vector, we select the x column of $W_y D$ which represents the transformation of D along dimension y since we have $\hat{D}_x(y) = W_y D(x, y)$ by the definition of standard multidimensional wavelet transformation (see Section 4). Thus, we have:

$$G(x) = \sum_y W_y D(x, y) \cdot \hat{Q}(y) \tag{5}$$

However, y dimension is selected on-the-fly (i.e. at the query submission) and we cannot pre-compute the data transformed along y dimension in advance. The following lemma, however, provides the opportunity of constructing $W_y D$ from the transformed data $\hat{D}$.

Lemma 7. *Given a wavelet-transformed datacube $\hat{D}$ and the set of dimensions y, the data transformed along y is computed by inverse transforming the data along other dimensions x:*

$$W_y D(x, y) = \sum_\alpha W_x^{-1}(x, \alpha) \cdot \hat{D}(\alpha, y) \tag{6}$$

Proof. Let the data D have two sets of dimensions x and y. Therefore, its wavelet transformation is defined as $\hat{D} = W_x W_y D$. By performing an inverse transformation along x on the both sides, we have: $W_y D = W_x^{-1} \hat{D}$.

5.2   Reconstruction Phase

Let us overview the process so far. First, we compute $W_y D$ by inverse transforming the $\hat{D}$ along x dimension. Then, after preparing $\hat{Q}$ on-the-fly, we perform a dot product between $\hat{Q}$ and the x column of $W_y D$ for each plot value $G(x)$. However, this process is not efficient yet since we must perform the costly operation of $W_x^{-1} \hat{D}$ at first and store a temporary large datacube $W_y D$. Toward addressing this inefficiency, now we propose to perform the aggregation before the inverse transformation to reduce the overall cost. In fact, we push the aggregation down to the wavelet domain and reconstruct the result from the wavelet-transformed temporary datacube. For this purpose, let us substitute Equation 6 into Equation 5 and interchange the linear operations as following:

$$G(x) = \sum_y (\sum_\alpha W_x^{-1}(x, \alpha) \hat{D}(\alpha, y)) \hat{Q}(y)$$
$$= \sum_\alpha W_x^{-1}(x, \alpha) (\sum_y \hat{D}(\alpha, y) \hat{Q}(y))$$

The equation above shows that we aggregate the data long y first, then we reconstruct the value by inverse transforming along x. We denote the second summation by $\hat{D}_G$ as it carries the G values of our plot query. In fact, $\hat{D}_G$ is the transformation of the aggregated data along y.

The following lemma summarizes the process of plot processing. In short, it states that $G(x)$ is computed by performing an inverse transform on $\hat{D}_G$ for all the plot points.

Lemma 8. *Given a wavelet-transformed datacube $\hat{D}$ and a wavelet-transformed query vector $\hat{Q}$ as the aggregation along y dimension, we compute the plot values with the following steps:*

$$\text{Step 1 (Aggregation)}: \quad \hat{D}_G(x) = \sum_x \hat{D}(x, y) \hat{Q}(y) \tag{7}$$

$$\text{Step 2 (Reconsutrction)}: \quad G(x) = \sum_y W_x^{-1}(x, y) \hat{D}_G(y) \tag{8}$$

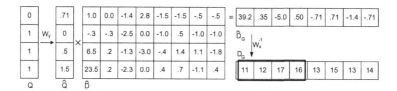

Fig. 4. Plot Query with Wavelets

Example 5. Figure 4 demonstrates the process of constructing $\hat{D}_G$ from $\hat{D}$. Here, every x element of $\hat{D}_G$ is computed by a dot product between $\hat{Q}$ and $\hat{D}_x$. Finally, we perform an inverse wavelet transform on $\hat{D}_G$ to compute D_G. The highlighted subset of D_G refers to our plot values G.   ❑

To conclude this section, we analyze the complexity of our algorithm by providing the following theorem.

Theorem 3. *Given a d-dimensional wavelet-transformed data $\hat{D}$ with the domain size of N per dimension, the range size of M per dimension, and the g grouping dimensions, the I/O complexity of plot query processing is:*

$$O((M + \log \frac{N}{M})^g \cdot \log^{d-g} N)$$

Proof. The aggregation complexity is $O(\log^{d-g} N)$ based on Theorem 1 for a $(d-g)$-dimensional range and the reconstruction complexity is $O((M+\log \frac{N}{M})^g)$ based on Theorem 2. Multiplying the two, we compute the overall complexity for plot query processing.

6   Approximation and Progressiveness

When execution time is limited, the accuracy of the plot result can be traded off for a better response time, that is, a fast less accurate result become preferred to an exact late result. Since the dominant factor for query processing is the database retrieval, we limit the retrievals to a certain number B, that is, we only retrieve the B most significant wavelet coefficients contributing to the query. Here, we adopt the two widely used methods, First-B and Highest-B, for selecting the B most significant wavelet coefficients. Using First-B, the most significant coefficients are the coefficients with the lowest frequencies. With Highest-B, the most significant coefficients are those that have the highest absolute values.

Let us recall that the process of plot query processing has two phases: 1) Aggregation phase $\hat{D} \rightarrow \hat{D}_G$, 2) Reconstruction phase $\hat{D}_G \rightarrow G$. Therefore, we can approximate either or both of these steps to approximate the plot output.

First, we intend to approximate the reconstruction process, that is, we need to select the best coefficients of $\hat{D}_G$ for reconstruction of G given a limited number of retrievals. The following example clarifies our purpose.

Example 6. Consider the wavelet data $\hat{D}_G$ illustrated in Figure 4. When the query is limited to retrieve only 2 coefficients, it is recommended to retrieve either of the following sets: $\{39.2, 0.35\}$ if we consider the First-B ordering (the least frequencies) or $\{39.2, -5.0\}$ if we consider the highest-B ordering (the highest absolute values).   ❏

Unfortunately, we cannot utilize the Highest-B method of the process mentioned above in practice because this requires knowing all the values of $\hat{D}_G$ in advance to determine the highest values. Therefore, we can only utilize the First-B method in this step, that is, we select the coefficients with the lowest frequencies.

In addition to the reconstruction phase, we can approximate the aggregated intermediate result $\hat{D}_G$ by selecting the most contributing coefficients, that is, the ones with the highest values of the pair of query $\hat{Q}$ and data $\hat{D}$ items (see Equation 7). However, the values of $\hat{D}$ are not known in advance and cannot be utilized for this process. Therefore, we advocate selecting the query coefficients at the query time to achieve good approximate results. Toward this end, the B most significant query coefficients are selected using Highest-B or First-B methods. We refer the reader to [20] for more information regarding various techniques used in wavelet query approximation.

Having the ability to approximate at both phases, aggregation and reconstruction, we are faced with this dilemma: either to compute the aggregated result in exact and then perform the approximate reconstruction phase, or to approximate both phases of aggregation and reconstruction together. The result of our empirical study shows that the latter outperforms the first one. We discuss this later in the experimental section.

By progressively increasing the term B, we can *order* coefficients based on their significants. We exploit this ordering to answer the plot query in a progressive manner so that each step produces a more precise evaluation of the actual answer. In fact, the real-world users of our technique have found its progressiveness the most appealing feature for processing large plot queries.

Let us end this section by emphasizing that we have studied the "query approximation" here. Adopting "data approximation" (use of compressed data) is straightforward. More specifically, we can compress the data with any of the two orderings, Highest-B or First-B. At the query time if the data coefficient is not stored, i.e. dropped previously due to the data compression, we assume this wavelet data coefficient is zero and continue the process. This assumption is basically the implementation of hard thresholding (see [13] for more information).

7   Experiments

In this section, we empirically examine our proposed method with three multidimensional datasets. We would like to emphasize that our experiments are performed on real datasets using our fully functional system (see [1,12] for further information).

We start the experiments by describing the datasets employed in our study. Next, we compare the query performance of our technique (Plot Query) with the individual queries in both original domain (Naive Batch Queries) and the wavelet domain (Batch Queries with Wavelets). Finally, we study the progressiveness and compare the different forms of approximation.

7.1 Experimental Datasets

We evaluate our framework with three real-world scientific datasets, namely *Precipitation*, *GPS*, and *AIRS*.

Precipitation is a 4-dimensional dataset that measures the daily precipitation for the Pacific NorthWest for 45 years. It consists of three dimension attributes, latitude, longitude, and time, and one measure attribute, precipitation. The size of this dataset is 5 MB.

GPS dataset contains profiles of atmospheric water vapor pressure with resolution of about a kilometer, derived from radio occultation data. This 5-dimensional dataset is provided by NASA/JPL and includes latitude, longitude, pressure level, and time as dimension attributes, and water vapor pressure as measure attribute. We obtained this data for 9 months and its size is 2 GB.

AIRS, standing for Atmospheric Infrared Sounder, collects the Earths atmospheric temperature profiles at a very high rate. This 5-dimensional dataset provided by NASA/JPL includes latitude, longitude, pressure level, and time as dimension attributes, and temperature as measure attributes. This data is gathered over a year and has a size of 320 GB.

We wavelet transformed the datacubes using our efficient transformation technique [11] and stored them into the disk using our efficient multidimensional tiling [21]. Each tile contains the wavelet coefficients that are related with each other under the particular access pattern of wavelets to minimize the number of disk I/Os needed to perform any operation in the wavelet domain. By reporting the number of retrieved "coefficients" in our experiments, we do not include the advantage of using this technique.

7.2 Performance Analysis

We generate 100 random plot queries (a random range for each query) and count the number of disk I/Os required to answer each query. We perform this experiment on our three datasets and used the three algorithm discussed in this paper; Naive Batch Queries, Batch Queries with Wavelets, and Plot Query (our proposed technique). The average number of I/Os across the queries is depicted in Figure 5.

Generally, Batch Queries with Wavelets outperforms Naive Batch Queries because we perform each aggregate queries in a less costly method ($O(N^d)$ is reduced to $O(\log^d N)$). More importantly, this figure shows that Plot Query dramatically outperforms both. Note that Y-axis is in logarithmic scale. The reason is that Plot Query is a one-pass algorithm which shares the coefficient among plot points whereas Batch Queries with Wavelets requires submitting a large set of individual aggregate queries.

Now, we study the effect of range size on the performance of plot queries. We generate 10 plot queries for varied range sizes, from 1% to 35% of the entire

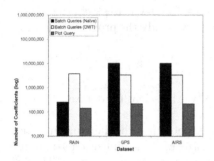

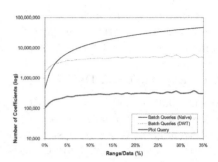

Fig. 5. Performance Analysis of Plot Processing

Fig. 6. Effect of Range Size

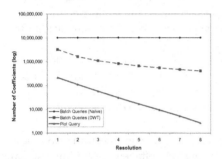

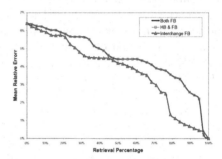

Fig. 7. Query Plots with different Resolutions

Fig. 8. Progressiveness

dataset. The result of our experiment is shown in Figure 6, the median of the 10 queries is reported for each point. Note that for this set of experiments and the following experiments, we report our results only on AIRS dataset because the trends and observations from other datasets were similar to those of AIRS.

Here, as the range size grows, the number of coefficients increases for Naive Batch Queries as we expected since its complexity was $O(M^d)$. However, the number of coefficients for the other two techniques, Batch Queries with Wavelets and Plot Query, are almost constant as the range grows. This is discussed earlier when we justified the use of wavelets in aggregate queries. We discussed that the complexity of range aggregate queries with wavelets is $O(\log^d N)$ which is independent of the range size. In addition, this figure shows the significant difference between the two wavelet methods, one more time this is due to the I/O sharing in the Plot Query method. Note that Y-axis is in logarithmic scale in this figure.

Next, we compare the three methods in dealing with multi-resolution parameter. We generate 100 random plot queries for each resolution level. A higher resolution refers to a coarser view of data. We show the average number of I/Os required to answer these queries versus resolution in Figure 7.

With Naive Batch Queries, generating a plot for a coarser view has the exact same cost as we generate it for the finer resolutions. This is because for coarser views, we submit less number of aggregate queries while the complexity to answer each query is higher. As the resolution grows (coarser views), Batch Query with Wavelets performs better. This is due to the fact that the number of range aggregate queries is reduced for coarser views. However, the steep of cost reduction is not as deep as the Plot Query method. The reason is that no matter what the resolution is, the cost of aggregate queries for Batch Query with Wavelets is $O(\log^N)$. On the contrary, Plot Query is a resolution-aware method which only retrieves the data up to the required level. The significant difference between the query processing methods is also illustrated in this figure. Note that Y-axis is in logarithmic scale.

We conclude our experiments with studying the progressiveness of our algorithm. We generate 100 random plot queries on AIRS and report the mean relative error on the values of the plots in its entirety. Here, we use three ordering schema. 1) Both F_B: First-B ordering for the reconstruction phase then First-B ordering for the aggregation phases, 2) H_B and F_B: Highest-B for the aggregation phase and First-B for the reconstruction phase, and 3) Interchangeable F_B's: First-B ordering for both reconstruction and aggregation at the same time.

Figure 8 shows that the first two algorithms perform similarly and both of them are inferior compared to the winning ordering which is the interchangeable F_B. The reason behind our observation is that the first two algorithms compute each coefficient of the intermediate datacube D_G exactly and then they move to computing the second coefficient. However, Interchangeable F_B estimates all coefficients of D_G of at the same time.

8   Conclusion

We have defined a plot as a single database query and propose a wavelet-based technique that exploits I/O sharing across plot points to evaluate the plot efficiently. Furthermore, we have extended our algorithm to progressive query processing by ordering the retrieval procedure. Our experimental results show that the approximate results produced by our progressive framework are very accurate long before the exact plot query is complete.

References

1. ProDA, http://infolab.usc.edu/projects/proda/
2. Chakrabarti, K., Garofalakis, M.N., Rastogi, R., Shim, K.: Approximate query processing using wavelets. In: Proc. of VLDB, pp. 111–122 (2000)
3. Chen, Z., Narasayya, V.: Efficient computation of multiple group by queries. In: SIGMOD 2005, pp. 263–274. ACM Press, New York (2005)
4. Garofalakis, M., Gibbons, P.B.: Wavelet synopses with error guarantees. In: Proc. of ACM SIGMOD (2002)

5. Geffner, S., Agrawal, D., Abbadi, A.E., Smith, T.: Relative prefix sums: An efficient approach for querying dynamic OLAP data cubes. In: Proc. of ICDE (1999)
6. Gibbons, P.B., Matias, Y.: New sampling-based summary statistics for improving approximate query answers. In: Proc. of SIGMOD, pp. 331–342 (1998)
7. Gilbert, A.C., Kotidis, Y., Muthukrishnan, S., Strauss, M.J.: Optimal and approximate computation of summary statistics for range aggregates. In: Proc. of PODS (2001)
8. Gunopulos, D., Kollios, G., Tsotras, V.J., Domeniconi, C.: Approximating multidimensional aggregate range queries over real attributes. In: Proc. of SIGMOD (2000)
9. Hellerstein, J.M., Haas, P.J., Wang, H.: Online aggregation. In: Proc. of SIGMOD, pp. 171–182. ACM Press, New York (1997)
10. Ho, C., Agrawal, R., Megiddo, N., Srikant, R.: Range queries in OLAP data cubes. In: Proc. of SIGMOD
11. Jahangiri, M., Sacharidis, D., Shahabi, C.: Shift-Split: I/O Efficient Maintenance of Wavelet-Transformed Multidimensional Data. In: Proc. of SIGMOD (2005)
12. Jahangiri, M., Shahabi, C.: ProDA: A Suite of WebServices for Progressive Data Analysis. In: Proc. of ACM SIGMOD (demonstration) (2005)
13. Jahangiri, M., Shahabi, C.: Wolap: Wavelet-based range aggregate query processing. In: Department of Computer Science Technical Reports. USC (2007)
14. Lazaridis, I., Mehrotra, S.: Progressive approximate aggregate queries with a multiresolution tree structure. In: Proc. of SIGMOD, pp. 401–412 (2001)
15. Nievergelt, Y.: Wavelets Made Easy. Springer, Heidelberg (1999)
16. Poosala, V., Ganti, V.: Fast approximate answers to aggregate queries on a data cube. In: Proc. of SSDBM, pp. 24–33. IEEE Computer Society, Los Alamitos (1999)
17. Riedewald, M., Agrawal, D., Abbadi, A.E.: pCube: Update-efficient online aggregation with progressive feedback. In: Proc. of SSDBM, pp. 95–108 (2000)
18. Schmidt, R., Shahabi, C.: How to evaluate multiple range-sum queries progressively. In: Proc. of ACM PODS, pp. 3–5 (2002)
19. Schmidt, R., Shahabi, C.: Propolyne: A fast wavelet based technique for progressive evaluation of polynomial range-sum queries. In: Jensen, C.S., Jeffery, K.G., Pokorný, J., Šaltenis, S., Bertino, E., Böhm, K., Jarke, M. (eds.) EDBT 2002. LNCS, vol. 2287. Springer, Heidelberg (2002)
20. Shahabi, C., Jahangiri, M., Sacharidis, D.: Hybrid Query and Data Ordering for Fast and Progressive Range-Aggregate Query Answering. International Journal of Data Warehousing and Mining 1(2), 49–69 (2005)
21. Shahabi, C., Schmidt, R.: Wavelet disk placement for efficient querying of large multidimensional data sets. Technical Reports, USC (2004)
22. Vitter, J.S., Wang, M.: Approximate computation of multidimensional aggregates of sparse data using wavelets. In: Proc. of SIGMOD, pp. 193–204 (1999)
23. Wu, Y.-L., Agrawal, D., Abbadi, A.E.: Using wavelet decomposition to support progressive and approximate range-sum queries over data cubes. In: Proc. of CIKM, pp. 414–421 (2000)
24. Zhao, Y., Deshpande, P.M., Naughton, J.F., Shukla, A.: Simultaneous optimization and evaluation of multiple dimensional queries. In: Proc. of SIGMOD, pp. 271–282. ACM Press, New York (1998)

Quality-Aware Probing of Uncertain Data with Resource Constraints

Jinchuan Chen and Reynold Cheng

Department of Computing,
Hong Kong Polytechnic University,
Hung Hom, Kowloon, Hong Kong
{csjcchen,csckcheng}@comp.polyu.edu.hk

Abstract. In applications like sensor network monitoring and location-based services, due to limited network bandwidth and battery power, a system cannot always acquire accurate and fresh data from the external environment. To capture data errors in these environments, recent researches have proposed to model uncertainty as a probability distribution function (pdf), as well as the notion of probabilistic queries, which provide statistical guarantees on answer correctness. In this paper, we present an entropy-based metric to quantify the degree of ambiguity of probabilistic query answers due to data uncertainty. Based on this metric, we develop a new method to improve the query answer quality. The main idea of this method is to acquire (or probe) data from a selected set of sensing devices, in order to reduce data uncertainty and improve the quality of a query answer. Given that a query is assigned a limited number of probing resources, we investigate how the quality of a query answer can attain an optimal improvement. To improve the efficiency of our solution, we further present heuristics which achieve near-to-optimal quality improvement. We generalize our solution to handle multiple queries. An experimental simulation over a realistic dataset is performed to validate our approaches.

1 Introduction

In many emerging and important applications like wireless sensor networks and location-based applications, the data obtained from the sensing devices are often imprecise [10,17,18]. Consider a monitoring application that employs a sensor network to obtain readings from external environments. Due to imperfection of physical devices, as well as limited battery power and network delay, it is often infeasible to obtain accurate readings. As a result, the data maintained in the monitoring applications are often contaminated with noises (e.g., sampling and measurement error). The *uncertainty* of these data should be modeled and handled carefully, or else the quality of the services or queries provided to users can be affected [4,10].

One commonly-used uncertainty model assumes that the exact value of a data item is located within a closed region, together with a probability distribution

B. Ludäscher and Nikos Mamoulis (Eds.): SSDBM 2008, LNCS 5069, pp. 491–508, 2008.

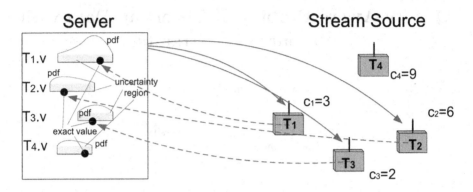

Fig. 1. Probing of Sensor Data for Uncertainty Reduction

function (*pdf*) of that value in the region [10,13,21]. An example is shown in Figure 1, where a monitoring server maintains the pdf of the temperature values acquired from four wireless sensors $(T_1, \ldots, T_4)$. Each of these pdf's is confined within a closed range of possible values, to model the fact that the data has not been updated for an extensive amount of time. These pdf's could be derived through techniques like time-series or model-based analysis [5,10]. In general, a pdf which spans over a larger uncertainty region is more vague (or uncertain) than the one with a smaller region.

To process uncertain data, *probabilistic queries* [4,21] have been proposed. These are the "probabilistic" counterparts of spatial queries, such as range queries and nearest neighbor queries. Probabilistic queries produce *imprecise* results, which are essentially answers that are augmented with probability values to indicate the likelihood of their occurrences. For example, a probabilistic range query, inquiring which of the four sensor data values in Figure 1 have non-zero probabilities of being inside a specified range $[10^\circ C, 20^\circ C]$, may produce an answer like: $\{(T_1, 0.9), (T_2, 0.5)\}$. This answer indicates that T_1 (T_2) has a chance of 0.9(respectively 0.5) for having a value between $[10^\circ C, 20^\circ C]$.

How can we interpret the probability values of these query answers? Intuitively, these values reflect the *ambiguity* of a query result, due to the impreciseness of the data being evaluated. In the previous example, since T_1 has a chance of 0.9 for satisfying the query, we know that T_1 is very likely to be located inside $[10^\circ C, 20^\circ C]$. The case of T_2 is more vague: it could either be inside or outside the specified range, with equal probabilities. In general, a query answer may consists of numerous probability values, making it hard for a query user to interpret the likelihood of their answers. A *quality metric* is desired, which computes a real-valued score for a probabilistic query answer [4,15]. This metric serves as a convenient indicator for the user to understand how vague his/her answer is, without the need of interpreting all the probabilities present in the answer. For example, if the score of his/her query answer is high, the user can immediately understand that the quality of his/her answer is good. In this paper, we define a quality score for a probabilistic range query based on the definition of entropy [20]. This metric quantifies the degree of query answer uncertainty by measuring the amount of information presented in a query.

More importantly, the quality score definition enables us to address the question: "how can the quality of my query answer be improved?" Let us consider the sensor network example in Figure 1 again. Suppose that the sensors have not reported their values for a long time. As a result, the sensor data kept in the server have a large degree of uncertainty. Consequently, the query answer quality is low (i.e., the query answers are vague), and a user may request the server to give him/her an answer with a higher quality. To satisfy the user's request, the system can acquire (or **probe**) the current data from the sensors, in order to obtain more precise information (i.e., possibly with a smaller uncertainty interval). A higher quality score for the query user's answer can then potentially be attained. In fact, if all the items $(T_1, \ldots, T_4)$ are probed, then the server will have up-to-date knowledge about the external world, thereby achieving the highest query quality.

In reality, it is unlikely that a system can always maintain an accurate state of the external environment, since probing a data item requires precious resources (e.g., network bandwidth and energy). It is thus not possible for the system to probe the data from all the sources in order to improve the quality of a query request. A more feasible assumption is that the system assigns to the user a certain amount of "resource budget", which limits the maximum amount of resources invested for a particular query. The question then becomes "how can the quality of a probabilistic query be maximized with probing under tight resource constraints?" To illustrate, let us consider Figure 1, where $c_1, \ldots, c_4$ are the respective costs for probing $T_1, \ldots, T_4$. The cost value of each sensor may represent the number of hops required to receive a data value from the sensor. Let us also assume that a query is associated with a resource budget of 8 units. If we want to improve the quality for this query, there are five *probing sets*, namely $\{T_1\}$, $\{T_2\}$, $\{T_3\}$, $\{T_1, T_2\}$ and $\{T_2, T_3\}$. Each of these sets describe the identities of the sensors to be probed. Moreover, the total sum of their probing costs is less than 8 units. Now, suppose the probing of T_2 and T_3 will yield the highest quality improvement. Then the system only needs to probe these two sensors, to ensure the maximum benefit.

Since testing the possible candidates in a brute-force manner requires an exponential-time complexity, we propose a polynomial-time solution based on dynamic programming. We also present a greedy solution to enhance scalability. Our experimental results show that the greedy solution achieves almost the same quality as the dynamic-programming solution. We study this problem for probabilistic range queries, which return the items within a user-defined region. This query is one of the most important queries commonly found in location-based services and sensor applications. Our solution can generally be applied to any multi-dimensional uncertain data, where the pdf's are arbitrary.

The problem studied in this paper addresses the balance between query quality and the amount of system resources consumed. A few related problems have been studied in [10,16], where probing plans are used to direct the server to acquire the least number of data items required to achieve the highest quality. However, these work do not consider the issue of maximizing quality under limited system resources allocated to a user. We further consider the scenario in which a group of query users share the same resource budget. This represents the case when a system allocates its resources to users with the same priority. We explain how our basic solution (tailored for a single query) can be extended to address this. To our understanding, this has not been studied before.

To summarize, our major contributions are:

1. We propose an entropy-based quality metric for probabilistic range queries.
2. We develop optimal and approximate solutions that maximize the quality of a probabilistic query under limited resource constraints.
3. We extend our solution to handle the case where multiple query users share the same resource budget.
4. We conduct extensive experiments with realistic datasets to validate the performance of our algorithms.

The rest of this paper is organized as follows. In Section 2, we present the related work. Section 3 illustrates the system architecture. We discuss the details of quality and resource budget for probabilistic range queries in Section 4. Then we give our solutions in Section 5. We report our experimental results in Section 6. Section 7 concludes the paper.

2   Related Work

In this section, we summarize the work done in probing and evaluation of probabilistic queries.

Probing Plans. In applications like sensor network monitoring, it is important for a system to generate a probing plan that only requests relevant sources to report their data values, in order to optimize the use of resources. In [19], efficient algorithms are derived to fetch remote data items in order to generate a satisfactory result quickly. Liu et al. [16] propose an optimal algorithm to find the exact result for minimum and maximum queries by probing the smallest set of data sources. The uncertainty model of a data item considered in these two work is simply a one-dimensional interval. Since the pdf of the value within the interval is not considered, the query results are "qualitative", i.e. they are not be augmented with probabilistic guarantees. Our paper, on the other hand, defines a quality metric for probabilistic query answers, and use this measure to devise probing plans. Although Madden et al. [8,10,11] consider the pdf of data values in their uncertainty models, their methods do not consider the strict resource constraints imposed on the system (e.g., the maximum amount of resources that can be spent on a query). The quality metric they consider is based on a simple probability threshold (e.g., the probability of the object should be higher than 95%). Our work proposes a feasible probing plan that achieves the highest quality under limited resource constraints. Our solutions can be applied to multi-dimensional uncertain data with artibtrary pdfs. We also use the amount of information gain (i.e. entropy) as the quality metric in our probing solutions, and this has not been studied before.

Probabilistic Queries. There are plenty of recent studies about efficient evaluation of probabilistic queries for large uncertain databases. In [3,4,5], efficient algorithms of evaluating probabilistic nearest-neighbor queries are proposed, which evaluate uncertain location data and provide probabilistic guarantees in answers. In [1], efficient methods for evaluating probabilistic location-dependent queries are studied. Indexing of probabilistic range queries is considered in [7], and the solution is extended to handle multi-dimensional uncertainty in [6,22]. The evaluation of probabilistic queries in sensor networks is considered in [2,10,14]. In this paper, we illustrate our probing techniques by using the probabilistic query evaluation methods in [4]. However, other advanced query evaluation or indexing techniques can also be used together with our probing algorithm.

3   System Architecture

Figure 2 describes the architecture of the system used in this paper. The *Data Manager* caches the value ranges and corresponding pdf of remote sensors. The *Query Register* receives queries from the users. The *Query Evaluator* evaluates the queries based on the information stored in the *Data Manager*. The *Probing Scheduler* is responsible for generating a *probing set* for each query – essentially the set of sensors to be probed. The benefits and costs of probing actions will be taken into account by the *Probing Scheduler* in deciding the what sensors to be consulted. More specifically, a user query is handled in four major steps:

- Step 1. The query is evaluated by the *Query Evaluator* based on the data cached in the *Data Manager*.
- Step 2. The *Probing Scheduler* decides the content of probing set.
- Step 3. The *Probing Scheduler* sends probing commands to the sensors defined in the probing set.

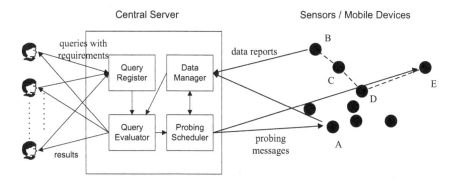

Fig. 2. System Architecture

– Step 4. The *Query Evaluator* reevaluates the query based on the refreshed data returned to the *Data Manager*, and returns results to the query issuer.

Uncertainty Model. We assume there are D data sources, namely $T_i (i = 1, 2, ..., D)$. Each data source T_i has an actual value, denoted by $T_i.v$, where $T_i.v \in R$. The uncertainty model of each T_i cached at the server consists of an uncertain region $[l_i, u_i]$, together with a pdf f_i. After probing, the value of the data item becomes "precise" (i.e., has a pdf value equal to one inside an infinitesimally-thin uncertain region). For simplicity, we illustrate our solution with an uncertainty model for one-dimensional data, but our methods can easily be extended to handle multi-dimensional data.

4   Quality and Resource Budget of Probabilistic Queries

In this section, we present the notion of *Quality Score* for probabilistic range query, the query that we extensively study in this paper. Section 4.1 details the definition and evaluation of probabilistic range queries. In Section 4.2 we present a quality metric for probabilistic range queries. Section 4.3 then discusses the metric of resource constraints, called *Resource Budget*, which is assigned to each query as the maximum amount of resources allowed in the process of query evaluation.

4.1   Probabilistic Range Query

The Probabilistic Range Query (PRQ)[4] returns a set of data objects, with the probabilities that their attribute values are in the specified range, called *qualification probabilities*:

Definition 1. Probabilistic Range Query (PRQ): *Given a closed interval* $[a, b]$, *where* $a, b \in R$ *and* $a \leq b$, *a PRQ (denoted by Q), returns a set of tuples* (T_i, p_i), *where* p_i *is the non-zero probability that* $T_i.v \in [a, b]$.

To illustrate, Figure 3 shows two PRQ's on data items A, B, C and D. The uncertainty region of each item is shown. For query Q_1, three items (A, B and C) are included in the result; item D is excluded since its uncertainty region does not overlap with the query range, yielding zero qualification probability. The result of Q_1 becomes: $(A, 0.25)$, $(B, 0.5)$, $(C, 0.75)$.

In general, the value of the qualification probability, i.e., p_i, can be calculated by using the Equation 1 [4].

$$p_i = \int_{R_i} f_i(x) dx \qquad (1)$$

where R_i is the overlapping region of the query range $[a, b]$ and $[l_i, u_i]$, and $f_i(x)$ is the uncertainty pdf of item T_i. In Figure 3, we shade the overlapping region of all the data items with the query range of Q_1 and Q_2.

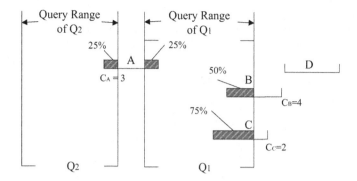

Fig. 3. An Example of Probabilistic Range Query

4.2 Quality Score

Let us now present a metric to measure the quality of the answer of a probabilistic range query. This metric is based on the notion of information entropy[20]. As a brief review, the information entropy measures the average number of bits required to encode a message, or the amount of information carried in the message:

Definition 2. Entropy: *Let* X_1, ..., X_n *be all possible messages, with respective probabilities* $p(X_1)$, ..., $p(X_n)$ *such that* $\sum_{i=1}^{n} p(X_i) = 1$. *The entropy of a message* $X \in \{X_1, ..., X_n\}$ *is:*

$$H(X) = -\sum_{i=1}^{n} p(X_i) log_2 p(X_i) \qquad (2)$$

Recall that in the answer of PRQ, each value p_i describes the probability that object T_i satisfies it. Thus there are two possible events: (1) T_i satisfies the PRQ with a probability as p_i; (2) T_i does not satisfy the PRQ with a probability of $1 - p_i$. Using Definition 2, the entropy of T_i for satisfying a PRQ is

$$g_i = -p_i log_2 p_i - (1 - p_i) log_2 (1 - p_i) \qquad (3)$$

We then use the sum of the entropy values for all the objects that satisfy the PRQ with non-zero probabilities as the quality metric. More specifically, for a result containing n answers $(T_1, p_1), (T_2, p_2), ..., (T_n, p_n)$, the *quality score*, denoted by H, of this result is defined by

$$H = -\sum_{i=1}^{n} (p_i log_2 p_i + (1 - p_i) log_2 (1 - p_i)) \qquad (4)$$

By substituting Equation 3 into Equation 4, we have

$$H = \sum_{i=1}^{n} g_i \qquad (5)$$

A larger value of H implies a lower quality. In particular, H is equal to zero if the result is precisely known, which happens when all the p_i's are equal to zero or one. The range of H is $[0, n]$. Notice that after probing item T_i, its uncertainty region shrinks to a point, and the server knows exactly whether T_i satisfies the range query. Thus, p_i equals to either zero or one. The corresponding ambiguity caused by the answer (T_i, p_i) is then "removed", and the entropy of the overall query result is reduced by an amount given by Equation 3 . We denote this amount of entropy reduction as the *gain* of probing T_i, denoted by g_i. As shown in Equation 3 the value of g_i only depends on the qualification probability of a single object T_i. Moreover, the gain is only non-zero for we choose items that have qualification probabilities in $(0,1)$, and the gain of probing a set of items is simply equal to the sum of their gains.

4.3   Resource Budget

We now present the *resource budget* model of a query, which limits the amount of resources that can be used to probe the sensing devices for this query.

In general, there are several types of important resources for a wireless sensor network, such as network bandwidth and the battery power used to transmit data. Here we use a single metric, namely the number of transmitted messages, to measure the cost. The number of transmitted messages for probing an item is the major source of consumption of the important resources. The more number of times the sensors are probed, the more amount of network bandwidth and battery power is required. Thus, we assume the server assigns to a query the maximum number of transmitted messages allowed as its *resource budget*, denoted as C.

The transmission cost of a data item can vary among the sensors. For example, a message generated from a sensor may need different number of hops to reach the base station. Figure 2 shows that four hops are required for probing item E (the dashed path), whereas only one hop is needed to probe item A. Thus probing E will cost more than A. We assume the server knows how many messages are

Table 1. Notations

Notation	Description
T	A remote stream source
$T.v$	The exact value of T
$[l, u]$	Lower and upper bounds of $T.v$
f	Probability distribution function of the $T.v$
Q	Probabilistic range query
C	Resource constraint assigned to Q
c	# of messages for probing T
H	Precision quality (entropy)
p	The probability that T satisfies Q
g	The benefit of probing T
n	# of items in the result set

needed for probing an item. We also use c_i to denote the number of messages for probing T_i. We list the notations used in this paper in Table 1.

5   Maximizing Quality with Limited Resources

As we have mentioned in Section 4.2, probing items that have non-zero qualification probabilities can often improve the quality of a query result. In general, there can be a tremendous number of objects present in the answer. Moreover, the amount of resource budget available probing is limited. In this section, we discuss how query quality can be maximized with limited resource budgets.

In Section 5.1 we present the *Single Query* (SQ) problem, where we explain how probing can be done efficiently for a query with limited resource budgets. We then extend our solution to support a more complicated and practical scenario, i.e. *Multiple Queries with Shared Budget* (MQSB), in Section 5.2. We give heuristics which provide close-to-optimal performance in Section 5.3.

5.1   Single Query (SQ)

In this scenario, only one query, Q, needs to be considered when choosing sensors. Suppose based on the cached data, the Query Evaluator has calculated the qualification probabilities, $\{p_1, p_2, ..., p_n\}$, of all the items $T_i (i = 1, ..., n)$ such that $p_i > 0$. The cost of probing T_i is c_i. Let the gain obtained by probing T_i be g_i (Equation 3). We formally define the *Single Query (SQ) problem* as follows.

$Maximize \sum_{i=1}^{n} x_i \cdot g_i$
$subject\ to \sum_{i=1}^{n} x_i \cdot c_i \leq C$
$x_i \in \{0, 1\}, i = 1, 2, ..., n$

Here we use an array $\overline{X} = x_1, x_2, ..., x_n$ to record the choices. Initially, all the values of x_i are zero. If item T_i is chosen for probing, we set x_i to 1.

To solve the SQ problem, we use dynamic programming. We observe that this problem has the *optimal substructure*, meaning that the optimal solutions of subproblems can be used to find optimal solutions of the SQ problem. Let us rewrite the SQ problem as $P(C, N)$, which is associated with a resource budget C and items $N = \{T_1, T_2, ..., T_n\}$, whose p_i's are all nonzero. Suppose we have found the optimal set $S = \{T_{\gamma_1}, T_{\gamma_2}, ..., T_{\gamma_m}\}$ $(m \leq n \wedge \gamma_i \in [1, n])$ for $P(C, N)$: among all the subsets of N whose costs are not larger than C, S is the one with the highest gain. Now we define a subproblem by randomly removing an item, e.g. T_{γ_1}, from the candidate item set, and reducing the budget to $C - c_{\gamma_1}$. That is, we consider a subproblem $P(C - c_{\gamma_1}, N/\{T_{\gamma_1}\})$. If $S_1 = S/\{T_{\gamma_1}\} = \{T_{\gamma_2}, ..., T_{\gamma_m}\}$, is the optimal set for this subproblem, the SQ problem can be solved by using the dynamic programming framework. Next we prove that S_1 must be the optimal set for $P(C - c_{\gamma_1}, N/\{T_{\gamma_1}\})$.

Proof. Suppose S_1 is not the optimal set for $P(C - c_{\gamma_1}, N/\{T_{\gamma_1}\})$, then we can find another set $S_1' \neq S_1$ which meets two requirements: (1) the cost of probing

S'_1 is not larger than $C - c_{\gamma_1}$ and (2) the gain of probing S'_1 is higher than that of probing S_1. Consider the set $S'_1 \cup \{T_{\gamma_1}\}$. Its cost is not larger than $C - c_{\gamma_1} + c_{\gamma_1} = C$. The gain of probing it is higher than that of probing the set $S_1 \cup \{T_{\gamma_1}\}$, or S. Thus $S'_1 \cup \{T_{\gamma_1}\}$ should be a better choice than S for the overall problem, which violates the condition that S is the optimal set. So S_1 must be the optimal set for $P(C - c_{\gamma_1}, N/\{T_{\gamma_1}\})$.   □

Algorithm DP. In this algorithm, we look for the optimal set for each subproblem denoted by $P(k, i)$, where the resource budget equals to k and the candidate item set is $\{T_1, ..., T_i\}$. There are totally $n \cdot C$ subproblems. For the subproblems with zero budget or empty candidate set, the optimal set is also an empty set. We use an array s to store the optimal sets for the subproblems, where $s[k, i]$ is the optimal set for the subproblem $P(k, i)$. Each element of s, e.g. $s[k, i]$, is also an array, where $s[k, i][j] = 1$ if T_j is chosen for probing, and zero otherwise. We also use an array v to store the gain by probing the optimal set $s[k, i]$. For each data item T_i, there are two possible choices. Either T_i is not chosen and $s[k, i - 1]$ is considered as the optimal set for $P(k, i)$, or this item is put into the solution set which contributes g_i to the solution gain but decrease the budget remaining for items $\{T_1, T_2, ..., T_{i-1}\}$ to $k - c_i$. The optimal set for the subproblem $P(k - c_i, i - 1)$ is $s[k - c_i, i - 1]$ with the gain $v[k - c_i, i - 1]$. Thus if T_i is chosen, the maximum possible gain is $v[k - c_i, i - 1] + g_i$. In Step 3, the gains of these two possible choices are compared, and the one with larger gain is taken as the optimal solution for current subproblem $P(k, i)$. Steps 4-5 handle the case that T_i is not chosen, while Steps 7-9 construct the optimal set and the corresponding gain if T_i is chosen. Another point to notice is, in order to put T_i into the solution set, the cost of probing T_i, i.e. c_i, must be not larger than the remaining budget k. Step 3 also tests whether this condition is satisfied.

Input An array of probing costs $c = (c_1, c_2, ..., c_n)$
      An array of gains $g = (g_1, g_2, ..., g_n)$
      The resource budget C
Output The optimal set

1. **for** $i := 1$ **to** n **do**
2. **for** $k := 1$ **to** C **do**
3.   **if** $c_i > k$ **or** $v[k, i - 1] > v[k - c_i, i - 1] + g_i$
4. $v[k, i] := v[k, i - 1]$
5. $s[k, i] := s[k, i - 1]$
6.   **else**
7. $v[k, i] := v[k - c_i, i - 1] + g_i$
8. $s[k, i] := s[k - c_i, i - 1]$
9. $s[k, i][i] := 1$
10. **return** $s[C, n]$

Fig. 4. Algorithm DP for SQ

Using Algorithm DP, we can find an optimal solution for the SQ problem. We will show soon that Algorithm DP can also be used to solve the MQSB problem, with little change to the calculation of *gain*.

Complexity. There are two for-loops in Algorithm DP. The computational complexity is thus $O(nC)$. The algorithm requires the storage of intermediate results, i.e. the optimal sets and corresponding gains for the subproblems. The variable s is a $3D$ array with space complexity of n^2C, while v is a $2D$ array with the size of nC. Thus the memory complexity of Algorithm DP is $O(n^2C)$.

5.2 Multiple Queries with Shared Budget (MQSB)

In many cases, more than one query are processed at the server simultaneously. A data item T_i may be involved in the results of multiple queries. By probing T_i, all queries containing it in their results will have a better quality. In order to apply Algorithm DP in this scenario, we need to change the method of calculating gain, i.e. Equation 3. Suppose there are m queries, $Q_1, Q_2, ..., Q_m$, we can have a set of m values for T_i, $p_{i1}, p_{i2}, ..., p_{im}$, where $p_{ij}(j = 1, 2, .., m)$ specifies the probability that T_i satisfies Q_j. After getting the exact value of T_i the result precision of these queries will be improved by H_{ij}. Here H_{ij}, the gain for Q_j obtained by probing T_i, is equal to $-p_{ij}log_2p_{ij} - (1 - p_{ij})log_2(1 - p_{ij})$, where $j = 1, 2, ..., m$(Equation 3). The gain of probing T_i is the sum of H_{ij}, or

$$G_i = \sum_{j=1}^{m} H_{ij} \tag{6}$$

For example, as in Figure 3, item A overlaps with the ranges of both Q_1 and Q_2, where $p_{A1} = p_{A2} = 0.25$. Thus $g_A = -2 \cdot (0.25 \cdot log_20.25 + 0.75 \cdot log_20.75) = 1.62$.

Suppose the server needs to process multiple queries in batches, and these queries share a single resource budget C. We denote this scenario as *Multiple Queries with Shared Budget, MQSB*. The formal definition of MQSB has the same form as that of SQ. The only difference is the use of G_i (Equation 6) to replace g_i (Equation 3). Therefore, Algorithm DP is also suitable for solving MQSB. Moreover, the approximate solutions, which will be discussed in Section 5.3, can also be used for MQSB.

Complexity of DP (MQSB). Compared with the SQ scenario, the inputed data size for the DP algorithm will be larger in the MQSB scenario. There are m queries evaluated concurrently in the MQSB scenario. If we let n to be the average size of the result sets for these m queries, the DP algorithm needs to process nm data items. Moreover, there would be extra cost of computing the gains by using Equation 6 in the MQSB scenario, which is $O(nm)$. Thus, the computational complexity of Algorithm DP would be $O(nmC + nm) = O(nmC)$ in the MQSB scenario, and the memory complexity is $O((nm)^2C)$.

5.3   Approximate Solutions

Greedy. The dynamic programming solution, Algorithm DP, can find the optimal sets. However, its complexity can be quite high. To enhance its scalability, we design a greedy algorithm. The general idea of Greedy is to make a locally optimal choice. Every unit of cost should be allocated to the items which can produce maximum benefit. To achieve this objective, we define a new metric to describe the amount of gain obtained by consuming a unit of resource. This metric is called *efficiency*, denoted by e_i. Equation 7 shows how to compute the value of e_i.

$$e_i = \frac{g_i}{c_i} \tag{7}$$

Input An array of probing costs $c = (c_1, c_2, ..., c_n)$
   An array of gains $g = (g_1, g_2, ..., g_n)$
   The resource budget C
Output The optimal set

1. $d := sort(c, g)$
2. $b := C$
3. **for** $i := 1$ **to** n **do**
4.   **if** $b \geq c_{d[i]}$
5. $s[d[i]] := 1$
6. $b := b - c_{d[i]}$
7. **return** s

Fig. 5. Algorithm Greedy

In Step 1 of the Greedy algorithm, the items are sorted by their efficiencies in descending order. The sorted indices are stored in an array d. Initially, the remaining budget, i.e. b, is set to the value of C. We then check the items sequentially in the order stored in d. If the remaining budget is not smaller than the cost of probing this item (Step 4), it is put into array s (Step 5) and the remaining budget is reduced by its cost (Step 6). Step 7 returns the probing set stored in s.

The Greedy algorithm has a time complexity of $O(n \log n)$ (to sort the items). The space requirement for Greedy is $O(n)$. It is thus more efficient than DP. However Greedy does not guarantee an optimal set can be found. We will compare the performance of these two algorithms in Section 6.

Random and MaxVal. We also develop two other simpler heuristics, called *Random* and *MaxVal*. The Random solution chooses items randomly until the resource budget is exhausted. The MaxVal heuristic probes items sequentially in descending order of their gains until the resource budget is exhausted.

Table 2 compares the complexities of the above algorithms in the SQ scenario.

Table 2. Complexity of Four Algorithms (SQ)

Algorithm	Computational Complexity	Space Complexity
DP	$O(nC)$	$O(n^2C)$
Greedy	$O(n \log n)$	$O(n)$
Random	$O(n)$	$O(n)$
MaxVal	$O(n \log n)$	$O(n)$

Table 3. Complexity of Four Algorithms (MQSB)

Algorithm	Computational Complexity	Space Complexity
DP	$O(nmC)$	$O((nm)^2C)$
Greedy	$O((nm) \log(nm))$	$O(nm)$
Random	$O(nm)$	$O(nm)$
MaxVal	$O((nm) \log(nm))$	$O(nm)$

For MQSB, the complexities of the optimal and approximate solutions are listed in Table 3. They are derived by substituting the value of n in Table 2 by nm.

6   Experimental Results

We have performed experimental evaluation on the effectiveness of our approaches. We first present our simulation model, followed by the detailed results.

6.1   Experiment Settings

We use a realistic data set, called Long Beach[1], which contains 53K rectangles, and each represents a region in the Long Beach country. The objects occupy a 2D space of $10,000 * 10,000$ units. We use the Long Beach data as an uncertain object database. We also assume that the uncertainty pdf of any uncertain object is a uniform distribution.

The cost of probing each item (i.e. c_i) are uniformly distributed in $[1, 10]$. The resource budget, C, ranges from 20 to 500. The performance metric is the result quality improved by probing a set of result items. Each data point is an average over 50 runs. Our experiments are run on a PC with 2.4GHz CPU and 512MB of main memory. Our simulation is written in j2sdk1.4.2_11.

6.2   Results

Effectiveness Analysis. Figure 6 compares the quality improvement using different probing strategies for the SQ problem. The x-axis is the value of resource budget which ranges from 20 to 500. The y-axis is the improved quality of query

---

[1] Available at http://www.census.gov/geo/www/tiger/

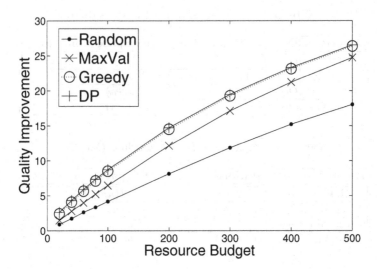

Fig. 6. Quality Improvement vs. Resource Budget (SQ)

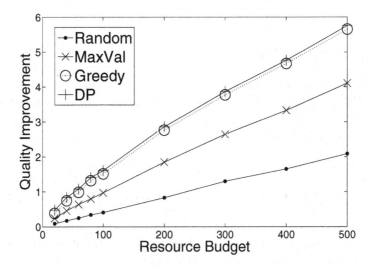

Fig. 7. Quality Improvement vs. Resource Budget (MQSB)

results. As shown in Figure 6, DP always outperforms MaxVal and Random. This is because DP derives the probing set with optimal resource utilization. The performance of Greedy is close to DP; in fact, DP performs about only 2% to 3% better than Greedy. This is because that the quality-aware probing problem is a variant of the knapsack problem [9], and it has been shown in [12] that the average performance of a greedy solution is close to the optimal one.

Figure 7 illustrates similar results for MQSB. In these experiments, 10 queries are executed concurrently in a batch.

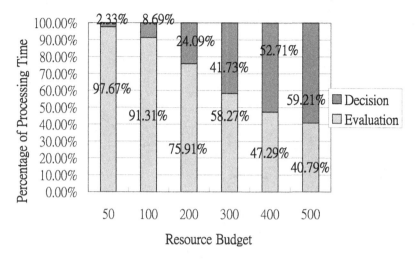

Fig. 8. Time Spent in Different Phases during Query Processing (SQ)

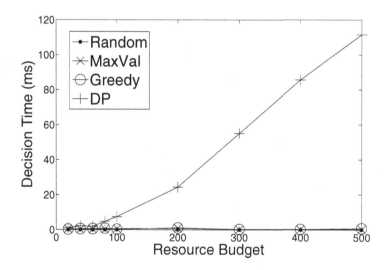

Fig. 9. Decision Time vs. Resource Budget (SQ)

Performance Analysis. Figure 8 shows a decomposition of the time spent in the server: (1) Evaluation - the time required by the Query Evaluator to compute the initial results (Step 1 in Section 3), and (2) Decision - the time for deciding the probing set contents (Step 2 in Section 3). We have ignored the processing time of Step 4 since after probing, the qualification probabilities will become either zero or one for the data items in the probing set, and no extra effort is needed to compute their qualification probabilities. Here, we use DP to find optimal probing set in the Decision step. As shown in Figure 8, the Decision

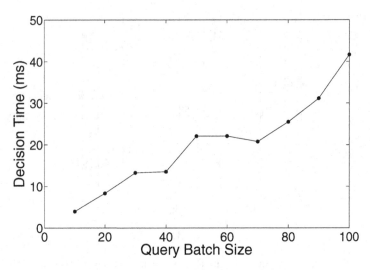

Fig. 10. Scalability of MQSB (Greedy)

step costs more time as the resource budget increases. The reason is that more choices are available with a larger resource budget.

Figure 9 shows the time spent in the Decision step for the SQ problem. DP uses more time to find the optimal probing sets, and the decision time of DP increases fast with the resource budget. The decision time for heuristics (i.e. Greedy, MaxVal and Random) are much less. The results in MQSB are similar and are omitted here.

Compared with DP, Greedy gets similar quality improvement with less time. In fact, Greedy performs very well under a large batch of queries in the MQSB scenario. Figure 10 illustrates the time required for finding the probing sets using the Greedy algorithm. The resource budget is set to 100. The number of queries evaluated in a batch varies from 10 to 100. As shown in the figure, the decision time increases gracefully with the query batch size.

7   Conclusions

The evaluation of probabilistic queries over uncertain data has attracted plenty of research interest in recent years. In this paper, we investigated the problem of optimizing the quality of probabilistic query answers with limited resources. We further extended our solution to handle the case where the resource budget is shared among multiple queries. While the DP algorithm provides an optimal solution in polynomial times, our experiments show that the Greedy heuristic can achieve close-to-optimal performance in less time. In the future, we will investigate this problem for other types of queries.

Acknowledgements

The work described in this paper was supported by the Research Grants Council of the Hong Kong SAR, China (Project No. PolyU 5133/07E), and the Germany/HK Joint Research Scheme (Project No. G_HK013/06). We would like to thank the anonymous reviewers for their insightful comments and suggestions.

References

1. Chen, J., Cheng, R.: Efficient evaluation of imprecise location-dependent queries. In: ICDE (2007)
2. Cheng, R., Chan, E., Lam, K.Y.: Efficient quality assurance of probabilistic queries in sensor networks. Sensor Network and Configuration: Fundamentals, Techniques, Platforms, and Experiments (May 2006)
3. Cheng, R., Chen, J., Mokbel, M., Chow, C.-Y.: Probabilistic verifiers: Evaluating constrained nearest-neighbor queries over uncertain data. In: ICDE (2008)
4. Cheng, R., Kalashnikov, D., Prabhakar, S.: Evaluating probabilistic queries over imprecise data. In: Proc. ACM SIGMOD (2003)
5. Cheng, R., Kalashnikov, D., Prabhakar, S.: Evaluation of probabilistic queries over imprecise data in constantly-evolving environments. Information Systems Journal (2006)
6. Cheng, R., Xia, Y., Prabhakar, S., Shah, R.: Change tolerant indexing on constantly evolving data. In: ICDE (2005)
7. Cheng, R., Xia, Y., Prabhakar, S., Shah, R., Vitter, J.S.: Efficient indexing methods for probabilistic threshold queries over uncertain data. In: Proc. VLDB (2004)
8. Chu, D., Deshpande, A., Hellerstein, J., Hong, W.: Approximate data collection in sensor networks using probabilistic models. In: ICDE (2006)
9. Cormen, T., Leiserson, C., Rivest, R., Stein, C.: Introduction to Algorithms. The MIT Press, Cambridge (2001)
10. Deshpande, A., Guestrin, C., Madden, S., Hellerstein, J., Hong, W.: Model-driven data acquisition in sensor networks. In: VLDB (2004)
11. Despande, A., Hong, W., Guestrin, C., Madden, S.R.: Exploiting correlated attributes in acquisitional query processing. In: ICDE (2005)
12. Diubin, G.: The average behaviour of greedy algorithms for the knapsack problem: general distributions. Mathematical Methods of Operations Research 57(3) (2003)
13. Pfoser, D., Jensen, C.: Capturing the uncertainty of moving-objects representations. In: Proc. SSDBM (1999)
14. Han, S., Chan, E., Cheng, R., Lam, K.Y.: A statistics-based sensor selection scheme for continuous probabilistic queries in sensor network. Real Time Systems Journal (RTS) (2006)
15. Lazaridis, I., Mehrotra, S.: Approximate selection queries over imprecise data. In: ICDE (2004)
16. Liu, Z., Sia, K.C., Cho, J.: Cost-efficient processing of min/max queries over distributed sensors with uncertainty. In: Preneel, B., Tavares, S. (eds.) SAC 2005. LNCS, vol. 3897. Springer, Heidelberg (2006)
17. Ljosa, V., Singh, A.: APLA: Indexing arbitrary probability distributions. In: ICDE (2007)

18. Olston, C., Jiang, J., Widom, J.: Adaptive filters for continuous queries over distributed data streams. In: SIGMOD (2003)
19. Olston, C., Widom, J.: Offering a precision-performance tradeoff for aggregation queries over replicated data. In: VLDB (2000)
20. Shannon, C.: The Mathematical Theory of Communication. University of Illinois Press (1949)
21. Sistla, P.A., Wolfson, O., Chamberlain, S., Dao, S.: Querying the uncertain position of moving objects. In: Temporal Databases: Research and Practice. Springer, Heidelberg (1998)
22. Tao, Y., Xiao, X., Cheng, R.: Range search on multidimensional uncertain data. ACM Transactions on Database Systems 32(15) (2007)

Efficient Computation of Statistical Significance of Query Results in Databases

Vishwakarma Singh[1], Arnab Bhattacharya[2], and Ambuj K. Singh[1]

[1] Department of Computer Science, University of California, Santa Barbara, USA
vsingh@cs.ucsb.edu, ambuj@cs.ucsb.edu
[2] Department of Computer Science and Engineering, Indian Institute of Technology (I.I.T.),
Kanpur, India
arnabb@iitk.ac.in

Abstract. Queries such as database similarity searches return results satisfying certain properties of distances or scores. For domain scientists, the absolute values of scores are seldom sufficient. Statistical significance or *p-value* of the result is a more useful criterion. This can be computed using an appropriate model of random objects. The problem of computing p-values becomes more acute when queries have multiple components. In this case, the returned score is an aggregate of individual scores. The simple way of calculating the p-value by enumerating all random possibilities fails for large database and query sizes. We propose an efficient method to calculate the approximate p-value of a multi-attribute result when the distribution of scores for the database objects is non-parametric. Experimental evaluation on large databases shows that our method is practical, runs 5 orders of magnitude faster than the basic approach, and has an error of less than 5% in p-value computation.

1 Motivation and Problem Statement

Many database systems retrieve results based on some distance or score measure between the query object and the database objects. Score is a quantitative measure of the similarity between objects based on multiple attributes. It has been widely used for ranking results in content-based multimedia retrieval systems. However, with the growing interest in analyzing the results of a database similarity query, computing rigorous statistical properties of the results is more meaningful.

Statistical significance helps the domain scientists in understanding the nature of the query and the statistical properties of the database objects. The most well known example is BLAST [1]. A standard measure of statistical significance is the *p-value*. The p-value of score s of a query result from a database is defined as the probability of randomly obtaining a result from the database with a score s or higher for the same query. It is the area under the probability distribution function (pdf) of the scores of random objects greater than s.

For a database management system (DBMS) serving single object queries, the score *pdf* can be characterized or calculated, and so, the p-value can be computed. However, there are database systems of complex objects where each object consist of multiple attributes or components. Such systems support queries with multiple attributes or objects and the score of a result is some aggregate function (e.g., sum) of the individual

B. Ludäscher and Nikos Mamoulis (Eds.): SSDBM 2008, LNCS 5069, pp. 509–516, 2008.

scores of each query component against its corresponding result component [5]. These queries are common for region based image retrieval (RBIR) systems [3] and information retrieval systems [9]. For example, in an RBIR system, a query region is composed of a number of sub-regions (e.g., tiles) [2,10]. The database images are also split into sub-regions. Each component sub-region has a corresponding score of its match with a query sub-region. The score of a result is the sum of the individual scores.

For a given query object Q of size r, a random database for computing the p-value can be modeled by considering all possible aggregates of size r composed of components from the database. To find the p-value, we need to calculate the score pdf for this random database. This simple method has a running time that grows exponentially with database size and query size and is, therefore, impractical. In this paper, we propose and solve the following problem: *"Given a query Q composed of r objects $Q_i, i = 1, \cdots, r$, database objects $D_j, j = 1, \cdots, n$, scoring functions $f_i : Q_i \times D \to \Re$, compute the p-value of obtaining a score s for a result $R = \cup_{i=1}^r R_i$, where $s = \sum_{i=1}^r f(Q_i, R_i)$, for a random database of objects, each having r component objects."*

Methods have been proposed for obtaining a single measure of statistical significance by combining the individual p-values. For example, the method in [4] requires finding the correlation among the attributes, which is done by sampling for large datasets. We adopt a more direct approach. We find the sum pdf of the individual pdfs of the components of the query. Then we calculate the p-value from this sum score pdf. Since score pdf of each component is independent of the other, this pdf is the *convolution* of all the individual pdfs. For most databases, the nature and the parameters of this pdf cannot be computed. We consider such cases where the probability distribution function of the cumulative scores is non-parametric.

2   Algorithm

For a multiple object query, the p-value can be found from the sum pdf of its components. The basic approach of calculating the sum pdf is to calculate the pdf of each query component and then find their convolution. Two score pdfs h_1 and h_2 can be convoluted to produce the sum score pdf h: the probability corresponding to score s considers all possible scores s_1 and s_2 from h_1 and h_2 such that $s = s_1 + s_2$. The cost of computing this convolution is, thus, *quadratic* in the number of distinct scores in the constituent pdfs. Hence, we can see that the convolution of multiple pdfs incurs a multiplicative cost on the size of the pdfs, and therefore, can be large. Assume that a query has r components, and each component has b distinct score values. The convolution of the first two components requires $b \times b = b^2$ operations and produces up to b^2 distinct scores. Convoluting this result with the third component requires $b^2 \times b = b^3$ operations, and so on. The total running time, therefore, is $b \times b \times \cdots \times b = O(b^r)$.

In order to speed up the p-value computation, we consider the two aspects of the problem—computing the score distribution for each query component and convoluting the distributions—separately. The first sub-problem is handled by pre-processing and maintaining a separate score pdf for each object component in the database. This can be done offline. For each component of the query, we approximate its score pdf by

---

Algorithm PRUNE
Input: Query $Q = \cup_{i=1}^{r} Q_i$, Score s, Database D, Number of bins b
Output: P-value p
1. **for** $i = 1$ to r
2. $D_i := 1\text{-NN}(Q_i, D)$
3. $h_i := \text{BinHistogram}(D_i, b)$
4. **end for**
/* σ_i is the sum pdf of bin histograms $1, \cdots, i$ */
5. $\sigma_1 := h_1$
6. **for** $i = 2$ to r
7. $B(\sigma_i) := s - \sum_{j=i+1}^{r} \max(h_j)$
8. $B(h_i) := B(\sigma_i) - \max(\sigma_{i-1})$
9. $B(\sigma_{i-1}) := B(\sigma_i) - \max(h_i)$
10. $\sigma_i := \text{Convolute(all bins } \sigma_{i-1,j} \geq B(\sigma_{i-1}), \text{ all bins } h_{i,k} \geq B(h_i))$
11. **end for**
12. $p := $ Sum of probabilities in all bins $\sigma_{r,j} \geq s$

Fig. 1. The PRUNE algorithm

the pdf corresponding to its nearest component in the database. The nearest database component can be retrieved very efficiently by indexing the feature vectors of the objects using R-trees [7].

To efficiently convolute the pdfs and compute the p-value, we developed an approximation technique PRUNE (Fig. 1). There are three main steps in the algorithm: (i) Use *histograms* to approximate the score probability distribution functions of each query object, (ii) Progressively *cascade* the convolution of query object histograms to obtain the score histogram for the entire query, and (iii) Use *bounds* to convolute the histograms. We next explain each step in detail.

2.1 Use of Histograms to Approximate Distributions

Since the cost of convoluting two pdfs is a quadratic function of the number of distinct values in the pdfs, instead of using an actual score pdf, we approximate it by a histogram with a fixed number of bins as shown in step 3 of Algorithm PRUNE (Fig. 1). For speed, simplicity, and convenience, we choose equi-width histograms. The whole score range is divided into a fixed number of equi-width bins. The accuracy of the approximation depends on the number of bins maintained. More bins have less error, but higher running time. Section 3 considers the effect of the number of bins on the running time and the error in calculating the p-value.

2.2 Cascaded Convolution of Histograms

As described earlier, the simple way of directly convoluting r histograms has a time complexity which is exponential in r. To avoid such high costs, we convolute the histograms in a progressive fashion. Initially, the histograms of two query component objects are convoluted to yield another score histogram, which is again binned into b bins.

Then, this histogram is convoluted with the next histogram and so on till all the r histograms have been convoluted.

Denoting the ith histogram by h_i and the convolution of histograms up to i components by σ_i, we compute $\sigma_i = \sigma_{i-1} \oplus h_i$ up to $i = r$. Each histogram convolution requires *quadratic* number of operations in terms of the number of bins in the histograms. The total time complexity, therefore, is $O(b^2 r)$. To make it even more efficient, we apply a bounding procedure which is described next.

2.3   Convolution of Bounded Histograms

The bounding method is based on the observation that computing the p-value for a score s requires counting only those scores that are greater than or equal to s. Scores in the histogram of a query object that cannot add up to s even when combined with the best scores of the histograms of other query objects need not be considered. Therefore, the bins in the histogram whose scores fall below this *threshold score* can be deleted. The bounding method achieves this pruning of histogram bins by evaluating the threshold score at each stage. This reduces the number of bins, and thus, the running time.

Fig. 2 shows an example of how such thresholds are computed. Assume that the histogram σ_{i-1} is convoluted with h_i to yield σ_i. Also, assume that the score s for which the p-value is being calculated is 100. If the maximum score in h_{i+1} is 40, then any score below $100 - 40 = 60$ in σ_i cannot add up to s. This is the threshold score for that histogram, and is highlighted in the figure. Thus, all scores below 60 can be deleted from σ_i. By analyzing this bounding behavior backwards for the histograms σ_{i-1} and h_i, it can be seen that such contributing pairs of scores need not be calculated at all. The maximum score in h_i is 55. Since we do not need any score in σ_i that is below 60, all scores below $60 - 55 = 5$ in σ_{i-1}, when added to any score in h_i will be less than 60, and hence, can be deleted. Continuing this reasoning, all scores below 35 in h_i can be deleted. The threshold scores are highlighted in the figure.

In this example, the number of bins in σ_{i-1} and h_i are reduced from 6 to 4 and 3 respectively. This translates to a saving of $6 \times 6 - 4 \times 3 = 24$ bin convolution operations.

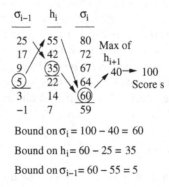

Bound on $\sigma_i = 100 - 40 = 60$

Bound on $h_i = 60 - 25 = 35$

Bound on $\sigma_{i-1} = 60 - 55 = 5$

Fig. 2. Efficient convolution of histograms. $\sigma_{i-1} \oplus h_i = \sigma_i$. The bins below the score thresholds (shown inside circles) can be pruned to save time.

Steps 7 to 10 of Algorithm PRUNE (Fig. 1) apply bounding to the cascaded convolution. As shown in the next section, the overall saving for r histogram convolutions is significant.

Note that the two sources of error in the p-value computation are the use of nearest neighbors and histogram binning. The bounding method does not introduce any error.

3   Experiments

In this section, we demonstrate the effectiveness of our PRUNE method over alternate approaches. We explain the empirical results in the context of region-based image retrieval (RBIR) system for a biomedical image database of fluorescent micrographs of feline retinas labeled with different antibodies [6]. The dataset consists of 805,272 tiles. The score between two tiles is a decreasing function of the L_1 distance between the color histogram features of the tiles. The tiles are the component objects in our system. The score of the alignment of a query region to a database region is the sum of the scores of the alignment of the individual tiles. The details of the dataset preparation, the features, the scoring function, and the retrieval system are explained in [10].

3.1   Running Time

The basic approach of online computation of score pdfs of each query tile and their convolution yields impractical time. Therefore, we do not consider it. Instead, we maintain a database of the pre-computed pdf of each database component. We use the following parameters for the analysis of running time: (i) the number of bins in the score histograms, (ii) the query score for which the p-value is computed, measured as a percentage of the maximum score that can be achieved by the query, and (iii) the query size, which is the number of tiles in the query image.

First, we compare the four different approaches of computing the p-value: (i) Using actual pdf without pruning, (ii) Using actual pdf with pruning, (iii) Using binned pdf

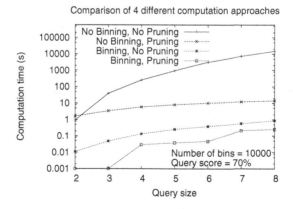

Fig. 3. Comparison of the various approaches of p-value computation

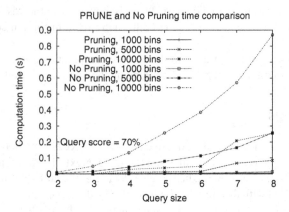

Fig. 4. The effect of pruning on the running time of p-value computation

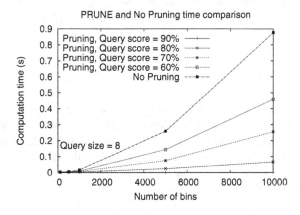

Fig. 5. The effect of query score and number of bins on the running time of p-value computation

without pruning, and (iv) Using binned pdf with pruning (PRUNE). Fig. 3 shows their running times for different query sizes. The pruning strategy shows a gain of about 10^3 for a query size of 8 without binning. Binning improves the computation time by 2 orders of magnitude with pruning and 5 orders of magnitude without pruning. In all cases, the PRUNE strategy finished in practical times—at most 255 ms.

Since the PRUNE strategy outperforms all other approaches we analyze it further with respect to other parameters. Fig. 4 shows that the efficiency of pruning increases with the increase in query size across varying number of bins in the histogram. Up to medium query sizes of 6, and number of bins 5000, the scalability is linear or better.

The next experiment (Fig. 5) shows that the pruning strategy performs better when the query score increases, across varying number of bins. When the query score is 80% of the maximum score, the pruning strategy is very effective for all histogram bin sizes.

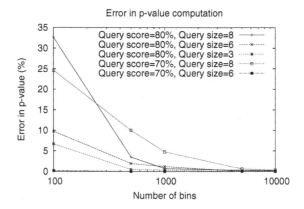

Fig. 6. The percentage error in p-value computation due to binning

The scalability is better for higher query scores. Thus, the empirical results strongly suggest that our PRUNE method is efficient and practical.

3.2 Error

We next performed experiments to measure the error in p-value computation induced by binning. Fig. 6 shows the error percentage across varying number of bins. When the query size is large, using less number of bins accumulates the error over more number of steps, resulting in more than 20% error. Increasing the number of bins to 1000 reduces the error to at most 5%, irrespective of the query size and the query score. This proves the effectiveness of our strategy.

4 Conclusion

In this paper, we defined the problem of efficiently computing the p-value for multi-object query results for non-parametric distributions. We proposed an approximate bounding procedure PRUNE and showed that it is faster than the alternate approaches by more than 5 orders of magnitude with the error in computation less than 5%. Possible future avenues of work include sampling to obtain the score histograms, computing bounds for other aggregate functions like max, and examining the order in which the component object histograms should be convoluted in order to minimize the number of operations.

We presented a $O(b^2 r)$ complexity algorithm for convoluting r histograms of b bins each. In future, we plan to examine the computation time and the possibility of optimization by utilizing the convolution theorem which states that the discrete Fourier transform (DFT) of the convolution of two equi-width histograms is equal to the product of the individual DFTs of the histograms [8]. This implies that the convolution of r histograms can be achieved in $O(rb \lg b)$ time using fast Fourier transform and its inverse [8].

References

1. Altschul, S.F., Gish, W., Miller, W., Myers, E.W., Lipman, D.J.: Basic Local Alignment Search Tool. J. Molecular Biology 215(3), 403–410 (1990)
2. Dagli, C., Huang, T.S.: A Framework for Grid-Based Image Retrieval. In: Proc. 17th Int. Conf. on Pattern Recognition (ICPR 2004), vol. 2, pp. 1021–1024 (2004)
3. Datta, R., Li, J., Wang, J.Z.: Content-Based Image Retrieval: Approaches and Trends of the New Age. In: MIR 2005: Proc. 7th ACM SIGMM Int. Workshop on Multimedia Information Retrieval, pp. 253–262 (2005)
4. Delongchamp, R., Lee, T., Velasco, C.: A Method for Computing the Overall Statistical Significance of a Treatment Effect Among a Group of Genes. BMC Bioinformatics (2006)
5. Fagin, R., Lotem, A., Naor, M.: Optimal Aggregation Algorithms for Middleware. J. Computer and System Sciences 66(4), 614–656 (2003)
6. Fisher, S.K., Lewis, G.P., Linberg, K.A., Verardo, M.R.: Cellular Remodeling in Mammalian Retina: Results from Studies of Experimental Retinal Detachment. Progress in Retinal and Eye Research 24(3), 395–431 (2005)
7. Guttman, A.: R-Trees: A Dynamic Index Structure for Spatial Searching. In: Special Interest Group on Management of Data (SIGMOD), pp. 47–57 (1984)
8. Press, W.H., Flannery, B.P., Teukolsky, S.A., Vetterling, W.T.: Numerical Recipes in C. Cambridge University Press, Cambridge (1992)
9. Salton, G.: Automatic Text Processing: The Transformation Analysis and Retrieval of Information by Computer. Addison-Wesley, Reading (1988)
10. Singh, V., Bhattacharya, A., Singh, A.K., Banna, C., Lewis, G.P., Fisher, S.K.: QUIP: Querying Significant Patterns from Image Databases. Technical report, Dept.of Computer Science, University of California, Santa Barbara (2007)

Analysis of Basic Data Reordering Techniques

Tan Apaydin[1], Ali Şaman Tosun[2,*], and Hakan Ferhatosmanoglu[1,**]

[1] The Ohio State University, Computer Science and Engineering
{apaydin,hakan}@cse.ohio-state.edu
[2] University of Texas at San Antonio, Computer Science
tosun@cs.utsa.edu

Abstract. Data reordering techniques are applied to improve the space and time efficiency of storage and query systems in various scientific and commercial applications. Run-length encoding is a prominent approach of compression in many areas, whose performance is significantly enhanced by achieving longer and fewer "runs" through data reordering. In this paper we theoretically study two reordering techniques, namely lexicographical order and Gray code order. We analyze these two methods in the context of bitmap indexes, which are known to have high query performances. We take into account the two commonly used bitmap encodings: equality and range. Our analysis indicates that, when we have all the possible data tuples, both ordering methods perform the same with equality encoding. However, Gray code achieves better compression with range encoding. Experimental results are provided to validate the theoretical analysis.

1 Introduction

Scientific data is mostly read-only and its volume can reach to the order of petabytes, e.g., astrophysics, genomic and proteomics, high energy physics. The techniques for maintaining the conventional databases usually do not apply to these applications, which brings up the need for effective indexing methods for efficient storage and retrieval. Bitmap indexes are compact index structures and they have been successfully applied to data warehouses and scientific databases by exploiting the property that scientific data are enumerated or numerical [6,11]. These structures have also been implemented in commercial Database Management Systems such as Oracle [1,2], Informix [4,7].

These indexes handle partial match and range queries very efficiently since they utilize the fast bitwise logical operations, which are directly supported by computer hardware. However, in order to maintain these advantages in large domains, effective compression schemes are applied. The most suited and widely used techniques are the adaptations of run-length based compression [10]. Besides reducing the data size, run-length compression has the benefit of partially avoiding the overhead of decompression in query processing for bitmap indexes [1,12]. Since run-length based compression schemes pack together the consecutive same-value-symbols, the compression ratio depends heavily on the occurrences of such patterns. Data can be reorganized/reordered to increase the length of the *runs* and improve the compression performance.

---

[*] Partially supported by US National Science Foundation (NSF) Grant CCF-0702728.
[**] Partially supported by US National Science Foundation (NSF) Grant IIS-0546713.

B. Ludäscher and Nikos Mamoulis (Eds.): SSDBM 2008, LNCS 5069, pp. 517–524, 2008.
© Springer-Verlag Berlin Heidelberg 2008

A common approach is lexicographical sorting, which is used in traditional databases to preserve locality and avoid disk seeks. Several other reordering techniques have been successfully applied to increase the performance of compression in different domains as well [5,9]. For a boolean matrix, the objective of finding an order of data that minimizes the total number of runs in the columns of the matrix was shown to be NP-hard through a reduction to Traveling Salesperson Problem (TSP) [5]. Gray codes have been proposed as an efficient alternative to simple lexicographic or expensive TSP heuristics, and shown to achieve comparable compression to TSP, while running significantly faster than these heuristics [9].

In this paper we examine the effectiveness of (re-)ordering methods on the data compression performances. We provide the theoretical foundations and performance analysis of lexicographic and Gray code order in the context of bitmap indexes. Comparatively, lexicographic order and Gray code order are investigated for two encoding techniques that are commonly used in bitmap indices, namely equality and range encodings, and their relative performances are studied with data reordering in consideration.

Compared to their own equality encoding versions, our study reveals that both Gray code and lexicographic order achieve greater compression performances for range encoding. On the other hand, comparison of the two ordering methods leads to the following outcomes. With equality encoding, when we have all the possible data tuples, lexicographic order and Gray code order perform the same. However, Gray code order achieves better compression than lexicographic order when range encoding is used. We also provide experimental results to validate the theoretical analysis.

The organization of the paper is as follows. In Section 2 we briefly cover the background information about the impact of reordering schemes on the compression performance, and provide the preliminaries for the rest of the paper. Section 3 provides the analysis for the equality encoding bitmap model. We provide the theoretical study on the range encoding model in Section 4. Experimental results are presented in Section 5, and finally we conclude in Section 6.

2   Background

In this section, the background information for the data ordering and compression approaches are provided, and the related work is discussed.

Efficient storage of large boolean tables are achieved by utilizing the run-length based compression [10], which is the process of replacing the consecutive occurrences of a symbol by a single instance and a count. We define a *run* as a sequence of 0's followed by (not including) a 1 or end symbol, or a sequence of 1's followed by (not including) a 0 or end symbol. For instance, in Figure 1(a) the first column has 2 runs and the second column has 4 runs. Variations of the run-length compression technique are utilized in different domains in the literature. For example, for bitmap indexes, the two most popular compression schemes are BBC [1] and WAH [12]. BBC stores the compressed data in bytes while WAH stores in words. They are designed not only to decrease the bitmap index size but also to speed up the query execution performance while running the queries over the compressed data.

$$
\begin{array}{c c}
\begin{array}{c}
t_1 \\ t_2 \\ t_3 \\ t_4 \\ t_5 \\ t_6 \\ t_7 \\ t_8
\end{array}
\begin{bmatrix}
0 \ 0 \ 0 \\
0 \ 0 \ 1 \\
0 \ 1 \ 0 \\
0 \ 1 \ 1 \\
1 \ 0 \ 0 \\
1 \ 0 \ 1 \\
1 \ 1 \ 0 \\
1 \ 1 \ 1
\end{bmatrix}
&
\begin{array}{c}
t_1 \\ t_2 \\ t_4 \\ t_3 \\ t_7 \\ t_8 \\ t_6 \\ t_5
\end{array}
\begin{bmatrix}
0 \ 0 \ 0 \\
0 \ 0 \ 1 \\
0 \ 1 \ 1 \\
0 \ 1 \ 0 \\
1 \ 1 \ 0 \\
1 \ 1 \ 1 \\
1 \ 0 \ 1 \\
1 \ 0 \ 0
\end{bmatrix}
\\[2pt]
\text{(a) Lexicographic} & \text{(b) Gray code}
\end{array}
$$

Fig. 1. Example of tuple reordering

For a boolean matrix, long runs of 0 or 1 blocks are necessary for run-length based compression to be effective. In other words, the performance depends on the number of runs, therefore reordering techniques are utilized for improvement by packing the same-value bits together. Adaptations of the traveling salesperson problem (TSP) solutions have been applied to the large boolean matrices in [5]. In order to improve the bitmap index compression, Gray code ordering (GCO) is proposed as a data reorganization technique in [9]. GCO based approaches are known to be faster than TSP-based solutions.

The original Gray code for binary numbers is an encoding such that two adjacent numbers differ only by one bit (Hamming distance is equal to 1). For instance (000, 001, 011, 010, 110, 111, 101, 100) is a *binary reflected* Gray code. One can achieve a Gray code recursively as follows: i) Let $S = (s_1, s_2, ..., s_n)$ be a Gray code. ii) First write S forwards and then reflect S by writing it backwards, so that we have $(s_1, s_2, ..., s_n, s_n, ..., s_2, s_1)$. iii) Append 0 to the beginning of first n numbers, and 1 to the beginning of last n numbers. For instance, take the Gray code (0, 1). Write it forwards and backwards, and we get: (0, 1, 1, 0). Then we add 0's and 1's to get: (00, 01, 11, 10).

Figure 1 illustrates the effect of running the GCO algorithm. On the left is the numeric (or lexicographic) order of a boolean matrix with 3 columns. In the rest of the paper, we refer to the lexicographic order shortly as *Lexico order*. GCO of the same matrix is presented on the right. As the figure illustrates, the aim of GCO is to produce longer and thus fewer runs than Lexico order. Figure 1(a) produces 14 runs (2 on the first column, 4 and 8 on the following columns) whereas Figure 1(b) has 10 runs (2 on the first column, 3 and 5 on the following columns).

We call a data set that has all the possible combinations of tuples as *full*. Table 1 is an example of a full data. Recall that the aim of GCO is to reorder the data so that the Hamming distances between the consecutive tuples will be 1. For a set of tuples there can be more than one order that have such property. Therefore, Gray codes are not unique. In this paper, to simplify the analysis, only the reflected GCO is taken into account.

Equality Encoding is the basic encoding scheme for bitmap indices, which is also known as Value-List index [8]. For an equality encoded bitmap index, data is partitioned into several bins, where the number of bins for each attribute could vary. If a value falls into a bin, this bin is marked "1", otherwise "0". Since a value can only fall into a single bin, only a single "1" can exist for each row of each attribute. Table 1 shows a two-attribute example such that the first attribute has 2 bins and the second attribute has

Table 1. Encoding example for two attributes with 2 and 3 bins

Tuple	Equality Encoding					Range Encoding				
	Attribute 1		Attribute 2			Attribute 1		Attribute 2		
	a	b	1	2	3	a	b	1	2	3
$t_1 = (b, 3)$	0	1	0	0	1	0	1	0	0	1
$t_2 = (a, 2)$	1	0	0	1	0	1	1	0	1	1
$t_3 = (a, 3)$	1	0	0	0	1	1	0	0	0	1
$t_4 = (b, 2)$	0	1	0	1	0	0	1	0	1	1
$t_5 = (b, 1)$	0	1	1	0	0	0	1	1	1	1
$t_6 = (a, 1)$	1	0	1	0	0	1	1	1	1	1

3 bins. The first tuple t_1 falls into the second bin of the first attribute and the third bin of the second attribute.

Another prominent encoding scheme is called *Range encoding* [3], which is also presented in Table 1. In this encoding, if a value falls into a bin b_i, all the greater bins and also b_i are marked "1"; and "0" otherwise. Range encoding performs better especially for single-sided range queries compared to equality encoding. For details we refer the reader to [3].

3   Equality Encoding

In this section, we investigate the behaviors of Lexico order and GCO schemes using equality encoding. Our main goal is to derive a formula for the total number of runs with full data. For the remaining of the paper, we use the terms *cardinality* and *number of bins* of an attribute interchangeably and they basically refer to the same value.

Define $F(x)$ as $F(x) = 3x - 2$. This function will be used to find the number of runs of an attribute as a function of its cardinality. The total number of attributes is denoted by A, and the cardinality of attribute i is denoted by C_i. For A attributes, where $A \geq 2$, following theorem presents the total number of runs for the full data using Lexico order.

Theorem 1. *For full data, number of runs in Lexico order using equality encoding is*

$$F(C_1) + \sum_{i=2}^{A} \left(F(C_i) \prod_{j=1}^{i-1} C_j - \left[(C_i - 2) \left(\left(\prod_{j=1}^{i-1} C_j \right) - 1 \right) \right] \right)$$

Proof. Number of runs for the first attribute is $F(C_1)$. With full data i^{th} attribute can be considered as $\prod_{j=1}^{i-1} C_j$ separate chunks where the tuples in a chunk have the same value for the attributes $A_1, ..., A_{i-1}$. An example is given in Figure 2. In the example first attribute has a single chunk, second attribute has 2 chunks ($C_1 = 2$) and third attribute has 4 chunks ($C_1 \cdot C_2 = 2 \cdot 2 = 4$). Since there are $\prod_{j=1}^{i-1} C_j$ chunks and an attribute with C_i produces $F(C_i)$ runs, there are $F(C_i) \prod_{j=1}^{i-1} C_j$ runs in attribute i. However, this assumes that runs finish and start at chunk boundaries and can not be combined. Runs for the first and last columns of an attribute can not be combined. However, runs for other columns can be combined since runs are of the form: 0's followed by 1 followed by 0's. Trailing runs of 0's for one chunk can be combined with leading 0's of next chunk. There are $C_i - 2$ inner columns in each attribute and there are $\prod_{j=1}^{i-1} C_j - 1$ run

0	1	0	1	0	1
0	1	0	1	1	0
0	1	1	0	0	1
0	1	1	0	1	0
1	0	0	1	0	1
1	0	0	1	1	0
1	0	1	0	0	1
1	0	1	0	1	0

Fig. 2. Example of chunks with 3 attributes each having 2 bins

merges for each inner column. Total number of run merges is $(C_i - 2)(\prod_{j=1}^{i-1} C_j - 1)$ and we subtract this to find the exact number of runs. □

For equality encoding with the full data, we next show that Lexico order and GCO produce the same number of runs.

Theorem 2. *For full data, the number of runs in GCO using equality encoding is equal to the number of runs in Lexico order, which is given in Theorem 1.*

Proof. Number of runs for the first attribute is $F(C_1)$. With full data, i^{th} attribute can be considered as $\prod_{j=1}^{i-1} C_j$ separate chunks where the tuples in a chunk have the same value for the attributes $A_1, ..., A_{i-1}$. Since there are $\prod_{j=1}^{i-1} C_j$ chunks and an attribute with C_i produces $F(C_i)$ runs, there are $F(C_i) \prod_{j=1}^{i-1} C_j$ runs in attribute i. For odd numbered attributes, binary numbers in a chunk appear in increasing order and for even numbered attributes, binary numbers in a chunk appear in decreasing order. In either case, number of runs for a column is the same. Above analysis assumes that runs finish and start at chunk boundaries and can not be combined. Rest of the proof is similar to the proof of Theorem 1. □

4   Range Encoding

In this section, we focus on range encoding and discuss the behaviors of Lexico order and GCO. Our main goal again includes deriving the total number of runs. In addition, we compare the compression performances of Lexico order and GCO both for equality and range encodings. Note that conversion of an equality encoded tuple T_i to its range encoded version $R(T_i)$ is a 1-1 transformation (see Table 1).

Total Runs for Lexico: Define function $E(x)$ as $E(x) = 2x - 1$, which will help deriving the number of runs of range encoding for both Lexico order and GCO. The formula for lexicographic order is given by the following theorem.

Theorem 3. *For full data, the number of runs in Lexico order of A attributes, where $A \geq 2$, using range encoding is*

$$E(C_1) + \sum_{i=2}^{A} \left(E(C_i) \prod_{j=1}^{i-1} C_j - \left[\left(\prod_{j=1}^{i-1} C_j \right) - 1 \right] \right)$$

Proof. The number of runs for the first attribute is $E(C_1)$. With full data, i^{th} attribute can be considered as $\prod_{j=1}^{i-1} C_j$ separate chunks where the tuples in a chunk have the same value for the attributes $A_1, \ldots, A_{i-1}$. Since there are $\prod_{j=1}^{i-1} C_j$ chunks and an attribute with C_i produces $E(C_i)$ runs, there are $E(C_i) \prod_{j=1}^{i-1} C_j$ runs in attribute i. However, this assumes that runs finish and start at chunk boundaries and can not be combined. Runs for the last column can be combined since all the entries are 1's. Runs for other columns can not be combined since they all have a number of 0's followed by a number of 1's. There are $\prod_{j=1}^{i-1} C_j - 1$ run merges for the last column of the attribute. We subtract the number of run merges $(\prod_{j=1}^{i-1} C_j - 1)$ to find the exact number of runs.   □

Equality Lexico vs. Range Lexico: For Lexico order, range encoding achieves better compression than equality encoding as shown by the following corollary.

Corollary 1. *For Lexico order of full data, range encoding produces fewer runs than equality encoding.*

Proof. Follows from the comparison of Theorems 1 and 3 using $E(C_i) < F(C_i)$ and $\left[(\prod_{j=1}^{i-1} C_j) - 1\right] < \left[(C_i - 2)\left((\prod_{j=1}^{i-1} C_j) - 1\right)\right]$.   □

Total Runs for GCO: The total number of runs is given below. Tricky part of the derivation is again to find out how many of the runs merge. Since runs can cross the chunk boundaries (see Figure 2), we should avoid overcounting.

Theorem 4. *For full data, the number of runs in GCO of A attributes, where $A \geq 2$, in range encoding is*

$$E(C_1) + \sum_{i=2}^{A} \left(E(C_i) \prod_{j=1}^{i-1} C_j - C_i \left[(\prod_{j=1}^{i-1} C_j) - 1 \right] \right)$$

Proof. Number of runs for the first attribute is $E(C_1)$. With full data, i^{th} attribute can be considered as $\prod_{j=1}^{i-1} C_j$ separate chunks where the tuples in a chunk have the same value for the attributes $A_1, \ldots, A_{i-1}$. Since there are $\prod_{j=1}^{i-1} C_j$ chunks and an attribute with C_i produces $E(C_i)$ runs, there are $E(C_i) \prod_{j=1}^{i-1} C_j$ runs in attribute i. This analysis assumes that runs finish and start at chunk boundaries and can not be combined. However, the runs for all the bins of an attribute can be combined. Since there are C_i bins in the attribute and there are $\prod_{j=1}^{i-1} C_j - 1$ run-merges for each inner column, we subtract total number of run-merges of the attribute given by $C_i(\prod_{j=1}^{i-1} C_j - 1)$ to find the exact number of runs.   □

Equality GCO vs. Range GCO: Range encoding using GCO produces fewer runs. In other words, the conversion from equality encoding into range encoding reduces the number of runs for full data. Since the conversion is 1-1, range encoding can be used as a way to achieve further compression, which is an open research question.

Corollary 2. *For GCO of full data, range encoding produces fewer runs than equality encoding.*

Proof. Follows from comparison of Theorems 2 and 4 using $E(C_i) < F(C_i)$ and $\left[(\prod_{j=1}^{i-1} C_j) - 1\right] < \left[C_i\left((\prod_{j=1}^{i-1} C_j) - 1\right)\right]$.   □

Range Lexico vs. Range GCO: For range encoding, finally we compare Lexico order and GCO in the following corollary.

Corollary 3. *GCO produces fewer number of runs than Lexico for range encoding.*

Proof. Follows from comparison of Theorems 3 and 4.   □

5   Experimental Results

For our experiments, we used full data sets with varying number and cardinality of attributes. In Figure 3(a), we present the total runs in log scale as a function of the attribute cardinality. The larger the cardinality, the higher the number of runs. (Recall that Lexico and GCO have the same number of runs for equality encoding.) We also repeated the experiment with different number of attributes (7, 8 and 9). Again, increasing the number of attributes leads to higher number of runs. For example, 7 attributes each with 5 bins produce 195,331 number of runs, 8 attributes each with 5 bins produce 976,584 runs, and 9 attributes produce 4,882,837 runs.

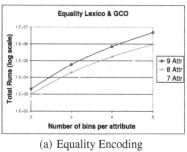
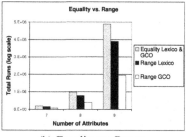

| (a) Equality Encoding | (b) Equality vs. Range |

Fig. 3. Experimental results with different number of attributes with different cardinalities

For a comparison between equality and range encodings, we present Figure 3(b) where each attribute has 5 bins. Since Lexico and GCO perform the same for equality encoding, we simply combined them and named that as *Equality Lexico & GCO*. Note that, among the three approaches (namely Equality Lexico & GCO, Range Lexico, and Range GCO), the best performance is achieved by Range GCO. For an example in Figure 3(b), the values for 9 attributes are as follows: Equality Lexico & GCO produces 4,882,837 runs. Range Lexico has 3,906,257 runs, and Range GCO produces 1,953,169 runs. For range encoding, note that the number of runs for Lexico is about twice the number of runs for GCO.

6   Conclusion

In this paper we studied the effectiveness of reordering methods that are applied for better compression performances in databases. High energy physics, astrophysics, genomic and proteomics are some of the applications that produce large data sets, which bring up the need for effective indexing techniques for efficient storage and querying. Bitmap indexes are practical structures that are prominently used for querying scientific data. In the literature, in order to reduce the sizes of these indexes, run-length based compression schemes are developed whose performances are improved by data reordering approaches.

We provided the theoretical foundations and performance analysis of lexicographic order and Gray code order in the context of bitmap indexes. Comparatively, lexicographic order and Gray code order are investigated for two encoding techniques that are commonly used in bitmap indices, namely *equality* and *range* encodings, and their relative performances are studied with data reordering in consideration.

Our study reveals that both Gray code and lexicographic order achieve greater compression performances for range encoding compared to their own equality encoding versions. On the other hand, comparison of the two ordering methods leads to the following observations. With equality encoding, when we have all the possible data tuples, lexicographic order and Gray code order perform the same. However, Gray code order achieves better compression than lexicographic order when range encoding is used. We also provided experimental results to validate the theoretical analysis.

References

1. Antoshenkov, G.: Byte-aligned bitmap compression. In: Data Compression Conference, Nashua, NH. Oracle Corp. (1995)
2. Antoshenkov, G., Ziauddin, M.: Query processing and optimization in oracle rdb. The VLDB Journal 5(4), 229–237 (1996)
3. Chan, C.Y., Ioannidis, Y.E.: Bitmap index design and evaluation. In: Proceedings of the 1998 ACM SIGMOD international conference on Management of data, pp. 355–366. ACM Press, New York (1998)
4. Informix. Decision support indexing for enterprise datawarehouse,
   http://www.informix.com/informix/corpinfo/-zines/whiteidx.htm
5. Johnson, D., Krishnan, S., Chhugani, J., Kumar, S., Venkatasubramanian, S.: Compressing large boolean matrices using reordering techniques. In: VLDB 2004 (2004)
6. Chen, J., Wu, K., Koegler, W., Shoshani, A.: Using bitmap index for interactive exploration of large datasets. In: Proceedings of SSDBM (2003)
7. O'Neil, P.: Informix and Indexing Support for Data Warehouses. Database Programming and Design 10, 38–43 (1997)
8. O'Neil, P., Quass, D.: Improved query performance with variant indexes. In: Proceedings of the 1997 ACM SIGMOD international conference on Management of data, pp. 38–49. ACM Press, New York (1997)
9. Pinar, A., Tao, T., Ferhatosmanoglu, H.: Compressing bitmap indices by data reorganization. In: ICDE, pp. 310–321 (2005)
10. Salomon, D.: Data Compression: The Complete Reference, 3rd edn (2004)
11. Stockinger, K., Shalf, J., Bethel, W., Wu, K.: Dex: Increasing the capability of scientific data analysis pipelines by using efficient bitmap indices to accelerate scientific visualization. In: Proceedings of SSDBM (2005)
12. Wu, K., Otoo, E.J., Shoshani, A.: Optimizing bitmap indices with efficient compression. ACM Trans. Database Syst. 31(1), 1–38 (2006)

Kriging for Localized Spatial Interpolation in Sensor Networks

Muhammad Umer, Lars Kulik, and Egemen Tanin

National ICT Australia, Department of Computer Science & Software Engineering, University of Melbourne, Victoria 3010, Australia
{mumer,lars,egemen}@csse.unimelb.edu.au

Abstract. The presence of coverage holes can adversely affect the accurate representation of natural phenomena being monitored by a Wireless Sensor Network (WSN). Current WSN research aims at solving the coverage holes problem by deploying new nodes to maximize the coverage. In this work, we take a fundamentally different approach and argue that it is not always possible to maintain exhaustive coverage in large scale WSNs and hence coverage strategies based solely on the deployment of new nodes may fail. We suggest spatial interpolation as an alternative to node deployment and present Distributed Kriging (DISK), a localized method to interpolate a spatial phenomenon inside a coverage hole using available nodal data. We test the accuracy and cost of our scheme with extensive simulations and show that it is significantly more efficient than global interpolations.

Keywords: Wireless Sensor Networks, Coverage Holes, Interpolation, Kriging.

1 Introduction

Monitoring physical phenomenon is an important application domain for wireless sensor networks (WSNs). The WSN data acquisition techniques accomplish a monitoring task by sampling data at different network locations over time. Currently, these techniques advocate the use of data suppression as a means of achieving energy efficiency [1]. The case for data suppression is based on the assumption that there is always an abundance of samples in a WSN and hence reporting accuracy is not compromised even if the data from a large part of the network is suppressed. However, experience with WSNs [2] and insights into future applications [3,4] reveal that situations may arise where the assumption of an abundance of samples does not hold. Such situations arise when node failures or the sparsity of a WSN trigger *gaps* or *coverage holes* in the reported data. We argue that for an accurate representation of a physical phenomenon in such scenarios *augmenting* the reported data as well as its *suppression* is a priority. In this paper, we investigate such situations and propose novel interpolation methods to augment the reported data in the presence of gaps.

Existence of coverage holes is an important reality for WSNs and has been studied extensively [2]. Current research has predominantly focused on the identification of holes in order to alert and restore the lost coverage of a WSN [5]. However, the replacement and restoration of nodes may not be sensible in hostile environments or not

B. Ludäscher and Nikos Mamoulis (Eds.): SSDBM 2008, LNCS 5069, pp. 525–532, 2008.

possible due to prohibitive costs. Therefore, our assertion is that coverage issues require data acquisition regimes that cope with missing data through other means than simply replacing or deploying more nodes. Thus, we need methods to interpolate the missing readings based on the available data and application specific expert knowledge.

Physical phenomenon are characterized by their spatial correlation, i.e., the fact that proximal locations have similar values and vary together. Spatial interpolation techniques can thus be used to estimate a phenomena in coverage holes. The challenge, however, is that typical interpolation techniques are not readily applicable to WSNs due to their reliance on global knowledge of the network [6]. Due to a WSNs' dynamic nature and large scale, such global information is prohibitively expensive to collect and maintain.

The key challenge that we address in this work is to perform accurate spatial interpolation for coverage holes with minimal power requirements. To maximize the use of available information we propose to first build a correlation model of the phenomenon under observation and then perform interpolation using this model. We first present the QS (Quad Suppress) algorithm, a distributed in-network aggregation algorithm for correlation modeling of a phenomenon. We then present the DISK (DIStributed Kriging) algorithm which utilizes the correlation model to perform interpolation in a fully distributed manner. With extensive simulations we show that QS and DISK are significantly more energy efficient than their global counterparts.

2   The Quad Suppress (QS) Algorithm

The first step towards localized spatial interpolation is to find an appropriate variogram model that best describes the spatial correlation in a dataset. The experimental variogram (EV) is a measure of spatial continuity in a spatial process defined as average squared difference between data values at a certain distance, called lag, h [7]. Assume a random variable Z represents a Gaussian spatial process and $Z(x)$ represent its realizations at location x then its EV can be given as:

$$2\gamma(h) = \frac{1}{N(h)} \sum_{N(h)} [Z(x) - Z(x + h)]^2 . \tag{1}$$

where, $N(h)$ is the number of data pairs at distance h. A variogram model is a curve fitted on the observed EV values.

Equation 1 shows that for a certain lag h, EV construction requires the difference of the value of each node from all nodes in its EV neighborhood, i.e., present within a distance of h units from itself. Thus, a simple global approach for EV construction could be to propagate all samples to a base station. However, to reduce communication costs we propose the QS algorithm, an alternative based on in-network aggregation. In-network aggregation algorithms organize the network in a tree like fashion such that each internal node aggregates all data coming from its child nodes and communicate only the partial aggregate to its parent. However, for EV construction there is an added constraint; an internal node can aggregate a child node, say K, only if it can ensure that it is also a parent of all nodes in K's EV neighborhood. The QS algorithm adopts a quadtree-like tree creation method to fulfill this constraint.

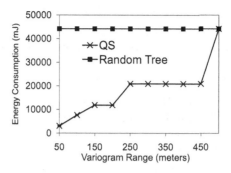

Fig. 1. Comparison of energy expenditure in QS and random tree algorithms for EV construction

The QS aggregation tree is built as follows. The base station partitions the space and chooses a cell-head for each quadrant. These cell-heads recursively partition their cells choosing new sub-cell heads. The process continues until a predetermined grid resolution is reached. A random aggregation tree is then built inside each of the unpartitioned cells rooted at the corresponding cell-head. Data aggregation is performed at each internal node of this tree. The benefit of creating the aggregation tree in a quadtree-like manner is that a cell-head can locally determine whether or not it covers the EV region of a child node. It can then accordingly aggregate or forward the node's value.

Figure 1 shows the considerable difference in energy expenditure of the QS algorithm and a random tree based global data collection using 2500 samples from Digital Elevation Model (DEM) dataset [8] covering a 500 m^2 area of the state of Colorado, US. The significant difference in the performance of the two algorithms can be explained by their data forwarding behavior. The global data collection scheme creates the aggregation tree randomly, thus an internal node cannot determine whether or not it covers the EV neighborhood of its child nodes. Consequently, all internal nodes propagate their data to the base station in unaggregated form. On the other hand, the QS algorithm reduces the communication costs significantly by performing the aggregation in the network as soon as it realizes that a node's EV neighborhood is covered.

3   Distributed Kriging (DISK) Algorithm

Once a variogram model is established for a given phenomenon, it can then be used for spatial interpolation using Kriging. Kriging is a well known geostatistical method used to estimate unknown values of a physical process using existing knowledge about the process and a model of its spatial variation, i.e., the variogram [7]. Assume that the values $Z(x_1), Z(x_2), \ldots, Z(x_N)$ represent realizations of a spatial process Z at locations $x_1, x_2, \ldots, x_N$, then the Kriging interpolator of Z at a point x_0 is given by [9]:

$$\hat{Z}(x_0) = \sum_{i=1}^{N} \lambda_i Z(x_i) . \tag{2}$$

where λ_i are the weights fulfilling the unbiasedness condition, i.e., $\sum_{i=1}^{N} \lambda_i = 1$ and the expected error is $E[\hat{Z}(x_0) - Z(x_0)] = 0$ [7]. It can be shown that optimal weights λ_i for the Kriging interpolator can be computed from the following system of linear equations (SLE)

$$\Lambda = A^{-1}b. \tag{3}$$

where Λ is a vector comprising of Kriging weights λ_i and a Lagrange multiplier (added for computational reasons), A is the covariance matrix of sample locations $x_1, x_2, \ldots, x_N$ and b is a vector whose elements represent the covariance between x_0 and each $x_i \in \{x_1, x_2, \ldots, x_N\}$. All covariances are based on an appropriate variogram model defined for the spatial process in question. We first use the QS algorithm to build a variogram model for the entire WSN and distribute this model in the network. In this way all nodes can autonomously compute their correlation with any other node in the network as required during the Kriging process.

In the following subsection we explain our interpolation approach, the DISK algorithm, which in essence is a distributed and localized form of the Kriging interpolation.

3.1 Iterative Formulation of Kriging System of Linear Equations

The basic building block of the DISK algorithm is an iterative approach towards the Gaussian elimination method. In terms of the Kriging SLE, the Gaussian elimination method can be interpreted as the process of finding a sequence of elementary row operations, or linear maps, that transforms matrix A to its reduced row-echelon form. The basic idea of our iterative elimination approach is presented in Figure 2. If the nodes performing a Kriging operation are assumed to be aligned along a chain, each node k adds a new variable, i.e., its Kriging weight (λ_k), and its corresponding linear combination ($\sum_{i=1}^{k} \lambda_i c_{ki} = b_k$) to the Kriging SLE. In matrix terms, each node in the chain adds a new row and column to the matrix A required to be inverted while not changing the original entries. We can then order the elimination process on the basis of the following recursive formulation of matrix A:

$$A_k = \begin{bmatrix} 1 & K^T \\ K & A_{k-1} \end{bmatrix} \quad K = \begin{bmatrix} c_{k(k-1)} \\ \ldots \\ c_{k1} \end{bmatrix} \quad A_2 = \begin{bmatrix} c_{22} & c_{21} \\ c_{12} & c_{11} \end{bmatrix} \tag{4}$$

where $k \geq 2$.

Now for A_k, we define $\Phi(T_k) : A_k \rightarrow A_k^{\Phi}$ as a group of all linear maps enumerated by the following recursive definition:

$$\Phi(T_{k-1}). \tag{5}$$

$$R_k \leftarrow R_k + (-k_{1i} \times R_{k-i}) \; \forall i \in \{1, 2 \ldots k-1\}. \tag{6}$$

$$R_k \leftarrow R_k \times \frac{1}{a_{11}}. \tag{7}$$

$$R_{k-i} \leftarrow R_{k-i} + (-k_{i1} \times R_k) \; \forall i \in \{1, 2 \ldots k-1\}. \tag{8}$$

where $k \geq 2$. R_i represents the ith row of A_k and $a_{11}, k_{i1}, k_{1i}, i = 1, 2 \ldots k-1$ represent elements of matrices A_k and its corresponding K and K^T constituent vectors,

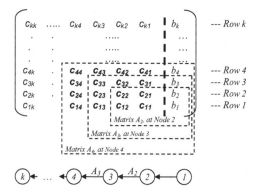

Fig. 2. Iteratively building Kriging SLE

respectively. $\Phi(T_{k-1})$ represents all linear maps defined for matrix A_{k-1}, while $\Phi(T_1)$ comprises one row operation defined as: $R_k \leftarrow R_k \times \frac{1}{a_{11}}$.

An immediate consequence of above formulation can be specified as the following Lemma:

Lemma 1. *If* $A_{k-1} = I$, *the identity matrix, and* A_k *defined in terms of Equation 4, then* $A_k^{\Phi} = I$.

Theorem 1. *Let* A_k *be an invertible matrix and* $\Phi(T_k) : A_k \rightarrow A_k^{\Phi} = I$ *be the group of linear maps corresponding to* A_k. *The solution of SLE* $\lambda = A_k^{-1}b$ *can be obtained by a recursive application of linear maps from* T^K *on* b.

The above formulation allows us to find the linear maps required to solve the Kriging SLE in an iterative manner. Consider the example in Figure 2. The first intermediate node (2) in the chain initiates the iterative process by computing the required transformations for A_2 and transmit the composite linear map, T_2, to the node above it (node 3). Node 3 computes the row and column entries for A_3 and according to the definition above, first applies the received linear map, T_2 on A_3 followed by linear maps that reduce all new row entries to 0, reduce the pivot element to 1 and reduce all new column entries to zero, i.e., applying Equations 6, 7 and 8 in that order. Node 3 then forwards the composite linear map T_3 to the next node in chain. The iterative application of the same procedure at each intermediate node in the chain results in a final composite linear map, T_k at root node of the chain. T_k can then be applied on vector b to compute the solution of the Kriging SLE resulting in the required Kriging weighting vector: Λ.

4   Experimental Study

We performed extensive simulations with the goal to evaluate the scalability of the DISK algorithm. Our simulations are based on two large datasets; a Digital Elevation Model (DEM) dataset from the state of Colorado, US [8] and simulated traffic data for the city of Melbourne, Australia. Along with DISK, we also simulate a Global Kriging

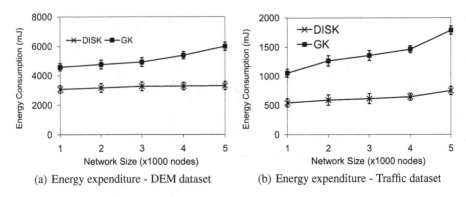

(a) Energy expenditure - DEM dataset   (b) Energy expenditure - Traffic dataset

Fig. 3. DISK and GK approaches for various network sizes

algorithm (GK) that assumes the full knowledge of node and coverage hole locations at the base station and performs its interpolation by propagating the required data to the base station. In the DEM dataset experiments, we estimate the altitude at various points while in the traffic dataset experiments, we estimate the number of cars at various locations. We measure the cost of a technique as the overall node energy used in data transmission. The accuracy is based on cross-validation of interpolation with the known values and is computed as the root mean square error (RMSE).

We define the Kriging neighborhood as the set of nodes located within a predefined distance to a point being interpolated. We measure the relationship between the network area and the Kriging neighborhood as a ratio between their sizes (in terms of number of nodes), referred to as the *network to neighborhood area ratio* (NNR).

The effect of network size. In this experiment we analyze the scalability of DISK and GK with increasing network size. In each step, we expand the deployment area and the Kriging neighborhood such that the NNR stays constant at 5%. It can be observed from Figure 3 that DISK scales well with increasing network size while GK increases in cost rapidly. For the network of 1000 nodes, DISK uses about 60% of energy used by GK while for the 5000 nodes network its energy usage reduces to about 48% of GK.

Figure 4 shows a comparison of the accuracy of DISK and GK algorithms. Although both techniques interpolate using the same Kriging neighborhood, the difference in accuracy can be attributed to the variogram model used in each case. DISK uses the localized variogram model built through the QS algorithm while GK creates the variogram model centrally after collecting all data. In the case of highly correlated DEM data the localized variogram model enables DISK to achieve slightly better accuracy than GK. In this case, lower estimation accuracy of GK can be attributed to the use of global information which adds noise to its estimation. On the other hand, DISK suffers in the traffic dataset due to low levels of spatial correlation in this data.

The effect of the Kriging neighborhood size. In this experiment we increase the NNR by increasing the Kriging neighborhood area while the network size is kept fixed at 5000 nodes. Figure 5 presents the result of these experiments for DEM data. Increasing

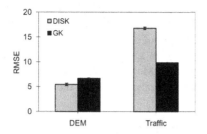

Fig. 4. Accuracy of Disk and GK. Root mean square error (RMSE) is computed in terms of meters and number of cars for DEM and traffic data respectively.

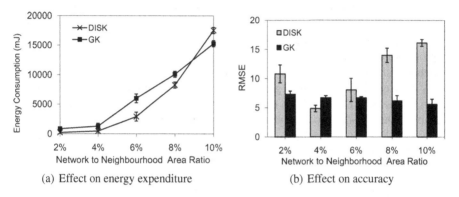

(a) Effect on energy expenditure   (b) Effect on accuracy

Fig. 5. Effect of size of the Kriging neighborhood

the neighborhood area results in an increase in the number of nodes in the neighborhood. Consequently, the communication cost of GK increases as there is more data to propagate to the base station. Similarly, the communication cost of DISK also rises due to the involvement of more nodes in the Kriging process. However, we observe that for accurate Kriging the NNR should not be increased beyond a certain value as further samples add noise to the estimation and lower the accuracy. For DEM data, the most accurate estimation results are obtained for 4% NNR where the cost of DISK is only about 40% of GK. A similar behavior with respect to energy expenditure was observed in our experiments with the traffic dataset (not shown here for the brevity of presentation).

5   Related Work

The problem of phenomenon estimation in WSNs for a region of interest not fully covered by a WSN is explored in [10]. The major limitations of [10] in comparison to DISK is its global nature. This method assumes complete knowledge of the network at the base station and pumps all of the data from sampled nodes to the sink. The Distributed Regression method [6] for global kernel regression closely compares to DISK. This method is based on the notion that a number of regional correlation structures can be identified inside a given deployment. It is thus suggested that message passing for a global regression task can be optimized by distributing the global computation among the constituent regions. Although we propose the idea of distributed interpolation on a regional basis as well, the concept of a region in our method is fundamentally different from [6]. For DISK, the interest in a region is not based upon its correlation pattern but its vicinity to a coverage hole. Moreover, the iterative Gaussian elimination step in DISK makes the computation more localized than the distributed regression method.

6   Conclusions and Future Work

Coverage holes are a reality for WSNs and it is often important to estimate the information within a coverage hole. Kriging is a well-established interpolation method that particularly suits this problem as spatial correlations in measurements is common in WSNs. The challenge, however, is to perform Kriging with minimal communication and computation and with high accuracy in a WSN. We address this challenge by proposing QS and DISK algorithms that enable distributed and localized variogram modeling and Kriging. In extensive simulations we show that our methods are significantly more energy efficient than global interpolations. In future, we plan to investigate the applications of DISK to form energy efficient node redeployment strategies where the movement of nodes can be guided by phenomenon estimation in a region of interest.

References

1. Silberstein, A., Braynard, R., Yang, J.: Constraint Chaining: On Energy-Efficient Continuous Monitoring in Sensor Networks. In: Proceedings of SIGMOD, pp. 157–168 (2006)
2. Ahmed, N., Kanhere, S.S., Jha, S.: The Holes Problem in Wireless Sensor Networks: A Survey. Mob. Comput. Commun. Rev. 9(2), 4–18 (2005)
3. Exploratory Research on Sensor-Based Infrastructure for Early Tsunami Detection (2006), http://www.cs.pitt.edu/s-citi/tsunami/
4. NASA Volcano Sensorweb (2007), http://sensorwebs.jpl.nasa.gov/
5. Wang, Y., Gao, J., Mitchell, J.S.: Boundary Recognition in Sensor Networks by Topological Methods. In: Proceedings of MobiCom, pp. 122–133 (2006)
6. Guestrin, C., Bodik, P., Thibaux, R., Paskin, M., Madden, S.: Distributed Regression: An Efficient Framework for Modeling Sensor Network Data. In: Proceedings of IPSN, pp. 1–10 (2004)
7. Isaaks, E., Srivatava, R.M.: An Introduction to Applied Geostatistics. Oxford Press, New York (1989)
8. Digital Elevation Models (USGS DEM) (2007), http://data.geocomm.com/dem/
9. Cressie, N.A.: Statistics for Spatial Data. Wileys, New York (1993)
10. Zhang, H., Moura, J.M.F., Krogh, B.: Estimation in Sensor Networks: A Graph Approach. In: Proceedings of IPSN, p. 27 (2005)

Scalable Ubiquitous Data Access
in Clustered Sensor Networks

Yueh-Hua Lee, Alex Thomo, Kui Wu, and Valerie King

University of Victoria, Victoria BC, Canada
yhl@uvic.ca,{thomo,wkui,val}@cs.uvic.edu

Abstract. Wireless sensor networks have drawn much attention due to
their ability to monitor ecosystems and wildlife habitats. In such systems,
the data should be intelligently collected to avoid human intervention. For
this, we propose a network infrastructure in which the sensor nodes are
designated as "data-generating" or "data-storage" nodes. Data-generating
nodes take measurements, whereas data-storage nodes make themselves
available to compute and store checksums of data received from nearby
data-generating nodes.

We propose a spatially-clustered architecture for our storage nodes
and a coding scheme that allows a data collector to recover all original
data by querying only a small random subset of storage nodes from
each cluster. The size of such a subset is equal to the number of data-
generating nodes that the cluster serves.

When the clustering structure of the storage nodes is unknown, we
show that recovering of the original data is still possible if a random
subset *of the right size* of storage nodes is selected for querying. We
determine this right size so as to have a successful decoding with a probability exceeding a given threshold.

1   Introduction

A wireless sensor network (WSN) consists of a large number of cheap, low-
power sensors with strictly limited resources. Most WSN applications collect
and process sensor readings such as temperature and humidity. In applications
such as the detection of fire and pollution, live data streams are delivered to
a data processing center through a connected network [7,8,1,5]. In some other
applications; however, the network may not be connected all the time or access
to live data streams may be costly and undesirable. To mention a few examples,
in ZebraNet [6] that tracks wild zebras in Africa, it is very hard to access live
data from zebras due to their spontaneous movement; in the habitat monitoring
system in Great Duck Island [9], some birds are notoriously sensitive to human
intervention, and as such, data are collected only occasionally. In these examples,
the data are stored temporarily (at storage nodes) for later data access.

A system infrastructure to achieve the above goal is to deploy sensors in the
field to form a distributed data storage network. The infrastructure contains
three different nodes: "data-generating nodes" (or data nodes for short), "data

B. Ludäscher and Nikos Mamoulis (Eds.): SSDBM 2008, LNCS 5069, pp. 533–540, 2008.

storage nodes" (or storage nodes for short), and "data collectors." The data nodes (e.g., the sensors on animals) takes measurements. When the data nodes are close to storage nodes, they upload their readings. The duty of the storage nodes is to encode and store the incoming data. Data collectors, which may be interested researchers using the system, will access the storage nodes at a later time. The storage nodes are usually stationary. They form an auto-configuration network, which is managed by an underlying network management scheme (e.g., clustering). The operations of the management scheme may be invisible to data nodes (zebras) and data collectors (human people), who may not be always on the monitored field.

It has been pointed out that naïve data storage without coding cannot achieve efficient data collection (cf. [3,12]). Suppose that the number of data nodes and storage nodes is K and N, respectively. The *ubiquitous access* property is

> *to recover all K data items by querying any K storage nodes.*

This is exactly the goal that we want to achieve with small system overhead.

Dimakis et al. in [3] proposed Decentralized Erasure Codes (DEC) for ubiquitous access to sensor data. Dimakis et al. showed that if each data node transmits the data to at least $5\frac{N}{K}\ln K$ storage nodes, then the ubiquitous access can be fulfilled. They assumed, however, a flat underlying network structure, and thus, the communication cost for the data distribution is prohibitive when N becomes large. Unfortunately, in many applications, we must deploy a great number of storage nodes to provide good coverage. In addition, the K data items are decodable only with a probability of $1 - \frac{K}{q}$, where q is the order of the Galois field used for encoding. To achieve a high decoding probability, q cannot be small, and this implies a significant overhead in calculations.

We remark that one could also achieve ubiquitous access by using Reed-Solomon (RS) codes. In such a case, the decoding is 100% certain. However, the RS code matrix is in general "denser" (i.e. with more non-zero elements) than the DEC's code-matrix, and this translates into more data messaging from the data nodes to storage nodes.

In this paper, we solve this problem by clustering storage nodes. Data items are propagated only within each cluster. As shown in Figure 1 (right), data distribution with our method only involves short-range intra-cluster communications. On the other hand, with a flat architecture, Figure 1 (left), the messages may be sent to faraway storage nodes. Notably, we have the "luxury" to adopt a deterministic code (such as RS codes) in a cluster because the number of storage nodes in a cluster is much smaller than their total number.

Next, we study the case when the data collectors are not fully aware of the clustering structure of storage nodes. This is motivated by the fact that the clusters may change dynamically and data collectors may not be always on the field. Most existing clustering algorithms (cf. [11]) dynamically adjust cluster heads or structure to balance energy consumption and prolong network lifetime. Therefore, for

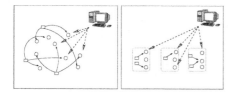

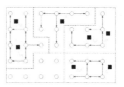

Fig. 1. A flat WSN (left) requires long-distance communication. Our scheme has intra-cluster communication only (right). Squares are data nodes; circles are storage nodes.

Fig. 2. A sensor netork of 5 clusters: The data nodes (squares) upload data to storage nodes (circles). The storage nodes are forwarding (indicated by arrows) the data among its cluster.

full generality, we assume that the cluster assignment is unknown to data collectors, who would like to decode all the data segments by simply querying a random set of storage nodes. With a theoretical analysis and numerical results, we show that the size of the random set can be close to K, the number of data nodes.

In summary, our contributions are:

1. We propose an architecture that reduces the cost to achieve ubiquitous access.
2. We propose a coding scheme which imposes no memory overhead.
3. We investigate the possibility of ubiquitous access in the case when a data collector does not know the cluster structure of the network. For this, we present a mathematical model for the decoding probability and show that, in practical cases, this probability is sufficiently high for moderate sample sizes.
4. We demonstrate that our coding scheme is more cost-efficient than DEC.

2   System Architecture and Encoding/Decoding Scheme

Our architecture is a clustered WSN (See Figure 2). In total, there are K data nodes and N storage nodes. We assume that the storage nodes are stationary but data nodes may move with animals as for example in the ZebraNet. An underlying clustering algorithm partitions the storage nodes into M clusters. Within a cluster, a cluster-head node maintains the full view of the cluster, including the number of data and storage nodes, node IDs, etc. Each data node takes measurements and send its readings to its nearest storage node. As the data nodes move, they transmit their data to different storage nodes belonging to different clusters. Suppose that the number of storage nodes in the mth cluster is N_m for $1 \leq m \leq M$. We assume that $K \leq N_m$ for each $1 \leq m \leq M$. This can be easily achieved by the underlying clustering algorithm as the common assumption is that $K \ll N$.

We denote by d_k, for $1 \leq k \leq K$, the content of such a data-message generated by data node k. When a storage node receives d_k, it propagates d_k in its cluster and updates its checksum and as follows. We denote by $s_{m,i}$, where $1 \leq m \leq M$ and $1 \leq i \leq N_m$, the code-checksums of the storage nodes of cluster m. Let H be the code-matrix of a *systematic* $(N_m + K, K)$ RS code over a $GF(2^w)$ field, where $2^w > N_m + K$. The encoding state of cluster m adheres to the following equation

$$H \cdot \begin{bmatrix} d_1 \\ \vdots \\ d_K \end{bmatrix} = \begin{bmatrix} 1 & \cdots & 0 \\ \vdots & & \vdots \\ 0 & \cdots & 1 \\ a_{1,1} & \cdots & a_{1,K} \\ \vdots & & \vdots \\ a_{N_m,1} & \cdots & a_{N_m,K} \end{bmatrix} \begin{bmatrix} d_1 \\ \vdots \\ d_K \end{bmatrix} = \begin{bmatrix} d_1 \\ \vdots \\ d_K \\ s_{m,1} \\ \vdots \\ s_{m,N_m} \end{bmatrix}.$$

Based on the above equation, a storage node checksum, say $s_{m,i}$, will be

$$s_{m,i} = a_{i,1}d_1 + a_{i,2}d_2 + \ldots + a_{i,K}d_K.$$

Of course, some data nodes may be served by other clusters. For those data nodes, we consider that they send the value of zero, which we assume is not a valid content. This coding scheme tolerates up to N_m *erasures*, which means that we can find out all of $d_1, \ldots, d_K$ values by selecting K storage nodes only.

Let K_m be the number of non-zero d_k's in the mth cluster. Clearly, $K_m \leq K$. The question is:

Can we recover the K_m non-zero d_k's by selecting only K_m (as opposed to K) storage nodes?

I.e. whether the K_m non-zero d_k's are calculated from a system of K_m equations. In order to achieve this, we modify our coding state. We observe that our erasure scheme is special in that all d_k's are always "erased." Hence, the problem boils down to solving (with respect to d_k's) a system of linear equations obtained by selecting any K_m equations from the following system of N_m equations represented in matrix form as

$$\begin{bmatrix} a_{1,1} & \cdots & a_{1,K} \\ \vdots & & \vdots \\ a_{N_m,1} & \cdots & a_{N_m,K} \end{bmatrix} \begin{bmatrix} d_1 \\ \vdots \\ d_K \end{bmatrix} = \begin{bmatrix} s_{m,1} \\ \vdots \\ s_{m,N_m} \end{bmatrix}.$$

The above translates into asking that any submatrix of the coefficients' matrix (on the left), obtained by selecting K_m rows, to be invertible.

In other words, because of our particular erasure model, we only need to find such a "nicely behaved" matrix for computing the $s_{m,i}$'s. Fortunately, an $N_m \times K$ Vandermonde matrix fits our needs. Such a Vandermonde matrix is

$$\begin{bmatrix} 1 & 1 & \cdots & 1 \\ 1 & b_1 & \cdots & b_{K-1} \\ \vdots & & \vdots & \\ 1 & b_1^{N_m-1} & \cdots & b_{K-1}^{N_m-1} \end{bmatrix}$$

where $1, b_1, \ldots, b_{K-1}$ are K different elements of the underlying $GF(2^w)$ field.

The desired property of such an $N_m \times K$ Vandermonde matrix is that every subset of K rows is guaranteed to be linearly independent. Furthermore, the west part, $(1 \ldots N_m) \times (1 \ldots K_m)$, for $1 \leq K_m \leq K$, is also a Vandermonde matrix with the property that any set of K_m rows is linearly independent.

Thus, by using the above $N_m \times K$ Vandermonde matrix as the lower part of our systematic code-matrix, and by consolidating the non-zero d_k's to be always the first in the data node vector, we guarantee that with only K_m equations (for $s_{m,i}$'s) we are able to calculate all the values of the non-zero d_k's. Hence, we need to query only K_m of the storage nodes in order to recover the contents of the K_m data nodes that are in the proximity of cluster m.

When using this scheme, we do not need to store the coefficients since they can be computed on the fly as powers of $b_1, \ldots b_{N_m}$, whereas these elements can be considered as consecutive powers of a field generator g.

Remark. Vandermonde matrices are commonly used in creating systematic Reed-Solomon codes for RAID schemes recovering from disk failures (see [10]). For this, one could start with an $(N_m + K) \times K$ Vandermonde matrix and then apply elementary matrix operations to bring it into a systematic form. The final matrix is, of course, not Vandermonde anymore.

We emphasize that, having an $(N_m + K) \times K$ code-matrix which has a $K \times K$ identity matrix as upper part and an $N_m \times K$ Vandermonde matrix as lower part (as we are proposing) would not (in general) allow us to fully decode having only K values out of d_k's and $s_{m,i}$'s. Such a matrix is not good for a RAID scheme as any disk, regular or redundant, can fail, and we need to recover the data using the remaining disks (whose number needs to be $\geq K$).

On the other hand, in our setting, we need instead to recover K_m of d_k's from $s_{m,i}$'s. This is to say that, the K_m data nodes in the vicinity of the cluster "always fail" whereas the storage nodes in the cluster are "always alive." For this, our proposed code-matrix allows us to decode the data of the K_m data nodes, by querying only K_m storage nodes (storing $s_{m,i}$'s).

3   Decoding and Sampling

The random sampling procedure is as follows. A data collector chooses a random subset of storage nodes and queries them. The storage nodes reply with their checksums. Then, the data collector is able to decode if it receives at least K_m checksums from cluster m, for $1 \leq m \leq M$.

As we mentioned before, the data collector may have no clue of the clustering structure. As such, when a data collector retrieves information from a WSN, it randomly queries a set S of sensors from the whole network. We call this set of sensors a *sample*. Let S_m be the intersection of S and m-th cluster. Therefore, there are $|S_m|$ selected storage nodes in the m-th cluster. Let $X_m = |S_m| - K_m$. Then, the data collector can fully decode if $|S_m| \geq K_m$ (or $X_m \geq 0$), for $m = 1, \ldots, M$. Clearly, the smallest sample size is $K = K_1 + \ldots + K_M$. We define $Pr(S)$ to be the probability that the user can decode using sample S. We can show that

Theorem 1.

$$Pr(S) = \frac{\sum_{X_1} \sum_{X_2} \cdots \sum_{X_M} \binom{N_1}{K_1+X_1}\binom{N_2}{K_2+X_2} \cdots \binom{N_M}{K_M+X_M}}{\binom{N}{|S|}}.$$

To illustrate, suppose that we have 3 clusters of 7 storage nodes each. Also suppose that $K_1 = K_2 = K_3 = 4$; and thus the smallest sample size is 12. If $|S|$ is 13, then the extra selected node can be in any cluster. There are three possible cases for the extra node. Therefore, $Pr(S) = \frac{\binom{7}{4+1}\binom{7}{4}\binom{7}{4}+\binom{7}{4}\binom{7}{4+1}\binom{7}{4}+\binom{7}{4}\binom{7}{4}\binom{7}{4+1}}{\binom{21}{13}}$
$=0.379$. Formally, we are solving the following problem.

Find minimum $|S|$

Subject to

$Pr(S) \geq 1 - \varepsilon$

$X_m, K_m, N_m \geq 0$, for $1 \leq m \leq M$

$X_1 + X_2 + \ldots + X_M = X$

$K_1 + K_2 + \ldots + K_M = K$

$N_1 + N_2 + \ldots + N_M = N$

$|S| = K + X$

$X \leq N - K$

We solve the above problem numerically by calculating $Pr(S)$ of different sample sizes ranging from K to N. This approach is inefficient since the calculation of $Pr(S)$ requires at least $O(\binom{M+X-1}{X})$ time, and the initial value of $|S|$ may be far from the optimum solution. The rest of this section approximates the $Pr(S)$ in linear time and provides a "safe" sample size that can be used directly by data collectors.

For the approximation, suppose that there are N balls in M bins. The balls are numbered from 1 to N, and the bins are numbered from 1 to M. The number of balls in the bin m is N_m. Besides, each bin has a threshold value K_m. We randomly pick up a ball, record the ball number and bin number, put it back, and pick up another ball. We consider the following question.

After we pick $|S|$ times, what is the probability that we pick at least K_m balls from bin m, for $m = 1, \ldots M$?

We denote this probability as $Pr'(S)$ and use $Pr'(S)$ to approximate $Pr(S)$. Next we show that $Pr'(S)$ is easy to calculate and provides a good lower bound for $Pr(S)$.

We firstly consider bin 1. Define an indicator random variable

$$Y_i = \begin{cases} 1 \text{ if the } i^{th} \text{ ball is picked from bin 1} \\ 0 \text{ otherwise} \end{cases}$$

Let random variable $Z_1 = \sum Y_i$ be the number of the ball that we pick from bin 1. And then, we denote E_1 as the event that $Z_1 \geq K_1$, and $\overline{E}_1$ as the event that $Z_1 < K_1$. Clearly, $Pr'(S) = 1 - Pr(\overline{E}_1 \cup \overline{E}_2 \cup \ldots \cup \overline{E}_M)$. We use the following lemma from [2] to analyze $Pr(\overline{E}_1)$.

Lemma 1. *Consider a sequence of n Bernoulli trials, where in the ith trial, success occurs with probability p_i and failure with probability $q_i = 1 - p_i$. Let Z be the random variable describing the total number of successes, and $\mu = E[Z]$. Then, for $r > 0$, $Pr(\mu - Z \geq r) \leq exp[\frac{-r^2}{2n}]$.*

Let $p_1 = N_1/N$ be the probability that we pick a ball from bin 1, and let r be $|S|p_1 - K_1$. Then, $Pr(\mu - Z_1 \geq r) = Pr(Z_1 \leq K_1) = Pr(\overline{E}_1)$ $\leq exp[\frac{-(|S|p_1 - K_1)^2}{2|S|}]$. If we take the union bound for all bins, we can find the probability that we fail to decode. We denote it as $Pr(\text{fail})$. $Pr(\text{fail}) = 1 - Pr'(S) \leq \sum_{m=1}^{M} Pr(\overline{E}_m) = \sum_{m=1}^{M} exp[\frac{-(|S|p_m - K_m)^2}{2|S|}]$. Assume that the $\widetilde{m}$-th bin has the maximum $exp[\frac{-(|S|p_m - K_m)^2}{2|S|}]$ value among all bins. Let $\widetilde{p}$ and $\widetilde{K}$ be the probability that we pick a ball from the $\widetilde{m}$-th bin and the threshold value of the $\widetilde{m}$-th bin respectively. We substitute $exp[\frac{-(|S|\widetilde{p} - \widetilde{K})^2}{2|S|}]$ for each $exp[\frac{-(|S|p_m - K_m)^2}{2|S|}]$ value in $Pr(\text{fail})$. Let the threshold be ε. Then, we have

$$Pr(\text{fail}) \leq M \cdot exp[\frac{-(|S|\widetilde{p} - \widetilde{K})^2}{2|S|}] < \varepsilon$$

In our sampling scheme, we sample the network without replacement. In other words, our scheme has better probability of success than the above ball and bin game. Finally, we solve the above equation and conclude with the following theorems.

Theorem 2. *The decoding probability $Pr(S)$ is no less than*

$$1 - \sum_{m=1}^{M} exp[\frac{-(|S|p_m - K_m)^2}{2|S|}].$$

Theorem 3. *We can choose a sample size*

$$|S| = \frac{2(\widetilde{p}\widetilde{K} + \ln \frac{M}{\varepsilon}) + \sqrt{4(\widetilde{p}\widetilde{K} + \ln \frac{M}{\varepsilon})^2 - 4\widetilde{p}^2\widetilde{K}^2}}{2\widetilde{p}^2}$$

to achieve more than $1 - \varepsilon$ probability of decoding.

4   Performance Evaluation and Conclusion

We use a grid network of N storage nodes as an example. The same analytic principle is applicable to other network topology. Suppose that the storage nodes are deployed as a grid in a unit square. Each data node moves around randomly and uploads its sensor reading to its nearest storage node. A clustering algorithm partitions the grid into M small squares. Regarding the data distribution, our coding scheme has a message complexity of $\sum_{m=1}^{M} K_m N_m (2/3)\sqrt{N_m}$, whereas the DEC has a message complexity of $5N \ln K \cdot (2/3)\sqrt{N}$. For an illustration, if $N_m \approx N/M$ and $K_m \approx K/M$ (for $1 \leq m \leq M$), then the condition for our scheme to be better than DEC is that

$$M \frac{K}{M} \frac{N}{M} \sqrt{\frac{N}{M}} \leq 5N \ln K \sqrt{N} \Rightarrow \left(\frac{K}{5 \ln K} \right)^{\frac{2}{3}} \leq M,$$

which is generally true. The experiments also confirm the same conclusion. We use 10,000 storage nodes and 100 data nodes in our simulation. The results show that our decoding probability is sufficiently high, and our data distribution cost is smaller than DEC's cost when $M > 4$. Detailed results and complexity analysis are in the full version of this paper at *http://web.uvic.ca/~yhl/ssdbm08full.pdf*.

In conclusion, we have proposed a cost-efficient coding scheme that fulfills the ubiquitous access to sensor data. The algorithm is easy to implement on any clustered WSN. Moreover, we give a mathematical model for the probability of decoding as well as analysis of our system cost. The experimental results demonstrate that our coding scheme outperforms DEC.

References

1. Bonnet, P., Gehrke, J.E., Seshadri, P.: Towards Sensor Database Systems. In: Proceedings of the Second International Conference on Mobile Data Management (2001)
2. Cormen, T., Leiserson, C., Rivest, R., Stein, C.: Introduction to Algorithms, 2nd edn. (1990)
3. Dimakis, A.G., Prabhakaran, V., Ramchandran, K.: Decentralized erasure codes for distributed networked storage. IEEE/ACM Transactions on Networking 14(SI) (June 2006)
4. Garcia-Molina, H., Ullman, J.D., Widom, J.D.: Database Systems: The Complete Book (2001)
5. Intanagonwiwat, C., Govindan, R., Estrin, D.: Directed diffusion: A scalable and robust communication paradigm for sensor networks. In: Proceedings of the Sixth Annual International Conference on Mobile Computing and Networking (Mobi-COM) (2000)
6. Juang, P., Oki, H., Wang, Y., Martonosi, M., Peh, L., Rubenstein, D.: Energy-Efficient Computing for Wildlife Tracking: Design Tradeoffs and Early Experiences with ZebraNet. In: Tenth International Conference on Architectural Support for Programming Languages and Operating Systems (ASPLOS-X) (2002)
7. Madden, S., Franklin, M., Hellerstein, J., Hong, W.: TAG: a tiny aggregation service for ad-hoc sensor networks. In: Proceedings of OSDI (2002)
8. Madden, S., Franklin, M., Hellerstein, J., Hong, W.: The design of an acquisitional query processor for sensor networks. In: Proceedings of ACM SIGMOD (2003)
9. Mainwaring, A., Polastre, J., Szewczyk, R., Culler, D., Anderson, J.: Wireless Sensor Networks for Habitat Monitoring. In: Proceedings of ACM WSNA 2002 (2002)
10. Plank, J.S.: A Tutorial on Reed-Solomon Coding for Fault-Tolerance in RAID-like Systems. Software – Practice & Experience 27(9), 995–1012 (1997)
11. Shan, X., Tan, J.: Mobile sensor deployment for a dynamic cluster-based target tracking sensor network. In: Proceedings of IEEE/RSJ International Conference on Intelligent Robots and Systems (2005)
12. Wang, D., Zhang, Q., Liu, J.: Partial Network Coding: Theory and Application for Continuous Sensor Data Collection. In: Proceedings of IWQOS 2006 (2006)

iJoin: Importance-Aware Join Approximation over Data Streams

Dhananjay Kulkarni[1] and Chinya V. Ravishankar[2]

[1] Boston University, USA
kulkarni@bu.edu
[2] University of California - Riverside, USA
ravi@cs.ucr.edu

Abstract. We address approximate join processing over data streams when memory limitations cause incoming tuples to overflow the available memory, precluding exact processing. Moreover, in many real-world applications such as for news-feeds and sensor-data, different tuples may have different *importance* levels. Current methods pay little attention to load-shedding when tuples bear such importance semantics, and perform poorly due to *premature tuple* drops and *unproductive tuple* retention. We propose a novel framework, called iJoin, which overcomes these drawbacks, maximizes result importance, and has the best performance compared to earlier work.

1 Introduction

Windowed Joins [1] are *stateful* operations, which means that all tuples in the window are required to compute the *exact* join result. However, when the memory is already full and a new tuple arrives, there is no recourse but to perform load-shedding [2]. The join result is said to be *approximate* when it is executed over the reduced window (after dropping tuples). Approximate join processing has been studied from different perspectives. When memory is full, [3,4,5,1] propose randomly evicting tuples. [6] argues that this scheme is likely to produce sub-optimal results, and proposes heuristics to maximize the output *size*. But, none of the previous work addresses tuple 'importance', even though many real-world applications demand it.

We present an example to motivate that load-shedding which does not address tuple-importance, such as random-drop [1], and semantic approximation [6] are sub-optimal for two reasons. First, some important tuples suffer *premature evictions* and fail to match with other tuples. Second, some *unproductive* tuples stay in memory too long without contributing much to result.

Let (`t`, `value`:`importance`) represent a stream tuple. Consider streams R and S tuples between time=1 to 8. Stream R's tuples are (1,a:1), (2,b:2), (3,c:3), (4,d:4), (5,d:4), (6,b:2), (7,a:1), (8,c:3) and S's tuples are (1,b:2), (2,a:1), (3,b:2), (4,b:2), (5,c:3), (6,c:3), (7,d:4), (8,a:1). Consider an equi-join over R and S, with a window of 8 time units. Formally, an importance-function F_i returns the importance of the tuple r(t) as

B. Ludäscher and Nikos Mamoulis (Eds.): SSDBM 2008, LNCS 5069, pp. 541–548, 2008.

a function of its attributes as shown in Equation 1, where F_i is the importance-function and $a_1, a_2, \ldots$ are the attributes.

$$r(t).imp = F_i(r(t).a_1, r(t).a_2, \ldots) \tag{1}$$

When a tuple $r(j)$ joins with a tuple $s(i)$, the output tuple $o(j)$ acquires importance based on $r(j).imp$ and $s(i).imp.$ as shown in Equation 2.

$$o(j).imp = \min\{r(j).imp, s(i).imp\} \tag{2}$$

Let $\Omega_q = \{o(i_1), \ldots, o(i_n)\}$ be the output tuples of a join query q. We define *join output quality* (or **total importance** IMP(q) of query q) as in Equation 3.

$$IMP(q) = \sum_{o(i) \in \Omega} o(i).imp \tag{3}$$

An *exact* join (Table 1(a)) in our example produces 16 outputs with a total importance of 36. Let M be the total amount of memory available, and for simplicity, let $M = 4$ be equally divided between R and S. If each tuple occupies 1 unit of memory, a maximum of 2 tuples from each stream can be buffered. Let the tuple $r(t)$ arrive when R's buffer is full. The DSMS has to make either of the following decisions, both resulting in an *approximate* result: (1) the DSMS can drop $r(t)$ and process the tuples already in the join buffer, or (2) the DSMS can evict some buffered tuple, clearing space for $r(t)$. Hence after time $t = 3$, we must evict 1 tuple per tuple arrival.

Table 1(b) illustrates a scheme where tuples are randomly dropped, producing only 3 output tuples with total importance of 5. Let $Str(v, t)$ denote a tuple with value v arriving in stream Str at time t. The total importance suffers because tuple $R(c, 3)$ is dropped prematurely, though it would have matched $S(c, 5)$ at $t = 6$. Also, $R(a, 1)$ produces no output tuple after $t = 2$, but occupies memory until $t = 5$. We call $R(c, 3)$ a *premature* tuple and $R(a, 1)$ an *unproductive* tuple. Our goal is to overcome these drawbacks.

We address the following problem: *Given the available memory M and a sliding-window join query $\langle \alpha, c, w \rangle$, where $\alpha = \{S_1, \ldots, S_n\}$ is the set of streams*

Table 1. Exact join and approximate join under RAND

(a) Exact (total importance = 36)

t	R	S	Output	Imp
2-3	a,b	b,a	b,a	3
3-4	a,b,c	b,a,b	b	2
4-5	a,b,c,d	b,a,b,b	b	2
5-6	a,b,c,d,d	b,a,b,b,c	c	3
6-7	a,b,c,d,d,b	b,a,b,b,c,c	b,b,b,c	9
7-8	a,b,c,d,d,b,a	b,a,b,b,c,c,d	a,d,d	5
8-9	a,b,c,d,d,b,a,c	b,a,b,b,c,c,d,a	c,c,a,a	8

(b) RAND (total imp. = 5)

t	R tuples	S tuples	Out	Imp
2-3	a,b	b,a	b,a	3
3-4	a,c	a,b	-	0
4-5	a,d	b,b	-	0
5-6	d,d	b,c	-	0
6-7	d,b	b,c	b	2
7-8	b,a	b,d	-	0
8-9	b,c	b,a	-	0

(with importance semantics), c is the join condition, and w is the time-window, compute the approximate join such that **total importance** *of the join output is maximized.*

Addressing tuple-importance and **fairness** simultaneously is challenging, because tuple characteristics are not know a priori. None of the previous work addresses fairness during load-shedding, so our contribution here is novel. Similar to Jain's fairness index [7], we measure of fairness as $\texttt{fairness} = \frac{(\sum L_i)^2}{(n \times \sum L_i^2)}$, where tuple lifetime L_i is the difference between time that tuple i gets evicted, and it's arrival time. Fairness ranges from $1/n$ (worst case) to 1 (best case), where n is the input size. As we will see later, our work is practical and allows each tuple a fair chance in being part of the join result.

2   The iJoin Approach

Our framework, which we call *iJoin* is outlined in Figure 1. As shown in Algorithm 1, various tuple-related information (or metadata) is also updated during join operation. If the buffer is full, we drop some tuples based on our load-shedding strategy, we accommodate $r(i)$, and then compute the join as in Algorithm 1.

We maintain the following metadata with each tuple: (1) arrival time (t_a) of the tuple, (2) tuple importance (imp) determined according to Equation 1, (3) number of tuples $(matches)$ that joined with the tuple, (4) timestamp of most recent matching tuple $(prevmatch)$. We use these statistics to determine a tuples 'worth' and choose which tuples to drop. Algorithm 1 shows how the metadata is updated when $r(i)$ finds a match to produce o(i). Function isMatch() in Algorithm 1 returns TRUE only when join condition c is satisfied.

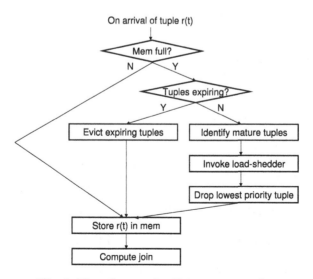

Fig. 1. Flow diagram for iJoin approximation

Algorithm 1. Join operation

Require: window size w, join condition c, $r(i), \gamma = \{s(j)\}$ such that $i - w \le j \le i, w$
Ensure: the output set Ω
1: **for all** $s(j) \in \gamma$ **do**
2:     **if** $isMatch(s(j), r(i), c) = TRUE$ **then**
3:         $o(i) = \{s(j), r(i)\}$, $o(i).imp = \min \{s(j).imp, r(i).imp\}$
4:         $\Omega \leftarrow \Omega \cup \{o(i)\}$
5:         $s(j).matches \leftarrow s(j).matches + 1$, $s(j).prevmatch \leftarrow i$
6:     **end if**
7: **end for**
8: **if** $\Omega \neq \emptyset$ **then**
9:     $r(i).matches \leftarrow |\Omega|$, $r(i).prevmatch \leftarrow i$
10: **end if**

Tuple Priority: A tuple's *residence priority (RP)* indicates how valuable the tuple is to the join process. As we will see later, tuples with the lowest RP get dropped from memory. A tuple $r(i)$ arriving at t_a is said to have a priority $RP(i, t_a)$, and is set to $RP_{init}(i)$ on arrival. We must construct a function F_p to determine the tuple priority at some time $t' > t_a$. Since we want to maximize the total importance of a query output, F_p (see Equation 4) is a function of importance, age, and matches. Residence priority at time t' is computed using Equation 5.

$$F_p(\mathtt{imp}, \mathtt{matches}, \mathtt{age}) = \frac{\mathtt{imp} \times \mathtt{matches}}{\mathtt{age}} \tag{4}$$

$$RP(r(i), t') = F_p(r(i).imp, r(i).matches, (t' - r(i).t_a)) \tag{5}$$

Establishing Tuple Maturity: To reduce premature drops, we place a residence threshold on the tuples before they are candidates for load shedding. A tuple is **mature** if it's age is greater than a certain threshold τ, which can be set higher (or lower) as per high (or low) memory availability.

Penalizing Unproductive Tuples: A tuple is *unproductive* if it does not produce a join output for a long time. Our goal is to identify such tuples, and penalize it for occupying memory so that they will quickly lose their residence priority and will be eventually evicted. For example, consider a output tuple produced at time $\mathtt{t}$. When a arriving tuple $s(t)$ matched with $r(i)$, we record this time by assigning $r(i).prevmatch = t$ and $s(t).prevmatch = t$. A tuple is considered unproductive at time t' if $t' - r(i).prevmatch \ge \Delta$, where Δ is a tunable parameter. If a tuple is identified as being unproductive, we reduce its priority by a penalty δ using Equation 6, where K is some constant.

$$\mathtt{Penalty}(\delta) = \mathtt{K} \times (t' - r(i).prevmatch) \tag{6}$$

We perform load shedding based on residence priority as shown in Algorithm 2. We first identify mature tuples as eviction candidates. Next, we examine

Algorithm 2. Load-shedding invoked at time (t)

Require: $\beta=\{r(i)\}$ such that t-w $\leq$ i $\leq$ t, Maturity threshold τ, Unproductivity
threshold Δ, Penalty δ, k .
1: **for all** r(i) $\in \beta$ **do**
2:    Apply Condition Maturity (τ) to determine if r(i) is MATURE
3:    **if** r(i) is NOT MATURE **then**
4:       $\beta \leftarrow \beta$ - {r(i)}
5:    **end if**
6: **end for**
7: **for all** r(i) $\in \beta$ **do**
8:    Determine the tuple-priority RP(r(i),t)
9:    Apply Condition Unproductivity (Δ) to determine if r(i) is UNPRODUCTIVE
10:   **if** r(i) is UNPRODUCTIVE **then**
11:      RP(r(i), t) $\leftarrow$ RP(r(i),t) - δ
12:   **end if**
13: **end for**
14: **for all** r(i) $\in \beta$ **do**
15:   Sort tuple by tuple priority RP(r(i), t)
16: **end for**
17: Drop r(j) such that RP(r(j),t) = min (RP(r,t)) $\forall$ r $\in \beta$

the `prevmatch` to check for unproductive tuples, apply a penalty, and compute
all new residence priorities. Finally, we drop the tuples with the lowest priority.
   We propose 3 schemes to recompute residence priorities.

(1) Successive: In this scheme, we compute tuple priority each time we need
to evict a tuple from memory. This scheme has high overhead, but provides the
best estimates of the 'worth' of tuples currently in memory.
(2) k-Successive: In this scheme, we compute tuple priorities after every k
load shedding decisions, reducing overhead by k, as compared to the successive
scheme. At relatively stable distributions, k could be set to a higher value.
(3) Adaptive: We monitor the *difference* between consecutive join-
approximation executions and recompute RP only if the difference in join quality
falls below a certain threshold ϵ. Increasing ϵ decreases the overhead.

3   Experiments

We used a synthetic dataset closely resembling real-world settings. The tuples
had categorical attributes, and arrived at the rates between 100–200 tuples per
second. Unless specified, we used the following default parameters: tuple arrival
rate (100–200 tuples per sec), tuple domain (1–100 categorical values), imp domain (1–100 decimal values), join memory (M=10 tuples), window size (w=25
secs), maturity threshold (τ=2 secs), unproductivity threshold (Δ=3 secs). We
used the successive recomputation policy. The importance of each tuple was
mapped to the range 1–100, and we allowed the importance function to be re-
defined no more than twice during execution.

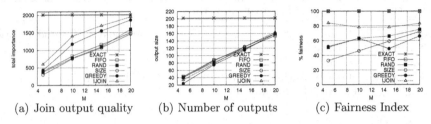

(a) Join output quality   (b) Number of outputs   (c) Fairness Index

Fig. 2. Performance when available memory is varied from 5 to 20

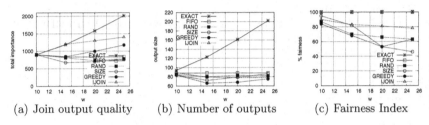

(a) Join output quality   (b) Number of outputs   (c) Fairness Index

Fig. 3. Performance when query window is varied from 10 to 25

We used an equi-join query over 2 streams as a test query. We allowed the query to run for 100 seconds on a Quad Xeon 550MHz. We measured the following: (1) **total importance**, (2) **output size**, and (3) **fairness**. We compared our work with the following load-shedding schemes: (1) **EXACT**: a the optimal scheme where memory is unlimited, (2) **FIFO**: a simple scheme with queue size M, (3) **RAND**: a random load-shedding scheme used in [1], (4) **SIZE**: a scheme similar to [6] that maximizes output size, and (5) **GREEDY**: a greedy scheme that evicts the least important tuple.

Figure 2 shows the effect of memory size. IJOIN has the best performance in meeting our approximation objective of maximizing total importance. EXACT, of course, performs best since it has unlimited memory. Though SIZE is designed to maximize the output-set, our heuristics did better, and also showed higher join quality. This might be due to SIZE's limited ability to detect correlation among joining streams. IJOIN also does the best with respect to fairness. Our scheme is developed to provide equal, or almost equal opportunity to each tuple to find a matching tuple, and our experiments show that IJOIN was more than 80% fair in doing so. FIFO obviously is 100% fair, but has the lowest quality.

Figure 3 shows the performance of various schemes when the window size varied from 10 to 25 seconds. Intuitively, a larger window places higher memory constraints on the join operation, as the available join memory was constant in this experiment. Only the IJOIN and GREEDY schemes seem to improve join quality when the window grows. This is due to IJOIN's ability to find better correlations as window grows, and use recent estimates to determine the more valuable tuples. IJOIN is between 80%-85% fair in retaining tuples in memory. In contrast, SIZE experiences a drop from 90% to 42%.

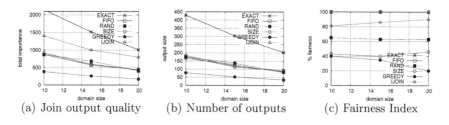

(a) Join output quality   (b) Number of outputs   (c) Fairness Index

Fig. 4. Performance when domain-size is varied from 10 to 20

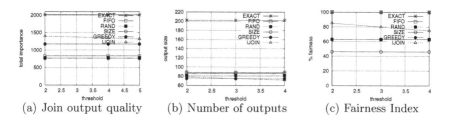

(a) Join output quality   (b) Number of outputs   (c) Fairness Index

Fig. 5. Performance when unproductivity threshold is increased from 2 to 5 seconds

Next, we varied the domain size of the join attribute from 10–20 distinct values and let all tuples have the same importance. Intuitively, when the domain size increases, the probability of the same value appearing in the window decreases. As shown in Figure 4, this affects the number of matching tuples, so we see a linear drop in the output size and total importance. IJOIN still outperforms all other approximation schemes, and is consistently fair between 80%-85%. Interestingly, GREEDY has the worst performance, because all tuples have the same importance-level which this leads to only a selected few tuples occupying the memory.

We also studied the effects of the unproductivity threshold Δ. As Figure 5 shows, IJOIN quality drops as Δ increases, since a tuple a can remain longer without contributing to any result. We recommend a finer threshold (low value) or applying higher penalty δ (see Algorithm 2) when tuples are identified as unproductive.

Table 2 shows that **successive** provides the best approximation, and yields the highest 'fairness' measure, at about 80%. However, **adaptive** has a lower overhead than **successive**, with only 600 tuple priority recomputations. **k-successive** is a

Table 2. Performance of re-computation schemes

	Successive	k-Successive	Adaptive
Total importance	1400	1200	1300
Output size (#tuples)	80	75	80
Fairness	81%	75%	80%
Overhead (#invocations)	high (1000)	low (100)	medium (600)

balance between these two schemes, with a low overhead and at about 75% fair. We suggest that **successive** be used in erratic environments, where data distribution and tuple correlations are likely to be volatile. **Adaptive** performs best when data distributions are relatively stable.

4   Conclusions

We have proposed a join-approximation framework, called *iJoin* to address tuple-importance. We have shown that limiting premature drops, and penalizing unproductive tuples is practical and very effective in improving quality of joins. We have also shown that fairness is important goal when memory is limited. Moreover, we have outlined 3 schemes to recompute tuple-priorities and shown that successive scheme is the best to use if data is erratic, and the adaptive scheme provides the best join quality vs. overhead trade-off.

Acknowledgements. This work was partially supported by grants from Tata Consultancy Services, Inc.

References

1. Kang, J., Naughton, J.F., Viglas, S.D.: Evaluating window joins over unbounded streams. In: Proceedings of ICDE (2003)
2. Tatbul, N., Cetintemel, U., Zdonik, S., Chemiack, M., Stonebraker, M.: Load shedding in a data stream manager. In: Proceedings of VLDB (2003)
3. Urhan, T., Franklin, M.J.: XJoin: A reactively-scheduled pipelined join operator. IEEE Data Engineering Bulletin 23(2), 27–33 (2000)
4. Viglas, S., Naughton, J.F., Burger, J.: Maximizing the output rate of multi-join queries over streaming information sources. In: Proceedings of VLDB (2003)
5. Viglas, S., Naughton, J.F.: Rate-based query optimization for streaming information sources. In: Proceedings of SIGMOD (2002)
6. Das, A.: Semantic approximation of data stream joins. IEEE Transactions on Knowledge and Data Engineering 17(1), 44–59 (2005) (Member-Johannes Gehrke and Member-Mirek Riedewald)
7. Jain, R., Chiu, D., Hawe, W.: A quantitative measure of fairness and discrimination for resource allocation in shared computer systems. DEC Research Report TR-301, Digital Equipment Corporation, Maynard, MA, USA (September 1984)

Efficient Continuous K-Nearest Neighbor Query Processing over Moving Objects with Uncertain Speed and Direction

Yuan-Ko Huang, Shi-Jei Liao, and Chiang Lee

Department of Computer Science and Information Engineering, National
Cheng-Kung University, Tainan, Taiwan, R.O.C.
{hyk,jayliau}@dblab.csie.ncku.edu.tw
leec@mail.ncku.edu.tw

Abstract. One of the important types of queries in spatio-temporal
databases is the Continuous K-Nearest Neighbor (CKNN) query, which
is to find among all moving objects the K-Nearest Neighbors (KNNs)
of a mobile user at each time instant within a user-given time interval
$[t_s, t_e]$. In this paper, we focus on how to process such a CKNN query
efficiently when the moving *speed* and *direction* of each moving object are
uncertain. We thoroughly analyze the complicated problems incurred by
this uncertainty and propose a *Continuous PKNN (CPKNN) algorithm*
to effectively tackle these problems.

1 Introduction

With the increasing number of real-world applications that involve large spatio-
temporal data sets, providing efficient query processing techniques for these applications becomes essential. One of the important types of queries in spatio-
temporal databases is the Continuous K-Nearest Neighbor (CKNN) query. A
CKNN query is defined as finding the K-nearest neighbors (KNNs) of a moving
user at each time instant within a user-given time interval $[t_s, t_e]$. In this paper,
we focus on how to process such a CKNN query efficiently when the moving
speed and *direction* of each moving object are uncertain. The uncertain ranges
of speed and of direction depend on factors such as the historical object information (e.g., location, speed, and direction), the traffic condition on the road,
and the location update frequency. We assume that the varying ranges of speed
and direction can be computed based on the above factors by the location server
and then is sent to each object. With the uncertain speed and direction, the
possible locations of each object would be bounded by a *sector region* (depicted
as a shaded region in Figure 1), which is formed by two segments and two arcs.
In other words, the exact location of an object should be somewhere inside this
sector region. As each moving object is aware of its sector region, a location
update is issued to the server only when it moves out of this region. In this way,
the update frequency for each object is reduced and the communication cost
between the object and the server and the update cost of the server are both
reduced.

B. Ludäscher and Nikos Mamoulis (Eds.): SSDBM 2008, LNCS 5069, pp. 549–557, 2008.

The problem caused by the release of the fixed speed and fixed direction assumption is that the uncertainty of moving speed and direction of an object makes it difficult to determine precisely the distance between a moving object and the query object. The uncertain distance between two such moving objects would lead to an uncertain (or possible) solution which complicates the process of evaluating a

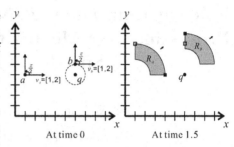

Fig. 1. Example of a $CKNN$ query

$CKNN$ query. For instance, Figure 1 shows an example of $CKNN$ query for moving objects with uncertain speed and direction. The start locations of objects a and b are at $(1,4)$ and $(6,5)$, respectively. Both of the two objects move with an uncertain speed, which lies in between 1 m/sec and 2 m/sec. Let the uncertain speed be denoted as within a range $[1,2]$. In addition, the uncertain moving directions of both a and b are within a range of angles, which is represented as $[0, \frac{\pi}{2}]$. For ease of illustration, query object q is stationary at location $(6,4)$ (note that this is however not required in our technique). Region R_a and region R_b in the figure represent the possible locations of the moving objects a and b, respectively. Apparently, R_a and R_b will grow as time passes. At time 1.5, if a and b are located at $(4,4)$ and $(6,8)$, respectively, then the 1NN of q is a. However, if a and b are at $(1,7)$ and $(6,7)$, respectively, which is as likely to occur as in the other case, then the 1NN of q is b. From this example, we see that all "possible" locations of an object should be taken into account in searching for the KNNs of q so as to guarantee that all answers will be included in the result. These possible answers are named the *possible KNNs* (or PKNNs for short). Efficiently identifying the PKNNs for every time instant is a complicated issue and will be thoroughly investigated in this paper.

The rest of this paper is organized as follows. In Section 2, we discuss some related works on processing $CKNN$ queries. In Section 3, we present our uncertain distance model. Section 4 describes the TPR$^{(s,d)}$-tree. The CPKNN algorithm is presented in Section 5. Finally, Section 6 concludes the paper.

2   Related Work

Past methods in the literature can be classified into three categories according to how the velocity of a moving object is treated. Papers in the first category assume the velocity (i.e., speed and direction) of a moving object is fixed [3,5,8,9]. Under this assumption, the trajectory of a moving object can be precisely determined. Tao et al. [8] proposed a repetitive query processing approach for answering $CKNN$ queries. Recently, they proposed another method [9] to overcome the high-cost problems of their repetitive query processing approach, but the method is only applicable to static objects. To adapt to moving objects and reduce computational cost, Raptopoulou et al. [5] proposed an efficient method that processes the $CKNN$

query by tracing the change of the K-th nearest neighbor. Iwerks et al. [3] answered the CKNN query through the technique of processing the *within query*. Papers of the second category release the fixed velocity assumption so that in their work the velocity of a moving object is uncertain. Under this circumstance, the exact moving trajectory is impossible to achieve as the location of every object can only be updated periodically so that we simply learn the approximate trajectory of a moving object [4,7,10,11]. Song *et al.* [7] proposed an approach in which the snapshot KNN query is re-evaluated whenever the location of the query object is updated. The data objects in their work are assumed to be static, which simplifies the location update problem. Mouratidis *et al.* [4], Xiong *et al.* [10], and Yu *et al.* [11] extended the approach presented in [7] to adapt to moving object datasets. Our previous work [1] represents an approach of the third category which remedies the shortcomings of the above categories of approaches by allowing an object to move with an uncertain speed and a fixed direction. It uplifts the practicality of the approaches in the first category in that the speed is allowed to vary within a range rather than fixed at an unchangeable speed. Also, it avoids query re-evaluation required in the approaches of the second category. However, the limitation of the fixed moving direction of an object restricts the applicability of the work to a realistic environment.

3   Uncertain Distance Model

The distance between two objects in our model is the Euclidean distance. We assume that the moving speed of an object is between a minimal and a maximal speed. Also, the moving direction lies in between a minimal and a maximal angle.

Note that in our model, each angle is represented as a *polar angle* ranging in $[0, 2\pi]$. For example, Figure 2 shows that the speed (and direction) of an object o varies within a range $[o.v, o.V]$ (and $[o.\theta, o.\Theta]$). In the case that object o moves with the minimal speed $o.v$ and the minimal angle $o.\theta$, its location at time t, denoted as $\vec{o}_\alpha(t)$, can be represented as $\vec{o}_\alpha(t) = o.\vec{s} + o.\vec{v}_\alpha \times (t - t_0)$, where $o.\vec{s} = (o.x_0, o.y_0)$ is the start location, $o.\vec{v}_\alpha = (o.v \times \cos(o.\theta), o.v \times \sin(o.\theta))$ is the velocity vector, and t_0 is the start time. In another case where object o's speed and di-

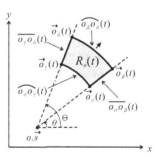

Fig. 2. The possible region

rection are $o.v$ and $o.\Theta$ respectively (i.e., o moves with the minimal speed and the maximal angle), the location of o at time t, denoted as $\vec{o}_\gamma(t)$, can be obtained from the above equation by substituting the velocity vector $o.\vec{v}_\gamma$ for $o.\vec{v}_\alpha$, where $o.\vec{v}_\gamma = (o.v \times \cos(o.\Theta), o.v \times \sin(o.\Theta))$. If the direction of o is uncertain and varies between $[o.\theta, o.\Theta]$, all of its possible locations at time t would form an *arc*, denoted as $o_\alpha \widehat{o}_\gamma(t)$, in which $\vec{o}_\alpha(t)$ and $\vec{o}_\gamma(t)$ are the two endpoints. That is, at time t object o is located at some point on the arc $o_\alpha \widehat{o}_\gamma(t)$.

Similarly, as object o's speed is $o.V$ (i.e., o moves with the maximal speed) and direction is within $[o.\theta, o.\Theta]$, the possible locations of o can be represented by another arc $o_\beta o_\delta(t)$ whose two endpoints are $o_\beta(t)$ and $o_\delta(t)$. Hence, when the speed and the direction are within $[o.v, o.V]$ and $[o.\theta, o.\Theta]$, respectively, all the possible locations of o form a *sector region* (as shown in Figure 2), which is enclosed by four endpoints $\vec{o}_\alpha(t)$, $\vec{o}_\beta(t)$, $\vec{o}_\gamma(t)$, and $\vec{o}_\delta(t)$, two segments $\overline{o_\alpha o_\beta}(t)$ and $\overline{o_\gamma o_\delta}(t)$, and two arcs $o_\alpha o_\gamma(t)$ and $o_\beta o_\delta(t)$. We term the sector region the *possible region* because the object can possibly locate in anywhere inside this region and denote it as $R_o(t)$. In addition, the set of the four endpoints, the set of the two segments, and the set of the two arcs of $R_o(t)$ are denoted as P_o, S_o, and A_o, respectively. As object o moves, $R_o(t)$ moves too and its scope will grow over time.

Given two moving objects o and q, the distance between these two objects is bounded by a minimal and a maximal distance. The minimal and the maximal distances between objects o and q vary over time because the two objects move continuously. We use two functions, $d_{o,q}(t)$ and $D_{o,q}(t)$, to represent at every time instant the minimal and the maximal distances, respectively. The two functions $d_{o,q}(t)$ and $D_{o,q}(t)$ can be obtained by using the relationship of the possible regions $R_o(t)$ and $R_q(t)$.

The possible regions $R_o(t)$ and $R_q(t)$ at an arbitrary instant t may or may not intersect each other. If $R_o(t)$ intersects $R_q(t)$, the minimal distance between $R_o(t)$ and $R_q(t)$ is equal to 0 (i.e., $d_{o,q}(t) = 0$). Otherwise, $d_{o,q}(t)$ would be equal to one of the following distances:

1. The distance between *an endpoint* $\vec{o}_i(t)$ of $R_o(t)$ and *an endpoint* $\vec{q}_j(t)$ of $R_q(t)$. Therefore, $d_{o,q}(t)$ is represented as $\{d(\vec{o}_i(t), \vec{q}_j(t))|\vec{o}_i(t) \in P_o, \vec{q}_j(t) \in P_q\}$.
2. The perpendicular distance from *an endpoint* $\vec{p}(t)$ of one possible region to *a segment* $\overline{p_i p_j}(t)$ of another possible region. That is, $d_{o,q}(t) = \{d(\vec{p}(t), \overline{p_i p_j}(t))| (\vec{p}(t) \in P_o, \overline{p_i p_j}(t) \in S_q) \vee (\vec{p}(t) \in P_q, \overline{p_i p_j}(t) \in S_o)\}$.
3. The minimal distance from *an endpoint* $\vec{p}(t)$ of one possible region to *an arc* $p_i p_j(t)$ of another possible region. Therefore, $d_{o,q}(t) = \{d(\vec{p}(t), p_i p_j(t))| (\vec{p}(t) \in P_o, p_i p_j(t) \in A_q) \vee (\vec{p}(t) \in P_q, p_i p_j(t) \in A_o)\}$.
4. The minimal distance from *a segment* $\overline{p_i p_j}(t)$ of one possible region to *an arc* $p_m p_n(t)$ of another possible region. Hence, $d_{o,q}(t) = \{d(\overline{p_i p_j}(t), p_m p_n(t))| (\overline{p_i p_j}(t) \in S_o, p_m p_n(t) \in A_q) \vee (\overline{p_i p_j}(t) \in S_q, p_m p_n(t) \in A_o)\}$.
5. The minimal distance between *an arc* $o_i o_j(t)$ of $R_o(t)$ and *an arc* $q_i q_j(t)$ of $R_q(t)$. That is, $d_{o,q}(t) = \{d(o_i o_j(t), q_i q_j(t))| o_i o_j(t) \in A_o, q_i q_j(t) \in A_q\}$.

As for the maximal distance between $R_o(t)$ and $R_q(t)$, it could only be equal to (1) the distance $D(\vec{o}_i(t), \vec{q}_j(t))$ between *two endpoints* $\vec{o}_i(t)$ and $\vec{q}_j(t)$, where $\vec{o}_i(t) \in P_o$ and $\vec{q}_j(t) \in P_q$ (i.e., $D_{o,q}(t) = D(\vec{o}_i(t), \vec{q}_j(t))$), or (2) the maximal distance $D(o_i o_j(t), q_i q_j(t))$ between *two arcs* $o_i o_j(t)$ and $q_i q_j(t)$, where $o_i o_j(t) \in A_o$ and $q_i q_j(t) \in A_q$ (i.e., $D_{o,q}(t) = D(o_i o_j(t), q_i q_j(t))$). With the minimal distance function $d_{o,q}(t)$ and the maximal distance function $D_{o,q}(t)$, the possible distances between two objects o and q within the time interval $[t_s, t_e]$ can be represented as a region which is bounded by $d_{o,q}(t)$ and $D_{o,q}(t)$. That is, each point in this region is a possible distance between o and q.

4   The $\text{TPR}^{(s,d)}$-tree

The $\text{TPR}^{(s,d)}$-tree is an enhanced TPR-tree [6], which is built to efficiently index moving objects with uncertain speed and direction. In a $\text{TPR}^{(s,d)}$-tree, objects are recursively grouped in a bottom-up manner according to their locations at the time when the index is built. For instance, Figure 3 gives a two-dimensional example where eleven objects a to k move with uncertain velocity. At time 0, these objects are grouped according to their spatial proximity into four leaf nodes E_3 to E_6. Then, nodes E_3 to E_6 are recursively grouped into nodes E_1 and E_2, that become the entries of the root. Each entry of a leaf node of a $\text{TPR}^{(s,d)}$-tree has the structure $(o.\vec{s}, o.v, o.V, o.\theta, o.\Theta, o.ptr)$, where $o.\vec{s}$ represents the start location, $o.v$ $(o.\theta)$ and $o.V$ $(o.\Theta)$ refer to the minimal and the

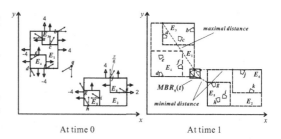

Fig. 3. A two-dimensional example

maximal moving speeds (angles), respectively, and $o.ptr$ is a pointer to the actual object tuple in the database. Each entry of an internal node has the structure $(MBR_E, E.\vec{v}, E.c, E.ptr)$, where MBR_E is the minimum bounding rectangle (MBR) that encloses all the objects in the child node E of this internal node, $E.\vec{v}$ represents the velocity vector of MBR_E, $E.c$ refers to the number of objects enclosed by MBR_E (i.e., the cardinality of node E), and $E.ptr$ is a pointer to node E. Let the velocity vector $E.\vec{v}$ be decomposed to four components $(E.v^{\vdash}, E.v^{\dashv}, E.v^{\perp}, E.v^{\top})$, which represent the velocities of *the left edge, the right edge, the lower edge,* and *the upper edge* of MBR_E, respectively. Then, $E.v^{\vdash} = \min\{o.v^{\vdash} | \forall o \in O_E\}$, where O_E is the set of objects enclosed by MBR_E and

$$o.v^{\vdash} = \begin{cases} -o.V & \text{if } \pi \in [o.\theta, o.\Theta], \\ \min\{o.V \times \cos(o.\theta), o.V \times \cos(o.\Theta)\} & \text{otherwise.} \end{cases}$$

Similarly, $E.v^{\dashv}$, $E.v^{\perp}$, and $E.v^{\top}$ can be obtained after minor revisions of the above equation. For the example shown in Figure 3, the moving speed of each object is assumed to be within the range $[2, 4]$, and the difference between the minimal angle and the maximal angle is $\frac{\pi}{6}$. As MBR_{E_3} encloses three objects a, b, and c (i.e., $E_3.c = 3$), its velocity vector $E_3.\vec{v}$ is determined by the speeds and the angles of the three objects and represented as $(-4, 4, 0, 4)$. Similarly, the velocity vectors of the other MBRs can be derived according to the speeds and the angles of the objects enclosed by those MBRs. With the velocity vector, the extent of each MBR at time t, defined as $MBR(t)$, need not be stored explicitly. Instead, $MBR(t)$ can be obtained from $MBR(0)$ (which represents its extent at time when the index is built) and the velocity vector. For instance, the extent of $MBR_{E_3}(t)$ at time 1 can be obtained from $MBR_{E_3}(0)$ and the velocity vector $(-4, 4, 0, 4)$, and is depicted as the dashed rectangle. From this example, we also see that the possible locations (which is enclosed by a sector region) of objects j and k are still inside the obtained $MBR_{E_6}(t)$. Therefore, we know that if a moving object o with uncertain speed and direction is grouped into a node E, then all the possible locations of o will be inside $MBR_E(t)$ at any time t.

5   CPKNN Algorithm

5.1   Filtering Step

Starting from the root, the traversal of the TPR$^{(s,d)}$-tree is based on the following principles. (1) When an entry o of the leaf node (i.e, an object) is encountered, its minimal distance function $d_{o,q}(t)$ and maximal distance function $D_{o,q}(t)$ are computed because it is possible to be a PKNN (i.e., a candidate). (2) An entry E of the internal node is visited only if its $MBR_E(t)$ contains any qualifying object within the query time interval $[t_s, t_e]$. Two parameters are utilized to determine whether an entry E needs to be visited or not. The first one is the *global minimal distance* between $MBR_E(t)$ and $MBR_q(t)$, denoted as d_E, within the time interval $[t_s, t_e]$. Note that $MBR_q(t)$ is the minimum bounding rectangle enclosing q's sector region at time t and is computed based on the start location, the moving speed, and the moving direction of q. At time t, as both $MBR_E(t)$ and $MBR_q(t)$ are rectangles, the minimal distance between them could be equal to either 0 (that is, $MBR_E(t)$ intersects $MBR_q(t)$) or the minimal distance from a corner of one rectangle to another rectangle. Therefore, if there exists a time instant $t \in [t_s, t_e]$ at which $MBR_E(t)$ intersects $MBR_q(t)$, the global minimal distance $d_E = 0$. Otherwise, $d_E = \min\{d_{c_i,E}(t) \cup d_{c_j,q}(t) | \forall c_i \in C_q, \forall c_j \in C_E, \forall t \in [t_s, t_e]\}$, where $d_{c_i,E}(t)$ (or $d_{c_j,q}(t)$) is the minimal distance from the corner c_i (or c_j) to $MBR_E(t)$ (or $MBR_q(t)$) at time t, and C_q and C_E are the sets of four corners of $MBR_q(t)$ and $MBR_E(t)$, respectively. Efficient methods for the computation of the minimal distance between *a moving point* and *a moving rectangle* have been discussed in previous work [9], which can be used to derive $d_{c_i,E}(t)$ and $d_{c_j,q}(t)$. The second parameter is the *global maximal distance* between $MBR_E(t)$ and $MBR_q(t)$, denoted as D_E, within the time interval $[t_s, t_e]$. Given two rectangles, at any time t the maximal distance between them would be equal to the maximum of the distances between four corners of one rectangle and four corners of another rectangle. As such, the global maximal distance D_E can be represented as $\max\{d_{c_i,c_j}(t) | \forall c_i \in C_q, \forall c_j \in C_E, \forall t \in [t_s, t_e]\}$, where $d_{c_i,c_j}(t)$ refers to the distance between the corner c_i of $MBR_q(t)$ and the corner c_j of $MBR_E(t)$ at time t.

During the traversal of the TPR$^{(s,d)}$-tree, a linked list $\mathcal{L}$ is utilized to keep the information about the entries of the nodes visited so far. Each element of $\mathcal{L}$ stores a node E's information, including the cardinality $E.c$ (i.e., the number of objects enclosed by $MBR_E(t)$), the global maximal distance D_E, and the global minimal distance d_E. Note that the elements of $\mathcal{L}$ are sorted in ascending order according to their D_E, and initially $\mathcal{L}$ only contains root node information of the TPR$^{(s,d)}$-tree. In each iteration, the first element of $\mathcal{L}$ (i.e., the element whose D_E is the smallest among the elements of $\mathcal{L}$) will be retrieved and the corresponding node of the TPR$^{(s,d)}$-tree is visited. If the visited node is an internal node, d_E and D_E of each entry of this internal node are computed and then inserted into $\mathcal{L}$. For each element E of $\mathcal{L}$, if its d_E is greater than D_{E_i} of the i-th element E_i which results in $\sum_{j=1}^{i-1} E_j.c < K \le \sum_{j=1}^{i} E_j.c$, then E can be deleted from $\mathcal{L}$. This is because at least K objects whose distances to the query object are less than d_E of the element E can be found so that the objects enclosed by $MBR_E(t)$ must not be the query result. When a leaf node is visited, each entry (i.e., an object) of this leaf node is possible to be a $PKNN$ (i.e., a candidate), and thus its $d_{o,q}(t)$ and $D_{o,q}(t)$ are computed. The above process proceeds until $\mathcal{L}$ is empty. The candidates will be further verified in the refinement step.

5.2 Refinement Step

The maximal distance function $D_{o,q}(t)$ and the minimal distance function $d_{o,q}(t)$ of each candidate that have been computed in the filtering step can now be shown in the *time-distance space*. For example, Figure 4 shows the maximal distance functions (depicted as solid curves) and the minimal distance functions (depicted as dotted curves) of four candidates a to d in the time-distance space.

Based on the relationship between the distance functions of candidates, we know that at each time instant $t \in [t_s, t_e]$, if there exist K candidates whose $D_{o,q}(t)$ are less than $d_{o_i,q}(t)$ of a candidate o_i, then o_i must not be a $PKNN$. Motivated by this, the candidate o_k whose maximal distance ranks at the K-th smallest among all candidates' maximal distances is used to generate an *answer curve* in the time-distance space, which is formed by the K-th smallest maximal distances (i.e., the values of $D_{o_k,q}(t)$) at all time instants within $[t_s, t_e]$. That is, at every time $t \in [t_s, t_e]$, there are exactly K candidates whose $D_{o,q}(t)$ are

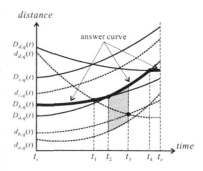

Fig. 4. The refinement step

less than or equal to $D_{o_k,q}(t)$. Continuing the example shown in Figure 4, assume that two NNs are to be found between $[t_s, t_e]$ (i.e., $K = 2$). Candidate b is the candidate o_k at t_s because it has the second smallest maximal distance $D_{b,q}(t_s)$. Then, candidate a replaces b to be the next o_k at t_2, and is replaced by d at t_4. Note that the time instant at which o_k is replaced by another candidate o (e.g., t_2 or t_4) can be obtained by solving the equation $D_{o_k,q}(t) = D_{o,q}(t)$. Finally, the answer curve is obtained and shown as the bold line in the figure.

Having generated the answer curve, the PKNNs can be found by examining the relationship between $d_{o,q}(t)$ of each candidate and the answer curve. There are three possible cases. The first one is that $d_{o,q}(t)$ of candidate o is *completely above* the answer curve (e.g., candidate c in Figure 4). It means that at least

K candidates are closer to the query object at each time instant than o is, so
that o can never become part of the query result. The second case is that $d_{o,q}(t)$
is *completely below* the answer curve (e.g., candidates a and b), and thus o is a
PKNN at each time instant $t \in [t_s, t_e]$. The last case is that $d_{o,q}(t)$ *intersects* the
answer curve at somewhere in $[t_s, t_e]$. It means that o cannot always be a PKNN
within $[t_s, t_e]$ (e.g., candidate d).

6   Conclusions

In this paper, we focused on processing the CKNN query for moving objects with
uncertain speed and direction. We proposed an uncertain distance model to formulate the uncertain distance between moving objects. Based on this model,
we developed the CPKNN algorithm to efficiently process a CKNN query. The
filtering step of the CPKNN algorithm employed a branch-and-bound traversal on the TPR$^{(s,d)}$-tree to prune non-qualifying objects. Then, the refinement
step is designed to determine the subintervals within which the query object
has the same PKNNs, and find the corresponding PKNNs within each of these
subintervals. Readers may refer to our technical report [2] for the details on the
performance of the system.

Acknowledgements. This work was supported by National Science Council of
Taiwan (R.O.C.) under Grants NSC95-2221-E-006-206-MY3.

References

1. Huang, Y.-K., Chen, C.-C., Lee, C.: Continuous k-nearest neighbor query for moving objects with uncertain velocity. Accepted and to appear in GeoInformatica
(available online)
2. Huang, Y.-K., Liao, S.-J., Lee, C.: Efficient continuous k-nearest neighbor query
processing over moving objects with uncertain speed and direction. Technical Report TR-03-27, Department of Computer Science and Information Engineering,
National Cheng Kung University
3. Iwerks, G., Samet, H., Smith, K.: Continuous k-nearest neighbor queries for continuously moving points with updates. In: Proceedings of the International Conference
on Very Large Data Bases, Berlin, Germany (2003)
4. Mouratidis, K., Hadjieleftheriou, M., Papadias, D.: Conceptual partitioning: An
efficient method for continuous nearest neighbor monitoring. In: Proceedings of
the ACM SIGMOD (2005)
5. Raptopoulou, K., Papadopoulos, A., Manolopoulos, Y.: Fast nearest-neighbor
query processing in moving-object databases. GeoInformatica 7(2), 113–137 (2003)
6. Saltenis, S., Jensen, C.S., Leutenegger, S.T., Lopez, M.A.: Indexing the positions
of continuously moving objects. In: Proceedings of the ACM SIGMOD (2000)
7. Song, Z., Roussopoulos, N.: K-nearest neighbor search for moving query point. In:
Proceedings of 7th International Symposium on Advances in Spatial and Temporal
Databases, Redondo Beach, CA, USA (2001)

8. Tao, Y., Papadias, D.: Time parameterized queries in spatio-temporal databases. In: Proceedings of the ACM SIGMOD, Madison, Wisconsin (2002)

9. Tao, Y., Papadias, D., Shen, Q.: Continuous nearest neighbor search. In: Proceedings of the International Conference on Very Large Data Bases, Hong Kong, China (2002)

10. Xiong, X., Mokbel, M.F., Aref, W.G.: Sea-cnn: Scalable processing of continuous k-nearest neighbor queries in spatio-temporal databases. In: Proceedings of the International Conference on Data Engineering (2005)

11. Yu, X., Pu, K.Q., Koudas, N.: Monitoring k-nearest neighbor queries over moving objects. In: Proceedings of the International Conference on Data Engineering (2005)

ProUD: Probabilistic Ranking in Uncertain Databases

Thomas Bernecker, Hans-Peter Kriegel, and Matthias Renz

Institute for Informatics, Ludwig-Maximilians-Universität München, Germany
{bernecker,kriegel,renz}@dbs.ifi.lmu.de

Abstract. There are a lot of application domains, e.g. sensor databases, traffic management or recognition systems, where objects have to be compared based on vague and uncertain data. Feature databases with uncertain data require special methods for effective similarity search. In this paper, we propose an effective and efficient probabilistic similarity ranking algorithm that exploits the full information given by inexact object representations. Thereby, we assume that the objects are given in form of discrete probabilistic object locations in particular several object snapshots with confidence values. Based on the given object representations, we suggest diverse variants of probabilistic ranking schemes. In a detailed experimental evaluation, we demonstrate the benefits of our probabilistic ranking approaches. The experiments show that we can achieve high quality query results while keeping the computational cost quite small.

1 Introduction

Similarity ranking is one of the most important query types in feature databases. A similarity ranking query iteratively reports objects in descending order of their similarity to a given query object. The iterative computation of the answers is very suitable for retrieving the results the user could have in mind. This is a big advantage of ranking queries against the most prominent similarity queries, the distance-range (ε-range) and the k-nearest neighbor query, in particular if the user does not know how to specify the query parameters ε and k.

Many modern applications have to cope with uncertain or imprecise data. Example applications are location determination and proximity detection of moving objects, similarity search and pattern matching in sensor databases or personal identification and recognition systems based on video images or scanned image data. The importance of this topic in the context of database systems is demonstrated by the increasing interest of the database research community in this subject matter. Several approaches coping with uncertain objects have been proposed [1,2,3,8]. All these approaches use continuous probability density functions (pdfs) for the description of the spatial uncertainty while the approaches proposed in [5,6] use discrete representations of uncertain objects. The approach proposed in [5] supports probabilistic distance range queries on uncertain objects. In [6] efficient methods for probabilistic nearest-neighbor queries are proposed. However, in fact only one-nearest neighbor queries are supported.

Similarity search in conjunction with multimedia data like images, music, or data from personal identification systems like face snapshots or fingerprints commonly involves distance computations within the feature space. If exact features cannot be generated from uncertain objects, we have to cope with *positionally* uncertain vectors in

B. Ludäscher and Nikos Mamoulis (Eds.): SSDBM 2008, LNCS 5069, pp. 558–565, 2008.

the feature space (i.e. objects are represented by ambiguous feature vectors). Basically, there exist two forms of representations of *positionally* uncertain data: Uncertain positions represented by a probability density function (pdf) or uncertain positions drawn by samples. In this paper we concentrate on uncertain objects represented by a set of sample positions, each associated with a confidence value. The confidence values indicate how well the corresponding sample matches the exact object. This form of representation is motivated by the fact that we often have only discrete but ambiguous object information as usually returned by common sensor devices, e.g. discrete snapshots of continuously moving objects.

A probabilistic ranking on uncertain objects computes for each object $o \in \mathcal{D}$ the probability that o is the K-th nearest neighbor $(1 \leq K \leq |\mathcal{D}|)$ of a given query object q. In the context of probabilistic ranking queries we propose diverse forms of ranking outputs which differ in the order the objects are reported to the user. Furthermore, we suggest diverse forms in which the results are reported (i.e. which kind of information is assigned to each result).

2 Problem Definition

In this section, we formally introduce the problem of probabilistic ranking queries on uncertain objects. We first start with the definition of (positionally) uncertain objects.

2.1 Positionally Uncertain Objects

Objects of a d-dimensional vector space $\mathbb{R}^d$ are called *positionally uncertain*, if they do not have a unique position in $\mathbb{R}^d$, but have multiple positions associated with a probability value. Thereby, the probability value assigned to a position $p \in \mathbb{R}^d$ of an object o indicates the likelihood that p is the best of all representations for o. A formal definition is given in the following:

Definition 1 (positionally uncertain object). *Let $\mathcal{D}$ be a database of objects located in a d-dimensional feature space $\mathbb{R}^d$. An object $o_i \in \mathcal{D}$ is called* positionally uncertain, *iff the object cannot be assigned to a unique position in $\mathbb{R}^d$. A positionally uncertain object o_i is represented by a set of M sample points $\mathcal{S}(o_i) = \{o_{i,1}, .., o_{i,M}\}$, where $o_{i,j} \in \mathbb{R}^d$ $(1 \leq j \leq M)$.*

Let us note that in many applications the positionally uncertain objects are already given in the discrete representation, i.e. by a set of sample points, in particular if the objects are derived from a sequence of sensor signals, e.g. in object tracking systems. Otherwise, we use the generally applicable concept of Monte-Carlo sampling to generate the set of samples according to a given continuous probability density function.

In the remainder, we call *positionally uncertain objects* simply *uncertain objects* and use both notions alternately.

2.2 Distance Computation for Uncertain Objects

Positionally uncertain objects involve uncertain distances between them. Like the uncertain position, the distance between two uncertain objects (or between two objects where at least one of them is an uncertain object) can be described by a probability density function (pdf) that reflects the probability for each possible distance value. However for uncertain objects with discrete uncertainty representations we need another form of distance.

Definition 2 (uncertain distance). *Let* $o_i = \{o_{i,1}, \ldots, o_{i,M}\} \in \mathcal{D}$ *and* $o_j = \{o_{j,1}, \ldots, o_{j,M}\} \in \mathcal{D}$ *be two uncertain objects, each represented by* M *sample points and let* $dist : \mathbb{R}^d \times \mathbb{R}^d \rightarrow \mathbb{R}_0^+$ *be a distance function. Then an* uncertain distance $d_{uncertain}$ *between two uncertain objects* o_i *and* o_j *is a collection of* M^2 *distance samples as defined below*

$$d_{uncertain}(o_i, o_j) = \{dist(o_{i,m}, o_{j,n}) | 1 \leq m \leq M, 1 \leq n \leq M\},$$

where $dist()$ is a L_p-norm based similarity distance.

The probability that the distance $d_{uncertain}(o_i, o_j)$ between two uncertain objects o_i and o_j is smaller than a given range $\varepsilon \in \mathbb{R}_0^+$ can be estimated by:

$$P(d_{uncertain}(o_i, o_j) \leq dist) = \frac{|\{d \in d_{uncertain}(o_i, o_j) | d \leq dist\}|}{|d_{uncertain}(o_i, o_j)|}.$$

Since distance computations between uncertain objects are very expensive, we need computationally inexpensive distance approximations to reduce the candidate set in a filter step. For this reason, we introduce distance approximations that lower and upper bound the uncertain distance between two uncertain objects.

Definition 3 (minimal object distance). *Let* $o_i = \{o_{i,1}, o_{i,2}, .., o_{i,M}\}$ *and* $o_j = \{o_{j,1}, o_{j,2}, .., o_{j,M'}\}$ *be two uncertain objects. Then the distance* $d_{min}(o_i, o_j) = \min_{s=1..M, s'=1..M'}\{dist(o_{i,s}, o_{j,s'})\}$ *is called minimal distance between the objects* o_i *and* o_j.

Likewise, we can define an upper distance bound for uncertain objects.

Definition 4 (maximal object distance). *Let* $o_i = \{o_{i,1}, o_{i,2}, .., o_{i,M}\}$ *and* $o_j = \{o_{j,1}, o_{j,2}, .., o_{j,M'}\}$ *be two uncertain objects. Then the distance* $d_{max}(o_i, o_j) = \max_{s=1..M, s'=1..M'}\{dist(o_{i,s}, o_{j,s'})\}$ *is called maximal distance between the objects* o_i *and* o_j.

2.3 Probabilistic Ranking on Uncertain Objects

The output of probabilistic queries is usually in form of a set of result objects, each associated with a probability value indicating the likelihood that the object fulfills the query predicate. However, in contrast to ε-range queries and k-nn queries, ranking queries do not have such an unique query predicate, since the query predicate changes with each ranking position. In case of a ranking query, a set of probability values is assigned to each result object, one for each ranking position. We call this form of ranking output *probabilistic ranking*.

Definition 5 (probabilistic ranking). *Let q be an uncertain query object and $\mathcal{D}$ be a database containing $N = |\mathcal{D}|$ uncertain objects. An* uncertain ranking *is a function $prob\_ranked_q : (\mathcal{D} \times \{1, .., N\}) \rightarrow [0..1]$ that reports for a database object $o \in \mathcal{D}$ and a ranking position $k \in \{1, .., N\}$ the probability which reflects the likelihood that o is at the kth ranking position according to the uncertain distance $d_{uncertain}(o, q)$ between o and the query object q in ascending order.*

If the result of the probabilistic ranking is reported to the user in its raw form, the user could be overstrained with ambiguous ranking results. For this reason, we suggest an unambiguous ranking based on the information given by the probabilistic ranking. The following proposed unambiguous ranking can be built in a post-processing step. Our unambiguous ranking *PRQ_MAC* assigns each object o a unique ranking position k by aggregating over the confidences of all prior ranking positions $i < k$ according to o.

Definition 6. *A probabilistic ranking query based on maximal aggregated confidence (PRQ_MAC) incrementally retrieves for the next ranking position $i \in \mathcal{I}_N$ a result tuple of the form $(o, \sum_{j=1..i} prob\_ranked_q(o, j))$, where $o \in \mathcal{D}$ has not been reported at previous ranking iterations (i.e. at ranking positions $j < i$) and $\forall p \in \mathcal{D}$ which have not been reported at previous ranking iterations, the following statement holds:*

$$\sum_{j=1..i} prob\_ranked_q(o, j) \geq \sum_{j=1..i} prob\_ranked_q(p, j).$$

3   Probabilistic Ranking Algorithm

The computation of the probabilistic ranking is very expensive and is the main bottleneck of the probabilistic ranking queries proposed in the previous section. In the following, we assume that each object is represented by M sample points. Furthermore, we assume that the object samples are stored in a spatial index structure like the R*-tree [7], in order to organize the uncertain objects such that proximity queries can be efficiently processed.

In the following, we concentrate on the computation of the probabilistic ranking query according to one sample point $q_j \in \mathbb{R}^d$ of the query object q. The computation is done for each sample point of the query object separately and, in a post-processing step, the results are then easily merged by building the average, to obtain the final result.

3.1   Iterative Probability Computation

Initially, an iterative computation of the nearest neighbors of q_j w.r.t. the sample points of all objects $o \in \mathcal{D}$ (sample point ranking $rank_s(q_j)$) is started using the ranking algorithm proposed in [4]. Then, we iteratively pick object samples from the sample point ranking $rank_s(q_j)$ according to the query sample point q_j. For each sample point $o_{i,s}$ ($1 \leq s \leq M$) returned from $rank_s(q_j)$, we immediately compute the probability that $o_{i,s}$ is the kth nearest neighbor of q_j for all k ($1 \leq k \leq i$). Thereby, all other samples $o_{i,t}$ ($t \neq s$) of object o_i have to be ignored due to the sample dependency within an object as mentioned above.

For the computation of the probabilistic ranking we need a table called *probability table* (PT) which is used to maintain the intermediate results w.r.t. $o_{i,s}$ and which finally contains the overall results of the probabilistic ranking.

Probability Table (PT). The *probability table* stores for each object o_i and each $k \in \mathbb{N}$ ($1 \leq k \leq N$) the actual probability that o_i is the kth-nearest neighbor of the query sample point q_s. The entries of PT according to the sth sample point of object o_i are defined as follows:

$$PT[k][i][s] =$$
$$P((k-1) \text{ objects } o \in \mathcal{D}(o \neq o_i) \text{ are closer to } q_j \text{ than the sample point } o_{i,s}).$$

We assume that object o_i is the ith object for which $rank_s(q_j)$ has reported at least one sample point. The same assumption is made for the sample points of an uncertain object (i.e., sample point $o_{i,s}$ is the sth-closest sample point of object o_i according to q_j). These assumptions hold for the remainder of this paper.

Now, we show how to compute an entry $PT[k][i][s]$ of the probability table using an additional structure called sample table (ST). The sample table stores for each accessed object l separately the portion of samples already returned from $rank_s(q_j)$ denoted by $ST[l][1]$, whereas $ST[l][0]$ denotes the portion of the remaining not yet returned samples, i.e. $ST[l][0] = 1 - ST[l][1]$. Let ST be a sample table of size N (i.e. ST stores the information corresponding to all N objects of the database $\mathcal{D}$). Let $\sigma_k(i) \subseteq \{o \in \mathcal{D} | o \neq o_i\}$ denote the set, called *k-set* of o_i, containing exactly $(k-1)$ objects. If we assume $k < N$, obviously $\binom{N}{k}$ different k-set permutations $\sigma_k(i)$ exist. For the computation of $PT[k][i][s]$, we have to consider the set S_k of all possible k-set permutations according to o_i. The probability that exactly $(k-1)$ objects are closer to the query-sample point q_j than the sample point $o_{i,s}$, can be computed as follows:

$$PT[k][i][s] = \sum_{\sigma_k(i) \in S_k} \prod_{\substack{l = 1..N \\ l \neq i}} \begin{cases} ST[l][1] & \text{,if } o_l \in \sigma_k(i) \\ ST[l][0] & \text{,if } o_l \notin \sigma_k(i) \end{cases}$$

Let us assume that we actually process the sample point $o_{i,s}$. Since the object samples are processed in ascending order according to their distance to q_j, the sample table entry $ST[l][1]$ reflects the probability, that object o_l is closer to q_j than the sample point $o_{i,s}$. On the other hand, $ST[l][0]$ reflects the probability that $o_{i,s}$ is closer to q_j than o_l.

In the following, we show how the entries of the probability table can be computed by fetching iteratively the sample points from $rank_s(q_j)$. Thereby, we assume that all entries of the probability table are initially set to zero. Then the iterative ranking process $rank_s(q_j)$ which reports one sample point of an uncertain object in each iteration, is started. Each reported sample point $o_{i,s}$ is used to compute for all k ($1 \leq k \leq N$) the probability value that corresponds to the table entry $PT[k][i][s]$. After filling the (i,s)-column of the probability table, we proceed with the next sample point fetched from $rank_s(q_j)$ in the same way as we did with $o_{i,s}$. This procedure is repeated until all sample points are fetched from $rank_s(q_j)$.

3.2 Accelerated Probability Computation

The computation of the probability table can be very costly in space and time. One reason is the size of the table that grows drastically with the number of objects and the number of samples for each object. The table size can be reduced as, in fact, we need only one value per object and ranking position which aggregates the results over the object samples. Another problem is the very expensive computation of the probability table entries PT[k][i][s]. In the following, we propose methods that reach a considerable reduction of the overall query cost.

In fact, at a time we explicitly have to maintain table entries for those objects from which at least one sample point has been reported from $rank_s(q_j)$, whereas we can skip those from which we already fetched all sample points.

The computational bottleneck of our probabilistic ranking algorithm is the computation of each table entry. For each computation of $PT[k][i][s]$ we have to compute the probabilities according to $\binom{N}{k}$ different k-set permutations which have to be summed up to the final probability value. For example, if $N = 100$ and $k = 20$ we need to consider about $1.73 \cdot 10^{13}$ k-set permutations.

In the case of subsequently fetching samples belonging to the same object, the ranking probabilities according to this object do not change. Hence, obviously only one computation of the probability value is required. However, the case where two adjacent sample points reported from the ranking belong to different objects often occurs. For this case we suggest a divide and conquer method which is able to drastically reduce the number of k-set permutations to be computed. Instead of considering all k of N permutations, we first split the k-set into two subsets of equal size. Then we only need to consider $(k-i)$ of $\frac{N}{2}$ permutations for $i = 1..k$ for the one subset, combined with the i of $\frac{N}{2}$ permutations of the other subset. As a consequence, instead of considering $\binom{N}{k}$ k-set permutations, the number of k-set permutations to be considered can be reduced to

```
ALGORITHM probability(ST,MIN,MAX,k)
   result = 0;
   N = MAX - MIN + 1;
   IF (k = 0) THEN result = ∏_{i=MIN..MAX} ST[i][0];
   ELSE IF (k ≥ N) THEN result = ∏_{i=MIN..MAX} ST[i][1];
   ELSE
      MID = ⌈(MIN + MAX)/2⌉;
      FOR (i = 0..min(⌈(MAX - MIN)/2⌉,k)) DO
         left = probability(ST,MIN,MID - 1,i);
         right = probability(ST,MID,MAX,(k - i));
         result = result + (left * right);
      END FOR
   END IF
   RETURN result;
```

Fig. 1. The sample point probability computation algorithm

$$\sum_{i=0..k} \left(\binom{\frac{N}{2}}{k-i} + \binom{\frac{N}{2}}{i} \right).$$

The k-set split can be recursively repeated for each subset. The recursive decomposition of a subset, from which we have to compute k ($0 < k < N$) out of N permutations stops if $k \geq N$. Otherwise, there exists only one permutation that can be immediately computed and reported to the calling function of the recursion. The algorithm for the computation of the sample point probability is depicted in Figure 1. The range of the k-set, that is currently worked on, is limited by the parameters MIN and MAX. The sample table, which is used for probability computation (cf. Section 3.1), is denoted by the additional parameter ST.

4 Experimental Evaluation

Due to space limitations, in this section we can only give a coarse summary of the experimental evaluation of our ranking methods. We applied our ranking methods on real world datasets as well as on artificial datasets. The artificial datasets which are used for the efficiency experiments contain 10 to 1000 3-dimensional uncertain objects. For the evaluation of the effectiveness of our methods we used three real-world datasets O_3, NSP_h and NSP_{frq}. The O_3 dataset is an environmental dataset consisting of 30 uncertain time series, each consisting of a series of measurements of O_3 concentration in the air measured within one month. The NSP datasets NSP_h and NSP_{frq} are chronobiologic datasets describing the cell activity of Neurospora[1] within sequences of day cycles. These datasets are used to investigate endogenous rhythms.

In the first experiments, we evaluated the quality of our probabilistic ranking query (*PRQ_MAC*) proposed in Section 2.3. We compare its quality with the quality of a non-probabilistic ranking (*MP, Mean Position*) which ranks the objects based on the distance between their mean positions. For these experiments, we used the three real-world datasets O_3, NSP_h and NSP_{frq}. The ranking quality is measured by the average precision over all recall values for each dataset. The average precisions for the dataset O_3 are prec(*PRQ_MAC*) = 0.65 and prec(*MP*) = 0.63, for the dataset NSP_h they are prec(*PRQ_MAC*) = 0.43 and prec(*MP*) = 0.35 and for the dataset NSP_{frq} they are prec(*PRQ_MAC*) = 0.70 and prec(*MP*) = 0.60. Obviously, the *PRQ_MAC* approach outperforms the non-probabilistic ranking approach.

In the next experiment, we evaluated the performance of our probabilistic ranking acceleration strategies proposed in Section 3.2 w.r.t. query processing time. The results of the experiments showed that the strategies are able to reduce the query cost by several orders of magnitude. Interestingly, the recursive computation of the probability permutations alone (i.e. without other strategies) yields a speed up of up to two orders of magnitude compared to the other strategies.

---

[1] Neurospora is the name of a fungal genus containing several distinct species. For further information see *The Neurospora Home Page*: http://www.fgsc.net/Neurospora/neurospora.html

5   Conclusions

In this paper, we proposed an approach that efficiently computes probabilistic ranking queries on uncertain objects represented by sets of sample points. In particular, we proposed methods that are able to break down the high computational complexity required to compute for an object o the probability, that o has the ranking position k ($1 \leq k \leq N$) according to the distance to a query object q. We theoretically and experimentally showed that against straightforward solutions our approach is able to speed-up the query by factors of several orders of magnitude. In the future we plan to apply probabilistic ranking queries to improve data mining applications.

Acknowledgments

We would like to thank Jan Remis and Roselyn Santos from the Institute of Medical Psychology (IMP) at the Ludwig-Maximilians University of Munich for making the neurospora dataset NSP available to us. Furthermore, we would like to thank U. Böllmann and M. Meindl for providing us with the environmental dataset TEMP from the Bavarian State Office for Environmental Protection, Augsburg, Germany.

References

1. Böhm, C., Pryakhin, A., Schubert, M.: Probabilistic Ranking Queries on Gaussians. In: Proc. of the 18th Int. Conf. on Scientific and Statistical Database Management (SSDBM 2006), pp. 169–178 (2006)
2. Cheng, R., Kalashnikov, D., Prabhakar, S.: Evaluating Probabilistic Queries over Imprecise Data. In: Proc. ACM SIGMOD Int. Conf. on Management of Data (SIGMOD 2003), San Diego, CA, pp. 551–562 (2003)
3. Cheng, R., Xia, Y., Prabhakar, S., Shah, R., Vitter, J.: Efficient Indexing Methods for Probabilistic Threshold Queries over Uncertain Data. In: Proc. 30th Int. Conf. on Very Large Databases (VLDB 2004), Toronto, Canada, pp. 876–887 (2004)
4. Hjaltason, G., Samet, H.: Ranking in Spatial Databases. In: Egenhofer, M.J., Herring, J.R. (eds.) SSD 1995. LNCS, vol. 951, pp. 83–95. Springer, Heidelberg (1995)
5. Kriegel, H.-P., Kunath, P., Pfeifle, M., Renz, M.: Probabilistic Similarity Join on Uncertain Data. In: Li Lee, M., Tan, K.-L., Wuwongse, V. (eds.) DASFAA 2006. LNCS, vol. 3882, pp. 295–309. Springer, Heidelberg (2006) (Best paper)
6. Kriegel, H.-P., Kunath, P., Renz, M.: Probabilistic Nearest-Neighbor Query on Uncertain Objects. In: Kotagiri, R., Radha Krishna, P., Mohania, M., Nantajeewarawat, E. (eds.) DASFAA 2007. LNCS, vol. 4443, pp. 337–348. Springer, Heidelberg (2007)
7. Kriegel, H.-P., Seeger, B., Schneider, R., Beckmann, N.: The R*-tree: An Efficient Access Method for Geographic Information System. In: Proc. Int. Conf. on Geographic Information Systems, Ottawa, Canada (1990)
8. Tao, Y., Cheng, R., Xiao, X., Ngai, W., Kao, B., Prabhakar, S.: Indexing Multi-Dimensional Uncertain Data with Arbitrary Probability Density Functions. In: Proc. 31th Int. Conf. on Very Large Data Bases (VLDB 2005), Trondheim, Norway, pp. 922–933 (2005)

Flexible Scientific Workflow Modeling Using Frames, Templates, and Dynamic Embedding*

Anne H.H. Ngu[1], Shawn Bowers[2], Nicholas Haasch[1], Timothy McPhillips[2], and Terence Critchlow[3]

[1] Texas State University, San Marcos, TX
[2] Genome Center, University of California, Davis, CA
[3] Pacific Northwest National Laboratory

Abstract. While most scientific workflows systems are based on dataflow, some amount of control-flow modeling is often necessary for engineering fault-tolerant, robust, and adaptive workflows. However, control-flow modeling within dataflow often results in workflow specifications that are hard to comprehend, reuse, and maintain. We describe new modeling constructs to address these issues that provide a structured approach for modeling control-flow within scientific workflows, and discuss their implementation within the Kepler scientific workflow system.

1 Introduction

Scientific workflow systems aim to provide end-to-end frameworks for automating and simplifying data processing tasks. These tasks often include data acquisition, transformation, integration, analysis, and visualization. Many existing scientific workflow systems (e.g., KEPLER [1], Taverna [2], and SCIRun [3]) are based on *dataflow* models of computation [4], where individual components (*actors* in KEPLER) are loosely coupled, communicate via streams of data objects, and are scheduled by the workflow system according to dataflow dependencies. An advantage of this approach is that actors, which may be native to the system or wrap external software components such as web services, scripts, or external applications, can become reusable components for use within multiple workflows. Because of the emphasis on data dependencies, these systems also provide a simple and intuitive model for scientific workflow designers [5].

However, while dataflow has become a standard model of computation in scientific workflow systems, control-flow modeling is often necessary for engineering fault-tolerant, robust, and adaptive workflows. Without mechanisms to control the scheduling and execution of actors, otherwise simple workflows quickly become hard to comprehend, reuse, and maintain. In KEPLER, e.g., special-purpose control-flow actors are often introduced into workflows for this purpose, which often lead to complex workflows with many of the above problems (*e.g.*, see Figure 1).

This paper describes a *structured* approach for introducing control-flow into dataflow-oriented scientific workflows. This work is based on our experiences developing workflows for a range of domains, including astrophysics, environmental monitoring, phylogenetics, and bioinformatics. These workflows typically involve generic

---

* This work supported in part through NSF grants DBI-0533368, IIS-0612326, IIS-0630033, and OCI-0722079.

B. Ludäscher and Nikos Mamoulis (Eds.): SSDBM 2008, LNCS 5069, pp. 566–572, 2008.

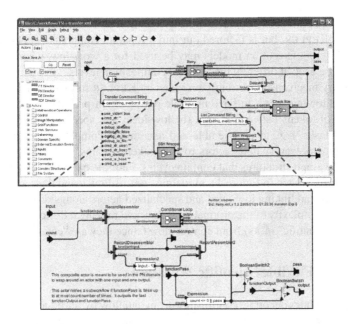

Fig. 1. Control-flow intensive workflow in KEPLER. "Retry," a composite actor for fault-tolerant data transfer (top), contains a subworkflow (bottom), which itself contains a "ConditionalLoop" subworkflow (not shown). Complex feedback loops and low-level actors demonstrate complexity of modeling control-flow using dataflow constructs.

processing steps related to submitting and running external jobs, transferring and processing data, and visualizing results. Each generic processing step typically has many different concrete implementations, where the desired implementation depends on the data input to the workflow, on the type of analysis being performed, or even on the state of the workflow environment at the time of execution.

Our approach is based on workflow *frames* and *templates* [6], which provide an abstraction for modeling control-flow issues surrounding the selection of concrete actor implementations among multiple alternatives. An advantage of our approach is that dataflow remains the primary model of scientific workflows, while allowing complex control-flow specifications to be embedded as subtasks or wrappers around existing actors. We also present an approach (extending [6]) in which the complete specification of frames is determined at workflow execution time through *dynamic embedding*.

2   The KEPLER Scientific Workflow System

KEPLER provides support for designing and executing scientific workflows. KEPLER workflows are created by selecting actors and wiring them together on a design canvas to form the desired workflow graph (Figure 1). Actors have input and output *ports* for communicating with other actors. Data from one actor is streamed asynchronously to another actor via data *tokens*. Composite actors encapsulate subworkflows, e.g., allowing one workflow to be reused in another. Actors typically have *parameter* ports for configuring default behavior. The overall execution of a workflow is *not* defined by actors within KEPLER, but is factored out into a separate component called a *director* [7]. Thus, different execution models may be used for a workflow, where a different director may be used at different hierarchical composition levels.

The primary mechanisms currently provided by KEPLER for managing control-flow include: (a) allowing actors to have multiple ports, which provides a mechanism for passing control-tokens between actors; (b) low-level and specialized actors for handling control tasks, *e.g.*, to disassemble and reassemble complex data types (such as records), and "Boolean-switch" actors to fork token streams; and (c) allowing complex workflow graph structures that contain cycles, multiple paths, etc. As described in [6,8] (and demonstrated in Figure 1), modeling control-flow using these constructs involves inserting and linking low-level and specialized actors alongside dataflow actors, increasing the complexity of the original workflow and making it difficult to distinguish dataflow (or "scientific" tasks) from control-flow (since they are "entangled").

3   Enabling Flexible Scientific Workflows

Frames and templates decouple control-flow from dataflow by introducing new abstractions: A *frame* wraps a set of alternative actor implementations, and a *template* specifies a subworkflow with "holes" that can be filled in at design time or runtime with actors or additional templates. We adopt and modify the finite-state transducer framework of [7] for specifying templates, which provides an intuitive language for encapsulating control-flow behavior within KEPLER. This approach allows workflow designers to change control-flow behavior by selecting and applying different templates and frames.

In [6], the specification of control-flow behavior using frames and templates is performed exclusively at workflow design time. For example, once a particular template is configured, it is not modified again regardless of changing runtime conditions (although the template may be specified to react to pre-specified runtime conditions). Further, once a frame or template has been embedded, different implementations cannot be re-selected for embedding at run time, unless this behavior is explicitly encoded within the template. *Dynamic embedding* extends the approach by allowing actors and control-flow behavior to be selected at workflow runtime. The rest of this section describes frames, templates, and dynamic embedding.

Actor Frames. Actors in KEPLER are always *concrete* in that they correspond to particular implementations that are directly executed in a workflow. As a simple example, a `gridftp` and a `sftp` (secure ftp) actor are tied to two different data-transfer implementations. A *frame* denotes a set of alternative actor implementations (or refinements) with similar, but not necessarily identical functionality. For workflow designers, frames are placeholders for actors that will be instantiated and specialized at runtime. Thus, a designer can place a frame on the design canvas, and connect it with other workflow components, without prematurely specifying which component is to be used. For actor developers, frames can be used as abstractions for a family of components with similar function, e.g., a `DataTransfer` frame can generalize the transfer of data without specifying whether the implementation is provided by `gridftp` or `sftp`.

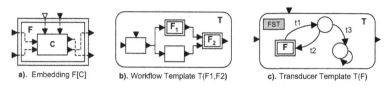

a). Embedding F[C]   b). Workflow Template T(F1,F2)   c). Transducer Template T(F)

Fig. 2. (a) Embedding of component C in frame F; (b) worfklow template $T(F_1, F_2)$; (c) finite state transducer template $T(F)$

A frame is a named entity F that acts as a placeholder for a component to be "plugged into" F (see Figure 2). When devising a frame F, a family of components $\mathbf{C}_F$ is envisioned, with each $C \in \mathbf{C}_F$ being a possible alternative for embedding into F. As with actors, frames have ports and structural types, which together form the *frame signature* Σ_F. This signature represents the common API of the family $\mathbf{C}_F$ of components that F abstracts. An *embedding* $F[C]$ of a component C into a frame F is a set of pairs associating (or "wiring") ports of C with ports of F. The embedded component C may also introduce new ports not in F; and an embedding $F[C]$ may not use all the ports of C. Parameter ports of F can also be connected to input ports of C and vice versa; however, other connection types are generally not allowed.

Workflow Templates. A frame F imposes constraints on its set of components $\mathbf{C}_F$ such that embeddings $F[C]$ should be well-formed and well-typed for any $C \in \mathbf{C}_F$. However, no assumptions can be made about the "inner workings" of C. A *workflow template* T provides a similar level of abstraction for a set of workflows $\mathbf{W}_T$. Unlike a frame, a template T (partially) specifies the behavior of the workflows it represents. In addition, a template includes an "inner" workflow graph W_T, where some of the components of W_T are not concrete actors, but frames (Figure 2b). Let $F_1, \ldots, F_n$ be the frames that occur in W_T, either directly, or indirectly through nested templates. We can view T as a partial workflow specification $T(F_1, \ldots, F_n)$, whose frames F_i can be independently specialized by embedded components (actors or templates) C_i. The resulting embedding $T(F_1[C_1], \ldots, F_n[C_n])$ is a concrete, executable workflow if no C_i has a frame; otherwise the embedding is a (more refined) template. The left diagram in Figure 3 shows an example data transfer frame embedded with a "retry" transducer and example state implementation.

Dynamic Embedding. Because frame and template embeddings are given at design time, they require all paths within a workflow to be completely bound and loaded prior to execution. In *dynamic embedding*, frames are embedded and instantiated on demand during workflow execution, which unlike static embedding, does not require the workflow system to load or manage the alternative implementations of a frame. Instead, alternatives are dynamically selected according to rules employed by the frame itself. Dynamic embedding is used when selection rules change often or are complex, and when deploying large workflows with many frames and alternative implementations.

Our implementation of dynamic embedding consists of: (1) waiting for data tokens to arrive at the frame's input ports, (2) selecting an embedded component using a set of selection criteria (given by the workflow designer), (3) transferring input tokens from the

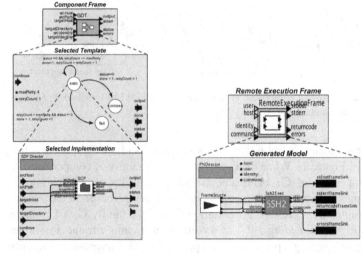

Fig. 3. A generic data transfer component statically embedded with a specific template and underlying state frame implementation (left), and a frame after being embedded dynamically (right)

frame to the embedded component, (4) automatic construction of an internal workflow to run the embedded component (with the appropriate director), (5) executing the constructed workflow, and (6) transferring output tokens from the embedded component to the frame actor. The right of Figure 3 shows a remote job-execution frame implemented via dynamic embedding. As the workflow executes, the frame selects the appropriate concrete actor based on given selection criteria. Also shown in Figure 3 (bottom, right) is one of the automatically generated internal workflows, which consists of a dataflow director, the selected actor, the source and sink actors for controlling input and output, and the workflow parameters.

Frames that employ dynamic embedding are implemented via higher-order actors, which invoke actors or subworkflows given as input. These frames configure subworkflows according to selection criteria and use higher-order actors to control their execution. The frames also use additional actors to mediate the flow of tokens to the underlying frame embeddings. The *select actor* implements the selection policies of the frame (given as rules based on runtime conditions and input tokens), and supplies the embedding. The *source actors* transfer input tokens from the frame actor to the selected actor. The *sink actors* transfer output tokens from the selected actor to the frame actor. Finally, a *port wiring* component is used to map ports and parameters of the selected actor to ports and parameters of the frame.

Implementation. We have developed a prototype implementation of dynamic embedding within KEPLER and have applied the approach to example workflows, including the Terascale Supernova Initiative [9]. This workflow, in particular, was developed to automate the repetitive and complex data transfer and monitoring tasks involved in running a supercomputer simulation from a user's local computer. We have also developed examples of frame-based workflows for inferring phylogenetic trees that select the appropriate implementations of actors (*e.g.*, for file conversion and tree inference) based on the type of input data provided to the workflow.

4   Summary and Related Work

Frames, templates, and dynamic embedding enable workflow designers to more easily specify control-flow tasks needed for fault-tolerant, reusable, and adaptive scientific workflows, while still supporting dataflow as the primary model of computation. Most scientific workflow systems are based on dataflow, as opposed to business workflow systems [10] and associated approaches (e.g., workflow patterns [11] and web-service composition [12,13]) that use control-based models such as Petri nets. A significant challenge in both fields is to seamlessly integrate control-flow *and* dataflow within a single model. Frames and templates are inspired by hierarchical finite state machines [14] and the nesting of heterogeneous computation models [15]. Our approach also extends adaptive workflow modeling [16] by supporting complex data and control structures. In future work, we will further explore dynamic embedding for designing and analyzing workflows, e.g., to automatically compose specifications of dynamically-embedded components and discover suitable actors for use within frames and templates.

References

1. Ludäscher, B., et al.: Scientific workflow management and the Kepler system. Concurrency and Computation: Practice & Experience (2006)
2. Oinn, T., et al.: Taverna: A tool for the composition and enactment of bioinformatics workflows. Bioinformatics 20 (2004)
3. Parker, S.G., Miller, M., Hansen, C.D., Johnson, C.R.: An integrated problem solving environment: The SCIRun computational steering system. In: HICSS (1998)
4. Lee, E.A., Parks, T.M.: Dataflow process networks. Proc. of the IEEE 83, 773–801 (1995)
5. Bowers, S., Ludäscher, B.: Actor-oriented design of scientific workflows. In: ER (2005)
6. Bowers, S., Ludäscher, B., Ngu, A.H.H., Critchlow, T.: Enabling scientific workflow reuse through structured composition of dataflow and control flow. In: IEEE SciFlow (2006)
7. Eker, J., et al.: Taming heterogeneity—The Ptolemy approach. Proc. of the IEEE 91 (2003)
8. Goderis, A., Goble, C., Sattler, U., Lord, P.: Seven bottlenecks to workflow reuse and repurposing. In: Gil, Y., Motta, E., Benjamins, V.R., Musen, M.A. (eds.) ISWC 2005. LNCS, vol. 3729, pp. 323–337. Springer, Heidelberg (2005)
9. Xin, X.: Case study: Terascale supernova initiative workflow (TSI-Swesty). LLNL Technical Note (2004)
10. Alonso, G., Mohan, C.: Workflow management systems: The next generation of distributed processing tools. In: Advanced Transaction Models and Architectures (1997)
11. van der Aalst, W.M.P., ter Hofstede, A.H.M., Kiepuszewski, B., Barros, A.P.: Workflow patterns. Distributed and Parallel Databases 14 (2003)
12. Curbera, F., et al.: Business Process Execution Language for Web Services (BPEL4WS), Version 1.0 (2002)
13. Martin, D.L., et al.: Bringing semantics to web services: The OWL-S approach. In: Intl. Workshop on Semantic Web Services and Web Process Composition (2004)

14. Girault, A., Lee, B., Lee, E.A.: Hierarchical finite state machines with multiple concurrency models. IEEE Transactions on CAD 18 (1999)
15. Lee, E.A., Neuendorffer, S.: Actor-oriented models for codesign: Balancing re-use and performance. In: Formal Methods and Models for System Design. Kluwer, Dordrecht (2004)
16. Ngu, A.H.H., et al.: Advanced process-based component integration in Telcordia's cable OSS. In: ICDE (2002)

Examining Statistics of Workflow Evolution Provenance: A First Study

Lauro Lins, David Koop, Erik W. Anderson, Steven P. Callahan,
Emanuele Santos, Carlos E. Scheidegger, Juliana Freire, and Cláudio T. Silva

SCI Institute & School of Computing, University of Utah

1   Introduction

Provenance (also referred to as audit trail, lineage, and pedigree) captures information about the steps used to generate a given data product. Such information provides documentation that is key to determining data quality and authorship, and necessary for preserving, reproducing, sharing and publishing the data. Workflow design, in particular for exploratory tasks (e.g., creating a visualization, mining a data set), requires an involved, trial-and-error process. To solve a problem, a user has to iteratively refine a workflow to experiment with different techniques and try different parameter values, as she formulates and test hypotheses. The maintenance of detailed provenance (or history) of this process has many benefits that go beyond documentation and result reproducibility. Notably, it supports several operations that facilitate exploration, including the ability to return to a previous workflow version in an intuitive way, to undo bad changes, to compare different workflows, and to be reminded of the actions that led to a particular result [2].

As provenance-enabled systems are deployed, and increasing volumes of provenance information are collected, there is a unique opportunity to leverage and obtain useful knowledge from this data. In this paper, we take a first step at analyzing this data. We present a preliminary analysis of workflow evolution provenance generated by thirty subjects who worked on six distinct exploratory tasks over the period of four months. This initial analysis shows that useful statistics can be extracted from this data that provide insights into how different people interact with workflow systems to solve problems.

2   Workflow Evolution Provenance: Background

Because scientific tasks evolve as users switch input data, vary parameters, and investigate different approaches, scientists often need to manage a large collection of workflows. The change-based provenance model [2] treats a workflow specification as a first-class data item and captures the provenance of its evolution by recording every change to the specification. As a user modifies a workflow (e.g., by adding a module, changing a parameter or deleting a connection), the provenance mechanism transparently records each change, akin to a database

B. Ludäscher and Nikos Mamoulis (Eds.): SSDBM 2008, LNCS 5069, pp. 573–579, 2008.

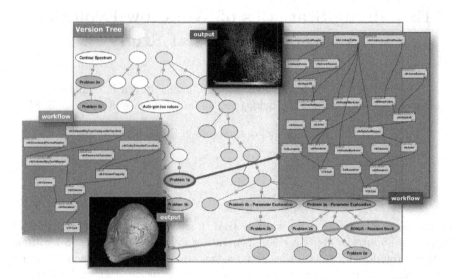

Fig. 1. A version tree with two workflow specifications and their outputs

transaction log. We can then reconstruct any workflow by replaying the sequence of captured changes from an empty specification to the desired version. In contrast to previous models which only capture provenance of data products (i.e., information about how a given data product was generated) [5], the change-based model captures both workflow and data provenance: it maintains a detailed record of the trail created by a user while solving a problem. In addition, this representation is concise and requires substantially less space than the alternative of storing multiple versions of a task specification.

Because the change-based provenance model captures the derivation of workflows, we can represent workflow evolution as a tree where each node is a *version* of the workflow specification and each edge coincides with an action. Given an edge from a parent node w_p to a child node w_c, its corresponding *action* is the sequence of changes necessary to transform w_p into w_c. Figure 1 shows an example of a workflow version tree, a couple of the workflow specifications, and the corresponding outputs from these workflows. Note that to reduce visual clutter, only important nodes of the tree are displayed by default, including those the user has tagged.

We have shown that maintaining detailed provenance of workflow evolution has many benefits and supports various activities that are crucial for performing reflective reasoning and obtaining insights, such as for example, following chains of reasoning backward and forward and comparing different results [3]. The tree-based view allows users to work collaboratively, to return to a previous version in an intuitive way, to undo bad changes, to reuse workflows and workflow fragments, to compare different workflows and their results, and to be reminded of the actions that led to a particular result [2,4].

The change-based model was originally implemented in the VisTrails system.[1] More recently, other workflow systems, including Taverna [6] and Kepler [1], have started to capture workflow evolution provenance.

3   Extracting Statistics from Workflow Evolution

Workflow evolution provenance makes it possible to analyze, in an unobtrusive manner, different aspects of workflow design. Furthermore, it provides a means to evaluate the utility of workflow systems and provenance to users, as they solve problems using workflows. In this section, we present an initial case study and discuss some statistics that can be extracted from this kind of provenance.

3.1   The Data

Our dataset was collected during a scientific visualization course.[2] A total of thirty students took the course. Throughout the semester, they were assigned six different tasks with fixed deadlines. Table 1 provides a short description as well as a subjective evaluation by the course instructor of the difficulty and open-endedness of each task.

Students used VisTrails to complete the tasks and for each task, they submitted a file containing all the actions they performed. These actions are transparently captured by VisTrails and stored according to the change-based model. Each action has a unique identifier; the identifier of its parent action; the user who performed the action; a timestamp indicating when the action took place; an optional tag; free-text annotations; and the required information to reproduce the action.

3.2   Analyzing Evolution Provenance at Different Levels

Because our provenance data encompasses a range of tasks completed by a set of users, it can be analyzed on different levels. Globally, we can observe trends across all tasks and users. At the task level, we can attempt to characterize tasks

Table 1. Description of the six tasks involved in the study with the instructor's expectation of difficulty and open-endedness on a scale from 1 to 5

Task	Description	Difficulty	Open-Endedness
Task 1	Introduction	1	1
Task 2	2D Visualization Techniques	3	2
Task 3	Scalar & Vector Field Visualization	3	2
Task 4	Isosurfacing & Volume Rendering	4	3
Task 5	Diffusion Tensor Imaging & InfoVis	4	4
Task 6	Open-Ended Visualization	5	5

[1] http://www.vistrails.org
[2] http://www.vistrails.org/index.php/SciVisFall2007

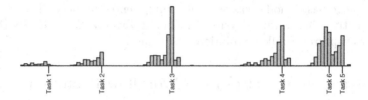

Fig. 2. Activity histogram binned by date with due dates indicated

by the types of actions involved. Finally, for a specific user, we can drill down to assess progress, work habits, and strategies used for different tasks.

Because we know exactly when each action occurred, it is possible to plot the total workload against time. The activity histogram in Figure 2 shows that, unsurprisingly, most work was condensed into the few days preceding the task deadlines. Besides that, the activity histogram also gives a good sense of which tasks required more effort. Although this measure may not match the assessment of the instructor, it gives a better measure of the effort the students put forth.

Global Analysis

One useful feature of workflow evolution provenance is that users can interact with this provenance as they work. For example, in VisTrails users can at any time access the version tree and select any existing workflow to execute it, to inspect its specification or to modify it. In this last case, a new *branch* with the modified workflow specification is created as a new leaf of the tree. In order to help users to identify workflow specifications, VisTrails allows them to *tag* the nodes in the tree. In our analysis, we found that the number of branches in the version tree is correlated with the number of tagged nodes, as shown in Figure 3. This indicates that,

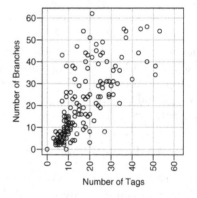

Fig. 3. The correlation between the number of branches and the number of tags per user-task

as users have to revisit a previously defined workflow, they would select a tagged node because it is easier to identify.

Analysis of Tasks. Workflow evolution information can also be helpful to characterize tasks. As noted in Table 1, the tasks assigned to the scientific visualization students varied in their goals, difficulty, due date, and how open-ended they were. To illustrate how workflow evolution data can be used to gain some insight into the types of work involved in a task, we classified the actions involved in workflow development into: *structural actions* (addition and deletion of modules and connections in the workflow); *parameter actions* (modification of parameter

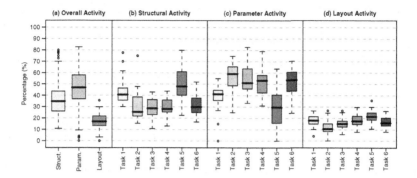

Fig. 4. Workflow Structural, Parameter and Layout Activity

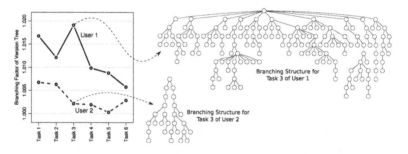

Fig. 5. Plot of Branching Factors for the six tasks from two different users. The branching structure for Task 3 is depicted on the right.

values in the workflow); and *layout actions* (changes to the locations of modules in visual programming interface).

Figure 4 shows an attempt to characterize tasks by the breakdown of actions involved. For all users, we calculated the overall percentage of actions that were structural, parameter and layout actions across all tasks (Figure 4(a)). In addition, we computed these percentages for each task, as shown in Figure 4(b), (c) and (d). The distributions of these percentages were plotted as boxplots. Note that the percentage of actions spent changing parameters has the greatest variance for most tasks. This should be expected as some users locate correct parameter values faster than others, and some will also expend more effort tweaking parameters than others. Another interesting feature of these plots is that Task 5 shows more structural activity than Tasks 2, 3, and 4. This is explained by the fact that students were given examples for the previous three tasks, and in Task 5, they were left to discover how to create workflows from scratch.

Analysis of Users. A useful application of workflow evolution provenance is to help in understanding how different users approach a problem. Figure 5 shows two trees created by different users for the same task. User 1 and User 2 clearly have different development styles: the tree derived by User 2 is both shorter and narrower than that of User 1. This figure also shows a plot of the branching factor of the version trees across the tasks for User 1 and User 2. A smaller branching factor indicates that a more direct path was used to obtain a solution. In contrast, a larger branching factor indicates that more trial-and-error steps were followed. There are many cases where branching can be useful, including when a user wishes to develop workflows that share a common sub-workflow: the user designs the first workflow, goes to the version tree, selects the node corresponding to the common sub-workflow and from there branches to the second workflow. We found a range of branching factors that varied across users and tasks.

Branching is just one variable from the workflow evolution provenance data that can be used to identify "user signatures", other variables, such as the time between actions and the umber of sessions may also lead to insights in this respect.

4   Discussion and Future Work

We have shown that workflow evolution provenance allows one to measure, summarize, and analyze new aspects of workflow specification and design. A detailed analysis of how time is spent in workflow design can help to provide an understanding of how users interact with workflow systems. In addition, these statistics can produce insights into the potential bottlenecks and how these systems can be improved. While our results represent only an initial examination, we have discovered a number of areas where comparative statistics offer a window into general workflow design patterns, task characterization, and exploratory styles.

Besides investigating additional measures and statistical analyses, there are several avenues we plan to pursue in future work. In the course of our study, we have identified some limitations of the VisTrails provenance capture mechanism. We plan to improve and augment the variables captured by the change-based model to allow for more accurate and detailed analyses. Specifically, while each change is time-stamped, it is difficult to determine the actual time involved in performing a single action. In addition, information about distinct sessions of work would be useful to better determine the actual time spent accomplishing the computational tasks. We also plan to cross quality or merit data about the workflow specifications with the provenance data to infer information about which practices led to good workflow specification and how time was used in these cases. For our initial analysis we considered only general actions for modifying workflows. In future work, we plan to perform analyses that take into account the semantics of the individual actions. For example, instead of looking at the addition and deletion of modules, for a visualization task, we could consider the addition of a volume renderer or of an isosurface extraction. By doing so, we could measure the effort involved in applying these two different visualization techniques.

References

1. Altintas, I., Barney, O., Jaeger-Frank, E.: Provenance collection support in the kepler scientific workflow system. In: Moreau, L., Foster, I. (eds.) IPAW 2006. LNCS, vol. 4145, pp. 118–132. Springer, Heidelberg (2006)
2. Freire, J., Silva, C.T., Callahan, S.P., Santos, E., Scheidegger, C.E., Vo, H.T.: Managing rapidly-evolving scientific workflows. In: Moreau, L., Foster, I. (eds.) IPAW 2006. LNCS, vol. 4145, pp. 10–18. Springer, Heidelberg (2006)
3. Norman, D.A.: Things That Make Us Smart: Defending Human Attributes in the Age of the Machine. Addison Wesley, Reading (1994)
4. Scheidegger, C.E., Vo, H.T., Koop, D., Freire, J., Silva, C.T.: Querying and creating visualizations by analogy. IEEE Transactions on Visualization and Computer Graphics 13(6), 1560–1567 (2007)
5. Simmhan, Y.L., Plale, B., Gannon, D.: A survey of data provenance in e-science. SIGMOD Record 34(3), 31–36 (2005)
6. Zhao, J., Goble, C., Stevens, R., Turi, D.: Mining taverna's semantic web of provenance. Concurrency and Computation: Practice and Experience (2007)

ELKI: A Software System for Evaluation of Subspace Clustering Algorithms

Elke Achtert, Hans-Peter Kriegel, and Arthur Zimek

Institute for Informatics, Ludwig-Maximilians-Universität München
{achtert,kriegel,zimek}@dbs.ifi.lmu.de
http://www.dbs.ifi.lmu.de

Abstract. In order to establish consolidated standards in novel data mining areas, newly proposed algorithms need to be evaluated thoroughly. Many publications compare a new proposition – if at all – with one or two competitors or even with a so called "naïve" *ad hoc* solution. For the prolific field of subspace clustering, we propose a software framework implementing many prominent algorithms and, thus, allowing for a fair and thorough evaluation. Furthermore, we describe how new algorithms for new applications can be incorporated in the framework easily.

1 Introduction

In an active research area like data mining, a plethora of algorithms is proposed every year. Most of them, however, are presented once and never heard about again. On the other hand, newly proposed algorithms are often evaluated in a sloppy way taking into account only one or two partners for comparison of efficiency and effectiveness, presumably because for most algorithms no implementation is at hand. And if an implementation is provided by the authors, a *fair* comparison is nonetheless all but impossible due to different performance properties of different programming languages, frameworks, and, last but not least, implementation details. Eventually, an evaluation based on implementations of different authors is more likely to be a comparison of the efforts of different authors in efficient programming rather than truly an evaluation of algorithmic merits.

Recently, an understanding for the need for consolidation of a maturing research area is rising in the community as illustrated by the discussions about the repeatability of results for SIGMOD 2008, the Panel on performance evaluation at VLDB 2007, and the tentative special topic of "Experiments and Analyses Papers" at VLDB 2008.

In the software system described in this paper, we try to facilitate a fair comparison of many subspace clustering algorithms based on experimental evaluation. The framework provides the data management independently of the tested algorithms. So all algorithms are comparable on equal conditions. The implementation aims at effectiveness in a balanced way for all algorithms. But even more important is an intuitive and easy-to-understand programming style to invite contributions in the future when the framework is made available open source.

B. Ludäscher and Nikos Mamoulis (Eds.): SSDBM 2008, LNCS 5069, pp. 580–585, 2008.

2   An Overview on the Software System

A wealth of data-mining approaches is provided by the almost "classical" open source machine learning framework Weka [1]. We consider Weka as the most prominent and popular environment for data mining algorithms. However, the focus and strength of Weka is mainly located in the area of classification, while clustering approaches are somewhat underrepresented.

The same holds true for another framework for data mining tasks: YALE [2]. This is a rather complex environment that completely incorporates Weka. The main focus of YALE is in supporting "rapid prototyping", i.e. to ease the definition of a specific data mining task as a combination of a broad range of available methods. While Weka is restricted to use numerical or nominal features (and in some cases strings), YALE does also extend the range of possible input data.

Although both, Weka and YALE, support the connection to external database sources, they are based on a flat internal data representation. Thus, experiments assessing the impact of an index structure on the performance of a data mining application are not possible using these frameworks.

On the other hand, frameworks for index structures, such as GiST [3], do not provide any precast connection to data mining applications.

To connect both worlds, we demonstrate the Java Software Framework ELKI (Environment for DeveLoping KDD-Applications Supported by Index Structures). ELKI comprises on the one hand a profound and easily extensible collection of algorithms for data mining applications, such as item-set mining, clustering, classification, and outlier-detection, and on the other hand ELKI incorporates and supports arbitrary index structures to support even large, high-dimensional data sets. But ELKI does also support the use of arbitrary data types, not only feature vectors of real or categorical values. Thus, it is a framework suitable to support the development and evaluation of new algorithms at the cutting edge of data mining as well as to incorporate experimental index structures to support complex data types.

ELKI intends to ease the development of new algorithms by providing a wealth of helper classes and methods for algebraic and analytic computations, and simulated database support for arbitrary data types using an index structure at will.

2.1   The Environment: A Flexible Framework

As a framework, our software system is flexible in a sense, that it allows to read arbitrary data types (provided there is a suitable parser for your data file or adapter for your database), and supports the use of any distance or similarity measure (like some kernel-function) appropriate for the given data type. So far, many implementations of data mining algorithms – especially subspace clustering algorithms – still rely on the numeric nature of feature vectors as underlying data structure. Our framework is already one step ahead and ready to work on complex data types. Generally, an algorithm needs to get provided a distance of some sort. Thus, distance functions connect arbitrary data types to arbitrary algorithms.

The architecture of the software system separates data types, data management, and data mining applications. So, different tasks can be implemented independently. A new data type can be implemented and used by many algorithms, given a suitable distance function is defined. An algorithm will perform its routine irrespectively of the data handling which is encapsulated in the database. A database may facilitate efficient data management via incorporated index structures.

2.2   Available Algorithms

While the framework is open to all kind of data mining applications, the main focus in the development of ELKI has been on clustering and especially subspace clustering (axis-parallel as well as arbitrarily oriented). Available general clustering algorithms are SLINK [4], k-means [5], EM-clustering [6], DBSCAN [7], Shared-Nearest-Neighbor-Clustering [8], OPTICS [9], and DeLiClu [10]. There are axis-parallel subspace and projected clustering approaches implemented like CLIQUE [11], PROCLUS [12], SUBCLU [13], PreDeCon [14], HiSC [15], and DiSH [16]. Furthermore, some biclustering or pattern-based clustering approaches are supported like δ-bicluster [17], FLOC [18] or p-cluster [19], and correlation clustering approaches are incorporated like ORCLUS [20], 4C [21], HiCO [22], COPAC [23], ERiC [24], and CASH [25]. The improvements on these algorithms described in [26] are also integrated in ELKI.

2.3   Development of Subspace Clustering Algorithms

Often, the main difference between clustering algorithms is the way to assess the distance or similarity between objects or clusters. So, while other well known and popular software systems like Weka [1] or YALE [2] predefine the Euclidean distance as only possible distance between different objects to use in clustering approaches (beside some kernel functions in classification approaches), ELKI allows the flexible definition of any distance measure. This way, subspace clustering approaches that differ mainly in the definition of distance between points (like e.g. COPAC and ERiC) can use the same algorithmic routine and become, thus, highly comparable in their performance.

Distance functions are used to perform range queries on a database object. Any implementation of an algorithm can rely on the database object to perform range queries with an arbitrary distance function and needs only to ask for k nearest neighbors not being concerned with the details of data handling.

A new data type is supposed to implement the interface DatabaseObject. A new algorithm class suitable to certain data types O needs to implement the Interface Algorithm<O extends DatabaseObject>. The central routine to implement the algorithmic behavior is void run(Database<O> database). Here, the algorithm is applied on an arbitrary database consisting of objects of a suitable data type. The database supports operations like

```
<D extends Distance<D>>
List<QueryResult<D>>
kNNQueryForObject(O queryObject,
                  int k,
                  DistanceFunction<O,D> distanceFunction)
```

performing a k-nearest neighbor query for a given object of a suitable data type O using a distance function that is suitable for this data type O and provides a distance of a certain type D. Such a query method returns a list of QueryResult<D> objects encapsulating the database id of the collected objects and their distance to the query object in terms of the specified distance function.

A new subspace clustering algorithm may therefore use a specialized distance function and implement a certain routine using this distance function on an arbitrary database.

2.4 Support of Arbitrary Index-Structures

As pointed out above, while existing frameworks for index-structures, such as GiST [3], do not provide any precast connection to data mining applications, well-known data-mining frameworks like Weka [1] or YALE [2] do not support the internal use of index structures.

Our software system ELKI supports the use of arbitrary index structures in combination with, e.g., a clustering algorithm. Already available within ELKI are metric index-structures like MTree [27], MkCoPTree and its variants MkTab-Tree and MkMaxTree [28], and MkAppTree [29] and spatial index-structures like RStarTree [30], DeLiCluTree [10], and RdkNNTree, an extension from [31] for $k \geq 1$.

Index structures are encapsulated in database objects. These database objects facilitate range queries using arbitrary distance functions. Algorithms operate on database objects irrespective of the underlying index structure. So the implementation of an algorithm, as pointed out above, is not concerned with the details of handling the data which can be supported by arbitrary efficient procedures.

This is interesting because the complexity of algorithms is often analyzed theoretically on the basis of index structures but often, if implementations are provided, an index structure is not included and cannot be incorporated in the particular implementation.

2.5 Setting Up Experiments

The integration of several algorithms into one software framework also allows for setting up complex experiments comparing different algorithms in an easy way and on equal terms. We plan to use the framework for extensive comparisons of a broad range of subspace clustering algorithms.

2.6   Availability and Documentation

The framework ELKI is available for download and use via

http://www.dbs.ifi.lmu.de/research/KDD/ELKI/.

There is provided an extensive documentation of the implementation and usage as well as examples to illustrate how to expand the framework by integrating new algorithms.

3   Conclusion

The software system ELKI presents a large collection of data mining applications (mainly clustering and – axis parallel or arbitrarily oriented – subspace and projected clustering approaches). Algorithms can be supported by arbitrary index structures and work on arbitrary data types given supporting data classes and distance functions. We therefore expect ELKI to facilitate broad experimental evaluations of algorithms – existing algorithms and newly developed ones alike.

References

1. Witten, I.H., Frank, E.: Data Mining: Practical machine learning tools and techniques, 2nd edn. Morgan Kaufmann, San Francisco (2005)
2. Mierswa, I., Wurst, M., Klinkenberg, R., Scholz, M., Euler, T.: YALE: Rapid prototyping for complex data mining tasks. In: Proc.KDD (2006)
3. Hellerstein, J.M., Naughton, J.F., Pfeffer, A.: Generalized search trees for database systems. In: Proc.VLDB (1995)
4. Sibson, R.: SLINK: An optimally efficient algorithm for the single-link cluster method. The Computer Journal 16(1), 30–34 (1973)
5. McQueen, J.: Some methods for classification and analysis of multivariate observations. In: 5th Berkeley Symposium on Mathematics, Statistics, and Probabilistics, vol. 1, pp. 281–297 (1967)
6. Dempster, A.P., Laird, N.M., Rubin, D.B.: Maximum likelihood from incomplete data via the EM algorithm. Journal of the Royal Statistical Society Series B 39(1), 1–31 (1977)
7. Ester, M., Kriegel, H.P., Sander, J., Xu, X.: A density-based algorithm for discovering clusters in large spatial databases with noise. In: Proc.KDD (1996)
8. Ertöz, L., Steinbach, M., Kumar, V.: Finding clusters of different sizes, shapes, and densities in noisy, high dimensional data. In: Proc.SDM (2003)
9. Ankerst, M., Breunig, M.M., Kriegel, H.P., Sander, J.: OPTICS: Ordering points to identify the clustering structure. In: Proc.SIGMOD (1999)
10. Achtert, E., Böhm, C., Kröger, P.: DeLiClu: Boosting robustness, completeness, usability, and efficiency of hierarchical clustering by a closest pair ranking. In: Ng, W.-K., Kitsuregawa, M., Li, J., Chang, K. (eds.) PAKDD 2006. LNCS (LNAI), vol. 3918, pp. 119–128. Springer, Heidelberg (2006)
11. Agrawal, R., Gehrke, J., Gunopulos, D., Raghavan, P.: Automatic subspace clustering of high dimensional data for data mining applications. In: Proc.SIGMOD (1998)

12. Aggarwal, C.C., Procopiuc, C.M., Wolf, J.L., Yu, P.S., Park, J.S.: Fast algorithms for projected clustering. In: Proc. SIGMOD (1999)
13. Kailing, K., Kriegel, H.P., Kröger, P.: Density-connected subspace clustering for high-dimensional data. In: Jonker, W., Petković, M. (eds.) SDM 2004. LNCS, vol. 3178. Springer, Heidelberg (2004)
14. Böhm, C., Kailing, K., Kriegel, H.P., Kröger, P.: Density connected clustering with local subspace preferences. In: Perner, P. (ed.) ICDM 2004. LNCS (LNAI), vol. 3275. Springer, Heidelberg (2004)
15. Achtert, E., Böhm, C., Kriegel, H.P., Kröger, P., Müller-Gorman, I., Zimek, A.: Finding hierarchies of subspace clusters. In: Fürnkranz, J., Scheffer, T., Spiliopoulou, M. (eds.) PKDD 2006. LNCS (LNAI), vol. 4213, pp. 446–453. Springer, Heidelberg (2006)
16. Achtert, E., Böhm, C., Kriegel, H.P., Kröger, P., Müller-Gorman, I., Zimek, A.: Detection and visualization of subspace cluster hierarchies. In: Kotagiri, R., Radha Krishna, P., Mohania, M., Nantajeewarawat, E. (eds.) DASFAA 2007. LNCS, vol. 4443, pp. 152–163. Springer, Heidelberg (2007)
17. Cheng, Y., Church, G.M.: Biclustering of expression data. In: Proc. ISMB (2000)
18. Yang, J., Wang, W., Wang, H., Yu, P.S.: δ-clusters: Capturing subspace correlation in a large data set. In: Proc. ICDE (2002)
19. Wang, H., Wang, W., Yang, J., Yu, P.S.: Clustering by pattern similarity in large data sets. In: Proc. SIGMOD (2002)
20. Aggarwal, C.C., Yu, P.S.: Finding generalized projected clusters in high dimensional space. In: Proc. SIGMOD (2000)
21. Böhm, C., Kailing, K., Kröger, P., Zimek, A.: Computing clusters of correlation connected objects. In: Proc. SIGMOD (2004)
22. Achtert, E., Böhm, C., Kröger, P., Zimek, A.: Mining hierarchies of correlation clusters. In: Proc. SSDBM (2006)
23. Achtert, E., Böhm, C., Kriegel, H.P., Kröger, P., Zimek, A.: Robust, complete, and efficient correlation clustering. In: Jonker, W., Petković, M. (eds.) SDM 2007. LNCS, vol. 4721. Springer, Heidelberg (2007)
24. Achtert, E., Böhm, C., Kriegel, H.P., Kröger, P., Zimek, A.: On exploring complex relationships of correlation clusters. In: Proc. SSDBM (2007)
25. Achtert, E., Böhm, C., David, J., Kröger, P., Zimek, A.: Robust clustering in arbitrarily oriented subspaces. In: Proc. SDM (2008)
26. Kriegel, H.P., Kröger, P., Schubert, E., Zimek, A.: A general framework for increasing the robustness of PCA-based correlation clustering algorithms. In: Proc. SSDBM (2008)
27. Ciaccia, P., Patella, M., Zezula, P.: M-Tree: an efficient access method for similarity search in metric spaces. In: Proc. VLDB (1997)
28. Achtert, E., Böhm, C., Kröger, P., Kunath, P., Pryakhin, A., Renz, M.: Efficient reverse k-nearest neighbor search in arbitrary metric spaces. In: Proc.SIGMOD (2006)
29. Achtert, E., BÖhm, C., Kröger, P., Kunath, P., Pryakhin, A., Renz, M.: Approximate reverse k-nearest neighbor search in general metric spaces. In: Proc. CIKM (2006)
30. Beckmann, N., Kriegel, H.P., Schneider, R., Seeger, B.: The R*-Tree: An efficient and robust access method for points and rectangles. In: Proc. SIGMOD, pp. 322–331 (1990)
31. Yang, C., Lin, K.I.: An index structure for efficient reverse nearest neighbor queries. In: Proc. ICDE (2001)

IVIP – A Scientific Workflow System to Support Experts in Spatial Planning of Crop Production

Christopher J. Tuot, Michael Sintek, and Andreas R. Dengel

German Research Center for Artificial Intelligence DFKI GmbH,
Trippstadter Straße 122, D-67663 Kaiserslautern, Germany
{christopher.tuot,michael.sintek,andreas.dengel}@dfki.de
http://www.dfki.de/km

Abstract. Decision making for crop production planning is essentially driven by location-based or more precisely by space-oriented information. Therefore, farmers and regional experts in the field mostly rely on new spatial-data-oriented decision making tools. IVIP[1] is a prototype for a Web-based Spatial Decision Support System (WSDSS) demonstrating the benefits of location-based decision making using digitalized geographic information about ground allocation and soil quality. We present how the library of potential models for the IVIP WSDSS has been realized by extending the Scientific Workflow Management System KEPLER that assists the collaboration of agricultural experts and computer scientists during model development. We first describe the requirements of our WSDSS, and then give a short introduction to the KEPLER platform and explain in detail which extensions have been realized: cascading client-server architecture, spatial operations support, and WSDL interface. Finally, we illustrate how the biomass yield model has been modeled in our system.

Keywords: Scientific workflow models, Scientific data integration, Spatial Decision Support System (SDSS), KEPLER, Workflow Management System (WMS), Web Service, WSDL, GIS, Forecast, Agriculture.

1 Motivation

Decision making for crop production planning is essentially driven by location-based or more precisely by space-oriented information. The required Global Positioning System (GPS) and sensor technology are becoming standard for agricultural machinery and most of the current Farm Management Information Systems (FMIS) already provide interfaces to import the acquired sensory information. With the standards for geospatial content and services developed by the Open Geospatial Consortium (OGC) the interoperability for geospatial technology has been highly facilitated. Nevertheless, the amount and the complexity of the spatial information becoming available is dramatically increasing and so is

---

[1] The IVIP project is funded by the Ministry for Economy, Transport, Agriculture and Viniculture (MWVLW) and is part of the EU program: "Ziel 2 Programm RLP."

B. Ludäscher and Nikos Mamoulis (Eds.): SSDBM 2008, LNCS 5069, pp. 586–591, 2008.

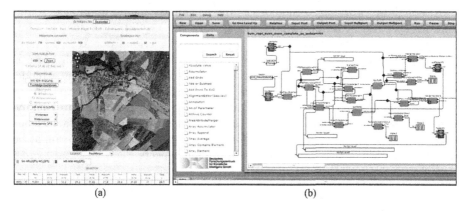

Fig. 1. Screenshots: (a) IVIP Spatial Decision Support System for Crop Production; (b) the bym Model in KEPLER Flex GUI

the demand for tools capable of handling such information. Therefore, farmers and regional experts in the field mostly rely on new spatial-data-oriented decision making tools. IVIP is a prototype (see Fig. 1 (a)) for a Web-based Spatial Decision Support System (WSDSS) demonstrating the benefits of location-based decision making using digitalized geographic information about ground allocation and soil quality.

2   WSDSS Requirements

Sprague [9] defines a Spatial Decision Support System (SDSS) as a standard decision support system plus some semi-structured spatial problem. A SDSS usually consists of three parts. First there must be a *Data-Base Management System* (DBMS) capable of handling geographical data like, e.g., a Geographical Information System (GIS). Then there must be a *library of potential models* that can be used to forecast the possible outcomes of decisions. Finally, an *interface* should aid the users to interact with the computer and should assist them in analyzing the outcomes. This defines the architecture of our WSDSS. In this paper, we focus on the library of models.

In the GIS world, geospatial information is based on *features*. A *feature* is an entity with a geographic location and some additional meta-information stored as key-value properties. The geographic location of a *feature* is described by a geometry based, e.g., on points, arcs, or polygons. Therefore, our WSDSS must be able to deal efficiently with **feature-data oriented models**.

Current SDSS solutions like, e.g., CommunityViz, an extension for ArcGIS, are definitively meant for GIS specialists. In IVIP, domain experts have **none or relatively limited programming skills and experience with GIS**. Nevertheless, those experts must be able to work with the WSDSS and in particular to develop new models.

Although networking technologies are constantly being improved, data-transfer speed is becoming a bottle-neck since the resolution of spatial information increases considerably at the same time. Therefore, our WSDSS must allow to **realize data-intensive operations** at the places where the data is actually located (in the network). Moreover, it is often the case that a model relies on other already existing models in the WSDSS which again can rely on other models. Consequently, a **cascading architecture for GIS operations** is needed which would not only solve the problem of efficient data-transfer but which would also define a good basis for further issues data providers might be concerned with like data-control access or billing aspects.

3   The KEPLER Platform

KEPLER[2] [1] is an open-source **scientific workflow system** that allows scientists to design scientific workflows and execute them efficiently. A *scientific workflow* is a **high level description** of a data-oriented process which can be used to solve a scientific problem. The dataflow consists of data being processed though parameterizable modules called *actors*. KEPLER is based on the PTOLEMY II [3] system for heterogeneous, concurrent modeling and design developed at UC Berkeley. Designing a scientific workflow in KEPLER is done following a visual programming principle and a simple workflow using existing actors can be realized with only a few mouse clicks. Depending on the size and the complexity (in terms of internal logic) of the workflows, basic programming skills can be helpful but are not mandatory. Advanced programming skills are only required for implementing new actors.

4   New Extensions for KEPLER

Spatial operations support: Possibilities to handle geospatial data in KEPLER have already been discussed in [5,8]. However, most of the propositions rely on externalizing the spatial operations, for example in Web Services or dedicated applications like the Geographic Resources Analysis Support System[4] (GRASS). Although delegating those complex operations to specialized applications makes sense, this has for the user enormous drawbacks in terms of workflow transparency. Therefore, we decided to handle geospatial operations directly in KEPLER. This tends to go back to the original idea of PTOLEMY however, high level workflows can still be designed on the feature level.

GeoTools[5] is an open-source Java library which provides OGC standards compliant methods for manipulating geospatial data. The **GeoTools** provide an efficient internal representation for features and all necessary basic spatial operations. We have implemented additional KEPLER actors using the GeoTools API.

---

[2] http://kepler-project.org/
[3] http://ptolemy.berkeley.edu/ptolemyII/
[4] http://grass.itc.it/
[5] http://geotools.codehaus.org/

Spatial information can now directly be imported in KEPLER using either the **WFS** actor, implementing the OGC Web Feature Service [10] protocol (WFS), or by using the GMLReader actor which imports data in the OGC Geography Markup Language [4] format (**GML**). We have also implemented several actors handling all common operations at the **feature level** (add, remove, update, merge). Finally, storing the resulting information can be done either using our **WFS-T** actor implementing the WFS transactional (WFS-T) protocol or by using the GMLWriter actor which exports GML data.

Client-Server architecture: Although KEPLER is currently distributed as a standalone application, its design actually made it possible to separate the GUI from the workflow engine itself. Our modified version of KEPLER can be deployed on a **standard application server** and be remotely controlled through the HTTP protocol. Besides defining the first step towards building a **cascaded architecture** to process spatial information, the client-server architecture also plays a major role in knowledge sharing.

One key feature of KEPLER is to allow scientists to exchange their workflows. Scientific workflows in KEPLER are stored using PTOLEMY's own Modeling Markup Language (MOML). MOML files only contain high level information about the workflow and the actors used are only referenced. Hence the definitions of the KEPLER actors themselves are not contained in the workflow but are literally part of the workflow engine. Therefore, if a workflow is using non-standard actors, it cannot be directly executed on another installation of KEPLER. Unfortunately, KEPLER still misses a proper management system to manage actors for tasks like adding, deleting, or updating actors. Those tasks remain too difficult for users without solid Java programming skills and really affect the possibilities of exchanging workflows. This is actually one of the issues the new KEPLER CORE initiative[6] will try to solve in the next years. Our idea is to further take advantage of our client-server architecture.

An application server is per definition capable of handling multiple-user requests. Let A and B be two scientists both running different instances of KEPLER (client-server architecture). A sharing a workflow with B could be done on the KEPLER installation of A. This insures that all necessary actors to run the workflow provided by A are present in the workflow engine. So the real question now is whether or not the standalone KEPLER GUI can **remotely execute a workflow** on a distant KEPLER. Unfortunately, although we managed to separate the GUI and the workflow engine, the GUI could in this state no more be used without spending a considerable amount of time to reinterface it with the KEPLER server. Alternatively, we went for a simplified version of the GUI supporting basic functionalities to create and execute workflows. We decided to develop a web-based KEPLER client using the new Adobe Flex 3[7] technology. Adobe Flex 3 is a cross platform open-source framework for creating rich applications. Flex applications are platform independent and are a good compromise between standard rich versus thin clients. Behind the pure software engineering

---

[6] http://www.kepler-project.org/Wiki.jsp?page=KeplerCORE
[7] http://labs.adobe.com/technologies/flex/

performance of implementing this client, downloading and installing the current 100 Megabytes KEPLER package is now no longer required to start working with scientific workflows. A single client-server KEPLER version can provide several scientists with the most core functionalities of KEPLER and allow them to efficiently share their workflows.

WSDL interface: Our client-server architecture of KEPLER was just the first step towards a cascaded architecture to process spatial information. Indeed the HTTP interface to remotely control KEPLER workflows is not suitable to allow two KEPLER instances to communicate with each other. The Web Service Description Language [3] (**WSDL**) is an XML-based language providing a model for describing Web services. Whereas Web Services can already be used in KEPLER workflows [6], our idea was to publish a KEPLER workflow itself as a Web service allowing any application supporting WSDL to use this workflow. Therefore we introduce two new actors: *WebServiceInput* and *webServiceOutput*. Those actors are only necessary for prototyping the input/output parameters of the workflow which unfortunately cannot yet be done automatically using the rest of the information contained in the workflow. A specific file folder is used as Web service pool and any workflow (properly designed with the WebServiceInput/Output actors) is directly available under the url:
http://serverIp:port/KFlexServer/wsdl/workflow=workflowName.

5   Example: The Biomass Yield Model in KEPLER

The Biomass Yield Model (**bym**) was developed at the University of applied sciences Eberswalde [2,7]. The original model was developed in Visual Basic using the software ArcGIS[8] from ESRI. Thanks to our custom GIS actors, we managed to design a KEPLER workflow computing the bym. The top level workflow in Fig. 1 (b) benefits from the composite actors architecture since the bym KEPLER workflow actually consists of eight encapsulated sub-workflows representing together more than 130 actors. Thus, the whole bym workflow has been broken into atomic operations which facilitates the work of the experts aiming at improving or optimizing this workflow.

The bym workflow can compute the yield for 16 different crops depending on the soil quality and the precipitation levels. Different kinds of scenarios are taken into consideration, e.g., conventional vs. ecological farming and three different levels of precipitation (low, normal, high). On the whole, six different scenarios are computed at once. The input of the workflow is a map containing the fields of a farmer, usually a GML file. The soil quality and the precipitation levels are on two remote GIS which are accessed using the WFS protocol. The results of the workflow are stored in a local GIS, i.e., in the same network. The storage is done using the WFS-T protocol. The data can then be visualized either directly in our GIS or with any mapping service application, e.g., OpenLayers[9] or Google Maps.

---

[8] http://www.esri.com/software/arcgis/
[9] http://www.openlayers.org/

The results delivered by this workflow represent the key information for our IVIP prototype, a Decision Support System to optimize crop production. All necessary parameters for the workflow can be entered using our GUI using the Web-Service interface of the bym KEPLER workflow to compute the results.

6 Conclusions and Future Work

In this paper we describe how we extended the Scientific Workflow Management System KEPLER to define the core of IVIP, a Web-based Spatial Decision Support System for crop production planning. The resulting cascading system for GIS operations enables non-GIS experts to design workflows for their spatial models. Next steps are to develop more GIS functionalities for KEPLER allowing to design more complex spatial models and to evaluate the benefits of our WSDSS.

References

1. Altintas, I., Berkley, C., Jaeger, E., Jones, M., Ludäscher, B., Mock, S.: Kepler: an extensible system for design and execution of scientific workflows. In: Proceedings. 16th International Conference on Scientific and Statistical Database Management, 2004, pp. 423–424 (2004)
2. Brozio, S., Piorr, H.-P., Müller, D., Torkler, F.: Potenziell nutzbare Biomasse – Modellierung mit GIS (2006)
3. Christensen, E., Curbera, F., Meredith, G., Weerawarana, S.: Web Services Description Language (WSDL). W3C Web Site (2001)
4. Cox, S., Daisey, P., Lake, R., Portele, C., Whiteside, A.: OpenGIS Geography Markup Language (GML) Implementation Specification. OpenGIS project document reference number OGC (2003)
5. Jaeger, E., Altintas, I., Zhang, J., Ludäscher, B., Pennington, D., Michener, W.: A Scientific Workflow Approach to Distributed Geospatial Data Processing using Web Services. In: 17th Intl. Conference on Scientific and Statistical Database Management (2005)
6. Perera, S., Gannon, D.: Enabling Web Service Extensions for Scientific Workflows (2006)
7. Pior, H.-P., Kersebaum, K.C., Koch, A.: Die Bedeutung von Extensivierung und ökologischem Landbau für Strukturwandel, Umweltentlastung und Ressourcenschonung in der Agrarlandschaft. Eberswalder Wissenschaftliche Schriften 3, 99–114 (1999)
8. Rueda, C., Gertz, M., Ludäscher, B., Hamann, B.: An Extensible Infrastructure for Processing Distributed Geospatial Data Streams. In: Proceedings of the 18th International Conference on Scientific and Statistical Database Management, pp. 285–290 (2006)
9. Sprague Jr., R., Carlson, E.: Building Effective Decision Support Systems. Prentice Hall Professional Technical Reference, Englewood Cliffs (1982)
10. Vretanos, P.: Web Feature Service Implementation Specification. OpenGIS project document: OGC, 2–58 (2002)

A FUSE-Based Tool for Accessing Meteorological Data in Remote Servers

Keiichirou Ui[1], Toshiyuki Amagasa[2], and Hiroyuki Kitagawa[2]

[1] College of Information Science
[2] Center for Computational Sciences
University of Tsukuba
1-1-1 Tennodai, Tsukuba 305–8573, Japan
kui@kde.cs.tsukuba.ac.jp, {amagasa, kitagawa}@cs.tsukuba.ac.jp

Abstract. This paper describes a tool for providing transparent access to online meteorological databases by way of local file system. The tool is based on FUSE, an implementation of usermode filesystem on Linux, and a user is allowed to deal with online meteorological data as if they were stored in his/her local file systems. The main target of this tool is scientists in meteorology those who are not familiar with computers and related technologies. For this reason, we attempt to provide easy ways for configuring, maintaining, and using the system. For speeding up the data transfer, we integrate a caching mechanism inside the system thereby making the number of HTTP requests smaller.

1 Introduction

Due to the rapid growth of Information and Comunication Technology, there is a growing demand for carrying out computationally intensive science in highly distributed network environments. In order to describe such an emerging new-style science, the term *e-Science* has been used. The term was created by John Taylor in 1999. Examples of the kind of science include social simulations, particle physics, earth sciences and bio-informatics.

When looking into the domain of earth science and meteorology, there is also an increasing demand for computationally intensive processing over large datasets like observational data. In fact, there are several databases available online. For example, Center for Climate System Research in the University of Tokyo [1], Disaster Prevention Research Institute in Kyoto University [2], National Centers for Environmental Prediction [3], and European Centre for Medium-Range Weather Forecasts [4] are providing online databases. We are also providing an online meteorological database, GPV/JMA Archive at http://gpvjma.ccs.hpcc.jp as a part of the activities at the Center for Computational Sciences

---

[1] http://www.ccsr.u-tokyo.ac.jp/
[2] http://www.dpri.kyoto-u.ac.jp/web_j/
[3] http://www.cdc.noaa.gov/cdc/reanalysis/
[4] http://data.ecmwf.int/data/

B. Ludäscher and Nikos Mamoulis (Eds.): SSDBM 2008, LNCS 5069, pp. 592–597, 2008.

in University of Tsukuba [1]. The archive is to provide the daily operational weather forecasting data provided by the Japan Meteorological Agency (JMA).

For accessing necessary datasets for daily research operations, domain scientists need to deal with such databases. In many cases, such a database provides a Web browser interface, but it is not so usable when dealing with many data and/or repeating routine tasks. Another access method might be to use grid middle-wares if supported. e-Science projects involve large teams managed, and they thus employ grid middlewares as the basis of their research activities due to the complexity of the software and the backend infrastructural requirements [2,3,4]. Grid middlewares provide efficient ways for sharing computational and information resources among participating scientists. However, it is often the case that setting up itself is a complicated task, and it might be overkill for those scientists who just want to access to desired data.

For the above reason, we have developed a light-weight tool for providing transparent access to online meteorological databases by way of local file system. The tool is based on FUSE [5], an implementation of usermode filesystem on Linux, which a user is allowed to deal with online meteorological data as if they were stored in his/her local file systems. Notice that the target of the tool is not limited to meteorology, and it can be used to integrate multidisciplinary information resources.

Actually, there have been several systems which enable us to deal with remote files by way of local filesystems with the help of user-level filesystems. Ufo [6] is an implementation of user-level filesystem on Solaris. SSHFS and HTTPFS are based on FUSE, and they allow us to access to remote data by respective network protocols. The main difference between our tool and them is that we focus on scientific data on the Web, and implement a caching mechanism by taking the features of those data into account.

2   The Proposed System

2.1   Filesystem in Userspace (FUSE)

We firstly introduce FUSE, an implementation of userspace filesystem in UNIX-like operating systems, on which we implement our proposed tool. Filesystem in Userspace (FUSE) is a UNIX kernel module, that allows users to create their own file systems without modifying the kernel code. This is achieved by the FUSE kernel module that bridges the file system code in user space to the kernel interfaces. So, when writing a filesystem code, all that a developer needs to do is to implement some necessary functions which basically correspond to system calls related to filesystem. As a consequence, he/she does not need to know much about kernel-module programming.

FUSE has been a part of the latest Linux kernel, and has been available in major distributions like RedHat and Debian. FUSE is also available for other

---

[5] http://fuse.sourceforge.net

UNIX-like operating systems, such as, FreeBSD, NetBSD (as PUFFS), OpenSo-
laris, and Mac OS X. FUSE supports major programming languages, such as C,
C++, Java, Perl, and so on. In this paper we adopt FuseFS [5], a Ruby binding
of FUSE that allows users to define a filesystem entirely by the Ruby language.

2.2   An Overview

MDFS (Meteorological Database File System) is
the tool that we have implemented. It works as a
client of FUSE, and a user is allowed to mount a
remote database as a subdirectory under the top
directory managed by MDFS. Once it is mounted,
he/she is allowed to get access to the database as
if it were a locally attached storage. Specifically,
Figure 1 depicts an overview of the system, and
the overall system works as follows:

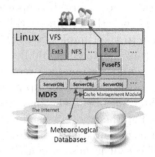

Fig. 1. A system overview

1) Suppose a user (or a client program) attempts
to access to a file under MDFS. It causes an invo-
cation of a system-call, which is caught by the VFS in the kernel. The message
is then passed to FUSE module and FuseFS. FuseFS translates the message to
an invocation of a method of MDFS.

2) MDFS attempts to extract information about the remote server to be accessed
by looking at the file path, and sends the request to the respective object which
is responsible for interacting with the remote server. We call such an object
"server object" hereafter.

3) The server object firstly query the cache module to check if there is a valid
cache entry. If so, it returns the cached file. Otherwise it tries to get the requested
data from the remote server by an appropriate protocol. Currently, only HTTP
is supported, but we plan to support other major protocols like FTP and SFTP.

4) Having received the requested data, the server object tries to transform the
data to the internal representation. Specifically, as for HTML data, it scrapes
the data to extract hyper links for subsequent generation of directory structures
under the MDFS filesystem. For other types of data, it is treated as a binary
data, and is conveyed to FuseFS. Also, the retrieved data is cached by the cache
module.

5) FuseFS translates the received data so that it can fit for the FUSE kernel
module, and sends the data back to FUSE. FUSE returns the data back to VFS,
and so on. Finally, the user (or the client) gets the response from the filesystem.

As described above, for each remote database, MDFS maintains a dedicated
object called "server object", which is responsible for interacting with its respective server. MDFS is allowed to interact with remote servers through server
objects. Consequently, a server object can be regarded as a network client from
the remote database side, while it is regarded as a filesystem from the MDFS
side.

```
1  url: http://remote.database.org/database/pub/
2  user: username
3  password: passphrase
4  delay: 0.7
```

Fig. 2. An example of server configuration (remotedb.yml)

2.3 Configuring Server Objects

Creating or configuring a server object is an easy task; all that a user needs to do is to write a configuration file in YAML format[6] specifying necessary information for accessing the database. Figure 2 depicts an example.

- The filename of a configuration, except for its suffix (.yml), is used as the root directory of the remote server. In this example, remotedb is used, that is, remotedb directory will appear just below the root directory managed by MDFS.
- url is mandatory, and is used to specify the network address of the remote server and the directory being accessed in the server.
- If the remote server requires an authentication, user and password are used to supply authentication information. These items are optional.
- Some databases do not permit continual access not only as a countermeasure against DoS attack, but also for supporting a large number of clients. To cope with such servers, delay can be used to specify intervals between successive accesses in seconds. This item is not mandatory.

```
==== Terminal 2 ====
$ ls
config/  lib/  mdfs.rb  tmp/

==== Terminal 1 ====
$ ruby mdfs.rb

==== Terminal 2 ====
$ ls
config/  lib/  mdfs.rb  mnt/  tmp/      ### "mnt" is the / of MDFS.
$ ls mnt/
gpvjma/  nws_noaa/  rish/              ### 3 servers are available.
$ ls mnt/gpvjma                        ### Looking into gpvjma.
2005/  2007/  ensemble_month_jma/  gsm_jma/  rsm_jma/
2006/  2008/  ensemble_week_jma/   msm_jma/  tmp/
$ ls mnt/gpvjma/gsm_jma/              ### Listing available GSM data.
GSM00X024  GSM00X084  GSM12X048  GSM12X180
GSM00X048  GSM12X024  GSM12X084  GSM12X192
$ ls -l mnt/gpvjma/gsm_jma/GSM00X024
-r--r--r-- 1 user group 1 Jan 23 23:14 mnt/gpvjma/gsm_jma/GSM00X024
```

Fig. 3. A session using MDFS

---

[6] YAML (YAML Ain't Markup Language) is a human-readable data serialization format for programming languages (http://yaml.org).

2.4   An Example Session

Figure 3 shows an example session using MDFS. In this session, two terminals are used; one is to invoke MDFS and the other is to access remote servers. After invoking the MDFS process, the mount point mnt appears. By looking into that directory, three directories appear, each of which corresponds to a remote database. In this session, GPV/JMA archive (http://gpvjma.ccs.hpcc.jp/~gpvjma/), RISH database (http://database.rish.kyoto-u.ac.jp/), and NOAA (http://www.noaa.gov/) are configured. By accessing those directories, a user can browse remote data as a part of his/her filesystem. Notice that most UNIX commands work fine even in MDFS. For example, one can use find to look for desired data by specifying filename. Since most users are accustomed to manipulate filesystems, MDFS can provide an easy and usable way to access to remote databases.

3   Preliminary Experiment

We have tested the feasibility of MDFS by a preliminary experiment. This section reports the experimental results.

We have compared the data transmission time of MDFS with an HTTP client (GNU wget). As for MDFS, we used the cp command to transfer a data item from the remote server to the localhost. In order to minimize the error caused by disk access, the null device (/dev/null) was used as the destination. To see the effectiveness of our cache mechanism, we repeated the same transmission for several times. Specifically, we copied the same data for three times for a session, and computed

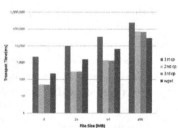

Fig. 4. Experimental results

the average over four sessions. Tested data sizes were 2MB, 16MB, 64MB, and 288MB. All data were copied from GPV/JMA Archive.

3.1   Experimental Results

Figure 4 shows the experimental results. Basically, the performance of MDFS is comparable to that of wget, but is slightly slower. This is due to the overhead caused by FuseFS, particularly for writing the accessed data to the cache in local storage. However, thanks to the caching mechanism, the 2nd and 3rd accesses are faster than wget.

An important observation here is that even for the 2nd and 3rd accesses, it is slower than wget for 288MB data. Probably, this owes to the nature of the Ruby language; it is an interpreter language, and it is therefore not so efficient at dealing with large data. Another fact might be related to the internal structure of FuseFS, that is, a data item being transmitted from FuseFS to FUSE have to be an Ruby object. So, when dealing with large data, the entire data is once loaded in main memory. This gives an significant impact on the entire system performance. In order to provide stable and robust performance, we need to develop additional mechanisms, because data sizes in many scientific databases are quite large.

4 Conclusions

In this paper we described a tool for providing transparent access to online meteorological databases by way of local file system. We employed FuseFS as the basis for developing a dedicated filesystem, and implemented a system to mount remote databases as directories in local filesystem. The system is easy to configure, and a scientist is allowed to deal with remote data as if they were stored in his/her local storage. The experimental results show the feasibility of the system.

As a part of our future work, we try to improve the performance particularly for large data. Another work is to integrate workflow mechanism in our tool so that a user is allowed to specify a set of regular tasks regarding data in remote servers.

Acknowledgments

This study has been supported by Grant-in-Aid for Scientific Research of JSPS (#18650018) and of MEXT (#19700083).

References

1. Amagasa, T., Kitagawa, H., Komano, T.: Constructing a Web Service System for Large-scale Meteorological Grid Data. In: 3rd IEEE Int'l Conf. on e-Science and Grid Computing (e-Science 2007), December 2007, pp. 118–124 (2007)
2. TeraGrid, http://www.teragrid.org/
3. LEAD Grid, https://portal.leadproject.org/gridsphere/gridsphere
4. GEO Grid, http://www.geogrid.org/
5. FuseFS, https://rubyforge.org/projects/fusefs/
6. Ufo, http://www.cs.ucsb.edu/projects/ufo/

$IndeGS^{RI}$: Efficient View-Dependent Ranking in CFD Post-processing Queries with RDBMS

Christoph Brochhaus and Thomas Seidl

Data Management and Exploration Group
RWTH Aachen University, Germany
{brochhaus,seidl}@informatik.rwth-aachen.de

Abstract. *Computational fluid dynamics* (CFD) linked with *virtual reality* (VR) visualization techniques offer comfortable means to explore interaction of fluids and gases with complex surfaces in the field of engineering or physics amongst others. Huge data sets, in the range of many gigabytes, require sophisticated storage schemes to enable efficient access during post-processing.

In this paper we introduce approximate geometric ranking methods for CFD data using off-the-shelf RDBMS by significantly extending the efficient indexing structure RI-tree. We further present preliminary, but very promising performance results of our ongoing research.

1 Introduction

Numerical simulations in the area of fluid dynamics became of growing importance by offering a very high level of accuracy and reproducibility of fluid behavior with complex surfaces and replacing tedious and expensive physical experiments. In both industrial development and research, CFD (*computational fluid dynamics*) [1] simulations are acknowledged methods in the field of physics or automotive engineering amongst others. During interactive post-processing, requested features of the CFD data sets are extracted by experts in the application domain. The results are commonly visualized in virtual reality environments, e.g. six-sided stereo projection systems, offering a high degree of interactivity by letting users fully immerse into the visualized objects. Common post-processing tasks include isosurface extraction (e.g.: "display regions with temperature = 125°C", cf. example in figure 1 in *Visualizer* window).

With CFD data sets up to many gigabytes in size, efficiency is one of the major requirements that VR frameworks have to satisfy, reducing expensive idle times until a result is presented and ready for visual inspection. In [2] we propose the **Index** based **Graphics data Server** *IndeGS*, utilizing novel secondary storage methods based on R-trees and supplying efficient dynamic view-dependent access methods. *IndeGS* offers quick and high-quality first impressions of result sets by streaming crucial parts of the solution, enabling the users to change view parameters as well as post-processing parameters "on the fly" with immediate system response.

Many post-processing tasks (isosurface extraction, geometrical selection, etc.) can be mapped to interval intersection queries on the CFD data to retrieve "active cells". The Relational Interval Tree (RI-tree) [3][4] offers very efficient query methods for

B. Ludäscher and Nikos Mamoulis (Eds.): SSDBM 2008, LNCS 5069, pp. 598–604, 2008.

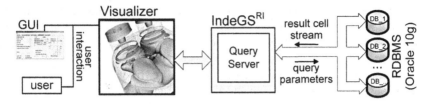

Fig. 1. IndeGS^{RI} infrastructure

intersection queries using a standard RDBMS, but does not provide a geometric ranking of the results aligned to the user's view position and direction during query execution. An erratic visualization without a ranking of the result stream does not support a good understanding of the result until the query execution, which is lengthy on very large data sets, is completely finished.

In this short paper we present our ongoing research on efficiently utilizing state-of-the-art RDBMS in the context of CFD post-processing. We introduce geometric ranking of data in the result stream during query processing and are thus able to visualize partial results depending on the user's view point and direction. We propose the extension of the RI-tree in combination with a standard RDBMS (here: Oracle 10g) via its extensible indexing interface. Figure 1 depicts the resulting infrastructure *IndeGS^{RI}*.

2   Relational Indexing of CFD Data

CFD data sets consist of collections of 3-dimensional cells of different structure (hexahedra, tetrahedra, etc.) defined by their respective corner points. These corner points carry geometric locality information and scalar values describing simulated properties (energy, density, temperature, etc.). The minimum and maximum of each scalar value of all corner points of one cell define a scalar range for this cell.

We next describe the Relational Interval Tree (*RI-tree*) for efficiently querying interval data. We substantially extend it in Section 2.2 for the use with CFD data and view dependent query processing. Our focused problem of CFD isosurfaces queries is mapped to interval intersection queries. To ensure a high-quality "first impression" of the result set, the user demands to be presented partial results during query execution depending on the user's standpoint. The RI-tree does not incorporate streaming of a geometric ranking of result data during query execution, thus catering for a random and erratic construction of the result set, irritating the user who is immersed into a virtual reality setup.

2.1   RI-Tree Basics

Interval trees have first been proposed by [5] for efficient isosurface extraction based on main memory access. Techniques presented in [6] enable the use of interval trees in conjunction with secondary storage by introducing the binary-blocked I/O interval tree. With the RI-tree [3][4], an efficient access method has been proposed to process interval intersection as well as stabbing queries on top of any existing relational database system. Data objects are managed by common built-in relational indexes following the paradigm of relational indexing. Its implementation is restricted to (procedural) SQL, and thus can be integrated easily into modern RDBMS (e.g. Oracle 10g, IBM DB2) through their extensible index interfaces.

The basic structure of the integer version of the RI-tree resembles a binary tree ("virtual backbone"), which is only arithmetically traversed and not materialized. Its node values are used as artificial keys for the stored intervals: upon insertion of an interval, the first node that hits the interval when descending the tree from the root node down to the interval location is assigned to that interval. An instance of the RI-tree consists of two relational indexes lowerIndex and upperIndex, both storing the artificial key value node, the bounds lower and upper, respectively, and the id of each interval. An interval is represented by a single entry in each of the two indexes. Primary keys based on B-tree indexes are generated on the attributes in lowerIndex and upperIndex. When querying the RI-tree, a candidate set of nodes is generated by traversing the virtual backbone. With these candidate sets, the lowerIndex and upperIndex are queried efficiently using their primary keys and intersecting entries are collected. For a detailed description we refer to [3] and [4], where the predominance of the RI-tree over competing interval indexing structures is shown. Minor modifications are applied to efficiently index intervals of floating point numbers.

2.2   RI-Tree Partitionings

During query processing on the standard RI-tree, the result stream is produced without respect to the geometric location of the result cells. To enable a "quick first impression", result cells close to the viewer are supposed to be presented first. We introduce a partitioning of the data space and integrate the partition information for the CFD cells into the RI-tree by creating new composite indexes lowerIndex(partition_ID, node, lower, cell_id) and upperIndex analogously. The corresponding B-trees are partitioned B-trees, as described in [7], and offer very efficient means to query sub-RI-trees (addressed by partition_ID).

Examples of axis-parallel partitionings of the data set based on different approaches are depicted in figure 2. In the case of a regular grid (figure 2a for two dimensions), we partition the data space into n^3 equi-sized cuboids, each of which holds the CFD cells located inside the cuboid. The density-based approach partitions the data space according to the n-quantiles for all geometrical dimensions (figure 2b) with the help of histograms describing the distribution of cells per dimension. The more advanced (*octree based*) approach recursively splits the data space into smaller (equi-depth) partitions (as when constructing an octree, cf. figure 2c). This step is repeated as long as the number of CFD cells in each new partition exceeds a predefined threshold t, leading to a finer grid in "dense" regions of the data set. To generate cuboids containing an equal number of cells, we developed an additional approach: the dimensions are split at the median point of the data set, thus resulting in $2^d = 8$ cuboids, each containing an equal number of cells. This process is repeated recursively on each of the cuboids until a desired partition size is achieved. CFD cells are not necessarily axis-parallel and might cross cuboid boundaries. To avoid complex splitting and storing of cells in numerous partitions, each cell is uniquely allocated to one partition defined by the cells center point. Other

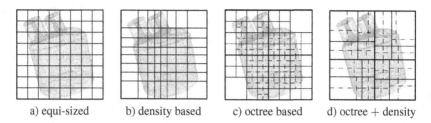

a) equi-sized   b) density based   c) octree based   d) octree + density

Fig. 2. Partitioning approaches

allocation schemes like "majority voting" on corner points or volume coverage can easily be realized. The partitioning information is maintained in separate straightforward tables and can be efficiently accessed during query execution.

3 Querying CFD Data in RDBMS

We focus on common queries that can be mapped to interval intersection queries on scalar values (e.g. "display isosurface for temperature = 125°C", "display regions with an energy between 4.5 and 4.7 J"). A non-sorted and erratic construction of the result set, as produced by a linear scan over the complete database, does not allow for a qualitative impression of the result until the data set is almost completely scanned (cf. results presented in [2]). Combining the partitioning of the data space with the extended RI-tree enables performing approximate view-dependent query execution. View-dependency is essential to achieve the perception of a "quick first impression" of the result set during query execution on many gigabytes of CFD data.

Figure 3 displays our approach to introduce view-dependency when querying the RI-tree: the view-point and view-direction (depicted by the arrow) define an order on the partitions by using appropriate distance functions. The closest partition, i.e. the partition the view-point lies in, is queried for active cells first, followed by the remaining cells. Figure 3 shows the ranking order induced by the Euclidean distance. Using the *hv*-distance [2], which is aligned to characteristics of human vision and ranks objects in the line of sight (direction of arrow) before objects in the peripheral field of vision, the partition order is like shown in figure 3b. The *hv*-distance speeds up the construction of

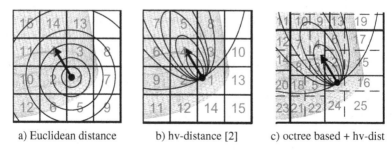

a) Euclidean distance   b) hv-distance [2]   c) octree based + hv-dist

Fig. 3. Example: Ranking of partitions

the result set in the user's field of vision when compared to traditional distance functions like Euclidean or Manhattan distance. Figure 3c shows an exemplary order for an octree-based partitioning in combination with the hv-distance.

The generic pseudo-code for the approximate ranking is quoted in Algorithm 1. "Generic" in this context means that partitioning methods, from which the partition candidates will be retrieved, can be arbitrarily defined (e.g. as presented in section 2.2). The distance function on which the ranking is based in line 2 can be any user-defined distance, e.g. Euclidean/Manhattan distance, hv-distance etc. A further reduction is possible on the level of partitions by pruning partitions from the candidate set (line 1) for which we can reliably state that they do not contain "active" cells by storing the range which is covered by the specified partition. If the queried scalar value is not contained in the range of a certain partition, this partition is removed from the candidate set without loss of results.

Algorithm 1. Retrieve "active" cells from database

Require: view point $q = (x, y, z)$, value $scalar = s$
1: PART $\Leftarrow$ retrieve partition candidates from DB
2: order partitions PART by distance to view point q
3: **while** PART not empty **do**
4:     remove first element of PART $\Rightarrow$ part_id
5:     query RI-tree index with (part_id, s)
6:     stream "active" cells to visualizer
7: **end while**

---

4   Preliminary Results

In this section we present preliminary results of experiments performed using $IndeGS^{RI}$. Our test data originates from a simulated fuel injection into a combustion engine cylinder consisting of $\approx$190,000 cells. During the continuation of our research we examine the expected scalability of our approaches on larger data sets (in the gigabyte range). We compared the performance of several partitioning schemes regarding runtime and correctness of the ranking, by running our experiments on Oracle 10g. We generated equi-distant and density-based partitionings with 8 partitions in each geometric dimension (resulting in 512 unique partitions). We performed isosurface queries on scalar "energy" of varying complexity (2,500 up to 10,000 result cells), with an average isosurface complexity of $\approx$5,200 cells.

First we measured the degree of disorder in the stream of result cells regarding the view-dependent ranking (here: by Euclidean distance). Figure 4a displays the average disorder, which is defined by the average displacement of all result cells in the measured stream compared to the correct ranking. The linear scan and standard RI-tree (without ranking functionality) yield the highest average disorder, with up to 3,100 for a result set size of 10,000 cells, i.e. each cell in the result stream is on average misplaced 3,100 positions compared to the correct ranking. The equi-distant and density-based partitioning variants show a significantly better quality of ranking. The disorder for the partitioned RI-tree almost grows linearly with the result set size, but the relative displacement is in the range of only $4.3 - 6.8\%$. The "sorted" variants order partial results after each partition query (cf. lines 3-7 in algorithm 3) before streaming the cells to the visualizer, thus adding marginal sorting costs to the overall execution time, but in return further reducing the average disorder to a small degree.

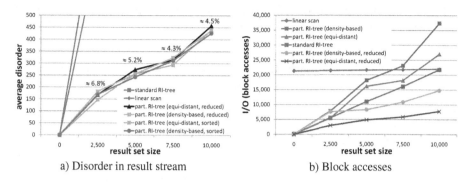

a) Disorder in result stream   b) Block accesses

Fig. 4. Evaluation of disorder and block accesses

Figure 4b displays the block accesses performed by the RDBMS. The "linear scan" requires a constant number of block reads as every database block is accessed. The RI-trees without prior exclusion of partitions perform unnecessary block reads when querying partitions not containing result cells, thus producing even more block reads than the standard RI-tree. Best performance is shown by the partitioned RI-tree variants which exclude partitions prior to query execution (titled "reduced" in figure 4b), which significantly outperform the standard RI-tree with factors of up to ≈ 3.

5 Future Work

Our preliminary experiments show promising results when applying the presented approaches of partitioning CFD data sets. We plan to investigate on other partitioning schemes, which are not restricted to an axis-parallel partitioning. Furthermore, one of the design goals is to further improve the "first impression" during query execution and we plan to extend our evaluation to various view-oriented distance functions.

In [2], we present dynamic query and result stream adaption when the user changes view-point and direction by moving in the VR environment during query execution. We plan to integrate handling of dynamic query adaption in *IndeGS*^{RI}.

References

1. Lomax, H., Pulliam, T.H., Wingg, D.W.: Fundamentals of Computational Fluid Dynamics. Springer, Heidelberg (2001)
2. Brochhaus, C., Seidl, T.: Efficient Index Support for View-Dependent Queries on CFD Data. In: Papadias, D., Zhang, D., Kollios, G. (eds.) SSTD 2007. LNCS, vol. 4605, pp. 57–74. Springer, Heidelberg (2007)
3. Kriegel, H.P., Pötke, M., Seidl, T.: Managing Intervals Efficiently in Object-Relational Databases. In: VLDB Conference, pp. 407–418 (2000)
4. Brochhaus, C., Enderle, J., Schlosser, A., Seidl, T., Stolze, K.: Efficient Interval Management Using Object-Relational Database Servers. Informatik - Forschung & Entwicklung 20(3), 121–137 (2005)

5. Cignoni, P., Marino, P., Montani, C., Puppo, E., Scopigno, R.: Speeding Up Isosurface Extraction Using Interval Trees. IEEE Trans. on Visualization and Comp. Graphics, 158–170 (1997)
6. Chiang, Y., Silva, C.: External Memory Techniques For Isosurface Extraction In Scientific Visualization. In: External Memory Algorithms and Visualization, vol. 50, pp. 247–277 (1999)
7. Graefe, G.: Partitioned B-trees - A User's Guide. In: BTW Conference, pp. 668–671 (2003)

Real-Time Integration of Geospatial Raster and Point Data Streams*

Carlos Rueda and Michael Gertz

Dept. of Computer Science, University of California, Davis, U.S.A.
{carueda,mgertz}@ucdavis.edu

Abstract. Sensor and network technology advances are increasingly placing an immense amount of real-time geospatial data streams at the scientist's disposal. The effective integration and assimilation of such datasets, however, is still a challenging goal. In this paper, we describe a computational framework that simplifies the design, execution, and visualization of processing workflows involving the integration of satellite raster and ground point data streams. The framework is enabled for interoperability by adhering to open sensor data standards, and demonstrated with the evaluation of key environmental inputs needed for the estimation of reference evapotranspiration over California.

1 Introduction

As geospatial sensor data becomes ubiquitous, there is an increasingly critical need for its effective integration in applications that require inputs from multiple sources. Moreover, as geospatial applications are themselves becoming increasingly sophisticated (variety of products, distributed, etc.), the real challenge, beyond data integration, is to accomplish a high degree of *interoperability* to better reuse and exploit the richness of available services and sensor data.

Integration and interoperability play a crucial role in the running of complex environmental simulation models that assist in forecasting and, more generally, in the estimation of meteorological variables that are very difficult or costly to obtain by direct means. For example, accurate evapotranspiration estimations over extended agricultural areas are important information for water agencies and individuals interested in delineating optimal irrigation plans, improving water quality, and increasing yield [1].

The main contribution of this paper consists of the development of a processing framework, which, founded on a precise geospatial data stream model, is enabled with the utilization of open geospatial standards especially targeted at sensor information and interoperability. By adopting a scientific workflow based realization, we also facilitate the usability of the system in scientific environments. We demonstrate our approach with a representative setting for the evaluation of key environmental inputs needed for the estimation of reference

* This work is in part supported by the NSF under awards ATM-0619139 (COMET), DBI-0619060 (REAP), and IIS-0326517 (GEOSTREAMS).

B. Ludäscher and Nikos Mamoulis (Eds.): SSDBM 2008, LNCS 5069, pp. 605–611, 2008.
© Springer-Verlag Berlin Heidelberg 2008

evapotranspiration over California. The evaluation is performed both quantitatively via regression analysis, and qualitatively via 3D visualizations. These operations are realized as reusable processing components that can be inserted and linked with other components in new scientific workflows.

2   Related Work

Although decades of research have seen several efforts on integrating static data sets (see, *e.g.*, [2]), only little work has been done on integrating heterogeneous data coming from diverse sensors and in a streaming fashion. Peng *et al.* [3] present a framework for publishing, browsing, and delivering real-time sensor data based on UDDI registries and a non-standard communication protocol. Moodley *et al.* [4] describe a satellite-based system to support wildfire disaster response management using Sensor Web technologies. Chu and Buyya [5] present a service oriented sensor Web architecture that integrates Sensor Web with Grid computing. Pursuing similar objectives in general, our work is founded on an especially defined stream model and uses a scientific workflow management system as the realization infrastructure.

Also, substantial research in the database community has been done toward the management of data streams (see, *e.g.*, [6]). Most of these efforts, however, adopt a relational approach (mainly in the context of event monitoring, financial market analysis, and network traffic analysis), but little research has been done for a systematic treatment in the case of streams of geospatial images. To represent geospatial image and point data streams in this paper, we use the model introduced for this case in our previous work [7,8]. Here, we use relevant concepts of such a model with a focus on integration and visualization of data with different spatio-temporal characteristics, and show how the underlying framework is extended to incorporate Web service interfaces in a fashion that is transparent to the components in the processing workflows.

3   Geospatial Data Stream Processing

Both images and point data from remote sensing instruments (*e.g.*, at ground stations or carried in satellites, aircraft, etc.) are regarded as *geospatial observations*. A geospatial observation $\mathbf{a} = (\mathbf{v}, \mathbf{x}, \tau)$ comprises the observed value $\mathbf{v}$ (scalar, vector, image) of a particular observable entity at a particular point location or area $\mathbf{x}$ and time of acquisition τ. A *geospatial data stream* is an unbounded, timestamp-ordered sequence of observations, $\langle \mathbf{a}_1, \mathbf{a}_2, \ldots \rangle$, where all values belong to a common set and all locations are given in reference to a common coordinate system. In the case of point data, the location is given in a suitable geographic reference and often associated as an explicit property of the originating sensor. In the case of sensor instruments able to generate single observations that cover extended geographical areas (*e.g.*, satellite imagery), these usually take the form of a raster (possibly multi-banded) image. In Sect. 4, we illustrate our approach with both point and raster data streams.

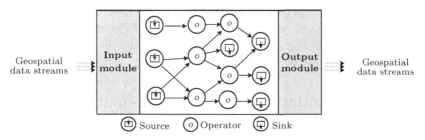

Fig. 1. Core architecture [7] and extensions developed in this work (represented in grayed boxes) for interoperability

We have extended our geospatial data stream processing framework [7] to incorporate communication components that conform to open standards for interoperability. See Fig. 1. This framework is coupled with the Kepler system [9], a comprehensive scientific workflow management system being developed by a community of contributors from several projects in diverse disciplines such as the geosciences, ecology, bioinformatics, and others. Kepler allows to design, execute, and deploy scientific workflows using Web- and Grid-based technologies.

To fully potentiate the interoperability of our framework with other data stream processing systems, we incorporate the utilization of open data and protocol standards. Of particular relevance are the activities recently taken by the Open Geospatial Consortium (OGC) Sensor Web Enablement (SWE) program [10], an initiative that seeks to provide interoperability between disparate sensors and sensor processing systems by establishing a set of standard protocols to enable a "Sensor Web," by which sensors of all types in the Web are discoverable, accessible, and taskable. SWE components include several models, XML Schemas, and Web service interfaces. In our prototype, we have included elements from the following SWE components:

- SensorML, *Sensor Model Language*: An XML Schema to describe sensors and sensor platforms.
- O&M, *Observations & Measurements*: A specification for encoding observations and measurements from sensors.
- SOS, *Sensor Observation Service*: A Web service interface to request and retrieve metadata about sensor systems as well as observation data.

The general sequence of steps to obtain sensor metadata is as follows. As an SOS service, the data provider first returns a *capabilities* document as a response to a *GetCapabilities* request by a client. This document includes the identification of the provider itself and the description of the available streams in the system, which are organized in the form of *observation offerings*. An offering includes information about the period of time for which observations can be requested, the phenomena being sensed, and the geographic region covered by the observations. Once a client is interested in a particular geospatial data stream, it will submit a *DescribeSensor* request to the provider. The corresponding response is a SensorML document describing the sensor that generates the stream. Next, the client will request the actual data from the sensor. This is done by submitting a *GetObservation* request, whose response is an O&M document [10]. The O&M response will include an `xlink:href` attribute that the client can use to connect to the actual data stream.

4   Application: Integration of Ground- and Satellite-Based Data Streams for Evaluation of an Environmental Model

The context of this application is the ETo Project [1]. This system calculates spatially distributed daily *reference evapotranspiration*, denoted ETo, and produces corresponding daily maps for the state of California at high spatial resolution (4km^2), using data from the CIMIS network of weather stations [11]. Evapotranspiration is the loss of water to the atmosphere caused by evaporation and plant transpiration. Accurate ETo estimations are important information for individuals and water agencies interested in delineating irrigation plans over extended areas. Several meteorological variables are required to calculate ETo, including temperature, relative humidity, and wind speed.

A main goal in the ETo system is to generate ETo maps for the whole state of California. To acomplish this goal, the measurements of some of the variables (*e.g.*, temperature) obtained at the CIMIS stations are interpolated to obtain the input rasters required to compute ETo for the state. A key step in the overall ETo process is the determination of the "best" interpolated surface to use for each variable. Experience has shown, however, that even when the interpolation techniques give similar quantitative errors (which are based on standard cross-validation methods), noticeable discrepancies in the spatial configuration of the generated surfaces are often revealed. Hence, combined visualizations of these interpolated surfaces over the same variable have been implemented as an important tool for inspection. Here, we include real-time GOES satellite data [12] as one more source for evaluating the output from the interpolation methods.

Several workflow steps are required to properly integrate and compare CIMIS and GOES raster data in this application. For instance, there are three geographical projections involved: Albers Equal Area (AEA) is the projection to be used for the generated ETo maps; latitude/longitude coordinates are used for the CIMIS stations; and an instrument specific coordinate system is used in GOES data. The required reprojections and other main steps carried out by the Kepler workflow shown in Fig. 2 are explained next.

Using the SOS interface, the `CimisReader` source actor reads data from the CIMIS system for a given date and hour, and provides the resulting geo-located point temperature observations into the workflow. The location of each observation is converted by the `L2A` actor from latitude/longitude coordinates to easting-northing coordinates (in meters) in the AEA projection. The resulting point data is read by the `Interpolator` actor, here instructed to use one of the available interpolation methods and, as the region of interest for the generated raster, the state of California (in AEA projection). The `Interpolator` actor makes a

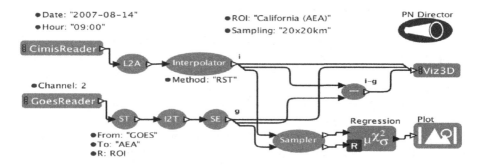

Fig. 2. Workflow for comparing CIMIS and GOES temperature raster images

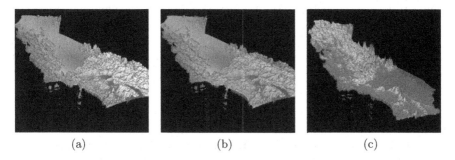

(a)   (b)   (c)

Fig. 3. Comparing interpolated and GOES temperature raster images

request to the ETo system back-end to perform the interpolation and receives the resulting raster as a Web Coverage Service response. The resulting interpolated temperature raster over the region is denoted **i** in Fig. 2.

While **L2A** applies a direct point-to-point reprojection, the **ST** actor in the workflow performs a full spatial transform of the incoming GOES infrared images to the region of interest given in the AEA projection. Finally, the **I2T** actor performs the pointwise conversion to temperature, and the stream extension **SE** actor spatially aggregates the incoming images to maintain a single composite over the covered region [7]. The resulting GOES-derived temperature raster over California is denoted **g** in Fig. 2.

An especially designed actor, **Viz3D**, allows the visualization of the input rasters for qualitative inspection. Complementary to side-by-side comparisons, the **Viz3D** actor can combine the raster surfaces in many ways to better appreciate the differences. Three examples are shown in Fig. 3. Raster **g** is mapped to gray-level in (a) and (b) to easily distinguish it from **i**, while in (b) **g** is also assigned a certain level of transparency to better appreciate the discrepancy with the interpolated raster. The difference **i** − **g** is directly displayed in (c) along with a semitransparent constant surface at $0°C$.

Quantitatively, a linear regression analysis is performed by the **Regression** actor on regularly-spaced selected points provided by the **Sampler** actor.

5 Conclusion and Future Work

We presented an extension to a computational framework for the processing of geospatial data streams with different spatio-temporal characteristics to allow the seamless integration of raster and point data, two common data formats in remote sensing imagery and ground based sensor systems. A main goal for interoperability is accomplished by adhering to open geospatial standards and service interfaces especially designed for sensor data. Our scenario highlighted the integration of point and raster data with different geographic projections, for the evaluation of temperature estimations over California. We described a regression analysis in the raster setting, which required the interpolation of the ground observations over the region of interest. Both qualitative and quantitative comparisons were presented as a final step in the overall integration exercise demonstrating the benefits of the framework in a scientific environment.

Our ongoing work includes the incorporation of sensor data registries, process protocols, and other SWE components into our prototype, as well as the evaluation of middleware stream system strategies toward the computational scalability of the framework in complex, demanding environments.

References

1. Hart, Q., Brugnach, M., Temesgen, B., Rueda, C., Ustin, S., Frame, K.: Daily reference evapotranspiration for California using satellite imagery and weather station measurement interpolation. Civil Eng. and Environ. Systems (to appear)
2. Ziegler, P., Dittrich, K.: Data Integration – Problems, Approaches, and Perspectives. In: Conceptual Modelling in Information Systems Engineering, pp. 39–58 (2007)
3. Peng, R., Hua, K.A., Hamza-Lup, G.L.: A Web services environment for Internet-scale sensor computing. In: Proc. IEEE International Conference on Services Computing, pp. 101–108 (2004)
4. Moodley, D., Terhorst, A., Simonis, I., McFerren, G., van den Bergh, F.: Using the Sensor Web to detect and monitor the spread of wild fires. In: 2nd International Symposium on Geo-information for Disaster Management (2006)
5. Chu, X., Buyya, R.: Service Oriented Sensor Web. In: Sensor Networks and Configuration, pp. 51–74 (2007)
6. Chaudhry, N., Shaw, K., Abdelguerfi, M.: Stream Data Management (Advances in Database Systems). Springer, Heidelberg (2005)
7. Rueda, C., Gertz, M.: Modeling satellite image streams for change analysis. In: Proceedings of the 15th annual ACM international Symposium on Advances in Geographic Information Systems (ACMGIS), pp. 43–50. ACM Press, New York (2007)
8. Gertz, M., Hart, Q., Rueda, C., Singhal, S., Zhang, J.: A data and query model for streaming geospatial image data. In: Grust, T., Höpfner, H., Illarramendi, A., Jablonski, S., Mesiti, M., Müller, S., Patranjan, P.-L., Sattler, K.-U., Spiliopoulou, M., Wijsen, J. (eds.) EDBT 2006. LNCS, vol. 4254, pp. 687–699. Springer, Heidelberg (2006)

9. Ludäscher, B., Altintas, I., Berkley, C., Higgins, D., Jaeger, E., Jones, M., Lee, E.A., Tao, J., Zhao, Y.: Scientific Workflow Management and the Kepler System. In: Concurrency and Computation: Practice & Experience (2005)
10. Open Geospatial Consortium. OpenGIS Sensor Web Enablement: Architecture Document, www.opengeospatial.org/pt/14140
11. California Irrigation Management Information System, http://wwwcimis.water.ca.gov
12. Geostationary Operational Environmental Satellite, www.goes.noaa.gov

IRMA: An Image Registration Meta-algorithm*
Evaluating Alternative Algorithms with Multiple Metrics

Kelvin T. Leung[1,2], D. Stott Parker[1,2], Alexandre Cunha[2,3],
Cornelius Hojatkashani[4], Ivo Dinov[2,4], and Arthur W. Toga[2,4]

[1] UCLA Computer Science Dept., Los Angeles CA 90095-1596
[2] UCLA Center for Computational Biology (CCB), Los Angeles CA 90095-1569
[3] Caltech Center for Advanced Computing Research (CACR), Pasadena CA 91125
[4] UCLA Laboratory of Neuro-Imaging (LONI), Los Angeles CA 90095-1569

Abstract. IRMA is a *meta-algorithm* for image registration (image alignment), evaluating results under multiple metrics using the LONI Pipeline workflow infrastructure, on the LONI/CCB grid computing facility. IRMA manages these results in a model base implemented with PostgreSQL. It permits scientists to catalog the results such as provenance information, and permits subsequent mining — exploring the space of alternatives in an organized fashion and building understanding about individual algorithms, and learn about strengths and weaknesses of algorithms over time.

1 Introduction

A common problem in scientific computing is to have to choose among a number of different metrics or scoring criteria. To minimize error, for example, one might choose among metrics like L_1 error, L_2 error, etc. These choices can be difficult to make, because each criterion is an objective that can be useful in its own right. Because they can differ significantly, however, ad hoc choices can have significant consequences.

It is sometimes said that a sign of successful diversity is that complete agreement becomes impossible — no single point of view is enough. Both science and statistics rest on diversity, encouraging different opinions as long as they add information, even when their objectives are inconsistent. Different methods for extracting information from data rest on different objectives or metrics as well.

This work grew out of investigations in data mining, where sophisticated sampling and meta-level methods for combining models have had enormous impact. Intuitively, just as using ensembles of methods can combat bias and reduce variance, using multiple methods should provide benefits in large-scale scientific computation. The results here report an effort spanning several years at UCLA, resulting in the development of a meta-algorithm for neuroimaging.

* This work supported by NIH grant 1U54RR021813 and the Center for Computational Biology, an NIH National Center of Biomedical Computing.

B. Ludäscher and Nikos Mamoulis (Eds.): SSDBM 2008, LNCS 5069, pp. 612–617, 2008.
© Springer-Verlag Berlin Heidelberg 2008

IRMA (Image Registration Meta-Algorithm) is a method for combining results of different image registration algorithms. Registration is effectively a problem of 'aligning brains', i.e., obtaining a correspondence between two different objects that finds as many points of similarity between them as possible.

In this paper we briefly review the problem of image registration, and present a sampler of metrics used in assessing the quality of a registration result. We then report on the development of IRMA and its application for this problem. We summarize the overall design of IRMA, its use of a database to manage metadata produced by use of multiple algorithms and metrics, and on the use of data mining methods to analyze this data.

2 Biomedical Image Registration

2.1 Essence of the Problem

Let $\mathcal{R}$ and $\mathcal{T}$ be, respectively, the reference and template images we want to register. In image registration we typically look for a transformation f such that under reasonable assumptions $D(\mathcal{R}, f(\mathcal{T})) \simeq 0$, where D is a measure of similarity (distance) between a pair of images. Thus, we want the transformed image $f(\mathcal{T})$ to be as close as possible to the target image $\mathcal{R}$. In general, if $D(\mathcal{A}, \mathcal{B}) \simeq 0$ then we say the images $\mathcal{A}$ and $\mathcal{B}$ are similar.

The similarity between images is commonly defined either as a function of the intensities (luminosities) of corresponding voxels across images and their distributions or is based on the morphology of the features present in both images. In practice, measuring similarity in an application depends on the application itself and on the modalities of the input images. Both intensity- and morphology-based metrics have been largely employed in the implementation of registration tools to attend different needs including comparing images with different modalities. We want the mapping f to be homeomorphic so that points close together in one image are carried over to points close together in the other image. Also, f must have a continuous inverse satisfying $D(f^{-1}(\mathcal{R}), \mathcal{T}) = D(\mathcal{R}, f(\mathcal{T}))$. In principle a metric (i.e., distance measure) should also satisfy properties such as: commutativity $(D(\mathcal{A}, \mathcal{B}) = D(\mathcal{B}, \mathcal{A}))$, identity $(D(\mathcal{A}, \mathcal{B}) = 0$ iff $\mathcal{A} = \mathcal{B})$, and scale invariance $(D(\alpha\mathcal{A}, \mathcal{B}) = D(\mathcal{A}, \alpha\mathcal{B}) = D(\mathcal{A}, \mathcal{B})$, for $\alpha > 0)$. When assessing registration, it is natural to investigate how the edges from the template image are mapped to the corresponding edges in the reference image. If the mapped and reference edges are perfectly sumperimposed, or very close in shape and space, then we say we have a good registration.

2.2 Diversity of Methods and Metrics

The metrics we consider here can be divided into two categories: *Intensity based metrics*, which solely rely on the luminosity of the voxels, and *Statistics metrics*, which are based on the distribution of intensities. There are strengths and weaknesses of each metric when applied to categories of image modalities. In fact, some of these metrics are designed and biased towards specific categories and therefore cannot encompass all possible images arising in practice.

We list here seven basic metrics that we have studied. (A survey covering the derivation and use of the entropy-related metrics is available in [1], and the Correlation and Woods metrics are summarized in [2].) Throughout this list, N is the size of the images (total number of voxels), and x ranges over the set of voxels in an image. For the problem of registration, the image $\mathcal{S}$ could be $f(\mathcal{T})$, so that for each of the following metrics $D(\mathcal{R}, \mathcal{S})$ is $D(\mathcal{R}, f(\mathcal{T}))$:

Mean Square Diff of Intensities	$\frac{1}{N}(\mathcal{R} - \mathcal{S})^2 = \frac{1}{N}\sum_x (\mathcal{R}(x) - \mathcal{S}(x))^2$
Absolute Difference of Intensities	$\frac{1}{N}\|\mathcal{R} - \mathcal{S}\| = \frac{1}{N}\sum_x \|\mathcal{R}(x) - \mathcal{S}(x)\|$
Entropy of Difference of Intensities	$\mathcal{H}(\mathcal{R} - \mathcal{S}) = \frac{1}{N}\sum_x p(\mathcal{R}(x) - \mathcal{S}(x)) \log p(\mathcal{R}(x) - \mathcal{S}(x))$
Mutual Information	$\mathcal{I}(\mathcal{R}, \mathcal{S}) = \mathcal{H}(\mathcal{R}) + \mathcal{H}(\mathcal{S}) - \mathcal{H}(\mathcal{R}, \mathcal{S})$
Normalized Mutual Information	$\mathcal{I}(\mathcal{R}, \mathcal{S})/\mathcal{H}(\mathcal{R}, \mathcal{S}) + 1 = (\mathcal{H}(\mathcal{R}) + \mathcal{H}(\mathcal{S}))/\mathcal{H}(\mathcal{R}, \mathcal{S})$
Correlation	$cor(\mathcal{S} \mid \mathcal{R}) = 1 - \frac{1}{N\,var}\sum_i N(i)\,var(i)$
Woods	$woo(\mathcal{S} \mid \mathcal{R}) = 1 - \frac{1}{N}\sum_i N(i)\,stddev(i)\,/\,mean(i)$

In the final two metrics, i ranges over intensity values, $N(i)$ is the number of voxels in $\mathcal{R}$ having value i, and *mean*, *var*, and *stddev* are statistics of the same voxel positions. This list of metrics illustrates the challenge: there is diversity of opinion about what constitutes a good registration. Some researchers go so far as to assert that high-quality registrations can only be discerned by experts, and then increase the challenge by noting that experts disagree. To help facilitate comparisons across metrics, the metric values were converted to uniform similarity 'scores' (values in $[0, 1]$, with 1 being good and 0 being bad), by normalizing and complementing them. as shown in Figure 1. An alternative view is that all metrics reflect some aspect of quality of a registration. There are many notions of quality. Since metrics can be computed automatically, evaluating a set of them gives us an inexpensive way of assessing multiple aspects of quality, a means for eliminating poor results, and a basis for machine learning.

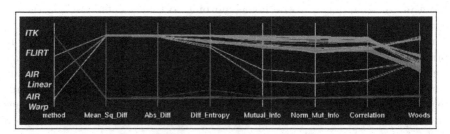

Fig. 1. A parallel coordinates plot, recording the performance of 4 algorithms under 7 metrics. The plot summarizes results for each of the 7 metrics (the columns) discussed in this paper, for 170 differently configured algorithm runs that made use of 4 different registration algorithms — AIR Warp, AIR Linear, FLIRT, ITK — the 'rows' or trajectories across this plot. Altogether there are 170 'rows' or 'trajectories', each giving the metric values obtained by one run. The plotted points have been jittered (perturbed) slightly so as to make visible the overlapping results for AIR.

2.3 Example

Figure 1 shows aggregate results of registering a set of 10 brain images (T2-weighted RARE 3D MRI volumes of dimension $256 \times 256 \times 256$, of wild-type mice, produced at UCLA), using 4 different method/parameter-setting runs per image: altogether a total of 170 different runs. The figure shows the results on the 7 different metrics just discussed for each of these runs.

The four algorithms here include two – Linear and Warp (nonlinear) – from UCLA's AIR registration package [2], the FLIRT tool from Oxford's FSL package [3], and the 3D Deformation Registration 7 program from NLM Insight Segmentation and Registration Toolkit (ITK) package [4]. In addition, a total of 17 different method/parameter combinations were used, i.e., there were a total of 17 different registration runs per image. These were configured as follows:

Method/Tool	Options/Parameters used
AIR Warp	same as for AIR Linear, default otherwise
AIR Linear	blur $\in$ {(11,11,11), (19,19,19) }, model $\in$ {6,7,9}
FLIRT	transformation $\in$ {6,7,8,12}
ITK	default

This gives a modest but representative sample of parameter settings. On the 10 brain images (NORM) we obtain a set of 170 registration runs, a nontrivial computational load requiring several hours to complete on the LONI/CCB grid.

The values for each metric have been rescaled independently, so that the spread in metric values covers the entire vertical scale. Thus the plot really only highlights patterns among the *relative ordering* among metric values for these different methods, and not patterns among the actual magnitudes of the metric values. Furthermore, of course, nothing about the relative merit of the 4 algorithms can be determined from one sample registration problem.

3 IRMA: An Image Registration Meta-algorithm

IRMA is a meta-algorithm for image registration developed with all of the considerations above in mind. We have implemented it as a set of programs that execute within the LONI Pipeline environment [5] at UCLA. Figure 2 shows a pipeline definition of IRMA for the 4 algorithms and 7 metrics described earlier. When executed it produced the results shown in Figure 1.

3.1 Essential Aspects of the IRMA Design

The basic idea behind IRMA is simple. Given a set of images $\mathcal{R} = \{\, \mathcal{R}_1, \ldots, \mathcal{R}_n \,\}$ produced by a set of registration algorithm runs $A = \{\, A_1, \ldots, A_m \,\}$ and their respective mappings f_i, $i = 1, \ldots, n$ (it is possible that a given algorithm gives more than one candidate mapping, thus $m \leq n$), we can automate the evaluation of each candidate mapping under a battery of metrics.

The implementation of IRMA goes beyond this basic idea in two ways: (1) IRMA uses a database system to store not only metric values obtained by each run, but also metadata about program execution, and (2) IRMA includes tools

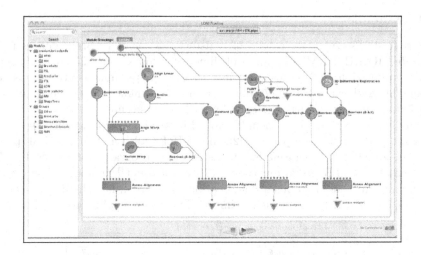

Fig. 2. Screen capture of the LONI Pipeline implementation of IRMA

for mining the resulting tables of metric values and execution information. Both of these design aspects are significant in achieving goals we have set for IRMA.

3.2   IRMA as a Scientific Database

As Figure 2 shows, an IRMA computation consists of a number of independent branches, each evaluating the results of a given algorithm and a choice of parameters. These evaluations are performed by the *Assess Alignment* modules shown in the figure, which in our example here produce the 7 metric values. These results are sent to a backend PostgreSQL database for later analysis. The LONI Pipeline also produces a log containing metadata about the execution of each IRMA step. IRMA extracts relevant parts of this metadata that also can be useful for subsequent analysis. Currently the database includes the following important fields: user name, execution directory, images and files used, commands/modules/algorithms used, and options/parameters/flags used.

IRMA is thus a system for managing data about program executions. Relying on database technoloyg provides three important benefits. First, it endows IRMA with the ACID properties (Atomicity, Consistency, Isolation, Durability) of database systems, permitting it to survive in the scientific world – a world in which tools are unreliable, as is unfortunately the case in neuroimaging. Second, IRMA can operate effectively on the grid: the LONI Pipeline is designed to create and manage independent processes, and IRMA provides the synchronization needed to aggregate their results. Third, IRMA is a model base, allowing ad hoc extraction and mining of execution log data. Although a given set of any given set of executions may not be large (170 runs in our example), a model base approach makes analysis of this data natural.

3.3   IRMA as a Data Mining Platform

IRMA's managing of information about execution provides interesting capabilities for data mining. For example, one can determine not only which algorithms and parameter settings give better results for brain volumes from a given source, but also analyze execution times and differences in performance by variants of a given algorithm. The introduction of mining capabilities should foster deeper understanding of registration algorithms, and permit corresponding advances in sophistication. It should also improve our understanding of the various metrics.

For example, *dimensionality reduction* methods based on eigendecompositions are successful at extracting dominant structural elements for this kind of dataset. Whereas the first eigenvector is weighted almost equally over all metrics, the second heavily emphasizes the Woods metric. We can reduce most of the performance data to a 2-dimensional space, capturing the variance among these metrics shown in Figure 1. This permits learning about both the quality of the individual image registration results and the algorithms and metrics used.

4   Conclusions

We have summarized the design and implementation of IRMA, an Image Registration Meta-Algorithm. IRMA permits the creation of LONI Pipeline modules that obtain the results of many image registration algorithms, and then evaluate these results under multiple metrics.

A key part of IRMA is a model base that permits management and analysis of these evaluation results. The model base, implemented with PostgreSQL, stores basic metadata about the execution of each IRMA process. These metadata include information about the execution of the process, and can be used in developing models about many aspects of algorithm performance over time.

References

1. Pluim, J.P.W., Maintz, J.B.A., Viergever, M.A.: Mutual information based registration of medical images. IEEE Trans. Med. Imaging 22(8), 986–1004 (2003)
2. Woods, R.P., et al.: Automated Image Registration: I. General Methods and Intrasubject, Intramodality Validation. J. Comp. Ass. Tomography 22, 139–152 (1998)
3. Smith, S.M., et al.: Advances in functional and structural MR image analysis and implementation as FSL. NeuroImage 23(S1), 208–219 (2004)
4. National Library of Medicine Insight Segmentation & Registration Toolkit (ITK), http://www.itk.org/index.htm
5. Dinov, I.D., Parker, D.S., Payan, A., Tam, J.-W., Cheung, C., Rajendiran, J., Hojatkashani, C., MacKenzie-Graham, A., Horn, J.V., Leung, K.T., Konstantinidis, F., Magsipoc, R., Woods, R.P., Toga, A.W.: Intelligent Graphical Workflow Pipeline Infrastructure for Automated Analys is of Neuroimaging Data, http://pipeline.loni.ucla.edu

Author Index

Lecture Notes in Computer Science

Sublibrary 3: Information Systems and Application, incl. Internet/Web and HCI

For information about Vols. 1– 4662
please contact your bookseller or Springer

Vol. 4853: F. Fonseca, M.A. Rodríguez, S. Levashkin (Eds.), GeoSpatial Semantics. X, 289 pages. 2007.

Vol. 4836: H. Ichikawa, W.-D. Cho, I. Satoh, H.Y. Youn (Eds.), Ubiquitous Computing Systems. XIII, 307 pages. 2007.

Vol. 4832: M. Weske, M.-S. Hacid, C. Godart (Eds.), Web Information Systems Engineering – WISE 2007 Workshops. XV, 518 pages. 2007.

Vol. 4831: B. Benatallah, F. Casati, D. Georgakopoulos, C. Bartolini, W. Sadiq, C. Godart (Eds.), Web Information Systems Engineering – WISE 2007. XVI, 675 pages. 2007.

Vol. 4825: K. Aberer, K.-S. Choi, N. Noy, D. Allemang, K.-I. Lee, L. Nixon, J. Golbeck, P. Mika, D. Maynard, R. Mizoguchi, G. Schreiber, P. Cudré-Mauroux (Eds.), The Semantic Web. XXVII, 973 pages. 2007.

Vol. 4823: H. Leung, F. Li, R. Lau, Q. Li (Eds.), Advances in Web Based Learning – ICWL 2007. XIV, 654 pages. 2008.

Vol. 4822: D.H.-L. Goh, T.H. Cao, I.T. Sølvberg, E. Rasmussen (Eds.), Asian Digital Libraries. XVII, 519 pages. 2007.

Vol. 4820: T.G. Wyeld, S. Kenderdine, M. Docherty (Eds.), Virtual Systems and Multimedia. XII, 215 pages. 2008.

Vol. 4816: B. Falcidieno, M. Spagnuolo, Y. Avrithis, I. Kompatsiaris, P. Buitelaar (Eds.), Semantic Multimedia. XII, 306 pages. 2007.

Vol. 4813: I. Oakley, S.A. Brewster (Eds.), Haptic and Audio Interaction Design. XIV, 145 pages. 2007.

Vol. 4810: H.H.-S. Ip, O.C. Au, H. Leung, M.-T. Sun, W.-Y. Ma, S.-M. Hu (Eds.), Advances in Multimedia Information Processing – PCM 2007. XXI, 834 pages. 2007.

Vol. 4809: M.K. Denko, C.-s. Shih, K.-C. Li, S.-L. Tsao, Q.-A. Zeng, S.H. Park, Y.-B. Ko, S.-H. Hung, J.-H. Park (Eds.), Emerging Directions in Embedded and Ubiquitous Computing. XXXV, 823 pages. 2007.

Vol. 4808: T.-W. Kuo, E. Sha, M. Guo, L.T. Yang, Z. Shao (Eds.), Embedded and Ubiquitous Computing. XXI, 769 pages. 2007.

Vol. 4806: R. Meersman, Z. Tari, P. Herrero (Eds.), On the Move to Meaningful Internet Systems 2007: OTM 2007 Workshops, Part II. XXXIV, 611 pages. 2007.

Vol. 4805: R. Meersman, Z. Tari, P. Herrero (Eds.), On the Move to Meaningful Internet Systems 2007: OTM 2007 Workshops, Part I. XXXIV, 757 pages. 2007.

Vol. 4804: R. Meersman, Z. Tari (Eds.), On the Move to Meaningful Internet Systems 2007: CoopIS, DOA, ODBASE, GADA, and IS, Part II. XXIX, 683 pages. 2007.

Vol. 4803: R. Meersman, Z. Tari (Eds.), On the Move to Meaningful Internet Systems 2007: CoopIS, DOA, ODBASE, GADA, and IS, Part I. XXIX, 1173 pages. 2007.

Vol. 4802: J.-L. Hainaut, E.A. Rundensteiner, M. Kirchberg, M. Bertolotto, M. Brochhausen, Y.-P.P. Chen, S.S.-S. Cherfi, M. Doerr, H. Han, S. Hartmann, J. Parsons, G. Poels, C. Rolland, J. Trujillo, E. Yu, E. Zimányie (Eds.), Advances in Conceptual Modeling – Foundations and Applications. XIX, 420 pages. 2007.

Vol. 4801: C. Parent, K.-D. Schewe, V.C. Storey, B. Thalheim (Eds.), Conceptual Modeling - ER 2007. XVI, 616 pages. 2007.

Vol. 4797: M. Arenas, M.I. Schwartzbach (Eds.), Database Programming Languages. VIII, 261 pages. 2007.

Vol. 4796: M. Lew, N. Sebe, T.S. Huang, E.M. Bakker (Eds.), Human–Computer Interaction. X, 157 pages. 2007.

Vol. 4794: B. Schiele, A.K. Dey, H. Gellersen, B. de Ruyter, M. Tscheligi, R. Wichert, E. Aarts, A. Buchmann (Eds.), Ambient Intelligence. XV, 375 pages. 2007.

Vol. 4777: S. Bhalla (Ed.), Databases in Networked Information Systems. X, 329 pages. 2007.

Vol. 4761: R. Obermaisser, Y. Nah, P. Puschner, F.J. Rammig (Eds.), Software Technologies for Embedded and Ubiquitous Systems. XIV, 563 pages. 2007.

Vol. 4747: S. Džeroski, J. Struyf (Eds.), Knowledge Discovery in Inductive Databases. X, 301 pages. 2007.

Vol. 4744: Y. de Kort, W. IJsselsteijn, C. Midden, B. Eggen, B.J. Fogg (Eds.), Persuasive Technology. XIV, 316 pages. 2007.

Vol. 4740: L. Ma, M. Rauterberg, R. Nakatsu (Eds.), Entertainment Computing – ICEC 2007. XXX, 480 pages. 2007.

Vol. 4730: C. Peters, P. Clough, F.C. Gey, J. Karlgren, B. Magnini, D.W. Oard, M. de Rijke, M. Stempfhuber (Eds.), Evaluation of Multilingual and Multi-modal Information Retrieval. XXIV, 998 pages. 2007.

Vol. 4723: M. R. Berthold, J. Shawe-Taylor, N. Lavrač (Eds.), Advances in Intelligent Data Analysis VII. XIV, 380 pages. 2007.

Vol. 4721: W. Jonker, M. Petković (Eds.), Secure Data Management. X, 213 pages. 2007.

Vol. 4718: J. Hightower, B. Schiele, T. Strang (Eds.), Location- and Context-Awareness. X, 297 pages. 2007.

Vol. 4717: J. Krumm, G.D. Abowd, A. Seneviratne, T. Strang (Eds.), UbiComp 2007: Ubiquitous Computing. XIX, 520 pages. 2007.

Vol. 4715: J.M. Haake, S.F. Ochoa, A. Cechich (Eds.), Groupware: Design, Implementation, and Use. XIII, 355 pages. 2007.

Vol. 4714: G. Alonso, P. Dadam, M. Rosemann (Eds.), Business Process Management. XIII, 418 pages. 2007.

Vol. 4704: D. Barbosa, A. Bonifati, Z. Bellahsène, E. Hunt, R. Unland (Eds.), Database and XML Technologies. X, 141 pages. 2007.

Vol. 4690: Y. Ioannidis, B. Novikov, B. Rachev (Eds.), Advances in Databases and Information Systems. XIII, 377 pages. 2007.

Vol. 4675: L. Kovács, N. Fuhr, C. Meghini (Eds.), Research and Advanced Technology for Digital Libraries. XVII, 585 pages. 2007.

Vol. 4674: Y. Luo (Ed.), Cooperative Design, Visualization, and Engineering. XIII, 431 pages. 2007.

Vol. 4663: C. Baranauskas, P. Palanque, J. Abascal, S.D.J. Barbosa (Eds.), Human-Computer Interaction – INTERACT 2007, Part II. XXXIII, 735 pages. 2007.